DSP First

Second Edition

Global Edition

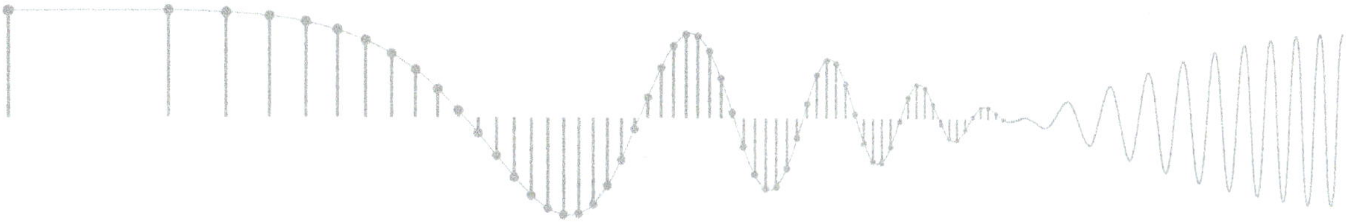

James H. McClellan
Georgia Institute of Technology

Ronald W. Schafer
Stanford University

Mark A. Yoder
Rose-Hulman Institute of Technology

PEARSON

Boston • Columbus • Indianapolis • New York • San Francisco
Hoboken • Amsterdam • Cape Town • Dubai • London • Madrid • Milan • Munich
Paris • Montreal • Toronto • Delhi • Mexico City • São Paulo
Sydney • Hong Kong • Seoul • Singapore • Taipei • Tokyo

Vice President and Editorial Director, ECS: Marcia J. Horton
Aquisitions Editor: Julie Bai
Editorial Assistant: Michelle Bayman
Senior Acquisitions Editor, Global Editions: Sandhya Ghoshal
VP of Marketing: Christy Lesko
Director of Field Marketing: Tim Galligan
Product Marketing Manager: Bram van Kempen
Field Marketing Manager: Demetrius Hall
Marketing Assistant: Jon Bryant
Senior Managing Editor: Scott Disanno
Program Manager: Joanne Manning
Project Editor, Global Editions: K.K. Neelakantan
Senior Manufacturing Controller, Global Editions: Trudy Kimber
Media Production Manager, Global Editions: Vikram Kumar
Global HE Director of Vendor Sourcing and Procurement: Diane Hynes
Operations Specialist: Maura Zaldivar-Garcia
Creative Director: Blair Brown
Senior Art Director: Janet Slowik
Cover Designer: Lumina Datamatics
Manager, Rights and Permissions: Rachel Youdelman
Full-Service Project Management: Pavithra Jayapaul, Jouve India

Image credits:
Cover photo: fotographic1980/Shutterstock.com
Image on page 22: James McClellan, Ronald Schafer, Mark Yoder

Pearson Education Limited
Edinburgh Gate
Harlow
Essex CM20 2JE
England

and Associated Companies throughout the world

Visit us on the World Wide Web at:
www.pearsonglobaleditions.com

British Library Cataloguing-in-Publication Data

A catalogue record for this book is available from the British Library

ISBN 10: 1292113863
ISBN 13: 9781292113869

Typeset in 10/12 Times LT Std Roman by Jouve India

Contents

5 FIR Filters 167

6 Frequency Response of FIR Filters 214

7 Discrete-Time Fourier Transform 259

B Programming in MATLAB 511

Preface

This book, entitled simply *DSP First*, is the second edition of the text *DSP First: A Multimedia Approach* (1998) which was packaged with a CD-ROM that provided many resources to extend the boundaries of a traditional textbook. In 2003, a second book entitled *Signal Processing First* was produced with a broader set of topics that included four new chapters on continuous-time signal processing and the Fourier transform, as well as updated versions of the first eight chapters of *DSP First*. New material was produced for the CD-ROM bundled with the 2003 textbook, and all the supporting resources have now moved to a website for easier access.

These three books and the Companion Website are the result of more than 20 years of work grounded on the premise that digital signal processing (DSP) is an ideal starting point for the study of both electrical engineering and computer engineering. In the summer of 1993, two of us (JHMc and RWS) began to develop a one-quarter course that was to become the required introductory course for Georgia Tech computer engineering (CmpE) students. We argued that the subject of digital signal processing had everything we wanted in a first course for computer engineers: it introduced the students to the use of mathematics as a language for thinking about and solving engineering problems; it laid useful groundwork for subsequent courses; it made a strong connection to digital computation as a means for implementing systems; and it provided the tools to discuss interesting applications that would motivate beginning engineers to do the hard work of connecting mathematics and computation to problem solving. Nothing has happened in the past 22 years to change our minds on this point. Indeed, our teaching experience with more than 6,000 students at Georgia Tech has only strengthened our conviction that digital signal processing, distilled to its essence, is an ideal introductory subject for *both* electrical and computer engineering students.[1] In fact, we have become firmly convinced that a course on DSP at the level of this text should be required of every engineering and computer science student.

From the beginning, we believed that "hands-on" experience with real signals was crucial, so we expended considerable effort on developing additional material for laboratory exercises and projects based on MATLAB. In the laboratory assignments, students can experience the effects of signal processing operations that they have implemented on sound and image signals. For example, they can synthesize music from sinusoids, but they can also see that those same sinusoids are the basis for the wireless systems that they use routinely to access the Internet. These experiences, available on the Companion Website, will augment and reinforce the mathematical concepts that form the basis of DSP.

In addition to the 25 detailed lab assignments, the Companion Website includes many resources that extend the printed textbook with material such as demonstrations and animations used in classes, and hundreds of solved homework problems. The impetus for having this website came from Mark Yoder who, in 1995, while on sabbatical leave at Georgia Tech from Rose-Hulman, had the idea to put all of this material into a form that other teachers (and students) could access easily. Interactive MATLAB demonstrations have been created for demonstrating specific topics such as convolution and frequency response, and most of these are now used as the basis for some of the laboratory exercises. As teachers, all this material has changed the way we present ideas, because it expands the ways to visualize a concept "beyond the equations." Over the years, the collection of resources on our website has continued to grow. In the future, we will explore new ideas for presenting the concepts of DSP, and hope to move beyond the printed page to an e-Text version that would truly integrate the narrative of the book with the visualizations of the companion website.

The distinguishing feature of this text (and its progenitors) is that it presents signal processing at a level consistent with an introductory ECE course, i.e., the sophomore level (second year) in a typical U.S. university. The list of topics in the book is not surprising given its emphasis on discrete-time signal processing, but since we want a course that is broadly accessible to sophomores, we feel that we must combine signal processing concepts with some introductory ideas. Part of the reason for this is that in many electrical engineering curriculums, signals and systems and DSP typically have been treated as junior- and senior-level courses, for which a traditional background of linear circuits and linear systems is assumed. Our approach, on the other hand, makes the subject much more accessible to students in other majors such as computer science

[1]In our development of these ideas, two books by Professor Ken Steiglitz of Princeton University had a major impact on our thinking: *An Introduction to Discrete Systems*, John Wiley & Sons, 1972, and *A Digital Signal Processing Primer: With Applications to Computer Music*, Addison-Wesley Publishing Company, 1996. Steiglitz's 1972 book was well ahead of its time, since DSP had few practical applications, and even simple simulations on then-available batch processing computers required significant programming effort. However, by 1993 when we began our work, easy-to-use software environments such as MATLAB were widely available for implementing DSP computations on powerful personal computers.

and other engineering fields. This point is increasingly important because non-specialists need to use DSP techniques routinely in many areas of science and technology.

Content of the New Edition. This new edition has an organization similar to the first edition of *DSP First*. A look at the table of contents shows that the book begins very simply (Chapter 2) with a detailed discussion of continuous-time sinusoidal signals and their representation by complex exponentials. This is a topic traditionally introduced in a linear circuits course, but including it here makes it immediately accessible for the rest of this book, especially for students who come from other backgrounds. If students have already studied linear circuits, this chapter can be skipped, or rapidly covered. We then proceed to introduce the spectrum concept (Chapter 3) by considering sums of sinusoidal signals, culminating with a brief introduction to Fourier series. Although Chapter 3 of the first edition covered the same basic ideas, this chapter has some new material.[2]

Next we make the transition to discrete-time signals by considering sampled sinusoidal signals (Chapter 4). We have found that it is not necessary to invoke the continuous-time Fourier transform to make the important issues in sampling clear. All that is needed is the simple trigonometric identity $\cos(\theta + 2\pi) = \cos(\theta)$. In fact, in Chapters 2–4 (with the exception of Fourier Series), we have only needed to rely on the simple mathematics of sine and cosine functions. The basic linear system concepts are then introduced with running average systems and other simple FIR filters (Chapter 5). Impulse sequences are introduced which leads to the impulse response characterizing a filter. Convolution is treated as a numerical operation in the first pass at this idea. The key concept of frequency response is derived and interpreted for FIR filters (Chapter 6). Sinusoids are the primary signals of interest, and we emphasize the magnitude and phase change experienced by a sinusoid when filtered by a linear time-invariant system.

At this point we depart significantly from the first edition by introducing (Chapter 7) the concept of discrete-time Fourier transform (DTFT), which arises naturally from the frequency response of a discrete-time system. The concept of the inverse DTFT completes the description of an invertible transform and also enables us to describe ideal filters. It is then natural to move from the DTFT to the discrete Fourier transform (DFT), which is simply a sampled version of the DTFT and thus computable through fast algorithms that are readily available (Chapter 8). Chapters 7 and 8 are completely new. They are a response to frequent requests from teachers who want to expose their students to the powerful concept of the Fourier transform, and we have found that sophomores are fully capable of understanding these concepts and putting them to use. These two chapters bring many of the ideas of practical spectrum analysis into focus with the goal of providing the knowledge to successfully employ the powerful spectrum analysis tools readily available in software environments such as MATLAB.

Finally, the last two chapters return to the flow of the first edition. We introduce z-transforms (Chapter 9) and IIR systems (Chapter 10). At this stage, a student who has faithfully read the text, worked homework problems, and done the laboratory assignments will be rewarded with the ability to understand applications involving the

[2]Furthermore, for instructors who prefer to dive deeper into Fourier analysis of periodic signals, Appendix C on Fourier series is essentially another entire chapter on that topic.

sampling theorem, discrete-time filtering, and spectrum analysis. Furthermore, they are well prepared to move on to courses in linear analog circuits, continuous-time signals and systems, and control systems. All of these courses can build on the foundation established through the study of this text.

Summary of What's New in This Edition

- New material on the Discrete-Time Fourier Transform (DTFT) has been developed and is presented in Chapter 7. The presentation makes an easy transition from the frequency response concept to begin the study of the general idea of a Fourier transform.
- New material on ideal filters and digital filter design is presented in Chapter 7 as a very useful application of the DTFT. The window method for FIR filter design is presented in detail.
- New material on the Discrete Fourier Transform (DFT) has been developed and is presented in Chapter 8. The presentation adopts the point of view that the DFT is a sampled version of the DTFT, and also develops the relationship of the DFT to the discrete Fourier series (DFS).
- New material on spectrum analysis and the spectrogram has been developed for the last sections of Chapter 8. This provides a solid foundation for understanding time-frequency analysis of signals as is commonly done with the FFT algorithm, as well as the role of windowing in frequency resolution.
- Chapters 7 and 8 are derived from Chapter 9 in the first edition and Chapter 13 in *Signal Processing First*. The new chapters are a significant rewrite to make this material accessible at the introductory level. The benefit is that students can learn the ideas of practical spectrum analysis which can then be reinforced with a lab experience where actual signals are processed with the tools available in MATLAB.
- The presentation of the spectrum in Chapter 3 has been expanded to include a formal discussion of properties of the spectrum (e.g., time-delay, frequency shifting). This sets the stage for later discussions of the DTFT and DFT.
- The material on Fourier Series which was part of Chapter 3 has been expanded, but most of it is now placed in Appendix C. Chapter 3 contains a sufficient description of the Fourier series to present the spectrum of one periodic signal, the full wave rectified sine. Appendix C provides an in-depth presentation for instructors who choose to emphasize the topic. Details of other periodic signals (square wave, triangular wave, and half-wave rectified sine) are given along with a derivation of Parseval's theorem and a heuristic discussion of convergence. Properties of the Fourier Series are also developed.
- Extensive changes have been made to the end-of-chapter problems. There are a total of 241 problems in the book: 83 are new, 86 are different from the first edition by virtue of changing the details, and 72 are the same as in the first edition.

- The Companion Website contains new material for labs, MATLAB visualizations, and solved homework problems. The Companion Website may be found at www.pearsonglobaleditions.com/McClellan.

At Georgia Tech, our sophomore-level, 3 credit course covers most of the content of Chapters 2–10 in a format involving two one-hour lectures, one 1.5 hour recitation, and one 1.5 hour laboratory period per week. As mentioned previously, we place considerable emphasis on the lab because we believe that it is essential for motivating our students to learn the mathematics of signal processing, and because it introduces our students to the use of powerful software in engineering analysis and design. At Rose-Hulman, we use *DSP First* in a freshman-level, 10-week course that covers Chapters 1–6, 9, and 10. The Rose format is 3 one-hour lectures per week and one three-hour lab. The students use MATLAB throughout the course. The entire content of the present text was used by RWS for a 10-week, four credit course at Stanford University. Since this course followed quarter-long courses in continuous-time signals and systems and linear circuits, it was possible to skip Chapters 2 and 3 and move immediately into a focus on discrete-time signals and systems using the remaining chapters. One credit was devoted to a weekly lab assignment which was done individually without a regularly scheduled laboratory period.

These examples from our own teaching experience show that the text and its associated supporting materials can be used in many different ways depending on instructor preference and number of course hours. As can be seen from the previous discussion, the second edition of *DSP First* is not a conventional signals and systems book. One difference is the inclusion of a significant amount of material on sinusoids and complex phasor representations. In a traditional electrical engineering curriculum, these basic notions are covered under the umbrella of linear circuits taken before studying signals and systems. Indeed, our choice of title for this book and the first edition is designed to emphasize this departure from tradition. An important point is that teaching signal processing first also opens up new approaches to teaching linear circuits, since there is much to build upon that will allow redirected emphasis in the circuits course.

A second difference from conventional signals and systems texts is that *DSP First* emphasizes topics that rely on "frequency domain" concepts. This means that, in an electrical engineering curriculum, topics like Laplace transforms, state space, and feedback control, would have to be covered in later courses such as linear circuits or an upper-level course on control systems. Although our text has clearly been shaped by a specific point of view, this does not mean that it and the associated website can be used in only one way. Indeed, as our own experience shows, by appropriate selection of topics, our text can be used for either a one-quarter or one-semester signals and systems course that emphasizes communications and signal processing applications from the frequency domain point of view. For most electrical engineering curricula, the control-oriented topics would be covered in another course.

In other disciplines such as computer science and computer engineering, *DSP First* emphasizes those topics that are most relevant to computing for signal analysis. This is also likely to be true in other engineering fields where data acquisition and frequency domain analysis play an important role in modern engineering analysis and design.

This text and its Companion Website represents an untold amount of work by the three authors, numerous colleagues, and many students. Fortunately, we have been able to motivate a number of extremely talented students to contribute MATLAB demos to this project. There are so many that to name them all would be impractical. We simply thank them all for their valuable contributions to our project. Greg Krudysz who authored several of the demos has now taken over the primary role of developing new demos and visualizations with GUIs and updating the existing ones. Since the beginning in 1993, many professors have participated in the sophomore course ECE-2025 (and now ECE-2026) at Georgia Tech as lecturers and recitation instructors. Once again, naming all the recitation instructors would be impractical, but their comments and feedback have given life to the course as it evolved during the past 12 years. For example, Pamela Bhatti developed a laboratory on simulating the filter bank of a Cochlear Implant hearing system. Recently, the lecturing and administration of the course has been shared by Russ Mersereau, Fred Juang, Chin Lee, Elliot Moore, Mark Clements, Chris Rozell, G. K. Chang, David Taylor, David Anderson, John Barry, Doug Williams, and Aaron Lanterman. We are indebted to them for their many suggestions that have made a positive impact on this second edition, especially the new material on the DFT and DTFT. We are also indebted to Wayne Padgett and Bruce Black, who have taught ECE-380 at Rose-Hulman and have contributed many good ideas, and we appreciate the work of Ed Doering who created a whole new set of labs for ECE-180, the new freshman-level DSP First. These labs start with traditional audio processing and end with video object tracking. A new first for freshman.

We also want to acknowledge the contributions of Tom Robbins (formerly at Pearson Prentice-Hall) who was an early supporter of our efforts to bring DSP to the fore in ECE education. Tom bought into our concept of *DSP First* from the beginning, and he encouraged us during the initial project, as well as the 2003 book. More recently, Andrew Gilfillan and Julie Bai have been the editors who helped make this second edition a reality.

Finally, we want to recognize the understanding and support of our wives (Carolyn McClellan, Dorothy Schafer, and Sarah Yoder). Carolyn's photo of the cat Kilby appears in Chapter 1. They have patiently supported us as this multi-year project continued to consume energy and time that might have been spent with them.

JHMc
RWS
MAY

Pearson would like to thank and acknowledge Ankita Pramanik, Indian Institute of Engineering Science and Technology, Shibpur, and Sanjay H.S., M.S. Ramaiah Institute of Technology, for contributing to the Global Edition, and Chih-Wei Liu, National Chiao Tung University, Siddhartha Bhattacharya, RCC Institute of Information Technology, and Neeraj Varshney, Indian Institute of Technology Kanpur, for reviewing the Global Edition.

CHAPTER

1

Introduction

This is a book about signals and systems. In this age of multimedia gaming computers, audio and video entertainment systems, and smartphones, it is almost certain that you, the reader of this text, have formed some impression of the meaning of the terms ***signal*** and ***system***, and you probably use the terms often in daily conversation.

It is likely that your usage and understanding of the terms are correct within some rather broad definitions. For example, you may think of a signal as "something" that carries information. Usually, that something is a pattern of variations of a physical quantity that can be manipulated, stored, or transmitted by physical processes. Examples include speech signals, audio signals, video or image signals, biomedical signals, radar signals, and seismic signals, to name just a few. An important point is that signals can take many equivalent forms or *representations.* For example, a speech signal is produced as an acoustic signal, but it can be converted to an electrical signal by a microphone, and then to a string of numbers as in digital audio recording.

The term ***system*** may be somewhat more ambiguous and subject to interpretation. For example, we often use "system" to refer to a large organization that administers or implements some process, such as the "Social Security system" or the "airline transportation system." However, we are interested in a much narrower definition that is very closely linked to signals. More specifically, a system, for our purposes, is something that can manipulate, change, record, or transmit signals. For example, a DVD recording stores or represents a movie or a music signal as a sequence of numbers. A DVD player is a system for converting the numbers stored on the disc (i.e., the numerical representation of the signal) to a video and/or acoustic signal. In general, systems ***operate*** on signals to produce new signals or new signal representations.

Our goal in this text is to develop a framework wherein it is possible to make precise statements about both signals and systems. Specifically, we want to show that mathematics is an appropriate language for describing and understanding signals and systems. We also want to show that the representation of signals and systems by mathematical equations allows us to understand how signals and systems interact and how we can design and implement systems that achieve a prescribed purpose.

1-1 Mathematical Representation of Signals

Signals are patterns of variations that represent or encode information. Many signals are naturally thought of as a pattern of variations in time. A familiar example is a speech signal, which initially arises as a pattern of changing air pressure in the vocal tract. This pattern, of course, evolves with time, creating what we often call a ***time waveform***. Figure 1-1 shows a plot of a recorded speech waveform. In this plot, the vertical axis represents microphone voltage (proportional to air pressure), and the horizontal axis represents time. Notice that there are four plots in the figure corresponding to four contiguous time segments of the speech waveform. The second plot is a continuation of the first, and so on, with each graph corresponding to a time interval of 50 milliseconds (ms).

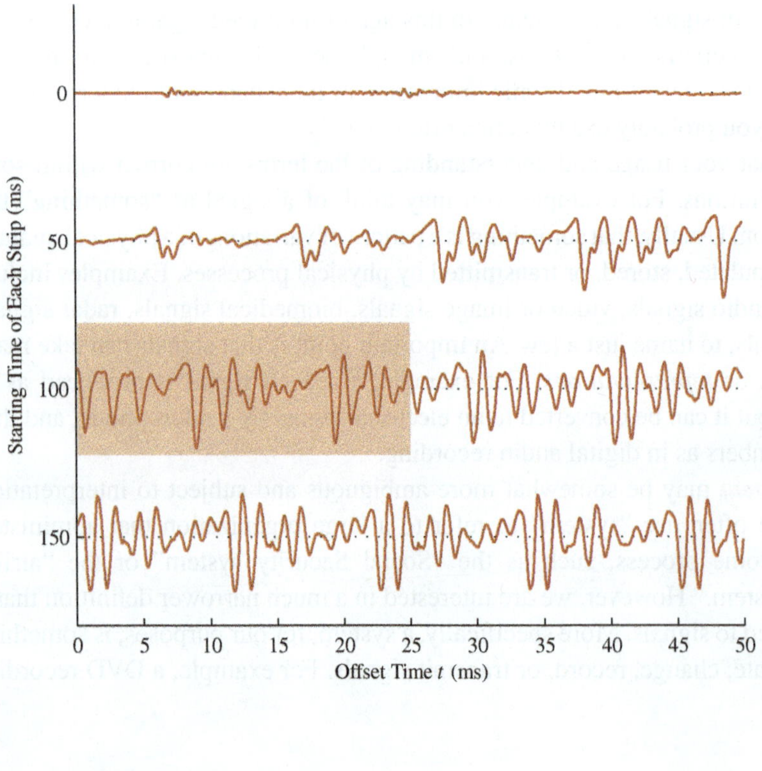

Figure 1-1 Strip plot of a speech signal where each row is a continuation of the row above. This signal $s(t)$ can be represented as a function of a single (time) variable. The shaded region is shown in more detail in Fig. 1-2.

The speech signal in Fig. 1-1 is an example of a one-dimensional ***continuous-time signal***. Such signals can be represented mathematically as a function of a single independent variable, which is normally called time and denoted t. Although in this particular case we cannot write a simple equation that describes the graph of Fig. 1-1 in terms of familiar mathematical functions, we can nevertheless associate a function $s(t)$ with the graph. Indeed, the graph itself can be taken as a definition of the function that assigns a number $s(t)$ to each instant of time (each value of t).

Many, if not most, signals originate as continuous-time signals. However, for reasons that will become increasingly obvious as we progress through this text, it is often desirable to obtain a discrete-time representation of a signal. This can be done by ***sampling*** a continuous-time signal at isolated, equally spaced points in time. The result is a sequence of numbers that can be represented as a function of an index variable that takes on only integer values. This can be represented mathematically as $s[n] = s(nT_s)$, where n is an integer (i.e., $\{\ldots, -2, -1, 0, 1, 2, \ldots\}$), and T_s is the ***sampling period***. Note that our convention is to use parentheses ( ) to enclose the independent variable of a continuous-variable function such as $s(t)$, and square brackets [ ] to enclose the independent variable of a discrete-variable function, e.g., the sequence $s[n]$. Sampling is, of course, exactly what we do when we plot values of a function on graph paper or on a computer screen. We cannot evaluate the function at every possible value of a continuous variable, but only at a set of discrete points. Intuitively, we know that the closer the spacing in time of the points, the more the sequence retains the shape of the original continuous-variable function. Figure 1-2 shows an example of a short segment of a discrete-time signal that was derived by sampling the speech waveform of Fig. 1-1 with a sampling period of $T_s = 1/8$ ms. In this case, the dots show the sample values for the sequence $s[n]$.

While many signals can be thought of as evolving patterns in time, many other signals are not time-varying patterns. For example, an image formed by focusing light through a lens is a spatial pattern, and thus is appropriately represented mathematically as a function of two spatial variables. Such a signal would be considered, in general, as a function of two independent variables [i.e., a picture might be denoted $p(x, y)$]. A photograph is another example, such as the gray-scale image shown in Fig. 1-3. In this case, the value $p(x_0, y_0)$ represents the shade of gray at position (x_0, y_0) in the image.

Images such as that in Fig. 1-3 are generally considered to be two-dimensional continuous-variable signals, since we normally consider space to be a continuum.

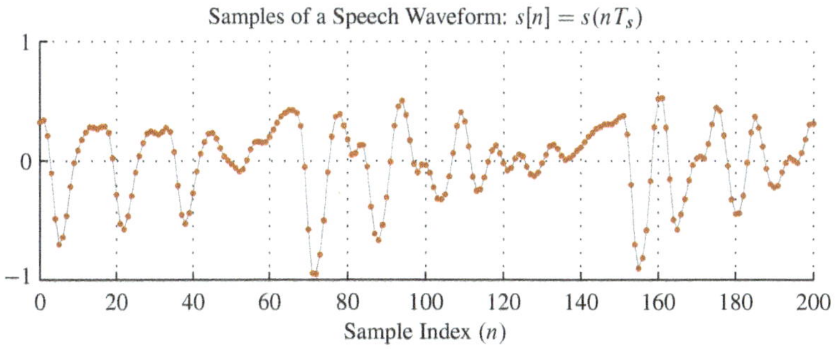

Figure 1-2 Discrete-time signal represented as a one-dimensional sequence which is a function of a discrete variable n. Signal samples are taken from the shaded region of Fig. 1-1. The continuous-time speech signal $s(t)$ is shown in gray.

Figure 1-3 Example of a signal that can be represented by a function of two spatial variables.

However, sampling can likewise be used to obtain a discrete-variable two-dimensional signal from a continuous-variable two-dimensional signal. In a digital camera, this sampling is done by recording light values which have been focused on a sensor array composed of millions of points, or mega-pixels. In a color camera, there would be three separate arrays for RGB: red, green, and blue. A two-dimensional gray-scale image like Fig. 1-3 would be represented by a two-dimensional discrete-variable sequence or an array of numbers, and would be denoted $p[m, n] = p(m\Delta_x, n\Delta_y)$, where both m and n would take on only integer values, and Δ_x and Δ_y are the horizontal and vertical sampling periods, respectively.

Two-dimensional functions are appropriate mathematical representations of still images that do not change with time; on the other hand, videos are time-varying images that would require a third independent variable for time, so a video signal would be denoted $v(x, y, t)$. In analog television, time is discrete (30 frames/s), each horizontal line (x) is continuous, but there are a finite number of horizontal lines, so y is discrete. In present day digital video, all three variables of the video signal $v(x, y, t)$ are discrete since the signal is a sequence of discrete images.

Our purpose in this section has been to introduce the idea that signals can be represented by mathematical functions. Although we will soon see that many familiar functions are quite valuable in the study of signals and systems, we have not even attempted to demonstrate that fact. Our sole concern is to make the connection between functions and signals, and, at this point, functions simply serve as abstract symbols for signals. Thus, for example, now we can refer to "the speech signal $s(t)$" or "the sampled image $p[m, n]$." Although this may not seem highly significant, we will see in the next section that it is indeed a very important step toward our goal of using mathematics to describe signals and systems in a systematic way.

1-2 Mathematical Representation of Systems

As we have already suggested, a system is something that transforms signals into new signals or different signal representations. This is a rather vague definition, but it is useful as a starting point. To be more specific, we say that a one-dimensional continuous-time system takes an input signal $x(t)$ and produces a corresponding output signal $y(t)$. This can be represented mathematically by

$$y(t) = \mathcal{T}\{x(t)\} \tag{1.1}$$

which means that the input signal (waveform, image, etc.) is operated on by the system (symbolized by the operator $\mathcal{T}$) to produce the output $y(t)$. While this sounds very abstract at first, a simple example shows that this need not be mysterious. Consider a system such that the output signal is the square of the input signal. The mathematical description of this system is simply

$$y(t) = [x(t)]^2 \tag{1.2}$$

which says that at each time instant the value of the output is equal to the square of the input signal value at that same time. Such a system would logically be termed a "squarer system." Figure 1-4 shows the output signal of the squarer for the input of Fig. 1-1. As would be expected from the properties of the squaring operation, we see that the output signal is always nonnegative and the larger signal values are emphasized relative to the smaller signal values.

The squarer system defined by (1.2) is a simple example of a ***continuous-time system*** (i.e., a system whose input and output are continuous-time signals). Can we build a physical system that acts like the squarer system? The answer is yes; the system of (1.2) can be approximated through appropriate connections of electronic circuits. On the other hand, if the input and output of the system are both discrete-time signals (sequences of numbers) related by

$$y[n] = (x[n])^2 \tag{1.3}$$

then the system would be a ***discrete-time system***. The implementation of the discrete-time squarer system would be trivial given a digital computer; one simply multiplies each discrete signal value by itself.

In thinking and writing about systems, it is often useful to have a visual representation of the system. For this purpose, engineers use ***block diagrams*** to represent operations performed in an implementation of a system and to show the interrelations among the many signals that may exist in an implementation of a complex system. An example of the general form of a block diagram is shown in Fig. 1-5. What this diagram shows is simply that the signal $y(t)$ is obtained from the signal $x(t)$ by the operation $\mathcal{T}\{\ \}$.

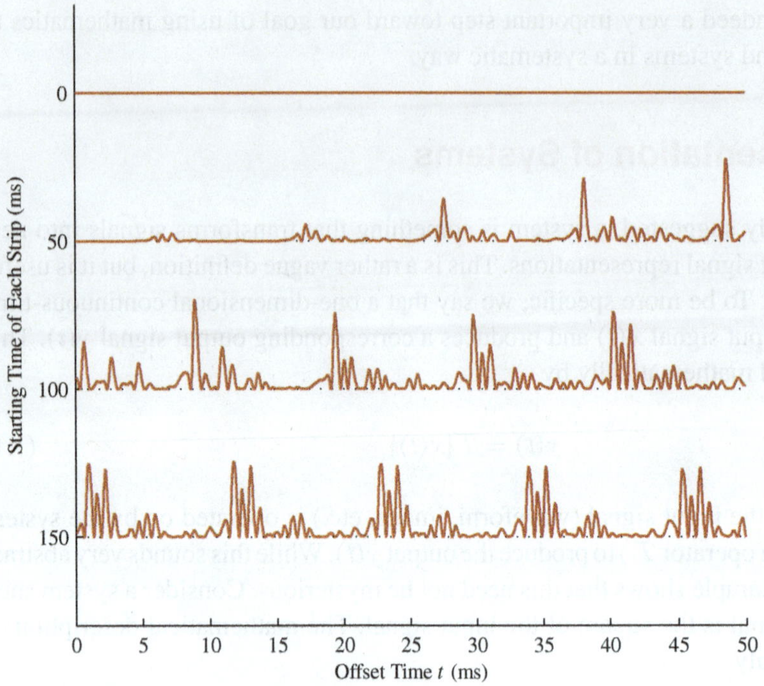

Figure 1-4 Output of a squarer system for the speech signal input of Fig. 1-1. The squarer system is defined by the equation $y(t) = [x(t)]^2$.

A specific example of a system was suggested earlier when we discussed the sampling relationship between continuous-time signals and discrete-time signals. A ***sampler*** is defined as a system whose input is a continuous-time signal $x(t)$ and whose output is the corresponding sequence of samples, defined by the equation

$$x[n] = x(nT_s) \tag{1.4}$$

which simply states that the sampler "takes an instantaneous snapshot" of the continuous-time input signal once every T_s s.[1] Thus, the operation of sampling fits our definition of a system, and it can be represented by the block diagram in Fig. 1-6. Often we will refer to the sampler system as an "ideal continuous-to-discrete converter" or ***ideal C-to-D converter***. In this case, as in the case of the squarer, the name that we give to the system is really just a description of what the system does.

Figure 1-5 Block diagram representation of a continuous-time system.

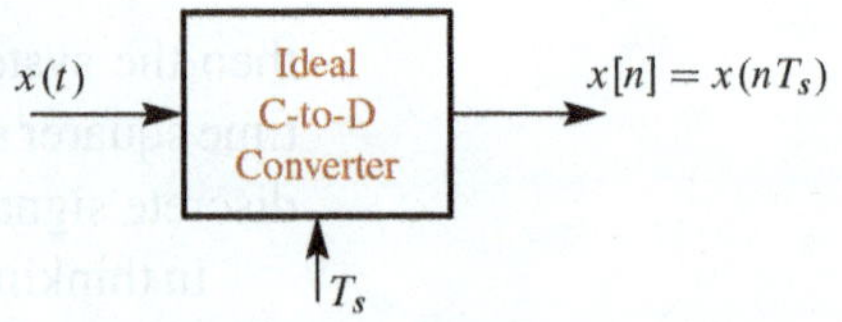

Figure 1-6 Block diagram representation of a sampler.

[1]The units of time in seconds are abbreviated as s.

1-3 Systems as Building Blocks

Block diagrams are useful for representing complex systems in terms of simpler systems, which are more easily understood. For example, Fig. 1-7 shows a block diagram representation of the process of recording and playback of music using MP3 compression. This block diagram breaks the operation down into four subsystems, each of which could be broken down further into smaller subsystems. The first operation is A-to-D (analog-to-digital) conversion to acquire the music waveform in digital form. The A-to-D system is a physical approximation to the ideal C-to-D converter defined in (1.4). An A-to-D converter produces finite-precision numbers as samples of the input signal (quantized to a limited number of bits), while the ideal C-to-D converter produces samples with infinite precision. For the high-accuracy A-to-D converters used in precision audio systems, the difference between an A-to-D converter and our idealized C-to-D converter is slight, but the distinction is very important—only finite-precision quantized sample values can be stored in digital memory of finite size.

Figure 1-7 shows that the output of the A-to-D converter is the input to a system that compresses the numbers $x[n]$ into a much smaller bit stream using the MP3 method. This is a complex process, but for our purposes it is sufficient to show it as a single operation. The output is a compressed digital representation that may be efficiently stored as data on a server or transmitted to a user. Once another user has the compressed data file, the MP3 compression must be reversed in order to listen to the audio signal. Since MP3 is a "lossy" compression scheme, the signal synthesized by the MP3 decoder is only an approximation to the original. The value of MP3 is that this approximation is audibly indistinguishable from the original because the MP3 encoding method exploits aspects of human hearing that render certain coding errors inaudible. Once the music waveform is reconstituted in digital form as $\hat{x}[n]$, the last block does the conversion of the signal from discrete-time form to continuous-time (acoustic) form using a system called a D-to-A (digital-to-analog) converter. This system takes finite-precision binary numbers in sequence and fills in a continuous-time function between the samples. The resulting continuous-time electrical signal could then be fed to other systems, such as amplifiers, loudspeakers, and headphones, for conversion to sound. In Chapter 4, we will discuss the ideal D-to-C converter, which is an idealization of the physical device called an D-to-A converter.

Systems like MP3 audio are all around us. For example, digital cameras use JPEG encoding to reduce digital image file sizes prior to storage, and JPEG decoding to view pictures. Most of the time we do not need to think about how such systems work, but this example illustrates the value of thinking about a complex system in a hierarchical form.

Figure 1-7 Simplified block diagram for MP3 audio compression and playback system.

In this way, we can first understand the individual parts, then the relationship among the parts, and finally the whole system. By looking at the MP3 audio system in this manner, we can discuss two things. First of all, the conversion from continuous-time to discrete-time and back to continuous-time can be considered separately from the other parts of the system. The effect of connecting these blocks to the system is then relatively easy to understand because they provide the input and output interface to real audio signals. Secondly, details of some parts can be hidden and left to experts who, for example, can develop more detailed breakdowns of the MP3 encoder and decoder subsystems. In fact, those systems involve many signal processing operations, and it is possible to specify their operations by connecting several canonical DSP blocks that we will study in this text.

1-4 The Next Step

The MP3 audio coding system is a good example of a relatively complicated discrete-time system. Buried inside the blocks of Fig. 1-7 are many discrete-time subsystems and signals. While we do not promise to explain all the details of MP3 coders or any other complex system, we do hope to establish the foundations for the understanding of discrete- and continuous-time signals and systems so that this knowledge can be applied to understanding components of more complicated systems. In Chapter 2, we will start at a basic mathematical level and show how the well-known sine and cosine functions from trigonometry play a fundamental role in signal and system theory. Next, we show how complex numbers can simplify the algebra of trigonometric functions. Subsequent chapters introduce the concept of the frequency spectrum of a signal and the concept of filtering with a linear time-invariant system. By the end of the book, if you have diligently worked the problems, experienced the demonstrations, and done the laboratory exercises on the Companion Website (which are marked with icons), you will be rewarded with a solid understanding of many of the key concepts underlying much of modern signal processing technology.

NOTE

Companion Website has many labs, demonstrations and homework problems with solutions

CHAPTER

2

Sinusoids

We begin our discussion by introducing a general class of signals that are commonly called *cosine signals* or, equivalently, *sine signals*, which are also commonly referred to as cosine or sine *waves*, particularly when speaking about acoustic or electrical signals. Collectively, such signals are called ***sinusoidal signals*** or, more concisely, ***sinusoids***. Sinusoidal signals are the basic building blocks in the theory of signals and systems, and it is important to become familiar with their properties. The most general mathematical formula for a sinusoid is

$$x(t) = A\cos(\omega_0 t + \varphi) \tag{2.1}$$

where $\cos(\cdot)$ denotes the cosine function that is familiar from the study of trigonometry. When defining a continuous-time signal, we typically use a function whose independent variable is t, a continuous real variable that represents time. From (2.1) it follows that $x(t)$ is a mathematical function in which the angle (or argument) of the cosine function is, in turn, a function of the variable t. Since we normally think of time as increasing uniformly, the angle of the cosine function likewise increases in proportion to time. The parameters A, ω_0, and φ are fixed numbers for a particular cosine signal. Specifically, A is called the ***amplitude***, ω_0 the ***radian frequency***, and φ the ***phase*** of the cosine signal.

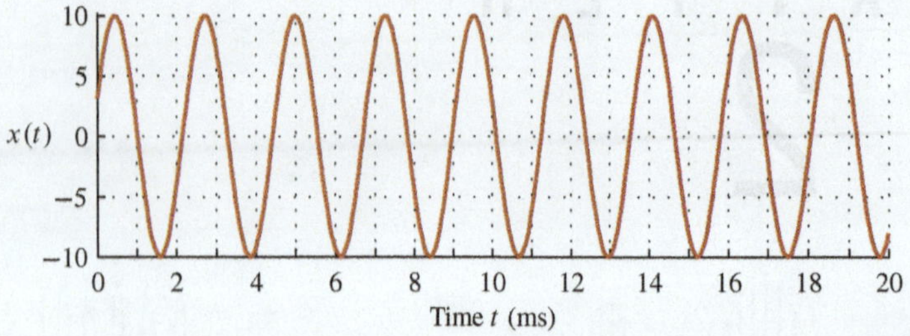

Figure 2-1 Sinusoidal signal generated from the formula: $x(t) = 10\cos(2\pi(440)t - 0.4\pi)$.

Figure 2-1 shows a plot of the continuous-time sinusoid

$$x(t) = 10\cos(2\pi(440)t - 0.4\pi)$$

where $A = 10$, $\omega_0 = 2\pi(440)$, and $\varphi = -0.4\pi$ in (2.1). Note that $x(t)$ oscillates between A and $-A$, and repeats the same pattern of oscillations every $1/440 = 0.00227$ s (approximately). This time interval is called the ***period*** of the sinusoid. We will show later in this chapter that most features of the sinusoidal waveform are directly dependent on the choice of the parameters A, ω_0, and φ.

2-1 Tuning-Fork Experiment

One of the reasons that cosine waves are so important is that many physical systems generate signals that can be modeled (i.e., represented mathematically) as sine or cosine functions versus time. Among the most prominent of these are signals that are audible to humans. The tones or notes produced by musical instruments are perceived as different pitches. Although it is an oversimplification to equate notes to sinusoids and pitch to frequency, the mathematics of sinusoids is an essential first step to understanding complicated sound signals and their perception by humans.

To provide some motivation for our study of sinusoids, we will begin by considering a very simple and familiar system for generating a sinusoidal signal. This system is a *tuning fork*, an example of which is shown in Fig. 2-2. When struck sharply, the tines of the tuning fork vibrate and emit a "pure" tone. This tone has a single frequency, which is usually stamped on the tuning fork. It is common to find "A–440" tuning forks, because 440 hertz (Hz) is the frequency of A above middle C on a musical scale, and is often used as the reference note for tuning a piano and other musical instruments. If you can obtain a tuning fork, perform the following experiment which is shown in a movie on the Companion Website.

DEMO
Tuning Fork

Strike the tuning fork against your knee, and then hold it close to your ear. You should hear a distinct "hum" at the frequency designated for the tuning fork. The sound will persist for a rather long time if you have struck the tuning fork properly; however, it is easy to do this experiment incorrectly. If you hit the tuning fork sharply on a hard surface such as a table, you will hear a high pitched metallic "ting" sound. This is *not* the characteristic sound

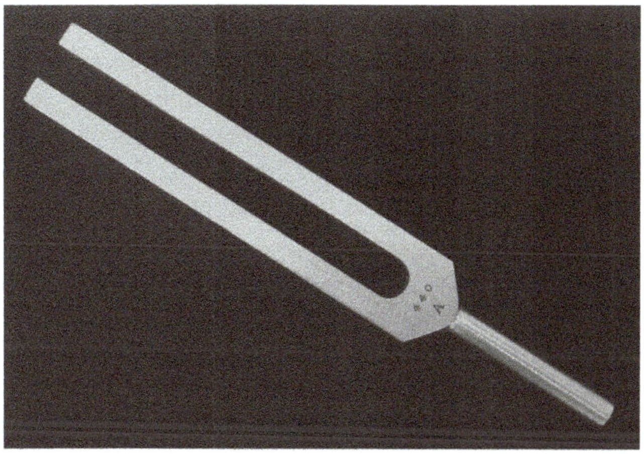

Figure 2-2 Picture of a tuning fork for 440 Hz.

that you are seeking. If you hold the tuning fork close to your ear, you will hear two tones: The higher-frequency "ting" will die away rapidly, and then the desired lower-frequency "hum" will be heard.

With a microphone and a computer equipped with an A-to-D converter, we can make a digital recording of the signal produced by the tuning fork. The microphone converts the sound into an electrical signal, which in turn is converted to a sequence of numbers stored in the computer. Then these numbers can be plotted as a waveform in MATLAB. A typical plot of a short section of the signal taken well after the initial strike of the tuning fork is shown in Fig. 2-3(a) for an A–440 tuning fork. In this case, the A-to-D converter sampled the output of the microphone at a rate of 10,000 samples/s. The signal

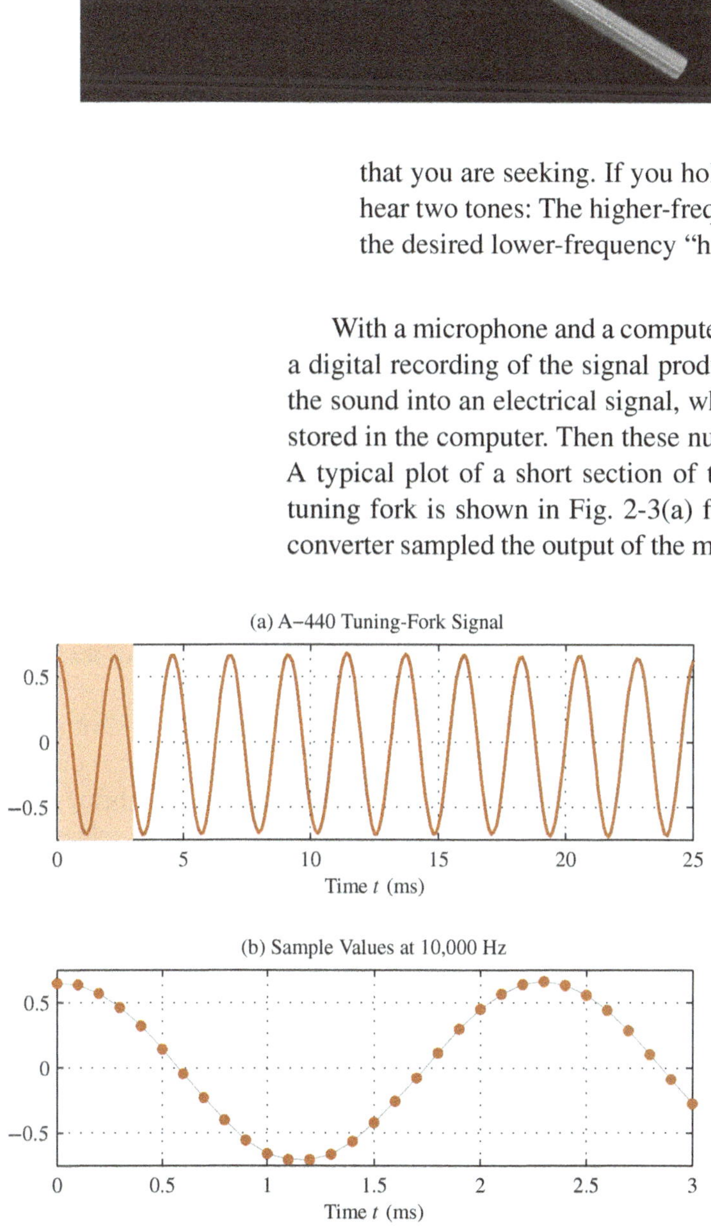

Figure 2-3 (a) Recording of an A–440 tuning-fork signal sampled at a sampling rate of 10,000 samples/s. (b) Zoom in to the first 3 ms taken from the top plot (shaded region), showing the individual sample values (connected by a thin gray line).

samples for the initial 3 ms of the signal are shown on a finer scale in Fig. 2-3(b). The "continuous looking" plot in Fig. 2-3(a) was constructed by connecting the sample values by straight lines. It appears that the signal generated by the tuning fork is very much like the cosine signal of Fig. 2-1. It oscillates between symmetric limits of amplitude and it also repeats periodically with a period of about 2.27 ms (0.00227 s). As we will see in Section 2-3.1, this period is proportional to the reciprocal of ω_0 (i.e., $2\pi/(2\pi(440)) \approx 0.00227$).

This experiment shows that common physical systems produce signals whose graphical representations look very much like cosine signals, as in the graphical plots of the mathematical functions defined in (2.1). Later, in Section 2-7, we will add further credence to the sinusoidal model for the tuning-fork sound by showing that cosine functions arise as solutions to the differential equation that (through the laws of physics) describes the motion of the tuning fork's tines. Before looking at the physics of the tuning fork, however, we should become more familiar with sinusoids and sinusoidal signals.

2-2 Review of Sine and Cosine Functions

Sinusoidal signals are defined in terms of the familiar sine and cosine functions of trigonometry. A brief review of the properties of these basic trigonometric functions is useful for understanding the properties of sinusoidal signals.

The sine and cosine functions are often introduced and defined through a triangle diagram like Fig. 2-4. The trigonometric functions sine and cosine take an angle as their argument. We often think of angles in degrees, but where sine and cosine functions are concerned, angles must be dimensionless. Angles are therefore specified in radians. If the angle θ is in the first quadrant ($0 \le \theta < \pi/2$ rad), then the sine of θ is the length y of the side of the triangle opposite the angle θ divided by the length r of the hypotenuse of the right triangle. Similarly, the cosine of θ is the ratio of the length of the adjacent side x to the length of the hypotenuse.

Note that as θ increases from 0 to $\pi/2$, $\cos\theta$ decreases from 1 to 0 and $\sin\theta$ increases from 0 to 1. When the angle is greater than $\pi/2$ radians, the algebraic signs of x and y come into play, x being negative in the second and third quadrants and y being negative in the third and fourth quadrants. This is most easily shown by plotting the values of $\sin\theta$

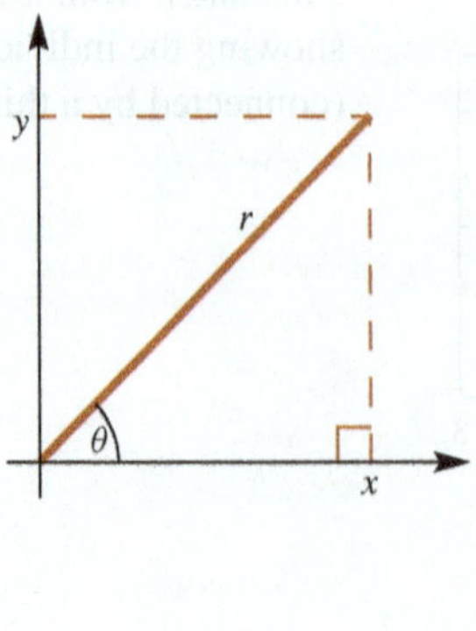

$$\sin\theta = \frac{y}{r} \implies y = r\sin\theta$$

$$\cos\theta = \frac{x}{r} \implies x = r\cos\theta$$

Figure 2-4 Definition of sine and cosine of an angle θ within a right triangle.

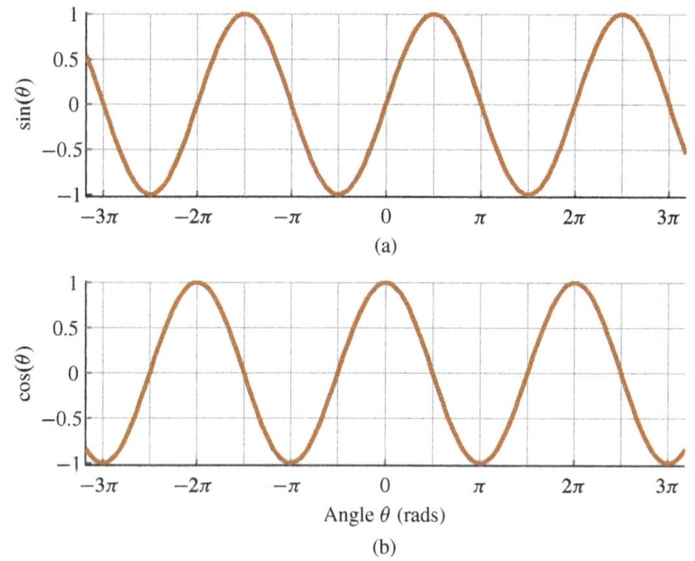

Figure 2-5 (a) Sine function and (b) cosine function plotted versus angle θ. Both functions have a period of 2π.

and $\cos\theta$ as a function of θ, as in Fig. 2-5. Several features of these plots[1] are worthy of comment. The two functions have exactly the same shape. Indeed, the sine function is just a cosine function that is shifted to the right by $\pi/2$ (i.e., $\sin\theta = \cos(\theta - \pi/2)$). Both functions oscillate between $+1$ and -1, and they repeat the same pattern periodically with period 2π. Furthermore, the sine function is an odd function of its argument, and the cosine is an even function. A summary of these and other properties is presented in Table 2-1.

Table 2-2 shows that the sine and cosine functions are very closely related. This often leads to opportunities for simplification of expressions involving both sine and cosine functions. In calculus, we have the interesting property that the sine and cosine functions are derivatives of each other:

$$\frac{d\sin\theta}{d\theta} = \cos\theta \quad \text{and} \quad \frac{d\cos\theta}{d\theta} = -\sin\theta$$

Table 2-1 Basic properties of the sine and cosine functions.

Property	Equation
Equivalence	$\sin\theta = \cos(\theta - \pi/2)$ or $\cos(\theta) = \sin(\theta + \pi/2)$
Periodicity	$\cos(\theta + 2\pi k) = \cos\theta$, when k is an integer
Evenness of cosine	$\cos(-\theta) = \cos\theta$
Oddness of sine	$\sin(-\theta) = -\sin\theta$
Zeros of sine	$\sin(\pi k) = 0$, when k is an integer
Ones of cosine	$\cos(2\pi k) = 1$, when k is an integer
Minus ones of cosine	$\cos[2\pi(k + \frac{1}{2})] = -1$, when k is an integer

[1] It is a good idea to memorize the form of these plots and be able to sketch them accurately.

Table 2-2 Some basic trigonometric identities.

Number	Equation
1	$\sin^2\theta + \cos^2\theta = 1$
2	$\cos 2\theta = \cos^2\theta - \sin^2\theta$
3	$\sin 2\theta = 2\sin\theta\cos\theta$
4	$\sin(\alpha \pm \beta) = \sin\alpha\cos\beta \pm \cos\alpha\sin\beta$
5	$\cos(\alpha \pm \beta) = \cos\alpha\cos\beta \mp \sin\alpha\sin\beta$

In trigonometry, there are many identities that can be used in simplifying expressions involving combinations of sinusoidal signals. Table 2-2 gives a brief list of *trigonometric identities* that will be useful. Recall from your study of trigonometry that these identities are not independent: e.g., identity #3 can be obtained from identity #4 by substituting $\alpha = \beta = \theta$. Also, these identities can be combined to derive other identities. For example, adding identity #5 with plus sign to identity #5 with minus sign leads to the identity

$$\cos\alpha\cos\beta = \tfrac{1}{2}\cos(\alpha+\beta) + \tfrac{1}{2}\cos(\alpha-\beta)$$

A more extensive table of trigonometric identities may be found online, or in any book of mathematical tables. However, once we introduce the complex phasor notation for sinusoids, there will be little or no need for these identities.

NOTE
Exercise solutions are available on Companion Website

EXERCISE 2.1 Use trigonometric identity #5 to derive an expression for $\cos 8\theta$ in terms of $\cos 9\theta$, $\cos 7\theta$, and $\cos\theta$.

2-3 Sinusoidal Signals

The most general mathematical formula for a sinusoidal time signal is obtained by making the argument (i.e., the angle) of the cosine function be a function of t. The following equation gives two equivalent forms:

$$x(t) = A\cos(\omega_0 t + \varphi) = A\cos(2\pi f_0 t + \varphi) \tag{2.2}$$

which are related by defining $\omega_0 = 2\pi f_0$. In either form, there are three independent parameters. The names and interpretations of these parameters are as follows:

(a) A is called the ***amplitude***. The amplitude is a scaling factor that determines how large the cosine signal will be. Since the function $\cos\theta$ oscillates between $+1$ and -1, the signal $x(t)$ in (2.2) oscillates between $+A$ and $-A$.

(b) φ is called the ***phase***. The units of phase must be radians, since the argument of the cosine must be in radians. We generally prefer to use the cosine function when defining the phase. If we happen to have a formula containing sine, e.g., $x(t) = A\sin(\omega_0 t + \varphi')$, then we can rewrite $x(t)$ in terms of cosine if we use the equivalence property in Table 2-1. The result is:

$$x(t) = A\sin(\omega_0 t + \varphi') = A\cos(\omega_0 t + \varphi' - \pi/2)$$

so we define the phase to be $\varphi = \varphi' - \pi/2$ in (2.2). For simplicity and to prevent confusion, we often avoid using the sine function.

(c) ω_0 is called the ***radian frequency***. Since the argument of the cosine function must be in radians, which is dimensionless, the quantity $\omega_0 t$ must likewise be dimensionless. Thus, ω_0 must have units of rad/s if t has units of seconds. Similarly, $f_0 = \omega_0/2\pi$ is called the ***cyclic frequency***, and f_0 must have units of s^{-1}, or hertz.

EXAMPLE 2-1 Plotting Sinusoids

Figure 2-6 shows a plot of the signal

$$x(t) = 20\cos(2\pi(40)t - 0.4\pi) \tag{2.3}$$

In terms of our definitions, the signal parameters are $A = 20$, $\omega_0 = 2\pi(40)$ rad/s, $f_0 = 40$ Hz, and $\varphi = -0.4\pi$ rad. The signal size depends on the amplitude parameter A; its maximum and minimum values are $+20$ and -20, respectively. In Fig. 2-6 the maxima occur at

$$t = \ldots, -0.02, 0.005, 0.03, \ldots$$

and the minima at

$$\ldots, -0.0325, -0.0075, 0.0175, \ldots$$

DEMO
Sinusoids

The time interval between successive maxima in Fig. 2-6 is 0.025 s, which is equal to $1/f_0$. To understand why the signal has these properties, we will need to do more analysis.

2-3.1 Relation of Frequency to Period

The sinusoid plotted in Fig. 2-6 is a periodic signal. The ***period*** of the sinusoid, denoted by T_0, is the time duration of one cycle of the sinusoid. In general, the frequency of the sinusoid determines its period, and the relationship can be found by applying the definition of periodicity $x(t + T_0) = x(t)$ as follows:

$$A\cos(\omega_0(t + T_0) + \varphi) = A\cos(\omega_0 t + \varphi)$$

$$\cos(\omega_0 t + \underbrace{\omega_0 T_0}_{=2\pi\,?} + \varphi) = \cos(\omega_0 t + \varphi)$$

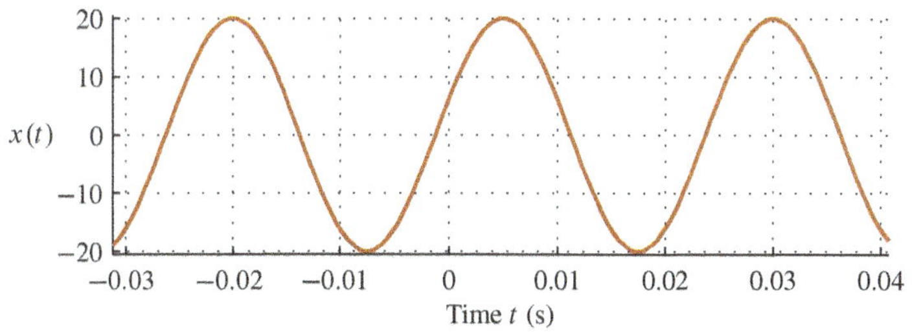

Figure 2-6 Sinusoidal signal with parameters $A = 20$, $\omega_0 = 2\pi(40)$, $f_0 = 40$ Hz, and $\varphi = -0.4\pi$ rad.

Since the cosine function has a period of 2π, the smallest positive value for $\omega_0 T_0$ is 2π, which gives a simple relationship between T_0 and the frequency f_0 in hertz,

$$T_0 = \frac{2\pi}{\omega_0} = \frac{2\pi}{2\pi f_0} = \frac{1}{f_0} \tag{2.4}$$

Since T_0 is the duration of one period of the signal, $f_0 = 1/T_0$ is the number of periods (cycles) per second. Therefore, cycles per second is an appropriate unit for f_0, and it was in general use until the 1960s when the unit hertz (abbreviated Hz) was adopted worldwide.[2] For ω_0, the radian frequency, the units are rad/s. The units of hertz, or cycles per second, for f_0 are often more convenient when describing the plot of a sinusoid, because the factor of 2π is not involved.

It is relatively easy to understand the effect of the frequency parameter because a higher value is more cycles per unit time. Figure 2-7 shows this effect for three choices of f_0 in the signal $x(t) = 5\cos(2\pi f_0 t)$. The sinusoid in Fig. 2-7(a) has a higher frequency than the one in Fig. 2-7(b). As we expect, the waveform shape is similar for both, but the higher frequency signal varies more rapidly with time (i.e., the cycle length (period) is a shorter time interval). We have already seen in (2.4) that this is true because the period of

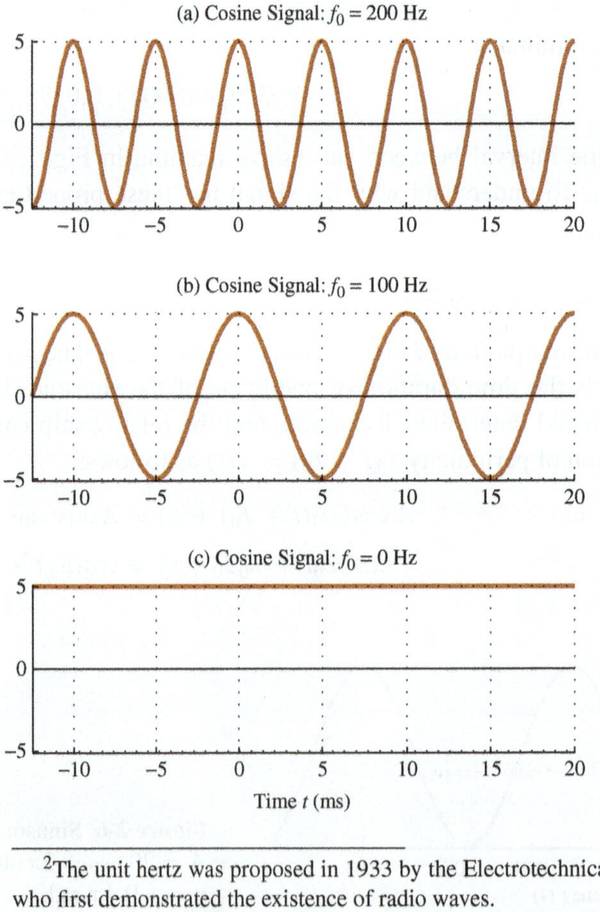

Figure 2-7 Cosine signals $x(t) = 5\cos(2\pi f_0 t)$ for several values of f_0: (a) $f_0 = 200$ Hz; (b) $f_0 = 100$ Hz; (c) $f_0 = 0$.

[2]The unit hertz was proposed in 1933 by the Electrotechnical Commission in honor of Heinrich Hertz, who first demonstrated the existence of radio waves.

a cosine signal is the reciprocal of the frequency. Note that when the frequency doubles (100 Hz → 200 Hz), the period is halved. This is an illustration of the general principle that the higher the frequency, the more rapid the signal waveform changes with time. Throughout this book, we will see more examples of this inverse relationship between time and frequency.

Finally, notice that $f_0 = 0$ is a perfectly acceptable frequency value, and when this value is used, the resulting signal is constant (Fig. 2-7(c)), since $5\cos(2\pi(0)t) = 5\cos(0) = 5$ for all values of t. Thus, the constant signal, often called DC,[3] is, in fact, a sinusoid of zero frequency.

2-3.2 Phase and Time Shift

The phase parameter φ (together with the frequency) determines the time locations of the maxima and minima of a cosine wave, as well as the zero crossings in-between. To be specific, notice that the sinusoid (2.2) with $\varphi = 0$ has a positive peak at $t = 0$. When $\varphi \neq 0$, the phase determines how much the maximum of the cosine signal is shifted away from $t = 0$.

Before we examine this point in detail for sinusoids, it is useful to become familiar with the general concept of ***time-shifting*** a signal. Suppose that a signal $s(t)$ is defined by a known formula or graph. A simple example is the following triangularly shaped function:

$$s(t) = \begin{cases} 2t & 0 \le t \le \frac{1}{2} \\ -\frac{2}{3}(t-2) & \frac{1}{2} \le t \le 2 \\ 0 & \text{elsewhere} \end{cases} \tag{2.5}$$

This simple function, shown in Fig. 2-8(a), has a slope of 2 for $0 \le t < \frac{1}{2}$ and a negative slope of $-\frac{2}{3}$ for $\frac{1}{2} < t \le 2$. Now consider the function $x_1(t) = s(t-2)$. From the definition of $s(t)$, it follows that $x_1(t)$ is nonzero for

$$0 \le (t-2) \le 2 \quad \Longrightarrow \quad 2 \le t \le 4$$

Within this time interval, the formula for the shifted signal is:

$$x_1(t) = \begin{cases} 2(t-2) & 2 \le t \le 2\frac{1}{2} \\ -\frac{2}{3}(t-4) & 2\frac{1}{2} \le t \le 4 \\ 0 & \text{elsewhere} \end{cases} \tag{2.6}$$

In other words, $x_1(t)$ is simply the $s(t)$ function with its origin (and peak) shifted to the *right* by 2 s; see Fig. 2-8(b). Similarly, $x_2(t) = s(t+1)$ is the $s(t)$ function shifted to the *left* by 1 s in Fig. 2-8(c); its nonzero portion is located in the interval $-1 \le t \le 1$.

[3]Electrical engineers use the abbreviation DC standing for direct current, which is a constant current.

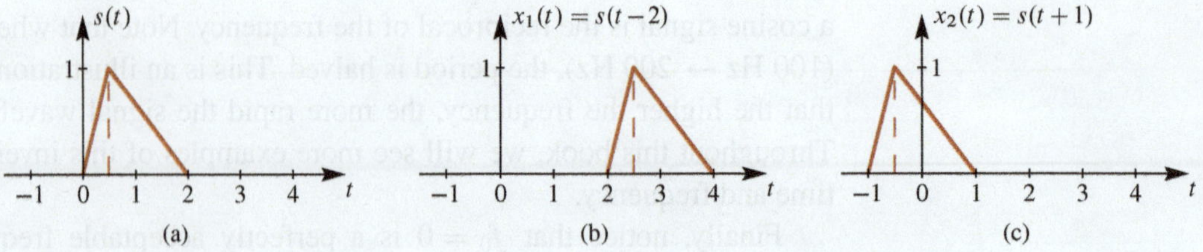

Figure 2-8 Illustration of time-shifting: (a) the triangular signal $s(t)$; (b) shifted to the right by 2 s, $x_1(t) = s(t-2)$; (c) shifted to the left by 1 s, $x_2(t) = s(t+1)$.

EXERCISE 2.2 Derive the equations for the shifted signal $x_2(t) = s(t+1)$.

We will have many occasions to consider time-shifted signals. Whenever a signal can be expressed in the form $x_1(t) = s(t - t_1)$, we say that $x_1(t)$ is a time-shifted version of $s(t)$. If t_1 is a positive number, then the shift is to the right, and we say that the signal $s(t)$ has been ***delayed*** in time. When t_1 is a negative number, then the shift is to the left, and we say that the signal $s(t)$ was ***advanced*** in time. In summary, time shifting is essentially a redefinition of the time origin of the signal. In general, any function of the form $s(t - t_1)$ has its time origin moved to the location $t = t_1$.

For a sinusoid, the time shift is defined with respect to a zero-phase cosine that has a positive peak at $t = 0$. Since a sinusoid has many positive peaks, we must pick one to define the time shift, so we pick the positive peak of the sinusoid that is closest to $t = 0$. In the plot of Fig. 2-6, the time where this positive peak occurs is $t_1 = 0.005$ s. Since the peak in this case occurs at a positive time (to the right of $t = 0$), we say that the time shift is a delay of the zero-phase cosine signal. Now we can relate the time delay to phase. Let $x_0(t) = A\cos(\omega_0 t)$ denote a cosine signal with zero phase. A delay of $x_0(t)$ can be converted to a phase φ by making the following comparison:

$$x_0(t - t_1) = A\cos\left(\omega_0(t - t_1)\right) = A\cos(\omega_0 t + \varphi)$$
$$\Rightarrow \cos(\omega_0 t - \omega_0 t_1) = \cos(\omega_0 t + \varphi)$$

Since the second equation must hold for all t, we conclude that $\varphi = -\omega_0 t_1$. Notice that the phase is negative when the time shift t_1 is positive (a delay). We can also express the phase in terms of the period ($T_0 = 1/f_0$) where we get the more intuitive formula

$$\varphi = -\omega_0 t_1 = -2\pi\left(\frac{t_1}{T_0}\right) \tag{2.7a}$$

which states that the phase is 2π times the fraction of a period given by the ratio of the time shift to the period. If we need to get the time shift from the phase, we solve for t_1

DEMO
Sine Drill

$$t_1 = -\frac{\varphi}{\omega_0} = -\frac{\varphi}{2\pi f_0} = -\frac{\varphi T_0}{2\pi} \tag{2.7b}$$

Since the positive peak nearest to $t = 0$ must always lie within the interval $[-\frac{1}{2}T_0, \frac{1}{2}T_0]$, the phase calculated in (2.7a) will always satisfy $-\pi < \varphi \le \pi$. However, the phase is also ambiguous because adding a multiple of 2π to the argument of a cosine function does not change the value of the cosine. This is a direct consequence of the fact that the cosine is periodic with period 2π. Each different multiple of 2π corresponds to picking a different peak of the periodic waveform. Thus, another way to compute the phase is to find any positive peak of the sinusoid and measure its corresponding time location. After the time location is converted to phase using (2.7a), an integer multiple of 2π can be added to or subtracted from the phase to produce a final result between $-\pi$ and $+\pi$. This gives a final result identical to locating the peak that is within half a period of $t = 0$. The operation of adding or subtracting multiples of 2π is referred to as ***reducing modulo 2π***, because it is similar to modulo reduction in mathematics, which amounts to dividing by 2π and taking the remainder. The value of phase that falls between $-\pi$ and $+\pi$ is called the ***principal value*** of the phase.

EXERCISE 2.3 In Fig. 2-6, it is possible to measure both a positive and a negative value of t_1 and then calculate the corresponding phases. Which phase is within the range $-\pi < \varphi \leq \pi$? Verify that the two phases differ by 2π.

EXERCISE 2.4 Starting with the plot in Fig. 2-6, sketch a plot of $x(t - t_1)$ when $t_1 = 0.0075$. Repeat for $t_1 = -0.01$. Make sure that you shift in the correct direction. For each case, compute the phase of the resulting time-shifted sinusoid.

EXERCISE 2.5 For the signal in Fig. 2-6, $x(t) = 20\cos(2\pi(40)t - 0.4\pi)$, find G and t_1 so that the signal $y(t) = Gx(t - t_1)$ is equal to $5\cos(2\pi(40)t)$.

2-4 Sampling and Plotting Sinusoids

All of the plots of sinusoids in this chapter were created using MATLAB. This had to be done with care, because MATLAB deals only with discrete signals represented by row or column vectors, but we actually want to plot a continuous function $x(t)$. For example, if we wish to plot a function such as

$$x(t) = 20\cos(2\pi(40)t - 0.4\pi) \tag{2.8a}$$

which is shown in Fig. 2-6, we must evaluate $x(t)$ at a discrete set of times. Usually we pick a *uniform* set $t_n = nT_s$, where n is an integer. Then from $x(t)$ we obtain the sequence of samples

$$x(nT_s) = 20\cos(80\pi nT_s - 0.4\pi) \tag{2.8b}$$

where T_s is called the *sample spacing* or ***sampling period***. When plotting the function $x(nT_s)$ using the `plot` function in MATLAB, we must provide a pair of row or column vectors, one containing the time values and the other the computed function values to be plotted. For example, the MATLAB statements

```
n = -6:9;
Ts = 0.005;
tn = n*Ts;
xn = 20*cos(80*pi*tn - 0.4*pi);
plot(tn,xn)
```

would create a row vector `tn` of 16 numbers between −0.03 and 0.045 spaced by the sampling period 0.005, and a row vector `xn` of samples of $x(t)$. Then the MATLAB function `plot` draws the corresponding points, connecting them with straight line segments. Constructing the curve between sample points in this way is called *linear interpolation*. The solid gray curve in Fig. 2-9(a) shows the result of linear interpolation when the sample spacing is $T_s = 0.005$. Intuitively, we realize that if the points are very close together, we will see a smooth curve. The important question is, "How small must we make the sample spacing, so that the cosine signal can be accurately reconstructed between samples by linear interpolation?" A qualitative answer to this question is provided by Fig. 2-9, which shows plots produced by three different sampling periods.

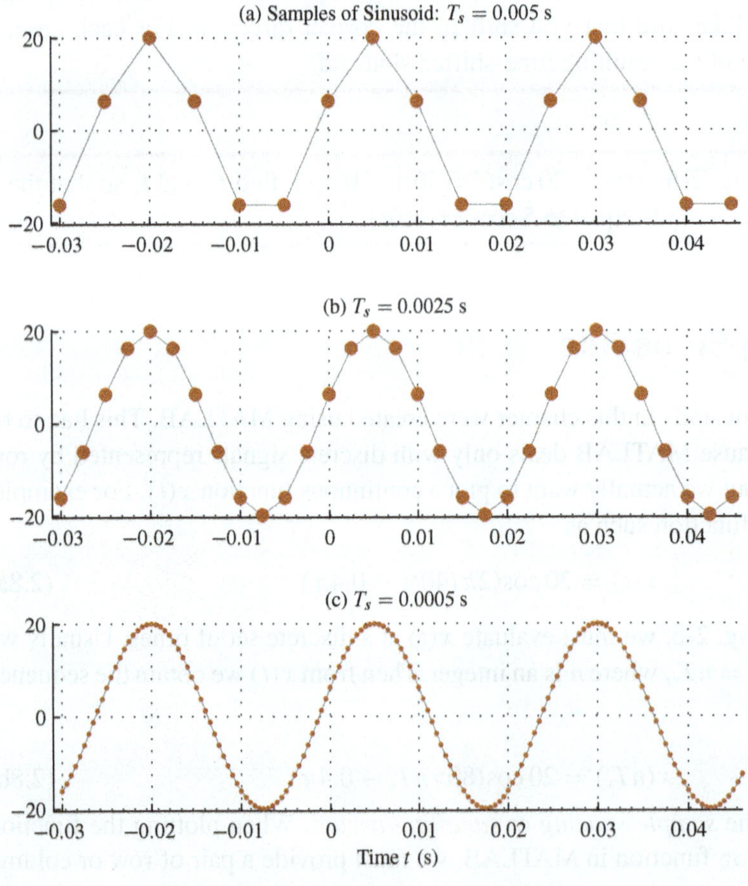

Figure 2-9 Plotting the 40-Hz sampled cosine 2.8(b) for (a) $T_s = 0.005$ s; (b) $T_s = 0.0025$ s; (c) $T_s = 0.0005$ s.

Obviously, the sample spacing of $T_s = 0.005$ in Fig. 2-9(a) is not sufficiently close to create an accurate plot when the sample points are connected by straight lines.[4] With a spacing of $T_s = 0.0025$ in Fig. 2-9(b), the plot starts to look more like a cosine, but it is still possible to see places where the points are connected by straight lines rather than the smooth cosine function. Only in the plot of Fig. 2-9(c), where the spacing is $T_s = 0.0005$, does the sampling spacing become so dense that our eye can easily visualize a curve that is a faithful representation of the cosine function. A precise answer to the question posed above would require a mathematical definition of accuracy; our subjective judgment would be too prone to variability among different observers. However, we learn from this example that as the sampling period decreases, more samples are taken during one period of the sinusoid. When $T_s = 0.005$, there are 5 samples per period; when $T_s = 0.0025$ there are 10 samples per period; and when $T_s = 0.0005$, there are 50 samples per period. It seems that 10 samples per period is not quite enough, and 50 samples per period is probably more than necessary, but, in general, the more samples per period, the smoother and more accurate is the linearly interpolated curve.

The choice of T_s also depends on the frequency of the cosine signal, because it is the number of samples per period that matters in plotting. For example, if the frequency of the cosine signal were 400 Hz instead of 40 Hz, then the period is ten times shorter and a sample spacing of $T_s = 0.0005$ would yield only 5 samples per period. The key to accurate reconstruction is to sample frequently enough so that the cosine signal does not change very much between sample points.

The problem of plotting a cosine signal from a set of discrete samples also depends on the interpolation method used. With MATLAB's built-in plotting function, linear interpolation is used to connect points by straight-line segments. This raises the question "If a more sophisticated interpolation method can be used, how *large* can the sample spacing be such that the cosine signal can be reconstructed accurately from the samples?" Surprisingly, the theoretical answer to this question is that the cosine signal can be reconstructed *exactly* from its samples if the sample spacing is less than half the period (i.e., the average number of samples per period need be only slightly more than two)! Linear interpolation certainly cannot achieve this result, but, in Chapter 4, where we examine the sampling process in more detail, we will explain how this remarkable result can be achieved. For now, our observation that smooth and accurate sinusoidal curves can be reconstructed from samples if the sampling period is "small enough" will be adequate for our purposes.

2-5 Complex Exponentials and Phasors

We have shown that cosine signals are useful mathematical representations for signals that arise in a practical setting, and that they are simple to define and interpret. However, it turns out that the analysis and manipulation of sinusoidal signals is often greatly simplified by

---

[4]The sample points are shown as dots in these three plots. This was achieved by adding a second plot using MATLAB's `hold` and `stem` functions.

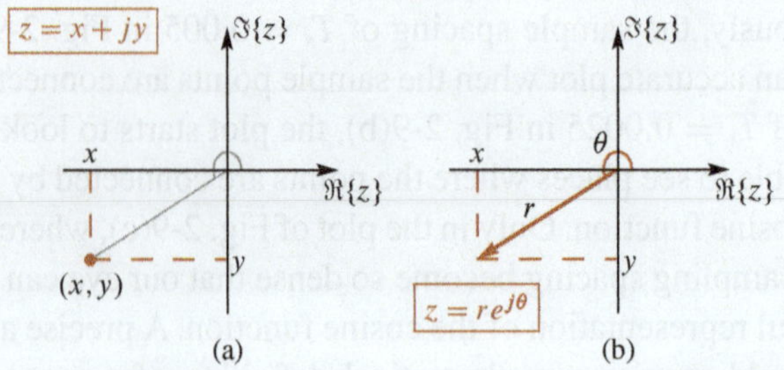

Figure 2-10 Representations of complex numbers in the complex plane. (a) Cartesian or rectangular form. (b) Polar form.

dealing with related signals called *complex exponential signals*. Although the introduction of the unfamiliar and seemingly artificial concept of complex exponential signals may at first seem to be making the problem more difficult, we will soon see the value in this new representation because we will avoid many tedious trigonometric manipulations that arise when using sinusoidal signals. Before introducing the complex exponential signal, we will first review some basic concepts concerning complex numbers.[5]

2-5.1 Review of Complex Numbers

A complex number z is an ordered pair of real numbers. Complex numbers may be represented by the notation $z = (x, y)$, where $x = \Re\{z\}$ is the ***real part*** and $y = \Im\{z\}$ is the ***imaginary part*** of z. Electrical engineers use the symbol j for $\sqrt{-1}$ instead of i, so we can also represent a complex number as $z = x + jy$. These two representations are called the ***Cartesian form*** of the complex number. Complex numbers are often represented as points in a ***complex plane***, where the real and imaginary parts are the horizontal and vertical coordinates, respectively, as shown in Fig. 2-10(a). With the Cartesian notation and the understanding that any number multiplied by j is included in the imaginary part, the operations of complex addition, complex subtraction, complex multiplication, and complex division can be defined in terms of real operations on the real and imaginary parts. For example, the sum of two complex numbers is defined as the complex number whose real part is the sum of the real parts and whose imaginary part is the sum of the imaginary parts.

Since complex numbers can be represented as points in a plane, it follows that complex numbers are analogous to vectors in a two-dimensional space. This leads to a useful geometric interpretation of a complex number as a vector, shown in Fig. 2-10(b). Vectors have length and direction, so another way to represent a complex number is the ***polar form*** in which the complex number is represented by r, its vector length, together with θ, its angle with respect to the positive real axis. The length of the vector is also called the ***magnitude*** of z (denoted $|z|$), and the angle is called the ***argument*** of z (denoted $\arg z$). This is indicated by the descriptive notation $z \leftrightarrow r\angle\theta$, which is interpreted to mean that the vector representing z has length r and makes an angle θ with the positive real axis.

[5]Appendix A provides a more detailed review of the fundamentals of complex numbers. Readers who know the basics of complex numbers and how to manipulate them can skim Appendix A for review and skip to Section 2-5.2.

It is important to be able to convert between the Cartesian and polar forms of complex numbers. Figure 2-10(b) shows a complex number z and the quantities involved in both the Cartesian and polar representations. Using this figure, as well as some simple trigonometry and the Pythagorean theorem, we can derive a method for computing the Cartesian coordinates (x, y) from the polar variables $r\angle\theta$:

$$x = r\cos\theta \qquad \text{and} \qquad y = r\sin\theta \tag{2.9}$$

and, likewise, for going from Cartesian to polar form

$$r = \sqrt{x^2 + y^2} \qquad \text{and} \qquad \theta = \arctan\left(\frac{y}{x}\right) \tag{2.10}$$

where the arctangent used must return angles in all four quadrants. Many calculators and computer programs have these two sets of equations built in, making the conversion between polar and Cartesian forms simple and convenient.

The $r\angle\theta$ notation is clumsy, and does not lend itself to ordinary algebraic rules. A much better polar form is given by using Euler's famous formula for the complex exponential

$$e^{j\theta} = \cos\theta + j\sin\theta \tag{2.11}$$

The Cartesian pair $(\cos\theta, \sin\theta)$ can represent any point on a circle of radius 1, so a slight generalization of (2.11) gives a representation valid for any complex number z

$$z = re^{j\theta} = r\cos\theta + jr\sin\theta \tag{2.12}$$

DEMO
zdrill

The complex exponential polar form of a complex number is most convenient when calculating a complex multiplication or division (see Appendix A for more details). It also serves as the basis for the complex exponential signal, which is introduced in the next section. A MATLAB GUI called *zdrill* is available (from the Companion Website) for building skills with basic complex number operations.

2-5.2 Complex Exponential Signals

The *complex exponential signal* is defined as

$$z(t) = A\,e^{j(\omega_0 t + \varphi)} \tag{2.13}$$

Observe that the signal $z(t)$ is a complex-valued function of t, but its magnitude is always constant (i.e., $|z(t)| = A$), and its angle is $\arg z(t) = (\omega_0 t + \varphi)$. Using Euler's formula (2.11), the complex exponential signal can be expressed in Cartesian form as

$$Ae^{j(\omega_0 t + \varphi)} = A\cos(\omega_0 t + \varphi) + jA\sin(\omega_0 t + \varphi) \tag{2.14}$$

As with the real sinusoid, A is the *amplitude*, and should be a positive real number; φ is the *phase*; and ω_0 is the *frequency* in rad/s. In (2.14) it is clear that the real part of the complex exponential signal is a real cosine signal as defined in (2.2), and that its imaginary part is a real sine signal. Figure 2-11 shows a plot of the following complex exponential signal:

$$\begin{aligned} z(t) &= 20\, e^{j(2\pi(40)t - 0.4\pi)} \\ &= 20\, e^{j(80\pi t - 0.4\pi)} \\ &= 20\cos(80\pi t - 0.4\pi) + j20\sin(80\pi t - 0.4\pi) \\ &= 20\cos(80\pi t - 0.4\pi) + j20\cos(80\pi t - 0.9\pi) \end{aligned}$$

Plotting a complex signal as a function of time requires two graphs, one for the real part and another for the imaginary part. Observe that the real and the imaginary parts of the complex exponential signal are both real sinusoidal signals, and only their phases differ by a phase shift of 0.5π rad.

The main reason that we are interested in the complex exponential signal is that it is an alternative *representation* for the real cosine signal. This is because we can always write

$$x(t) = \Re\left\{Ae^{j(\omega_0 t + \varphi)}\right\} = A\cos(\omega_0 t + \varphi) \tag{2.15}$$

In fact, the real part of the complex exponential signal shown in Fig. 2-11 is identical to the cosine signal plotted in Fig. 2-6. Although it may seem that we have complicated things by first introducing the imaginary part to obtain the complex exponential signal and then throwing it away by taking only the real part, we will see that many calculations are simplified by using properties of the complex exponential representation. It is possible, for example, to replace most trigonometric manipulations with algebraic operations on the exponents.

EXERCISE 2.6

Demonstrate that equating the real parts of both sides of the equation $e^{j(\alpha+\beta)} = e^{j\alpha}e^{j\beta}$ will lead to identity #5 in Table 2-2. Also, show that identity #4 is obtained from equating the imaginary parts.

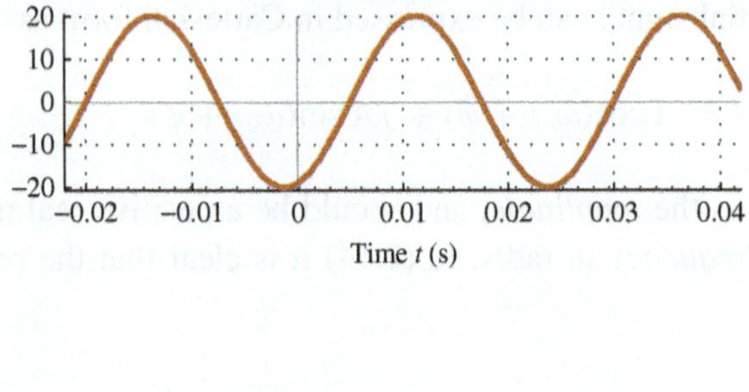

Figure 2-11 Real and imaginary parts of the complex exponential signal $z(t) = 20e^{j(2\pi(40)t - 0.4\pi)}$. The phase difference between the two sinusoids is 90° or $\pi/2$ rad.

2-5.3 The Rotating Phasor Interpretation

When two complex numbers are multiplied, it is best to use polar form for both numbers because we just multiply the magnitudes and add the angles. To justify this statement, consider $z_3 = z_1 z_2$, where $z_1 = r_1 e^{j\theta_1}$ and $z_2 = r_2 e^{j\theta_2}$. Then the product is

$$z_3 = r_1 e^{j\theta_1} r_2 e^{j\theta_2} = r_1 r_2 e^{j\theta_1} e^{j\theta_2} = \overbrace{r_1 r_2}^{r_3} e^{j\overbrace{(\theta_1 + \theta_2)}^{\theta_3}}$$

where we have used the law of exponents to combine the two complex exponentials. The vectors illustrated in Fig. 2-12 show that complex multiplication involves rotation and scaling.

This geometric view of complex multiplication leads to a useful interpretation of the complex exponential signal as a complex vector that rotates as time increases. If we use A and φ to define a complex number:

$$X = Ae^{j\varphi} \tag{2.16}$$

then the complex exponential (2.13) can be expressed as

$$z(t) = Xe^{j\omega_0 t} \tag{2.17}$$

(i.e., $z(t)$ is the product of the complex constant X and the complex-valued time function $e^{j\omega_0 t}$). The complex number X, which is aptly called the ***complex amplitude***, has a polar representation created from the amplitude and the phase of the complex exponential signal. Taken together, the complex amplitude $X = Ae^{j\varphi}$ and the frequency ω_0 are sufficient to represent $z(t)$, as well as the real cosine signal, $x(t) = A\cos(\omega_0 t + \varphi)$, using (2.15). The complex amplitude is also called a *phasor*. The *phasor* terminology is common in electrical circuit theory, where complex exponential signals are used to greatly simplify the analysis and design of circuits. Since it is a complex number, X can be represented graphically as a vector in the complex plane, where the vector's magnitude ($|X| = A$) is the amplitude, and its angle ($\angle X = \varphi$) is the phase of a complex exponential signal defined by (2.17). In the remainder of the text, the term *complex amplitude* will be used more often than *phasor*, but the two terms can be used interchangeably because they refer to the same quantity defined in (2.16).

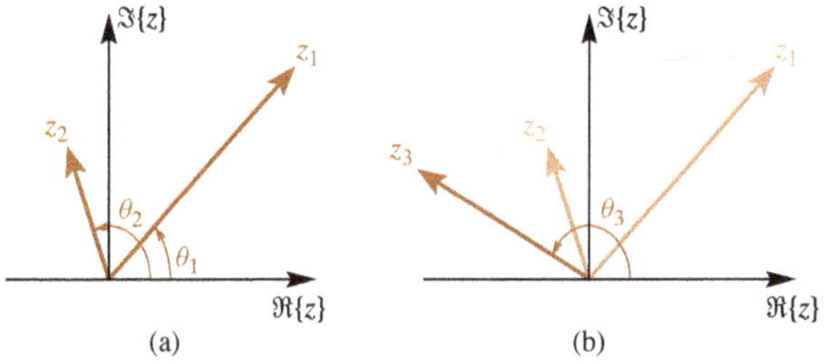

Figure 2-12 Geometric view of complex multiplication $z_3 = z_1 z_2$. The vector z_3 is a rotated and scaled version of z_2 because $\angle z_3 = \angle z_2 + \theta_1$, and the length of z_3 is r_1 times the length of z_2. Here, $r_1 = 1.72$, $r_2 = 0.77$, and $r_3 = 1.32$.

The complex exponential signal defined in (2.17) can also be written as

$$z(t) = Xe^{j\omega_0 t} = Ae^{j\varphi}e^{j\omega_0 t} = Ae^{j\theta(t)}$$

where $\theta(t)$ is a time-varying angle function

$$\theta(t) = \omega_0 t + \varphi \qquad \text{(radians)}$$

At a given instant in time, t, the value of the complex exponential signal, $z(t)$, is a complex number whose magnitude is A and whose angle is $\theta(t)$. Like any complex number, $z(t)$ can be represented as a vector in the complex plane. In this case, the tip of the vector always lies on the perimeter of a circle of radius A. Now, if t increases, the complex vector $z(t)$ will simply rotate at a constant rate, determined by the radian frequency ω_0. In other words, multiplying the phasor X by $e^{j\omega_0 t}$ as in (2.17) causes the phasor X to rotate by an angular amount of $\omega_0 t$. Since $|e^{j\omega_0 t}| = 1$, no scaling occurs. Thus, another name for the complex exponential signal is *rotating phasor*.

If the frequency ω_0 is positive, the direction of rotation is counterclockwise, because $\theta(t)$ will increase with increasing time. Similarly, when ω_0 is negative, the angle $\theta(t)$ changes in the negative direction as time increases, so the complex phasor rotates clockwise. Thus, rotating phasors are said to have *positive frequency* if they rotate counterclockwise, and *negative frequency* if they rotate clockwise.

A rotating phasor makes one complete revolution every time the angle $\theta(t)$ changes by 2π radians. The time it takes to make one revolution is also equal to the period, T_0, of the complex exponential signal, so

$$\omega_0 T_0 = (2\pi f_0)T_0 = 2\pi \qquad \Longrightarrow \qquad T_0 = \frac{1}{f_0}$$

Notice that the phase φ defines where the phasor is pointing when $t = 0$. For example, if $\varphi = \pi/2$, then the phasor is pointing straight up when $t = 0$, whereas if $\varphi = 0$, the phasor is pointing to the right when $t = 0$.

The plots in Fig. 2-13(a) illustrate the relationship between a single complex rotating phasor and the cosine signal waveform. The upper left plot shows the complex plane with two vectors. The vector at an angle in the third quadrant represents the signal

$$z(t) = e^{j(t-\pi/4)}$$

at the specific time $t = 1.5\pi$. The horizontal vector pointing to the left is equal to the real part of the vector $z(t)$ at the particular time $t = 1.5\pi$ which is

$$x(1.5\pi) = \Re\{z(1.5\pi)\} = \cos(1.5\pi - \pi/4) = -\frac{\sqrt{2}}{2}$$

As t increases, the rotating phasor $z(t)$ rotates in the counterclockwise direction, and its real part $x(t)$ oscillates left and right along the real axis. This is shown in the lower left plot, which shows how the real part of the rotating phasor has varied over one period (i.e., $0 \le t \le 2\pi$). We will return to discuss Fig. 2-13(b) in Sect. 2-5.4.

DEMO
Rotating Phasors

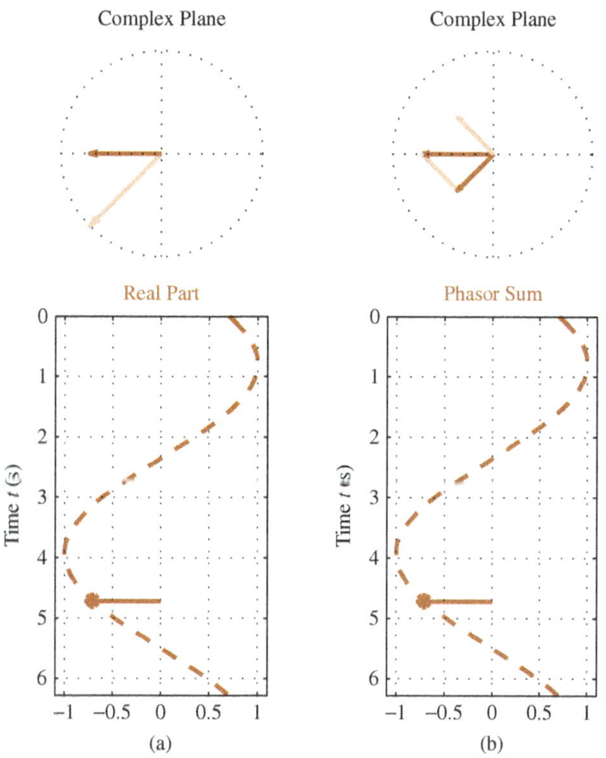

Figure 2-13 Rotating phasors: (a) single phasor rotating counterclockwise; (b) complex conjugate rotating phasors.

2-5.4 Inverse Euler Formulas

The inverse Euler formulas allow us to write the cosine function in terms of complex exponentials as

$$\cos\theta = \frac{e^{j\theta} + e^{-j\theta}}{2} \tag{2.18}$$

and also the sine function can be expressed as

$$\sin\theta = \frac{e^{j\theta} - e^{-j\theta}}{2j} \tag{2.19}$$

(See Appendix A for more details.)

Equation (2.18) can be used to express $\cos(\omega_0 t + \varphi)$ in terms of two complex exponentials as follows:

$$\begin{aligned} A\cos(\omega_0 t + \varphi) &= A\left(\frac{e^{j(\omega_0 t+\varphi)} + e^{-j(\omega_0 t+\varphi)}}{2}\right) \\ &= \tfrac{1}{2}Xe^{j\omega_0 t} + \tfrac{1}{2}X^*e^{-j\omega_0 t} \\ &= \tfrac{1}{2}z(t) + \tfrac{1}{2}z^*(t) \\ &= \Re\{z(t)\} \end{aligned}$$

where * denotes complex conjugation. This formula has an interesting interpretation. The real cosine signal with frequency ω_0 is actually composed of a positive frequency (ω_0) complex exponential and a negative frequency ($-\omega_0$) complex exponential. The complex amplitude of the positive frequency complex exponential signal is $\frac{1}{2}X = \frac{1}{2}Ae^{j\varphi}$, and the complex amplitude of the negative frequency complex exponential is $\frac{1}{2}X^* = \frac{1}{2}Ae^{-j\varphi}$. In other words, the real cosine signal can be represented as the sum of two complex rotating phasors that are rotating in opposite directions starting from phasors that are complex conjugates of each other.

DEMO
Rotating Phasors

Figure 2-13(b) illustrates how the sum of the two half-amplitude complex conjugate rotating phasors becomes the real cosine signal. In this case, the vector at an angle in the third quadrant (heavy-orange line) is the complex rotating phasor $\frac{1}{2}z(t)$ at time $t = 1.5\pi$. As t increases after that time, the angle would increase in the counterclockwise direction. Similarly, the vector in the second quadrant (plotted with a light-orange line) is the complex rotating phasor $\frac{1}{2}z^*(t)$ at time $t = 1.5\pi$. As t increases after that time, the angle of $\frac{1}{2}z^*(t)$ will increase in the clockwise direction. The horizontal vector pointing to the left is the sum of these two complex conjugate rotating phasors. The result is the same as the real vector in the plot on the left, and therefore the real cosine wave traced out as a function of time is the same in both cases. The lower right shows the variation of the real values of $\cos(t - \pi/4)$ for $0 \le t \le 2\pi$.

This representation of real sinusoidal signals in terms of their positive- and negative-frequency components is a remarkably useful concept. The negative frequencies, which arise due to the complex exponential representation, turn out to lead to many simplifications in the analysis of signals and systems problems. We will develop this representation further in Chapter 3, where we introduce the idea of the spectrum of a signal.

EXERCISE 2.7

Show that the following representation can be derived for the real sine signal:

$$A\sin(\omega_0 t + \varphi) = \tfrac{1}{2}Xe^{-j\pi/2}e^{j\omega_0 t} + \tfrac{1}{2}X^*e^{j\pi/2}e^{-j\omega_0 t}$$

where $X = Ae^{j\varphi}$. In other words, the sine signal is composed of two complex exponentials with the same positive and negative frequencies, but the complex coefficients multiplying the terms are different from those of the cosine signal. Specifically, the sine signal requires additional phases of $\mp\pi/2$ applied to the complex amplitude X and X^*, respectively.

2-6 Phasor Addition

There are many situations in which it is necessary to add two or more sinusoidal signals. When *all signals have the same frequency*, the addition simplifies. This addition problem arises in electrical circuit analysis, and it will arise again in Chapters 5 and 6, where we introduce the concepts of discrete-time filtering and frequency response. Thus it is useful to develop a short cut for adding several sinusoids having the same frequency, *but with different amplitudes and phases*. Our goal is to prove that the following statement is true:

$$\sum_{k=1}^{N} A_k \cos(\omega_0 t + \varphi_k) = A \cos(\omega_0 t + \varphi) \tag{2.20}$$

by showing how to get A and φ from the sum. Equation (2.20) states that a sum of N cosine signals of differing amplitudes and phases, but with the same frequency, can always be reduced to a single cosine signal of the same frequency. A proof of (2.20) could be accomplished by using trigonometric identities such as

$$A_k \cos(\omega_0 t + \varphi_k) = A_k \cos\varphi_k \cos(\omega_0 t) - A_k \sin\varphi_k \sin(\omega_0 t) \tag{2.21}$$

but this approach is exceedingly tedious to carry out. As we will see, a much simpler approach can be based on the complex exponential representation of the cosine signals.

2-6.1 Addition of Complex Numbers

When two complex numbers are added, it is necessary to use the Cartesian form. If $z_1 = x_1 + jy_1$ and $z_2 = x_2 + jy_2$, then $z_3 = z_1 + z_2 = (x_1 + x_2) + j(y_1 + y_2)$ (i.e., the real and imaginary parts of the sum are the sums of the real and imaginary parts, respectively). Using the vector interpretation of complex numbers, where both z_1 and z_2 are viewed as vectors with their tails at the origin, the sum z_3 is the result of vector addition, and is constructed as follows:

(a) Draw a copy of z_1 with its tail at the head of z_2. Call this displaced vector $\hat{z}_1$.

(b) Draw the vector from the origin to the head of $\hat{z}_1$. This is the sum z_3.

This process is depicted in Fig. 2-14 for the case $z_1 = 4 - j3$ and $z_2 = 2 + j5$.

2-6.2 Phasor Addition Rule

The phasor representation of cosine signals can be used to show that the following sum of sinusoids:

$$x(t) = \sum_{k=1}^{N} A_k \cos(\omega_0 t + \varphi_k) = A \cos(\omega_0 t + \varphi) \tag{2.22}$$

can be carried out by adding complex amplitudes to get A and φ:

$$\sum_{k=1}^{N} A_k e^{j\varphi_k} = Ae^{j\varphi} \tag{2.23}$$

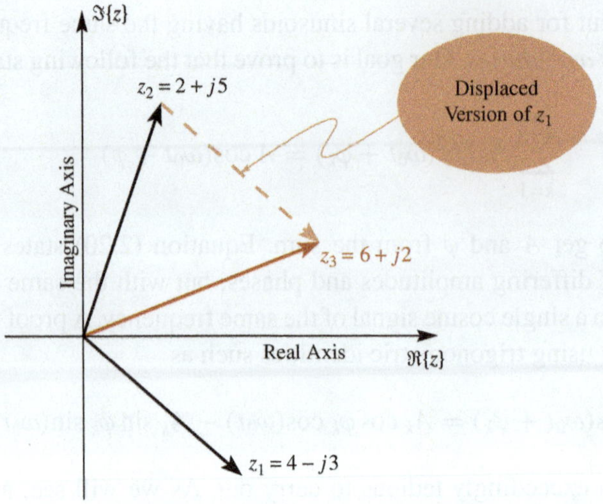

Figure 2-14 Graphical construction of complex number addition, $z_3 = z_1 + z_2$, shows that the process is the same as vector addition.

That is, the sum of two or more cosine signals each having the same frequency (2.22), but having different amplitudes and phases, can be expressed as a single equivalent cosine signal. The resulting amplitude (A) and phase (φ) of the term on the right-hand side of (2.22) can be computed from the individual amplitudes (A_k) and phases (φ_k) on the left-hand side by doing the complex addition in (2.23). This method is called the ***phasor addition rule***. Proof of the phasor addition rule requires the following two pieces of information:

(a) Any sinusoid can be written in the form:

$$A\cos(\omega_0 t + \varphi) = \Re\{Ae^{j(\omega_0 t + \varphi)}\} = \Re\{Ae^{j\varphi}\, e^{j\omega_0 t}\}$$

(b) For any set of complex numbers $\{X_k\}$, the sum of the real parts is equal to the real part of the sum, so we have

$$\Re\left\{\sum_{k=1}^{N} X_k\right\} = \sum_{k=1}^{N} \Re\{X_k\}$$

Proof of the phasor addition rule involves the following algebraic manipulations:

$$\begin{aligned}
\sum_{k=1}^{N} A_k\cos(\omega_0 t + \varphi_k) &= \sum_{k=1}^{N} \Re\left\{A_k e^{j(\omega_0 t + \varphi_k)}\right\} \\
&= \Re\left\{\sum_{k=1}^{N} A_k e^{j\varphi_k} e^{j\omega_0 t}\right\} \\
&= \Re\left\{\left(\sum_{k=1}^{N} A_k e^{j\varphi_k}\right) e^{j\omega_0 t}\right\} \\
&= \Re\left\{\left(Ae^{j\varphi}\right) e^{j\omega_0 t}\right\} = \Re\left\{Ae^{j(\omega_0 t + \varphi)}\right\} = A\cos(\omega_0 t + \varphi)
\end{aligned}$$

Two steps (shown in color) are important in this proof. In the third line, the complex exponential $e^{j\omega_0 t}$ is factored out of the summation because all the sinusoids have the same frequency. In going from the third line to the fourth, the crucial step is replacing the summation term in parentheses with a single complex number, $Ae^{j\varphi}$, as defined in (2.23), because we are adding N complex constants.

2-6.3 Phasor Addition Rule: Example

We now consider an example of adding two sinusoids, where

$$x_1(t) = 1.7\cos(20\pi t + 70\pi/180)$$
$$x_2(t) = 1.9\cos(20\pi t + 200\pi/180)$$

The frequency of both sinusoids is 10 Hz, so the period is $T_0 = 0.1$ s. The sum $x_1(t)+x_2(t)$ is done via phasor addition (2.23) of the complex amplitudes which requires four steps:

1. Represent $x_1(t)$ and $x_2(t)$ by the phasors:

$$X_1 = A_1 e^{j\varphi_1} = 1.7e^{j70\pi/180}$$
$$X_2 = A_2 e^{j\varphi_2} = 1.9e^{j200\pi/180}$$

2. Convert both phasors to rectangular form:

$$X_1 = 0.5814 + j1.5975$$
$$X_2 = -1.7854 - j0.6498$$

3. Add the real parts and the imaginary parts:

$$\begin{aligned} X_3 &= X_1 + X_2 \\ &= (0.5814 + j1.5975) + (-1.7854 - j0.6498) \\ &= -1.204 + j0.9477 \end{aligned}$$

4. Convert back to polar form, obtaining[6]

$$X_3 = 1.5322e^{j141.79\pi/180}$$

[6]With a modern scientific calculator, step 1 is data entry, and then a single button would be pushed to do steps 2–4.

The resultant phasor X_3 is converted back to a 10-Hz sinusoid, so the final formula for $x_3(t)$ is

$$x_3(t) = 1.5322\cos(20\pi t + 141.79\pi/180)$$
$$\text{or} \quad x_3(t) = 1.5322\cos(20\pi(t + 0.0394))$$

The waveforms of the three signals are shown Fig. 2-15(b) and the phasors used to solve the problem are shown on the left in Fig. 2-15(a). Notice that the times where the maximum of each cosine signal occurs can be derived from the phase through the formula

$$t_m = -\frac{\varphi T_0}{2\pi}$$

which gives

$$t_{m1} = -0.0194, \quad t_{m2} = -0.0556, \quad t_{m3} = -0.0394 \text{ s}$$

These times are marked with vertical dashed lines in the corresponding waveform plots in Fig. 2-15(b).

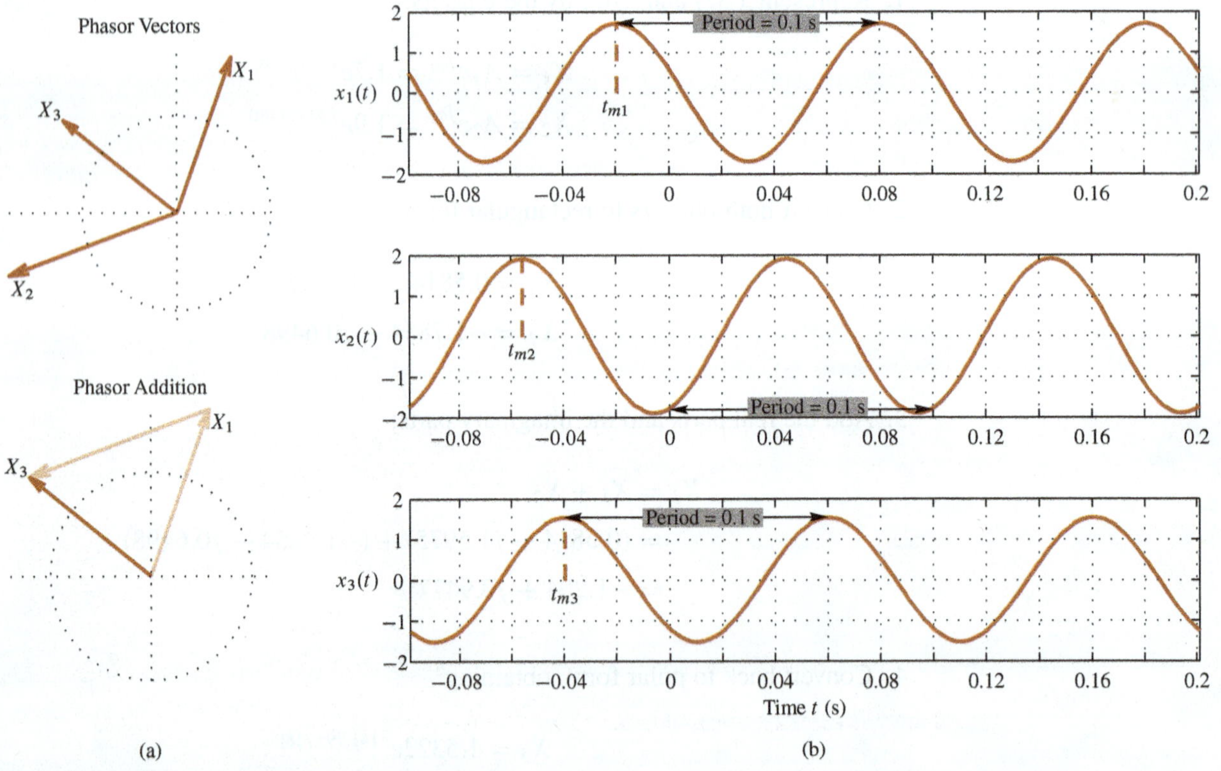

Figure 2-15 (a) Adding sinusoids by doing a phasor addition, which is actually a graphical vector sum. (b) The time of the signal maximum is marked on each $x_i(t)$ plot.

Table 2-3 Phasor addition example.

Z	= X	+ jY	Magnitude	Phase	Ph/pi	Ph(deg)
Z1	0.5814	1.5975	1.7	1.222	0.389	70.00
Z2	-1.7854	-0.6498	1.9	-2.793	-0.889	-160.00
Z3	-1.204	0.9477	1.5322	2.475	0.788	141.79

2-6.4 MATLAB Demo of Phasors

DEMO
ZDrill

LAB
DSP-First Toolbox

The process of phasor addition can be accomplished easily using MATLAB. The answer generated by MATLAB and printed with the special function `zprint` (provided on the website) for this particular phasor addition is given in Table 2-3. Help on `zprint` gives

```
ZPRINT   print out complex # in rect and polar form
  usage:    zprint(z)
   z = vector of complex numbers; each one will be printed
       in a format showing real, imag, mag and phase
```

The MATLAB code that generates the plots in Fig. 2-15 uses the special MATLAB functions (provided on the Companion Website) `zprint`, `zvect`, `zcat`, `ucplot`, and `zcoords` to make the vector plots.

2-6.5 Summary of the Phasor Addition Rule

In this section, we have shown how a real cosine signal can be represented as the real part of a complex exponential signal (complex rotating phasor), and we have applied this representation to show how to simplify the process of adding several cosine signals of the same frequency.

$$x(t) = \sum_{k=1}^{N} A_k \cos(\omega_0 t + \varphi_k) = A\cos(\omega_0 t + \varphi)$$

In summary, all we have to do to get the cosine signal representation of the sum is:

(a) Obtain the phasor representation $X_k = A_k e^{j\varphi_k}$ of each of the individual signals.

(b) Add the phasors of the individual signals to get $X = X_1 + X_2 + \cdots = Ae^{j\varphi}$. This requires polar-to-Cartesian-to-polar format conversions.

(c) Multiply X by $e^{j\omega_0 t}$ to get $z(t) = Ae^{j\varphi}e^{j\omega_0 t}$.

(d) Take the real part to get $x(t) = \Re\{Ae^{j\varphi}e^{j\omega_0 t}\} = A\cos(\omega_0 t + \varphi) = x_1(t) + x_2(t) + \cdots$.

In other words, A and φ must be calculated by doing a vector sum of all the X_k phasors.

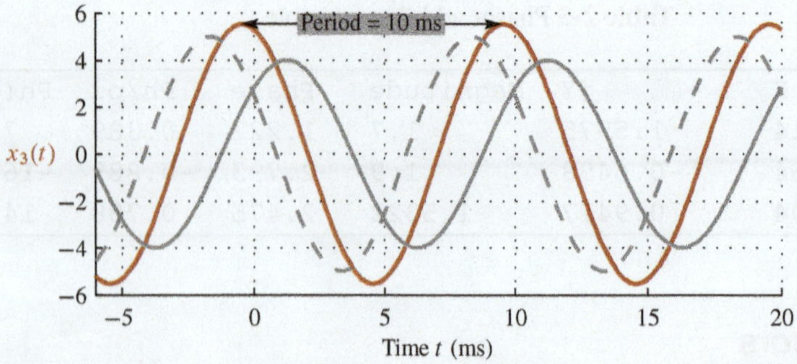

Figure 2-16 Two sinusoids and their sum, $x_3(t) = x_1(t) + x_2(t)$.

EXERCISE 2.8 Consider the two sinusoids,

$$x_1(t) = 5\cos(2\pi(100)t + \pi/3)$$

$$x_2(t) = 4\cos(2\pi(100)t - \pi/4)$$

Obtain the phasor representations of these two signals, add the phasors, plot the two phasors and their sum in the complex plane, and show that the sum of the two signals is

$$x_3(t) = 5.536\cos(2\pi(100)t + 0.2747)$$

In degrees the phase should be 15.74°. Examine the plots in Fig. 2-16 to see whether you can identify the cosine waves $x_1(t)$, $x_2(t)$, and $x_3(t) = x_1(t) + x_2(t)$.

2-7 Physics of the Tuning Fork

In Section 2-1, we described a simple experiment in which a tuning fork was seen to generate a signal whose waveform looked very much like that of a sinusoidal signal. Now that we know a lot more about sinusoidal signals, it is worthwhile to take up this issue again. Is it a coincidence that the tuning-fork signal looks like a sinusoid, or is there a deeper connection between vibrations and sinusoids? In this section, we present a simple analysis of the tuning-fork system that shows that the tuning fork does indeed vibrate sinusoidally when given a displacement from its equilibrium position. The sinusoidal motion of the tuning-fork tines is transferred to the surrounding air particles, thereby producing the acoustic signal that we hear. This simple example illustrates how mathematical models of physical systems that are derived from fundamental physical principles can lead to concise mathematical descriptions of physical phenomena and of signals that result.

2-7.1 Equations from Laws of Physics

A simplified drawing of the tuning fork and its motion is shown in Fig. 2-17. As we have seen experimentally, when struck against a firm surface, the tines of the tuning fork vibrate

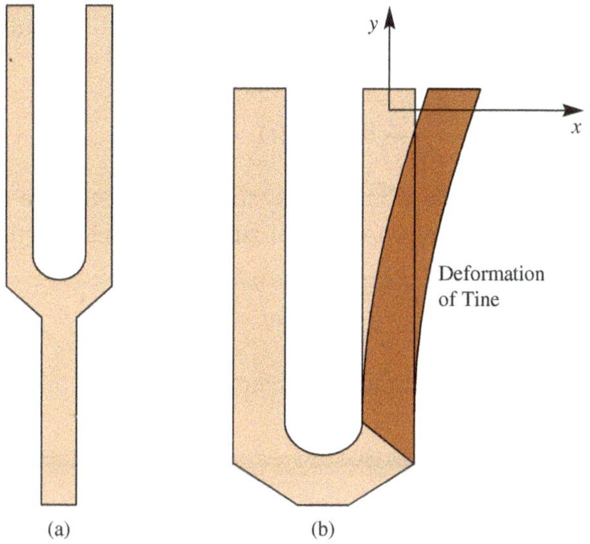

Figure 2-17 (a) Tuning fork. (b) The coordinate system needed to write equations for the vibration of the tine.

and produce a "pure" tone. We are interested in deriving the equations that describe the physical behavior of the tuning fork so that we can understand the basic mechanism by which the sound is produced.[7] Newton's second law, $F = ma$, will lead to a differential equation whose solution is a sine or cosine function, or a complex exponential.

When the tuning fork is struck, one of the tines is deformed slightly from its rest position, as depicted in Fig. 2-17(b). We know from experience that unless the deformation was so large as to break or bend the metal, there would be a tendency for the tine to return to its original rest position. The physical law that governs this movement is Hooke's law. Although the tuning fork is made of a very stiff metal, we can think of it as an elastic material when the deformation is tiny. Hooke's law states that the restoring force is directly proportional to the amount of deformation. If we set up a coordinate system as in Fig. 2-17(b), the deformation is along the x-axis, and we can write

$$F = -kx$$

where the parameter k is the elastic constant of the material (i.e., its stiffness). The minus sign indicates that this restoring force acts in the negative direction when the displacement of the tine is in the positive x direction (i.e., it acts to pull the tine back toward the neutral position).

Now this restoring force due to stiffness produces an acceleration as dictated by Newton's second law, which states

$$F = ma = m\,\frac{d^2x}{dt^2}$$

where m is the mass of the tine, and the second derivative with respect to time of position x is the acceleration of the mass along the x-axis. Since these two forces must balance

[7] The generation and propagation of sound are treated in any general college physics text.

each other (i.e., the sum of the forces is zero), we get a second-order differential equation that describes the motion $x(t)$ of the tine for all values of time t

$$m \frac{d^2x(t)}{dt^2} = -k\, x(t) \tag{2.24}$$

This particular differential equation is rather easy to solve, because we can, in fact, guess the solution. From the derivative properties of sine and cosine functions, we are motivated to try as a solution the function

$$x(t) = A \cos \omega_0 t$$

where the parameter ω_0 is a constant that must be determined. The second derivative of $x(t)$ is

$$\begin{aligned}\frac{d^2x(t)}{dt^2} &= \frac{d^2}{dt^2}(A \cos \omega_0 t)\\ &= \frac{d}{dt}(-\omega_0 A \sin \omega_0 t)\\ &= -\omega_0^2 A \cos \omega_0 t\end{aligned}$$

Notice that the second derivative of the cosine function is the same cosine function multiplied by a constant $(-\omega_0^2)$. Therefore, when we substitute $x(t)$ into (2.24), we get

$$\begin{aligned} m \frac{d^2x(t)}{dt^2} &= -k\, x(t)\\ -m\, \omega_0^2 A \cos \omega_0 t &= -k\, A \cos \omega_0 t\end{aligned}$$

Since this equation must be satisfied for all t, it follows that the coefficients of $\cos \omega_0 t$ on both sides of the equation must be equal, which leads to the following algebraic equation

$$-m\, \omega_0^2 = -k$$

This equation can be solved for ω_0, obtaining

$$\omega_0 = \pm\sqrt{\frac{k}{m}} \tag{2.25}$$

Therefore, we conclude that one solution of the differential equation is

$$x(t) = A \cos\left(\sqrt{\frac{k}{m}}\, t\right)$$

From our model, $x(t)$ describes the motion of the tuning-fork tine. Therefore, we conclude that the tines oscillate sinusoidally. This motion is, in turn, transferred to the particles of air in the locality of the tines, thereby producing the tiny sinusoidal variations in air pressure that make up an acoustic wave. The formula for the frequency lets us draw two conclusions:

(a) Of two tuning forks having the same mass, the stiffer one will produce a higher frequency. This is because the frequency is proportional to $\sqrt{k}$, which is in the numerator of (2.25).

DEMO
Tuning Fork

(b) Of two tuning forks having the same stiffness, the heavier one will produce a lower frequency. This is because the frequency is inversely proportional to the square root of the mass, $\sqrt{m}$, which is in the denominator of (2.25).

2-7.2 General Solution to the Differential Equation

There are many possible solutions to the tuning-fork differential equation (2.24). We can prove that the following function

$$x(t) = A\cos(\omega_0 t + \varphi)$$

will satisfy the differential equation (2.24) by substituting back into (2.24), and taking derivatives. Once again the frequency must be $\omega_0 = \sqrt{k/m}$. Only the frequency ω_0 is constrained by our simple model; the specific values of the parameters A and φ are not important. From this we can conclude that any scaled or time-shifted sinusoid with the correct frequency will satisfy the differential equation that describes the motion of the tuning fork's tine. This implies that an infinite number of different sinusoidal waveforms can be produced in the tuning-fork experiment. For any particular experiment, A and φ would be determined by the exact strength and timing of the sharp force that gave the tine its initial displacement. However, the frequency of all these sinusoids will be determined only by the mass and stiffness of the tuning-fork metal.

EXERCISE 2.9

Demonstrate that a complex exponential signal can also be a solution to the tuning-fork differential equation:

$$\frac{d^2x}{dt^2} = -\frac{k}{m}x(t)$$

By substituting $z(t)$ and $z^*(t)$ into both sides of the differential equation, show that the equation is satisfied for all t by both of the signals

$$z(t) = Xe^{j\omega_0 t} \quad \text{and} \quad z^*(t) = X^*e^{-j\omega_0 t}$$

Determine the value of ω_0 for which the differential equation is satisfied.

The tuning fork and its oscillatory behavior is just one example of *simple harmonic motion* where a sinusoid is an excellent description of motion in a physical system. The same mathematics comes up in the description of a mass on a spring, a simple pendulum, and even molecular vibration and many other physical phenomena.

2-7.3 Listening to Tones

The observer is an important part of any physical experiment. This is particularly true when the experiment involves listening to the sound produced. In the tuning-fork experiment, we perceive a tone with a certain pitch (related to the frequency) and loudness (related to the amplitude). The human ear and neural processing system respond to the frequency and amplitude of a sustained sound like that produced by the tuning fork, but the phase is not perceptible. This is because phase depends on an arbitrary definition of the starting time of the sinusoid, for example, a sustained tone sounds the same now as it did 5 minutes ago. On the other hand, we could pick up the sound with a microphone and sample or display the signal on an oscilloscope. In this case, it would be possible to make precise measurements of frequency and amplitude, but phase would be measured accurately only with respect to the time base of the sampler or the oscilloscope.

2-8 Time Signals: More Than Formulas

The purpose of this chapter has been to introduce the *concept of a sinusoid* and to illustrate how sinusoidal signals can arise in real situations. Signals, as we have defined them, are varying patterns that convey or represent information, usually about the state or behavior of a physical system. We have seen by both theory and observation that a tuning fork generates a signal that can be represented mathematically as a sinusoid. In the context of the tuning fork, the sinusoidal wave conveys and represents information about the state of the tuning fork. Encoded in the sinusoidal waveform is information such as whether the tuning fork is vibrating or at rest, and, if it is vibrating, the frequency and the amplitude of its vibrations. This information can be extracted from the signal by human listeners, or it can be recorded for later processing by either humans or computers. Another simple physical system that also generates nearly sinusoidal waveforms is shown in the ***Clay Whistle*** demo on the web site.

DEMO
Clay Whistle

Although the solution to the differential equation of the tuning fork (2.24) is a sinusoidal function, the resulting mathematical formula is simply a model that results from an idealization of the tuning fork. It is important to recall that the sound signal is a separate entity from the formula. The actual waveform produced by a tuning fork is probably not exactly sinusoidal. The signal is *represented* by the mathematical formula $x(t) = A\cos(\omega_0 t + \varphi)$, which can be derived from an idealized model based on physical principles. This model is a good approximation of reality, but an approximation nevertheless. Even so, this model is extremely useful since it leads directly to a useful mathematical representation of the signal produced by the tuning fork.

In the case of complicated signals generated by several musical instruments, the signal cannot be so easily reduced to a mathematical formula. Figure 2-18 shows a short segment of a recording of orchestra music. Just the appearance of the waveform suggests a much more complex situation. Although it oscillates like a sinusoid, it clearly is not periodic (at least, as far as we can see from the given segment). Orchestra music consists of many instruments sounding different notes in unison. If each instrument

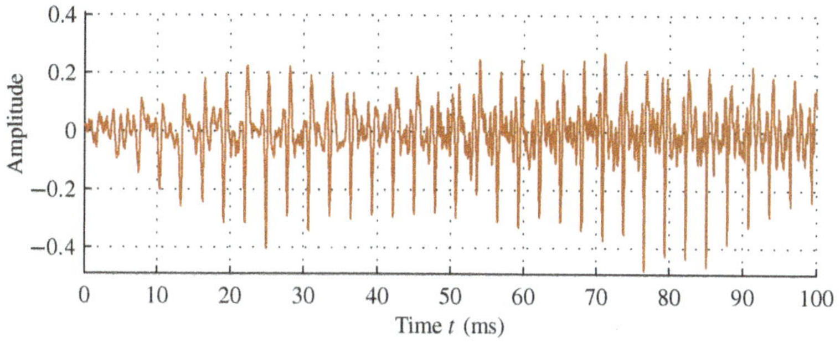

Figure 2-18 A short segment of an orchestra music signal.

produced a pure sinusoidal tone at the frequency of the note that is assigned to it, then the composite orchestra signal would be simply a sum of sinusoids with different frequencies, amplitudes, and phases. While this is far from being a correct model for most instruments, it is actually a highly appropriate way to think about the orchestra signal. In fact, we will see very soon that sums of sinusoids of different frequencies, amplitudes, and phases can result in an infinite variety of waveforms. Indeed, it is true that almost any signal can be represented as a sum of sinusoidal signals. When this concept was first introduced by Jean-Baptiste Joseph Fourier in 1807, it was received with great skepticism by the famous mathematicians of the world. Nowadays, this notion is commonplace (although no less remarkable). The mathematical and computational techniques of Fourier analysis underlie the frequency-domain analysis tools used extensively in electrical engineering and other areas of science and engineering.

2-9 Summary and Links

We have introduced sinusoidal signals in this chapter. We have attempted to show that they arise naturally as a result of simple physical processes and that they can be represented by familiar mathematical functions, and also by complex exponentials. The value of the mathematical representation of a signal is twofold. First, the mathematical representation provides a convenient formula to consistently describe the signal. For example, the cosine signal is completely described in terms of just three parameters. Second, by representing both signals and systems through mathematical expressions, we can make precise statements about the interaction between signals and systems.

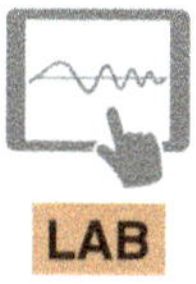

Introduction to MATLAB

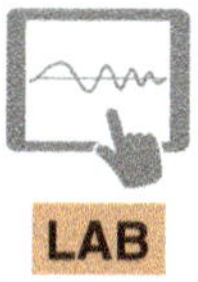

Complex Exponentials

In connection with this chapter, two laboratories are found on the Companion Website. The first lab involves some introductory exercises on the basic elements of the MATLAB programming environment, and its use for manipulating complex numbers and plotting sinusoids. Appendix B is also available for a quick overview of essential ideas about MATLAB. Labs on Complex Exponentials deal with sinusoids and phasor addition. In these labs, students must re-create a phasor addition demonstration similar to Fig. 2-15. Copies of the lab write-ups are also found on the website.

DEMO
Tuning Fork and Clay Whistle

On the Companion Website one can also find the following resources:

(a) A tuning-fork movie that shows the experiment of striking the tuning fork and recording its sound. Several recorded sounds from different tuning forks are available, as well as sounds from clay whistles.

DEMO
Sine Drill

(b) The drill program called `sindrill` in MATLAB can be used to practice with the amplitude, phase, and frequency of sinusoidal plots.

(c) A set of movies showing rotating phasor(s) and how they generate sinusoids through the real part operator.

2-10 Problems

NOTE
Problems with Solutions

The problems in this section address the major concepts of this chapter. On the Companion Website there are many related homework problems which can be used to review and practice. Most of them have detailed solutions.

P-2.1 Define $x(t)$ as

$$x(t) = 3\cos(0.2\pi t - 0.25\pi)$$

Make a plot of $x(t)$ over the range $-10 \le t \le 20$.

P-2.2 Figure P-2.2 is a plot of a sinusoidal wave. From the plot, determine values for the amplitude (A), phase (ϕ), and frequency (ω_0) needed in the representation:

$$x(t) = A\cos(\omega_0 t + \phi)$$

Give the answer as numerical values, *including the units* where applicable.

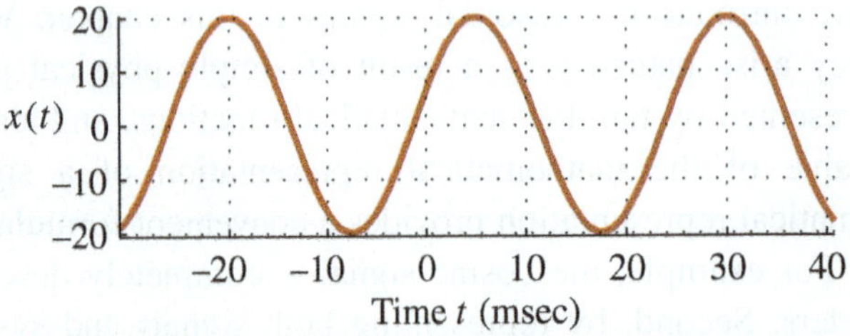

Figure P-2.2

P-2.3 Make a carefully labeled sketch for each of the following functions:

(a) Sketch $\cos\theta$ for values of θ in the range $0 \le \theta \le 6\pi$.

(b) Sketch $\cos(0.2\pi t)$ for values of t such that three periods of the function are shown.

(c) Sketch $\cos(2\pi t/T_0 + \pi/2)$ for values of t such that three periods of the function are shown. Label the horizontal axis in terms of the parameter T_0.

P-2.4 Use the series expansions for e^x, $\cos(\theta)$, and $\sin(\theta)$ given here to verify Euler's formula.

$$e^x = 1 + x + \frac{x^2}{2!} + \frac{x^3}{3!} + \cdots$$

$$\cos(\theta) = 1 - \frac{\theta^2}{2!} + \frac{\theta^4}{4!} + \cdots$$

$$\sin(\theta) = \theta - \frac{\theta^3}{3!} + \frac{\theta^5}{5!} + \cdots$$

P-2.5 Use complex exponentials (i.e., phasors) to show the following trigonometric identities:

(a) $\cos(\theta_1 + \theta_2) = \cos(\theta_1)\cos(\theta_2) - \sin(\theta_1)\sin(\theta_2)$

(b) $\cos(\theta_1 - \theta_2) = \cos(\theta_1)\cos(\theta_2) + \sin(\theta_1)\sin(\theta_2)$

Hint: Write the left-hand side of each equation as the real part of a complex exponential.

P-2.6 Use Euler's formula for the complex exponential to prove DeMoivre's formula

$$(\cos\theta + j\sin\theta)^n = \cos n\theta + j\sin n\theta$$

Use it to evaluate $\left(\frac{3}{5} + j\frac{4}{5}\right)^{100}$.

P-2.7 Simplify the following expressions:

(a) $3e^{j\pi/3} + 4e^{-j\pi/6}$

(b) $\left(\sqrt{3} - j3\right)^{10}$

(c) $\left(\sqrt{3} - j3\right)^{-1}$

(d) $\left(\sqrt{3} - j3\right)^{1/3}$

(e) $\Re\left\{je^{-j\pi/3}\right\}$

Give the answers in *both* Cartesian form $(x + jy)$ and polar form $(re^{j\theta})$.

P-2.8 Suppose that MATLAB is used to plot a sinusoidal signal. The following MATLAB code generates the signal and makes the plot. Derive a formula for the signal; then draw a sketch of the plot that will be done by MATLAB.

```
dt = 1/100;
tt = -1 : dt : 1;
Fo = 2;
zz = 300*exp(j*(2*pi*Fo*(tt - 0.75)));
xx = real( zz );
%
plot( tt, xx ), grid on
title( 'SECTION of a SINUSOID' )
xlabel('TIME   (sec)')
```

P-2.9 For the sinusoidal waveform shown in Fig. P-2.9, determine the complex phasor representation

$$X = A e^{j\varphi}$$

by finding values of the parameters ω_0, φ, and A such that the waveform is represented by

$$x(t) = \Re\{X e^{j\omega_0 t}\}$$

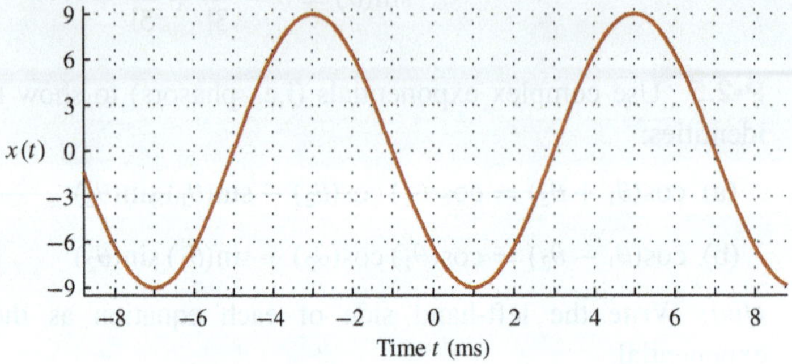

Figure P-2.9

P-2.10 Define $x(t)$ as

$$x(t) = 2\sin(\omega_0 t + \pi/4) + \cos(\omega_0 t)$$

(a) Express $x(t)$ in the form $x(t) = A\cos(\omega_0 t + \phi)$.

(b) Find a complex-valued signal $z(t)$ such that $x(t) = \Re e\{z(t)\}$.

P-2.11 Define $x(t)$ as

$$x(t) = 5\cos(\omega t) + 5\cos(\omega t + 120°) + 5\cos(\omega t - 120°)$$

Simplify $x(t)$ into the standard sinusoidal form: $x(t) = A\cos(\omega t + \phi)$. Use phasors to do the algebra, but also provide a plot of the vectors representing each of the three phasors.

P-2.12 Solve the following equation for θ:

$$\Re e\{(1 + j)e^{j\theta}\} = -1$$

Give the answers in radians. Make sure that you find *all* possible answers.

P-2.13 Define the following complex exponential signal:

$$(t) = 5e^{j\pi/3}e^{j10\pi t}$$

DEMO
Rotating Phasors

(a) Make a plot of $r(t) = \Re\{js(t)\}$. Pick a range of values for t that will include exactly three periods of the signal.

(b) Make a plot of $q(t) = \Re\{\frac{d}{dt}s(t)\}$, showing plot three cycles of the signal.

P-2.14 Two sinusoidal signals are defined as

$$x_1(t) = \sqrt{5}\cos(7t - \pi/3) \qquad x_2(t) = \sqrt{5}\cos(7t + \pi)$$

and their sum is denoted by $x(t) = x_1(t) + x_2(t)$.

(a) Find a complex-valued signal $z_1(t)$ such that $x_1(t) = \Re\{z_1(t)\}$.

(b) Find a complex-valued signal $z_2(t)$ such that $x_2(t) = \Re\{z_2(t)\}$.

(c) Find a complex-valued signal $z(t)$ such that $x(t) = \Re\{z(t)\}$.

P-2.15 Define $x(t)$ as

$$x(t) = 5\cos(\omega t + 3) + 7\cos(\omega t - 5) + 3\cos(\omega t)$$

where the phases have units of radians. Express $x(t)$ in the form $x(t) = A\cos(\omega t + \phi)$. Use complex phasor manipulations to obtain the answer. Explain your answer by giving a phasor diagram.

P-2.16 The phase of a sinusoid can be related to time shift as follows:

$$x(t) = A\cos(2\pi f_0 t + \varphi) = A\cos(2\pi f_0(t - t_1))$$

In the following parts, assume that the period of the sinusoidal wave is $T_0 = 8$ s.

(a) "When $t_1 = -2$ s, the value of the phase is $\varphi = \pi/2$." Explain whether this is True or False.

(b) "When $t_1 = 3$ s, the value of the phase is $\varphi = 3\pi/4$." Explain whether this is True or False.

(c) "When $t_1 = 7$ s, the value of the phase is $\varphi = \pi/4$." Explain whether this is True or False.

P-2.17 Define a signal as $x(t) = \sum_{k=0}^{4} \cos(\omega t + \frac{2}{5}\pi k)$

(a) Convert the sum to its phasor representation and then plot the phasors as vectors in the complex plane. Notice that vectors are distributed in a fashion similar to the roots of unity (see Appendix A), but one is missing.

(b) Express $x(t)$ in the form $x(t) = A\cos(\omega t + \varphi)$ by finding the numerical values of A and φ.

P-2.18 Determine the values of the parameters $\{\omega, \varphi, A\}$ in the equations below.

(a) $9\cos(\omega t + \varphi) = Ae^{(j8t+j\varphi)} + Ae^{(-j8t+j2\pi/3)}$

(b) $10\cos(9t - \pi/3) = A\cos(\omega t - \pi/2) + 5\cos(\omega t + \varphi)$

P-2.19 Sinusoidal equations can be solved as algebraic equations if the sinusoids are converted to phasors. For example, given the equation

$$9\cos(2t + \tfrac{1}{2}) = 3A\cos(2(t-1) + \varphi) + 4\cos(2t)$$

solve for A and φ. Here are the steps:

(a) Determine the phasor representation for each term in the above equation.

(b) Let $z = Ae^{j\varphi}$, and rewrite the phasor equation in part (a) as

$$9e^{j0.5} = (3e^{-j2})z + 4$$

(c) Use algebra to solve for z, which will be complex.

(d) Finally, express the answer z in *polar form* in order to determine the values for A and φ.

P-2.20 Simultaneous sinusoidal equations can be solved as simultaneous algebraic equations once the sinusoids are converted to phasors. Consider the following two equations with unknowns A_1, φ_1, A_2, and φ_2:

$$\cos(7t) = A_1\cos(7t + \varphi_1) + A_2\cos(7t + \varphi_2)$$
$$\sin(7t) = 2A_1\cos(7t + \varphi_1) + A_2\cos(7t + \varphi_2)$$

(a) Determine the phasor representation for each term in the above equations.

(b) Let $z_1 = A_1e^{j\varphi_1}$ and $z_2 = A_2e^{j\varphi_2}$, and then rewrite the two phasor equations in part (a) as

$$1 = z_1 + z_2$$
$$-j = 2z_1 + z_2$$

(c) Use algebra to solve for z_1 and z_2, which will be complex numbers.

(d) Finally, express the answers z_1 and z_2 in *polar form* in order to determine the values for A_1, φ_1, A_2, and φ_2.

P-2.21 Solve the following two simultaneous sinusoidal equations with unknowns A_1, φ_1, A_2, and φ_2:

$$\cos(3(t-1)) = 4A_1\sin(3t + \varphi_1) + A_2\cos(3t + \varphi_2)$$
$$\sin(3t + 2) = 3A_1\sin(3t + \varphi_1) + A_2\cos(3t + \varphi_2)$$

(a) Convert to phasors, z_1 and z_2, and determine the answers for A_1, φ_1, A_2, and φ_2 from the polar forms for the phasors z_1 and z_2.

(b) Provide two geometrical diagrams to explain the answer by plotting the vectors $4z_1$, z_2, and $4z_1 + z_2$, as well as the vectors $3z_1$, z_2, as $3z_1 + z_2$.

P-2.22 Solve the following equation for M and ψ. Obtain *all* possible answers. Use the phasor method, and provide a geometrical diagram to explain the answer.

$$5\cos(\omega_0 t) = M\cos(\omega_0 t - \pi/6) + 5\cos(\omega_0 t + \psi)$$

Hint: Describe the figure in the z-plane given by the set $\{z : z = 5e^{j\psi} - 5\}$ where $0 \le \psi \le 2\pi$.

P-2.23 Complex exponentials obey a simple rule for the operation of time shifting: the complex amplitude experiences a phase change. Consider the complex signal $z(t) = Ze^{j10\pi t}$ where $Z = 7e^{j0.3\pi}$.

(a) The time-shifted version of $z(t)$ can be represented as a new complex exponential with complex amplitude W. Thus, $z(t - t_d) = w(t) = We^{j10\pi t}$. Determine the complex amplitude W when $t_d = 0.24$ s.

(b) Suppose that $y(t)$ is a different time-shifted version of $z(t)$, and its complex amplitude is $Y = 7e^{-j0.1\pi}$. Determine the smallest positive value for the time shift t_d such that $y(t) = z(t - t_d)$.

P-2.24 Let $x[n]$ be a sequence of complex numbers defined as

$$x[n] = 7e^{j(0.22\pi n - 0.25\pi)}$$

for integers n such that $-\infty < n < \infty$. If we define a new sequence $y[n]$ as follows

$$y[n] = x[n+1] - 2x[n] + x[n-1] \quad \text{for all } n$$

it is possible to express $y[n]$ in the form $y[n] = Ae^{j\varphi}e^{j\hat{\omega}_0 n}$.

(a) Determine the frequency $\hat{\omega}_0$ in the expression for $y[n]$.

(b) Determine the values of A and φ.

P-2.25 Complex exponentials obey the expected rules of calculus and algebra when doing operations such as integrals and derivatives. Consider the complex signal $z(t) = Ze^{j2\pi t}$ where $Z = e^{j\pi/4}$.

(a) Show that the first derivative of $z(t)$ with respect to time can be represented as a new complex exponential $Qe^{j2\pi t}$, such that $\frac{d}{dt}z(t) = Qe^{j2\pi t}$. Determine the value for the complex amplitude Q.

(b) Plot both Z and Q in the complex plane. How much greater (or smaller) is the angle of Q than the angle of Z?

(c) Compare $\Re\{\frac{d}{dt}z(t)\}$ to $\frac{d}{dt}\Re\{z(t)\}$ for the given signal $z(t)$. Do you think that this would be true for any complex exponential signal?

(d) Evaluate the definite integral of $z(t)$ over the range $-0.5 \le t \le 0.5$:

$$\int_{-0.5}^{0.5} z(t)dt = ?$$

Express your answer in polar form. Note that integrating a complex quantity follows the expected rules of algebra: You could integrate the real and imaginary parts separately, but it is much easier to *use the integration formula for an exponential* directly on $z(t)$.

P-2.26 Determine two different frequencies for the complex-valued solutions $x(t) = Xe^{j\omega t}$ of the following differential equation:

$$\frac{d^2x(t)}{dt^2} = -289\,x(t)$$

P-2.27 As noted in Section 2-7, the tuning fork is one example of simple harmonic motion, because the vibration is governed by a second-order differential equation (2.24). There are many other physical phenomena that can be modeled by simple harmonic motion, and one important case is the electric circuit comprised of an inductor and a capacitor as shown in Fig. P-2.27. To derive the differential equation, it will be necessary to apply a bit of electric circuit theory. Observe that the inductor and capacitor are connected in what is called a parallel connection with the same voltage $v(t)$ across both elements. Also, note that the current $i(t)$ is (arbitrarily) defined to flow in the clockwise direction around the loop going through the terminals of the two elements. Thus, for the definitions of $v(t)$ and $i(t)$ in Fig. P-2.27, the capacitor (C) and inductor (L) are characterized by the following relations:

$$i(t) = C\frac{dv}{dt} \quad \text{and} \quad v(t) = -L\frac{di}{dt}$$

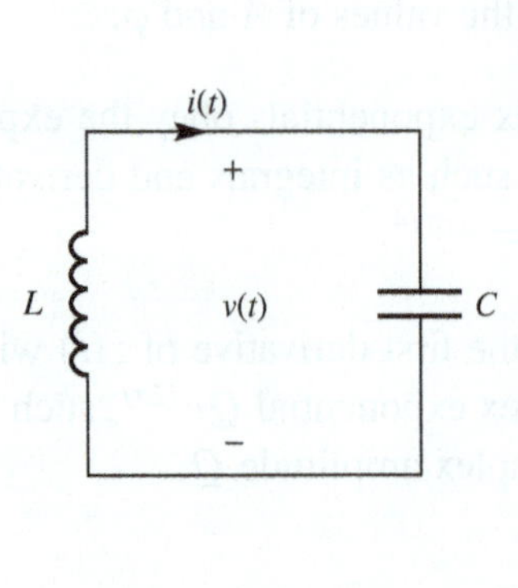

Figure P-2.27

The minus sign is needed for the inductor equation because the direction of the assumed current flow is from negative to positive voltage through L.

(a) Use the fact that the voltage $v(t)$ is the same across both elements as shown in Fig. P-2.27 to combine the two equations above and show that the voltage satisfies the differential equation

$$\frac{d^2 v(t)}{dt^2} = -\frac{1}{LC} v(t) \tag{2.26}$$

(b) Compare (2.26) to the differential equation (2.24) for the tuning fork. Since the form is the same, we should expect the voltage in the circuit of Fig. P-2.27 to display simple harmonic motion behavior. After comparing terms with (2.25), determine the frequency of the harmonic oscillation of the voltage $v(t)$.

(c) By substitution, show that $v(t) = A\cos(\omega_0 t + \varphi)$ is a solution to (2.26), and in the process, determine ω_0 in terms of the element values L and C. In general, the coefficients A and φ can be chosen to match initial conditions in the circuit.

(d) Suppose that $A = 5$ and $\varphi = \pi/3$ for the voltage $v(t)$ in the previous part. Use the capacitor relation above to obtain the current $i(t)$ from $v(t)$, and then determine the phase difference between $v(t)$ and $i(t)$.

(e) For the current obtained in the previous part, show that $i(t)$ satisfies the same harmonic differential equation as $v(t)$, that is,

$$\frac{d^2 i(t)}{dt^2} = -\frac{1}{LC} i(t) \tag{2.27}$$

P-2.28 In a mobile radio system (e.g., cell phones), there is one type of degradation that can be modeled easily with sinusoids. This is the case of *multipath fading* caused by reflections of the radio waves interfering destructively at some locations. Suppose that a transmitting tower sends a sinusoidal signal, and a mobile user receives not one but two copies of the transmitted signal: a direct-path transmission and a reflected-path signal (e.g., from a large building) as depicted in Fig. P-2.28.

The received signal is the sum of the two copies, and since they travel different distances they have different time delays. If the transmitted signal is $s(t)$, then the received signal[8] is

$$r(t) = s(t - t_1) + s(t - t_2)$$

[8]For simplicity we are ignoring propagation losses: When a radio signal propagates over a distance R, its amplitude will be reduced by an amount that is proportional to $1/R^2$.

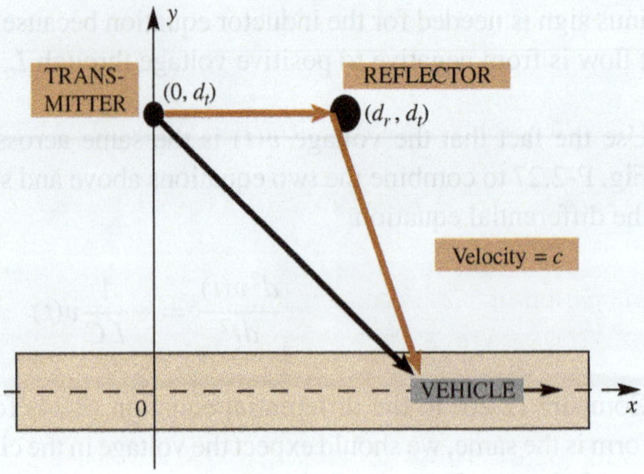

Figure P-2.28

In a mobile phone scenario, the distance between the mobile user and the cell tower is always changing. If $d_t = 1000$ m, then the direct-path distance is

$$d_1 = \sqrt{x^2 + 10^6} \quad \text{(meters)}$$

where x is the position of a mobile user who is moving along the x-axis. Assume that the reflector is at $d_r = 55$ m, so the reflected-path distance is

$$d_2 = \sqrt{(x - 55)^2 + 10^6} + 55 \quad \text{(meters)}$$

(a) The amount of the delay (in seconds) can be computed for both propagation paths, using the fact that the time delay is the distance divided by the speed of light (3×10^8 m/sec). Determine t_1 and t_2 as a function of the mobile's position (x).

(b) Assume that the transmitted signal is

$$s(t) = \cos(300 \times 10^6 \pi t)$$

Determine the received signal when $x = 0$. Prove that the received signal is a sinusoid and find its amplitude, phase, and frequency when $x = 0$.

(c) The amplitude of the received signal is a measure of its strength. Show that as the mobile user moves, it is possible to find positions where the signal strength is zero. Find one such location.

(d) If you have access to MATLAB, write a script that will plot signal strength versus position x, thus demonstrating that there are numerous locations where no signal is received. Use x in the range $-100 \le x \le 100$.

CHAPTER

3

Spectrum Representation

This chapter introduces the concept of the ***spectrum*** of a signal—a compact representation of the frequency content of a signal that can be represented as a sum of sinusoids. In Chapter 2, we learned about the properties of sinusoidal waveforms that can be represented by

$$x(t) = A\cos(2\pi f_0 t + \varphi) = \Re\left\{Xe^{j2\pi f_0 t}\right\} \tag{3.1}$$

Note that because we know the functional form, the waveform $x(t)$ in (3.1) is defined for all t by only three numbers: the amplitude, A, the frequency, f_0, and the phase, φ. In Chapter 2 we defined the complex amplitude $X = Ae^{j\varphi}$ and called it a phasor. The sinusoidal waveform of (3.1) is a good mathematical model for the voltages and currents found in electrical power networks. In the study of electric circuits, phasors are useful since they can simplify the addition of sinusoids having the *same* frequency. However, the phasor concept can be generalized and provides the basis for representing many other types of signals besides the electrical voltage at the wall plug.

In this chapter, we will show how more complicated waveforms can be represented as sums of sinusoids having *different* frequencies and corresponding different amplitudes and phases (or phasors). The spectrum of such a multi-frequency signal is defined simply as the collection of amplitude, phase, and frequency information that allows us to represent a signal in such a way.[1] We will show that it is useful to depict the corresponding spectrum information in a graphical representation with complex amplitude plotted as a function of frequency. This visual representation of the spectrum allows us quickly and easily to see interrelationships among spectral components of different frequencies. As a result, this visual representation of the signal is invaluable in understanding how signals can be modified by operations on signals.

3-1 The Spectrum of a Sum of Sinusoids

One of the reasons that sinusoids are so important for our study is that they are the basic building blocks for representing more complicated signals. In this chapter we will show some extraordinarily complicated-looking waveforms that can be constructed from rather simple combinations of basic cosine waves. The most general and powerful method for producing new signals from sinusoids is the *additive linear combination*, where a real signal is created by adding together a constant and N sinusoids, each with a different frequency, amplitude, and phase. If the signal is real, it may be represented by the right-hand side in

$$x(t) = A_0 + \sum_{k=1}^{N} A_k \cos(2\pi f_k t + \varphi_k) \tag{3.2}$$

where, in the most general case, each amplitude, phase, and frequency[2] may be chosen independently. Such a signal may also be represented using complex exponentials and complex amplitudes for the individual sinusoidal components, that is,

$$x(t) = X_0 + \sum_{k=1}^{N} \Re\left\{X_k e^{j2\pi f_k t}\right\} \tag{3.3}$$

where $X_0 = A_0$ represents a real constant component, and each complex amplitude

$$X_k = A_k e^{j\varphi_k}$$

represents the magnitude and phase of a rotating phasor whose frequency is f_k.

The *inverse Euler formula* gives a way to represent $x(t)$ in (3.3) in yet another alternative form

$$x(t) = X_0 + \sum_{k=1}^{N} \left\{\frac{X_k}{2} e^{j2\pi f_k t} + \frac{X_k^*}{2} e^{-j2\pi f_k t}\right\} \tag{3.4}$$

[1]The term **spectrum** is used in many ways—often without precise definition. Our use of the term is clearly defined in terms of sums of sinusoids or complex exponentials. Other usage of the term may be less precise, but still appropriate when signals are thought of as being composed of sinusoidal components.

[2]For this chapter, we prefer cyclic frequency f_k to radian frequency $\omega_k = 2\pi f_k$, because it is easier and more natural to describe physical quantities such as musical notes in Hz.

As in the case of individual sinusoids, this form follows from the fact that the real part of a complex number is equal to one-half the sum of that number and its complex conjugate. Equation (3.4) also shows that each sinusoid in the sum decomposes into two rotating phasors, one with positive frequency, f_k, rotating in the counter-clockwise direction, and the other with negative frequency, $-f_k$, rotating in the clockwise direction.

The signal representation in (3.4) is called the ***two-sided spectrum*** because it uses $2N+1$ *positive and negative frequencies* along with the corresponding $2N+1$ complex amplitudes to specify a signal composed of sinusoids (3.2). To be specific, our definition of the spectrum is the set of pairs

$$\left\{(0, X_0),\ \left(f_1, \tfrac{1}{2}X_1\right),\ \left(-f_1, \tfrac{1}{2}X_1^*\right), \ldots,\ \left(f_k, \tfrac{1}{2}X_k\right),\ \left(-f_k, \tfrac{1}{2}X_k^*\right), \ldots\right\} \tag{3.5}$$

Each pair $(f_k, \frac{1}{2}X_k)$ indicates the size and relative phase of the complex exponential component contributing at frequency f_k. It is common to refer to the spectrum as the ***frequency-domain representation*** of the signal. In contrast, the ***time-domain representation*** gives the values of the time waveform itself, whereas the frequency-domain representation simply gives the information required to synthesize it with (3.4).

When the number of components is small and the frequencies are spaced widely, the spectrum is a compact representation of the signal $x(t)$. In such cases, it is common to refer to the spectrum as being ***sparse***, meaning consisting of only a few components. In Chapter 7 we will generalize the spectrum concept to include a dense continuum of frequencies. In this case, the spectrum is called the discrete-time Fourier transform.

EXAMPLE 3-1 Two-Sided Spectrum

To determine the spectrum of the following signal,

$$x(t) = 10 + 14\cos(200\pi t - \pi/3) + 8\cos(500\pi t + \pi/2)$$

which is the sum of a constant and two sinusoids, we must convert from the general form in (3.2) to the two-sided form in (3.4). After we apply the inverse Euler formula, we get the following five terms:

$$\begin{aligned} x(t) = {} & 10 + 7e^{-j\pi/3}e^{j2\pi(100)t} + 7e^{j\pi/3}e^{-j2\pi(100)t} \\ & + 4e^{j\pi/2}e^{j2\pi(250)t} + 4e^{-j\pi/2}e^{-j2\pi(250)t} \end{aligned} \tag{3.6}$$

Note that the constant component of the signal, often called the ***DC component***,[3] can be expressed as a complex exponential signal with zero frequency (i.e., $10e^{j0t} = 10$). Therefore, in the list form suggested in (3.5), the spectrum of this signal is the set of five rotating phasors represented by the frequency/complex amplitude pairs

$$\{(0,\ 10),\ (100,\ 7e^{-j\pi/3}),\ (-100,\ 7e^{j\pi/3}),\ (250,\ 4e^{j\pi/2}),\ (-250,\ 4e^{-j\pi/2})\}$$

[3]The terminology "DC" comes from electric circuits, where a constant value of current is called direct current, or DC. It is common to call $X_0 = A_0$ the DC component of the spectrum. Since the DC component is constant, its frequency is $f = 0$.

3-1.1 Notation Change

The relationship between X_k and the spectrum involves special cases because the factor of $\frac{1}{2}$ multiplies every X_k in the spectrum (3.5), except for X_0. The companion formulas such as (3.4) are cumbersome. Therefore, we introduce a_k as a new symbol for the complex amplitude in the spectrum, and define it as follows:

$$a_k = \begin{cases} A_0 & \text{for } k = 0 \\ \frac{1}{2}A_k e^{j\varphi_k} & \text{for } k \neq 0 \end{cases} \tag{3.7}$$

This allows us to say that the spectrum is the set of $(f_k,\ a_k)$ pairs. The primary motivation for this notational change is to simplify the formulas for the Fourier series coefficients developed later in Section 3-5. For example, (3.4) can now be written as a single compact summation

$$x(t) = \sum_{k=-N}^{N} a_k e^{j2\pi f_k t} \tag{3.8}$$

where we have defined $f_0 = 0$.

3-1.2 Graphical Plot of the Spectrum

A plot of the spectrum is much more revealing than the list of $(f_k,\ a_k)$ pairs. Each frequency component can be represented by a vertical line at the appropriate frequency, and the length of the line can be drawn proportional to the magnitude, $|a_k|$. This is shown in Fig. 3-1 for the signal in (3.6). Each ***spectral line*** is labeled with the value of a_k to complete the information needed to define the spectrum.

This simple but effective plot makes it easy to see two things: the relative location of the frequencies, and the relative amplitudes of the sinusoidal components. This is why the spectrum plot is widely used as a graphical representation of the signal. As we will see in Chapters 4 and 6, the frequency-domain representation is exceedingly useful because it is often very easy to see how systems affect a signal by determining what happens to the signal spectrum as it is transmitted through the system. This is why the spectrum is the key to understanding sophisticated signal processing systems such as radios, televisions, cell phones, MP3 players, and the like.

DEMO
Rotating Phasors

In a spectrum like Fig. 3-1, the complex amplitude of each negative frequency component is the complex conjugate of the complex amplitude at the corresponding positive frequency component. This property is called *conjugate symmetry.* The spectrum is conjugate symmetric whenever $x(t)$ is a real signal, because the complex rotating phasors at f_k and $-f_k$ in (3.4) rotate in opposite directions and must combine to form a real signal (see Fig. 2-13(b) and the ***Rotating Phasors*** movie found on the website).

Spectra of signals comprised of individual sinusoids are often called "line spectra." The term "line" seems appropriate for us here because we plot the components as vertical lines positioned at the individual frequencies. However, the term originated

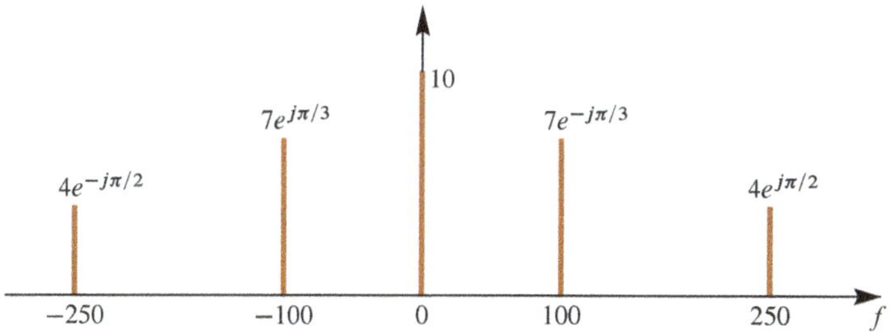

Figure 3-1 Spectrum plot for the signal $x(t) = 10 + 14\cos(200\pi t - \pi/3) + 8\cos(500\pi t + \pi/2)$. Units of frequency (f) are Hz. Negative frequency components should be included for completeness, even though for real signals, they are conjugates of the corresponding positive frequency components.

in physics where lines are observed in emission or absorption spectra formed with optical prisms (which serve as optical spectrum analyzers). These lines correspond to energy being emitted or absorbed at wavelengths that are characteristic of different atoms or ions. This is illustrated in Fig. 3-2, which shows how the frequency spectrum of white light from a star (assumed to contain all frequencies with essentially the same amplitude) is modified by transmission through the atmosphere of an exoplanet. The plot shows that the spectrum of white light is modified by absorption of certain frequencies. This example illustrates how modification of signals by an operation on that signal can be easily represented and understood in terms of the spectrum of that signal.

3-1.3 Analysis versus Synthesis

In the context of the spectrum, synthesis is the process of using the frequency-domain representation (i.e., the spectrum) to create the time-domain representation (i.e., the waveform $x(t)$). Analysis, on the other hand, is an operation on the time-domain signal that produces the frequency-domain spectrum. As described so far, both operations seem relatively easy, but a general procedure for computing and plotting the spectrum for an arbitrarily chosen signal requires the study of Fourier analysis, for example, Fourier Series in Section 3-5 and Appendix C, as well as Fourier transforms of discrete signals in Chapters 7 and 8.

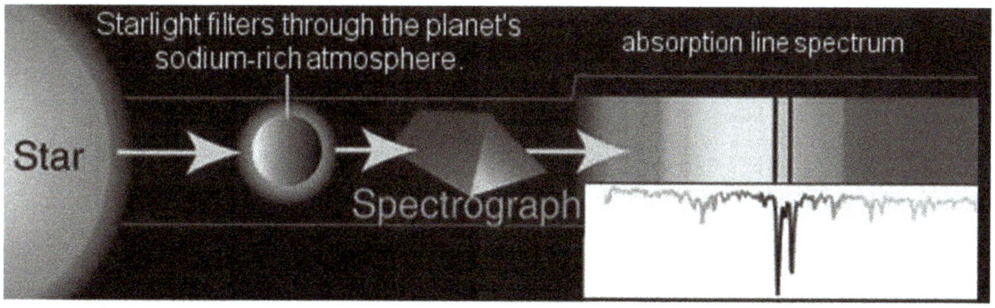

Figure 3-2 Illustration of spectrum analysis of white light transmitted through the atmosphere of exoplanet HD 209458. (NASA photo). For a color version of this picture, see http://science.nasa.gov/science-news/science-at-nasa/2001/ast27nov_1/

Up to this point, we have treated the special case where we can assume that the signal of interest is the sum of a constant and a *finite number* of sinusoids. The *synthesis procedure* is relatively easy when starting from a plot like Fig. 3-1. The values of (f_k, a_k) can be extracted from the spectrum plot, and (3.8) can be used to calculate the signal values in the time domain. When the signal is defined by a mathematical formula consisting of a sum of sinusoids as in (3.2), the *analysis procedure* is also straightforward. Each sinusoid in the mathematical formula (3.2) is expressed as two complex exponentials by using the inverse Euler relation to get a form like (3.4). Then the spectrum values can be extracted by noting the frequencies and their corresponding complex amplitudes. Such a simple analysis procedure rarely applies in practice. For example, if the signal were a recording of human speech, there is no mathematical formula for the signal values and the analysis problem becomes much more difficult. However, the basic concepts of analysis to obtain the spectrum and synthesis to construct the time signal are very general, and they will show up repeatedly throughout this book.

Before tackling the case of Fourier series for periodic signals, we show examples where the sum of just a few sinusoids can be used to synthesize audio signals whose waveforms are much different from sinusoids and are interesting to hear. The relationship between the time waveform and the spectrum will be emphasized. In some examples that we will show, the time waveform appears to be very complicated while the spectrum is sparse and relatively simple and therefore easy to understand.

3-2 Sinusoidal Amplitude Modulation

LAB

AM and FM Sinusoidal Signals

So far we have considered signals that can be represented as sums of sinusoids of different frequencies, but another useful mathematical signal model is the product of sinusoids. When we multiply two sinusoids having different frequencies, we can create an interesting audio effect called a ***beat note***. The phenomenon, which may sound like a warble, is best heard by picking one of the frequencies to be very small (e.g., 10 Hz), and the other around 1 kHz. Some musical instruments naturally produce beating tones. Another common situation where sinusoidal multiplication is used is radio broadcasting. In this case, the multiplication is called ***amplitude modulation*** (AM), hence the name AM radio for some broadcast stations.

3-2.1 Multiplication of Sinusoids

A signal produced by multiplying two sinusoids must be rewritten as a sum in order to obtain its spectrum, because our spectrum is a graphical representation of an *additive* linear combination of complex exponential signals. Other mathematical combinations of sinusoids would also have to be rewritten in the additive form in order to display their spectrum representation. This is illustrated by the following example.

EXAMPLE 3-2 Spectrum of a Product Signal

For the special case of a signal formed as the product of two sinusoids with frequencies $\frac{1}{2}$ Hz and 5 Hz

$$x(t) = \cos(\pi t)\sin(10\pi t) \tag{3.9}$$

it is necessary to rewrite $x(t)$ as a sum before its spectrum can be defined. One approach is to use the inverse Euler formula as follows:

$$x(t) = \left(\frac{e^{j\pi t} + e^{-j\pi t}}{2}\right)\left(\frac{e^{j10\pi t} - e^{-j10\pi t}}{2j}\right) \tag{3.10a}$$

$$= \tfrac{1}{4}e^{-j\pi/2}e^{j11\pi t} + \tfrac{1}{4}e^{-j\pi/2}e^{j9\pi t} + \tfrac{1}{4}e^{j\pi/2}e^{-j9\pi t} + \tfrac{1}{4}e^{j\pi/2}e^{-j11\pi t} \tag{3.10b}$$

$$= \tfrac{1}{2}\cos(11\pi t - \pi/2) + \tfrac{1}{2}\cos(9\pi t - \pi/2) \tag{3.10c}$$

In this derivation, we see four terms in the additive combination (3.10b), so there are four spectrum components at frequencies $\pm 11\pi$ and $\pm 9\pi$ rad/s, which convert to hertz as 5.5, 4.5, −4.5, and −5.5 Hz. The magnitude is the same ($\frac{1}{4}$) for all four components. It is also worth noting that neither of the original frequencies (5 Hz and $\frac{1}{2}$Hz) used to define $x(t)$ in (3.9) appear in the spectrum.

EXERCISE 3.1 Let $x(t) = \sin^2(10\pi t)$. Find an additive combination in the form of (3.8) for $x(t)$, and then plot the spectrum. Count the number of frequency components in the spectrum. What is the highest frequency contained in $x(t)$? Use the inverse Euler formula rather than a trigonometric identity to obtain the additive form.

3-2.2 Beat Note Waveform

Beat notes[4] are produced by adding two equal-amplitude sinusoids with nearly identical frequencies. The example in (3.9) and (3.10c) suggests that the product of two sinusoids is equivalent to a sum of two equal-amplitude sinusoids. Thus we can derive a general relationship for the spectrum of the product form if we start with an additive combination of two equal-amplitude sinusoids:

$$x(t) = \cos(2\pi f_1 t) + \cos(2\pi f_2 t) \tag{3.11}$$

DEMO
Spectrograms of Simple Sounds

with $f_2 > f_1$. The two frequencies can be expressed as $f_1 = f_c - f_\Delta$ and $f_2 = f_c + f_\Delta$, where we have defined a *center frequency* $f_c = \frac{1}{2}(f_1 + f_2)$ and a *deviation frequency* $f_\Delta = \frac{1}{2}(f_2 - f_1)$, which is smaller than f_c. The additive form (3.11) gives the spectrum of the beat note signal which is plotted in Fig. 3-3. Examples of beat notes and similar signals can be found on the Companion Website.

[4]We use the term *beat note* since the phenomenon is most often observed as an auditory effect when two musical notes (often idealized as sinusoids) are sounded together, for example, by playing two neighboring piano keys simultaneously.

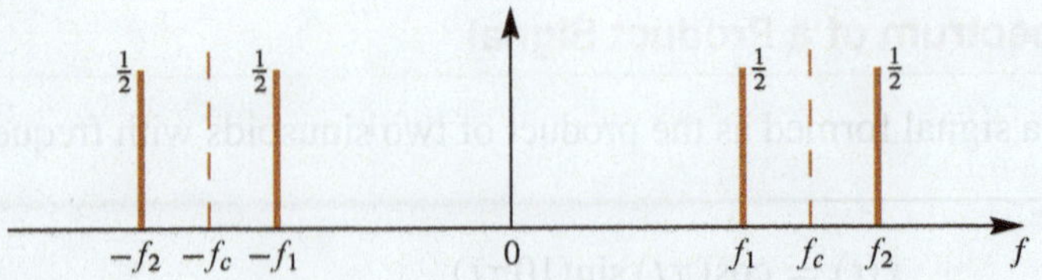

Figure 3-3 Spectrum of the sum of two equal-amplitude sinusoids. When $f_\Delta = \frac{1}{2}(f_2 - f_1)$ is small, this is the beat signal in (3.11) and (3.12). The negative-frequency lines are centered around $f = -f_c$.

Using the complex exponential representation of the two cosines, we can go from the additive form to the product form. After we rewrite $x(t)$ as a product of two cosines, we have a form that is easier to plot in the time domain, and from the plot we are able to see why (3.11) is perceived as a beat note when the two frequencies are almost the same. The analysis proceeds as follows:

$$\begin{aligned} x(t) &= \cos(2\pi f_1 t) + \cos(2\pi f_2 t) \\ &= \Re\left\{e^{j2\pi f_1 t}\right\} + \Re\left\{e^{j2\pi f_2 t}\right\} \\ &= \Re\left\{e^{j2\pi(f_c - f_\Delta)t} + e^{j2\pi(f_c + f_\Delta)t}\right\} \\ &= \Re\left\{e^{j2\pi f_c t}\left(e^{-j2\pi f_\Delta t} + e^{j2\pi f_\Delta t}\right)\right\} \\ &= \Re\left\{e^{j2\pi f_c t}\left(2\cos(2\pi f_\Delta t)\right)\right\} \\ &= 2\cos(2\pi f_\Delta t)\cos(2\pi f_c t) \end{aligned} \tag{3.12}$$

EXERCISE 3.2 If the two sinusoids being added were to have nonzero phases, that is,

$$x(t) = \cos(2\pi f_1 t + \varphi_1) + \cos(2\pi f_2 t + \varphi_2)$$

the frequencies, f_c and f_Δ, would remain the same, but the phases, φ_Δ and φ_c, would change.

$$x(t) = 2\cos(2\pi f_\Delta t + \varphi_\Delta)\cos(2\pi f_c t + \varphi_c)$$

Determine the phases needed in the product representation.

For a numerical example, we take $f_c = 200$ Hz and $f_\Delta = 20$ Hz in (3.12), so that

$$x(t) = 2\cos(2\pi(20)t)\cos(2\pi(200)t) \tag{3.13}$$

NOTE
Time-Domain Plot of a Beat Note

MATLAB was used to make the *time-domain* plots of the two cosine wave components and the product signal $x(t)$. Figure 3-4(a) shows the two sinusoidal components, $2\cos(2\pi(20)t)$ and $\cos(2\pi(200)t)$, that make up the product in (3.13). The plot of the beat note is constructed by first plotting $2\cos(2\pi(20)t)$ and its negative version $-2\cos(2\pi(20)t)$ to define boundaries inside of which we then draw the higher frequency signal. These boundaries are called the signal's ***envelope***. One signal varies much more

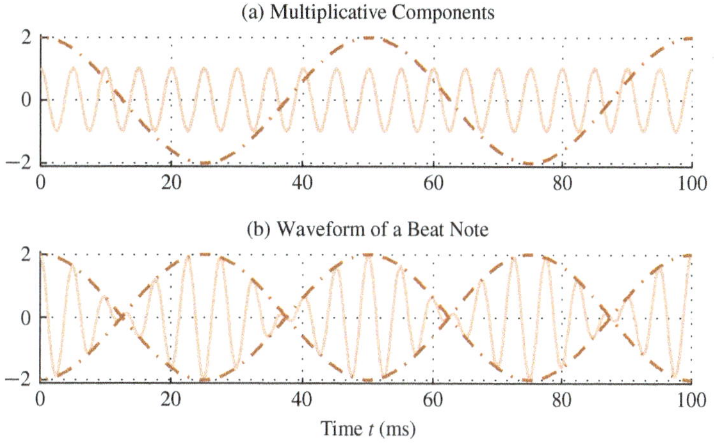

Figure 3-4 (a) Multiplicative components of a beat note with $f_c = 200$ Hz and $f_\Delta = 20$ Hz. (b) The corresponding product signal. The time interval between nulls (zeros) of the envelope is $\frac{1}{2}(1/f_\Delta) = 25$ ms, which is dictated by the frequency difference, f_Δ.

DEMO
beatcon

rapidly than the other, so the slow signal can be viewed as a slowly time-varying amplitude of the faster signal. The resulting beat note is plotted in Fig. 3-4(b), where it can be seen that the effect of multiplying the higher-frequency sinusoid (200 Hz) by the lower-frequency sinusoid (at 20 Hz) is to change the amplitude of the peaks of the higher-frequency waveform. If we play such a signal through a loudspeaker or create it with a musical instrument, it will appear to fade in and out because the signal envelope is rising and falling periodically with period $\frac{1}{2}(1/f_\Delta) = 25$ ms, as in Fig. 3-4(b). This results in the audible phenomenon called "beating" of tones in music. The amplitude goes up and down twice during one period of the lower-frequency sinusoid, so when we listen to such a signal $x(t)$, we can hear the amplitude variation with frequency $2f_\Delta$ Hz. When f_c is much higher in frequency than f_Δ, we also hear a tone near f_c Hz that seems to be changing up and down in frequency. Musicians can use this beating phenomenon as an aid in tuning two instruments to the same pitch. When two notes are close but not identical in frequency, the beating phenomenon is heard. As one pitch is changed to become closer to the other, the effect disappears, and the two instruments are then "in tune."

EXAMPLE 3-3 Decreasing the Frequency Difference, f_Δ

If f_Δ is decreased from 20 Hz to 9 Hz, we see in Fig. 3-5(a,b) that the envelope of the 200-Hz tone changes much more slowly. The time interval between nulls (zeros) of the envelope is one-half of the period of the f_Δ sinusoid, $\frac{1}{2}(1/f_\Delta)$, so the more closely spaced the frequencies of the sinusoids in (3.11), the slower the envelope variation. Figures 3-4 and 3-5 are simplified somewhat by using zero-phase cosines for both terms in (3.11), but other phase relationships would give similar patterns. Finally, note that for $x(t)$ in Fig. 3-4 the spectrum contains frequency components at ± 220 Hz and ± 180 Hz, while for the signal in Fig. 3-5 the spectrum has frequencies ± 209 Hz and ± 191 Hz.

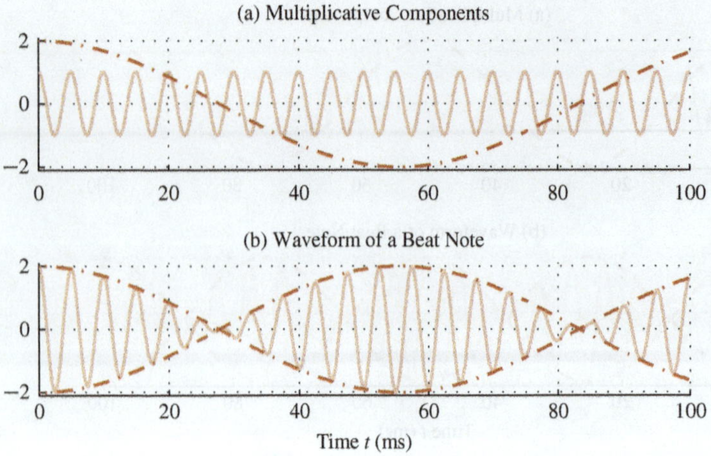

Figure 3-5 Beat note with $f_c = 200$ Hz and $f_\Delta = 9$ Hz. Nulls are now $\frac{1}{2}(1/f_\Delta) = 55.6$ ms apart.

3-2.3 Amplitude Modulation

Multiplying sinusoids is commonplace in communication systems where modulation of the envelope of a high frequency sinusoid is used to transmit information signals such as data, voice, or music. ***Amplitude modulation*** is the process of multiplying a high-frequency sinusoidal signal by a low-frequency signal, such as a voice or music signal. It is the technique used in broadcast AM radio: In fact "AM" is just the abbreviation for amplitude modulation. The AM signal is a product of the form

$$x(t) = v(t)\cos(2\pi f_c t) \tag{3.14}$$

where it is assumed that the frequency of the cosine term (f_c Hz) is much higher than any frequencies contained in the spectrum of $v(t)$. In radio, $v(t)$ represents the voice or music signal to be transmitted. The cosine wave $\cos(2\pi f_c t)$ in (3.14) is called the *carrier signal*, and its frequency f_c is called the *carrier frequency*.

LAB

AM and FM Sinusoidal Signals

With our limited knowledge of the spectrum at this point, the form of $v(t)$ in (3.14) must be a finite sum of sinusoids, but that case is sufficient to understand that the modulation process (3.14) involves *frequency shifting* of the spectrum of $v(t)$.

EXAMPLE 3-4 Amplitude Modulation

If we let $v(t) = 5+4\cos(40\pi t)$ and $f_c = 200$ Hz, then the AM signal is a multiplication similar to the beat signal:

$$x(t) = \left[5 + 4\cos(40\pi t)\right]\cos(400\pi t) \tag{3.15}$$

The signal $v(t)$ has a DC term large enough to make $v(t)$ positive for all t. A plot of $x(t)$ is given in Fig. 3-6(b), where it can be seen that the effect of multiplying the

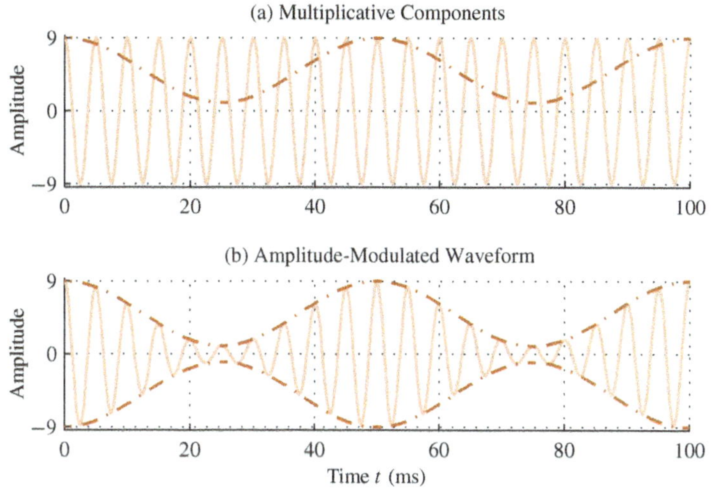

Figure 3-6 AM signal for $f_c = 200$ Hz and $f_\Delta = 20$ Hz. (a) The signal $5 + 4\cos(40\pi t)$ shown as a dark-orange line, along with a 200 Hz sinusoid $\cos(400\pi t)$ (light color orange, with 9× expanded scale for convenience in plotting). (b) AM signal (light orange) formed as the product of the two signals in (a) oscillates at 200 Hz, but has amplitude variations, which are referred to as its envelope (dark orange).

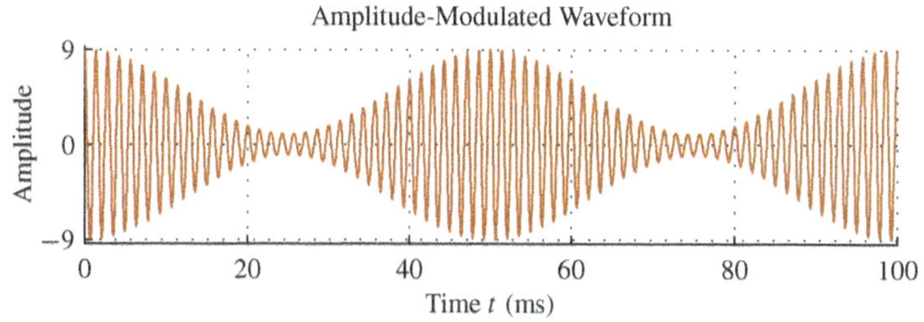

Figure 3-7 AM signal for $f_c = 700$ Hz and $f_\Delta = 20$ Hz. The higher carrier frequency makes it possible to see the outline of the modulating cosine without drawing the envelope.

higher-frequency carrier sinusoid (200 Hz) by the lower-frequency sinusoid (at 20 Hz) is to "modulate" (or change) the amplitude envelope of the carrier waveform—hence the name amplitude modulation for a signal like $x(t)$ in Fig. 3-7.

The primary difference between the AM signals of (3.14) and (3.15) and the beat note signals is that the AM envelope is usually set up to have no zero crossings. When the carrier frequency becomes very high compared with the frequencies in $v(t)$ as in Fig. 3-7, it is possible to see the outline of the modulating signal $v(t)$ without drawing the envelope signal explicitly. This characteristic simplifies the implementation of a detector circuit in AM broadcast receivers where the objective is to recover the envelope $v(t)$ from the AM signal $x(t)$.

3-2.4 AM Spectrum

In the frequency domain, the AM signal spectrum is easy to describe: it consists of scaled copies of the spectrum of $v(t)$ centered at $f = \pm f_c$, as shown in Fig. 3-8. The same statement applies to the spectrum of a beat note signal, which is the product of two sinusoids as discussed in Example 3-2 and Section 3-2.2. The spectrum of the AM

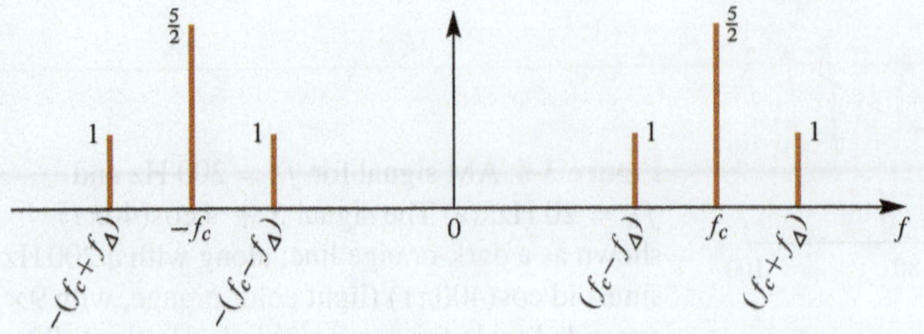

Figure 3-8 Spectrum of the AM signal in (3.15), (3.16), (3.17), and Fig. 3-6, where $f_c = 200$ Hz and $f_\Delta = 20$ Hz for $v(t)$ defined in (3.15).

signal $x(t)$ in (3.15) can be derived by writing $v(t)$ and $\cos(400\pi t)$ in terms of complex exponentials

$$\begin{aligned} x(t) &= v(t)(\tfrac{1}{2}e^{j400\pi t} + \tfrac{1}{2}e^{-j400\pi t}) \\ &= (5 + 2^{j40\pi t} + 2^{-j40\pi t})\tfrac{1}{2}e^{j400\pi t} + (5 + 2^{j40\pi t} + 2^{-j40\pi t})\tfrac{1}{2}e^{-j400\pi t} \end{aligned} \tag{3.16}$$

and then multiplying out to get the following additive combination for the spectrum:

$$x(t) = \tfrac{5}{2}e^{j400\pi t} + e^{j440\pi t} + e^{j360\pi t} + \tfrac{5}{2}e^{-j400\pi t} + e^{-j440\pi t} + e^{-j360\pi t} \tag{3.17}$$

The spectrum of $v(t)$ has three spectrum lines at $f = 0,\ 20,\ -20$ Hz, but now there are six spectral components for $x(t)$: at the frequencies $f = 200,\ 220,\ 180$ Hz and $f = -200,\ -180,\ -220$ Hz. Thus the spectrum (Fig. 3-8) for $x(t)$ contains two similar subsets, one centered at $f = f_c$ and the other at $f = -f_c$. Each subset is a frequency-shifted (and scaled) version of the two-sided spectrum of $v(t)$. The foregoing example is generalized in Section 3-3.5.

EXERCISE 3.3 Derive the spectrum of $v(t) = 5 + 4\cos(2\pi(20)t)$ and plot it as a function of frequency. Compare this result with the spectral plot for the AM signal in Fig. 3-8.

3-2.5 The Concept of Bandwidth

The concept of bandwidth is closely tied to the spectrum representation of signals, and it is of considerable importance in almost all applications where signals are transmitted, manipulated, or stored. Oftentimes we only want a coarse characterization of the spectrum of a signal. We can, for example, use the highest frequency for which the spectrum is nonzero as a measure of spectrum content. This is an appropriate measure when we consider sampling in Chapter 4. In the case of the AM signal whose spectrum is shown in Fig. 3-8, this highest frequency would be $f_c + f_\Delta$. Alternatively, it may be more important and more informative to emphasize the group (or band) of frequencies containing both the lowest and highest positive frequencies. Such measures of the spectrum as these are often referred to as the *bandwidth of the spectrum* or, by implication, the *bandwidth of the signal.*

In the case of the signal of Fig. 3-8, we might say that the bandwidth of the signal is $(f_c + f_\Delta) - (f_c - f_\Delta) = 2f_\Delta$, and we might be even more specific by noting that this band of frequencies is centered on frequency f_c.[5] The bandwidth characterization is important in the AM radio broadcast system because each AM radio station is assigned a specific carrier frequency f_c and a bandwidth of 10 kHz centered at the carrier frequency. In this way, many local stations can share the (wide) bandwidth of free space without interfering with each other.[6] This sharing of the free-space communication channel is called *frequency-division multiplexing*.

3-3 Operations on the Spectrum

In signal processing, we usually manipulate signals with time-domain operations. One of the benefits of the spectrum representation is that time-domain operations on the signal $x(t)$ often result in very simple changes to the spectrum. Since the spectrum consists of a set of frequency/complex amplitude pairs, $\mathcal{S} = \{(f_k, a_k)\}$, when we operate on a signal $x(t)$ in the time domain, we might change the frequencies $\{f_k\}$, or the complex amplitudes $\{a_k\}$, or both. The correspondence between a time-domain operation and spectrum changes is often referred to as a *property* of the spectrum representation. This section gives several examples of spectrum properties.

3-3.1 Scaling or Adding a Constant

Multiplying a signal by a scale factor (i.e., $\gamma x(t)$) multiplies all of the complex amplitudes in its spectrum by the same scale factor (γ)

$$\gamma x(t) = \gamma \sum_{k=-M}^{M} a_k e^{j2\pi f_k t} = \sum_{k=-M}^{M} (\gamma a_k) e^{j2\pi f_k t}$$

but leaves all the frequencies unchanged. On the other hand, adding a constant to a signal (i.e., $x(t) + c$) changes the complex amplitude at only one frequency $(f = 0)$. The following equation

$$x(t) + c = \sum_{f_k \neq 0} a_k e^{j2\pi f_k t} + \underbrace{a_0 e^{j2\pi(0)t} + c}_{\text{new DC}}$$

[5] If the signal is real, we know that the spectrum has complex conjugate symmetry so we can focus on only positive frequencies.

[6] AM broadcast stations have carrier frequencies ranging from 540 Khz to 1700 Khz. Other types of radio communication use other portions of what we refer to collectively as the *electromagnetic spectrum*.

shows that the new DC value is $a_0 + c$. If the spectrum of $x(t)$ has no DC component ($a_0 = 0$), then adding c would produce a new spectrum line at $f = 0$, so the pair $(0, c)$ would now be included in the spectrum set.

Figure 3-9 illustrates both operations for the case $y(t) = 2x(t) + 6$ when the signal is

$$x(t) = \tfrac{3}{2} + 6\cos(6\pi t - \pi/3) + 4\cos(14\pi t + \pi/4)$$

The spectrum of $x(t)$ is shown in Fig. 3-9(a), and the spectrum of $y(t)$ in Fig. 3-9(c). The complex amplitudes at $f = \pm 3$ and ± 7 are doubled when going from Fig. 3-9(a) to Fig. 3-9(c). At DC ($f = 0$) the constant 6 is added because it is a DC-only signal whose spectrum is shown in Fig. 3-9(b). Thus the DC component in the spectrum of $y(t)$ is $2(1.5) + 6 = 9$.

3-3.2 Adding Signals

Another simple operation is addition of two signals, $x_1(t) + x_2(t)$, but there are two cases depending on the frequencies in the spectra. In principle, the spectrum of $x_1(t)$ is added to the spectrum of $x_2(t)$, because

$$x_1(t) + x_2(t) = \sum_{f_{1k}} a_{1k} e^{j2\pi f_{1k} t} + \sum_{f_{2\ell}} a_{2\ell} e^{j2\pi f_{2\ell} t}$$

where (f_{1k}, a_{1k}) and $(f_{2\ell}, a_{2\ell})$ are the frequencies and complex amplitudes for $x_1(t)$ and $x_2(t)$, respectively. If a spectrum line is present in both spectra at $f = f_{1k} = f_{2\ell}$, then the corresponding complex amplitudes, a_{1k} and $a_{2\ell} = a_{2k}$, must be summed by phasor

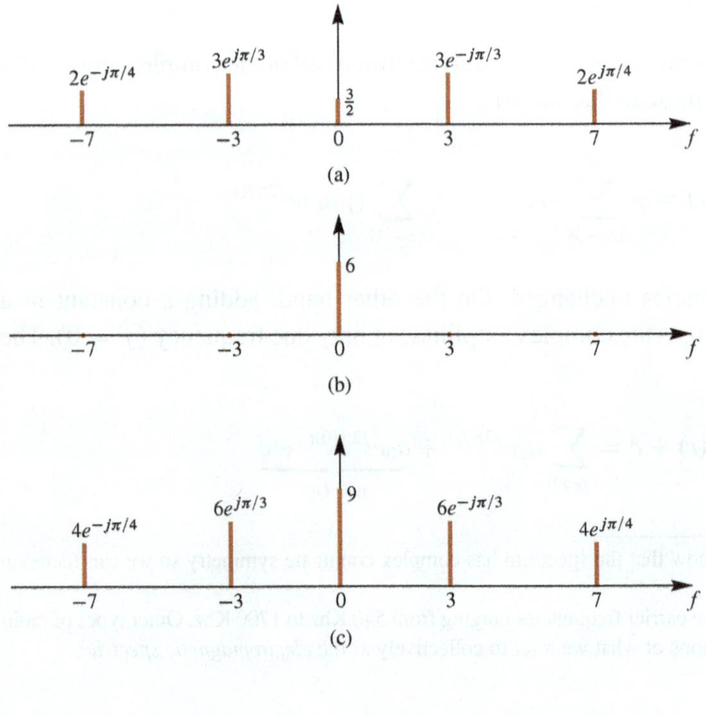

Figure 3-9 Spectrum plots for adding a constant to a scaled signal. (a) Spectrum of $x(t)$, (b) spectrum of constant signal, and (c) spectrum of $y(t) = 2x(t) + 6$.

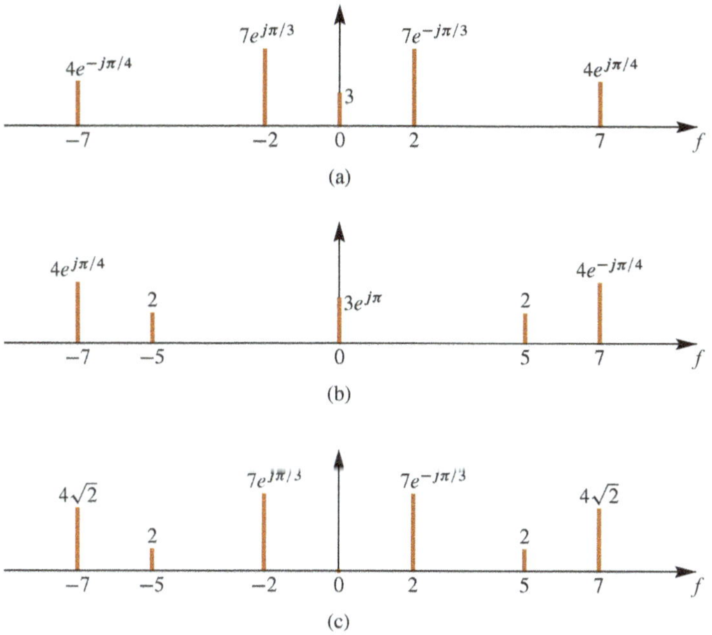

Figure 3-10 Spectrum plots for the sum of two signals. (a) Spectrum of $x_1(t)$, (b) spectrum of $x_2(t)$, and (c) spectrum of the sum signal, $x_1(t) + x_2(t)$.

addition.[7] The other case is a spectrum line that is present in only one spectrum. In this case, there is nothing to add so that the spectrum line becomes part of the spectrum of the sum signal without any change.

EXAMPLE 3-5 Adding Spectra

The plots in Fig. 3-10 illustrate several points about the spectrum for the sum of two signals. The spectrum of $x_1(t)$ is shown in Fig. 3-10(a), and the spectrum of $x_2(t)$ in Fig. 3-10(b). When adding the spectra, we see that both spectra have components at the frequencies $f = \pm 7$ and 0, so three phasor additions must be done.

$$\begin{aligned} 3 + 3e^{j\pi} &= 0 && \text{(at } f = 0) \\ 4e^{-j\pi/4} + 4e^{j\pi/4} &= 4\sqrt{2} && \text{(at } f = -7) \\ 4e^{j\pi/4} + 4e^{-j\pi/4} &= 4\sqrt{2} && \text{(at } f = 7) \end{aligned}$$

Since the complex amplitudes at $f = 0$ cancel out, there is no DC component in the sum spectrum of Fig. 3-10(c). The spectrum for $x_1(t)$ has components at $f = \pm 2$, but the spectrum for $x_2(t)$ does not, so those components are copied directly to the sum spectrum. Likewise, the components at $f = \pm 5$ in the spectrum for $x_2(t)$ are also copied to the sum spectrum.

[7]After the phasor addition it might happen that a complex amplitude becomes zero, in which case that frequency would drop out of the spectrum set.

3-3.3 Time-Shifting $x(t)$ Multiplies a_k by a Complex Exponential

If we form a new signal $y(t)$ by time-shifting $x(t)$, then the frequencies remain the same, but there is a phase change in the complex amplitudes of the spectrum.

$$y(t) = x(t - \tau_d) \quad \longleftrightarrow \quad b_k = a_k e^{-j2\pi f_k \tau_d}$$

because

$$y(t) = x(t - \tau_d) = \sum_k a_k e^{j2\pi f_k (t-\tau_d)} = \sum_k \underbrace{\left(a_k e^{-j2\pi f_k \tau_d}\right)}_{b_k} e^{j2\pi f_k t}$$

EXAMPLE 3-6 Delayed Cosine

Consider a cosine wave $x(t) = 6\cos(250\pi t)$ that is delayed by $\tau_d = 2\,\text{ms}$, which is one quarter of its period. The spectrum set of $x(t)$ is $\mathcal{S} = \{(125, 3), (-125, 3)\}$. In the spectrum of $x(t - 0.002)$ the phase change for the $f = +125$-Hz term is

$$e^{-j250\pi(0.002)} = e^{-j0.5\pi} = -j$$

Thus the spectrum of $x(t - 0.002)$ can be written as the spectrum set $\mathcal{S} = \{(125, -3j), (-125, 3j)\}$. It should be easy to verify that

$$x(t - 0.002) = 6\sin(250\pi t)$$

This result is consistent with the fact that a quarter cycle delay of a cosine wave is a sine wave.

3-3.4 Differentiating $x(t)$ Multiplies a_k by $(j2\pi f_k)$

If we form a new signal $y(t)$ by taking the derivative of $x(t)$, then the frequencies remain the same, while the complex amplitudes of the spectrum of $y(t)$ can be obtained as follows:

$$y(t) = \tfrac{d}{dt}x(t) \quad \longleftrightarrow \quad b_k = (j2\pi f_k)a_k$$

Since the derivative of a sum is the sum of the individual derivatives, we can apply the derivative operator to each term in the spectrum of $x(t)$ to obtain

$$y(t) = \tfrac{d}{dt}x(t) = \sum_k a_k \frac{d}{dt} e^{j2\pi f_k t} = \sum_k \underbrace{(j2\pi f_k)a_k}_{b_k} e^{j2\pi f_k t}$$

EXAMPLE 3-7 Derivative of Sine plus DC

Consider taking the derivative of a DC-offset sine wave $x(t) = 7 + 6\sin(250\pi t)$. The spectrum set for $x(t)$ is $\mathcal{S} = \{(0, 7), (125, -3j), (-125, 3j)\}$. In the spectrum of $\frac{d}{dt}x(t)$, the $f = 0$ term is 7 multiplied by $(j2\pi(0)) = 0$, so it is eliminated; the $f = +125$-Hz term is multiplied by $(j2\pi(125))$

$$(j2\pi(125))(-3j) = 750\pi$$

Thus the spectrum set of $\frac{d}{dt}x(t)$ is $\mathcal{S} = \{(125, 750\pi),\ (-125, 750\pi)\}$. From the spectrum it should be easy to verify that

$$\tfrac{d}{dt}x(t) = 1500\pi\cos(250\pi t)$$

which coincides with the fact that the derivative of a sine function is a cosine and the derivative of a constant is zero.

3-3.5 Frequency Shifting

If you multiply a signal by a sinusoid or a complex exponential, the frequencies in the spectrum are changed in a simple predictable fashion. Here is the derivation for a signal $x(t)$ multiplied by the complex exponential $Ae^{j\varphi}e^{j2\pi f_c t}$ whose frequency is f_c Hz.

$$y(t) = Ae^{j\varphi}e^{j2\pi f_c t}\underbrace{\sum_k a_k e^{j2\pi f_k t}}_{x(t)} \tag{3.18a}$$

$$y(t) = \sum_k (a_k Ae^{j\varphi})e^{j2\pi(f_k+f_c)t} \tag{3.18b}$$

The interpretation of (3.18b) is that the spectrum set of $y(t)$ is $\mathcal{S}_y = \{(f_k + f_c, a_k Ae^{j\varphi})\}$ which means that every frequency in $x(t)$ is shifted to the right by f_c, assuming $f_c > 0$, and the complex amplitudes are all multiplied by the complex amplitude of the complex exponential.

It is easy to plot this frequency shifting behavior for the spectrum. Figure 3-11(b) shows the spectrum for a signal that has been multiplied by a complex exponential with $f_c = 9$, $A = 1$, and $\varphi = 0$. The spectrum of $x(t)$ in Fig. 3-11(a) has seven spectral lines, and since there is conjugate symmetry, $x(t)$ is real. The signal $x(t)e^{j2\pi(9)t}$ has a spectrum that consists of the same seven spectral lines, but shifted up in frequency to be centered around $f = 9$ Hz. Note that this signal is no longer real, and its spectrum does not obey the conjugate symmetry property.

When the signal $x(t)$ is multiplied by a real sinusoid $A\cos(2\pi f_c t + \varphi)$, we need to apply the frequency shifting concept twice.

$$y(t) = x(t)A\cos(2\pi f_c t + \varphi) = x(t)\tfrac{1}{2}Ae^{j\varphi}e^{j2\pi f_c t} + x(t)\tfrac{1}{2}Ae^{-j\varphi}e^{-j2\pi f_c t} \tag{3.19}$$

The first term in (3.19) raises all the frequencies in $x(t)$ by f_c, which is a frequency shift to the right as shown in Fig. 3-11(b); the second term in (3.19) lowers all the

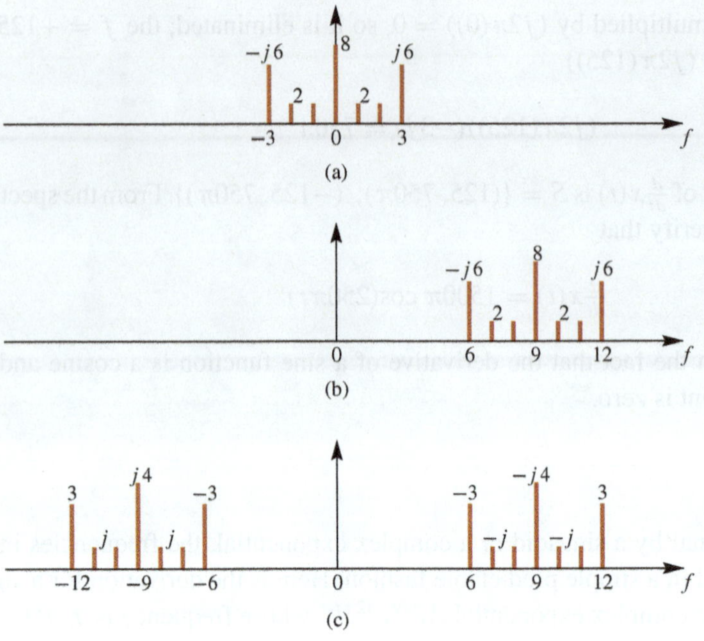

Figure 3-11 Spectrum plots for frequency shifting of a signal. (a) Spectrum of $x(t)$, (b) spectrum of $x(t)e^{j2\pi(9)t}$, and (c) spectrum of $x(t)\sin(2\pi(9)t)$. Note that the shifting frequency f_c is large enough so that the two shifted copies of the spectrum of $x(t)$ do not overlap, which simplifies the picture.

frequencies in $x(t)$ by f_c, which is a frequency shift to the left. The complex amplitudes are changed as follows: for the spectral lines shifted up by f_c, the new complex amplitudes are $a_k(\frac{1}{2}Ae^{j\varphi})$; for those shifted down, the new complex amplitudes are $a_k(\frac{1}{2}Ae^{-j\varphi})$, i.e., the phase change is opposite.

These two frequency shifts are illustrated in Fig. 3-11(c) for the signal $x(t)\sin(2\pi(9)t)$ which is the same as $x(t)\cos(2\pi(9)t - \frac{1}{2}\pi)$. When the spectrum of $x(t)$ in Fig. 3-11(a) is shifted up by 9 Hz, the complex amplitudes are multiplied by $\frac{1}{2}e^{-j\pi/2} = -j\frac{1}{2}$. For example, the spectral line at $f = 3$ in Fig. 3-11(a), which has a complex amplitude of $j6$, is moved to $f = 12$ in Fig. 3-11(c) and has a complex amplitude of $(j6)(-j\frac{1}{2}) = 3$. When the spectrum of $x(t)$ in Fig. 3-11(a) is shifted down by 9 Hz, the complex amplitudes are multiplied by $\frac{1}{2}e^{+j\pi/2} = j\frac{1}{2}$. Then the spectral line at $f = 3$ in Fig. 3-11(a) is moved to $f = -6$ in Fig. 3-11(c) and has a complex amplitude of $(j6)(j\frac{1}{2}) = -3$. After all the spectral lines are frequency shifted and their complex amplitudes are multiplied by $\frac{1}{2}Ae^{\pm j\varphi}$, the spectrum can be examined for conjugate symmetry. Since the signal $x(t)\sin(2\pi(9)t)$ is real, the conjugate symmetry property must be true in Fig. 3-11(c). As a final comment, we note that when f_c is greater than the largest frequency in the spectrum of $x(t)$, the upshifted and downshifted spectra do not overlap. This is the case in Fig. 3-11(c) which makes the plot easy to draw.

The frequency shifting property is based on the fact that multiplying two complex exponentials is equivalent to adding their exponents. In Chapter 7 we will show that this frequency shifting property is true for a much broader class of signals than just sums of sinusoids.

3-4 Periodic Waveforms

A periodic signal satisfies the condition that $x(t + T_0) = x(t)$ for all t, which states that the signal repeats its values every T_0 s. The time interval T_0 is called the ***period*** of $x(t)$, and, if it is the smallest such repetition interval, it is called the ***fundamental period***. In this section, we study how a sum of sinusoids can be used to synthesize a periodic signal, and we show that the summed sinusoids must have ***harmonically*** related frequencies; frequencies that are integer multiples of one frequency F_0. In other words, the signal would be synthesized as the sum of $N + 1$ sinusoids[8]

$$x(t) = A_0 + \sum_{k=1}^{N} A_k \cos(2\pi k F_0 t + \varphi_k) \tag{3.20}$$

where the frequency, f_k, of the kth cosine component in (3.20) is

$$f_k = kF_0 \qquad \text{(harmonic frequencies)}$$

The frequency f_k is called the kth ***harmonic*** of F_0 because it is an integer multiple of the basic frequency F_0, which is called the ***fundamental frequency*** if it is the largest such F_0.

Does the sum in (3.20) give a periodic signal, and, if so, what is the period of $x(t)$? The kth harmonic sinusoid has a shortest period of $1/(kF_0)$, but it also repeats k times every $1/F_0$ s. Since each cosine in (3.20) repeats with a period of $1/F_0$, the sum must have exactly the same repetition period, so, $x(t + 1/F_0) = x(t)$. It is easy to prove that this period, $T_0 = 1/F_0$, is the shortest repetition interval, so it is called the ***fundamental period***.[9] In other words, the fundamental period is the reciprocal of the fundamental frequency.

EXAMPLE 3-8 Fundamental Period T_0

Sometimes the fundamental period is not obvious. For example, the signal $x(t) = \cos^2(4\pi t)$ has a period of 0.5 s, but its fundamental period is actually $T_0 = 0.25$ s; because $\cos^2(4\pi(t + \frac{1}{4})) = \cos^2(4\pi t + \pi) = \cos^2(4\pi t)$.

EXAMPLE 3-9 Calculating F_0

The fundamental frequency is the *largest* F_0 such that $f_k = kF_0$ where k is an integer. In mathematical terms, it is related to the *greatest common divisor (gcd)* of a set of integers. If all the frequencies f_k are integers, then we can state

$$F_0 = \gcd\{f_k\} \qquad k = 1, 2, \ldots, N$$

[8] One of the sinusoids is the DC component A_0 which is a sinusoid with zero frequency.

[9] The fundamental period can be found using the least common multiple of the periods of all the sinusoids in (3.20).

For example, if the signal is the sum of sinusoids with frequencies 12, 20, and 60 Hz, then $F_0 = 4$ Hz, because 12 Hz is the 3rd harmonic, 20 Hz is the 5th harmonic, and 60 Hz is the 15th harmonic.

EXERCISE 3.4 Consider the case where the harmonic frequencies are not integers, for example, $\{f_k\} = \{3.5,\ 5.6,\ 9.8\}$ (in Hz). It might be possible to find a scaling constant λ such that λf_k is an integer for all k. In this particular case, $\lambda = 10$. Now we can use the gcd on the scaled frequencies as follows:

$$F_0 = (1/\lambda)\, \text{gcd}\, \{\lambda f_k\}$$

Apply this result to find the fundamental frequency for $\{3.5,\ 5.6,\ 9.8\}$ Hz.

Using the complex exponential representation of the cosines, we can also write (3.20) as

$$x(t) = \sum_{k=-N}^{N} a_k e^{j2\pi k F_0 t} \tag{3.21}$$

where a_k was defined by (3.7) in Section 3-1.1.

EXERCISE 3.5 Show that one possible period of the complex exponential signal $v_k(t) = e^{j2\pi k F_0 t}$ is $T_0 = 1/F_0$. Also, show that the fundamental period of $v_k(t)$ is $1/(kF_0)$.

3-4.1 Synthetic Vowel

As an example of synthesizing a periodic signal, consider a case where the fundamental frequency F_0 is 100 Hz, and where the sum in (3.21) contains nonzero terms for $\{a_{\pm 2},\ a_{\pm 4},\ a_{\pm 5},\ a_{\pm 16},\ a_{\pm 17}\}$. The numerical values of the complex amplitudes are listed in Table 3-1. The coefficients of the complex exponential signals were chosen so that the synthesized signal approximates the waveform produced by a male speaking the vowel sound "ah." The two-sided spectrum of this signal is plotted in Fig. 3-12 as two plots, Fig. 3-12(a) for the magnitudes, and Fig. 3-12(b) for the phases. This is different from Fig. 3-1, where it was convenient to make just one plot and label the nonzero spectral components with their complex amplitudes. Note that all the frequencies are integer multiples of 100 Hz, even though there is no spectral component at 100 Hz itself. Since the gcd of the frequencies is 100, the fundamental frequency is 100 Hz and the fundamental period is $1/100 = 0.01$ s. Also the negative frequency components in Fig. 3-12(b) have phase angles that are the negative of the phase angles of the corresponding positive frequency components, because the complex amplitudes for the negative frequencies are the complex conjugates of the complex amplitudes for the corresponding positive frequencies.

The synthetic vowel signal has ten nonzero spectral components, but only five sinusoidal terms after the positive and negative frequency terms are combined. It is

Table 3-1 Complex amplitudes for the periodic signal that approximates a complicated waveform like a vowel, such as "ah." The a_k coefficients are given for positive indices k, but the values for negative k are the conjugates, $a_{-k} = a_k^*$.

k	f_k (Hz)	a_k	Mag	Phase
1	100	0	0	0
2	200	$0.00772 + j0.122$	0.1223	1.508
3	300	0	0	0
4	400	$-0.08866 + j0.2805$	0.2942	1.877
5	500	$0.48 - j0.08996$	0.4884	−0.185
6	600	0	0	0
⋮	⋮	⋮	⋮	⋮
15	1500	0	0	0
16	1600	$0.01656 - j0.1352$	0.1362	−1.449
17	1700	$0.04724 + j0$	0.04724	0

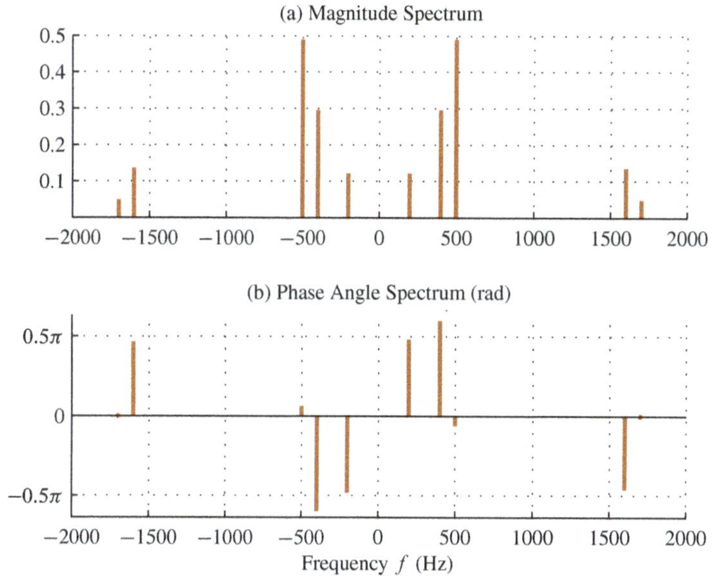

Figure 3-12 Spectrum of signal defined in Table 3-1. The magnitude (a) is an even function with respect to $f = 0$; the phase (b) is odd since $a_{-k} = a_k^*$.

interesting to examine the contribution of each real sinusoidal component separately. We can do this by successively plotting the waveforms corresponding to only one sinusoid, then two sinusoids, then three, etc. Figure 3-13(a) shows a plot of only the $k = 2$ sinusoidal term in Table 3-1. Note that since the frequency of this single component is $2F_0 = 200$ Hz, the waveform is periodic with period $1/200 = 5$ ms. Next, Fig. 3-13(b) shows $x_4(t)$, which is a plot of the sum of the $k = 2$ and $k = 4$ terms. Since these two frequencies are multiples of 200 Hz, the period of $x_4(t)$ is still 5 ms.

Figure 3-13(c) shows a plot of the sum of the first three terms, $x_5(t)$. Now we see that the period of the waveform is increased to 10 ms. This is because the three frequencies, 200, 400, and 500 Hz are integer multiples of 100 Hz; so, the fundamental

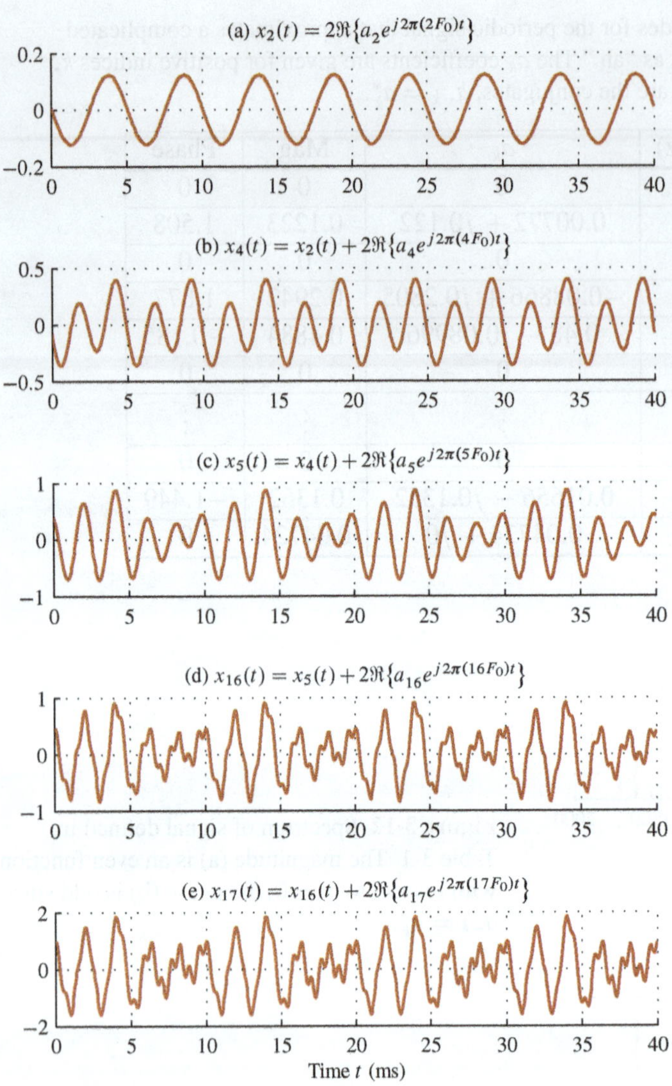

Figure 3-13 Summing all five terms in Table 3-1. (a) The 200-Hz term alone. (b) Sum of the 400-Hz and 200-Hz terms. Additional terms are added one at a time until the entire synthetic vowel signal is created in (e). (c) Adding the 500-Hz term, which changes the fundamental period, (d) adding the 1600-Hz term, and (e) adding the 1700-Hz term.

frequency is now 100 Hz. Figure 3-13(d) shows that the addition of the component with frequency 1600 Hz does not change the period. Finally, Fig. 3-13(e) shows $x_{17}(t)$, the sum of all the terms in Table 3-1. Note that the period of $x_{17}(t)$ is $T_0 = 10$ ms, which equals $1/F_0$, even though there is no component with frequency F_0. The signal $x_{17}(t)$ is typical of waveforms for vowel sounds in speech although many more frequencies might be needed to accurately represent the waveform of a natural speech signal. The high frequencies in the signal contribute the fine detail in the waveform. This is evident in Fig. 3-13 as the waveform becomes increasingly complicated and more rapidly varying as higher-frequency components such as the 16th and 17th harmonics are added. The Companion Website has a more detailed presentation of this *Vowel Synthesis* example.

DEMO
Vowel Synthesis

3-4.2 Sine-Cubed Signal

In Section 3-2 we determined the spectra of products of sine waves by expanding each sine wave into complex exponentials using the inverse Euler relation. After multiplying the complex exponential forms and simplifying exponents, it was possible to obtain a representation of the product as a sum of complex exponentials. The spectrum was then obtained by associating each complex amplitude with the corresponding frequency. This same approach can be used with other combinations of sine waves including those involving sine waves raised to a power. Raising a periodic function to a power results in a periodic function with the same or possibly shorter period. A simple example that illustrates this point is the sine-cubed signal

$$x(t) = \sin^3(4\pi t) \tag{3.22}$$

which is plotted in Fig. 3-14. Our objective is to obtain a representation of $x(t)$ as a sum of complex exponentials in the form (3.21), so that we may easily determine the spectrum of $x(t)$.

Using the inverse Euler formula for $\sin(\cdot)$ along with the binomial expansion for $(a-b)^3$, we can rewrite the sine-cubed function as a sum:

$$\begin{aligned} x(t) &= \left(\frac{e^{j4\pi t} - e^{-j4\pi t}}{2j}\right)^3 \\ &= \tfrac{1}{-8j}\left(e^{j12\pi t} - 3e^{j8\pi t}e^{-j4\pi t} + 3e^{j4\pi t}e^{-j8\pi t} - e^{-j12\pi t}\right) \\ &= \tfrac{j}{8}e^{j12\pi t} + \tfrac{-3j}{8}e^{j4\pi t} + \tfrac{3j}{8}e^{-j4\pi t} + \tfrac{-j}{8}e^{-j12\pi t} \end{aligned} \tag{3.23}$$

We see that (3.23) is the *sum of four terms* with radian frequencies: $\omega = \pm 4\pi$ and $\omega = \pm 12\pi$ rad/s. Since $\gcd(4, 12)\pi = 4\pi$, the fundamental frequency is $\omega_0 = 4\pi$ rad/s (or $F_0 = 2$ Hz). The frequencies in (3.23) are the first and third harmonics of the fundamental frequency. Therefore, $N = 3$ and $T_0 = 0.5$ s in (3.21). The coefficients in (3.21) are indexed in terms of the harmonic number k, so

$$a_k = \begin{cases} 0 & \text{for } k = 0 \\ \mp j\frac{3}{8} & \text{for } k = \pm 1 \\ 0 & \text{for } k = \pm 2 \\ \pm j\frac{1}{8} & \text{for } k = \pm 3 \end{cases} \tag{3.24}$$

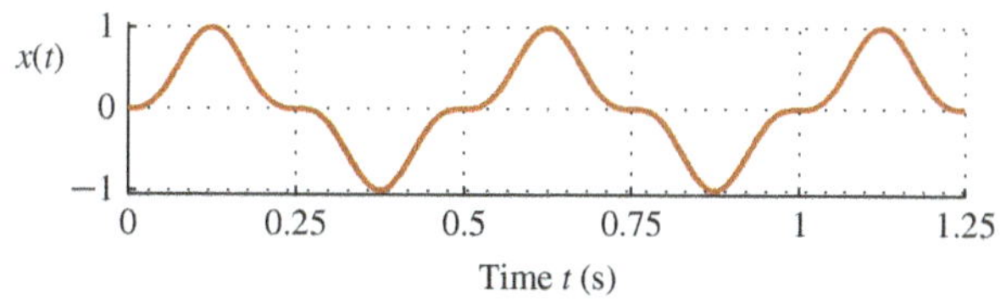

Figure 3-14 Waveform of the signal $x(t) = \sin^3(4\pi t)$.

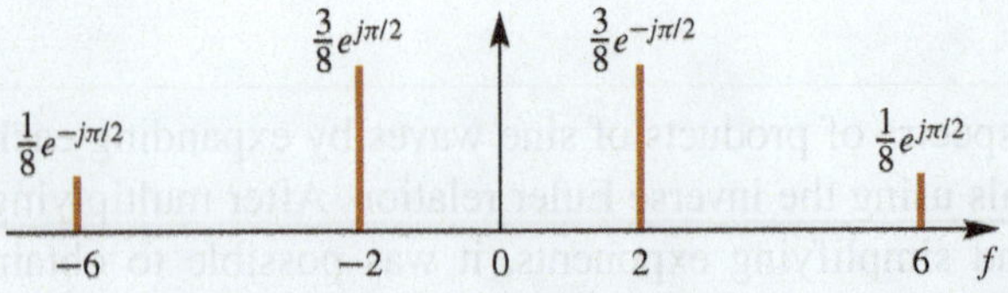

Figure 3-15 Spectrum of the "sine-cubed" signal derived from its representation as a sum of complex exponentials. Only the range from −6 to +6 Hz is shown. The spectrum is zero outside this interval. The complex amplitude of each spectrum line is equal to the coefficient a_k for that frequency, kF_0.

Now we can plot the spectrum because we know that we have four nonzero a_k components located at the four frequencies: $f = \{-6, -2, 2, 6\}$ Hz. In Fig. 3-15 the spectrum is plotted versus frequency in hertz, so the third harmonic is at 6 Hz, and the second harmonic is missing.

3-4.3 Example of a Non-Periodic Signal

When we add harmonically related complex exponentials, we get a periodic signal, but what happens when the frequencies have no harmonic relation to one another? The sinusoidal synthesis formula

$$x(t) = a_0 + \sum_{k=1}^{N} 2|a_k| \cos(2\pi f_k t + \angle a_k)$$

$$= a_0 + \sum_{k=1}^{N} \left(a_k e^{j2\pi f_k t} + a_k^* e^{-j2\pi f_k t}\right)$$

is still valid, but we want to demonstrate that periodicity is tied to harmonic frequencies, so we present a simple example.

First, we make a harmonic signal $x_h(t)$ using $a_1 = 1$, $a_3 = -1/3$, and $a_5 = 1/5$ with fundamental frequency $F_0 = 10$ Hz:

$$x_h(t) = 2\cos(20\pi t) - \tfrac{2}{3}\cos(20\pi(3)t) + \tfrac{2}{5}\cos(20\pi(5)t) \tag{3.25}$$

A plot of $x_h(t)$ is shown in Fig. 3-16 using a "strip chart" format. The plot consists of three lines, each one containing 2 s of the signal. The first line starts at $t = 0$, the second at $t = 2$ s, and the third at 4 s. This lets us view a long section of the signal, and we can see that $x_h(t)$ is clearly periodic, with fundamental period equal to 0.1 s. The amplitudes in (3.25) are proportional to those of a periodic square wave, which is discussed in Appendix C. For our purpose here, it is sufficient to note that because the frequencies are harmonic, the signal is periodic with period 0.1 s.

DEMO
Spectrograms: Simple Sounds: Square Waves

To illustrate what happens when the frequencies are not harmonic, we create a second signal where the complex amplitudes are the same but the frequencies are slightly perturbed from those of $x_h(t)$. Defining $x_{nh}(t)$ to be the sum of three sinusoids:

$$x_{nh}(t) = 2\cos(20\pi t) - \tfrac{2}{3}\cos(20\pi(2\sqrt{2})t) + \tfrac{2}{5}\cos(20\pi(3\sqrt{3})t)$$

The complex amplitudes are the same as in (3.25), but the frequencies have been changed slightly to irrational numbers (in Hz). Hence, there is no gcd and no fundamental

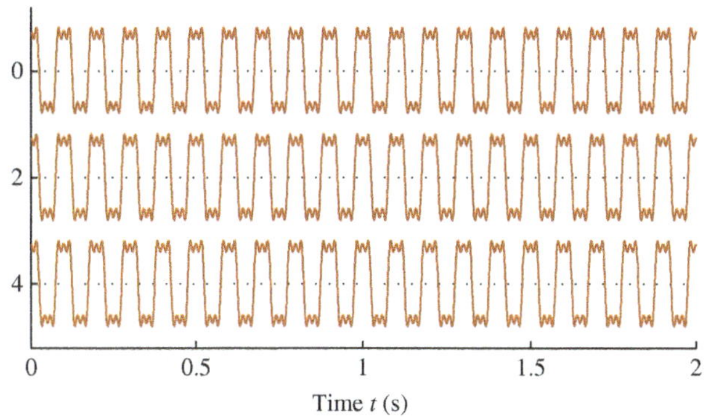

Figure 3-16 Sum of three cosine waves with harmonic frequencies. The spectrum is shown in Fig. 3-18(a), and the fundamental frequency of $x_h(t)$ is 10 Hz.

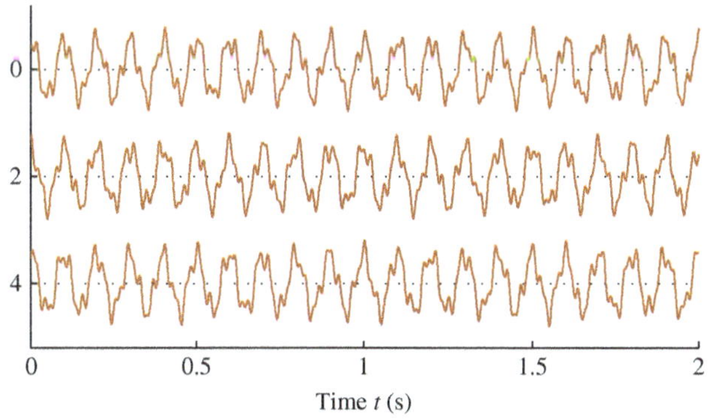

Figure 3-17 Sum of three cosine waves with nonharmonic frequencies, $x_{nh}(t)$. The spectrum is shown in Fig. 3-18(b).

frequency. The plot of $x_{nh}(t)$ in Fig. 3-17 is certainly not periodic, although the plot hints at a strong periodic component with period 0.1 s.[10] No matter how hard you try, you cannot find an exact repetition in this signal, yet the complex amplitudes are the same and the three frequencies are only slightly different from those of the waveform of Fig. 3-16.

The spectrum plots in Fig. 3-18 help explain the difference between the signals in Figs. 3-16 and 3-17. In Fig. 3-18(a), the frequencies are integer multiples of a common frequency, $F_0 = 10$ Hz, so the waveform of Fig. 3-16 is periodic with period $T_0 = 1/10$ s. The waveform of Fig. 3-17 is *nonperiodic*, because the frequencies are irrational numbers and are slightly different: $20\sqrt{2} \approx 28.28$ instead of 30, and $30\sqrt{3} \approx 51.96$ instead of 50. These slight shifts of frequency make a dramatic difference in the time waveform. For periodicity, we would need a fundamental frequency in the "frequency domain" of Fig. 3-18(b). To see that there is no fundamental, let us assume that $20\sqrt{2}$ is the pth harmonic, and $30\sqrt{3}$ is the qth harmonic. Then the ratio $(20\sqrt{2})/(30\sqrt{3})$ has to be the rational number p/q. Since that is impossible, we conclude that there is no fundamental frequency.

[10] There are 20 "major" oscillations on the first line, which corresponds to 2 s of time. Thus, the "period" of the waveform appears to be approximately $2/20 = 0.1$ s, but in fact, the signal never repeats exactly. The dominance of the 0.1-second period occurs because the sinusoidal component with frequency 10 Hz has an amplitude that is several times greater than the other two components.

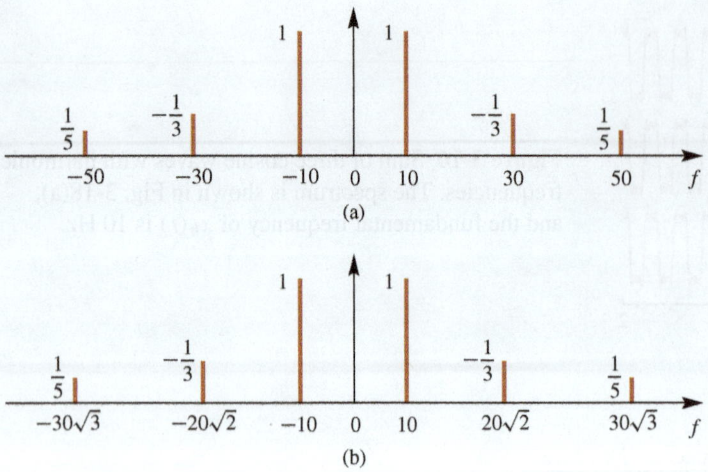

Figure 3-18 Spectrum of (a) the harmonic waveform $x_h(t)$ in Fig. 3-16, which has a period of 0.1 s, and (b) the nonharmonic waveform $x_{nh}(t)$ in Fig. 3-17 that is not periodic.

3-5 Fourier Series

The examples in Section 3-4 show that we can synthesize a variety of *periodic* waveforms by using a sum of *harmonically related* sinusoids. Now, we want to explore the question: *Can every periodic signal be synthesized as a sum of harmonically related sinusoids?* So far, we have only shown examples for N finite when we used (3.21) to obtain the plots of waveforms shown in Section 3-4. However, as we now show, if we let N be infinite, virtually all periodic waveforms can be synthesized with a sum of complex exponential signals.

Henceforth, when we mention sums of complex exponentials, we assume that N can be either finite or infinite as needed. This is expressed in general by using what we call the ***Fourier synthesis summation***:[11]

Fourier Synthesis Summation

$$x(t) = \sum_{k=-\infty}^{\infty} a_k e^{j(2\pi/T_0)kt} \tag{3.26}$$

This notation includes both the finite and infinite cases, since in the finite case, we can simply assume that $a_k = 0$ for $|k| > N$. Perhaps a more common name for the mathematical expression in (3.26) is *Fourier series*, but we prefer the more descriptive term Fourier synthesis summation. The kth complex exponential in (3.26) has a frequency equal to $f_k = k/T_0$ Hz, so all the frequencies are integer multiples of the fundamental frequency $F_0 = 1/T_0$ Hz.[12] Now we want to describe a general theory that shows how

[11] Jean-Baptiste Joseph Fourier (1768–1830) was a French mathematician and physicist who is credited with introducing the idea of representing functions by sums of sinusoids. This contribution is recognized today as one of the most important methods in science and engineering, and the field of mathematics that grew out of his work is justifiably named in his honor.

[12] There are three ways to refer to the fundamental frequency: radian frequency ω_0 in rad/s, cyclic frequency F_0 in Hz, or period T_0 in seconds. Each one has its merits in certain situations. The relationship among these is $\omega_0 = 2\pi F_0 = 2\pi/T_0$. Note that $f_0 = 0/T_0$ is not the fundamental frequency; rather, it is $f_1 = 1/T_0 = F_0$.

to find the coefficients a_k so that *any periodic signal can be synthesized with a sum of harmonically related sinusoids*, even when the sum may need an infinite number of terms. This is the mathematical theory of ***Fourier series***.

There are two aspects of the Fourier theory: analysis and synthesis. Starting from $x(t)$ and calculating $\{a_k\}$ is called ***Fourier analysis***. The reverse process of starting from $\{a_k\}$ and generating $x(t)$ is called ***Fourier synthesis***, which was the focus of Section 3-4. In order to answer the question about synthesizing every periodic signal via (3.26), we must learn how Fourier analysis is done.

3-5.1 Fourier Series: Analysis

The Fourier coefficients $\{a_k\}$ for the harmonic sum in (3.26) can be derived from the signal $x(t)$ by using the ***Fourier analysis integral***.

Fourier Analysis Integral

$$a_k = \frac{1}{T_0}\int_0^{T_0} x(t)e^{-j(2\pi/T_0)kt}\,dt \tag{3.27}$$

Only one period of $x(t)$ is required for the integrand, so T_0 should be the fundamental period of $x(t)$. Note that t plays the role in (3.27) of a "dummy variable of integration" that disappears from the result after substituting the limits. It would perhaps be less confusing if we used a different symbol, since t is also the symbol used for the independent variable in (3.26). However, it is generally not necessary to make such a careful distinction when we use (3.27) and (3.26) as a pair that together comprise the Fourier representation of a periodic signal. An important point, however, is to note that after (3.27) is evaluated, the variable t is integrated out and the result depends upon k alone, and time only re-enters through (3.26). Thus, it is common to refer to the function $x(t)$ as the *time-domain representation* of a periodic signal, and the coefficients a_k as the *frequency-domain representation* or the *spectrum representation* of the signal.

The Fourier integral (3.27) can be evaluated if we have a formula that defines $x(t)$ over one period. As the example in Section 3-5.2 shows, the definite integral would be evaluated to obtain a formula for a_k as a function of k. Several more examples are presented in Appendix C to illustrate the mechanics of evaluating the Fourier analysis integral. On the other hand, if $x(t)$ is known only as a recording, then numerical methods such as those discussed in Chapter 8 will be needed.

3-5.2 Analysis of a Full-Wave Rectified Sine Wave

In this section, we carry out the Fourier analysis of a signal called the full-wave rectified sine (FWRS), an example of which is shown in Fig. 3-19(b). This signal is found in DC power supplies as well as AC to DC power converters—the ubiquitous devices that are needed to recharge batteries in cell phones, laptops, and so on. Battery charging requires constant DC current, but plugging into a wall socket gets AC (alternating current) which,

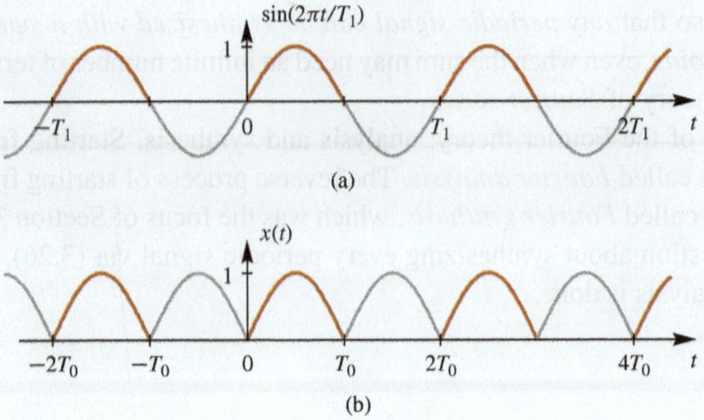

Figure 3-19 (a) Sinusoid with period T_1. (b) Periodic FWRS signal with period T_0, which is created by taking the absolute value of the sinusoid in (a).

in our terms, is a pure sinusoid with zero DC value. A simple electronic circuit built with diodes will turn a sinusoid into a signal with nonzero DC component which can then be cleaned up by filtering and used as a DC power source for recharging.

The mathematical formula for the FWRS signal is the absolute value of a sinusoid.

$$x(t) = |\sin(2\pi t/T_1)| \tag{3.28}$$

The fundamental period T_0 is equal to $\frac{1}{2}T_1$ because the negative lobe of the sinusoid is identical to the positive lobe after being flipped by the absolute value operator, as illustrated by Fig. 3-19. The FWRS shown in Fig. 3-19(b) is a continuous signal, but its first derivative is discontinuous, resulting in the sharp points at locations where the signal is zero (at $t = 0,\ T_0,\ 2T_0, \ldots$). For 50 Hz AC power[13] the period of the sinusoid is $T_1 = 0.02$ s but after "rectification" the period would be halved as shown in Fig. 3-19(b), so the fundamental frequency of the FWRS would be $F_0 = 100$ Hz.

The Fourier integral for this case is the following definite integral

$$a_k = \frac{1}{T_0}\int_0^{T_0} |\sin(2\pi t/T_1)|\, e^{-j(2\pi/T_0)kt}\, dt \tag{3.29a}$$

where $T_0 = \frac{1}{2}T_1$. Since $\sin(2\pi t/T_1) \geq 0$ over the interval specified by the limits on t from 0 to T_0, we can write

$$a_k = \frac{1}{T_0}\int_0^{T_0} \sin(2\pi t/T_1) e^{-j(2\pi/T_0)kt}\, dt \tag{3.29b}$$

The integration in (3.29b) can be carried out efficiently after using the inverse Euler formula for sine to write (3.29b) as the sum of two integrals

[13] In most of the world, power companies distribute AC power at 50 hertz, but, in the Americas, 60 hertz is common.

$$a_k = \frac{1}{2jT_0}\int_0^{T_0} e^{j(2\pi/T_1)t}e^{-j(2\pi/T_0)kt}\,dt + \frac{-1}{2jT_0}\int_0^{T_0} e^{-j(2\pi/T_1)t}e^{-j(2\pi/T_0)kt}\,dt \tag{3.30}$$

Each integrand can be combined into a single complex exponential which is then easy to integrate. Since $T_1 = 2T_0$, the integrands in (3.30) are the complex exponentials

$$e^{\pm j(2\pi/2T_0)t}e^{-j(2\pi/T_0)kt} = e^{j(\pm 1-2k)(\pi/T_0)t}$$

The definite integrals can be carried out and simplified as follows:

$$\int_0^{T_0} e^{j(\pm 1-2k)(\pi/T_0)t}dt = \left.\frac{e^{j(\pm 1-2k)(\pi/T_0)t}}{j(\pm 1-2k)(\pi/T_0)}\right|_0^{T_0} = \frac{e^{j(\pm 1-2k)((\pi/T_0)T_0)}-1}{j(\pm 1-2k)(\pi/T_0)} \tag{3.31}$$

The term in the numerator can be simplified as

$$e^{j(\pm 1-2k)((\pi/T_0)T_0)} - 1 = \underbrace{e^{j(\pm 1-2k)\pi}}_{\text{always }=-1} -1 = -2 \tag{3.32}$$

The complex exponential is always equal to -1 because its angle is an odd multiple of π, since $(\pm 1 - 2k)$ is always an odd integer for any integer k.

Now we apply (3.31) and (3.32) to the two integrals in (3.30) to obtain a simple expression for a_k

$$a_k = \frac{1}{2jT_0}\left(\frac{-2}{j(1-2k)(\pi/T_0)} - \frac{-2}{j(-1-2k)(\pi/T_0)}\right) \tag{3.33a}$$

$$= \frac{-2}{-2\pi(1-2k)} - \frac{-2}{-2\pi(-1-2k)} \tag{3.33b}$$

$$= \frac{1}{\pi(1-2k)} + \frac{1}{\pi(1+2k)} \tag{3.33c}$$

After combining the two terms in (3.33c) over a common denominator, the final result for a_k can be written as

$$a_k = \frac{2}{\pi(1-4k^2)} \tag{3.34}$$

Thus, we have shown that the FWRS wave can be represented as the Fourier series

$$x(t) = |\sin(2\pi t/T_1)| = \sum_{k=-\infty}^{\infty} \frac{2}{\pi(1-4k^2)}e^{j(2\pi/T_0)kt} \tag{3.35a}$$

$$= \frac{2}{\pi} + \sum_{k=1}^{\infty}\frac{4}{\pi(1-4k^2)}\cos(2\pi kt/T_0) \tag{3.35b}$$

where $T_0 = T_1/2$.

3-5.3 Spectrum of the FWRS Fourier Series

When we discussed the spectrum in Section 3-1, we described a procedure for plotting the spectrum when $x(t)$ is composed of a sum of complex exponentials. By virtue of the synthesis formula (3.26), the Fourier series coefficients a_k are, in fact, the complex amplitudes that define the spectrum of $x(t)$. Therefore, a spectrum plot of a periodic signal has lines only at multiples of the fundamental frequency $F_0 = 1/T_0$. Usually it is convenient to draw the spectrum line at frequency kF_0 with length equal to $|a_k|$, but in Fig. 3-20 the lengths are not to scale because the magnitude of a_k in (3.34) drops off very quickly and the lines would barely be visible for $|k| \geq 3$, e.g., $|a_3| = 2/(35\pi) \approx 0.018$. The spectrum plot is a complete representation if we also label each spectrum line with the corresponding complex value of a_k, but note that the FWRS spectrum is purely real. In Fig. 3-20, the spectrum is plotted only for $-4F_0 \leq f \leq 4F_0$, but the symbol $\cdots$ indicates that the spectrum continues for higher frequencies which all have nonzero coefficients as $kF_0 \to \infty$.

3-5.3.1 DC Value of Fourier Series

The Fourier series coefficient for $k = 0$ has a special interpretation as the *average value* of the signal $x(t)$. If we set $k = 0$ in the analysis integral (3.27), then we have

DC Component

$$a_0 = \frac{1}{T_0}\int_0^{T_0} x(t)\,dt \tag{3.36}$$

The integral is the area under the function $x(t)$ for one period. If we think of the area as a sum and realize that dividing by T_0 is akin to dividing by the number of elements in the sum, we can interpret (3.36) as the average value of the signal. The a_0 coefficient is the amplitude of a complex exponential of zero frequency (i.e., $0F_0 = 0$ Hz). Thus, it is a constant signal component. Hence, in electrical engineering it is also called the "DC component" of the signal. For the FWRS example, the DC value is nonzero

$$a_0 = \left.\frac{2}{\pi(1-4k^2)}\right|_{k=0} = 2/\pi \approx 0.6366$$

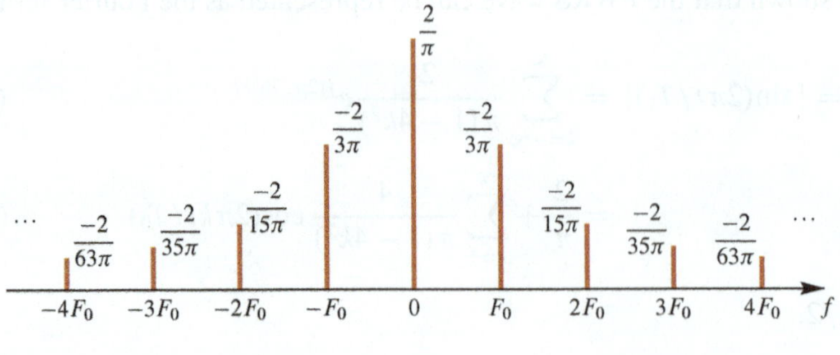

Figure 3-20 Spectrum of the periodic FWRS signal represented in terms of its Fourier series coefficients a_k and their corresponding frequencies kF_0. *Note:* Vertical lines are not to scale.

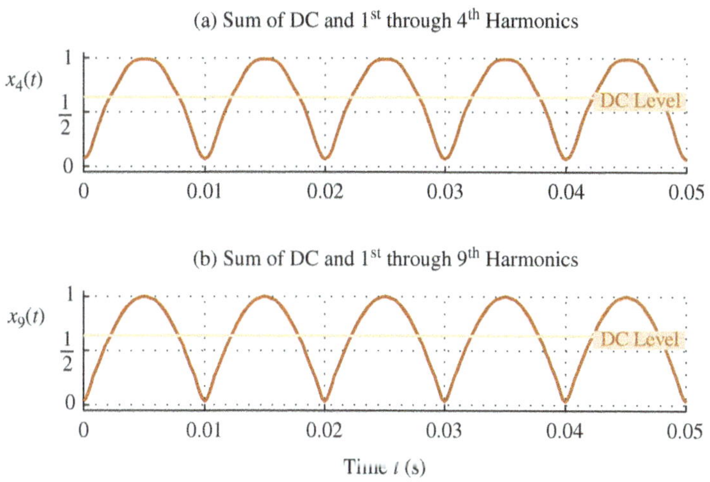

Figure 3-21 Summing harmonic components for the FWRS signal via (3.37); (a) $N = 4$ and (b) $N = 9$.

which is 63.66% of the maximum value of the rectified sine as shown in Fig. 3-21. For the AC-DC converter application, we can observe that the operation of taking the absolute value has produced an output signal with nonzero DC from an AC input that has zero DC, which is exactly what is needed.

3-5.3.2 Finite Synthesis of a Full-Wave Rectified Sine

DEMO
Fourier Series

The fact that the spectrum of a FWRS signal has an infinite number of spectral lines means that it has infinite bandwidth. To be specific, the millionth harmonic in the FWRS spectrum is nonzero, so is the billionth harmonic, and so on. Fortunately, the magnitude of these very high harmonics is extremely small—for example, the magnitude of the millionth harmonic is $\approx 10^{-13}$. In fact, (3.34) says that the magnitude falls off as k^{-2} as $k \to \infty$. A bandlimited signal could be obtained by dropping all the very high harmonics, and it might be an almost perfect representation of the FWRS signal. Therefore, it is useful to study the approximate resynthesis of a FWRS signal from a finite number of its spectral lines, and ask questions like "How good is the approximation when using $N = 4$, or $N = 9$, or $N = 20$ spectral lines?"

An approximate signal can be synthesized according to the *finite* Fourier synthesis summation formula[14]

$$x_N(t) = \sum_{k=-N}^{N} a_k e^{j(2\pi/T_0)kt} \tag{3.37}$$

The synthesis formula (3.37) usually can be simplified to a cosine form. For the particular case of the FWRS coefficients (3.34), when we take the DC value plus the first and second

[14] If you have access to MATLAB, it is straightforward to write a *Fourier Synthesis Program* that implements (3.37). This MATLAB programming exercise is described in more detail in the Music Synthesis Lab.

harmonic terms (i.e., $N = 2$), we get the sum of two sinusoids plus the constant (DC) level derived in (3.38).

$$\begin{aligned} x_2(t) &= a_{-2}e^{-j(4\pi/T_0)t} + a_{-1}e^{-j(2\pi/T_0)t} + a_0 + a_1 e^{j(2\pi/T_0)t} + a_2 e^{j(4\pi/T_0)t} \\ &= \frac{-2}{15\pi}e^{-j(4\pi/T_0)t} + \frac{-2}{3\pi}e^{-j(2\pi/T_0)t} + \frac{2}{\pi} + \frac{-2}{3\pi}e^{j(2\pi/T_0)t} + \frac{-2}{15\pi}e^{j(4\pi/T_0)t} \\ &= \frac{2}{\pi} + \frac{-2}{3\pi}(e^{j(2\pi/T_0)t} + e^{-j(2\pi/T_0)t}) + \frac{-2}{15\pi}(e^{j(4\pi/T_0)t} + e^{-j(4\pi/T_0)t}) \\ &= \frac{2}{\pi} - \frac{4}{3\pi}\cos(2\pi t/T_0) - \frac{4}{15\pi}\cos(4\pi t/T_0) \end{aligned} \tag{3.38}$$

LAB

Synthesis of Sinusoidal Signals

Using a simple MATLAB M-file, a synthesis was done via (3.37) with a fundamental frequency of $F_0 = 100$ Hz. Since the fundamental period is $T_0 = 1/100 = 0.01$ s, the harmonic frequencies are $(2\pi/T_0)k = (200\pi)k$ rad/s. Two cases are shown in Fig. 3-21(a) for $N = 4$ and Fig. 3-21(b) for $N = 9$. Notice that the period of the synthesized waveform is always the same, because it is determined by the fundamental frequency. After comparing the waveforms in Fig. 3-21 to the ideal continuous FWRS signal in Fig. 3-19, we see that a finite Fourier sum approximation (3.37) is roughly correct even with only the first four harmonics.

As more harmonic terms are added, the approximating formula is still a finite sum of sinusoids but higher frequency terms are included. In Fig. 3-21 the worst approximation is at the integer multiples of T_0 (i.e., 0, 0.1, 0.2, ...) where the FWRS should be zero. At these locations the derivative of the FWRS is discontinuous so higher frequency terms are needed. In the $N = 9$ case, the approximation is very close to the original waveform. Adding harmonics for $N > 9$ will improve the synthesis very little, mostly by sharpening the dips at $t = 0.01, 0.02, \ldots$. The $N = 4$ case is a poorer approximation to the true FWRS signal, because it is not using enough sinusoidal terms. It is possible to quantify the approximation error by plotting the spectrum as in Fig. 3-20, and examining the relative magnitudes of the higher frequency components being dropped for $k > 4$ (see Section C-4.3). The `fourierseries` MATLAB demo, described in Appendix C, provides an interactive GUI where the number of terms can be varied to study how the quality of the approximation improves with more terms.

As a final comment, we return to the application of AC-DC converters where the purpose of full wave rectification of a power line sine wave is to obtain a waveform whose DC value is nonzero. After rectification, additional processing is used to remove the fundamental and all higher harmonics of the original sine wave. The result is a signal that is only the DC level in Fig. 3-21. In a practical system comprised of resistor and capacitor circuit elements, the higher frequencies can be reduced in magnitude, but not completely removed, so a small "ripple" with frequency F_0 remains on the (approximate) DC signal.

3-6 Time–Frequency Spectrum

We have seen that a wide range of interesting waveforms can be synthesized by a sum of sinusoids

$$x(t) = A_0 + \sum_{k=1}^{N} A_k \cos(2\pi f_k t + \varphi_k) \tag{3.39}$$

These waveforms range from constants, to cosine signals, to general periodic signals, to complicated-looking signals that are not periodic. One assumption we have made so far is that the amplitudes, phases, and frequencies in (3.39) *do not change with time*. However, most real-world signals exhibit frequency changes over time. Music is a good example. For very short time intervals, the music may have a "constant" spectrum, but over the long term, the frequency content of a musical piece changes dramatically. Indeed, the changing frequency spectrum is the very essence of music. Human speech is another good example. Vowel sounds, if held for a long time, exhibit a "constant" nature because the vocal tract resonates with its characteristic frequencies, and the vocal cords vibrate with a more or less constant fundamental frequency. However, as we speak different words, the frequency content is continually changing. In any event, most interesting signals can be modelled as a sum of sinusoids if we let the frequencies, amplitudes, and phases vary with time. Therefore, we need a way to describe such time–frequency variations which leads us to study the concept of a time–frequency spectrum. The most common time–frequency representation is the ***spectrogram***.

The mathematical concept of a time–frequency spectrum is a sophisticated theoretical idea, but the intuitive notion of a time-varying spectrum is supported by common, everyday examples. A non-engineering example to cite is musical notation (Fig. 3-22). A musical score specifies how a piece is to be played not only by giving the notes to be played, but also the time duration of each note, and the starting time of each. The notation itself is not completely obvious, but many students have seen it while learning a musical instrument. In Fig. 3-22 the horizontal "axis" is time, while the vertical axis is frequency. The time duration for each note is encoded in the notehead, which might be open or filled, the stem, which is usually present, and flags, which might be attached to the stem. The encoded duration is sufficient to define a whole note, half note, quarter note, eighth, sixteenth, etc. In Fig. 3-22 most of the notes are sixteenth notes, indicating that the piece should be played briskly. If we assign a time duration to a sixteenth note, then all sixteenth notes should have the same duration. An eighth note would have twice

Figure 3-22 Sheet music notation is a time–frequency diagram.

the duration of a sixteenth note, and a quarter note would be four times longer than a sixteenth note, and so on.

The vertical (frequency) axis uses a much more complicated notation to define frequency. If you look carefully at Fig. 3-22, you can see that the black dots that mark the notes lie either on one of the horizontal lines or in the space between two lines. Each of these positions denotes a white key on the piano keyboard depicted in Fig 3-23, and each key produces a tone of a different frequency. The black keys on the piano require additional notation which is done by using "sharps" (♯) or "flats" (♭). Figure 3-22 has a few notes sharped, e.g., $D_5^\sharp$. The musical score is divided into a treble section (the top five lines) and a bass section (the bottom five lines). The vertical reference point for the notes is "middle C," which lies on an invisible horizontal line between the treble and bass sections (key number 40 in Fig. 3-23). Thus the bottom horizontal line in the treble section represents the white key (E_4) that is two above middle C, that is, key number 44 in Fig. 3-23.

Once the mapping from the musical score to the piano keys has been made, we can write a mathematical formula for the frequency of each note. A piano keyboard with 88 keys is divided into octaves containing twelve keys each. The frequencies of keys in an ***octave*** are twice those in the next lower octave. Within an octave, the neighboring keys maintain a constant frequency ratio (r). Since there are twelve keys per octave, the ratio (r) can be found as follows:

LAB

Synthesis of Sinusoidal Signals

$$r^{12} = 2 \quad \Longrightarrow \quad r = 2^{1/12} = 1.0595$$

With this ratio, we can compute the frequencies of all keys if we have one reference. The convention is that the A key above middle C, called A-440, has frequency 440 Hz. Since A-440 is key number 49 and middle C is key number 40, the frequency of middle C is

$$f_{\text{middle C}} = 440 \times 2^{(40-49)/12} \approx 261.6 \text{ Hz}$$

LAB

FM Synthesis for Musical Instruments

It is not our objective to explain how to read sheet music, although two of the lab projects on the Companion Website investigate methods for synthesizing waveforms to create songs and musical sounds. What is interesting about musical notation is that it uses a two-dimensional display to indicate frequency content that changes with time. If we adopt a similar notation, we can specify how to synthesize sinusoids with time-varying frequency content. A very simple notation is illustrated in Fig. 3-24. This time-frequency plot shows the frequencies of each note and their durations (0.2 s).

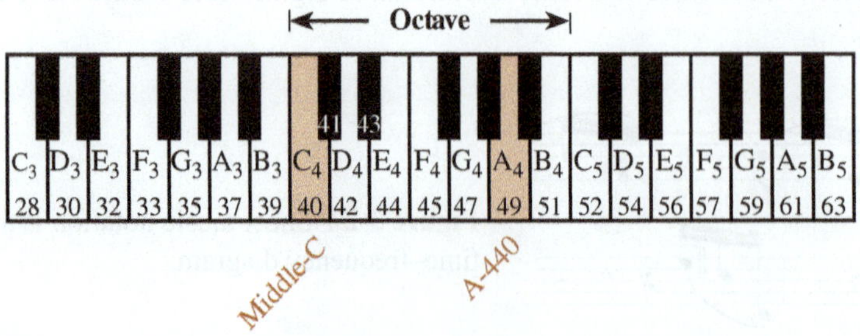

Figure 3-23 Piano keys can be numbered from 1 to 88. Three octaves are shown. Middle C is key 40. A-440 is key 49. The subscripts on the note names denote octave numbers.

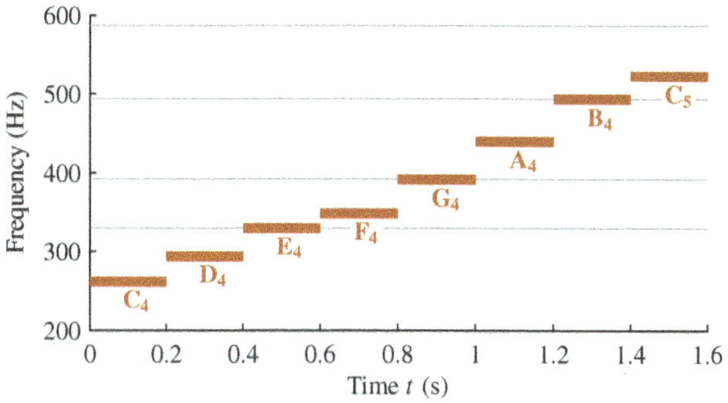

Figure 3-24 Ideal time–frequency diagram for playing the C-major scale. The horizontal dotted lines correspond to the first four horizontal lines in the treble staff of sheet music (Fig. 3-22).

3-6.1 Stepped Frequency

The simplest example of time varying frequency content is to make a waveform whose frequency stays constant for a short duration and then steps to a higher (or lower) frequency. Such stepped-frequency waveforms are also used in radar transmitters where frequency coverage and high power are needed with simple transmit hardware. In the case of music, an example of a simple stepped-frequency song would be a ***scale*** which is a succession of notes progressing over one octave. For example, the C-major scale consists of playing the notes $\{C_4, D_4, E_4, F_4, G_4, A_4, B_4, C_5\}$ one after another, starting at C_4 which is middle C. This scale is played completely on the white keys. The frequencies of these notes are:

Middle C	D_4	E_4	F_4	G_4	A_4	B_4	C_5
262 Hz	294	330	349	392	440	494	523

A graphical presentation of the C-major scale is shown in the time-frequency plot of Fig. 3-24. It should be interpreted as follows: Synthesize the frequency 262 Hz as $\cos(2\pi(262)t)$ for 200 ms, then the frequency 294 Hz as $\cos(2\pi(294)t)$ during the next 200 ms, and so on. The total waveform duration is 1.6 s. In music notation, the notes would be written as in Fig. 3-25(a), where each note is a quarter note.

3-6.2 Spectrogram Analysis

The frequency content of a signal can be considered from two points of view: analysis or synthesis. For example, the ideal time–frequency diagram in Figure 3-24 specifies a rule for synthesizing the C-major scale. Analysis is a more challenging problem, as we saw in Section 3-5, where the Fourier series analysis integral (3.27) was given.

Analysis for time-varying frequencies is an important topic of both research and application in most signal processing courses even up to advanced graduate courses. One reason is that there are several ways to write a mathematical formula like the Fourier series integral to do the analysis. On the other hand, excellent numerical routines are now available for time–frequency analysis. Specifically, it is easy to compute a ***spectrogram***, which is a two-dimensional function of time and frequency that displays a measurement of the time variation of the spectral content of a signal. The Companion Website has a demonstration of real sounds and their spectrograms, including piano sounds.

DEMO

Spectrograms of Real Sounds

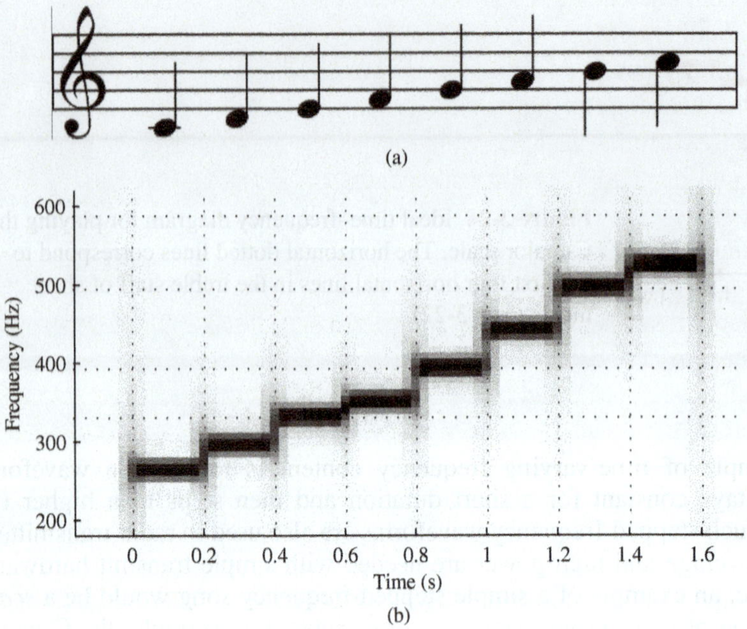

Figure 3-25 (a) Musical notation for the C-major scale, and (b) corresponding spectrogram computed using MATLAB's `spectrogram` function. The horizontal dotted lines denote the frequencies of the first four lines of the musical staff.

Chapter 8 presents the theory that underlies the spectrogram and explains the trade-offs that must be made between frequency resolution and time resolution. In MATLAB, the function `spectrogram` computes the spectrogram, and its default values work well for most signals.[15] Therefore, it is reasonable to see what sort of output can be produced by the `specgram` function. Figure 3-25(b) shows the results of applying `specgram` to the stepped-frequency time waveform that makes up the C-major scale represented in musical notation by Fig. 3-25(a). The calculation is performed by doing a numerical frequency analysis on short segments of the sampled signal and plotting the results at the specific time at which the analysis is done. By repeating this process with slight displacements in time, a two-dimensional array is created whose magnitude can be displayed as a grayscale image whose horizontal axis is time and whose vertical axis is frequency. The darkness of the marking is proportional to the amplitude of the signal at each time-frequency point. The time axis must be interpreted as the "time of analysis" because the frequency calculation is not instantaneous: rather, it is based on a finite segment of the signal—in this case, 25.6 ms.

It is quite easy to identify the frequency content due to each note, but there are also some interfering artifacts that make the spectrogram in Fig. 3-25(b) less than ideal. In Section 8-7 we will undertake a discussion of time–frequency *analysis* in order to explain how the spectrogram is calculated and how one should choose the analysis parameters to get a good result. Even though the spectrogram is a highly advanced idea in signal analysis, its application is relatively easy and intuitive, especially for

[15]The MATLAB syntax is `[S,F,T] = spectrogram(X,WINDOW,NOVERLAP,NFFT,Fs,'yaxis')`. The ***DSP First*** toolbox contains equivalent capability with the functions called `plotspec` and `spectgr`.

music signals, which are described symbolically by a notation that is very much like a spectrogram.

3-7 Frequency Modulation: Chirp Signals

Section 3-6 revealed the possibility that interesting sounds can be created when the frequency varies as a function of time. In this section, we use a different mathematical formula to create signals whose frequency is *continuously* time-varying.

3-7.1 Chirp or Linearly Swept Frequency

LAB

FM Synthesis for Musical Instruments

Suppose that we want to create a signal that covers, or *sweeps* over, an interval of frequencies. For example, in the audible region, we might begin at 300 Hz and *sweep* up to 800 Hz. One method for producing such a signal would be to use a stepped-frequency signal that concatenates a large number of short constant-frequency sinusoids, starting at 300 Hz and ending at 800 Hz. Figure 3-26 shows a time waveform where the frequency is being stepped. Notice the jumps at $t = 4, 8, 12, 16,$ and 20 ms, which, in this case, are caused by using $\varphi = 0$ for each short sinusoidal segment. The frequency stepping is not continuous and would not sound smooth even with many steps. In addition, this approach is tedious to implement because the boundary between the short sinusoids is discontinuous unless we are careful to adjust the initial phase of each short sinusoid.

A "chirp" signal is a swept-frequency signal whose frequency continuously changes linearly from some starting value to an ending frequency. The chirp formula can be derived from the complex-exponential point of view. First of all, recall that the constant-frequency sinusoid is the real part of a complex (rotating) phasor

$$x(t) = \Re\{Ae^{j\overbrace{(\omega_0 t + \varphi)}^{\text{angle}}}\} = A\cos(\omega_0 t + \varphi) \tag{3.40}$$

and notice that the angle of the complex exponential is equal to the argument of the cosine. When we generalize the sinusoid to a wider class of signals written as

$$x(t) = \Re\{Ae^{j\psi(t)}\} = A\cos(\psi(t)) \tag{3.41}$$

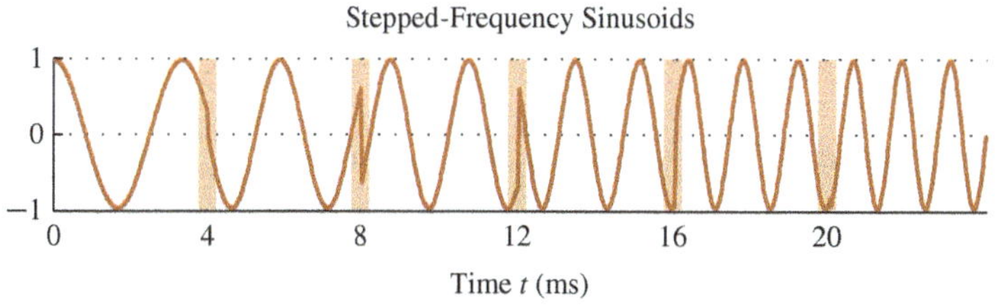

Figure 3-26 Stepped-frequency sinusoid, with six frequencies changing from 300 Hz to 800 Hz in 100 Hz increments. The frequency changes every four milliseconds.

where $\psi(t)$ denotes any function of time, we refer to $\psi(t)$ as the ***angle function*** of the cosine.[16] For example, we would create a signal with quadratic angle function by defining

$$\psi(t) = 2\pi\mu t^2 + 2\pi f_0 t + \varphi \tag{3.42}$$

Now we consider how to get frequency variation from the angle function. For the constant-frequency sinusoid in (3.40), the constant ω_0 is the *rate of change of the angle function* which can be obtained by taking the time derivative of the angle function. This notion of taking the derivative generalizes if we define the ***instantaneous frequency*** for signals in (3.41) as the derivative of the angle function,

$$\omega_i(t) = \frac{d}{dt}\psi(t) \qquad \text{(rad/s)} \tag{3.43}$$

which is the instantaneous slope of $\psi(t)$. The units of $\psi(t)$ are rad, and therefore, the units of $\omega_i(t)$ are rad/s, but we often want the frequency in Hz, which requires dividing by 2π,

$$f_i(t) = \frac{1}{2\pi}\frac{d}{dt}\psi(t) \qquad \text{(Hz)} \tag{3.44}$$

If the angle function of $x(t)$ is quadratic as in (3.42), then taking the derivative (3.44) gives an instantaneous frequency that changes *linearly* with time, that is,

$$f_i(t) = 2\mu t + f_0 \tag{3.45}$$

DEMO

Spectrograms: Chirp Sounds

The frequency variation produced by the time-varying angle function such as (3.42) is called frequency modulation (FM), and signals of this class are called ***linear FM signals***. Figure 3-27 shows an example of the ideal instantaneous frequency for linear-FM. Finally, since the linear variation of the frequency (3.45) can produce an audible sound similar to a bird chirp, linear FM signals are also called ***chirp*** signals, or simply *chirps*.

The process can be reversed to synthesize a time-domain signal with a desired linear frequency sweep. Since (3.43) states that the instantaneous frequency is the derivative of the angle function $\psi(t)$, the actual angle function needed in (3.41) is obtained from the integral of $\omega_i(t)$.

EXAMPLE 3-10 Synthesize a Chirp Formula

Suppose we want to synthesize a frequency sweep from $f_1 = 300$ Hz to $f_2 = 800$ Hz over a 0.5 second time interval, where the beginning and ending times are $t = 0$ and $t = T_2 = 0.5$ s as in Fig. 3-27. First of all, it is necessary to create a formula for the desired instantaneous frequency

$$f_i(t) = \frac{f_2 - f_1}{T_2}t + f_1 = \frac{800 - 300}{0.5}t + 300$$

[16] The *angle function* of the constant-frequency cosine wave would be $\psi(t) = \omega_0 t + \varphi$.

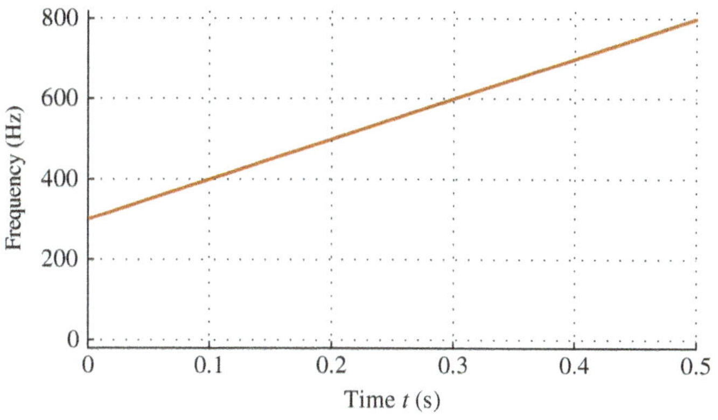

Figure 3-27 Instantaneous frequency of a linear-FM signal that sweeps continuously from 300 Hz to 800 Hz in 0.5 s.

which is the sloping line in Fig. 3-27. Then we must integrate $\omega_i(t) = 2\pi f_i(t)$ to get the angle function:

$$\begin{aligned}\psi(t) &= \int_0^t 2\pi f_i(u)\,du \\ &= \int_0^t 2\pi\,(1000u + 300)\,du \\ &= 1000\pi t^2 + 600\pi t + \varphi\end{aligned}$$

where the phase, $\varphi = \psi(0)$, is an arbitrary constant of integration. The time-domain chirp signal corresponding to Fig. 3-27 is $x(t) = \cos(1000\pi t^2 + 600\pi t + \varphi)$.

3-7.2 A Closer Look at Instantaneous Frequency

It may be difficult to see why the derivative of the angle function would be the instantaneous frequency. The following experiment provides a clue.

(a) Define a "chirp" signal using the method in Example 3-10 with the following parameters:

$$\begin{aligned}f_1 &= 300 \text{ Hz} \\ f_2 &= 800 \text{ Hz} \\ T_2 &= 20 \text{ ms}\end{aligned}$$

In other words, determine μ and f_0 in (3.42) to define $x(t)$ so that it sweeps the specified 500 Hz frequency range in 20 ms.

(b) Now make a plot of the signal synthesized in part (a). In Fig. 3-28, this plot is Fig. 3-28(b). It is difficult to verify whether or not this chirp signal has the correct frequency content. However, the rest of this experiment demonstrates that the derivative of the angle function is the "correct" definition of instantaneous frequency at two different times, picked as 5.45 ms and 16 ms (approximately). These are locations of two positive peaks of the chirp as marked in Fig. 3-28(b).

(c) First, plot a 450 Hz sinusoid, $x_1(t) = \cos(900\pi(t - 0.00545))$ which should have a positive peak at $t = 5.45$ ms. This constant frequency sinusoid is shown in Fig. 3-28(a).

(d) Next, generate and plot a 700 Hz sinusoid, $x_2(t) = \cos(1400\pi(t - 0.016))$ which should have a positive peak at $t = 16$ ms. This constant frequency sinusoid is shown in Fig. 3-28(c).

(e) Now compare the three signals in Fig. 3-28 with respect to the frequency content of the chirp. Concentrate on the frequency of the chirp in the time range 4 to 7 ms. Notice that the 450 Hz sinusoid nearly matches the chirp in this time region. Evaluate the theoretical $f_i(t)$ in this region.

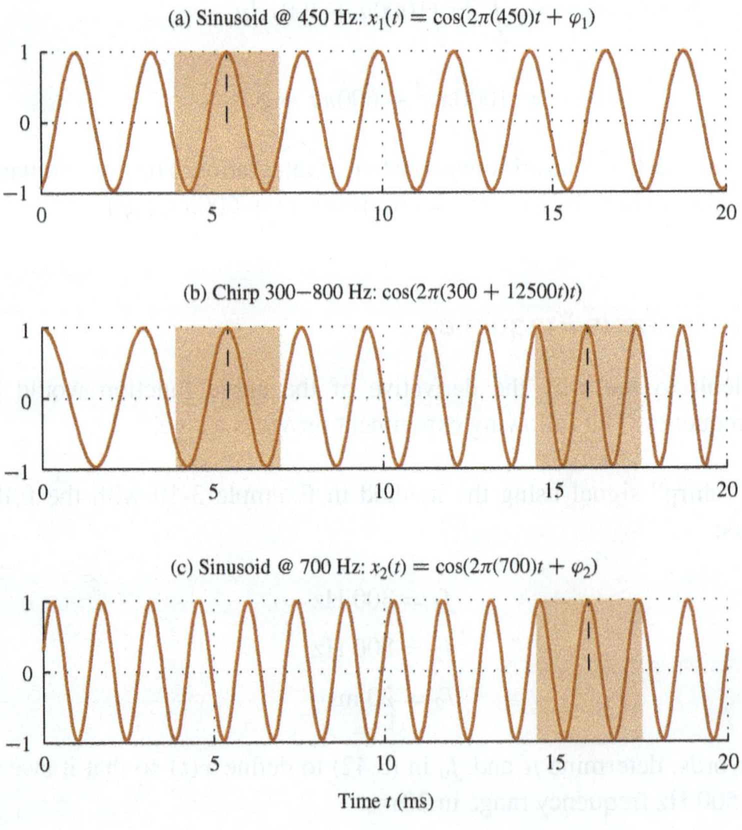

Figure 3-28 Comparing a chirp signal (b) with constant-frequency sinusoids (a) 450 Hz and (c) 700 Hz. Notice where the local frequency of the chirp is equal to one of the sinusoids.

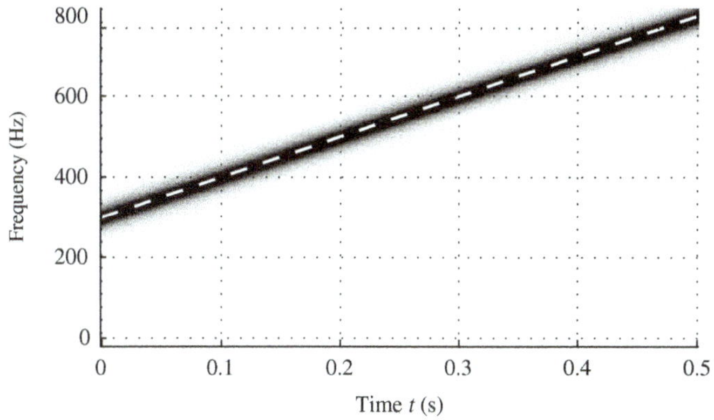

Figure 3-29 Spectrogram of a linear-FM signal that sweeps continuously from 300 Hz to 800 Hz in 0.5 s. The ideal instantaneous frequency, which changes at a rate of 1000 Hz/s, is shown as a dashed white line.

(f) Likewise, it is possible to find another region near 16 ms (14.5 to 17.5 ms) where the chirp frequency is nearly equal to 700 Hz.

The foregoing example provides justification for using the spectrogram to analyze signals whose frequency content is changing. Over a short time interval the chirp signal in Fig. 3-28(b) can be approximated by a constant frequency sinusoid. Thus, a Fourier analysis integral could be applied to such short time intervals to determine the *local* frequency content of the signal. This is exactly what is done by the spectrogram. Figure 3-29 shows an example of the spectrogram applied to the signal whose ideal instantaneous frequency was given in Fig. 3-27. For comparison, the ideal instantaneous frequency is shown superimposed on the spectrogram.

DEMO
FM Synthesis

We have seen that for signals of the form $x(t) = A\cos(\psi(t))$, the instantaneous frequency of the signal is the derivative of the angle function $\psi(t)$. If $\psi(t)$ is constant, the frequency is zero. If $\psi(t)$ is linear, $x(t)$ is a sinusoid at some fixed frequency. If $\psi(t)$ is quadratic, $x(t)$ is a chirp signal whose frequency changes linearly versus time. More complicated variations of $\psi(t)$ can produce a wide variety of signals. An example is the sound made by a police or ambulance siren. Another important application of these FM signals is in music synthesis. This application is illustrated with demos and a lab on the Companion Website.

3-8 Summary and Links

This chapter introduced the concept of the ***spectrum***, in which we represent a signal by its sinusoidal components. The spectrum is a graphical presentation of the complex amplitude for each frequency component in the signal. We showed how complicated signals can be formed from relatively simple spectra, and we presented the essential concepts of the Fourier series so that we could form the spectrum of arbitrary periodic signals. Finally, we ended with a discussion of how the spectrum can vary with time.

At this point, so many different demonstrations and projects can be done that we must limit the list somewhat. Among the laboratory projects on the website are three devoted to different aspects of the spectrum. The first contains exercises on the Fourier

LAB
Synthesis of Sinusoidal Signals

LAB
FM Synthesis for Musical Instruments

series representation of the square wave and a sawtooth wave. The second requires students to develop a music synthesis program to play a piece such as Bach's "Jesu, Joy of Man's Desiring." This synthesis must be done with sinusoids, but can be refined with extras such as a tapered amplitude envelope. Finally, the third lab deals with beat notes, chirp signals, and spectrograms. The second part of this lab involves a music synthesis method based on frequency modulation. The FM synthesis algorithm can produce realistic sounds for instruments such as a clarinet or a drum. In addition, there is one more lab that involves some practical systems that work with sinusoidal signals, such as the dual-tone multi-frequency (DTMF) telephone dialing system. This lab, however, requires some knowledge of filtering so it is reserved for Chapter 7. Write-ups of the labs can be found on the Companion Website.

The website also contains many demonstrations of sounds and their spectrograms:

DEMO
Links to Many Demos

(a) Spectrograms of simple sounds such as sine waves, square waves, and other harmonics.

(b) Spectrograms of realistic sounds, including a piano recording, a synthetic scale, and a synthesized music passage done by one of the students who took an early version of a course that included the material of this chapter.

(c) Spectrograms of chirp signals that show how the rate of change of the instantaneous frequency affects the sound you hear.

(d) An explanation of the FM synthesis method for emulating musical instruments. Several example sounds are included for listening.

(e) Rotating phasors.

(f) Vowel synthesis.

3-9 Problems

NOTE
Problems with Solutions

The problems in this section address the major concepts of this chapter. On the Companion Website there are many related homework problems which can be used to review and practice. Most of them have detailed solutions.

P-3.1 A signal $x(t)$ has the two-sided spectrum representation shown in Fig. P-3.1.

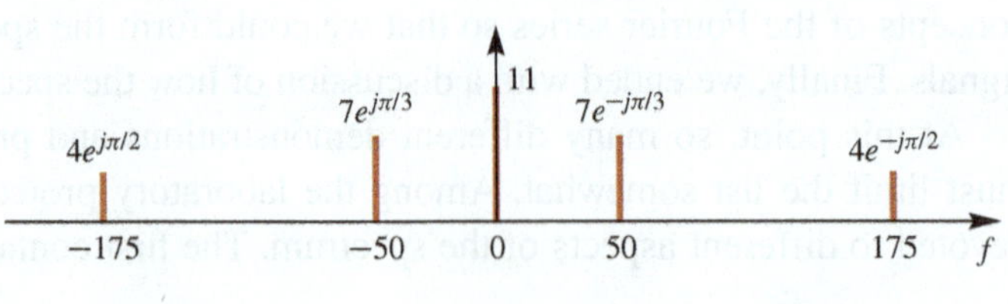

Figure P-3.1

(a) Write an equation for $x(t)$ as a sum of cosines.

(b) Is $x(t)$ a periodic signal? If so, determine its fundamental period and its fundamental frequency.

(c) Explain why negative frequencies are needed in the spectrum.

P-3.2 A signal composed of sinusoids is given by the equation

$$x(t) = 16\cos(500\pi t + \pi/4) + 9\cos(1000\pi t - \pi/3) - 5\cos(750\pi t)$$

(a) Sketch the spectrum of this signal, indicating the complex amplitude of each frequency component.

(b) Is $x(t)$ periodic? If so, what is the period?

(c) Now consider a new signal defined as $y(t) = 12 - 5\sqrt{3}\cos(750\pi t + \pi/2) + 5\cos(1500\pi t + \pi/2)$. Draw the spectrum of this new signal. Is $y(t)$ periodic? If so, what is the period?

P-3.3 The following MATLAB code defines a vector `zz`, which corresponds to the signal $z(t)$:

```
tt = -10:0.01:10;     %- in seconds
kk=1; zz = 2 + kk*cos(kk*tt);
kk=2; zz = zz + kk*cos(kk*tt);
kk=3; zz = zz + real(2*exp(j*pi/3)*exp(k*kk*tt) );
```

(a) Write a formula for $z(t)$ in terms of sinusoids in standard form, that is, $z(t) = A_0 + \sum_k A_k \cos(\omega_k t + \varphi_k)$.

(b) Plot the spectrum for $z(t)$ with labels for all of the frequencies and complex amplitudes.

P-3.4 Let $x(t) = \sin^3(54\pi t) + \sin^2(36\pi t)$.

(a) Determine a formula for $x(t)$ as the real part of a sum of complex exponentials.

(b) Determine the fundamental period for $x(t)$.

(c) Plot the *spectrum* for $x(t)$.

P-3.5 Make a sketch of the spectrum (vs. ω) of the signal defined by:

$$x(t) = \sum_{k=-3}^{3} \frac{1}{1 + jk} e^{jkt}$$

Label each spectrum line with the corresponding complex amplitude in polar form.

P-3.6 Shown in Fig. P-3.6 is a spectrum plot for the periodic signal $x(t)$. The frequency axis has units of rad/s.

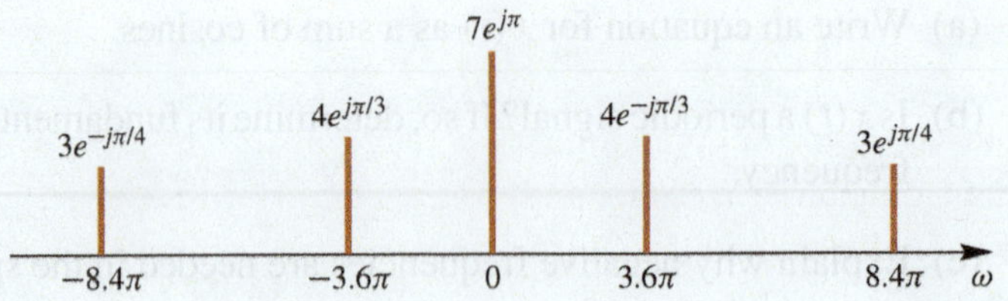

Figure P-3.6

(a) Determine the fundamental frequency ω_0 of this signal.

(b) Determine the (fundamental) period T_0 of $x(t)$, which is the shortest possible period.

(c) Determine the DC value of this signal.

(d) A periodic signal of this type can be represented as a Fourier series of the form

$$x(t) = \sum_{k=-\infty}^{\infty} a_k e^{j\omega_0 kt}.$$

If the Fourier series coefficients of $x(t)$ are denoted by a_k, $k = 0, \pm1, \pm2, \pm3, \ldots$, determine which coefficients are nonzero. List these nonzero Fourier series coefficients and their values in a table.

P-3.7 In Section 3-2.2, we discussed a simple example of the "beating" of one cosine wave against another. In this problem, we consider a more general case. Let

$$x(t) = A\cos(2\pi(f_c - f_\Delta)t) + B\cos(2\pi(f_c + f_\Delta)t)$$

For the case discussed in Section 3-2.2, the amplitudes are equal, $A = B = 1$.

(a) Use phasors to obtain a complex signal $z(t)$ such that

$$x(t) = \Re\{z(t)\}$$

(b) By manipulating the expression for $z(t)$ and then taking the real part, show that in the more general case above, $x(t)$ can be expressed in the form

$$x(t) = C\cos(2\pi f_\Delta t)\cos(2\pi f_c t) + D\sin(2\pi f_\Delta t)\sin(2\pi f_c t)$$

and find expressions for C and D in terms of A and B. Check your answer by substituting $A = B = 1$.

(c) Find values for A and B so that

$$x(t) = 2\sin(2\pi f_\Delta t)\sin(2\pi f_c t)$$

Plot the spectrum of this signal.

P-3.8 Consider the signal

$$x(t) = 8 + 10\cos(2\pi(100)t + \tfrac{1}{4}\pi) + 20\cos(2\pi(250)t)$$

(a) Using Euler's relation, the signal $x(t)$ defined above can be expressed as a sum of complex exponential signals using the finite Fourier synthesis summation (3.37)

Determine values for f_0, N, and all the complex amplitudes, a_k. *It is not necessary to evaluate any integrals to obtain* a_k.

(b) Is the signal $x(t)$ periodic? If so, what is the fundamental period?

(c) Plot the spectrum of this signal versus f in Hz.

P-3.9 An amplitude-modulated cosine wave is represented by the formula

$$x(t) = [12 + 9\cos(\pi t - \tfrac{1}{3}\pi)]\sin(14\pi t)$$

(a) Use phasors to show that $x(t)$ can be expressed in the form

$$x(t) = A_1\cos(\omega_1 t + \varphi_1) + A_2\cos(\omega_2 t + \varphi_2) + A_3\cos(\omega_3 t + \varphi_3)$$

where $\omega_1 < \omega_2 < \omega_3$; that is, find values of the parameters A_1, A_2, A_3, φ_1, φ_2, φ_3, ω_1, ω_2, and ω_3.

(b) Sketch the two-sided spectrum of this signal on a frequency axis. Be sure to label important features of the plot. Label your plot in terms of the numerical values of A_i, φ_i, and ω_i.

P-3.10 Consider a signal $x(t)$ such that

$$x(t) = 2\cos(\omega_1 t)\cos(\omega_2 t) = \cos[(\omega_2 + \omega_1)t] + \cos[(\omega_2 - \omega_1)t]$$

where $0 < \omega_1 < \omega_2$.

(a) What is the general condition that must be satisfied by $\omega_2 - \omega_1$ and $\omega_2 + \omega_1$ so that $x(t)$ is periodic with period T_0? Determine T_0 and verify that $x(t) = x(t + T_0)$.

(b) What does the result of (a) imply about ω_1 and ω_2? For example, must ω_2 be an integer multiple of ω_1?

P-3.11 A periodic signal is given by the equation

$$x(t) = 4 + 6\cos(60\pi t - \tfrac{1}{4}\pi) + 3\sin(90\pi) - 6\cos(150\pi t + \tfrac{1}{6}\pi)$$

(a) Determine the fundamental frequency ω_0, the fundamental period T_0, the number of terms N, and the coefficients a_k in the finite Fourier representation (3.37) for the signal $x(t)$ above. It is possible to do this *without evaluating any integrals.*

(b) Sketch the spectrum of this signal indicating the complex amplitude of each frequency component. Indicate the complex amplitude values at the appropriate frequencies. *There is no need to make separate plots for real/imaginary parts or magnitude/phase.*

(c) Now consider a new signal defined by adding one more sinusoid $y(t) = x(t) + 10 + 12\cos(50\pi t - \pi/6)$. How is the spectrum changed? Is $y(t)$ still periodic? If so, what is the fundamental period?

P-3.12 We have seen that musical tones can be modelled mathematically by sinusoidal signals. If you read music or play the piano, you know that the piano keyboard is divided into octaves, with the tones in each octave being twice the frequency of the corresponding tones in the next lower octave. To calibrate the frequency scale, the reference tone is the A above middle C, denoted A_4, which is often called A–440, since its frequency is 440 Hz. Each octave contains 12 tones, and the ratio between the frequencies of successive tones is constant. Thus, the ratio must be $2^{1/12}$. Since middle C is nine tones below A–440, its frequency is approximately $(440)2^{-9/12} \approx 261.6$ Hz. The names of the tones (notes) of the octave starting with middle C and ending with high C are:

Note name	C_4	$C_4^\sharp$	D_4	$E_4^\flat$	E_4	F_4	$F_4^\sharp$
Note number	40	41	42	43	44	45	46
Frequency							
Note name	$F_4^\sharp$	G_4	$G_4^\sharp$	A_4	$B_4^\flat$	B_4	C_5
Note number	46	47	48	49	50	51	52
Frequency				440			

(a) Make a table of the frequencies of the tones in octave 4, the octave beginning with middle C, assuming that A_4 above middle C is tuned to 440 Hz.

(b) The above notes on a piano are numbered 40 through 52. If n denotes the note number and f denotes the frequency of the corresponding tone, give a formula for the frequency of the tone as a function of the note number.

(c) A *chord* is a combination of musical notes sounded simultaneously. A *triad* is a three-note chord. The D-major chord is composed of the tones of D_4, $F_4^\sharp$ and A_4 sounded simultaneously. From the set of corresponding frequencies determined in (a), make a sketch of the essential features of the spectrum of the D-major chord assuming that each note is realized by a pure sinusoidal tone. *Do not specify the complex phasors precisely.*

P-3.13 The signal $x(t)$ in Fig. P-3.13 consists of a DC component plus a cosine signal.

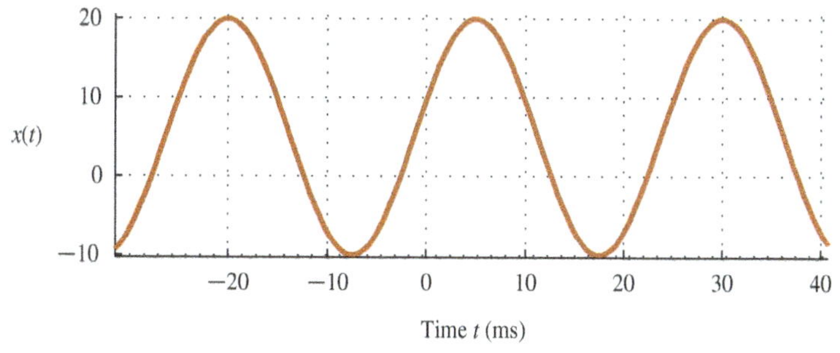

Figure P-3.13

(a) Determine the frequency of the DC component, as well as the frequency of the cosine component.

(b) Write an equation for the signal $x(t)$. You should be able to determine numerical values for all the amplitudes, frequencies, and phases in your equation by inspection of the above graph.

(c) Expand the equation obtained in the previous part into a sum of complex exponential signals with positive and negative frequency components.

(d) Plot the two-sided spectrum of the signal $x(t)$. Label the complex amplitudes for each positive and negative frequency component in the spectrum.

P-3.14 The two-sided spectrum of a signal $x(t)$ is given in the following table, but some of the values are not yet known.

Frequency (rad/s)	ω_2	-70π	0	ω_1	100π
Complex Amplitude	$8e^{-j\pi/3}$	X_{-1}	20	$\sqrt{2}+j\sqrt{2}$	X_2

(a) If $x(t)$ is a *real* signal, use symmetry properties of the spectrum to determine the numerical values of the parameters: X_{-1}, X_2, ω_1, and ω_2.

(b) Write an expression for $x(t)$ involving only real numbers and cosine functions.

(c) This signal is periodic; determine its fundamental period.

(d) Without plotting the waveform, show that $x(t) \geq 0$ for all t.

P-3.15 Determine the spectrum for each of the following signals. Give your answer as a plot.

(a) $x(t) = \cos^2(7\pi t - 0.1\pi)$

(b) $y(t) = \cos^2(7\pi t - 0.1\pi)\cos(77\pi t + 0.1\pi)$

P-3.16 In AM radio, the transmitted signal is voice (or music) combined with a *carrier signal.* The carrier is a sinusoid at the assigned broadcast frequency of the AM station. For example, WCNN in Atlanta has a *carrier frequency* of 680 kHz. If we use the notation $v(t)$ to denote the voice/music signal, then the actual transmitted signal for WCNN might be:

$$x(t) = (v(t) + A)\cos(2\pi(680 \times 10^3)t)$$

where A is a constant. The constant A is introduced to make $(A + v(t))$ positive, which simplifies the design of a receiver for AM—thus, $A > \min v(t)$.

(a) Voice-band signals tend to contain frequencies less than 4000 Hz (4 kHz). Suppose that $v(t)$ is a 3000 Hz sinusoid, $v(t) = \cos(2\pi(3000)t + 0.3\pi)$. Draw the spectrum for $v(t)$.

(b) Now draw the spectrum for $x(t)$, assuming a carrier at 680 kHz with zero phase. Use $v(t)$ from part (a) and assume that $A = 1.5$.
Hint: Substitute for $v(t)$ and expand $x(t)$ into a sum of cosine terms of three different frequencies.

P-3.17 The two-sided spectrum representation of a real-valued signal $x(t)$ is shown below (frequency in Hz):

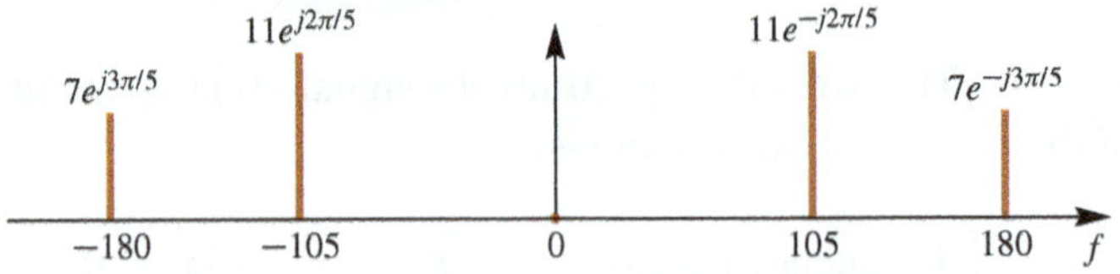

Figure P-3.17

(a) Determine the *fundamental frequency* (in Hz) for $x(t)$.

(b) Write the formula for $x(t)$ as a sum of real-valued sinusoids.

(c) The signal $x_2(t) = x(t - 0.05)$ is a time-delayed version of $x(t)$. Determine the spectrum of $x_2(t)$. Simplify the numerical values for the complex amplitudes, so that the phases lie in $(-\pi, \pi]$.

(d) A third signal is defined as $x_3(t) = x(t)\, e^{j210\pi t}$. This new signal is *complex-valued,* and it has a nonzero DC component. Determine the complex amplitude for the DC component of $x_3(t)$.

P-3.18 A real signal $x(t)$ has the two-sided spectrum shown in Fig. P-3.17.

(a) Plot the spectrum of the signal $y(t) = x(t) + 8$.

(b) Plot the spectrum of the signal $z(t) = x(t) + 18\cos(80\pi t + 0.8\pi)$.

(c) Determine the fundamental frequency of the signal $z(t)$.

(d) Define $q(t)$ as the derivative of $y(t)$ from part (a). Plot the spectrum of the signal $q(t) = \frac{d}{dt}y(t)$.

P-3.19 The two-sided spectrum of a signal $x(t)$ consists of two components given in the following table:

Frequency (rad/s)	ω_1	10π
Complex Amplitude	$8e^{j\pi/3}$	X_1

(a) If $x(t)$ is a *real* signal, determine the numerical values of the parameters X_1 and ω_1.

(b) Make a plot of the spectrum of $x(t)$; label both spectral lines.

(c) Define a new signal $y(t)$ as $y(t) = x(t)\cos(\omega_c t + \varphi)$. The two-sided spectrum of the signal $y(t)$ consists of three components given in the following table:

Frequency (rad/s)	-20π	ω_a	ω_b
Complex Amplitude	B	A	$-4j$

If $y(t)$ is a *real* signal, use the frequency shifting property to determine the numerical values of the parameters A, B, ω_a, ω_b, ω_c, and φ.

P-3.20 A real signal $x(t)$ has the two-sided spectrum shown in Fig. P-3.20. The frequency axis has units of rad/s.

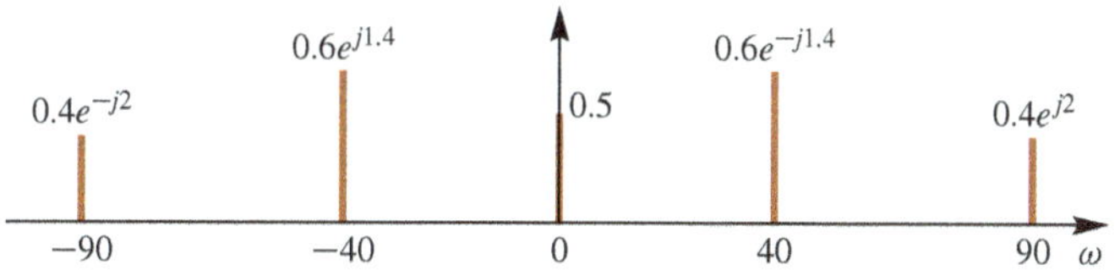

Figure P-3.20

(a) Determine the fundamental frequency ω_0 of this signal.

(b) Determine the fundamental period T_0 of $x(t)$, which is the shortest possible period.

(c) Determine the DC value of this signal.

(d) A periodic signal of this type can be represented as a Fourier series of the form

$$x(t) = \sum_{k=-\infty}^{\infty} a_k e^{j\omega_0 kt}$$

The kth term in the series is called the kth *harmonic*. Determine which harmonics (positive and negative) are present. In other words, for the Fourier series coefficients, a_k, determine which coefficients are nonzero. List the indices of the nonzero Fourier series coefficients and their values in a table.

P-3.21 A chirp signal $x(t)$ is one that sweeps in frequency from $\omega_1 = 2\pi f_1$ to $\omega_2 = 2\pi f_2$ as time goes from $t = 0$ to $t = T_2$. The general formula for a chirp is

$$x(t) = A\cos(\alpha t^2 + \beta t + \varphi) = \cos(\psi(t)) \tag{3.46}$$

where

$$\psi(t) = \alpha t^2 + \beta t + \varphi$$

The derivative of $\psi(t)$ is the ***instantaneous frequency*** (in rad/s), which is also the frequency heard if the frequencies are in the audible range.

(a) For the chirp in (3.46), determine formulas for the beginning frequency (ω_1) and the ending frequency (ω_2) in terms of α, β, and T_2.

(b) For the chirp signal

$$x(t) = \Re\{e^{j(57t^2+26t+17)}\}$$

Make a plot of the instantaneous frequency (in Hz) versus time over the range $0 \le t \le 1$s.

(c) If

$$x(t) = \Re\{e^{j(2\pi(5/2.0794)(8^t-1)+\varphi)}\}$$

derive the instantaneous frequency and plot it versus time.

P-3.22 A linear-FM "chirp" signal is one that sweeps in frequency from $\omega_1 = 2\pi f_1$ to $\omega_2 = 2\pi f_2$ during a finite time interval.

(a) Determine the mathematical formula for a chirp whose instantaneous frequency sweeps from $f_1 = 4800$ Hz down to $f_2 = 800$ Hz as time goes from $t = 0$ to $t = 2$ s.

(b) In the following chirp

$$y(t) = \cos(400\pi t^2 + 500\pi t - \pi/4) \qquad 0 \le t \le 3\,s$$

determine the starting and ending instantaneous frequencies.

P-3.23 To see why the derivative of the angle function would be the instantaneous frequency, repeat the experiment of Section 3-7.2.

(a) Use the following parameters to define a chirp signal:

$$f_1 = 1\text{ Hz}, \quad f_2 = 9\text{ Hz}, \quad T_2 = 2\text{ s}$$

Determine α and β in (3.46) to define $x(t)$ so that it sweeps the specified frequency range.

(b) The rest of this problem is devoted to a MATLAB experiment that demonstrates why the derivative of the angle function is the "correct" definition of instantaneous frequency. First, make a plot of the instantaneous frequency $f_i(t)$ (in Hz) versus time.

(c) Make a plot of the signal synthesized in (a). Pick a time-sampling interval that is small enough so that the plot is very smooth. Put this plot in the middle panel of a 3×1 by using subplot, `subplot(3,1,2)`.

(d) Generate a 4 Hz sinusoid, and plot it in the upper panel of a 3×1 by using subplot, `subplot(3,1,1)`.

(e) Generate an 8 Hz sinusoid, and then plot it in the lower panel of a 3×1 by using subplot, `subplot(3,1,3)`.

(f) Compare the three signals and comment on the apparent frequency of the chirp in the time range $1.6 \le t \le 2$ s. Which sinusoid matches the chirp in this time region? Compare the expected $f_i(t)$ in this region with 4 Hz and 8 Hz.

P-3.24 If $x(t) = \cos(\psi(t))$ then the instantaneous frequency (in Hz) is defined as the derivative of its argument $\psi(t)$,

$$f_i(t) = \frac{1}{2\pi}\frac{d\psi(t)}{dt}$$

For each of the sinusoids below, find an expression for the instantaneous frequency in Hz and sketch $f_i(t)$ for the specified range of t.

(a) $x_1(t) = \cos(\sqrt{t}), \text{ for } 0.01 \le t \le 3.6$

(b) $x_2(t) = \cos(2t^2 + \pi/4)^2, \text{ for } -2 \le t \le 2$

(c) $x_3(t) = \cos(e^{2.5t}), \text{ for } 0 \le t \le 3$

(d) $x_4(t) = \cos(\sin(3\pi t)\cos(5\pi t)), \text{ for } 0 \le t \le 4$

P-3.25 It is easy to determine the value of $x(t)$ at $t = 0$ directly from the complex ampltiudes of the spectrum.

(a) When $x(t)$ has a spectrum with lines at $f = f_k$, and complex amplitudes a_k, prove that $x(0)$ is the sum of all the complex amplitudes (i.e., $\sum a_k$).

Note that this result says that $x(0)$ does not depend on the frequencies.

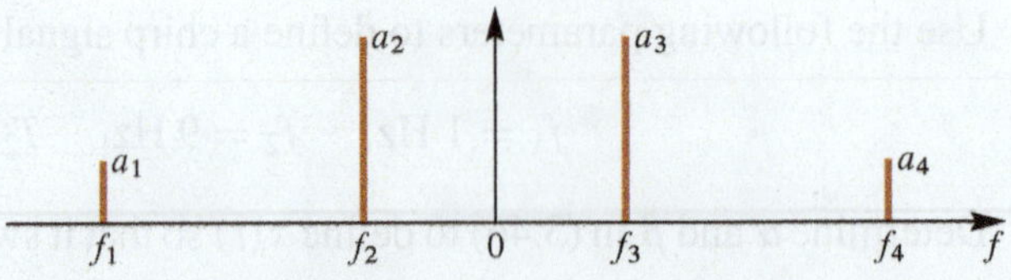

Figure P-3.25

(b) Test this result on the graphical spectrum given in Fig. P-3.26. Assume that $a_1^* = a_4 = 2e^{j3\pi/4}$ and $a_2^* = a_3 = 6e^{j\pi/4}$, and then determine the value of $x(0)$.

P-3.26 The plots in Fig. P-3.26 show waveforms on the left and spectra on the right. Match the waveform letter with its corresponding spectrum number. In each case, write the formula for the signal as a sum of sinusoids.

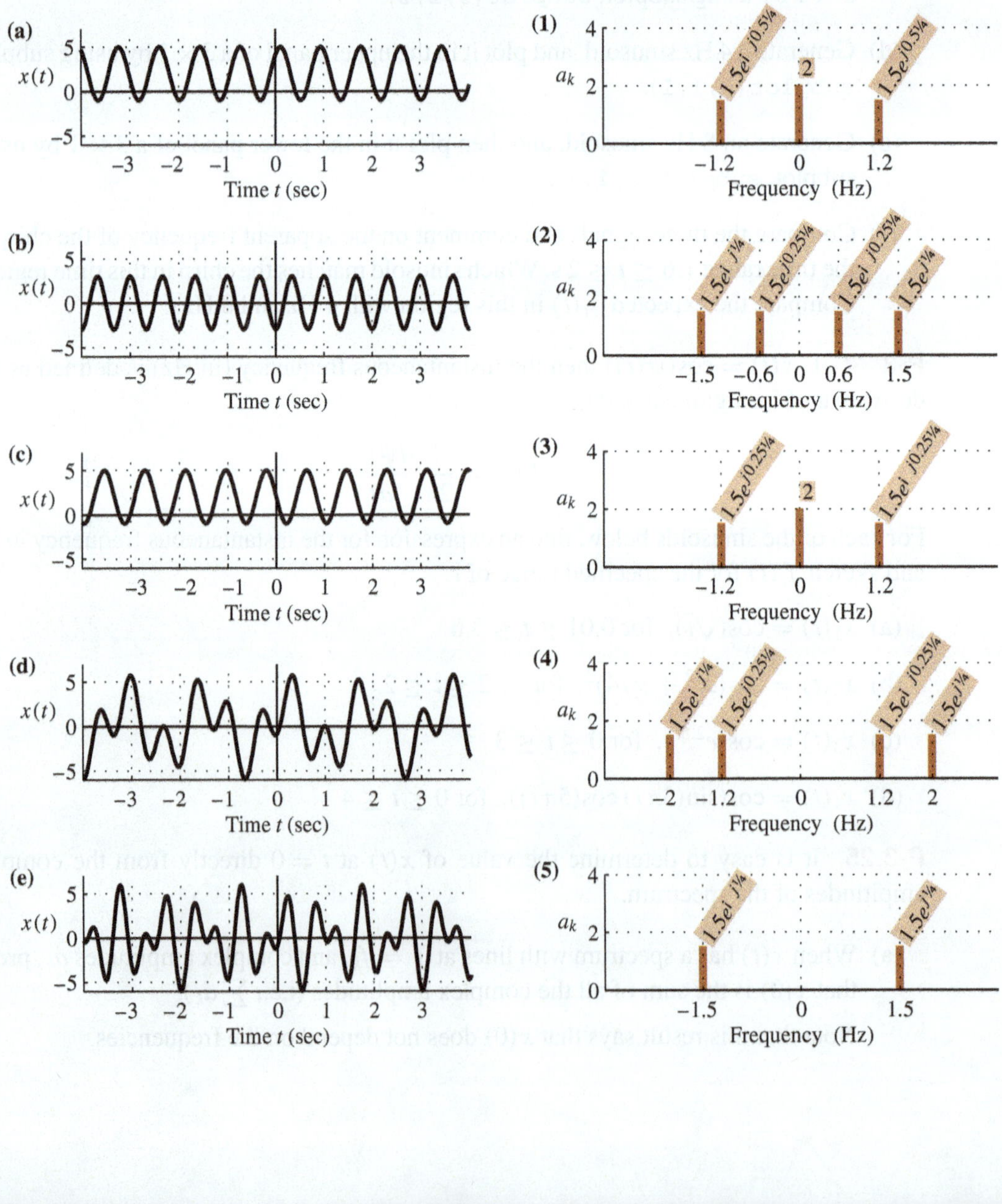

Figure P-3.26

P-3.27 Shown in Fig. P-3.25 are spectrograms (labeled with numbers 1–6) for six signals over the time period $0 \le t \le 2$ s. The frequency axis for each plot has units of Hz. For each signal description below, identify the corresponding spectrogram.

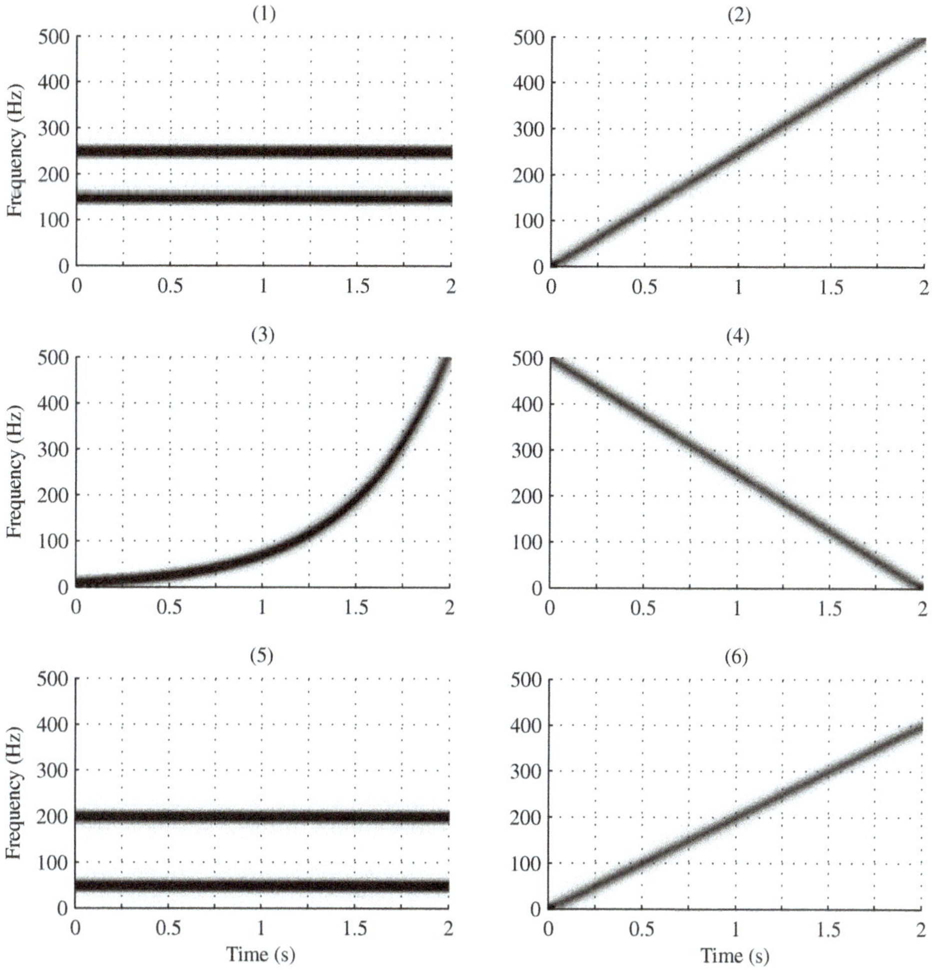

Figure P-3.27

(a) $x(t) = \cos(-250\pi t^2)$

(b) $x(t) = \cos(100\pi t - \pi/4) + \cos(400\pi t)$

(c) $x(t) = \cos(1000\pi t - 250\pi t^2)$

(d) $x(t) = \cos(100\pi t)\cos(400\pi t)$

(f) $x(t) = \cos(200\pi t^2)$

(e) $x(t) = \cos(30e^{2t})$

CHAPTER

4

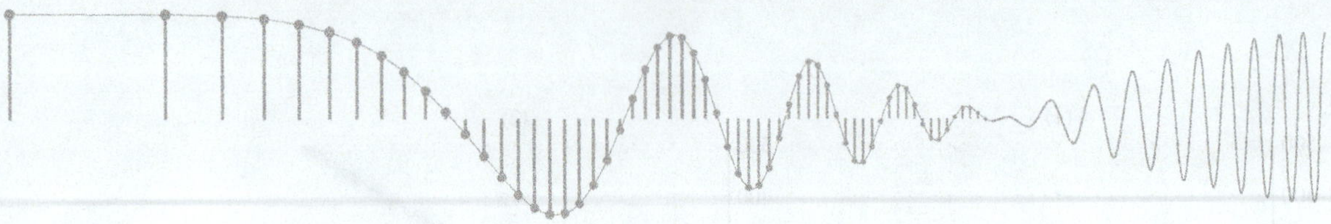

Sampling and Aliasing

Computers process numbers or sets of numbers, so continuous waveforms like sinusoidal signals must be turned into vectors, or streams of numbers for digital signal processing. The common way to get the numbers is to sample values of the continuous-time signal at a constant rate such as, e.g., 48,000 samples/s. A key question is how many numbers per second are needed to adequately represent a continuous-time signal. For constant-rate sampling, this question boils down to finding the minimum rate. This chapter is concerned with the conversion of signals between the analog (continuous-time) domain and the digital (discrete-time) domain. The primary objective of our presentation is an understanding of the ***sampling theorem***, which states that when the sampling rate is greater than twice the highest frequency contained in the spectrum of the analog signal, the original signal can be reconstructed exactly from the samples. As we noted in Chapter 2, a signal whose spectrum has a finite highest frequency is called a ***bandlimited*** signal, and, theoretically, such a signal can be sampled and reconstructed without error.

The process of converting from digital back to analog is called ***reconstruction***. A common example is playback of audio or video from a CD or DVD. For example, music on a CD or DVD is sampled at 44,100 times/s and stored in a digital form from which a player reconstructs the continuous (analog) waveform that we hear. The reconstruction process must "fill in" the missing signal values between the sample times t_n by constructing a smooth curve through the discrete-time sample values $x(t_n)$. Mathematicians call this process *interpolation* because it may be represented as time-domain interpolation formula. On the other hand, we will see that a spectrum or frequency-domain point of view is an alternative way to understand important issues such as bandwidth and reconstructing a signal from its samples.

4-1 Sampling

In previous chapters we have considered several examples of ***continuous-time*** signals, such as sinusoids and periodic signals. Continuous-time signals are represented mathematically by functions of time, $x(t)$, where t is a real number, and we assume that any value of $x(t)$ is also a real number. It is common to call *continuous-time* signals ***analog*** signals, meaning that both the signal amplitude and the time variable are assumed to be real numbers (not discrete values). Recall the discussion in Section 2-4 on p. 39, where we plotted waveforms using MATLAB, but actually did not plot the true continuous-time waveform. Instead, we actually plotted values of the waveform only at isolated (discrete) points in time and then connected those points with straight lines. It might have gone unnoticed, but any value of the signal was represented with a finite number of bits. The key point is that a signal represented in a digital computer is *discrete* in both time and amplitude. However, we will study discrete-time signals where the time axis is discrete, but the signal values are considered to be real numbers.[1]

A ***discrete-time signal*** is represented mathematically by an indexed sequence of numbers. When stored digitally, the signal values are held in memory locations, so they would be indexed by memory addresses.[2] We denote the values of the discrete-time signal as $x[n]$, where n is the integer index indicating the order of the values in the sequence. The square brackets [ ] enclosing the argument n provide a notation that distinguishes between the continuous-time signal $x(t)$ and a corresponding discrete-time signal $x[n]$.[3]

We can obtain discrete-time signals in either of the following ways:

(a) We can ***sample*** a continuous-time signal at equally spaced time instants, $t_n = nT_s$, that is,

$$x[n] = x(nT_s) \qquad -\infty < n < \infty \tag{4.1}$$

where $x(t)$ represents any continuously varying signal such as speech or audio, which are intrinsically analog signals since they arise in the natural world. The individual values of $x[n]$ are called *samples* of the continuous-time signal.

---

[1]Quantization of the signal values is rather difficult to analyze and is beyond the scope of this basic text.

[2]Furthermore, memory registers, being of finite size, only allow the samples to be represented by a finite set of numbers. This quantization is often inconsequential, and we neglect it in our subsequent discussions.

[3]The terminology for discrete-time signals is not universal, so we may occasionally use the word *sequence* in place of signal, or the adjective *digital* in place of discrete-time, to refer to $x[n]$.

The fixed time interval between samples, T_s, can also be expressed as a fixed ***sampling rate***, f_s, in samples/s:

$$f_s = \frac{1}{T_s} \text{ samples/s}$$

Therefore, an alternative way to write the sequence in (4.1) is $x[n] = x(n/f_s)$.

Sampling can be viewed as a transformation or operation that acts on a continuous-time signal $x(t)$ to produce an output, which is a corresponding discrete-time signal $x[n]$. In engineering, it is common to call such a transformation a *system,* and to represent it graphically with a block diagram that shows the input and output signals along with a name that describes the system operation. The sampling operation is an example of a system whose input is a continuous-time signal and whose output is a discrete-time signal as shown in Fig. 4-1. The system block diagram of Fig. 4-1 represents the mathematical operation in (4.1) and is called an ***ideal continuous-to-discrete (C-to-D) converter***. Its idealized mathematical form is useful for analysis, but an actual hardware system for doing sampling, called an analog-to-digital (A-to-D) converter, can only approximate the perfect sampling of the C-to-D converter. Hardware A-to-D converters differ from ideal C-to-D converters because of real-world problems such as amplitude quantization to 12- or 16-bit digital word length, jitter in the sampling times, and other factors that are difficult to express in a mathematical model. Since these factors can be made negligible with careful design, we can henceforth safely confine our presentation to the ideal C-to-D system.

(b) We can also *compute* the values of a discrete-time signal directly from a formula. A simple example is

$$w[n] = n^2 - 5n + 3$$

which yields the sequence of values $\{3, -1, -3, -3, -1, 3, 9, \dots\}$ corresponding to the indices $n = 0, 1, 2, 3, 4, 5, 6, \dots$. The sequence $w[n]$ could have been obtained by sampling $w(t) = t^2 - 5t + 3$ with a sampling period $T_s = 1$ s, but in fact there might not be any such signal in the physical world that was sampled. Nevertheless, we still refer to the individual values of the sequence as *samples.* Discrete-time signals described by formulas are very common in our study of discrete-time signals and systems.

When we plot discrete-time signals, we use the stem format shown in Fig. 4-2, which shows eight values (samples) of the sequence $w[n]$ defined previously. Stem

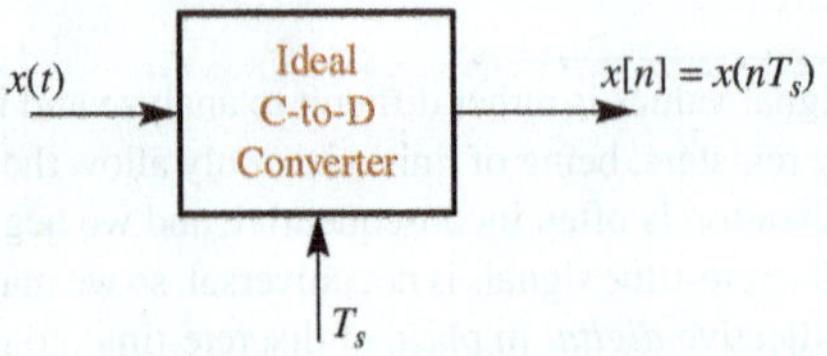

Figure 4-1 Block diagram representation of the ideal continuous-to-discrete (C-to-D) converter. The parameter T_s specifies uniform sampling of the input signal every T_s seconds.

plots show clearly that the discrete-time signal has values only for integer indices: in between, the discrete-time signal is *undefined.*

4-1.1 Sampling Sinusoidal Signals

Sinusoidal signals are examples of continuous-time signals that exist in the "real world" outside the computer, and for which we can also write a simple mathematical formula. Since the effects of sampling are easily understood for sinusoids, we use them as the basis for our study of sampling.

If we sample a sinusoid of the form $A\cos(\omega t + \varphi)$, we obtain

$$\begin{aligned} x[n] &= x(nT_s) \\ &= A\cos(\omega n T_s + \varphi) \\ &= A\cos(\hat{\omega} n + \varphi) \end{aligned} \tag{4.2}$$

where we have defined $\hat{\omega}$ to be

Normalized Radian Frequency

$$\hat{\omega} \stackrel{\text{def}}{=} \omega T_s = \frac{\omega}{f_s} \tag{4.3}$$

The signal $x[n]$ in (4.2) is a ***discrete-time cosine signal***, and $\hat{\omega}$ is its ***discrete-time frequency***. We use a "hat" over ω to denote that this is a new frequency variable. It is a normalized version of the continuous-time radian frequency with respect to the sampling frequency. Since ω has units of rad/s, the units of $\hat{\omega} = \omega T_s$ are *radians*, that is, $\hat{\omega}$ is a dimensionless quantity. This is entirely consistent with the fact that the index n in $x[n]$ is dimensionless. Once the samples are taken from $x(t)$, the time scale information is no longer part of the signal. The discrete-time signal $x[n]$ is just a sequence of numbers, and these numbers alone carry no information about the sampling period, T_s, used in obtaining them. An immediate implication of this observation is that an infinite number of continuous-time sinusoidal signals can be transformed into the same discrete-time sinusoid by sampling. All we need to do is change the sampling period inversely proportional to the input frequency of the continuous-time sinusoid. For

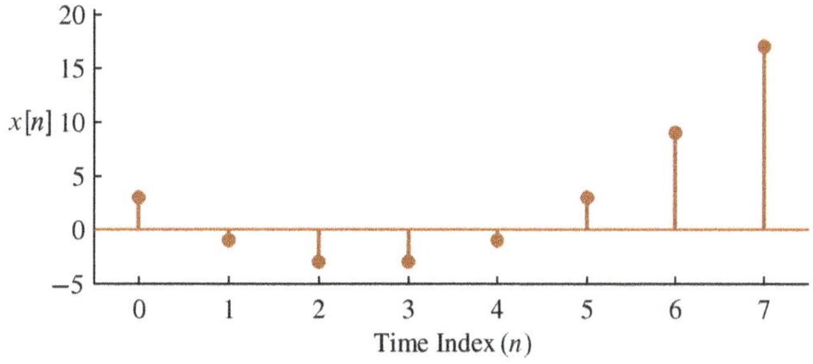

Figure 4-2 Plotting format for discrete-time signals, called a stem plot. In MATLAB, the function `stem` produces this plot. Some students also refer to the stem plot as a "lollypop" plot.

example, if $\omega = 200\pi$ rad/s and $T_s = 0.5$ ms, then $\hat{\omega} = 0.1\pi$ rad. On the other hand, if $\omega = 1000\pi$ rad/s and $T_s = 0.1$ ms, $\hat{\omega}$ is still equal to 0.1π rad.

Figure 4-3(a) shows $x(t) = \cos(200\pi t)$, a continuous-time sinusoid with frequency $f_0 = 100$ Hz. Figure 4-3(b) shows the samples taken with sampling period $T_s = 0.5$ ms. The sequence is given by the formula $x[n] = x(nT_s) = \cos(0.1\pi n)$, so the discrete-time radian frequency is $\hat{\omega}_0 = 0.1\pi$. (Since $\hat{\omega}_0$ is dimensionless, it is redundant to specify its units as rad.) The sampling rate in this example is $f_s = 1/T_s = 2000$ samples/s. The sample values are plotted as discrete points as in Fig. 4-3(b). The points are not connected by a continuous curve because we do not have any *direct* information about the value of the signal between the sample values. In this case, there are 20 sample values per period of the signal, because the sampling frequency (2000 samples/s) is 20 times higher than the frequency of the continuous signal (100 Hz). From this discrete-time plot, it appears that the sample values alone are sufficient to visually reconstruct a continuous-time cosine wave, but without knowledge of the sampling rate, we cannot tell what the analog frequency ω should be.

Another example of sampling is shown in Fig. 4-3(c). In this case, the 100 Hz sinusoid is sampled at a lower rate ($f_s = 500$ samples/s) resulting in the sequence of samples $x[n] = \cos(0.4\pi n)$. In this case, the discrete-time radian frequency is $\hat{\omega} = 0.4\pi$. The time between samples is $T_s = 1/f_s = 2$ ms, so there are only five samples per period

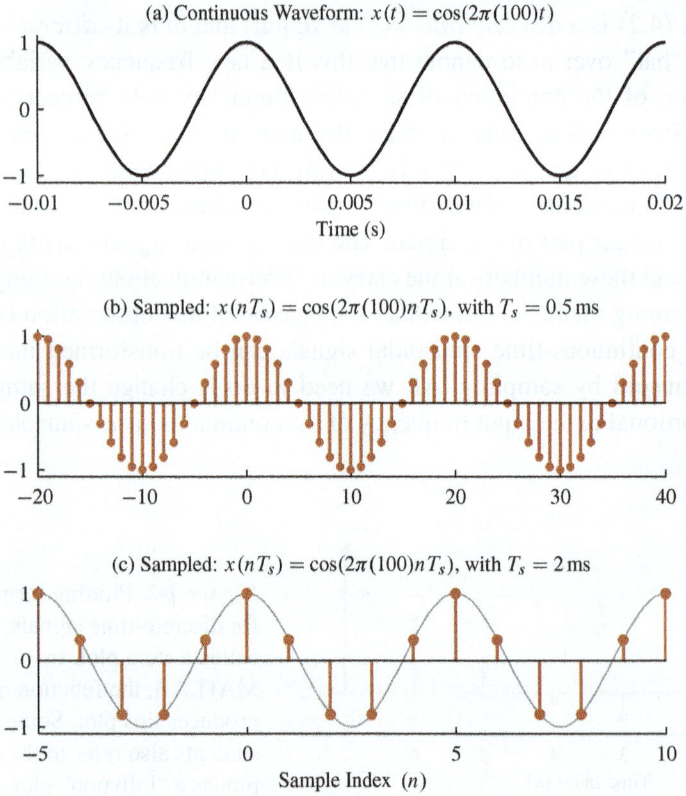

Figure 4-3 A continuous-time 100 Hz sinusoid (a) and two discrete-time sinusoids formed by sampling at $f_s = 2000$ samples/s (b) and at $f_s = 500$ samples/s (c).

of the continuous-time signal. We see that without the original waveform superimposed, it would be difficult to discern the precise waveshape of the original continuous-time sinusoid.

EXERCISE 4.1

If the sampling rate is $f_s = 1000$ samples/s, and the continuous-time signal is $x(t) = \cos(\omega t)$, what value of ω gives a sequence of samples identical to the discrete-time signal shown in Fig. 4-3(c)?

At first glance, Exercise 4.1 appears to ask a rather simple question because the formula in (4.3) can be solved for ω in terms of f_s and $\hat{\omega}$. Then it is easy to obtain an answer for Exercise 4.1 as $\omega = (0.4\pi) f_s = 400\pi$ rad/s. But is this the only possible answer? In order to understand this question and find many other answers to Exercise 4.1, we need to introduce the concept of an ***alias***.

4-1.2 The Concept of Aliases

A simple definition of the word ***alias*** would involve something like "two names for the same person, or thing." When a mathematical formula defines a signal, that formula can act as a name for the signal. Hence we introduce the concept of an alias by showing that two different discrete-time sinusoid formulas can define the same signal values. The sinusoid $x_1[n] = \cos(0.4\pi n)$ is the mathematical formula for the plot shown in Fig. 4-3(c), so it is one "name" that identifies that signal as having a frequency of 0.4π. Now consider another sinusoid formula, $x_2[n] = \cos(2.4\pi n)$, which is a different name that specifies a different frequency, 2.4π. To see how a plot of $x_2[n]$ would look, we invoke the simple trigonometric identity $\cos(\theta + 2\pi) = \cos(\theta)$ to obtain

$$x_2[n] = \cos(2.4\pi n) = \cos(0.4\pi n + 2\pi n) = \cos(0.4\pi n)$$

DEMO
Sampling Theory Tutorial

because $2\pi n$ is an integer number of periods of the cosine function. In other words, the plot of $x_2[n]$ is identical to the plot of $x_1[n]$, and this phenomenon that we are going to call aliasing is solely due to the fact that cosine functions are periodic with period 2π.

Figure 4-4 shows that the stem plots of these two signals, $x_1[n] = \cos(0.4\pi n)$ and $x_2[n] = \cos(2.4\pi n)$, are identical. Figure 4-4 also shows continuous plots of the signals $\cos(0.4\pi t)$ and $\cos(2.4\pi t)$, which when sampled with $T_s = 1$ would give $x_1[n]$ and $x_2[n]$. It is clear from Fig. 4-4 that the values of these continuous cosine signals are equal at integer values, n. Since $x_2[n] = x_1[n]$ for all integers n, we see that $x_2[n] = \cos(2.4\pi n)$ is another name for the same discrete-time signal (that is, it is an *alias* of 0.4π).

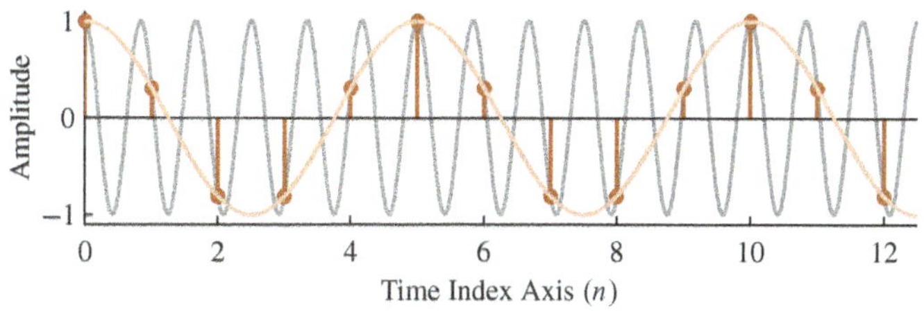

Figure 4-4 Illustration of aliasing: two continuous-time signals drawn through the same samples. The samples belong to two different cosine signals with different frequencies, but the cosine functions have the same values at $n = 0, 1, 2, 3, \ldots$. *Aliasing* occurs when a high frequency is indistinguishable from a low frequency after sampling.

The frequency of $x_2[n]$ is $\hat{\omega} = 2.4\pi$, while the frequency of $x_1[n]$ is $\hat{\omega} = 0.4\pi$. When speaking about the frequencies, we say that 2.4π is an *alias* of 0.4π. There are many more frequency aliases as the following exercise suggests.

EXERCISE 4.2 Show that $7\cos(8.4\pi n - 0.2\pi)$ is an alias of $7\cos(0.4\pi n - 0.2\pi)$. In addition, find two more frequencies that are aliases of 0.4π rad.

In the previous exercise, it should be easy to see that adding any integer multiple of 2π to 0.4π gives an alias, so the following general formula holds for the frequency aliases of 0.4π:[4]

$$\hat{\omega}_\ell = 0.4\pi + 2\pi\ell \qquad \ell = 0, 1, 2, 3, \ldots \tag{4.4}$$

The ***principal alias*** is defined to be the unique alias frequency in the interval $-\pi < \hat{\omega} \leq \pi$. In this case, the principal alias is 0.4π.

However, we are not finished yet. There are even more aliases. Another trigonometric identity states that $\cos(2\pi - \theta) = \cos(\theta)$, so we can generate another alias for $x_1[n] = \cos(0.4\pi n)$ as follows:

$$\begin{aligned} x_3[n] &= \cos(1.6\pi n) \\ &= \cos(2\pi n - 0.4\pi n) \\ &= \cos(0.4\pi n) \end{aligned}$$

The frequency of $x_3[n]$ is 1.6π. A general form for all the alias frequencies of this type would be

$$\hat{\omega}_\ell = -0.4\pi + 2\pi\ell \qquad \ell = 1, 2, 3, \ldots \tag{4.5}$$

For reasons that will become clear when discussing Fig. 4-12, these aliases of a negative frequency are called ***folded aliases***. In the folded case, the principal alias is -0.4π.

If we examine the aliases for the general discrete-time sinusoid (4.2), an extra complication arises for the folded case as illustrated by the following analysis:

$$\begin{aligned} A\cos((2\pi - \hat{\omega})n - \varphi) &= A\cos(2\pi n - \hat{\omega}n - \varphi) \\ &= A\cos(-\hat{\omega}n - \varphi) \\ &= A\cos(\hat{\omega}n + \varphi) \end{aligned} \tag{4.6}$$

Notice that the algebraic sign of the phase angles of the folded aliases must be opposite to the sign of the phase angle of the principal alias.

EXERCISE 4.3 Show that the signal $7\cos(9.6\pi n + 0.2\pi)$ is an alias of the signal $7\cos(0.4\pi n - 0.2\pi)$. It might be instructive to make MATLAB stem plots of these two signals to verify that the phase must change sign to have identical plots.

[4] The integer ℓ could be negative if we allow negative frequencies, but we prefer to avoid that case for the time being.

In summary, we can write the following general formulas for all aliases of a sinusoid with frequency $\hat{\omega}_0$:

$$\hat{\omega}_0, \ \hat{\omega}_0 + 2\pi\ell, \ 2\pi\ell - \hat{\omega}_0 \qquad (\ell = \text{integer}) \tag{4.7}$$

because the following signals have exactly the same values for all n:

$$\begin{aligned} A\cos(\hat{\omega}_0 n + \varphi) &= A\cos((\hat{\omega}_0 + 2\pi\ell)n + \varphi) \\ &= A\cos((2\pi\ell - \hat{\omega}_0)n - \varphi) \end{aligned} \tag{4.8}$$

If we were to make a stem plot of all of these signals (with specific numbers for A, $\hat{\omega}$, φ, and ℓ), we would not be able to tell them apart, as was shown in the example of Fig. 4-4.

4-1.3 Sampling and Aliasing

The concept of aliases might seem to be merely a mathematical curiosity, but it is crucial to the understanding of sampling and reconstruction of analog signals. Recall that if we sample a continuous-time cosine signal $x(t) = \cos(\omega_0 t)$, we get a discrete-time cosine signal $x[n] = x(nT_s) = \cos(\hat{\omega}_0 n)$, where $\hat{\omega}_0 = \omega_0 T_s$. As we will show in the following sections, if we hope to reconstruct the original analog signal, it is necessary that the normalized frequency $\hat{\omega}_0$ be the principal alias, that is,

$$-\pi < \hat{\omega}_0 = \omega_0 T_s < \pi \tag{4.9}$$

When (4.9) is ***not*** satisfied, we say that ***"aliasing has occurred."*** Henceforth, when we use the term ***aliasing***, we mean that when a signal is sampled, the resulting samples are identical to those obtained by sampling a ***lower frequency signal*** (corresponding to the principal alias). Since the original identity of the signal is ambiguous when aliasing occurs in sampling, it is not straightforward to reconstruct the original analog signal from its samples. Thus, it is generally desirable to avoid aliasing when designing a system that samples continuous-time signals.

4-1.4 Spectrum of a Discrete-Time Signal

We have seen that it is sometimes very useful to represent a continuous-time sinusoidal signal by a spectrum plot consisting of spectrum lines. We should expect to see something similar in a spectrum plot for a discrete-time signal. The spectrum of a continuous-time sinusoid exhibits two spectrum lines at the frequencies $\pm\omega$ rad/s. The alias phenomenon changes the spectrum plot because a given discrete-time sinusoidal sequence could correspond to an infinite number of different frequencies $\hat{\omega}$. Our approach is to take this into account by making a spectrum plot that explicitly shows that there are many different discrete-time sinusoids that have the same samples. Figure 4-5 shows that we can do this by drawing the spectrum representation of the principal alias along with several more of the other aliases.[5] In Fig. 4-5, the spectrum plot includes a representation of the principal alias

[5] We should draw an infinite number of aliases, but we cannot show them all.

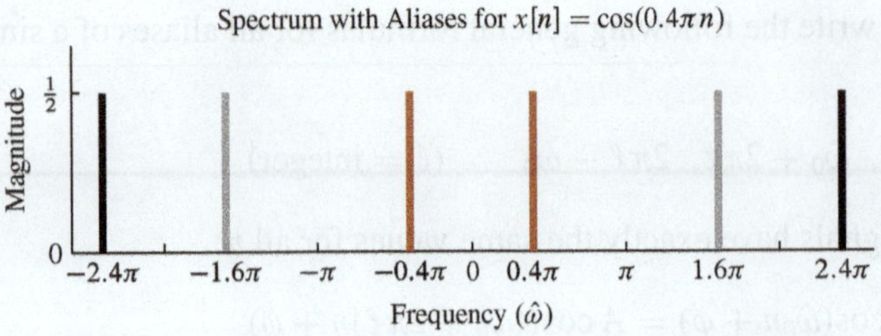

Figure 4-5 Spectrum of a discrete-time sinusoid. The principal alias components are at $\hat{\omega} = \pm 0.4\pi$.

$x_1[n] = \cos(0.4\pi n)$, and two of the aliases $x_2[n] = \cos(2.4\pi n)$ and $x_3[n] = \cos(1.6\pi n)$. Like the two-sided representation in (3.4) on p. 70, the spectrum of each discrete-time alias signal consists of a positive frequency component and a corresponding component at negative frequency. Thus, $x_1[n]$ is represented by the components at $\pm 0.4\pi$, and $x_2[n]$ is represented by the black components at $\pm 2.4\pi$, and the first folded alias is represented by the gray components. Another way of thinking about constructing the plot in Fig. 4-5 is that the spectrum of the principal alias signal consisting of components at $\pm 0.4\pi$ was moved up by 2π and down by 2π. In other words, the principal (or base) band of the spectrum was shifted by integer multiples of 2π.

In the case of the spectrum representation of a continuous-time signal, the assumption was that *all* of the spectrum components were added together to synthesize the continuous-time signal. This is *not* so for the discrete-time case because there are an infinite number of aliases for each principal alias component and they represent the same thing. In a plot like Fig. 4-5, we show spectrum lines for several of the aliases simply to emphasize the fact that many different frequencies could represent the same time-domain sequence. In the case of a sum of discrete-time sinusoids, such as

$$y_1[n] = 2\cos(0.4\pi n) + \cos(0.6\pi n) \tag{4.10}$$

we would make a plot like Fig. 4-5 with many aliases for each of the cosine signals and superimpose them on a single plot. To synthesize the time-domain signal corresponding to a given spectrum representation, we simply need to select one signal from each of the distinct alias sets. For example, (4.10) includes only the principal aliases since 0.4π and 0.6π are both less than π.

EXERCISE 4.4 Make a spectrum plot for the signal of (4.10) similar to Fig. 4-5 showing the principal alias and two other alias frequencies for each cosine signal. How would your plot change if the signal was $y_2[n] = 2\cos(0.4\pi n) + \cos(2.6\pi n)$?

4-1.5 The Sampling Theorem

The plots shown in Fig. 4-3 naturally raise the question of how frequently we must sample in order to retain enough information to reconstruct the original continuous-time signal from its samples. The amazingly simple answer is given by the *Shannon sampling theorem*, a theoretical pillar of modern digital communications, digital control, and digital signal processing.

> **Shannon Sampling Theorem**
> A continuous-time signal $x(t)$ with frequencies no higher than f_{max} can be reconstructed exactly from its samples $x[n] = x(nT_s)$, if the samples are taken at a rate $f_s = 1/T_s$ that is greater than $2f_{max}$.

Notice that the sampling theorem involves two issues. First, it talks about reconstructing the signal from its samples, although it never specifies the process for doing the reconstruction. Second, it gives a minimum sampling rate that is dependent on the frequencies that are contained in the spectrum of the continuous-time signal $x(t)$.

The minimum sampling rate of $2f_{max}$ is called the ***Nyquist rate***.[6] We can see examples of Nyquist rate sampling in many commercial products. For example, audio CDs use a sampling rate of 44.1 kHz for storing music signals in a digital format. This number is slightly more than two times 20 kHz, which is the generally accepted upper limit for human hearing and perception of musical sounds. In all applications, the Nyquist rate is significant because we are usually motivated to use the lowest possible sampling rate in order to minimize system cost in terms of storage and processing speed per sample.

The Shannon theorem states that reconstruction of a sinusoid is possible if we have more than two samples per period. What happens when we don't sample fast enough? The simple answer is that aliasing occurs and reconstruction becomes difficult or impossible. Section 4-2 delves into this issue by using a spectrum view of the C-to-D conversion process.

4-1.6 Ideal Reconstruction

The sampling theorem suggests that a process exists for reconstructing a continuous-time signal from its samples. This reconstruction process would undo the C-to-D conversion so it is called ***D-to-C conversion***. The ideal D-to-C converter block depicted in Fig. 4-6 shows the input and output signals. Since the sampling process of the ideal C-to-D converter is defined by the *mathematical* substitution $t = n/f_s$ as in (4.1), we would expect the same relationship to govern the ideal D-to-C converter, that is,

$$y(t) = y[n]\Big|_{n=f_s t} \qquad -\infty < t < \infty \tag{4.11}$$

However, this substitution is not practical because it is *only* true when $y[n]$ is expressed as a sum of sinusoids formula. In the special case where $y[n]$ was produced by sampling one sinusoid above the Nyquist rate to obtain $y[n] = A\cos(2\pi f_0 n/f_s + \varphi)$, we can use

[6] Harry Nyquist and Claude Shannon were researchers at Bell Telephone Laboratories who each made fundamental contributions to the theory of sampling and digital communication during the period from 1920 to 1950.

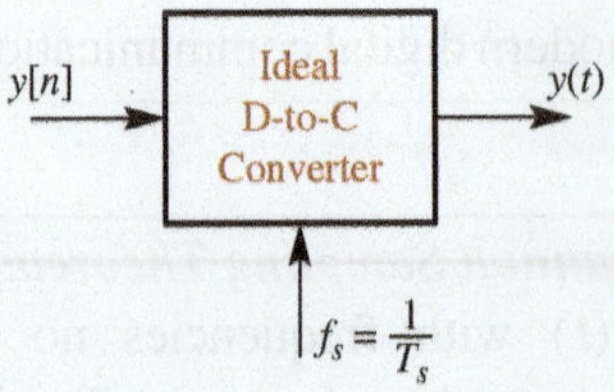

Figure 4-6 Block diagram of the ideal discrete-to-continuous (D-to-C) converter when the sampling rate is $f_s = 1/T_s$. The output signal must satisfy $y(nT_s) = y[n]$.

the substitution in (4.11) to produce $y(t) = A\cos(2\pi f_0 t + \varphi)$ at the D-to-C output. This is convenient because we obtain a mathematical formula for $y(t)$. In practice, we only have a sequence of numbers for $y[n]$ that was obtained by sampling, and we do not know the formula for $y[n]$, or there is no simple formula for the signal, and so things are not that easy. An actual hardware digital-to-analog (D-to-A) converter built with electronic components cannot perform a mathematical substitution like (4.11). Instead, it must "fill in" the signal values between the sampling times, $t_n = nT_s$. In Section 4-3, we will see how ***interpolation*** can be used to build an D-to-A converter that *approximates* the behavior of the ideal D-to-C converter.

If the ideal C-to-D converter samples a cosine signal without the confusion of aliasing, then we can describe its operation as ***frequency scaling***. For $y[n]$ above, the discrete-time frequency is $\hat{\omega} = 2\pi f_0/f_s$, and the continuous-time frequency of $y(t)$ is $\omega = 2\pi f_0$. Thus the relationship appears to be $\omega = \hat{\omega} f_s = \hat{\omega}/T_s$. We can use the same result to describe the D-to-C converter if we solve the frequency scaling equation (4.3) for ω in terms of $\hat{\omega}$,

$$\omega = \hat{\omega} f_s \tag{4.12}$$

Thus each sinusoidal component in a discrete-time signal could be mapped to a sinusoidal component of the continuous-time output signal. But there is one more issue: the discrete-time signal has aliases—an infinite number of them given by (4.7). Which discrete-time frequency should be used in (4.12)? Theoretically, the selection rule is arbitrary, but as we define the ideal D-to-C converter, it always selects the lowest possible frequency components (the principal aliases). These frequencies are guaranteed to lie in the range $-\pi < \hat{\omega} \le \pi$, so when D-to-C converting from $\hat{\omega}$ to analog frequency, the output frequency ***always lies between*** $-f_s/2$ ***and*** $+f_s/2$.

DEMO
Continuous-Discrete Sampling

In summary, the Shannon sampling theorem guarantees alias-free sampling if $x(t)$ contains no frequencies higher than $f_{\max}$, and if $f_s > 2f_{\max}$. Furthermore, when aliasing did not occur in sampling, the output signal $y(t)$ of the ideal D-to-C converter is *identical* to the signal $x(t)$, the original input to the ideal C-to-D converter in Fig. 4-7.

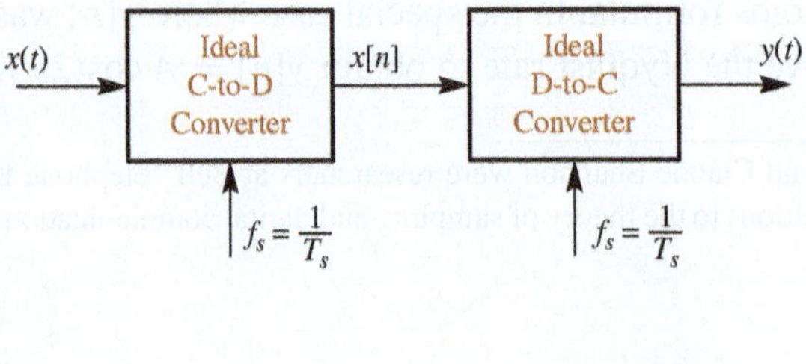

Figure 4-7 Sampling and reconstruction system where the rates of the C-to-D and D-to-C converters are the same.

4-2 Spectrum View of Sampling and Reconstruction

DEMO
Aliasing and Folding

We now present a detailed spectrum explanation of the sampling and reconstruction process. Because the output of the C-to-D converter is a discrete-time signal with an infinite number of spectral aliases, we need to use the spectrum diagram that includes *all* of the aliases as discussed in Section 4-1.4. Then the possible aliasing of frequencies in Fig. 4-7 can be tracked via the movement of spectrum lines in the frequency domain.

4-2.1 Spectrum of a Discrete-Time Signal Obtained by Sampling

If a discrete-time sinusoid (4.2) is obtained by sampling, its frequency is given by (4.3). After we include all the aliases as predicted by (4.7), it is necessary to plot an infinite number of spectrum lines to have a thorough representation of the spectrum. Of course, we can only plot a few of them and indicate that more exist outside our plotting range.

Suppose that we start with a continuous-time sinusoid, $x(t) = A\cos(\omega_0 t + \varphi)$, whose spectrum consists of *two* spectrum lines at $\pm\omega_0$ with complex amplitudes of $\frac{1}{2}Ae^{\pm j\varphi}$. The spectrum of the sampled discrete-time signal

$$\begin{aligned} x[n] &= x(n/f_s) = A\cos((\omega_0/f_s)n + \varphi) \\ &= \tfrac{1}{2}Ae^{j\varphi}e^{j(\omega_0/f_s)n} + \tfrac{1}{2}Ae^{-j\varphi}e^{j(-\omega_0/f_s)n} \end{aligned}$$

also has two spectrum lines at $\hat{\omega} = \pm\omega_0/f_s$, *but it also must contain all the aliases at the following discrete-time frequencies:*

$$\begin{aligned} \hat{\omega} &= \omega_0/f_s + 2\pi\ell \qquad \ell = 0, \pm1, \pm2, \ldots \\ \hat{\omega} &= -\omega_0/f_s + 2\pi\ell \quad \ell = 0, \pm1, \pm2, \ldots \end{aligned} \tag{4.13}$$

This illustrates the important fact that when a discrete-time sinusoid is derived by sampling, the alias frequencies all are based on the *normalized* value, ω_0/f_s, of the frequency of the continuous-time signal. The alias frequencies are obtained by adding integer multiples of 2π radians to the frequency $\hat{\omega} = \omega_0/f_s$.

EXERCISE 4.5 Plot the spectrum of the continuous-time signal $x(t) = \cos(2\pi(100)t + \pi/3)$.

The next sections show examples of sampling a continuous-time 100 Hz sinusoid of the form $x(t) = \cos(2\pi(100)t + \pi/3)$. The sampling frequency is varied to show the discrete-time spectrum for different cases, where f_s is above or below the Nyquist rate in Figs. 4-8 through 4-11. All were constructed with a MATLAB M-file that plots the location of spectrum components involved in the sampling process.

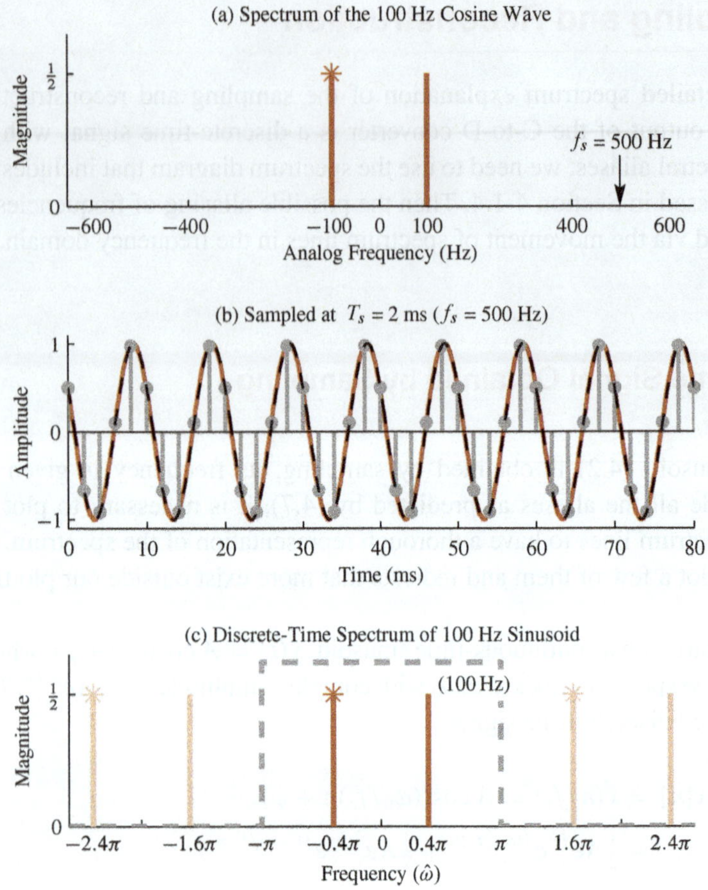

Figure 4-8 *Over-sampling* a 100 Hz sinusoid at $f_s = 500$ samples/s. (a) Continuous-time spectrum; (b) time-domain plot, showing the samples $x[n]$ as gray dots, the original signal $x(t)$ as a continuous orange line, and the reconstructed signal $y(t)$ as a dashed black line, which is the 100 Hz sinusoid passing through the same sample points; and (c) discrete-time spectrum plot, showing the positive and negative frequency components of the original sinusoid at $\hat{\omega} = \pm 0.4\pi$ rad, along with two aliases.

4-2.2 Over-Sampling

In most applications, we try to obey the constraint of the sampling theorem by sampling at a rate higher than twice the highest frequency in order to avoid the problems of aliasing. This is called *over-sampling*. For example, when we sample the 100 Hz sinusoid, $x(t) = \cos(2\pi(100)t + \pi/3)$, at a sampling rate of $f_s = 500$ samples/s, we are sampling at two and a half times the Nyquist rate. The time- and frequency-domain plots are shown in Fig. 4-8. With reference to the C-to-D and D-to-C blocks in Fig. 4-7, Fig. 4-8(a) shows the spectrum of the input signal $x(t)$, while Fig. 4-8(c) shows the spectrum of $x[n] = \cos(0.4\pi n + \pi/3)$, which is the output of the C-to-D converter. The time-domain plots of $x(t)$, $x[n]$, and the reconstructed output $y(t)$ are all in Fig. 4-8(b).

For the continuous-time spectrum in Fig. 4-8(a), the frequency axis is measured in hertz. Note that the negative-frequency complex exponential component of the cosine is denoted with a * at the top of its spectrum line to indicate that its complex value is the conjugate of the value of the positive frequency component. Figure 4-8(b) shows $x(t)$, $x[n]$, and $y(t)$ together, but its horizontal axis is labeled with time (t) in milliseconds, so

$x[n]$ is plotted versus nT_s. Figure 4-8(c) contains the spectrum of the discrete-time signal versus the normalized frequency $\hat{\omega}$. According to the frequency scaling equation (4.3), the input analog frequency of 100 Hz maps to $\hat{\omega} = 2\pi(100)/f_s = 2\pi(100)/500 = 0.4\pi$, so in Fig. 4-8(c) we plot spectrum lines at $\hat{\omega} = \pm 0.4\pi$. Then we also draw all the aliases at

$$\hat{\omega} = 0.4\pi + 2\pi\ell \qquad \ell = 0, \pm 1, \pm 2, \ldots$$
$$\hat{\omega} = -0.4\pi + 2\pi\ell \qquad \ell = 0, \pm 1, \pm 2, \ldots$$

The D-to-C converter transforms the discrete-time spectrum to the continuous-time output spectrum. But there is one complication: the D-to-C converter must select just one pair of spectrum lines from all the possibilities given by (4.13). The selection rule is arbitrary, but in order to be consistent with real D-to-A converter operation, we must assert that the ideal D-to-C converter always selects the lowest possible frequency for each set of aliases. These are what we have called the principal alias frequencies, and in Fig. 4-8(c) these are the frequency components that fall inside the dashed box, that is, in the range $|\hat{\omega}| < \pi$. Since the frequency components inside the dashed box satisfy $|\hat{\omega}| < \pi$, the spectral lines for the output ***always lie between*** $-f_s/2$ ***and*** $+f_s/2$. When the principal alias frequencies in Fig. 4-8(c) are converted from $\hat{\omega}$ to analog frequency, the result is $f = \pm 0.4\pi(f_s/2\pi) = \pm 100$ Hz which is identical to Fig. 4-8(a).

In summary, for the over-sampling case where the original frequency f_0 is less than $f_s/2$, the original waveform is reconstructed exactly. In the present example, $f_0 = 100$ Hz and $f_s = 500$, so the Nyquist condition of the sampling theorem is satisfied, and the output $y(t)$ equals the input $x(t)$ as shown in Fig. 4-8(b).

4-2.3 Aliasing Due to Under-Sampling

When $f_s < 2f_0$, the signal is *under-sampled* and we say that aliasing has occurred. For example, if $f_s = 80$ Hz and $f_0 = 100$ Hz, the sampling rate is below the Nyquist rate and we can show that aliasing confusion occurs. In Fig. 4-9(a), the spectrum of the analog input signal $x(t)$ is shown, along with an arrow indicating the sampling rate at $f_s = 80$ Hz. The spectrum of the discrete-time signal (Fig. 4-9(c)) contains lines at $\hat{\omega} = \pm 2\pi(100)/f_s = \pm 2\pi(100)/80 = \pm 2.5\pi$, as predicted by the frequency scaling equation (4.3). In order to complete the discrete-time spectrum we must also draw all the aliases at

$$\hat{\omega} = 2.5\pi + 2\pi\ell \qquad \ell = 0, \pm 1, \pm 2, \ldots$$
$$\hat{\omega} = -2.5\pi + 2\pi\ell \qquad \ell = 0, \pm 1, \pm 2, \ldots$$

In Fig. 4-9(b), the 100 Hz sinusoid (solid orange line) is sampled too infrequently to be recognized as the original 100 Hz sinusoid. When we examine the D-to-C process for this case, we use the lowest frequency spectrum lines from the discrete-time spectrum (Fig. 4-9(c)). These are the principal aliases at $\hat{\omega} = \pm 0.5\pi$, so the output spectrum lines are at $f = \pm 0.5\pi(f_s/2\pi) = \pm 80/4 = \pm 20$ Hz. Alternatively, we can state this result in the time domain by observing that the same samples would have been obtained from a

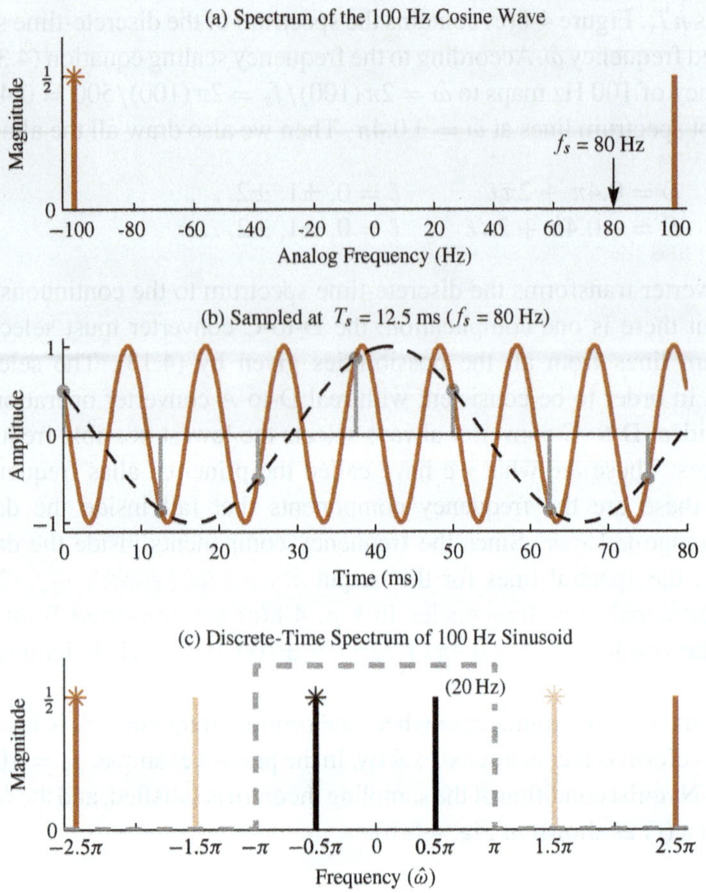

Figure 4-9 *Under-sampling* a 100 Hz sinusoid at $f_s = 80$ samples/s. (a) Continuous-time spectrum; (b) time-domain plot, showing the samples $x[n]$ as gray dots, the original signal $x(t)$ as a continuous orange line, and the reconstructed signal $y(t)$ as a dashed black line, which is a 20 Hz sinusoid passing through the same sample points; and (c) discrete-time spectrum plot, showing the positive and negative frequency components of the original sinusoid at $\hat{\omega} = \pm 2.5\pi$ rad, along with two sets of alias components.

20 Hz sinusoid. The reconstructed signal is that 20 Hz sinusoid shown as the black dashed line in Fig. 4-9(b). Notice that the alias frequency of 20 Hz can be found by subtracting f_s from 100 Hz.

The aliasing of sinusoidal components can have some dramatic effects. Figure 4-10 shows the case where the sampling rate and the frequency of the sinusoid are the same. In Fig. 4-10(b), we see that one sample is taken per period, and all the samples are the same value. In other words, $x[n]$ is DC, which is the same as a sinusoid with zero frequency. We can justify this result in the frequency domain (Fig. 4-10(c)) by noting that the discrete-time spectrum contains lines at $\hat{\omega} = \pm 2\pi(100/f_s) = \pm 2\pi$, and also at the aliases separated by $2\pi\ell$. Thus, two of the aliases land at $\hat{\omega} = 0$, and the sum of these two is the value that is reconstructed by the D-to-C converter.

4-2.4 Folding Due to Under-Sampling

Figure 4-11 shows the case where *under-sampling* leads to the type of aliasing called folding. Here the sampling rate is $f_s = 125$ samples/s, which is between f_0 and $2f_0$. Once again, Fig. 4-11(a) shows the spectrum of the continuous-time signal with spectrum

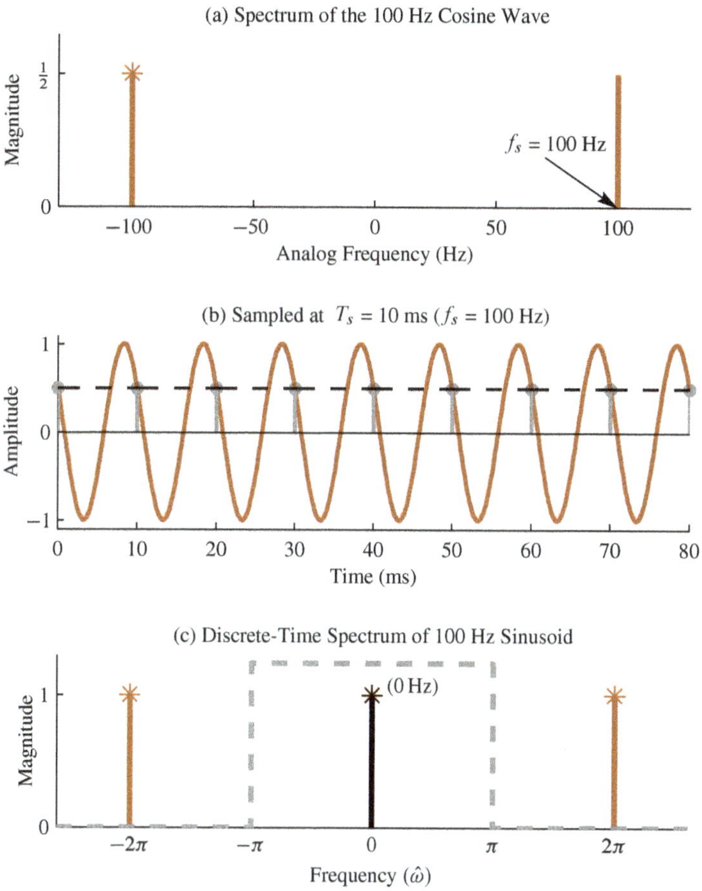

Figure 4-10 *Under-sampling* a 100 Hz sinusoid at $f_s = 100$ samples/s. (a) Continuous-time spectrum; (b) time-domain plot, showing the samples $x[n]$ as gray dots, the original signal $x(t)$ as a continuous orange line, and the reconstructed signal $y(t)$ as a dashed black line, which is a 0 Hz sinusoid passing through the same sample points; and (c) discrete-time spectrum plot, showing the positive and negative frequency components of the original sinusoid at $\hat{\omega} = \pm 2\pi$ rad, along with parts of two sets of alias components that are superimposed at $\hat{\omega} = 0$ to give a DC value of $\frac{1}{2}e^{j\pi/3} + \frac{1}{2}e^{-j\pi/3} = \cos(\pi/3)$.

lines at ± 100 Hz. The discrete-time spectrum (Fig. 4-11(c)) is constructed by mapping ± 100 Hz to the two spectrum lines at $\hat{\omega} = \pm 2\pi(100/f_s) = \pm 1.6\pi$, and then including all the aliases to get lines at

$$\hat{\omega} = 1.6\pi + 2\pi\ell \qquad \ell = 0, \pm 1, \pm 2, \ldots$$
$$\hat{\omega} = -1.6\pi + 2\pi\ell \qquad \ell = 0, \pm 1, \pm 2, \ldots$$

In this case, an interesting thing happens. The two frequency components between $\pm\pi$ are $\hat{\omega} = \pm 0.4\pi$, but the one at $\hat{\omega} = +0.4\pi$ is an alias of the negative frequency component at -1.6π, which is why this situation is called folding. The analog frequency of the reconstructed output is $f = 0.4\pi(f_s/2\pi) = f_s/5 = 25$ Hz. An additional fact about folding is that phase of the reconstructed analog signal changes sign. If the original 100 Hz sinusoid had a phase of $\varphi = +\pi/3$, then the phase of the component at $\hat{\omega} = -1.6\pi$ would be $-\pi/3$, and it follows that the phase of the aliased component at $\hat{\omega} = +0.4\pi$ would also be $-\pi/3$. After reconstruction, the phase of $y(t)$ would be $-\pi/3$. It is possible to observe this "phase-reversal" in the time-domain plot of Fig. 4-11(b). The input signal (solid orange line) is going down at $t = 0$, while the reconstructed output (black dashed line) is going up. This means that when we sample a 100 Hz sinusoid at a sampling rate

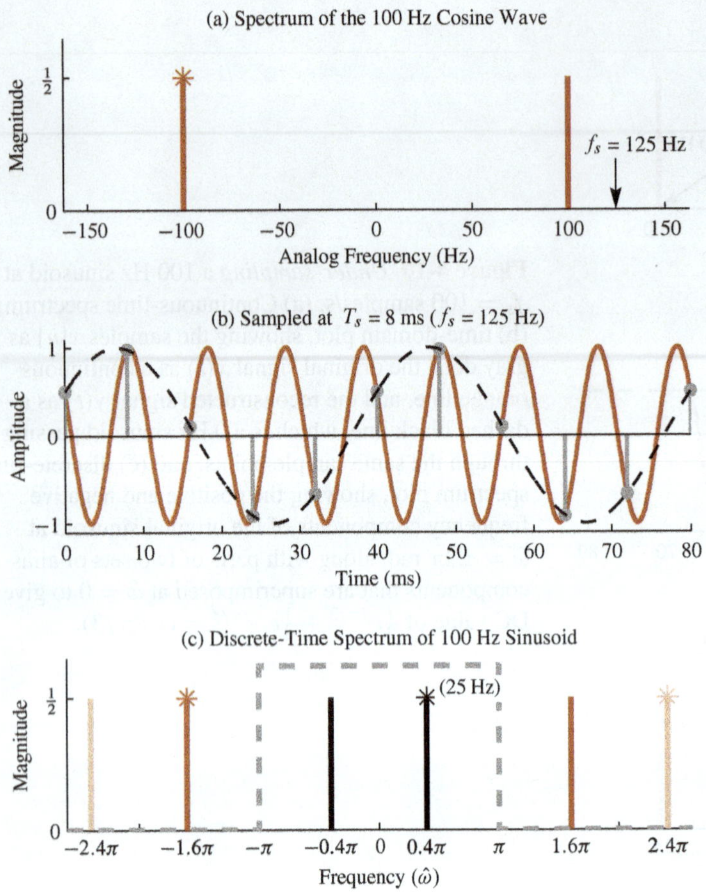

Figure 4-11 *Under-sampling* a 100 Hz sinusoid at $f_s = 125$ samples/s. (a) Continuous-time spectrum; (b) time-domain plot, showing the samples $x[n]$ as gray dots, the original signal $x(t)$ as a continuous orange line, and the reconstructed signal $y(t)$ as a dashed black line, which is a 25 Hz sinusoid passing through the same sample points; and (c) discrete-time spectrum plot, showing the positive and negative frequency components of the original sinusoid at $\hat{\omega} = \pm 1.6\pi$ rad, along with two sets of aliases. Note how the phase is negated in this example of aliasing by folding.

of 125 samples/s, we get the same samples that we would have gotten by sampling a 25 Hz sinusoid, but with opposite phase. This is depicted in Fig. 4-11 by showing that the complex amplitude of the positive frequency component at 0.4π is due to the negative frequency alias $(-1.6\pi + 2\pi)$.

4-2.5 Maximum Reconstructed Frequency

The preceding examples have one thing in common: the output frequency is always less than $f_s/2$. For a sampled sinusoid, the ideal D-to-C converter picks the alias frequency closest to $\hat{\omega} = 0$ and maps it to the output analog frequency via $f = \hat{\omega}(f_s/2\pi)$. Since the principal alias is guaranteed to lie between $-\pi$ and $+\pi$, the output frequency always lies between $-f_s/2$ and $+f_s/2$.

One striking demonstration of this fact can be made by implementing the system in Fig. 4-7 with a linear FM chirp signal as the input, and then listening to the reconstructed

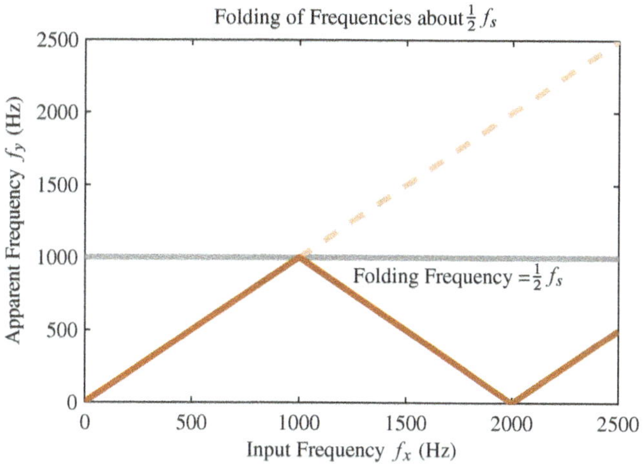

Figure 4-12 Folding of a sinusoid sampled at $f_s = 2000$ samples/s. The apparent frequency is the lowest frequency of a sinusoid that has exactly the same samples as the input sinusoid.

DEMO
Chirp Synthesis

output signal, or computing its spectrogram.[7] This ***Chirp Synthesis*** demo is available on the Companion Website, and is part a lab entitled AM and FM Sinusoidal Signals.

To understand the Chirp Synthesis demo, suppose that the instantaneous frequency of the input chirp increases according to the formula $f_x(t) = \alpha t$ Hz, where α is the chirp rate. Then the input signal is $x(t) = \cos(\pi \alpha t^2)$, and after sampling at a rate f_s, we obtain

$$x[n] = \cos(\pi \alpha n^2 / f_s^2)$$

LAB
AM and FM Sinusoidal Signals

When $y(t)$ is reconstructed from $x[n]$, what would you hear? It turns out that the output frequency goes up and down as shown in Fig. 4-12. We suspect that something like this will happen because we know that the output cannot have a frequency higher than $f_s/2$, even though the instantaneous frequency of the input continually increases above $f_s/2$.

Explaining the up and down behavior of the output frequency takes more thought. If we concentrate on a very short time interval surrounding the time $t = t_0$, then the concept of instantaneous frequency implies that input signal can be assumed to have a constant frequency equal to $f_x(t_0) = \alpha t_0$ near $t = t_0$. The spectrum of this constant-frequency approximation to $x(t)$ near $t = t_0$ has two spectrum lines at $\pm \alpha t_0$ which, after sampling, map to the normalized discrete radian frequencies, $\pm \hat{\omega}_x = \pm 2\pi(\alpha t_0)/f_s$. Then we must determine all of the aliases, and use the principal alias to determine the reconstructed output frequency. There is no easy formula, so we just consider a few separate cases. First of all, when the input frequency $f_x(t) = \alpha t$ goes from 0 to $f_s/2$, $\hat{\omega}_x$ increases from 0 to π and is equal to the principal alias, so the output frequency will be equal to the input frequency. Next consider the case when the input frequency $f_x(t) = \alpha t$ increases from $f_s/2$ to f_s. The corresponding frequencies for $x[n]$ increase from $\hat{\omega}_x = \pi$ to $\hat{\omega}_x = 2\pi$ for the positive frequency, while its negative frequency companion goes from $-\hat{\omega}_x = -\pi$ down to $-\hat{\omega}_x = -2\pi$. The principal alias of this negative frequency component is $-\hat{\omega}_x + 2\pi$ which goes from $\hat{\omega}_x = +\pi$ down to $\hat{\omega}_x = 0$, while the principal

[7]The spectrogram of a chirp without aliasing was shown in Fig. 3-29 in Section 3-7.2. An in-depth discussion of the spectrogram can be found in Section 8-7.

alias of the positive frequency component goes from $\hat{\omega} = -\pi$ up to $\hat{\omega}_x = 0$. Therefore, the output signal which is reconstructed from the principal alias has a frequency going from $f = f_s/2$ down to $f = 0$.

Figure 4-12 shows a plot of the output frequency $f_y(t)$ versus the input frequency $f_x(t)$. As f_x goes from $f_s/2$ to f_s, the output frequency f_y decreases from $f_s/2$ to 0 and then starts increasing when $\hat{\omega}_x$ becomes greater than 2π. The terminology *folded* frequency comes from the fact that the input and output frequencies are mirror images with respect to the line at $f_s/2$, and would lie on top of one another if the graph were folded about the $f_s/2$ line in Fig. 4-12.

4-2.6 Sampling and Reconstruction GUI

To develop greater understanding of the sampling and reconstruction process discussed previously and shown in Fig. 4-7, we can use the MATLAB GUI called `con2dis` whose interface is shown in Fig. 4-13. The GUI display is a 2×3 array of panels. The top row shows the time-domain signals and from left to right follows the signal flow in Fig. 4-7. The leftmost signal is the input, which is a continuous-time sinusoid; the middle is the discrete-time signal after sampling; and the right is the output of the D-to-C converter, which is again a continuous-time sinusoid. The bottom row shows the frequency-domain spectrum of the

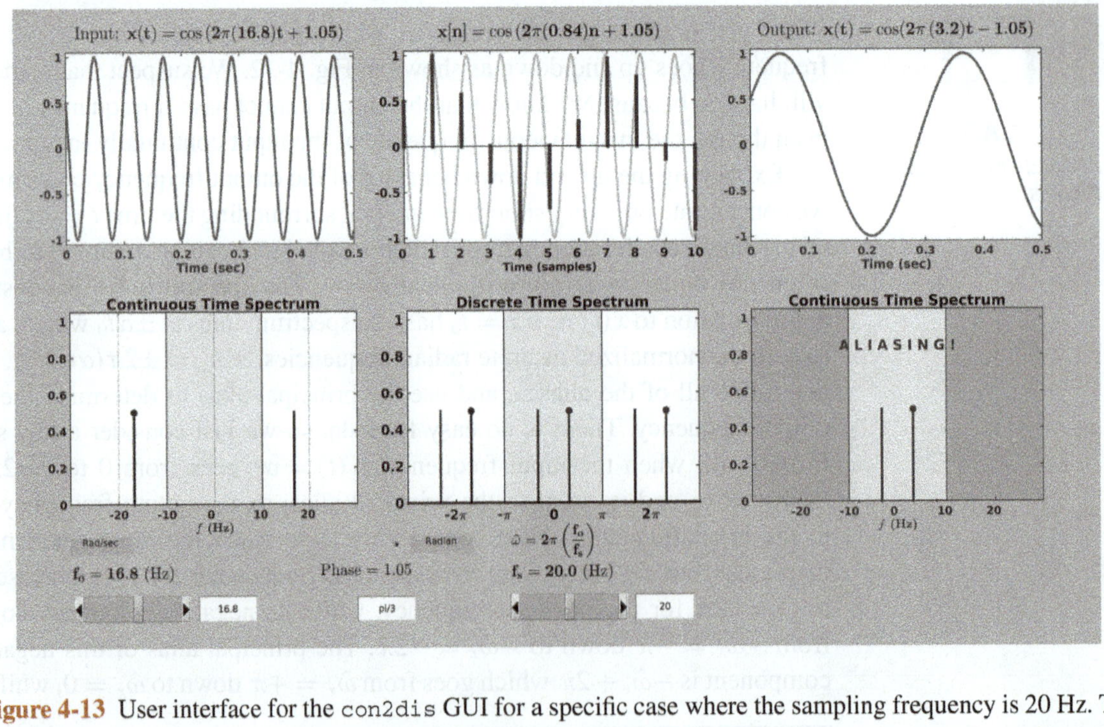

Figure 4-13 User interface for the `con2dis` GUI for a specific case where the sampling frequency is 20 Hz. The frequency of the input signal being sampling is 16.8 Hz which is between $f_s/2$ and f_s, so there is a folded alias. The frequency of the output signal must be less than $f_s/2$ which is 3.2 Hz as seen in the upper right panel.

DEMO
Continuous-Discrete Sampling

time-domain signals above. The leftmost spectrum shows the two spectrum lines[8] for the input continuous-time sinusoid; the middle spectrum is the spectrum for the discrete-time signal and exhibits the 2π periodicity of the $\hat{\omega}$ domain; and the rightmost spectrum again has two spectrum lines which are the lowest frequency pair of lines obtained by frequency scaling ($f = \hat{\omega}/2\pi f_s$) from the two spectrum lines in the $\hat{\omega}$ domain which are between $-\pi$ and $+\pi$. The frequency axis of the GUI display can be toggled between radian and cyclic frequency units.

The GUI can be controlled in two ways: the input frequency and phase can be varied while the sampling rate is constant, or the sampling rate can be varied for a fixed input sinusoid. By using the frequency slider in the lower left to continuously raise the input frequency, the folding phenomenon described in Fig. 4-12 can be observed by watching the output frequency go up and down.

4-3 Discrete-to-Continuous Conversion

The purpose of the ideal D-to-C converter is to interpolate a smooth continuous-time function through the given discrete-time samples $y[n]$. Thus, in the special case when $y[n] = A\cos(2\pi f_0 n T_s + \varphi)$, and if $f_0 < f_s/2$, then according to the sampling theorem, the ideal converter should produce

$$y(t) = A\cos(2\pi f_0 t + \varphi) \tag{4.14}$$

For sampled sinusoidal signals *only*, the ideal D-to-C converter in effect replaces n by $f_s t$ in the cosine formula. On the other hand, if $f_0 > f_s/2$, then we know that aliasing or folding confusion has occurred, and the ideal D-to-C converter reconstructs a cosine wave with frequency equal to the principal alias frequency which is less than $f_s/2$.

4-3.1 Interpolation with Pulses

How does an actual D-to-C converter work? In this section, we explain how the D-to-C converter does interpolation through the given discrete-time samples $y[n]$, and then describe a practical system that is nearly the same as the ideal D-to-C converter. These actual hardware systems, called D-to-A converters, approximate the behavior of the ideal D-to-C system as in (4.14).

A general formula that describes a broad class of D-to-C converters is given by the equation

$$y(t) = \sum_{n=-\infty}^{\infty} y[n]p(t - nT_s) \tag{4.15}$$

where $p(t)$ is the characteristic pulse shape of the converter. Equation (4.15) states that the output signal is produced by superimposing scaled and time-shifted pulses. In other words, at each sample time $t_n = nT_s$, a shifted pulse $p(t - nT_s)$ is emitted with an amplitude

[8]Note that the negative-frequency spectral line has a dot at the top.

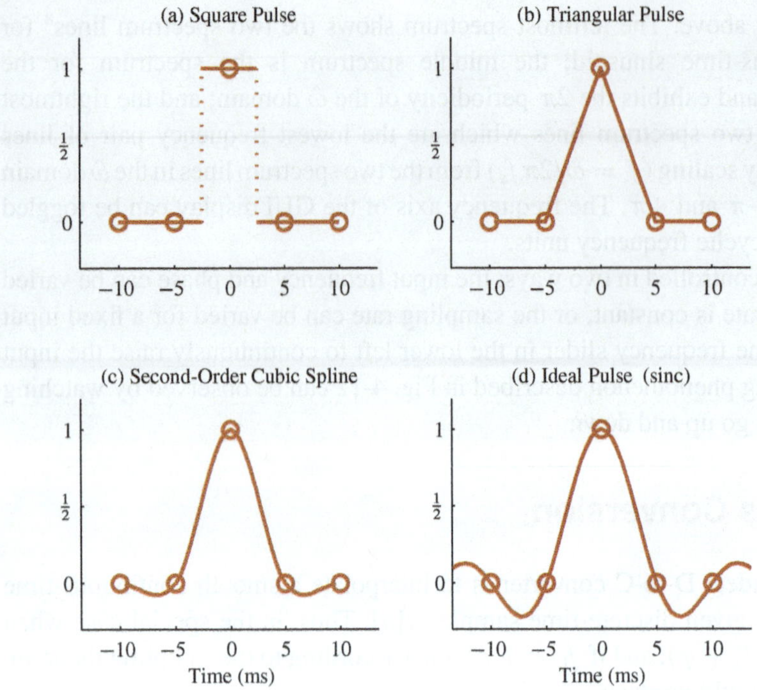

Figure 4-14 Four different pulses for D-to-C conversion. The sampling period is $T_s = 5$ ms, or equivalently, $f_s = 200$ Hz. The duration of the pulses is (a) T_s for the square pulse, (b) $2T_s$ for the triangular pulse, (c) $4T_s$ for the cubic spline, and (d) infinite for the ideal pulse.

proportional to the sample value $y[n]$ corresponding to that time instant.[9] Note that all the pulses have the same waveshape $p(t)$. If the pulse duration is greater than or equal to T_s, then the gaps between samples are filled by adding overlapped pulses.

Equation (4.15) requires that we choose the pulse waveform $p(t)$. Unfortunately, we have not yet studied the mathematical tools needed to derive the optimal pulse shape for exact reconstruction of a waveform $y(t)$ from its samples $y[n] = y(nT_s)$ as is predicted to be possible in the Shannon sampling theorem. This optimal pulse shape can be constructed during a mathematical proof of the sampling theorem. Nonetheless, we will demonstrate the plausibility of (4.15) by considering some simple examples that use suboptimal pulse shapes. Figure 4-14 shows four possible pulse waveforms for D-to-C conversion when $f_s = 200$ Hz.

DEMO
Reconstruction Movies

4-3.2 Zero-Order Hold Interpolation

The simplest pulse shape that we might propose is a symmetric square pulse of the form

$$p(t) = \begin{cases} 1 & -\frac{1}{2}T_s < t \le \frac{1}{2}T_s \\ 0 & \text{otherwise} \end{cases} \tag{4.16}$$

The total width of the square pulse plotted in Fig. 4-14(a) is $T_s = 5$ msec and its amplitude is 1. Therefore, each term $y[n]p(t - nT_s)$ in the sum (4.15) creates a flat region of amplitude $y[n]$ centered on $t = nT_s$. This is shown in Fig. 4-15(a), which shows the

[9]A *pulse* is a continuous-time waveform that is concentrated in time.

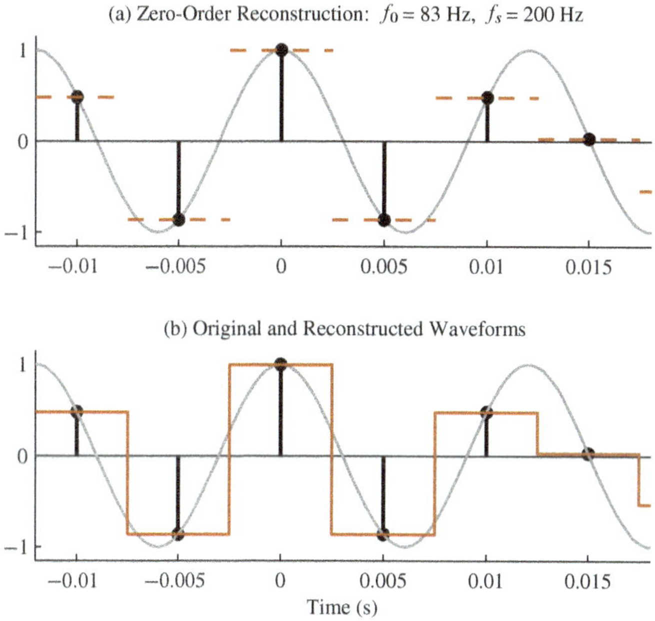

Figure 4-15 D-to-C conversion using a square pulse. (a) Shifted square pulses (dashed orange); (b) sum using (4.15) (solid orange). Flat regions are characteristic of the zero-order hold.

original 83 Hz cosine wave (solid gray line), its samples taken at a sampling rate of 200 samples/s, and the individual shifted and scaled pulses, $y[n]p(t-nT_s)$ (dashed line). The sum of all the shifted and scaled pulses creates a "stairstep" waveform, as shown in Fig. 4-15(b), where the gray curve is the original cosine wave $x(t)$, and the solid orange curve shows the reconstructed waveform using the square pulse.[10]

The space between samples has indeed been filled with a continuous-time waveform—however, it is clear from Fig. 4-15(b) that the reconstructed waveform for the square pulse is a poor approximation of the original cosine wave. Thus, using a square pulse in (4.15) is D-to-C conversion, but not *ideal D-to-C conversion*. Even so, this is a useful model, since many physically realizable D-to-A converters produce outputs that look exactly like this!

4-3.3 Linear Interpolation

The triangular pulse plotted in Fig. 4-14(b) is defined as a pulse consisting of the first-order polynomial (straight line) segments

$$p(t) = \begin{cases} 1 - |t|/T_s & -T_s \le t \le T_s \\ 0 & \text{otherwise} \end{cases} \tag{4.17}$$

Figure 4-16(a) shows the original 83 Hz cosine wave (solid gray curve) and its samples, together with the scaled pulses $y[n]p(t-nT_s)$ (dashed orange) for the triangular pulse shape. In this case, the output $y(t)$ of the D-to-C converter at any time t is the sum of

[10] Since a constant is a polynomial of zero order, and since the effect of the flat pulse is to "hold" or replicate each sample for T_s s, the use of a flat pulse is called ***zero-order hold reconstruction***.

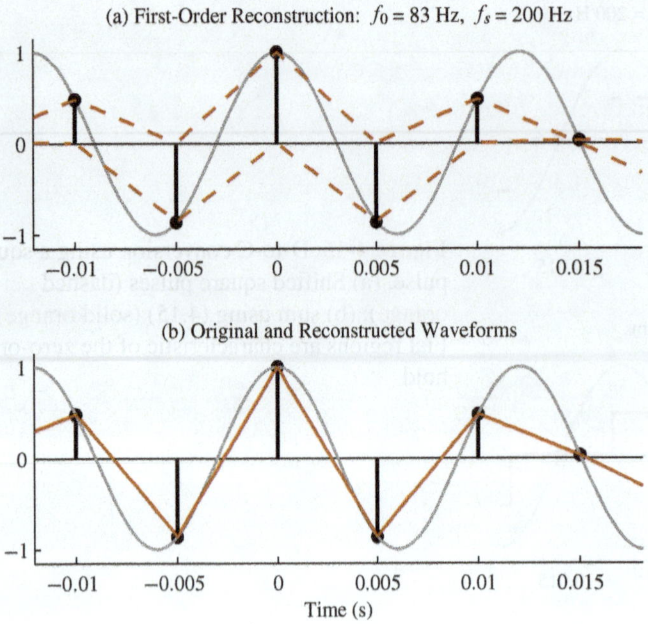

Figure 4-16 D-to-C conversion using a triangular pulse. (a) Shifted triangular pulses (dashed orange); (b) sum using (4.15) (solid orange). The result is reconstruction by linear interpolation.

two scaled triangular pulses that overlap at that time instant. Since the duration of the triangular pulse is $2T_s$, and neighboring pulses are shifted by T_s, no more than two pulses can overlap at any given time. The resulting output is shown as the solid orange curve in Fig. 4-16(b). Note that the result is that samples are connected by straight lines. Also note that the values at $t = nT_s$ are exactly correct. This is because the triangular pulses are zero at $\pm T_s$, and only one scaled pulse (with value $y[n]$ at $t = nT_s$) contributes to the value at $t = nT_s$. In this case, we say that the continuous-time waveform has been obtained by *linear interpolation* between the samples. This is a smoother and better approximation of the original waveform (thin gray curve), but there is still significant reconstruction error for this signal.

4-3.4 Cubic-Spline Interpolation

A third pulse shape is shown in Fig. 4-14(c). This pulse consists of four cubic spline (third-order polynomial) segments. Note that its duration is twice that of the triangular pulse and four times that of the square pulse. Also, note that this pulse is zero at the key locations:

$$p(t) = 0 \qquad \text{for } t = \pm T_s,\ \pm 2T_s$$

and its derivative is smooth at the sample locations. The reconstruction with the cubic-spline pulses is shown in Fig. 4-17. Figure 4-17(a) shows the original 83 Hz waveform (solid gray), its samples ($f_s = 200$ samples/s), and the shifted and scaled pulses $y[n]p(t - nT_s)$ (dashed orange). Now note that, for values of t between two adjacent sample times, four pulses overlap and must be added together in the sum (4.15). This means that the reconstructed signal at a particular time instant, which is the sum

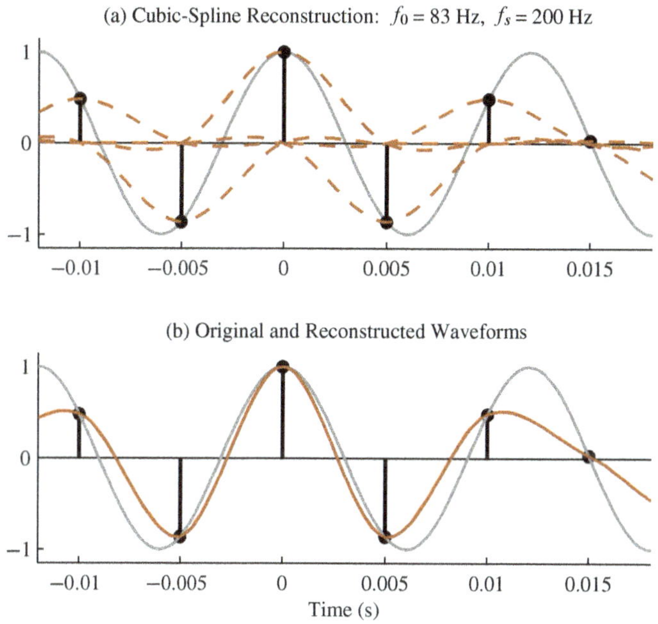

Figure 4-17 D-to-C conversion using a cubic-spline pulse. (a) Shifted cubic-spline pulses (dashed orange); (b) sum using (4.15) (solid orange).

of these overlapping pulses, depends on the two samples preceding and the two samples following that time instant. Figure 4-17(b) shows the original waveform (light gray), the samples, and the output of the D-to-C converter with "cubic-spline pulse" interpolation (solid orange curve). Now we see that the approximation is getting smoother and better, but sometimes it is far from perfect. In the next section, we will see that this is because the sampling rate is only 2.4 times the highest frequency.

4-3.5 Over-Sampling Aids Interpolation

From the previous three examples, it seems that one way to make a smooth reconstruction of a waveform such as a sinusoid is to use a pulse $p(t)$ that is smooth and has a long duration. Then the interpolation formula involves several neighboring samples in the computation of each output value $y(t)$—however, the sampling rate is another important factor. If the original waveform does not vary much over the duration of $p(t)$, then we can expect a good reconstruction. One way to ensure this is by over-sampling (that is, using a sampling rate that is much greater than the frequency of the cosine wave). This is illustrated in Figs. 4-18, 4-19, and 4-20, where the frequency of the cosine wave is still $f_0 = 83$ Hz, but the sampling frequency has been raised to $f_s = 500$ samples/s. Now the reconstruction pulses are the same shape as in Fig. 4-14, but they are much shorter, since $T_s = 2$ ms instead of 5 ms. The signal changes much less over the duration of a single pulse, so the waveform appears "smoother" and is much easier to reconstruct accurately using only a few samples. Note that even for the case of the square pulse (Fig. 4-18) the reconstruction is better, but still discontinuous; the triangular pulse (Fig. 4-19) gives an excellent approximation; and the cubic-spline pulse gives a reconstruction that is indistinguishable from the original signal on the plotting scale of Fig. 4-20.

DEMO
Reconstruction Movies

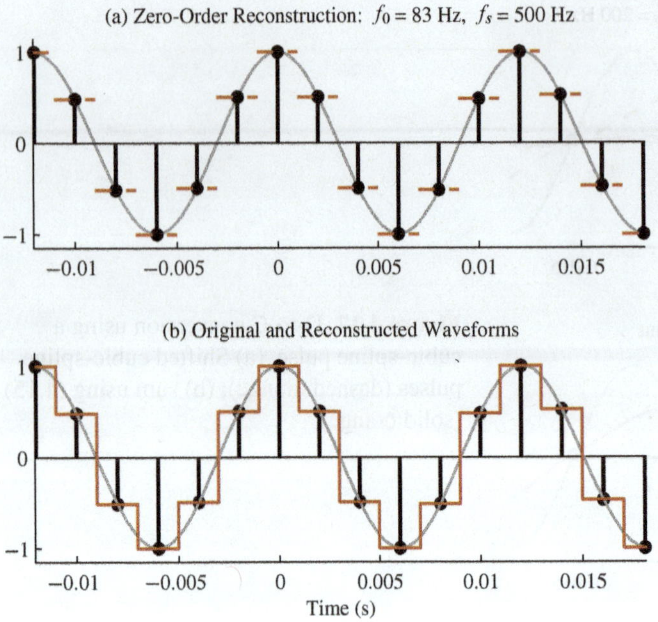

Figure 4-18 D-to-C conversion using a square pulse. The original 83 Hz sinusoid was over-sampled at $f_s = 500$ samples/s.

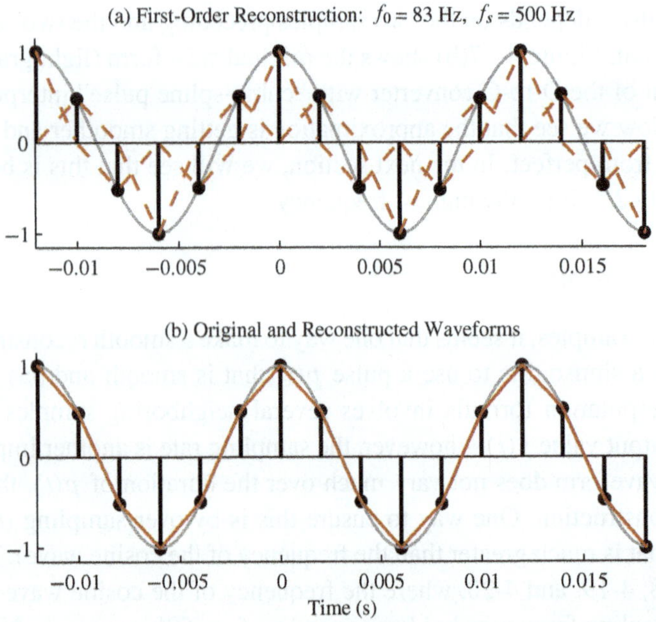

Figure 4-19 D-to-C conversion using a triangular pulse at $f_s = 500$ samples/s.

Figures 4-18, 4-19, and 4-20 show that over-sampling can make it easier to reconstruct a waveform from its samples. Indeed, this is why CD and DVD audio players have over-sampled D-to-A converters! In these systems, 4× or 2× over-sampling is used to increase the sampling rate before sending the samples to the D-to-A converter. This makes it possible to use a simpler (and therefore less expensive) D-to-A converter to reconstruct an accurate output from a CD or DVD player.

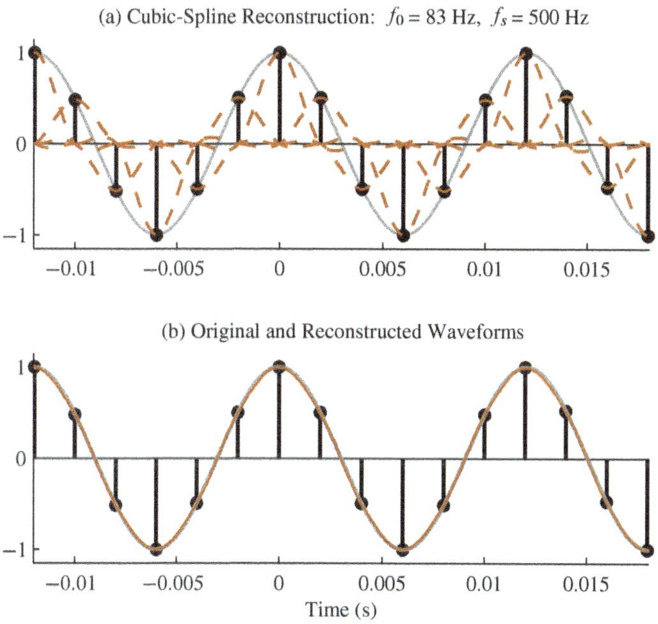

Figure 4-20 D-to-C conversion using a cubic-spline pulse at $f_s = 500$ samples/s.

4-3.6 Ideal Bandlimited Interpolation

So far in this section, we have demonstrated the basic principles of discrete-to-continuous conversion. We have shown that this process can be approximated very well by a sum of shifted pulses (4.15). One question remains: What is the pulse shape that gives "ideal D-to-C conversion"? The answer, which relies on the continuous-time Fourier Transform, is a *sinc function* which is given by the following equation:

$$p(t) = \text{sinc}(t/T_s) = \frac{\sin(\pi t/T_s)}{(\pi t/T_s)} \qquad \text{for } -\infty < t < \infty \tag{4.18}$$

The infinite length of this sinc pulse implies that to reconstruct a signal at time t exactly from its samples requires *all* the samples, not just those near time t. Figure 4-14(d) shows this pulse over the interval $-2.6T_s < t < 2.6T_s$. It continues to decay outside this interval, but never does reach and stay at zero. Since $p(0) = 1$ and $p(nT_s) = 0$ for $n = \pm1, \pm2, \ldots$, this type of reconstruction is still an interpolation process, and it is called *bandlimited interpolation*, because it can be shown that using (4.18) in the interpolation formula (4.15) is equivalent to selecting the principal alias components inside the gray boxes in Figs. 4-8 through 4-11. Using (4.18) to reconstruct from samples of a cosine wave always reproduces a cosine wave exactly. If the sampling rate satisfies the conditions of the sampling theorem, the reconstructed cosine wave is identical to the original signal that was sampled. If aliasing occurred in sampling, the ideal D-to-C converter reconstructs a cosine wave with the alias frequency that is less than $f_s/2$.

4-4 The Sampling Theorem

This chapter has discussed the issues that arise in sampling continuous-time signals. Using the example of the continuous-time cosine signal, we have illustrated the phenomenon of aliasing, and we have shown how the original continuous-time cosine signal can be reconstructed by interpolation. All of the discussion of this chapter has been aimed at the goal of establishing confidence in the Shannon sampling theorem, which, because of its central importance to our study of digital signal processing, is repeated here.

> ***Shannon Sampling Theorem***
>
> A *bandlimited* continuous-time signal $x(t)$ with *no spectral components at frequencies above* $f_{\max}$ can be reconstructed exactly from its samples $x(nT_s)$, if the samples are taken at a rate $f_s = 1/T_s$ that is greater than $2f_{\max}$.

A block diagram representation of the sampling theorem is shown in Fig. 4-21 in terms of the ideal C-to-D and D-to-C converters that we have defined in this chapter. The sampling theorem states that if the input is composed of sinusoidal signals with frequencies $\{f_k\}$ and all those frequencies are in the range $0 \le f_k \le f_{\max}$, then the reconstructed signal is equal to the original signal that was sampled (that is, $y(t) = x(t)$, if $f_s > 2f_{\max}$).

Signals composed of sums of sinusoids such that all frequencies are limited to a "band of frequencies" of the form $0 \le f \le f_{\max}$ are called *bandlimited signals*.[11] Such signals could be represented as

$$x(t) = \sum_{k=0}^{N} x_k(t) \tag{4.19}$$

where each of the individual signals is of the form $x_k(t) = A_k \cos(2\pi f_k t + \varphi_k)$ with $0 \le f_k \le f_{\max}$. As we have seen in Chapter 3 and Appendix C, such an additive combination of sinusoids can produce an infinite variety of both periodic and nonperiodic signal waveforms. If we sample the signal represented by (4.19) we obtain

$$x[n] = x(nT_s) = \sum_{k=0}^{N} x_k(nT_s) = \sum_{k=0}^{N} x_k[n] \tag{4.20}$$

where $x_k[n] = A_k \cos(\hat{\omega}_k n + \varphi_k)$ with $\hat{\omega}_k = 2\pi f_k/f_s$. Thus, if we sample a sum of continuous-time cosines, we obtain a sum of sampled cosines, and if we sample above the Nyquist rate, then $f_s > 2f_{\max} \ge 2f_k$, and $|\hat{\omega}_k| \le \pi$ for each $x_k[n]$.

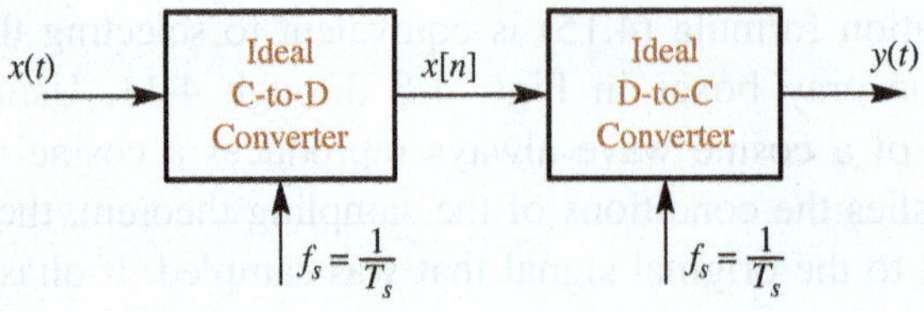

Figure 4-21 Sampling and reconstruction system with identical rates for the C-to-D and D-to-C converters.

[11]The corresponding complex exponential signals would be limited to the band $-f_{\max} \le f \le f_{\max}$.

The final step in the process of sampling followed by reconstruction in Fig. 4-21 is D-to-C conversion by interpolation with

$$y(t) = \sum_{n=-\infty}^{\infty} x[n]p(t - nT_s) \tag{4.21}$$

where for perfect reconstruction, $p(t)$ would be given by (4.18). This expression for the reconstructed output is a linear operation on the samples $x[n]$. This means that the total output $y(t)$ consists of the sum of the outputs due to each of the different components $x_k[n]$. We can see this by substituting (4.20) into (4.21) as in

$$y(t) = \sum_{n=-\infty}^{\infty} \left(\sum_{k=0}^{N} x_k[n] \right) p(t - nT_s) = \sum_{k=0}^{N} \left(\sum_{n=-\infty}^{\infty} x_k[n]p(t - nT_s) \right) \tag{4.22}$$

Now since each individual sinusoid is assumed to satisfy the conditions of the sampling theorem, it follows that the D-to-C converter reconstructs each component perfectly, and therefore we conclude that

$$y(t) = \sum_{k=0}^{N} x_k(t) = x(t)$$

Thus, we have shown that the Shannon sampling theorem applies to *any* signal that can be represented as a bandlimited (finite) sum of sinusoids, and since it can be shown that most real-world signals can be represented as bandlimited signals, it follows that the sampling theorem is a very general guide to sampling all kinds of signals.

4-5 Strobe Demonstration

One effective means for demonstrating aliasing is to use a strobe light to illuminate a spinning object. Prior to computer-controlled ignition, this stroboscopic effect was the routine method to set the ignition timing of automobile engines and it gives a practical example where aliasing is a desirable effect. In this section, we consider a disk attached to the shaft of an electric motor that rotates at a constant angular speed (see Fig. 4-22).

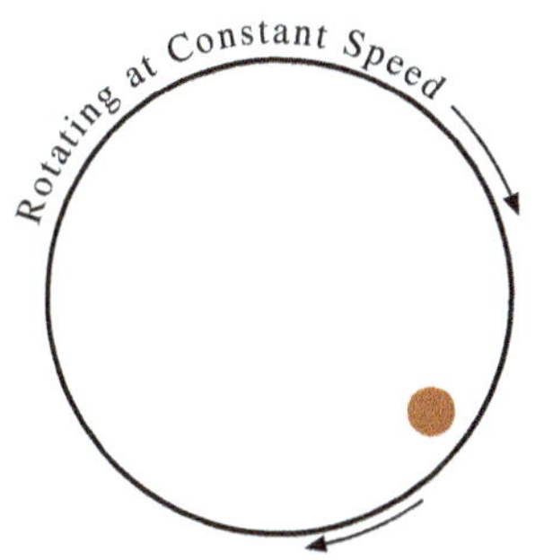

Figure 4-22 The disk attached to the shaft of a motor rotates clockwise at a constant speed.

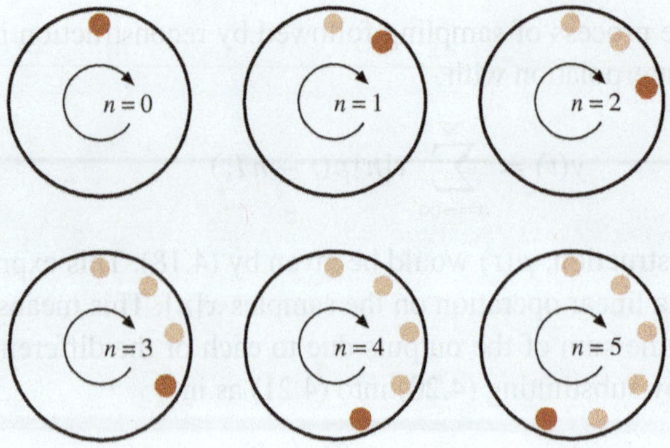

Figure 4-23 Six successive positions of the spot for a very high flashing rate. Light spots indicate previous locations of the dark spot. The disk is rotating clockwise, and the spot appears to move in the same direction. Angular change is $-40°$ per flash.

DEMO
Strobe Movies

On this white disk is painted a spot that is easily seen whenever the strobe light flashes. Initially, we assume that the rotation speed of the motor is constant at 750 rpm, and the flash rate of the strobe light is variable over a wide range.

Suppose that the flashing rate is very high, say nine times the rotation rate (that is, $9 \times 750 = 6750$ rpm). In this case, the disk does not rotate very far between flashes. In fact, for the $9\times$ case, it moves only $360°/9 = 40°$ per flash, as in Fig. 4-23. Since the movement is clockwise, the angular change from one flash to the next is $-40°$. Therefore, a higher rate of rotation is converted into a lower rate.

If the flash rate of the strobe is set equal to 750 flashes/min, the rotation rate of the disk, then the spot appears to stand still. This happens because the spot makes exactly one revolution between flashes and therefore is always at the same position when illuminated by the strobe. It is exactly the same situation illustrated in Fig. 4-10 where the frequency of the sampled sinusoid aliased to zero. This is not the only flash rate for which the spot appears to stand still. In fact, a slower flash rate that permits two or three or any integer number of complete revolutions between flashes will create the same effect. In our case of a 750 rpm motor, flash rates of 375, 250, $187\frac{1}{2}$, 150, and 125 flashes/min also work.

By using flashing rates that are close to these numbers, we can make the spot move slowly, and we can also control its direction of motion (clockwise or counterclockwise). For example, if we set the strobe for a flashing rate that is just slightly higher than the rotation rate, we observe another aliasing effect. Suppose that the flashing rate is 806 flashes/min then the disk rotates slightly less than one full revolution between flashes. We can calculate the movement if we recognize that one complete revolution takes $1/750$ min. Therefore,

$$\begin{aligned}\Delta\theta &= -360° \times \frac{1/806}{1/750} \\ &= -360° \times \frac{750}{806} = -335° = +25°\end{aligned}$$

Once again, the minus sign indicates rotation in a clockwise direction, but since the angular change is almost $-360°$, we would observe a small positive angular change instead, and

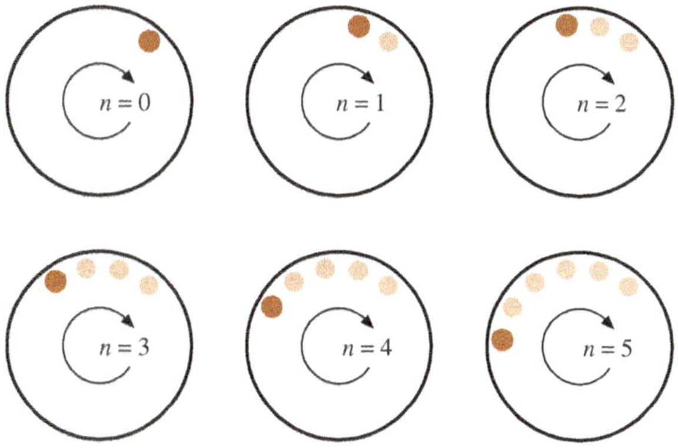

Figure 4-24 Six successive positions of the spot for a flashing rate that aliases the spot motion. Light spots indicate previous locations of the dark spot. The disk is rotating clockwise, but the spot appears to move counterclockwise. Angular change is $+25^\circ$ per flash.

the spot would appear to move in the counterclockwise direction, as shown in Fig. 4-24. The fact that one can distinguish between clockwise rotation and counterclockwise rotation is equivalent to saying that positive and negative frequencies have separate physical meanings. Figures 4-23 and 4-24 are reminiscent of an effect often observed on television where the wheels of a moving car sometimes appear to rotate opposite to the car's direction of motion—or, in old Western movies, where stage coach wheels seem to rotate in strange ways. The effect is due to the sampling inherent in the frame rate of video: 30 frames/s for analog TV, and 24 frames/s for movie film.

In order to analyze the strobe experiment mathematically, we need a notation for the motion of the spot as a function of time. The spot moves in an x–y coordinate system, so a succinct notation is given by a complex number whose real part is x and whose imaginary part is y. The position of the spot is

$$p(t) = x(t) + jy(t)$$

Furthermore, since the motion of the spot is on a circle of radius r, a better formula for $p(t)$ is a complex exponential with constant frequency, amplitude and phase:

$$p(t) = re^{-j(2\pi f_m t - \varphi)} \tag{4.23}$$

The minus sign in the exponent indicates clockwise rotation, and the initial phase φ specifies the location of the spot at $t = 0$ (see Fig. 4-25). The frequency of the motor rotation f_m is constant, as is the radius r.

The effect of the strobe light is to sample $p(t)$ at a fixed rate given by the flashing rate f_s. Thus the position of the spot at the nth flash can be expressed as the discrete-time signal

$$p[n] = p(t)\Big|_{t=nT_s} = p(nT_s) = p(n/f_s)$$

Substituting into the complex exponential (4.23) we get

$$p[n] = re^{-j(2\pi(f_m/f_s)n - \varphi)} \tag{4.24}$$

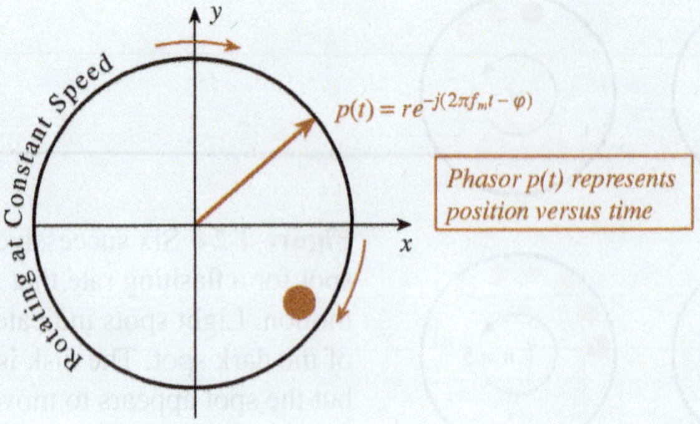

Figure 4-25 The position of the spot on the disk can be expressed as a rotating phasor $p(t) = x(t) + jy(t)$ versus time. The frequency f_m is the rotation rate of the motor in rpm.

If the constraint of the sampling theorem is met (i.e., $f_s > 2|f_m|$), then there is no aliasing in the experiment. In fact, the angular change from one sample time to the next is between $-180°$ and $0°$, so the spot will appear to rotate clockwise.

The more interesting cases occur when the flashing rate drops below $2|f_m|$. Then the disk makes one or more revolutions between flashes, which introduces the aliasing phenomenon. Using the sampled position formula (4.24), we can solve the following type of problem: Find *all* possible flashing rates so that the spot moves counterclockwise at a rate of 25° per flash, which is equivalent to one revolution every 14.4 flashes. Assume a constant motor speed of f_m rpm. One twist to this problem is that the flashing rate should be expressed as flashes/min to match the motor speed in rpm.

A systematic approach is possible if we use the property $e^{j2\pi\ell} = 1$, whenever ℓ is an integer. The desired rotation of the spot can be expressed as

$$d[n] = re^{+j(2\pi(25/360)n+\psi)}$$

where the term $2\pi(25/360)$ equals 25° converted to radians. The initial phase ψ is set equal to φ in $p[n]$, so that $d[0] = p[0]$. Then we can equate $p[n]$ and $d[n]$, but we include the factor $e^{j2\pi\ell n}$, which is just multiplying by one

$$p[n] = d[n]e^{j2\pi\ell n}$$

Now with the following analysis we can generate an equation that can be solved for the flashing rates:

$$re^{-j(2\pi(f_m/f_s)n-\varphi)} = re^{+j(2\pi(25/360)n+\varphi)}\, e^{j2\pi\ell n}$$

$$\Rightarrow \quad -\left(2\pi\frac{f_m}{f_s}n - \varphi\right) = +\left(2\pi\frac{25}{360}n + \varphi\right) + 2\pi\ell n$$

$$-\frac{f_m}{f_s} = \frac{25}{360} + \ell$$

$$\Rightarrow \quad -f_m = f_s\left(\frac{25}{360} + \ell\right)$$

So, finally we can solve for the flashing rate (per minute)

$$f_s = \frac{-f_m}{(5/72) + \ell} \tag{4.25}$$

where ℓ is any integer. Since we want positive values for the flashing rate, and since there is a minus sign associated with the clockwise rotation rate of the motor $(-f_m)$, we choose $\ell = -1, -2, -3, \ldots$ to generate the different solutions. For example, when the motor rpm is 750, the following flashing rates (in flashes/min) give the desired spot movement of 25° per flash:

ℓ	-1	-2	-3	-4
f_s	805.97	388.49	255.92	190.81

EXERCISE 4.6 To test your knowledge of this concept, try solving for *all* possible flashing rates so that the spot moves clockwise at a rate of one revolution every nine flashes. Assume the same motor rotation speed of 750 rpm clockwise. What is the maximum flashing rate for this case?

4-5.1 Spectrum Interpretation

The strobe demo involves the sampling of complex exponentials, so we can present the results in the form of a spectrum diagram rather than using time-domain equations, as in the foregoing section. The rotating disk has an analog spectrum given by a single frequency component at $f = -f_m$ cycles/min (see Fig. 4-26).

When the strobe light is applied at a rate of f_s flashes per minute, the resulting spectrum of the discrete-time signal $p[n]$ contains an infinite number of frequency lines at the normalized frequencies

$$\hat{\omega}_\ell = 2\pi\frac{-f_m}{f_s} + 2\pi\ell = -\hat{\omega}_m + 2\pi\ell \tag{4.26}$$

for $\ell = 0, \pm1, \pm2, \pm3, \ldots$. Equation (4.26) reminds us that there are two steps needed to create the spectrum (Fig. 4-27) of the discrete-time signal representing the sampled motion:

(a) Compute a *normalized motor frequency* $\hat{\omega}_m$ by dividing by f_s. Then plot a spectrum line at $\hat{\omega} = -\hat{\omega}_m$. The minus sign represents the clockwise rotation.

(b) Replicate the normalized spectral line for the motor by shifting it by all integer multiples of 2π. This accounts for the aliases of the discrete-time exponential, $p[n]$ in (4.24).

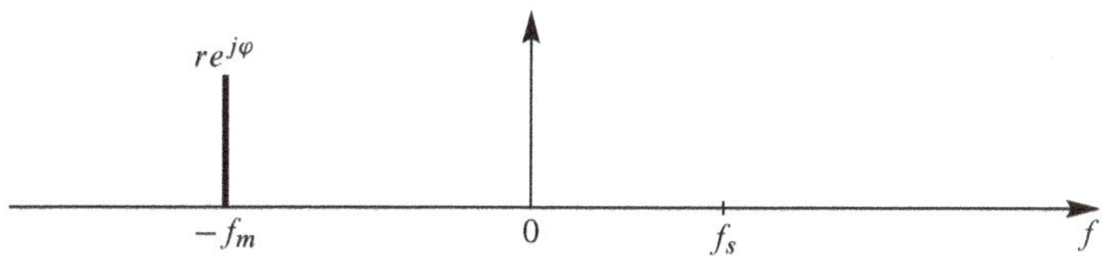

Figure 4-26 Analog spectrum representing the disk spinning clockwise (negative frequency) at f_m rpm. The units for f are cycles/min.

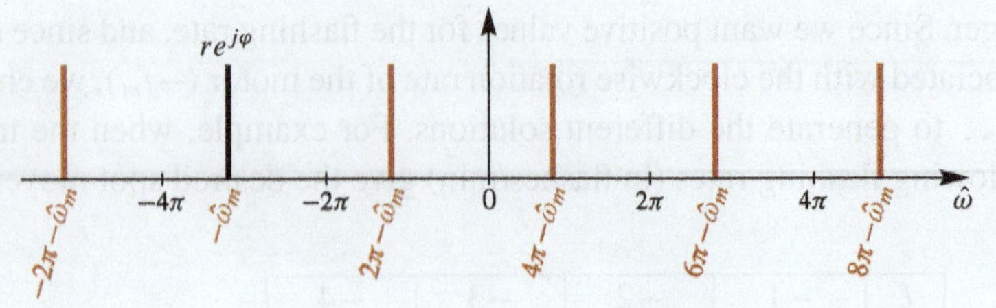

Figure 4-27 Discrete-time spectrum representing the disk spinning clockwise at f_m rpm, but strobed at f_s flashes per minute. The horizontal axis is normalized frequency: $\hat{\omega} = 2\pi f/f_s$. The normalized motor frequency $-\hat{\omega}_m = 2\pi(-f_m/f_s)$ and its aliases are shown.

How do we use Fig. 4-27 to explain what we see in the strobe-light experiment? Figure 4-27 contains an infinite number of spectrum lines, so it offers an infinite number of possibilities for what we will observe. Our visual system, however, must pick one and it is reasonable that it picks the slowest one. In effect, the human visual system acts as a D-to-C converter. Thus, in the discrete-time spectrum, only the frequency component closest to $\hat{\omega} = 0$ counts in D-to-C reconstruction, so the strobed signal $p[n]$ appears to be rotating at that lowest normalized frequency. However, one last conversion must be made to give the perceived analog rotation rate in rpm. The discrete-time frequency ($\hat{\omega}$) must be converted back to analog frequency via $f = \hat{\omega} f_s$. In Fig. 4-27, the alias for $\ell = 2$ is the one closest to zero frequency, so the corresponding analog frequency is

$$f_{\text{spot}} = \frac{1}{2\pi}(\hat{\omega}_2) f_s = \frac{1}{2\pi}(4\pi - \hat{\omega}_m) f_s = 2 f_s - f_m$$

This may seem like a roundabout way to say that the observed rotation rate of the spot differs from $-f_m$ by an integer multiple of the sampling rate, but it does provide a systematic method to draw the graphical picture of the relative frequency locations. Finally, the spectrum picture makes it easy to identify the lowest discrete-time frequency as the one that is "reconstructed." The discussion in Section 4-3 explains why this is true.

The case where the sampling rate is variable and f_m is fixed is a bit harder to solve, but this is the most common situation for stroboscopic applications. A graphical approach is still possible because the desired spot frequency defines a line in the discrete-time spectrum, say $\hat{\omega}_d$. The units of $\hat{\omega}_d$ are radians per flash. This line is the one closest to the origin, so we must add an integer multiple $2\pi\ell$ to $\hat{\omega}_d$ to match the normalized motor rotation frequency.

$$\hat{\omega}_d + 2\pi\ell = \hat{\omega}_m = \frac{-2\pi f_m}{f_s}$$

This equation can be solved for the flashing rate f_s, but the final answer depends on the integer ℓ, which predicts, by the way, that there are many answers for f_s, as we already saw in (4.25). The result is

$$f_s = \frac{-2\pi f_m}{\hat{\omega}_d + 2\pi\ell}$$

4-6 Summary and Links

LAB
Digital Images: A/D and D/A

This chapter introduced the concept of sampling and the companion operation of reconstruction. With sampling, the possibility of aliasing always exists, and we have created the strobe demo to illustrate that concept in a direct and intuitive way.

The lab entitled Digital Images: A/D and D/A shows how sampling applies to digital images. Some aspects of sampling have already been used in the music synthesis labs that are associated with Chapter 3, because the sounds must be produced by creating their samples in a computer before playing them out through a D-to-A converter.

The website also contains many demonstrations of sampling and aliasing:

DEMO
Synthetic Strobe Movies

DEMO
Reconstruction Movies

(a) Strobe movie filmed using the natural strobing of a video camera at 30 frames/s.

(b) Synthetic strobe demos produced as MATLAB movies.

(c) Reconstruction movies that show the interpolation process for different pulse shapes and different sampling rates.

(d) A sampling and aliasing MATLAB GUI called `con2dis`, shown in Fig. 4-13.

(e) Aliasing and folding movies that show how the spectrum changes for different sampling rates.

4-7 Problems

NOTE
Problems with Solutions

The problems in this section address the major concepts of sampling and aliasing in this chapter. On the Companion Website there are many related homework problems which can be used to review and practice. Most of them have detailed solutions.

P-4.1 Let $x(t) = 10\cos(9\pi t - \pi/5)$. In each of the following parts, the discrete-time signal $x[n]$ is obtained by sampling $x(t)$ at a rate f_s samples/s, and the resultant $x[n]$ can be written as

$$x[n] = A\cos(\hat{\omega}_1 n + \varphi)$$

For each part below, determine the values of A, φ, and $\hat{\omega}_1$ such that $0 \le \hat{\omega}_1 \le \pi$. In addition, state whether or not the signal has been over-sampled or under-sampled.

(a) Sampling frequency is $f_s = 11$ samples/s.

(b) Sampling frequency is $f_s = 7$ samples/s.

(c) Sampling frequency is $f_s = 4$ samples/s.

P-4.2 Discrete-time signals with different frequencies can be equal; this is one manifestation of aliasing and folded aliases. Determine which of the following discrete-time cosine signals are equal (i.e., plots of $x_i[n]$ versus n will be identical):

$$x_1[n] = \cos(0.8\pi n + 0.2\pi)$$

$$x_2[n] = \cos(-2.8\pi n + 0.2\pi)$$

$$x_3[n] = x_1[n] + x_2[n]$$

$$x_4[n] = \cos(5.2\pi n - 0.2\pi)$$

$$x_5[n] = x_1[n] + \cos(-7.1\pi n - 0.1\pi)$$

P-4.3 An ideal D-to-C converter produces an output signal whose $\hat{\omega}$ frequency content is the same as the input signal's discrete-time frequency content over the range $-\pi < \hat{\omega} \leq \pi$. However, the continuous-time frequency axis (ω) is a scaled version of $\hat{\omega}$. For each of the discrete-time signals $y[n]$ below, determine the continuous-time output signal $y(t)$ from an ideal D-to-C converter whose sampling rate is $f_s = 3200$ Hz.

(a) $y[n] = 8\cos(\frac{\pi}{4}n - \pi/3)$

(b) $y[n] = \cos(1.4\pi n)$

(c) $y[n] = 2\cos(2.5\pi n - \pi/4) + \sin(1.4\pi n)$

P-4.4 The spectrum diagram, which gives the frequency content of a continuous-time signal, helps in determining the Nyquist rate for sampling that signal.

(a) Given the signal $x(t) = \cos(3000\pi t)\cos(6000\pi t)$, draw a sketch of its spectrum. Label the frequencies and complex amplitudes of each component spectral line.

(b) Determine the minimum sampling rate that can be used to sample $x(t)$ without aliasing for any of its components.

(c) Given the signal $r(t) = \cos(2 \times 10^6 \pi t)\sin(5 \times 10^6 \pi t)\sin(9 \times 10^6 \pi t)$, determine the minimum sampling rate to avoid aliasing for any of its components.

(d) Given the signal $v(t) = \cos(2 \times 10^6 \pi t) + \sin(5 \times 10^6 \pi t) + \sin(9 \times 10^6 \pi t)$, determine the minimum sampling rate to avoid aliasing for any of its components.

P-4.5 The Fourier Series representation of a signal is a summation of complex exponential signals, so by focusing on the highest frequency, we can determine the Nyquist rate for sampling a continuous-time signal such as

$$x(t) = \sum_{k=-10}^{10} (j8\pi k)e^{j2\pi kt} \tag{4.27}$$

(a) For $x(t)$ determine the minimum (Nyquist) sampling rate to guarantee no aliasing.

(b) Define a new signal $r(t) = x(t)\cos(24\pi t)$. Determine the Nyquist sampling rate for $r(t)$.

(c) Define another new signal $v(t) = r(t) + r(t - 0.02)$. Determine the Nyquist sampling rate for $v(t)$.

P-4.6 Refer to Fig. 4-21 for the system with ideal C-to-D and D-to-C converters. If the sampling rate is high enough to avoid aliasing, and the D-to-C rate equals the C-to-D rate, then the reconstructed output signal is equal to the input signal (this is the gist of the Sampling Theorem). However, when the D-to-C rate is different, a careful analysis of the spectra is needed. For example, if the output rate is twice the input rate, the output signal is played out faster and we would expect the frequency content of the output signal to be higher.

(a) If the input $x(t)$ is given by the two-sided spectrum representation in Fig. P-4.6, and the sampling rate of the C-to-D converter is $f_s = 2500$ samples/s, sketch the spectrum of the discrete-time signal $x[n]$.

(b) Use the spectrum for $x[n]$ found in the previous part to determine a simple formula for the output signal $y(t)$ when the rate of the D-to-C converter is $f_s = 1500$ samples/s (i.e., different from the C-to-D rate).

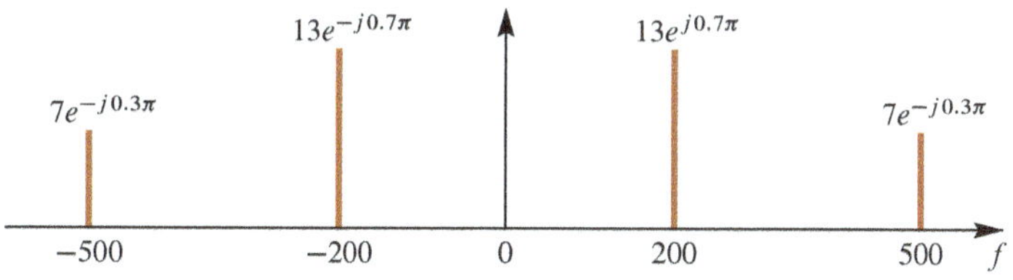

Figure P-4.6

P-4.7 In the system with ideal C-to-D and D-to-C converters (Fig. 4-21), assume that the sampling rate of both converters is $f_s = 475$ samples/s, and the input signal $x(t)$ is given by the two-sided spectrum representation in Fig. P-4.6.

(a) Is the output signal $y(t)$ equal to the input signal $x(t)$?

(b) For the case where the sampling rate of the C-to-D converter is $f_s = 475$ Hz, sketch the spectrum of the discrete-time signal $x[n]$. Label the frequencies and complex amplitudes of all the spectrum lines.

(c) Use the spectrum for $x[n]$ found in the previous part to determine a simple formula for the output signal $y(t)$ when $f_s = 475$ Hz (for both the C-to-D and D-to-C converters).

P-4.8 In the system with ideal C-to-D and D-to-C converters (Fig. 4-21), assume that the sampling rate of both converters is $f_s = 700$ samples/s, and the input signal $x(t)$ is given by the two-sided spectrum representation in Fig. P-4.6.

(a) State the Nyquist rate for $x(t)$ defined in Fig. P-4.6. Is the input signal aliased?

(b) For the case where the sampling rate of the C-to-D converter is $f_s = 700$ Hz, sketch the spectrum of the discrete-time signal $x[n]$. Label the frequencies and complex amplitudes of all the spectrum lines.

(c) Use the spectrum for $x[n]$ found in the previous part to determine a simple formula for the output signal $y(t)$ when $f_s = 700$ Hz *for both converters.*

P-4.9 Refer to Fig. 4-21 for the system with ideal C-to-D and D-to-C converters.

(a) Suppose that the output from the C-to-D converter is $x[n] = \cos(0.4\pi n)$, and the sampling rate of the C-to-D converter is $f_s = 7000$ samples/s. Determine a formula for the continuous-time sinusoidal input $x(t)$ using the smallest frequency greater than 9000 Hz.

(b) Suppose the output from the C-to-D converter is $x[n] = \cos(0.2\pi n)$, the input signal is $x(t) = \cos(525\pi t)$, and the sampling rate (f_s) of the C-to-D converter is less than 135 samples/s. Determine the largest possible sampling rate satisfying these three conditions.

(c) Suppose the output from the D-to-C converter is $y(t) = \cos(325\pi t)$, and the input to the C-to-D converter is $x(t) = \cos(75\pi t)$. If both converters are operating at the same rate (f_s), determine the smallest possible sampling rate f_s satisfying these conditions.

P-4.10 A continuous-time sinusoid $a_1(t) = \cos(\omega_1 t + 0.1\pi)$ is sampled at $f_{s_1} = 40$ Hz to give $a_1[n]$, and a second continuous-time sinusoid $a_2(t) = \cos(\omega_2 t + 0.1\pi)$ is sampled at $f_{s_2} = 50$ Hz to give $a_2[n]$. Equality of the discrete-time sampled signals (i.e., $a_1[n] = a_2[n]$), may or may not be equivalent to equality of the continuous-time signals (i.e., $a_1(t) = a_2(t)$).

(a) If $\omega_2 = 30\pi$ rad/s, determine ω_1 so that $a_1[n] = a_2[n]$. *Assume there is no aliasing when sampling* $a_1(t)$.

(b) For the previous part, determine the common frequency $\hat{\omega}$ of the discrete-time signals $a_i[n] = \cos(\hat{\omega}n + \varphi_i)$, where $i = 1, 2$.

(c) Now assume that $\omega_1 = \omega_2 = 2\pi(44\frac{4}{9})$ rad/s. Determine the frequencies $\hat{\omega}_1$ and $\hat{\omega}_2$ of the discrete-time signals $a_1[n]$ and $a_2[n]$. *Hint:* Use aliasing to obtain frequencies that satisfy $0 \le \hat{\omega}_1,\ \hat{\omega}_2 \le \pi$.

(d) Are the discrete-time signals found in the previous part equal? If yes, find the phases; otherwise, explain why not.

P-4.11 When the cosine waveform $x(t) = 3\cos(500\pi t)$ is sampled at a rate f_s, we obtain a discrete-time signal $x[n]$ given by

$$x[n] = x(n/f_s) = 3\cos(800\pi n/f_s)$$

for $-\infty < n < \infty$. In the following parts, assume that $f_s = 3600$ Hz.

(a) Determine how many samples are taken in one period of the cosine wave $x(t)$. This answer is the average number of samples per period, which is an integer in this case.

(b) Now consider another cosine waveform $y(t)$ with a different frequency ω_0

$$y(t) = 3\cos(\omega_0 t)$$

Find a value for ω_0, between 7000π and 9999π rad/s, such that the signal samples are identical to $x[n]$ above, that is, $y[n] = y(n/f_s) = x(n/f_s)$ for all n. *Hint:* The solution involves aliasing.

(c) For the frequency found in (b), determine the average number of samples taken in one period of $y(t)$. This answer is unlikely to be an integer.

P-4.12 Suppose that a discrete-time signal $x[n]$ given by the formula

$$x[n] = 11.3\cos(0.3\pi n + 0.6\pi)$$

was obtained by sampling a continuous-time signal $x(t) = A\cos(2\pi f_0 t + \varphi)$ at a sampling rate of $f_s = 4000$ samples/s. Determine three different continuous-time signals that could have produced $x[n]$. All these continuous-time signals have different frequencies which should be positive and less than 7000 Hz. Write the mathematical formula for all three continuous-time signals.

P-4.13 When operating on a signal in MATLAB, sampling is always being done because MATLAB stores signals as vectors which are, in effect, discrete-time signals. Furthermore, MATLAB performs ideal C-to-D conversion when operating on signals that are defined by mathematical formulas. As a result, there are cases where aliasing might occur unintentionally in a MATLAB program. Discrete-time MATLAB signals can be converted to continuous-time with the function `soundsc` which plays out an analog signal for listening.

(a) In the following MATLAB code, a discrete-time sinusoid is generated and then played out with `soundsc`:

```
nn = 0:2190099;
xx = (7/pi)*cos(1.8*pi*nn + 2.03);
soundsc(xx,16000)
```

Determine the analog frequency (in hertz) that will be heard.

(b) The following MATLAB code generates samples of a sinusoidal waveform:

```
tt = 0:1/2400:10000;
xx = cos(2*pi*1800*tt + pi/3);
soundsc(xx,fsamp);
```

Although the sinusoid was not written to have a frequency of 2400 Hz, it is possible to play out the vector `xx` so that it sounds like a 2400 Hz tone. Determine the value of `fsamp` that should be used in `soundsc` to play the vector `xx` as a 2400 Hz tone.

(c) The following snippet of MATLAB code generates a very long sinusoid:

```
tt = 0:(1/8000):64;
xx = 1.23*cos(2*pi*440*tt);
soundsc(xx,40000);
```

Determine the frequency (in Hz) and the duration (in seconds) of the tone played by `soundsc`. *Note:* Assume that MATLAB has an infinite amount of memory so that the very long vector `xx` can be stored.

P-4.14 An amplitude-modulated (AM) cosine wave is represented by the formula

$$x(t) = [4 + \sin(6600\,t)]\cos(2000\pi t)$$

(a) Sketch the two-sided spectrum of this signal. Be sure to label important features of the plot.

(b) Explain why $x(t)$ is not periodic.

(c) What relation must the sampling rate f_s satisfy so that $y(t) = x(t)$ in Fig. 4-21?

P-4.15 Suppose that a discrete-time signal $x[n]$ is given by the formula

$$x[n] = 9\cos(0.6\pi n + 0.4\pi)$$

and that it was obtained by sampling a continuous-time signal at a sampling rate of $f_s = 1/T_s = 9600$ samples/s.

(a) Determine two *different* continuous-time signals $x_1(t)$ and $x_2(t)$ whose samples are equal to $x[n]$; in other words, find $x_1(t)$ and $x_2(t)$ such that $x[n] = x_1(nT_s) = x_2(nT_s)$ if $T_s = 1/f_s$. Both of these signals should have a frequency less than 10,000 Hz. Give a formula for each signal.

(b) If $x[n]$ is given by the equation above, what signal is reconstructed by an ideal D-to-C converter operating at sampling rate of 6000 samples/s? That is, determine the output $y(t)$ in Fig. 4-21 when the sampling rates are different.

P-4.16 Bandpass sampling refers to sampling a signal that is known to lie in a restricted frequency range, for example, all frequencies in $x(t)$ are between 7000 and 8000 Hz. The usual Sampling Theorem requires $f_s > 16000$ Hz because the reconstruction is performed with the lowest frequency possible. However, in this case, the available bandwidth of the signal is only 1000 Hz, so it turns out that $x(t)$ can be sampled at $f_s = 2000$ Hz and recovered perfectly from those samples *because the range of recovered frequencies is known*. Suppose that a discrete-time signal $x[n]$ is given by the formula

$$x[n] = \cos(0.7\pi n + 0.2\pi)$$

and that it was obtained by sampling a continuous-time signal at a sampling rate of $f_s = 2000$ samples/s.

(a) When $f_s = 2000$ Hz and the possible bandwidth is also 2000 Hz, reconstruction from the sampled signal is ambiguous. Determine two different continuous-time signals $x_1(t)$ and $x_2(t)$ with frequencies between 7000 and 9000 Hz, and whose samples are equal to $x[n]$.

(b) Determine the continuous-time signal $x(t)$ whose frequency lies between 7000 and 8000 Hz and whose samples are equal to $x[n]$.

(c) Suppose that $x[n] = \cos(\hat{\omega} n + 0.2\pi)$, where $\hat{\omega} \in [0, \pi]$ is the discrete-time signal obtained by sampling $x(t) = \cos(2\pi f t + \varphi)$ at a sampling rate of $f_s = 2000$ Hz. Determine the general formula for the frequency f in terms of $\hat{\omega}$ such that the frequency satisfies $f \in [7000, 8000]$ Hz. Also, determine the phase φ.

Comment: The answer in the last part is simple only when one end of the frequency interval is an integer multiple of f_s (that is, $8000 = 4f_s$). The reader is encouraged to go further with this problem by reworking the last part when $f \in [8500, 9500]$ Hz.

P-4.17 A nonideal D-to-C converter takes a discrete-time signal (sequence) $y[n]$ as input and produces a continuous-time output $y(t)$ according to the relation

$$y(t) = \sum_{n=-\infty}^{\infty} y[n]p(t - nT_s)$$

where $T_s = 0.4$ s. Assume the input sequence is given by the formula

$$y[n] = \begin{cases} 0.1(-2)^n & 0 \le n \le 4 \\ 0 & \text{otherwise} \end{cases}$$

(a) For the pulse shape

$$p(t) = \begin{cases} 1 & -0.2 \le t \le 0.2 \\ 0 & \text{otherwise} \end{cases}$$

carefully sketch the output waveform $y(t)$.

(b) For the pulse shape

$$p(t) = \begin{cases} 1 - 2.5|t| & -0.4 \le t \le 0.4 \\ 0 & \text{otherwise} \end{cases}$$

carefully sketch the output waveform $y(t)$.

P-4.18 A continuous-time sinusoid $a(t) = 5\cos(\omega_a t)$ is sampled at $f_{s_1} = 25$ Hz to give $a[n]$, and a second continuous-time sinusoid $b(t) = 5\cos(\omega_b t)$ is sampled at $f_{s_2} = 65$ Hz to give $b[n]$.

(a) If $\omega_b = 26\pi$ rad/s, determine ω_a so that $a[n] = b[n] = 5\cos(\hat{\omega}n)$. *Assume there is no aliasing when sampling $a(t)$.* Determine the frequency $\hat{\omega}$ of the discrete-time signal.

(b) Now assume that both ω_a and ω_b are unknown, but they are equal ($\omega_a = \omega_b$). Determine the common frequency ω_a, so that the discrete-time signals are equal (that is, $a[n] = b[n]$).

(c) Among the many answers possible in the previous part, only two satisfy $0 < \omega_a < 40\pi$ rad/s. Determine those answers.

(d) If $\omega_a = 30\pi$ and $\omega_b = 60\pi$, and $a(t)$ and $b(t)$ are sampled at the same sampling frequency f_s, such that $30 < f_s < 60$, determine f_s for which $a[n] = b[n]$.

P-4.19 Often MATLAB is used to plot a sinusoidal signal and this requires that the time axis be defined because the `plot` command operates on vectors for horizontal and vertical axes. The following MATLAB code generates a signal $x[n]$ and plots $x[n]$ versus n:

```
Ts = 0.01;    % units are seconds
Duration = 0.3;
tt = 0 : Ts : Duration;
Fo = 380;    % units are Hz
xn = 2*cos( 2*pi*Fo*tt  + 0.6*pi );
stem( 0:length(xn)-1, xn )
```

(a) Make the `stem` plot of the signal $x[n]$ with n as the horizontal axis. Either sketch it or plot it using MATLAB. Assume that the n index starts at zero.

(b) In the plot, measure the period of the discrete-time signal $x[n]$.

(c) Determine the correct formula for the discrete-time signal in the form

$$x[n] = A\cos(\hat{\omega}_0 n + \varphi)$$

with $\hat{\omega}_0 \in [0, \pi]$.

(d) Explain how aliasing affects the formula and the stem plot that you made.

P-4.20 The intention of the following MATLAB program is to plot 17 cycles of a 15 Hz sinusoid, but it has a bug:

```
Fo = 15;
To = 1/Fo;      %-- Period
Ts = 0.05;
tt = 0 : Ts : (17*To);
xx = real((1+j*0.1)*exp(j*2*pi*Fo*tt));
%-- The  xx vector has samples of the 15-Hz sinusoid
plot( tt, xx, '-o' )
xlabel('TIME   (sec)'),  grid on
```

(a) First of all, if you plot 17 periods of a sinusoid how many zero crossings would you observe in the plot? How many positive peaks and how many negative peaks?

(b) Make the plot in MATLAB, and count the number of zero crossings and peaks.

(c) Measure the *period in the plot* (in s), and then calculate the frequency (in Hz) of a sinusoid with that period. Explain how aliasing or folding predicts this frequency.

(d) The samples of the signal `xx` form a discrete-time signal, $x[n]$. Determine the period of $x[n]$ and then write a formula for $x[n]$.

(e) Determine an acceptable value of `Ts` to get a very smooth plot of the desired 15 Hz signal.

P-4.21 An amplitude-modulated (AM) wave is represented by the formula

$$x(t) = [7 + \cos(\pi t)]\sin(15\pi t + \pi/2)$$

(a) Use *phasors* to show that $x(t)$ can be expressed in the form

$$x(t) = A_1\cos(\omega_1 t + \varphi_1) + A_2\cos(\omega_2 t + \varphi_2) + A_3\cos(\omega_3 t + \varphi_3),$$

where $\omega_1 < \omega_2 < \omega_3$. In addition, find values for the parameters $A_1,\ A_2,\ A_3,\ \varphi_1,\ \varphi_2,\ \varphi_3,\ \omega_1,\ \omega_2,\ \omega_3$.

(b) Sketch the two-sided spectrum of this signal on a frequency axis. Be sure to label important features of the plot, and include the numerical values of the parameters A_i, φ_i, and ω_i for $i = 1,\ 2,\ 3$.

(c) Determine the minimum sampling rate that can be used to sample $x(t)$ without aliasing of any of the components.

P-4.22 Refer to Fig. 4-21 for the sampling and reconstruction system. Assume that the sampling rates of the C-to-D and D-to-C converters are equal, and the input to the ideal C-to-D converter is

$$x(t) = 2\cos(2\pi(50)t + \pi/2) + \cos(2\pi(150)t)$$

(a) If the output of the ideal D-to-C converter is equal to the input $x(t)$, that is,

$$y(t) = 2\cos(2\pi(50)t + \pi/2) + \cos(2\pi(150)t)$$

what general statement can you make about the sampling frequency f_s in this case?

(b) If the sampling rate is $f_s = 250$ samples/s, determine the discrete-time signal $x[n]$, and give an expression for $x[n]$ as a sum of cosines. *Make sure that all frequencies in your answer are positive and less than π radians.*

(c) Plot the spectrum of the signal in part (b) over the range of frequencies $-\pi \leq \hat{\omega} \leq \pi$. Make a plot for your answer, labeling the frequency, amplitude, and phase of each spectral component.

(d) If the output of the ideal D-to-C converter is

$$y(t) = 2\cos(2\pi(50)t + \pi/2) + 1$$

determine the value of the sampling frequency f_s. (Remember that the input $x(t)$ is as defined above.)

P-4.23 In the rotating disk and strobe demo described in Section 4-5, we observed that different flashing rates of the strobe light would make the spot on the disk stand still or move in different directions.

(a) Assume that the disk is rotating *clockwise* at a constant speed of 13 rev/s. If the flashing rate is 15 flashes/s, express the movement of the spot on the disk as a rotating complex phasor, $p[n]$, that gives the position of the spot at the nth flash. Assume that the spot is at the top when $n = 0$ (the first flash).

(b) For the conditions in (a), determine the apparent speed (in rev/s) and direction of movement of the "strobed" spot.

(c) *Now assume that the rotation speed of the disk is unknown.* If the flashing rate is 13 flashes/s, and the spot on the disk moves *counterclockwise* by 15° with each flash, determine the rotation speed of the disk (in rev/s). If the answer is not unique, give *all* possible rotation speeds.

P-4.24 In the rotating disk and strobe demo, we observed that different flashing rates of the strobe light would make the spot on the disk stand still.

(a) Assume that the disk is rotating in the counterclockwise direction at a constant speed of 12 rev/s. Express the movement of the spot on the disk as a rotating complex phasor.

(b) If the strobe light can be flashed at a rate of n flashes/s, where n is an integer greater than zero, determine all possible flashing rates such that the disk can be made to stand still.

Note: The only possible flashing rates are 1 per second, 2 per second, 3 per second, etc.

(c) If the flashing rate is 13 times/s, explain how the spot moves, and write a complex rotating phasor that gives the position of the spot at each flash.

P-4.25 The following complex-valued signal is a phasor:

$$z[n] = e^{j\theta[n]}$$

where $\theta[n]$ is the phase.

(a) When the phase changes by a constant amount versus n, the phasor rotates at a constant speed. For the following phasor:

$$z[n] = e^{j(0.08\pi n - 0.25\pi)}$$

make a plot of the phasor locations for $n = 0, 1, 2, 7, 10, 17, 20, 33, 50$ and 99.

(b) What is the period of $z[n]$?

(c) Repeat for the complex phasor that corresponds to the following chirp signal:

$$c[n] = e^{j0.1n^2}$$

In this case, plot the phasor locations for $n = 0, 1, 2, 3, 4,$ and 7.

(d) It might be surprising to learn that the chirp signal $c[n]$ defined in (c) is periodic. Determine its period.

P-4.26 A digital chirp signal is synthesized according to the following formula:

$$x[n] = \Re\{e^{j\theta[n]}\} = \cos(\pi(0.7 \times 10^{-3})n^2)$$

for $n = 0, 1, 2, \ldots, 200$.

(a) Make a plot of the rotating phasor $e^{j\theta[n]}$ for $n = 10, 50,$ and 100.

(b) If this signal is the input to a D-to-A converter whose sampling rate is 8 kHz, make a plot of the instantaneous analog frequency (in Hz) versus time for the analog signal.

(c) If the *constant frequency* digital signal $v[n] = \cos(0.7\pi n)$ is the input to a D-to-A converter whose sampling rate is 8 kHz, what (analog) frequency will be heard?

P-4.27 In the MATLAB code below, determine the values of `ALPHA`, `BETA`, and `fs` in the definition of the FM signal whose spectrogram shown in Fig. P-4.27 is produced by the `plotspec` function, which is available on the website. Notice that both the positive and negative frequency regions are shown by adding a tiny imaginary part to the real signal `xn`.

NOTE
DSP-First M-Files

```
%
fs = ????;
tt = 0:1/fs:2;
psi = ALPHA*pi*tt.^4 + BETA*pi*tt;
xn = cos(psi);
plotspec(xn+j*1e-11, fs, 128), grid on %-- make a spectrogram
```

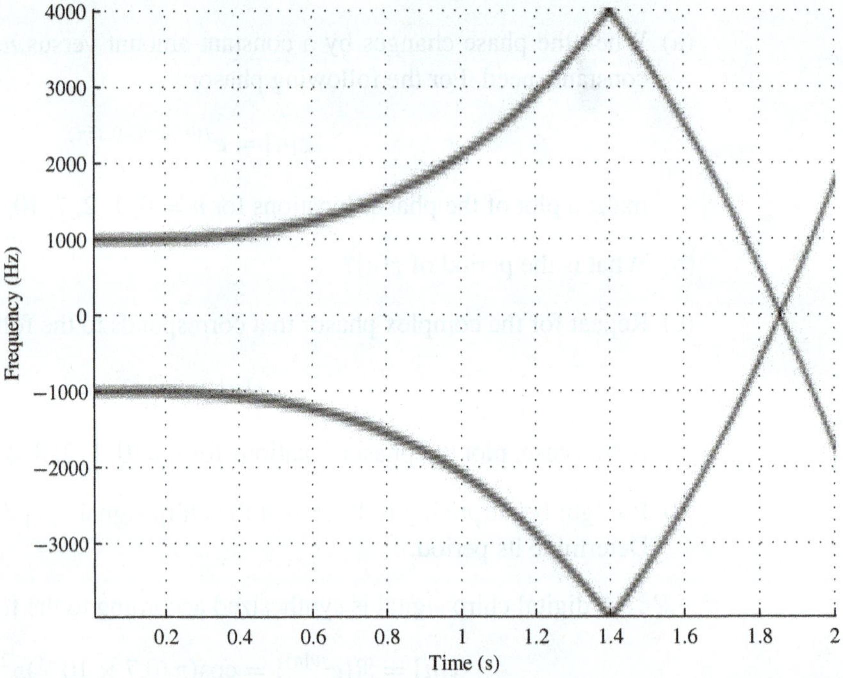

Figure P-4.27

CHAPTER

5

FIR Filters

Up to this point, we have focused our attention on signals and their mathematical representations. In this chapter, we begin to emphasize *systems* or ***filters***. Strictly speaking, a filter is a system that is designed to remove some component or modify some characteristic of a signal, but often the two terms are used interchangeably. In this chapter, we introduce the class of ***finite impulse response (FIR)*** systems, or, as we often refer to them, *FIR filters*. These filters are systems for which each output value is the sum of a finite number of weighted values of the input sequence. We define the basic input–output structure of the FIR filter as a time-domain computation based upon what is often called a ***difference equation***. The unit impulse response of the filter is defined and shown to completely describe the filter via the operation of convolution. The general concepts of linearity and time invariance will also be presented. These two properties completely characterize a wide class of filters that are exceedingly important in both the continuous-time and the discrete-time cases.

Our purpose in this chapter is to introduce the basic ideas of discrete-time systems and to provide a starting point for further study. The analysis of both discrete-time and continuous-time systems is a rich subject of study that is necessarily based on

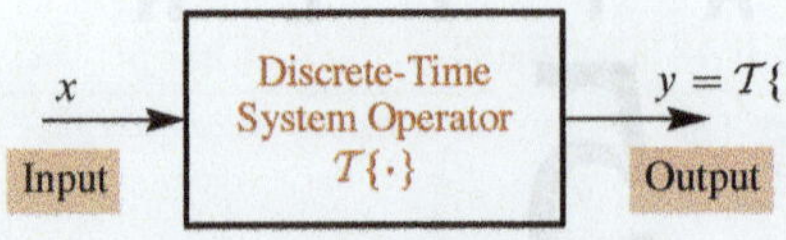

Figure 5-1 Block-diagram representation of a discrete-time system.

mathematical representations and manipulations.[1] The systems that we introduce in this chapter are the simplest to analyze. The remainder of the text is concerned with extending the ideas of this chapter to other classes of systems and with developing tools for the analysis of these systems.

5-1 Discrete-Time Systems

A discrete-time system is a computational process for transforming one sequence, called the *input signal*, into another sequence called the *output signal*. As we have already mentioned, systems are often depicted by block diagrams such as the one in Fig. 5-1. In Chapter 4, we used similar block diagrams to represent the operations of sampling and reconstruction. In the case of sampling, the input signal is a continuous-time signal and the output is a discrete-time signal, while for reconstruction the opposite is true. Now we want to begin to study discrete-time systems where the input and output are both discrete-time signals. Such systems are important because they can be implemented with digital computation and because they can be designed to modify signals in many useful ways.

In general, we represent the operation of a system by the notation

$$x[n] \stackrel{\mathcal{T}}{\longmapsto} y[n]$$

which states concisely that the output sequence[2] y is related to the input sequence x by a computational process (or mapping) that can be described mathematically by an operator $\mathcal{T}$.[3] Since a discrete-time signal is a sequence of values, such operators can be described by giving a formula for computing the values of the output sequence from the values of the input sequence. For example, the relation

$$y[n] = (x[n])^2$$

defines a very simple system for which the output sequence values are the square of the corresponding input sequence values. In this case the output at time index n depends

[1] Systems can be analyzed effectively by mathematical methods called transforms, such as the Fourier transform and the z-transform, which are introduced in Chapters 7, 9, and 10, and are covered extensively in more advanced signals and systems texts.

[2] The term sequence is equivalent to a discrete-time signal, so the two terms will be used interchangeably.

[3] An operator is a mathematical concept that is analogous to the concept of a function. Instead of mapping a number into another number, a discrete-time operator $\mathcal{T}$ maps an entire sequence (the input) into another sequence (the output).

only on the input value at the same time n. A more complicated example would be the following definition of a maximum-of-three system:

$$y[n] = \max\{x[n], x[n-1], x[n-2]\}$$

In this case, the output value depends on three consecutive input values, the "current value," indexed by n, and two "previous values" indexed by $n-1$ and $n-2$. As these two simple examples illustrate, a system definition is often given as a formula or computational process for transforming a sequence $x[n]$ into another sequence $y[n]$.

Since infinite possibilities exist for defining discrete-time systems, it is necessary to limit the range of possibilities by placing some restrictions on the properties of the systems that we study. Therefore, we begin our study of discrete-time systems in this chapter by introducing a very important class of discrete-time systems called ***FIR filters***. Specifically, we will discuss the representation, implementation, and analysis of discrete-time FIR systems, and illustrate how such systems can be used to modify signals.

5-2 The Running-Average Filter

A simple but useful transformation of a discrete-time signal is to compute a ***running average*** of two or more consecutive values of the sequence, thereby forming a new sequence of the average values.[4] Averaging is commonly used whenever data fluctuate and must be smoothed prior to interpretation. For example, stock-market prices fluctuate noticeably from day to day, or hour to hour. Therefore, one might take an average of the stock price over several days before looking for any trend. Another everyday example concerns credit-card balances where interest is charged on the *average* daily balance. The FIR filter is a generalization of the idea of a running average.

In order to motivate the general definition of the class of FIR systems, let us consider the simple running average as an example of a system that processes an input sequence to produce an output sequence. To be specific, consider a 3-point running average where each sample of the output sequence is the sum of three consecutive input sequence samples divided by three.[5] If we apply this algorithm to the triangularly shaped sequence shown in Fig. 5-2(a), we can compute a new sequence called $y[n]$, which is the output of the averaging operator. The sequence in Fig. 5-2(a) is an example of a ***finite-length*** signal. The ***support*** of such a sequence is the set of indices over which the sequence is nonzero; in this case, the support of the sequence is the finite interval $0 \le n \le 4$. A 3-point average of the values $\{x[0], x[1], x[2]\} = \{2, 4, 6\}$ gives the answer $\frac{1}{3}(2+4+6) = 4$. This result defines one of the output values. The next output value is obtained by averaging $\{x[1], x[2], x[3]\} = \{4, 6, 4\}$, which yields an average value of 14/3. Before going any further, we should decide on the output indexing. For example, the values 4 and 14/3

[4]The running average is often called a *moving average*.

[5]It is common practice to refer to signal values as points or samples. The term samples is short for sample values which are signal values since many signals arise from sampling.

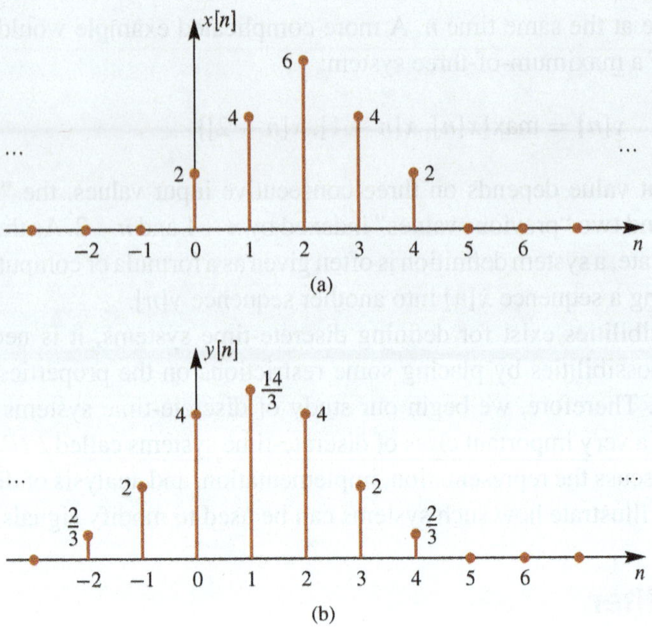

Figure 5-2 (a) Finite-length input signal, $x[n]$. (b) Output of running-average filter, $y[n]$.

could be assigned to $y[0]$ and $y[1]$, *but this is only one of many possibilities.* With this indexing, the equations for computing the output from the input are

$$y[0] = \tfrac{1}{3}(x[0] + x[1] + x[2])$$
$$y[1] = \tfrac{1}{3}(x[1] + x[2] + x[3])$$

which generalizes to the following input–output equation:

$$y[n] = \tfrac{1}{3}(x[n] + x[n+1] + x[n+2]) \tag{5.1}$$

The equation given in (5.1) is called a ***difference equation***. It is a complete description of the FIR system because we can use (5.1) to compute the entire output signal for all index values $-\infty < n < \infty$. For the input of Fig. 5-2(a), the output from (5.1) is the signal $y[n]$ tabulated as follows:

n	$n<-2$	-2	-1	0	1	2	3	4	5	$n>5$
$x[n]$	0	0	0	2	4	6	4	2	0	0
$y[n]$	0	$\frac{2}{3}$	2	4	$\frac{14}{3}$	4	2	$\frac{2}{3}$	0	0

Note that the values in orange type in the $x[n]$ row are the numbers involved in the computation of $y[2]$. Also note that $y[n] = 0$ outside of the finite interval $-2 \leq n \leq 4$; so, the output also has finite support. The output sequence is also plotted in Fig. 5-2(b). Observe that the support of the output sequence is longer (has more nonzero values) than the input sequence, which is typical for an FIR filter.

The choice of the output indexing is arbitrary, but it does matter when speaking about properties of the filter because n is often a time index. For example, the filter defined

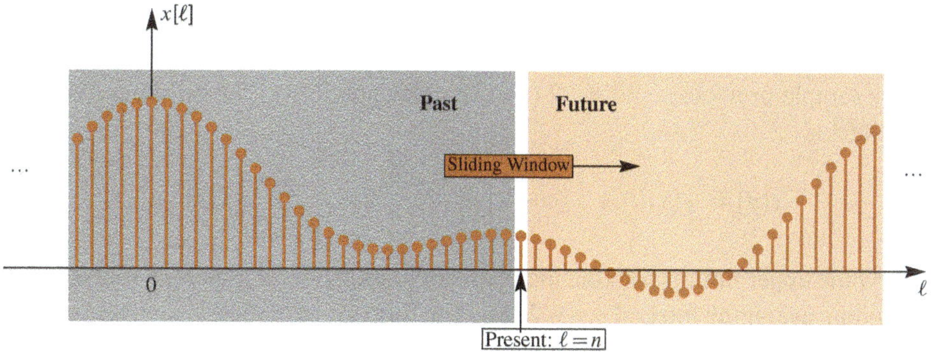

Figure 5-3 Filter calculation at the *present* time ($\ell = n$) uses values within a sliding window. Gray shading indicates the past ($\ell < n$); orange shading, the future ($\ell > n$). Here, the sliding window encompasses values from both the future and the past.

in (5.1) has the property that its output starts (becomes nonzero) before the input starts. This would certainly be undesirable if the input signal values came directly from an A-to-D converter, as is common in audio signal-processing applications. In this case, n would stand for time, and we can interpret $y[n]$ in (5.1) as the computation of the ***present*** value of the output based on three input values. Since these inputs are indexed as $n, n+1$, and $n+2$, two of them are "in the future." This form of the 3-point running-average filter can be represented compactly by the summation notation on the right-hand side of the following equation:

$$y[n] = \tfrac{1}{3}(x[n] + x[n+1] + x[n+2]) = \tfrac{1}{3}\sum_{\ell=n}^{n+2} x[\ell] \tag{5.2}$$

In this representation, ℓ is a "dummy" counting index for the sum and n denotes the index of the nth sample of the output sequence. When we represent the input sequence as $x[\ell]$, with $-\infty < \ell < \infty$, we are considering the sequence in its entirety. If we denote n as the current time,[6] then (5.2) states that the output at n is the average of the input at n (i.e., $x[n]$) and two future values $x[n+1]$ and $x[n+2]$.

In general, sample values from either the ***past*** or the ***future*** or both may be used in the computation of the running average. Figure 5-3 depicts a case when both past and future samples of the input are involved in computing the present value of the output sequence. In all cases of a 3-point running average, a ***sliding window*** of three samples determines which three samples are used in the computation of $y[n]$. Figure 5-3 also depicts the case when more than three samples are averaged to produce the output sample.

A filter that uses only the present and past values of the input is called a ***causal*** filter, implying that the cause does not precede the corresponding effect. Therefore, a filter that uses future values of the input is called ***noncausal***. Noncausal systems cannot be implemented in a real-time application because the input is not yet available when the output has to be computed. In other cases, where stored data blocks are manipulated inside a computer, the issue of causality may not be a crucial limitation.

[6]Oftentimes, we will not be so careful about our notation as in this discussion. When the meaning is clear, we may simply refer to "the sequence $x[n]$" even when we mean the entire sequence, not just the value of the sequence at n.

An alternative output indexing scheme can produce a 3-point averaging filter that is causal. In this case, the output value $y[n]$ is the average of input values at n (the present), $n-1$ (one sample previous), and $n-2$ (two samples previous). The difference equation for this filter[7] is

$$y[n] = \tfrac{1}{3}(x[n] + x[n-1] + x[n-2]) = \sum_{\ell=n-2}^{n} \tfrac{1}{3}x[\ell] \tag{5.3}$$

where now the upper limit is n so the index of summation does not include future values, and the sliding averaging interval is $n-2$ to n. The form given in (5.3) is a ***causal running averager***, or it may well be called a *backward average*. By a straightforward manipulation of the sum, (5.3) can also be expressed as

$$y[n] = \tfrac{1}{3}(x[n] + x[n-1] + x[n-2]) = \sum_{k=0}^{2} \tfrac{1}{3}x[n-k] \tag{5.4}$$

In (5.4), the counting index of the sum is denoted k to distinguish it from ℓ in (5.3), even though this is really not necessary. The indices ℓ and k are "dummy" indices in the sense that they are used only to count through the terms being summed—hence, they are replaced by numbers as in the expressions on the left in (5.3) and (5.4). Therefore, both ℓ and k could be replaced by any convenient symbol distinct from n.

Using the causal difference equation (5.3) or (5.4), we can make a table of all output values.

n	$n<-2$	-2	-1	0	1	2	3	4	5	6	7	$n>7$
$x[n]$	0	0	0	2	4	6	4	2	0	0	0	0
$y[n]$	0	0	0	$\frac{2}{3}$	2	4	$\frac{14}{3}$	4	2	$\frac{2}{3}$	0	0

Notice that now the orange-colored values of $x[n]$ are used to compute $y[4]$ instead of $y[2]$. The resulting signal $y[n]$ has the same values as before, but its support is now the index interval $0 \le n \le 6$. Observe that the output of the causal filter is simply a shifted version of the output of the previous noncausal filter. This filter is causal because the output depends on only the present and two previous (i.e., past) values of the input. Therefore, the output does not change from zero until the input changes from zero.

EXERCISE 5.1 Determine the output of a *centralized running averager*

$$y[n] = \tfrac{1}{3}(x[n+1] + x[n] + x[n-1])$$

for the input in Fig. 5-2(a). Is this filter causal or noncausal? What is the support of the output for this input? How would the plot of the output compare to Fig. 5-2(b)?

[7] The divide by 3 needed for the 3-pt running averager has been incorporated into a coefficient of 1/3 for each term in the sums (5.3) and (5.4).

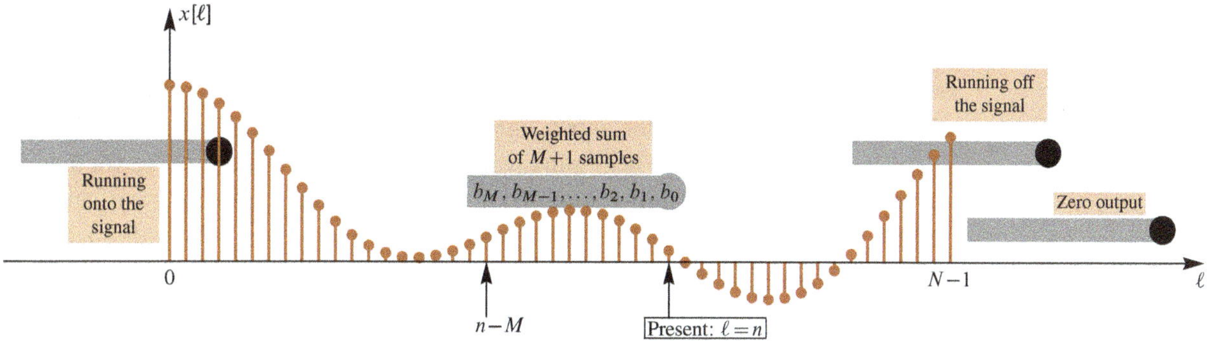

Figure 5-4 Operation of the Mth-order causal FIR filter showing various positions of the (gray) sliding window of $M+1$ samples over which the weighted average is calculated. When the input signal $x[\ell]$ has finite length (N samples), the sliding window runs onto and off of the input data, so the resulting output signal also has finite length.

5-3 The General FIR Filter

The causal running average (5.4) is a special case of the general causal difference equation

$$y[n] = \sum_{k=0}^{M} b_k\, x[n-k] = b_0 x[n] + b_1 x[n-1] + \cdots + b_M x[n-M] \qquad (5.5)$$

where the coefficients b_k are fixed numbers. Usually the b_k coefficients are not all the same, and then we say that (5.5) defines a ***weighted running average*** of $M+1$ samples. The 3-point running average (5.4) is the case where $M=2$ and $b_k = 1/3$ in (5.5). It follows from (5.5) that the computation of $y[n]$ involves the samples $x[\ell]$ for $\ell = n, n-1, n-2, \ldots, n-M$ (i.e., $x[n], x[n-1], x[n-2]$, etc). Figure 5-4 illustrates how the causal FIR filter uses $x[n]$ and the past M samples to compute the output. Since the sum in (5.5) does not involve future samples of the input, the system is causal, and, therefore, the output cannot start until the input becomes nonzero.[8]

A second general form of the FIR filter that reduces to (5.3) when $M=2$ and $b_k = 1/3$ for $k = 0, 1, 2$ can be obtained from (5.5) by making the substitution $\ell = n-k$ (or equivalently, $k = n-\ell$) and adjusting the limits on the sum to obtain

$$y[n] = \sum_{\ell=n-M}^{n} b_{n-\ell}\, x[\ell] = b_M x[n-M] + b_{M-1} x[n-M+1] + \cdots + b_0 x[n] \qquad (5.6)$$

Equations (5.5) and (5.6) both show that the subscript index of the filter coefficient sequence b_k is reversed with respect to the indexing of the input sequence. In (5.6) b_0 multiplies the present value of the input $x[n]$ while b_M multiplies the input value farthest in the past, $x[n-M]$. This form gives rise to the sliding window visualization of Fig. 5-4, where each gray bar contains the filter coefficients listed in reverse order.

[8]Note that a noncausal system can be represented by altering (5.5) to include negative values of the summation index k.

Figure 5-4 also shows what happens when the input signal is finite length with a beginning and end. In this case, the input is zero for $\ell < 0$ and for $N \leq \ell$ so it can be nonzero only when $0 \leq \ell \leq N-1$. This possibly nonzero region is called the *support* of the signal $x[n]$. When the input signal has finite support of length N, there is an interval of M samples at the beginning, where the computation involves fewer than $M+1$ nonzero samples as the sliding window of the filter engages with the input, and another interval of M samples at the end where the sliding window of the filter disengages from the input sequence. Furthermore, Fig. 5-4 shows that when the filter runs off the input signal, additional output values are created, so the output sequence can have a support that is as much as M samples longer than the input support. Thus, the length of the output sequence would be $N+M$ samples, where N is the length of the input signal. These general statements about the input and output of an FIR filter are illustrated by the following example.

EXAMPLE 5-1 Pulse Input to 3-Point Running Average

Consider a 3-point running average system expressed in the sliding window form of (5.6)

$$y[n] = \sum_{\ell=n-2}^{n} \tfrac{1}{3} x[\ell]$$

with input

$$x[n] = \begin{cases} 1 & 0 \leq n \leq 10 \\ 0 & \text{otherwise} \end{cases}$$

as plotted in Fig. 5-5(a). We refer to this $x[n]$ as a "pulse input" because it is nonzero and constant over a short interval of length 11. The shaded regions in Fig. 5-5(a) highlight

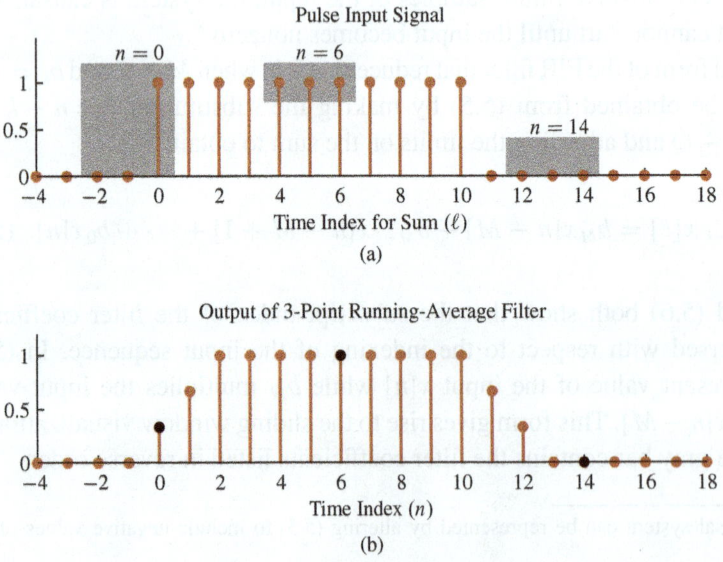

Figure 5-5 Computing the output of 3-point running-average filter. (a) Input signal $x[n]$ is a length-11 pulse. Shown in gray are three positions of the length-3 averaging interval which is the sliding window of this FIR filter. (b) Corresponding output of 3-point running averager with values at $n = 0, 6, 13$ highlighted (black dots).

the three samples of $x[\ell]$ involved in the computation of the output at $n = 0$, $n = 6$, and $n = 14$. These specific cases illustrate how the causal *sliding window* interval of length three scans across the input sequence moving sample-by-sample to produce the output signal. Note the three highlighted samples of the output in Fig. 5-5(b) which correspond to the three positions of the averaging interval in Fig. 5-5(a). Finally, note that the output is zero when $n < 0$ and $n > 12$ because for these values of n, the averaging interval includes only zero values of the input. Also note that the output for $n = 0, 1, 2$ follows a straight line from 0 to 1 because the averaging interval includes an additional unit sample as it moves to the right over the interval $0 \leq n \leq 2$. A similar taper occurs at the end of the pulse interval as the averaging interval includes fewer nonzero samples as it moves to the right. The output is constant and equal to one in the interval $2 \leq n \leq 10$, where the averaging interval includes three samples of $x[n]$ where $x[n] = 1$.

Example 5-1 illustrates the previous assertion about the support of the input and output signals: If $x[\ell]$ is nonzero only in the interval $0 \leq \ell \leq N - 1$, then for a causal FIR filter having $M + 1$ coefficients $\{b_k\}$ for $k = 0, 1, \ldots, M$, the corresponding output $y[n]$ can be nonzero only when $n \geq 0$ and $n - M \leq N - 1$. Thus the support of the output signal is the interval $0 \leq n \leq N + M - 1$, and the length of the output support is $N + M$ samples. This is true regardless of the values of the $M + 1$ coefficients. In Example 5-1, $N = 11$ and $M = 2$ so the output is nonzero in the interval $0 \leq n \leq 12$ and the length of the output sequence is 13 samples. Problem P-5.6 considers the general case where the signal $x[n]$ begins at $n = N_1$ and ends at $n = N_2$.

EXAMPLE 5-2 FIR Filter Coefficients

In general, the FIR filter is completely defined once the set of filter coefficients $\{b_k\}$ is known. For example, if the coefficients of a causal filter are

$$\{b_k\} = \{3, \ -1, \ 2, \ 1\}$$

then we have a length-4 filter with $M = 3$, and (5.5) expands into a 4-point difference equation:

$$y[n] = \sum_{k=0}^{3} b_k x[n-k] = 3x[n] - x[n-1] + 2x[n-2] + x[n-3]$$

The parameter M is the ***order*** of the FIR filter. The number of filter coefficients is also called the filter ***length*** (L). Usually, the length is one greater than the order, i.e., $L = M+1$. The terminology "order" will make more sense after we have introduced the z-transform in Chapter 9.

EXERCISE 5.2 Compute the output $y[n]$ for the length-4 filter whose coefficients are $\{b_k\} = \{3, -1, 2, 1\}$. Use the input signal given in Fig. 5-2(a). Verify that the partial answer tabulated here is correct, then fill in the missing values.

n	$n<0$	0	1	2	3	4	5	6	7	8	$n>8$
$x[n]$	0	2	4	6	4	2	0	0	0	0	0
$y[n]$	0	6	10	18	?	?	?	8	2	0	0

5-3.1 An Illustration of FIR Filtering

To illustrate some of the things that we have learned so far, and to show how FIR filters can modify sequences that vary in interesting ways, consider the signal

$$x[n] = \begin{cases} (1.02)^n + \frac{1}{2}\cos(2\pi n/8 + \pi/4) & 0 \le n \le 40 \\ 0 & \text{otherwise} \end{cases}$$

This signal, which is the sum of two components, is shown as the orange dots in Fig. 5-6(a). We often have real signals of this form; where one component is the signal of interest which, in this case, is the slowly varying exponential component $(1.02)^n$ and the other component is often considered *noise* or *interference* that degrades the observation of the desired signal. In this case, we consider the sinusoidal component $\frac{1}{2}\cos(2\pi n/8+\pi/4)$ to be an interfering signal that we wish to remove. The solid, exponentially growing curve shown superimposed in each of the plots in Fig. 5-6 simply connects the sample values of the desired signal $(1.02)^n$ by straight lines for reference in all three plots.

Now suppose that $x[n]$ is the input to a causal 3-point running averager, that is,

$$y_3[n] = \sum_{k=0}^{2} \tfrac{1}{3} x[n-k] \tag{5.7}$$

In this case, $M = 2$ and all the coefficients are equal to 1/3. The output of this filter is shown in Fig. 5-6(b). We can notice several things about these plots.

(a) Observe that the input sequence $x[n]$ is zero prior to $n = 0$, and from (5.7) it follows that the output must be zero for $n < 0$.

(b) The output becomes nonzero at $n = 0$, and the shaded interval[9] of length $M = 2$ samples at the beginning of the nonzero part of the output sequence is the interval where the 3-point averager "runs onto" the input sequence. For $2 \le n \le 40$, the input samples within the 3-point averaging window are all nonzero.

[9]In Example 5-1, shading was used to highlight the moving averaging interval relative to the input sequence. In Fig. 5-6(b) and (c) shading is used on the output sequence to highlight the fixed intervals at the beginning and end of the output where the averaging interval engages with and disengages from the finite-length input sequence.

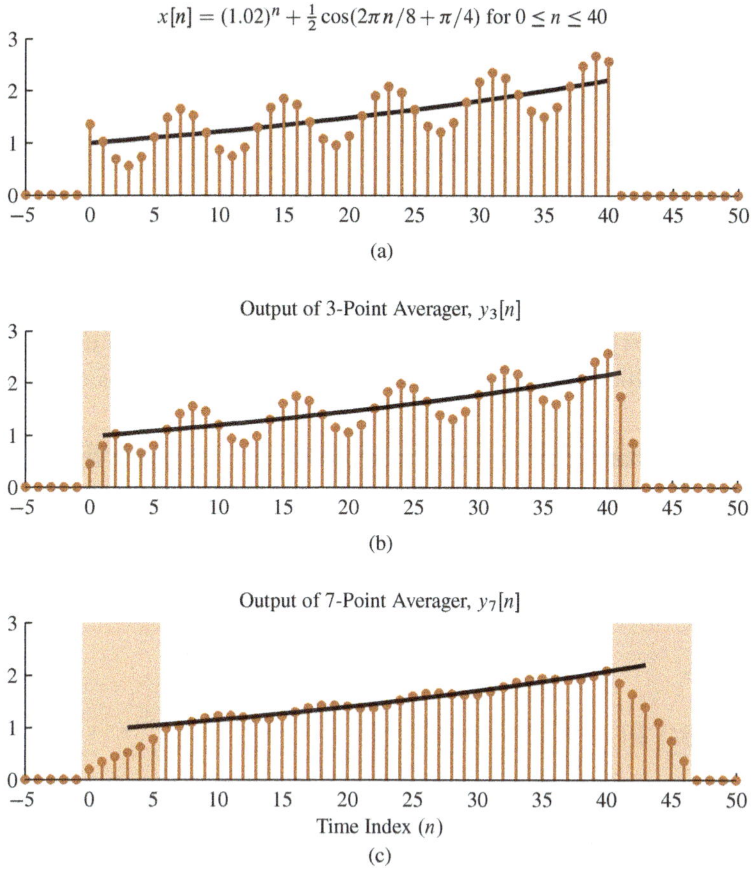

Figure 5-6 Illustration of running-average filtering. (a) Input signal; (b) output of 3-point averager; (c) output of 7-point averager.

(c) There is another shaded interval of length $M = 2$ samples at the end (after sample 40), where the filter window "runs off of" the input sequence.

(d) Observe that the size of the sinusoidal component in $y_3[n]$ has been reduced slightly, but that the component is not eliminated by the filter. The solid line showing the values of the exponential component has been shifted to the right by $M/2 = 1$ sample to account for the shift introduced by the causal filter.

The 3-point running averager has barely reduced the size of the fluctuations in the input signal, so we have not recovered the desired component. Intuitively, we might think that averaging over a longer interval might produce better results. The plot in Fig. 5-6(c) shows the output of a 7-point running averager as defined by

$$y_7[n] = \sum_{k=0}^{6} \tfrac{1}{7}x[n-k] \tag{5.8}$$

In this case, since $M = 6$ and all the coefficients are equal to $^1/_7$, we observe the following:

(a) The shaded regions at the beginning and end of the output are now $M = 6$ samples long.

(b) Now the size of the sinusoidal component is greatly reduced relative to the input sinusoid, and the component of the output is very close to the exponential component of the input (after a shift of $M/2 = 3$ samples).

What can we conclude from this example? First, it appears that FIR filtering can modify signals in ways that may be useful. Second, the length of the averaging interval seems to have a big effect on the resulting output. Third, the running-average filters appear to introduce a shift equal to $M/2$ samples. All of these observations can be shown to apply to more general FIR filters defined by (5.5). However, before we can fully appreciate the details of this example, we must explore the properties of FIR filters in greater detail. We will gain full appreciation of this example only upon the completion of Chapter 6.

EXERCISE 5.3 Use MATLAB to implement an 8-point averager and process the composite signal in Fig. 5-6(a). The output should have no sinusoidal interference in the region where the length-8 sliding window completely overlaps the input signal. Verify that this is true in the MATLAB output and then explain why this is true by considering the effect of the 8-point averager on the sinusoid alone.

5-4 The Unit Impulse Response and Convolution

In this section, we introduce three new ideas: the unit impulse sequence, the unit impulse response, and the convolution sum. We show that the *impulse response provides a complete characterization* of the FIR filter, because the convolution sum gives a formula for computing the output from the input when the unit impulse response is known.

5-4.1 Unit Impulse Sequence

The ***unit impulse*** is perhaps the simplest sequence because it has only one nonzero value, which occurs at $n = 0$. The mathematical notation is that of the Kronecker ***delta function***

$$\delta[n] = \begin{cases} 1 & n = 0 \\ 0 & n \neq 0 \end{cases} \tag{5.9}$$

It is tabulated in the second row of this table:

n	$\ldots$	-2	-1	0	1	2	3	4	5	6	$\ldots$
$\delta[n]$	0	0	0	1	0	0	0	0	0	0	0
$\delta[n-2]$	0	0	0	0	0	1	0	0	0	0	0

A shifted impulse such as $\delta[n-2]$ is nonzero when its argument is zero, that is, when $n - 2 = 0$, or equivalently $n = 2$. The third row of the previous table gives the values of the shifted impulse $\delta[n-2]$, and Fig. 5-7 shows a plot of that sequence.

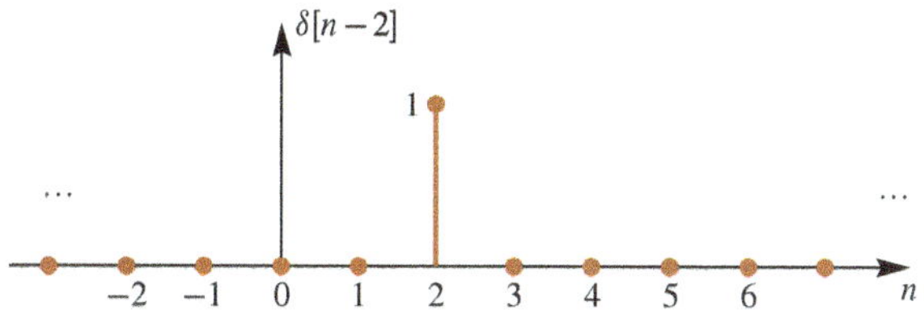

Figure 5-7 Shifted impulse sequence, $\delta[n-2]$.

The shifted impulse is a concept that is very useful in representing signals and systems. For example, we can show that the formula

$$x[n] = 2\delta[n] + 4\delta[n-1] + 6\delta[n-2] + 4\delta[n-3] + 2\delta[n-4] \tag{5.10}$$

is equivalent to defining $x[n]$ by tabulating its five nonzero values. To interpret (5.10), we must observe that the appropriate definition of multiplying a sequence by a number is to multiply each value of the sequence by that number; likewise, adding two or more sequences is defined as adding the sequence values at corresponding positions (times). The following table shows the individual sequences in (5.10) and their sum:

n	...	−2	−1	0	1	2	3	4	5	6	...
$2\delta[n]$	0	0	0	2	0	0	0	0	0	0	0
$4\delta[n-1]$	0	0	0	0	4	0	0	0	0	0	0
$6\delta[n-2]$	0	0	0	0	0	6	0	0	0	0	0
$4\delta[n-3]$	0	0	0	0	0	0	4	0	0	0	0
$2\delta[n-4]$	0	0	0	0	0	0	0	2	0	0	0
$x[n]$	0	0	0	2	4	6	4	2	0	0	0

Equation (5.10) is a compact mathematical representation of the signal in Fig. 5-2(a). It turns out that any sequence can be represented in this way. The equation

$$\begin{aligned} x[n] &= \sum_k x[k]\delta[n-k] \\ &= \cdots + x[-1]\delta[n+1] + x[0]\delta[n] + x[1]\delta[n-1] + x[2]\delta[n-2] + \cdots \end{aligned} \tag{5.11}$$

is true if k ranges over all the nonzero values of the sequence $x[n]$. Equation (5.11) states the obvious: The sequence is formed by using scaled shifted impulses to place samples of the right size at the correct index positions.

5-4.2 Unit Impulse Response Sequence

The output from a filter is often called the *response* to the input, so when the input is the unit impulse, $\delta[n]$, the output is called the ***unit impulse response***.[10] We reserve the notation $h[n]$ for this special output signal. To emphasize the impulse response in

[10] We usually shorten this to ***impulse response***, with *unit* being understood.

block diagrams of systems, we show the impulse sequence $\delta[n]$ at the input and the impulse response $h[n]$ at a corresponding position on the output side. This is illustrated in the block diagram of Fig. 5-8. In other words, it is always true that when the input to the FIR filter (5.5) is a unit impulse sequence, $x[n] = \delta[n]$, the output is the (unit) impulse response $h[n]$. Substituting $x[n] = \delta[n]$ in (5.5) gives the output $y[n] = h[n]$:

$$h[n] = \sum_{k=0}^{M} b_k\,\delta[n-k] = \begin{cases} b_n & n = 0, 1, 2, \ldots, M \\ 0 & \text{otherwise} \end{cases}$$

As we have observed, the sum evaluates to a single term for each value of n because each $\delta[n-k]$ is nonzero only when $n-k=0$, or $n=k$. In tabular form, the impulse response is

n	$n<0$	0	1	2	3	...	M	$M+1$	$n>M$
$x[n] = \delta[n]$	0	1	0	0	0	0	0	0	0
$y[n] = h[n]$	0	b_0	b_1	b_2	b_3	...	b_M	0	0

In other words, the impulse response $h[n]$ of the FIR filter is simply the sequence of difference equation coefficients. Since $h[n] = 0$ for $n < 0$ and for $n > M$, the length of the impulse response sequence $h[n]$ is finite. This is why the system (5.5) is called a ***FIR*** system. Figure 5-9 illustrates a plot of the impulse response for the case of the causal 3-point running-average filter. Since the impulse response is identical to the sequence of difference equation coefficients, it follows that the FIR filter is completely defined by the impulse response. In Section 5-7, we show that this characterization is also true for the much broader class of linear time-invariant (LTI) systems.

EXERCISE 5.4 Determine and plot the impulse response of the FIR system

$$y[n] = \sum_{k=0}^{10} (k-3)x[n-k]$$

Figure 5-8 Block diagram showing definition of the impulse response. When the input signal is $x[n] = \delta[n]$, then by definition, the output signal is called the impulse response, $y[n] = h[n]$.

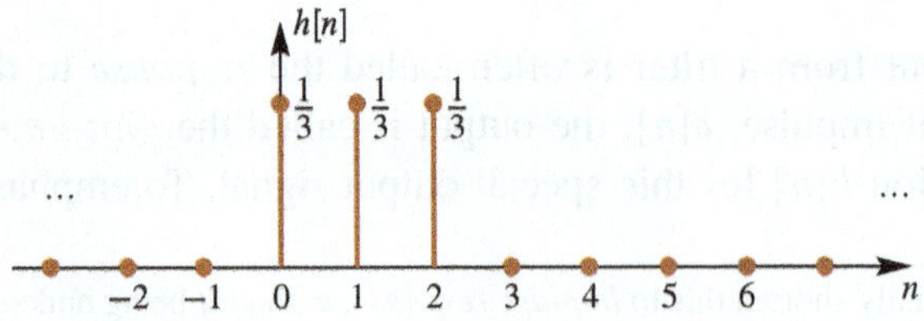

Figure 5-9 Impulse response of a 3-point running-average filter, $h[n]$.

5-4.2.1 The Unit-Delay System

One important system is the operator that performs a delay or shift by an amount n_0

$$y[n] = x[n - n_0] \tag{5.12}$$

When $n_0 = 1$, the system is called a ***unit delay***. The output of the unit delay is particularly easy to visualize. In a plot, the values of $x[n]$ are moved to the right by one time index. For example, $y[4]$ takes on the value of $x[3]$, $y[5]$ takes on the value of $x[4]$, $y[6]$ is $x[5]$, and so on.

The delay system is actually the simplest of FIR filters; it has only one nonzero coefficient. For example, a system that produces a *delay of 2* has filter coefficients $\{b_k\} = \{0,\ 0,\ 1\}$. The order of this FIR filter is $M = 2$, and its difference equation is

$$\begin{aligned} y[n] &= b_0 x[n] + b_1 x[n-1] + b_2 x[n-2] \\ &= (0)x[n] + (0)x[n-1] + (1)x[n-2] = x[n-2] \end{aligned}$$

Note: The length of the delay-by-2 filter can be either $L = 3$ or $L = 1$, depending on the situation. In MATLAB a length-3 vector `[0,0,1]` is needed to implement the filter. On the other hand, there is only one nonzero filter coefficient and, hence, only one term in the difference equation.

Figure 5-10 shows the output of the delay-by-2 system for the input of Fig. 5-2(a) on p. 170. The impulse response of the delay system is obtained by substituting $\delta[n]$ for $x[n]$ in (5.12). For the delay-by-2 case,

$$h[n] = \delta[n - n_0] = \delta[n-2] = \begin{cases} 1 & n = 2 \\ 0 & n \neq 2 \end{cases}$$

A plot of this impulse response would be the shifted impulse shown previously in Fig. 5-7.

5-4.3 FIR Filters and Convolution

A general expression for the FIR filter's output (5.5) can be derived in terms of the impulse response. Since the filter coefficients in (5.5) are identical to the impulse response values, we can replace b_k in (5.5) by $h[k]$ to obtain

$$y[n] = \sum_{k=0}^{M} h[k]\,x[n-k] \tag{5.13}$$

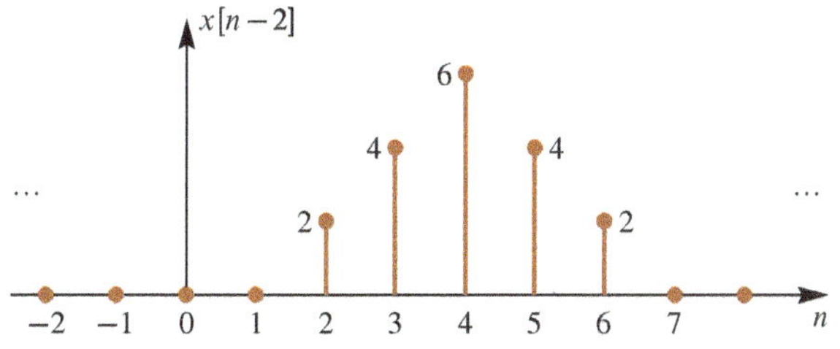

Figure 5-10 Delayed finite-length input signal, $y[n] = x[n-2]$ for the input of Fig. 5-2(a).

Now the relation between the input and the output of the FIR filter is expressed in terms of the input and the impulse response. The sum in (5.13) is called a finite ***convolution sum***, and we say that the output is obtained by ***convolving*** the sequences $h[n]$ and $x[n]$. This terminology emphasizes that convolution is an operation between two sequences. This point of view is formalized by introducing some new notation for the "convolution operator." We use a star ($*$) to represent the operation of evaluating (5.13) for $-\infty < n < \infty$ by writing

$$y[n] = h[n] * x[n] \tag{5.14}$$

which is read as "the output sequence $y[n]$ is obtained by convolving the impulse response $h[n]$ with the input sequence $x[n]$."

Later, in Section 5-7, we will prove that convolution is the fundamental input–output algorithm for a large class of very useful filters that includes FIR filters as a special case. We will show that a general form of convolution that also applies to infinite-length signals is

$$y[n] = h[n] * x[n] = \sum_{k=-\infty}^{\infty} h[k]\,x[n-k] \tag{5.15}$$

This convolution sum (5.15) has infinite limits to accomodate impulse responses of infinite length, but reduces to (5.13) when $h[n] = 0$ for $n < 0$ and $n > M$.

5-4.3.1 Computing the Output of a Convolution

In Section 5-2, the output of the running average FIR filter was constructed by tabulating values. This method works for short signals, but lacks the generality needed in more complicated problems—however, there is a simple interpretation of (5.13) that leads to an algorithm for doing *numerical convolution.* First of all, we describe the algorithm which can be implemented using the table in Fig. 5-11 that tracks the relative position of the signal values. The example in Fig. 5-11 shows how to convolve $x[n] = \{2,\ 4,\ 6,\ 4,\ 2\}$ with $h[n] = \{3,\ -1,\ 2,\ 1\}$. The first step is to write out the signal values of $x[n]$ and $h[n]$ on separate rows. Then we use a method similar to what we call[11] "synthetic polynomial

n	$n<0$	0	1	2	3	4	5	6	7	$n>7$
$x[n]$	0	2	4	6	4	2	0	0	0	0
$h[n]$	0	3	−1	2	1					
$h[0]x[n]$	0	6	12	18	12	6	0	0	0	0
$h[1]x[n-1]$	0	0	−2	−4	−6	−4	−2	0	0	0
$h[2]x[n-2]$	0	0	0	4	8	12	8	4	0	0
$h[3]x[n-3]$	0	0	0	0	2	4	6	4	2	0
$y[n]$	0	6	10	18	16	18	12	8	2	0

Figure 5-11 Numerical convolution of finite-length signals via synthetic polynomial multiplication.

[11] Our use of this term is motivated by analogy to the more well-known process of synthetic polynomial division.

multiplication" to form the output as the sum of shifted rows. Each shifted row is produced by multiplying the $x[n]$ row by one of the $h[k]$ values and shifting the result to the right so that it lines up with the $h[k]$ position. The final answer is obtained by summing down the columns.

The justification of this algorithm for evaluating the convolution sum (5.13) comes from examining individual terms in the sum

$$y[n] = h[0]x[n] + h[1]x[n-1] + h[2]x[n-2] + \cdots$$

A term such as $x[n-2]$ is the $x[n]$ signal with its values shifted two columns to the right. The multiplier $h[2]$ scales the shifted signal $x[n-2]$ to produce the contribution $h[2]x[n-2]$, which is the highlighted row in the table. All the individual terms must be added which is accomplished by summing down the columns.

EXERCISE 5.5 Use the "synthetic multiplication" convolution algorithm to compute the output $y[n]$ for the length-4 filter whose impulse response is $h[n] = \delta[n] - 2\delta[n-1] + 2\delta[n-2] - \delta[n-3]$. Use the input signal given in Fig. 5-2(a) on p. 170.

5-4.3.2 The Length of a Convolution

Using our new terminology, an FIR filter is a system whose output is the convolution of the finite-length impulse response of the system with the input to the system. From this view point, Example 5-1 can be reconsidered as the convolution of the 3-point impulse response

$$h[n] = \tfrac{1}{3}\delta[n] + \tfrac{1}{3}\delta[n-1] + \tfrac{1}{3}\delta[n-2]$$

with a length-11 pulse signal. The filter's output $y[n] = h[n] * x[n]$ shown in Fig. 5-5(b) is nonzero for $0 \le n \le 12$. This example illustrates the general principle that if $x[n]$ is nonzero only in the interval $0 \le n \le L_x - 1$, then for a causal FIR filter having an impulse response of length $L_h = M + 1$ samples, the corresponding output $y[n]$, which is the convolution of $h[n]$ with $x[n]$, can be nonzero only when $n \ge 0$ and $n - M \le L_x - 1$, or equivalently, the support of $y[n]$ is the interval $0 \le n \le L_x + M - 1$. Thus the length of $y[n]$ is

$$L_y = L_x + L_h - 1, \tag{5.16}$$

which is $L_y = 11 + 3 - 1 = 13$ in Fig. 5-5(b).

Another example is the convolution in Fig. 5-11 which is the convolution of a length-5 signal, $x[n]$, with a length-4 signal, $h[n]$. The result $y[n]$ starts at $n = 0$ and ends at $n = 7$, so it is a length-8 signal. The same result is obtained by using the general formula (5.16), because $L_y = 5 + 4 - 1 = 8$ in Fig. 5-11.

When the signals start at $n = 0$, it is possible to use the synthetic multiplication table to prove that the formula (5.16) is true. Since the length of $x[n]$ is L_x, the last nonzero signal value is $x[L_x - 1]$. Likewise, for $h[n]$ where $h[L_h - 1]$ is the last nonzero value. In the figure, the row for $h[0]x[n]$ starts at $n = 0$ and ends at $n = L_x - 1$, while the row for $h[L_h - 1]x[n - (L_h - 1)]$ ends at $n = (L_x - 1) + (L_h - 1)$. After summing down the columns, we see that the convolution result $y[n]$ starts at $n = 0$ and ends at $n = L_h + L_x - 2$, so the length of $y[n]$ is $L_x + L_h - 1$. Problem P-5.6 considers the general case where the signal $x[n]$ begins at sample N_1 and ends at sample N_2.

5-4.3.3 Convolution in MATLAB

In MATLAB, FIR systems can be implemented by using the `conv` function.[12] For example, the following MATLAB statements

```
xn = sin(0.07*pi*(0:50));
hn = ones(11,1)/11;
yn = conv(hn, xn);
```

evaluate the convolution of the 11-point sequence `hn` with the 51-point sinusoidal sequence `xn`. The particular choice for the MATLAB vector `hn` is actually the impulse response of an 11-point running average system:

$$h[n] = \begin{cases} 1/11 & n = 0, 1, 2, \dots, 10 \\ 0 & \text{otherwise} \end{cases}$$

That is, all 11 filter coefficients are the same and equal to $1/11$.

EXERCISE 5.6 In MATLAB, we can compute only the convolution of finite-length signals. Determine the length of the output sequence computed by the MATLAB convolution above.

EXERCISE 5.7 Use MATLAB to duplicate the plots of Fig. 5-5 (without the shading) of Example 5-1.

5-4.3.4 Polynomial Multiplication in MATLAB

We have already hinted that this operation called ***convolution*** is equivalent to polynomial multiplication. In Section 9-4 on z-transforms, we will prove that this correspondence is true. At this point, we note that in MATLAB there is no function for multiplying polynomials. Instead, we must know that convolution is equivalent to polynomial multiplication. Then we can represent the polynomials by sequences of their coefficients and use the `conv` function to convolve them, thereby doing polynomial multiplication.

[12] Note that MATLAB indexes all data vectors starting at 1, while our mathematical formulas index sequences such as $x[n]$ starting at $n = 0$. This difference must be accounted for in MATLAB programs.

EXERCISE 5.8 Use MATLAB's `conv` function to compute the following product of polynomials:

$$P(x)=(1+2x+3x^2+5x^4)(1-3x-x^2+x^3+3x^4)$$

5-4.3.5 Filtering the Unit-Step Signal

In previous sections, we have described the FIR filtering of finite-length signals, but there is no reason that the input signal cannot be infinite duration. An example is the ***unit-step*** signal which is zero for $n < 0$ and "turns on" at $n = 0$

$$u[n] = \begin{cases} 0 & n < 0 \\ 1 & n \geq 0 \end{cases} \tag{5.17}$$

Figure 5-12(a) shows a plot of $u[n]$. The symbol $u[n]$ is reserved for the unit step, so when we want the input signal to be the unit step we write $x[n] = u[n]$, and the output of an Mth-order FIR filter would be

$$y[n] = \sum_{\ell=n-M}^{n} b_{n-\ell}u[\ell] \tag{5.18}$$

if we use the form in (5.6). Since the FIR filter is causal and the unit step starts at $n = 0$, the output is zero for $n < 0$. Determining the output for $n \geq 0$ can be done by writing out terms

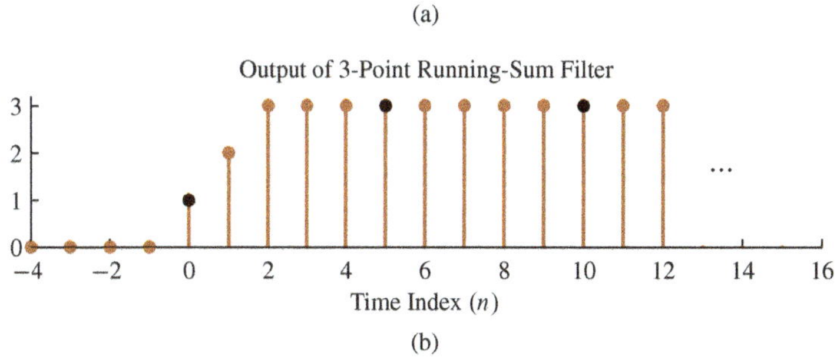

Figure 5-12 Filtering the unit-step signal with a 3-point running sum filter. (a) Input signal is a unit step plotted versus ℓ, the dummy time index of the convolution sum (5.18). Shown in gray are three positions of the length-3 summing interval which is the sliding window of this FIR filter. (b) Corresponding output of 3-point running sum with values at $n = 0, 5, 10$ highlighted (black dots).

$$y[0] = b_0 u[0] = b_0$$
$$y[1] = b_1 u[0] + b_0 u[1] = b_0 + b_1$$
$$y[2] = b_2 u[0] + b_1 u[1] + b_0 u[2] = \sum_{k=0}^{2} b_k$$
$$\vdots \quad \vdots$$
$$y[M] = \sum_{k=0}^{M} b_k$$
$$y[n] = \sum_{k=0}^{M} b_k \quad \text{for } n > M$$

Thus there are three cases for the output: (1) when $n < 0$, $y[n] = 0$; (2) when $0 \le n < M$, $y[n]$ is the sum of the first $n + 1$ filter coefficients; and (3) when $M \le n$, $y[n]$ is the sum of all $M + 1$ filter coefficients.

When the FIR filter is an L-point running sum all the filter coefficients are equal to one, so the output signal for $n \ge 0$ has the values

$$y[n] = \{\underbrace{1, 2, 3, \ldots, L-1}_{\text{rising}}, \underbrace{L, L, L, \ldots}_{\text{flat plateau}}\}$$

Figure 5-12(b) shows the result for $L = 3$.

EXAMPLE 5-3 Pulse as the Difference of Two Shifted Unit Steps

Show that a length-5 pulse can be written as

$$x[n] = u[n - 100] - u[n - 105]$$

when the starting index of the pulse is at $n = 100$.

5-4.3.6 Commutative Property of Convolution

The convolution operator point of view introduced in (5.14) turns out to be a very useful way to think about FIR filters and the more general class of LTI systems to be discussed in Sections 5-6 and 5-7. When thought of as an operator, convolution has many interesting properties that are useful in the analysis and design of linear systems. For example, convolution is a ***commutative*** operation, that is,

$$y[n] = h[n] * x[n] = x[n] * h[n] \tag{5.19}$$

This property is suggested by the equivalence of (5.5) and (5.6) for the FIR filter. Since the impulse response of an FIR filter is the sequence of filter coefficients, (5.6) can be rewritten as a convolution

$$y[n] = \sum_{k=n-M}^{n} x[k]\,h[n-k] = x[n] * h[n] \tag{5.20}$$

where $h[n]$ is now flipped within the summation. When $x[n]$ is finite and nonzero only for $0 \le n \le M_x$, the sum in (5.20) can be simplified further to

$$y[n] = \sum_{k=0}^{M_x} x[k]\,h[n-k] = x[n] * h[n] \tag{5.21}$$

EXERCISE 5.9 Demonstrate that convolution is commutative by interchanging the $x[n]$ and $h[n]$ rows in Fig. 5-11 and then carrying out the synthetic polynomial multiplication.

The commutative property of convolution is summarized as follows:

Convolution Is Commutative

$$y[n] = h[n] * x[n] = x[n] * h[n] \tag{5.22}$$

In Section 5-6, it is shown that the operation of convolution is also associative and that convolution distributes across addition. Thus, convolution for sequences displays the same properties as multiplication of numbers.

EXERCISE 5.10 Demonstrate that convolution in MATLAB is commutative by adding to the program in Section 5-4.3.3 the statement `ync = conv(xn, hn);`, running it again, and comparing `yn` to `ync`.

5-4.3.7 MATLAB GUI for Convolution

DEMO
Discrete Convolution

Figure 5-13 shows a screenshot of a MATLAB GUI that can generate examples of convolution with a variety of simple finite-length signals. The specific case shown in the GUI is the convolution of a length-11 square pulse of amplitude 0.5 with an impulse response defined by

$$h[n] = \delta[n] - \delta[n-1] \tag{5.23}$$

When the length-2 impulse response values are used in the convolution equation (5.20) with $M = 1$ and $h[n]$ given by (5.23), we obtain

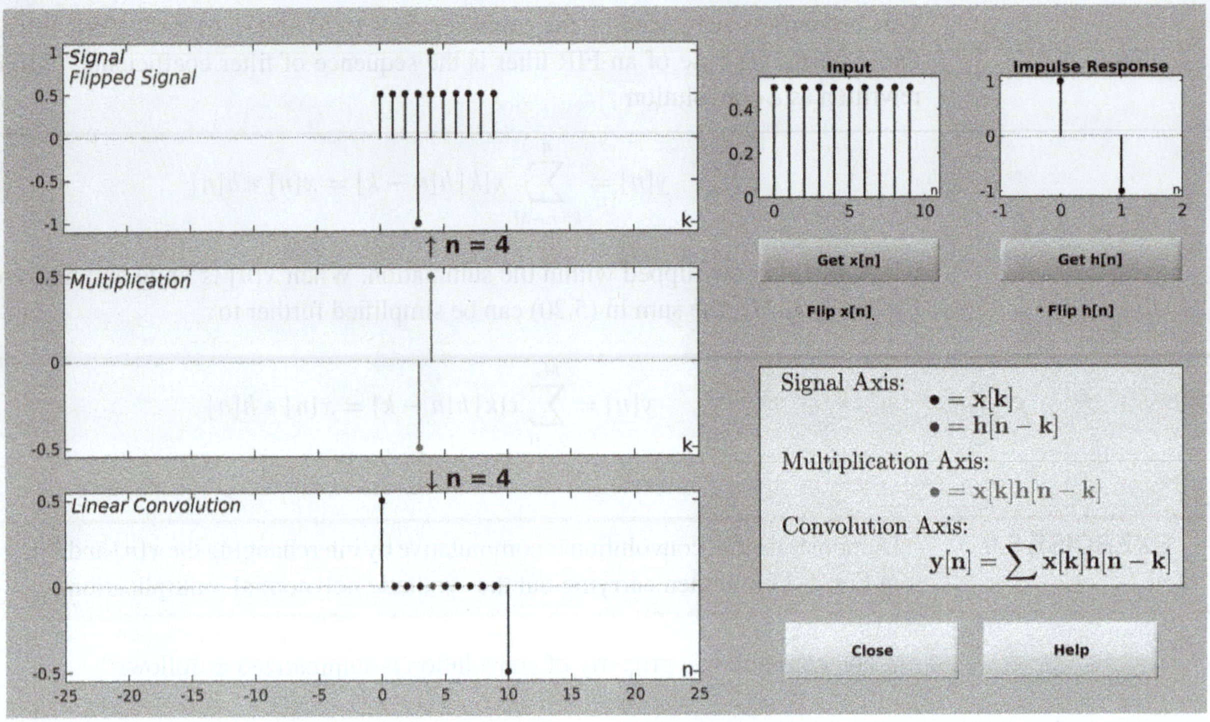

Figure 5-13 Graphical user interface for `dconvdemo` which illustrates the *sliding window* nature of FIR filtering (i.e., convolution). The impulse response $h[n]$ is flipped before sliding. The user can pick the input signal and the impulse response, and then slide the filter over the input signal to observe how the output is constructed.

$$\begin{aligned} y[n] &= \sum_{k=n-1}^{n} x[k]h[n-k] \\ &= x[n-1]h[n-(n-1)] + x[n]h[n-n] \\ &= x[n-1]h[1] + x[n]h[0] \end{aligned}$$

Therefore, the result can be expressed as the difference equation

$$y[n] = x[n] - x[n-1]$$

The impulse response is "flipped" because $h[1] = -1$ multiplies $x[n-1]$ which is in the past, while $h[0] = 1$ multiplies $x[0]$ which is the present value of the signal. Therefore, the "flip $h[n]$" button has to be selected in order to visualize the sliding window interpretation of FIR filtering. In Fig. 5-13, the pointer to the output index n can be moved back and forth along the horizontal axis in order to animate the sliding window concept. In the screen shot, the pointer is at $n = 4$ and the amplitude of the pulse is $1/2$ at both $n = 4$ and $n = 3$, so the computation of $y[4]$ is

$$y[4] = 1x[4]^{\,\frac{1}{2}} + (-1)x[3]^{\,-\frac{1}{2}} = 0$$

In this GUI it is possible to flip either $x[n]$ or $h[n]$. This is an expression of the commutativity property of convolution as discussed in Section 5-4.3.6. For the convolution $x[n] * h[n]$ in (5.20), $h[n]$ is flipped in the GUI while for $h[n] * x[n]$ as in (5.13), $x[n]$ is flipped in the GUI. Other properties of convolution are discussed in Section 5-7.2.

EXERCISE 5.11 Use the `dconvdemo` GUI to repeat the example of Fig. 5-13 with the input "flipped" and thereby verify the commutativity of convolution. Pay close attention to how the flipped input now slides past the impulse response which is fixed at indices 0 and 1.

5-5 Implementation of FIR Filters

Recall that the general definition of an FIR filter is

$$y[n] = \sum_{k=0}^{M} b_k x[n-k] \tag{5.24}$$

In order to use (5.24) to compute the output of the FIR filter, we need the following: (1) a means for multiplying delayed-input signal values by the filter coefficients; (2) a means for adding the scaled sequence values; and (3) a means for obtaining delayed versions of the input sequence. We find it useful to represent the operations of (5.24) as a block diagram. Such representations lead to new insights about the properties of the system and about alternative ways to implement the system.

5-5.1 Building Blocks

The three basic building-block systems we need are the multiplier, the adder, and the unit-delay operator. These are depicted in Fig. 5-14.

5-5.1.1 Multiplier

The first elementary system performs multiplication of a signal by a constant (see Fig. 5-14(a)). The output signal $y[n]$ is given by the rule

$$y[n] = \beta x[n]$$

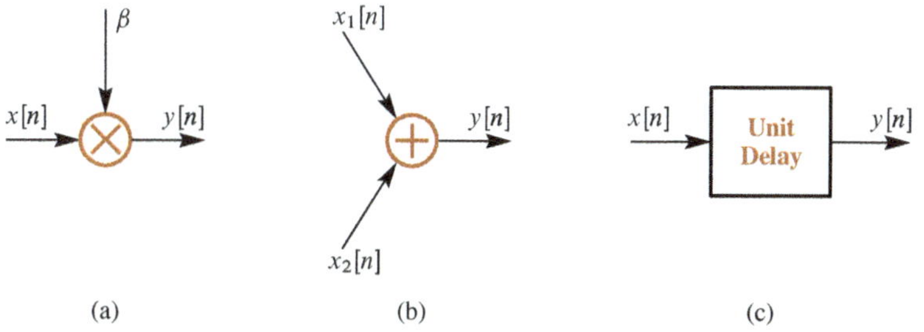

Figure 5-14 Building blocks for making any LTI discrete-time system: (a) multiplier, $y[n] = \beta x[n]$; (b) adder, $y[n] = x_1[n] + x_2[n]$; and (c) unit-delay, $y[n] = x[n-1]$.

where the coefficient β is a constant. This system can be the hardware multiplier unit in a computer. For a DSP microprocessor, the speed of this multiplier is one of the fundamental limits on the throughput of the digital filtering process. In applications such as image convolution, billions of multiplications per second may have to be performed to implement a good filter, so quite a bit of engineering work has been directed at designing fast multipliers for DSP applications. Furthermore, since many filters require the same sequence of multiplications over and over, pipelining the multiplier also results in a dramatic speed up of the filtering process.

Notice, by the way, that the simple multiplier is also an FIR filter, with $M = 0$, and $b_0 = \beta$ in (5.5). The impulse response of the multiplier system is simply $h[n] = \beta\delta[n]$.

5-5.1.2 Adder

The second elementary system in Fig. 5-14(b) performs the addition of two signals. This is a different sort of system because it has two inputs and one output. In hardware, the adder is simply the hardware adder unit in a computer. Since many DSP operations require a multiplication followed immediately by an addition, it is common in DSP microprocessors to build a special multiply-accumulate unit, often called a "MADD" or "MAC" unit.[13]

Notice that the adder is a pointwise combination of the values of the two input sequences. It is not an FIR filter, because it has more than one input—however, it is a crucial building block of FIR filters. With many inputs, the adder could be drawn as a multi-input adder, but, in digital hardware, the additions are typically done two inputs at a time.

5-5.1.3 Unit Delay

The third elementary system performs a delay by one unit of time. It is represented by the block diagram in Fig. 5-14(c). In the case of discrete-time filters, the time dimension is indexed by integers, so this delay is by one "count" of the system clock. The hardware implementation of the unit delay is actually performed by acquiring a sample value, storing it in memory for one clock cycle, and then releasing it to the output. The delays by more than one time unit that are needed to implement (5.24) can be implemented (from the block-diagram point of view) by cascading several unit delays in a row. Therefore, an M-unit delay requires M memory cells configured as a shift register, which can be implemented as a circular buffer in computer memory.

5-5.2 Block Diagrams

In order to create a graphical representation that is useful for hardware structures, we use ***block-diagram notation***, which defines the interconnection of the three basic building blocks to make more complex structures.[14] In such a diagram, the nodes (i.e., junction

[13]For example, the Texas Instruments C667x DSP can perform up to 320 billion 16-bit fixed-point and 160 billion 32-bit floating-point multiply-adds per second (ca. 2013).

[14]Such interconnections are called ***directed graphs***.

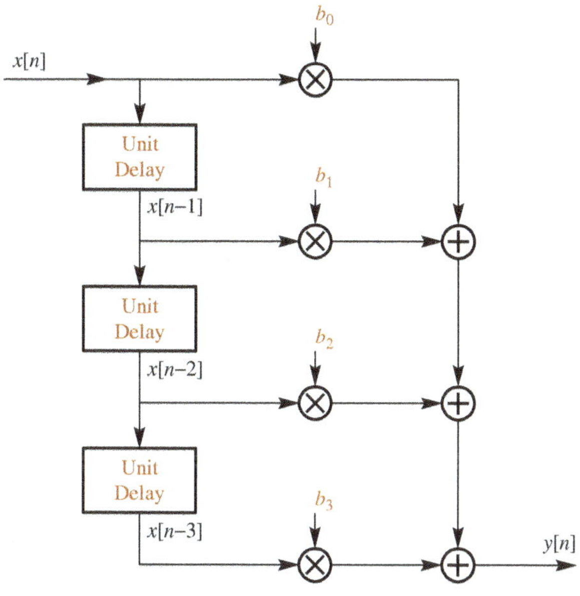

Figure 5-15 Block-diagram structure for the third-order FIR filter.

points) are either summing nodes, splitting nodes, or input–output nodes. The connections between nodes are either delay branches or multiplier branches. Figure 5-15 shows the general block diagram for a third-order FIR digital filter ($M = 3$). This structure shows why the FIR filter is also called a ***feed-forward difference equation***, since all paths lead forward from the input to the output. There are no closed-loop paths in the block diagram. In Chapter 10, we discuss filters with *feedback*, where both input and past output values are involved in the computation of the output.

Strictly speaking, the structure of Fig. 5-15 is a block-diagram representation derived from grouping the filter computation as follows:

$$y[n] = (((b_0x[n] + b_1x[n-1]) + b_2x[n-2]) + b_3x[n-3])$$

which is equivalent to (5.24) with $M = 3$. The input signal is delayed by the cascaded unit delays, each delayed signal is multiplied by a filter coefficient, and the products are accumulated to form the sum. Thus, it is easy to see that there is a one-to-one correspondence between the block diagram and the difference equation (5.24) of the FIR filter, because both are defined by the filter coefficients $\{b_k\}$. A useful skill is to start with one representation and then produce the other.

The structure in Fig. 5-15 displays a regularity that makes it simple to define longer filters; the number of cascaded delay elements is increased to M, and then the filter coefficients $\{b_k\}$ are substituted into the diagram. This standard structure is called the ***direct form***. Going from the block diagram back to the difference equation is just as easy, as long as we stick to direct form. Here is a simple exercise to make the correspondence.

EXERCISE 5.12 Determine the difference equation for the block diagram of Fig. 5-16.

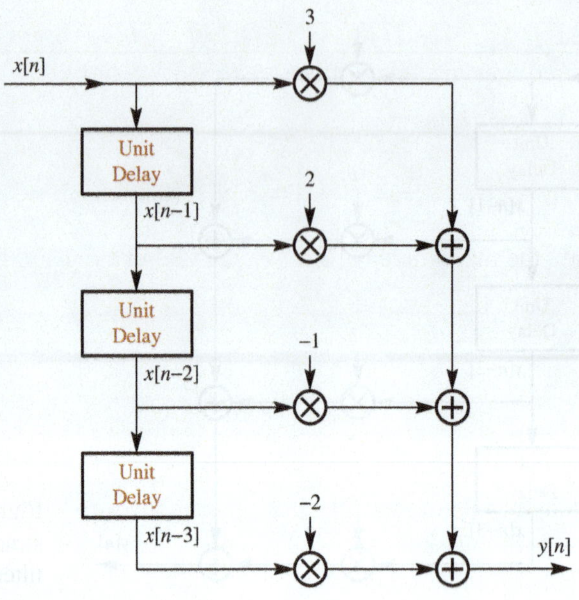

Figure 5-16 Third-order FIR filter in direct form.

5-5.2.1 Other Block Diagrams

A block diagram of a discrete-time system is really just a pictorial representation of a particular way to organize the computation of the output from the input. Many different block diagrams *implement* the same FIR filter, in the sense that the external behavior from input to output is the same, although the details of the internal computations may vary. Direct form is just one possibility. Other block diagrams would represent a different internal computation, or a different order of computation. In some cases, the internal multipliers might use different coefficients. After we have studied the z-transform in Chapters 9 and 10, we will have the tools to produce many different implementations.

When faced with an arbitrary block diagram, the following four-step procedure may be used to derive the difference equation from the block diagram.

(a) Give a unique signal name to the input of each unit-delay block.

(b) Notice that the output of a unit delay can be written in terms of its input.

(c) At each summing node of the structure, write a signal equation. Use the signal names introduced in steps 1 and 2.

(d) At this point, you have several equations involving $x[n]$, $y[n]$, and the internal signal names. These can be reduced to one equation involving only $x[n]$ and $y[n]$ by eliminating variables, as is done with simultaneous equations.

Let us try this procedure on the simple but useful example shown in Fig. 5-17. First of all, we observe that the internal signal variables $\{v_1[n],\ v_2[n],\ v_3[n]\}$ have been defined in Fig. 5-17 as the inputs to the three delays. Then the delay outputs are, from top to bottom, $v_3[n-1]$, $v_2[n-1]$, and $v_1[n-1]$. We also notice that $v_3[n]$ is a scaled version

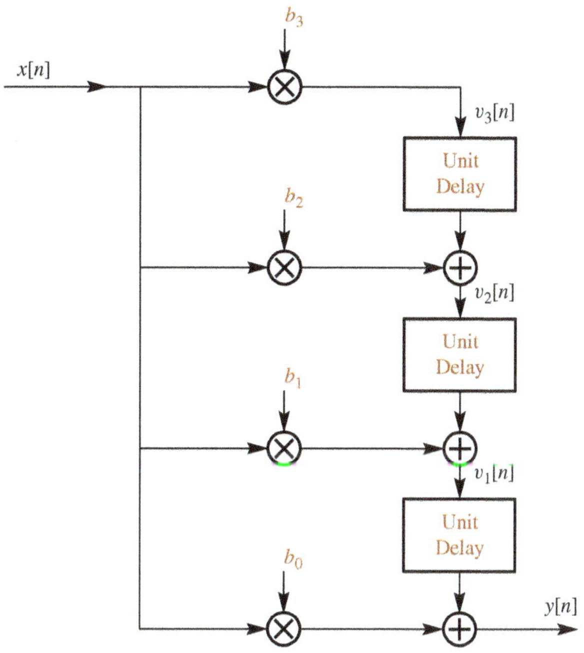

Figure 5-17 Transposed form block diagram structure for the third-order FIR filter.

of the input $x[n]$. In addition, we write the three equations at the output of the three summing nodes:

$$y[n] = b_0 x[n] + v_1[n-1]$$
$$v_1[n] = b_1 x[n] + v_2[n-1]$$
$$v_2[n] = b_2 x[n] + v_3[n-1]$$
$$v_3[n] = b_3 x[n]$$

Now we have four equations in five "unknowns." To show that this set of equations is equivalent to the direct form, we can eliminate the $v_i[n]$ by combining the equations in a pairwise fashion.

$$v_2[n] = b_2 x[n] + b_3 x[n-1]$$
$$y[n] = b_0 x[n] + b_1 x[n-1] + v_2[n-2]$$
$$= b_0 x[n] + b_1 x[n-1] + b_2 x[n-2] + b_3 x[n-3]$$

Thus, we have derived the same difference equation as before, so this new structure must be a different way of computing the same thing. In fact, it is widely used and is called the ***transposed form***[15] for the FIR filter.

EXERCISE 5.13 For the filter in Fig. 5-16, draw the *transposed form* block diagram.

[15]This terminology is suggested by a comparison of Figs. 5-15 and 5-17. Transposition of a single-input-single-output block diagram is explored in more detail in Section 10-4.3.

5-5.2.2 Internal Hardware Details

A block diagram shows dependencies among the different signal variables. Therefore, different block diagrams that implement the same input–output operation may have dramatically different characteristics as far as their internal behavior goes. Several issues come to mind:

(a) The order of computation is specified by the block diagram. In high-speed applications where parallelism or pipelining must be exploited, dependencies in the block diagram represent constraints on the computation.

(b) Partitioning a filter for an ASIC or FPGA implementation would be done (at the highest level) in terms of the block diagram. Likewise, algorithms that must be mapped onto special DSP architectures can be managed by using the block diagram with special compilers that translate the block diagram into optimized code for the DSP chip.

(c) Finite word-length effects are important if the filter is constructed using fixed-point arithmetic in order to speed up computation and reduce power consumption. In such cases, round-off error and overflow are important real-world problems that depend on the internal order of computation.

Now that we know something about convolution and the implementation of FIR filters, it is time to consider discrete-time systems in a more general way. In the next section, we show that FIR filters are a special case of the general class of LTI systems. Much of what we have learned about FIR filters will apply to this more general class of systems.

5-6 Linear Time-Invariant (LTI) Systems

In this section, we discuss two general properties of systems. These properties, ***linearity*** and ***time invariance***, lead to greater insight and understanding of system behavior. To facilitate the discussion of these general properties, it is useful to recall the block-diagram representation of a general discrete-time system shown in Fig. 5-1 on p. 168. This block diagram used $y[n] = \mathcal{T}\{x[n]\}$ to mean that the system processes $x[n]$ to produce $y[n]$. Alternatively, we can say that $x[n]$ goes through the system $\mathcal{T}$ and $y[n]$ comes out, which is captured by the notation

$$x[n] \stackrel{\mathcal{T}}{\longmapsto} y[n] \tag{5.25}$$

In a specific case, the system is defined by giving a formula or algorithm for computing all values of the output sequence from the values of the input sequence. A specific example is the ***square-law*** system ($\mathcal{S}$) defined by the rule:

$$x[n] \stackrel{\mathcal{S}}{\longmapsto} (x[n])^2 \tag{5.26}$$

Another example is (5.12), which defines the general delay system—still another is (5.24), which defines the general FIR filter and includes the delay system as a special case. These FIR filters are both linear and time-invariant, while the square-law system is not linear, but is time-invariant.

5-6.1 Time Invariance

A discrete-time system is said to be *time-invariant* if, when an input is delayed (shifted) by n_0, the output is delayed by the same amount. The condition must be true for any signal $x[n]$ and for any choice of n_0, the integer that determines the amount of time shift. A block-diagram view of the time-invariance property is given in Fig. 5-18. In the upper branch, the input is shifted prior to the system; in the lower branch, the output is shifted. Thus, a system can be tested for time-invariance by checking whether or not $w[n] = y[n - n_0]$ in Fig. 5-18.

Consider the example of the square-law system ($\mathcal{S}$) defined by (5.26). In the top branch of Fig. 5-18, we use the delayed input as the input to the square-law system, so we obtain

$$x[n] \overset{\mathcal{D}}{\longmapsto} x[n - n_0] \overset{\mathcal{S}}{\longmapsto} (x[n - n_0])^2$$

where $\mathcal{D}$ denotes the delay operator. In the bottom branch, $x[n]$ is the input to the square-law system, so $y[n] = (x[n])^2$ and

$$x[n] \overset{\mathcal{S}}{\longmapsto} (x[n])^2 \overset{\mathcal{D}}{\longmapsto} (x[n - n_0])^2$$

Since the two outputs are the same, the square-law system is time-invariant.

A second simple example is the time-flip system ($\mathcal{F}$), defined by the mapping

$$x[n] \overset{\mathcal{F}}{\longmapsto} x[-n]$$

The flip system simply reverses the order of the input sequence about the origin. If we delay the input and then reverse the order about the origin, we obtain

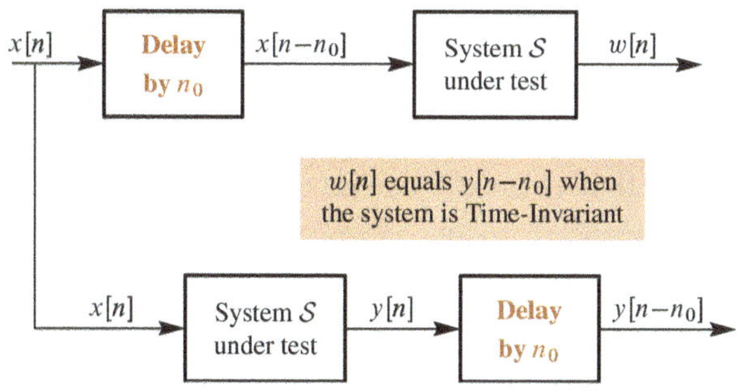

Figure 5-18 Testing the time-invariance property of system $\mathcal{S}$ by checking the interchange of operations.

$$x[n] \stackrel{\mathcal{D}}{\longmapsto} x[n-n_0] \stackrel{\mathcal{F}}{\longmapsto} x[(-n)-n_0] = w[n]$$

However, if we first flip the input sequence and then delay it, we obtain a different sequence from $w[n]$. Since $y[n] = x[-n]$, the output after delaying $y[n]$ is

$$x[n] \stackrel{\mathcal{F}}{\longmapsto} x[-n] \stackrel{\mathcal{D}}{\longmapsto} x[-(n-n_0)] = x[-n+n_0]$$

Thus, the time-flip system is *not* time-invariant.

EXERCISE 5.14 Test the system defined by $x[n] \stackrel{\mathcal{T}}{\longmapsto} nx[n]$ to discover that it is not a time-invariant system.

5-6.2 Linearity

Linear systems have the property that if $x_1[n] \longmapsto y_1[n]$ and $x_2[n] \longmapsto y_2[n]$, then

$$x[n] = \alpha x_1[n] + \beta x_2[n] \longmapsto y[n] = \alpha y_1[n] + \beta y_2[n] \tag{5.27}$$

where, with no name specified, $\longmapsto$ denotes the system under test. This mathematical condition must be true for any choice of the constants α and β, and for all signals $x_1[n]$ and $x_2[n]$. Equation (5.27) states that if the input consists of a sum of scaled sequences, then the corresponding output is a sum of scaled outputs corresponding to the individual input sequences. A block-diagram view of the linearity property is given in Fig. 5-19, which shows that a system can be tested for the linearity property by checking whether or not $w[n] = y[n]$.

The linearity condition in (5.27) is also called the ***principle of superposition***: If the input is the sum (superposition) of two or more scaled sequences, we can find the output due to each sequence acting alone and then add (superimpose) the separate scaled outputs. Sometimes it is useful to separate (5.27) into two properties. Setting $\alpha = \beta = 1$ we get the *additive* property

$$x[n] = x_1[n] + x_2[n] \longmapsto y[n] = y_1[n] + y_2[n] \tag{5.28a}$$

and using $\beta = 0$ there is only one input, which gives the *scaling* property

$$x[n] = \alpha x_1[n] \longmapsto y[n] = \alpha y_1[n] \tag{5.28b}$$

Both (5.28a) and (5.28b) must be true in order for (5.27) to be true.

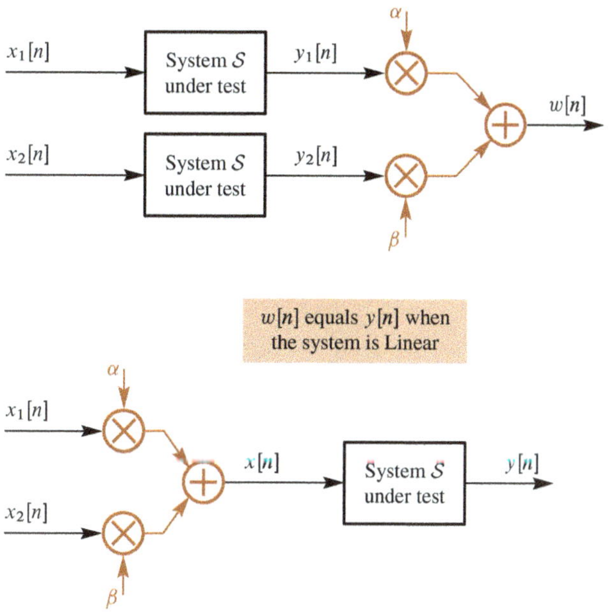

Figure 5-19 Testing linearity of system $\mathcal{S}$ by checking the interchange of operations.

EXERCISE 5.15 Use (5.28b) to show that for linear systems, if $x[n] = 0$ for all n, then $y[n] = 0$ for all n.

Reconsider the example of the square-law system defined by (5.26). The output $w[n]$ in the upper part of Fig. 5-19 for this system is

$$w[n] = \alpha(x_1[n])^2 + \beta(x_2[n])^2$$

while, in the lower part, the output is

$$\begin{aligned} y[n] &= (\alpha x_1[n] + \beta x_2[n])^2 \\ &= \alpha^2(x_1[n])^2 + 2\alpha\beta x_1[n]x_2[n] + \beta^2(x_2[n])^2 \end{aligned}$$

Thus, $w[n] \neq y[n]$, and the square-law system has been shown not to be linear. Systems that are not linear are called *nonlinear* systems.

EXERCISE 5.16 Show that the time-flip system $y[n] = x[-n]$ is a linear system.

5-6.3 The FIR Case

FIR systems described by (5.5) satisfy both the linearity and time invariance conditions. A mathematical proof of time-invariance can be constructed using the procedure in Fig. 5-18.

If we define the signal $v[n]$ to be $x[n-n_0]$, then the difference equation relating $v[n]$ to $w[n]$ in the upper branch of Fig. 5-18 is

$$\begin{aligned} w[n] &= \sum_{k=0}^{M} b_k v[n-k] \\ &= \sum_{k=0}^{M} b_k x[(n-k)-n_0] \\ &= \sum_{k=0}^{M} b_k x[(n-n_0)-k] \end{aligned}$$

For comparison, we construct $y[n-n_0]$ in the lower branch

$$y[n] = \sum_{k=0}^{M} b_k x[n-k]$$

$$\Rightarrow \quad y[n-n_0] = \sum_{k=0}^{M} b_k x[n-n_0-k]$$

These two expressions are identical, so the output $w[n]$ is equal to $y[n-n_0]$, and we have proven that the FIR filter is time-invariant.

The linearity condition is simpler to prove. Substitute into the difference equation (5.5) and collect terms with $x_1[n-k]$ and $x_2[n-k]$:

$$\begin{aligned} y[n] &= \sum_{k=0}^{M} b_k x[n-k] \\ &= \sum_{k=0}^{M} b_k(\alpha x_1[n-k] + \beta x_2[n-k]) \\ &= \alpha \sum_{k=0}^{M} b_k x_1[n-k] + \beta \sum_{k=0}^{M} b_k x_2[n-k] \\ &= \alpha y_1[n] + \beta y_2[n] \end{aligned}$$

The FIR filter obeys the principle of superposition; therefore, it is a linear system.

A system that satisfies both properties is called a ***linear time-invariant*** system, or simply ***LTI***. It should be emphasized that the LTI condition is a general condition. The FIR filter is an example of an LTI system. Not all LTI systems are described by (5.5), but all systems described by (5.5) are LTI systems.

5-7 Convolution and LTI Systems

Consider an LTI discrete-time system as depicted in the block diagram of Fig. 5-8. The impulse response $h[n]$ of the LTI system is simply the output when the input is the unit impulse sequence $\delta[n]$. In this section, we will show that the impulse response is a complete characterization for any LTI system, and that convolution is the *general formula* that allows us to compute the output from the input *for any LTI system*. In our initial discussion of convolution, we considered only finite-length input sequences and FIR filters. Now, we give a completely general presentation that allows infinite-length signals and impulse responses.

5-7.1 Derivation of the Convolution Sum

We begin by recalling from the discussion in Section 5-4 that *any* signal $x[n]$ can be represented as a sum of scaled and shifted impulse signals. Each nonzero sample of the signal $x[n]$ multiplies an impulse signal that is shifted to the index of that sample. Specifically, we can write $x[n]$ as follows:

$$x[n] = \sum_{\ell} x[\ell]\delta[n-\ell] = \cdots + x[-2]\delta[n+2] + x[-1]\delta[n+1] \\ + x[0]\delta[n] + x[1]\delta[n-1] + x[2]\delta[n-2] + \cdots \tag{5.29}$$

In the most general case, the range of summation in (5.29) could be from $-\infty$ to $+\infty$. In (5.29) we have left the range indefinite, realizing that the sum would include all nonzero samples of the input sequence.

A sum of scaled sequences such as (5.29) is commonly referred to as a "linear combination" or superposition of scaled sequences. Thus, (5.29) is a representation of the sequence $x[n]$ as a linear combination of scaled, shifted impulses. Since LTI systems respond in simple and predictable ways to sums of signals and to shifted signals, this representation is particularly useful for our purpose of deriving a general formula for the output of an LTI system.

Figure 5-8 reminds us that the output due to the input $\delta[n]$ is, by definition, the impulse response $h[n]$. Time invariance gives us additional information, specifically, the response due to $\delta[n-1]$ is $h[n-1]$. In fact, we can write a family of input–output pairs as follows:

$$\begin{aligned}
&\delta[n] &&\longmapsto h[n] \\
\Rightarrow\ &\delta[n-1] &&\longmapsto h[n-1] \\
\Rightarrow\ &\delta[n-2] &&\longmapsto h[n-2] \\
\Rightarrow\ &\delta[n-(-1)] &&\longmapsto h[n-(-1)] = h[n+1] \\
\Rightarrow\ &\delta[n-\ell] &&\longmapsto h[n-\ell] \quad \text{for any integer } \ell
\end{aligned}$$

Now we are in a position to use linearity, because (5.29) expresses a general input signal as a linear combination of shifted impulse signals. We can write out a few of the cases:

$$\begin{aligned}
x[0]\delta[n] &\longmapsto x[0]h[n] \\
x[1]\delta[n-1] &\longmapsto x[1]h[n-1] \\
x[2]\delta[n-2] &\longmapsto x[2]h[n-2] \\
x[\ell]\delta[n-\ell] &\longmapsto x[\ell]h[n-\ell] \quad \text{for any integer } \ell
\end{aligned}$$

Then we use superposition to put it all together as

$$x[n] = \sum_{\ell} x[\ell]\delta[n-\ell] \longmapsto y[n] = \sum_{\ell} x[\ell]h[n-\ell] \tag{5.30}$$

The derivation of (5.30) did not assume that either $x[n]$ or $h[n]$ was of finite duration, so, in general, we may write the convolution sum with infinite limits as

The Convolution Sum Formula

$$y[n] = x[n] * h[n] = \sum_{\ell=-\infty}^{\infty} x[\ell]h[n-\ell] \tag{5.31}$$

This expression represents the convolution operation in the most general sense, so we have proved for discrete-time systems that ***all LTI systems can be represented by a convolution sum***. The infinite limits take care of all possibilities, including the special cases where either or both of the sequences are finite in length.

5-7.2 Some Properties of LTI Systems

The properties of convolution are the properties of LTI systems. Here we examine some properties of convolution that can be developed from well-known concepts in algebra. As discussed in Section 5-4.3, convolution is a *binary operation between two signals,* so it has an algebraic nature (like multiplication and addition). The convolution of two sequences $x_1[n]$ and $x_2[n]$ is an operation that is denoted by $*$, that is,

$$y[n] = x_1[n] * x_2[n] = \sum_{\ell=-\infty}^{\infty} x_1[\ell]x_2[n-\ell] \tag{5.32}$$

We say that the first sequence $x_1[n]$ is *convolved* with the second sequence $x_2[n]$ to produce the result $y[n]$. In Section 5-4.3.6, we showed that convolution with a finite-impulse response is a commutative operation, and this is true for infinite sequences as well. We now show that in the general case of infinite-length sequences, the convolution operator is commutative, has an identity element, is associative, and has a distributive property, and then we will relate each of these to properties of LTI systems.

Commutative Property of Convolution An operator is commutative if order does not matter; in this case, convolving $x_2[n]$ with $x_1[n]$ must give the same result as convolving $x_1[n]$ with $x_2[n]$.

$$x_2[n] * x_1[n] = x_1[n] * x_2[n] \tag{5.33}$$

Justifying (5.33) involves a simple algebraic manipulation of the convolution sum, where we make the change of variables $k = n - \ell$ and sum on the new dummy variable k:

$$y[n] = x_2[n] * x_1[n] \tag{5.34a}$$

$$= \sum_{\ell=-\infty}^{\infty} x_2[\ell]x_1[n-\ell] \tag{5.34b}$$

$$= \sum_{k=+\infty}^{-\infty} x_2[n-k]x_1[k] \qquad \textit{(Note: } k = n - \ell\textit{)} \tag{5.34c}$$

$$= \sum_{k=-\infty}^{\infty} x_1[k]x_2[n-k] = x_1[n] * x_2[n] \tag{5.34d}$$

On the third line (5.34c), the limits on the sum can be swapped without consequence, because a set of numbers can be added in any order. When convolution represents an LTI system, the commutative property (5.33) becomes

$$y[n] = x[n] * h[n] = h[n] * x[n] \tag{5.35}$$

which means that the input signal is convolved with the impulse response, or vice versa. As we have seen, the FIR filter can be written in two ways: the summation in (5.13) or in (5.20). As pointed out in Exercise 5.9, both FIR summations are consistent with the computational algorithm in Fig. 5-11 on p. 182, because either $x[n]$ or $h[n]$ could be written on the first row, so $x[n]$ and $h[n]$ are interchangeable.

Identity Element of Convolution The unit impulse sequence $\delta[n]$ is the identity element of the convolution operator, that is, a signal is unchanged after convolving with an impulse, $x[n] * \delta[n] = x[n]$. Justification comes from considering the general case of convolving with a shifted impulse.

Convolution with an Impulse

$$x[n] * \delta[n - n_0] = x[n - n_0] \tag{5.36}$$

To show this, recall that convolution is the representation of an LTI system, where we convolve the input signal $x[n]$ with the impulse response $h[n]$, $x[n] * h[n]$. The delay by n_0 system is LTI, and its impulse response is known to be $h[n] = \delta[n - n_0]$, as previously discussed in Section 5-4.2.1. Thus we can write the input–output behavior of the LTI delay system in terms of a convolution as

$$y[n] = x[n - n_0] = x[n] * h[n] = x[n] * \delta[n - n_0]$$

This is a very useful result because (5.36) states that "to convolve any sequence $x[n]$ with an impulse located at $n = n_0$, all we need to do is translate the origin of $x[n]$ to n_0." For example, if $x[n]$ starts at $n = 0$, then $x[n - n_0]$ starts at $n = n_0$. When $n_0 = 0$, we get the special case $x[n] * \delta[n] = x[n]$ which can be stated succinctly as "$\delta[n]$ is the identity element of convolution."

Associative Property of Convolution The associative property states that, when convolving three signals,

$$(x_1[n] * x_2[n]) * x_3[n] = x_1[n] * (x_2[n] * x_3[n]) \tag{5.37}$$

we can convolve any two of them and then convolve that result with the third signal. Furthermore, the commutative property allows these convolutions to be performed in any order. The justification of the associative property will be given when discussing cascade systems in the next section.

Distributive Property of Convolution The distributive property states that convolution distributes over addition

$$(x_1[n] + x_2[n]) * h[n] = x_1[n] * h[n] + x_2[n] * h[n] \tag{5.38}$$

The convolution written in (5.38) with $h[n]$ is, in fact, a restatement of the additive property of superposition for an LTI system whose impulse response is $h[n]$.

5-8 Cascaded LTI Systems

LAB

Sampling, Convolution, and FIR Filtering

In a cascade connection of two systems, the output of the first system is the input to the second system, and the overall output of the cascade system is taken to be the output of the second system. Figure 5-20(a) shows two LTI systems (LTI 1 and LTI 2) in cascade.

LTI systems have the remarkable property that *two LTI systems in cascade can be implemented in either order*. This property is a direct consequence of the commutative property applied to the impulse responses of LTI systems. In order to show that the order of cascaded LTI systems doesn't matter, we first determine the impulse response of two systems in cascade. In Fig. 5-20(a), the impulse input and the corresponding outputs are

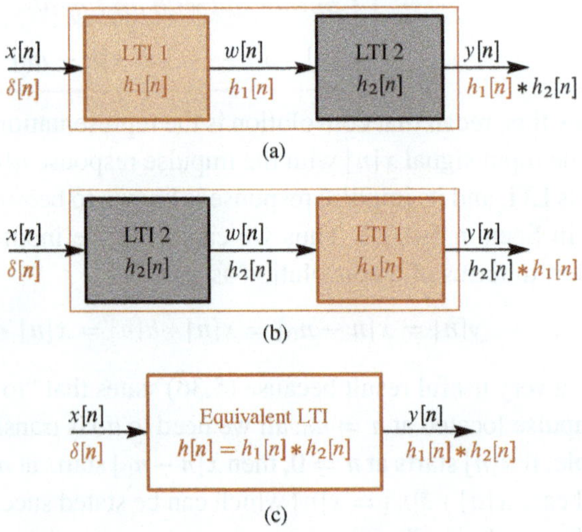

Figure 5-20 Switching the order of cascaded LTI systems. The single LTI system in (c) is equivalent to the cascade connection of the two systems in (a), and also to the cascade in (b).

shown below the arrows. When the input to the first system is an impulse, the output of LTI 1 is by definition its impulse response, $h_1[n]$, which then becomes the input to LTI 2. Next, the output of LTI 2 is the convolution of LTI 2's input $h_1[n]$ with its impulse response $h_2[n]$. Therefore, the overall impulse response of Fig. 5-20(a) is $h_1[n] * h_2[n]$ according to (5.31). In the same way, we can easily show that the overall impulse response of the other cascade system in Fig. 5-20(b) is $h_2[n] * h_1[n]$. Since convolution is commutative, the two cascade systems have the same impulse response

$$h[n] = h_1[n] * h_2[n] = h_2[n] * h_1[n] \tag{5.39}$$

which could be used as the impulse response $h[n]$ of a single system that is equivalent to either cascade. Stated another way, if the same signal $x[n]$ is the input to each of the three systems in Fig. 5-20, the output is the same for all three systems, because the overall impulse response is the same.

A by-product of this analysis of cascade systems is a proof of the associative property of convolution. For the cascade in Fig. 5-20(a) the output $y[n]$ for an input $x[n]$ is

$$y[n] = v[n] * h_2[n] = (x[n] * h_1[n]) * h_2[n]$$

and for the equivalent system in Fig. 5-20(c)

$$y[n] = x[n] * h[n] = x[n] * (h_1[n] * h_2[n])$$

The equality of these two expressions is precisely the associative property in (5.37).

EXAMPLE 5-4 Impulse Response of Cascaded Systems

To illustrate the utility of the results that we have obtained for cascaded LTI systems, we derive a single system that is equivalent to the cascade of two systems defined by

$$h_1[n] = \begin{cases} 1 & 0 \le n \le 3 \\ 0 & \text{otherwise} \end{cases} \qquad h_2[n] = \begin{cases} 1 & 1 \le n \le 3 \\ 0 & \text{otherwise} \end{cases}$$

The equivalent system, as in Fig. 5-20(c), has an impulse response equal to $h[n] = h_1[n] * h_2[n]$. Therefore, to find the overall impulse response $h[n]$ we must convolve $h_1[n]$ with $h_2[n]$. This can be done by using the synthetic polynomial multiplication algorithm of Section 5-4.3.1. In this case, the computation is as follows:

n	$n < 0$	0	1	2	3	4	5	6	$n > 6$
$h_1[n]$	0	1	1	1	1	0	0	0	0
$h_2[n]$	0	0	1	1	1				
$h_1[0]h_2[n]$	0	0	0	0	0	0	0	0	0
$h_1[1]h_2[n-1]$	0	0	1	1	1	1	0	0	0
$h_1[2]h_2[n-2]$	0	0	0	1	1	1	1	0	0
$h_1[3]h_2[n-3]$	0	0	0	0	1	1	1	1	0
$h[n]$	0	0	1	2	3	3	2	1	0

The equivalent impulse response in the bottom row also defines the filter coefficients $\{b_k\}$, and we can express $h[n]$ as a sum of weighted shifted impulses

$$h[n] = \sum_{k=0}^{6} b_k \delta[n-k]$$

where $\{b_k\}$ is the sequence $\{0, 1, 2, 3, 3, 2, 1\}$. The system with this impulse response $h[n]$ can be implemented either by the single difference equation

$$y[n] = \sum_{k=0}^{6} b_k x[n-k] \tag{5.40}$$

where $\{b_k\}$ is the above sequence, or by the pair of cascaded difference equations

$$\begin{aligned} v[n] &= \sum_{k=0}^{3} x[n-k] \\ y[n] &= \sum_{k=1}^{3} v[n-k] \end{aligned} \tag{5.41}$$

Example 5-4 illustrates an important point. There is a difference between (5.40) and (5.41) when doing computation. It can be seen that the cascade implementation in (5.41) requires a total of only five additions to compute each value of the output sequence, while (5.40) requires five additions and an additional four multiplications by coefficients that are not equal to unity. On a larger scale (longer filters), such differences in the amount and type of computation can be very important in practical applications of FIR filters. Thus, the existence of alternative equivalent implementations of the same filter provides flexibility in design choices.

EXERCISE 5.17 The two difference equations in (5.40) and (5.41) would be represented by two distinctly different block diagrams of the type shown in Fig. 5-15. Construct the detailed block diagrams of the direct form (5.40) and the cascade form (5.41) showing all multiplications and additions. (Multiplications by 1 do not need to be explicitly shown as multiply operations.)

5-9 Example of FIR Filtering

We conclude this chapter with an example of the use of FIR filtering on a real signal. An example of sampled data that can be viewed as a signal is the Dow-Jones Industrial Average (DJIA). The DJIA is a sequence of numbers obtained by averaging the closing prices of a selected number of representative stocks. It has been computed since 1897,

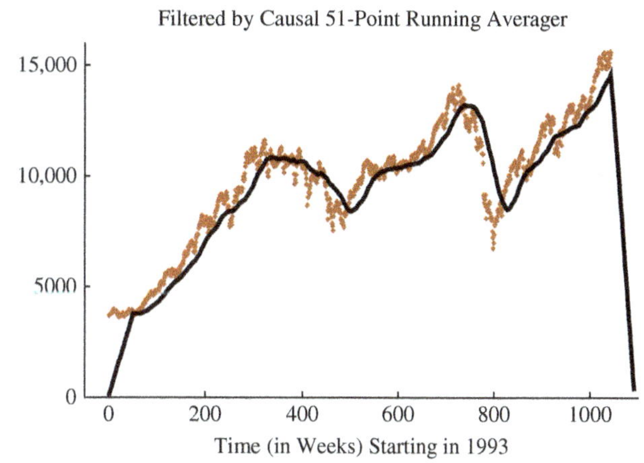

Figure 5-21 DJIA weekly closings (December 1993 to October 2013) filtered by a 51-point running-average FIR filter. Input (orange) and output (black). The output samples are connected to help distinguish them from the input samples.

and is used in many ways by investors and economists. The entire sequence dating back to 1897 makes up a signal that has positive values and is exponentially growing over the long term. In order to obtain an example where fine detail is visible, we have selected the segment from the end of 1993 to 2013 of the weekly closing price average to be our signal $x[n]$.[16] The raw DJIA signal is the one showing high variability in Fig. 5-21, where each weekly value is plotted as a distinct point. The smoother curve in Fig. 5-21 is the output of a 51-point causal running averager, that is,

$$y[n] = \frac{1}{51}\sum_{k=0}^{50} x[n-k] = h[n] * x[n] \qquad (5.42)$$

where $h[n]$ is the impulse response of the causal 51-point running averager.

Notice that, as we observed in Fig. 5-5 of Section 5-3.1, there are regions at the beginning and end (50 samples each, in this case) where the filter is engaging and disengaging the input signal. The left-hand region extends up to $n = 50$, which is the smallest value of n such that $n - k$ in (5.42) is never a negative index—specifically, for $n \geq 50$ no zero samples are included in the sum in (5.42). On the scale required to show the input and output signals over the selected time interval, the output in this region appears to rise at a constant rate. This is because the input signal varies only slightly around a value of approximately 3800 over the interval $0 \leq n \leq 50$, so the output of the running-average filter increases approximately linearly as more and more samples of the input are included in the 51 sample averaging interval. Similarly, the output tapers down to zero when n goes past the last sample index in the input as fewer and fewer of the input samples in the interval $n - 50$ to n are nonzero. The nonzero portion of the output is 50 points longer than the input because the filter eventually gives zero output after the index $n - 50$ goes past the end of the input.

Also notice that much of the fine-scale week-to-week variation has been removed by the filter. Finally, notice that the output is shifted relative to the input. In

[16] The complete set of weekly sampled data for the period 1897–2014 would involve more than 6000 samples and would range in value from about 40 to above 17,000. A plot of the entire sequence would not show the effects that we wish to illustrate.

Chapter 6, we will develop techniques that will explain why the shift introduced by this filter is exactly $M/2 = 50/2 = 25$ samples. Because the sampling period is 1 week, the 51-point averaging is over 51 weeks, and the shift corresponds to a delay of 25 weeks.[17]

In this example, it is informative to compare the input and output without the 25-sample shift, which can be done if we use the noncausal *centralized* running averager

$$\tilde{y}[n] = \frac{1}{51} \sum_{k=-25}^{25} x[n-k] = x[n] * \tilde{h}[n]$$

where $\tilde{h}[n]$ is the impulse response of the centralized running averager. The output $\tilde{y}[n]$ can be obtained by shifting (advancing) the output of the causal running averager $y[n]$ by 25 samples to the left. In terms of our previous discussion of cascaded FIR systems, we could think of the centralized system as a cascade of the causal system with a system whose impulse response is $\delta[n+25]$, because (5.36) says that convolution of $y[n]$ with an impulse shifts the origin of the sequence to the location of the impulse (in this case, to $n = -25$).

$$\begin{aligned}\tilde{y}[n] &= y[n] * \delta[n+25] \\ &= (x[n] * h[n]) * \delta[n+25] \\ &= x[n] * (h[n] * \delta[n+25]) \\ &= x[n] * h[n+25]\end{aligned}$$

LAB
Sampling, Convolution, and FIR Filtering

Thus, we find that the impulse response of the centralized running averager, $\tilde{h}[n] = h[n+25]$, is a shifted version of the impulse response of the causal running averager. By shifting the impulse response, we remove the delay introduced by the causal system.

EXERCISE 5.18 Determine the impulse response $h[n]$ of the 51-point causal running averager and determine the impulse response $\tilde{h}[n]$ for the 51-point centralized running averager.

The cascade representation of the centralized running averager is depicted in Fig. 5-22. One way to describe the system in Fig. 5-22 is to say that the second system *compensates* for the delay of the first, that is, the centralized running averager is a *zero-delay* running-average filter.

Figure 5-23 shows the input $x[n]$ and output $\tilde{y}[n]$ for the delay-compensated running-average filter. It now can be seen that corresponding features of the input and output are well aligned.

[17] We use 51 weeks instead of 52, because a 52-week running average has a delay of $51/2 = 25.5$. The interpretation of a half-sample delay as a shift is not as straightforward as when the delay is an integer number of samples.

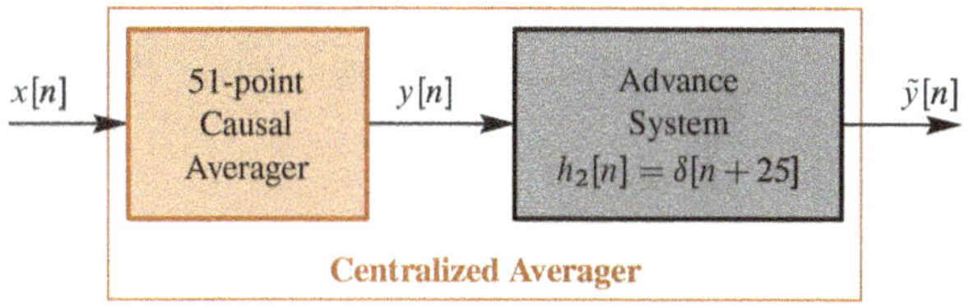

Figure 5-22 Cascade interpretation of centralized running averager in terms of causal running averager with delay compensation.

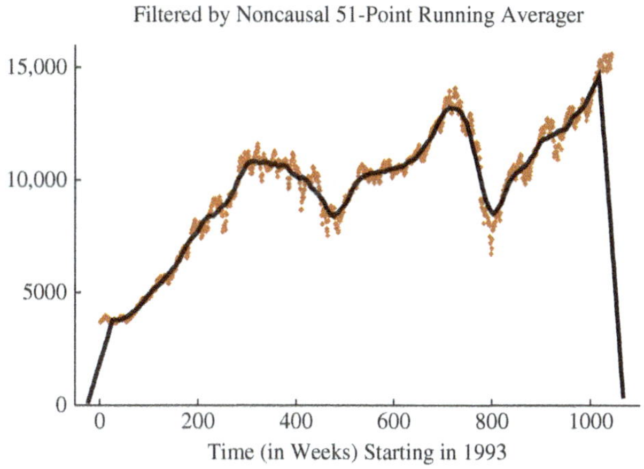

Figure 5-23 Input (orange) and output (black) for delay-compensated 51-point running averager applied to DJIA weekly closings.

5-10 Summary and Links

DEMO
Discrete Convolution

This chapter introduced the concept of FIR filtering along with the general concepts of convolution, the impulse response, and linear time-invariant systems. Among the laboratory projects on the Companion Website, the one entitled *Sampling, Convolution, and FIR Filtering* deals with discrete convolution and the effects of sampling. This lab also requires the use of two MATLAB GUIs, one for sampling, `con2dis` from Chapter 4, the other for discrete convolution, `dconvdemo`.

5-11 Problems

NOTE
Problems with Solutions

The problems in this section address the major concepts of FIR filtering, LTI systems and convolution in this chapter. On the Companion Website there are many related homework problems which can be used to review and practice. Most of them have detailed solutions.

P-5.1 If the impulse response $h[n]$ of an FIR filter is

$$h[n] = 7\delta[n] + \delta[n-2] - 6\delta[n-5]$$

write the difference equation for the FIR filter.

P-5.2 Evaluate the expression $G = \sum_{n=0}^{13} (4\delta[n-3] - 4\delta[n-8])\, e^{-j(\frac{2\pi}{3})n}$. Express the numerical answer for G as a complex number in *polar form.*

P-5.3 The length-L running sum is defined as

$$y[n] = \frac{1}{L} \sum_{k=0}^{L-1} x[n-k]$$

When the input $x[n]$ is the *unit-step* signal, determine the output signal as follows:

(a) Compute the numerical values of $y[n]$ over the time interval $-5 \le n \le 10$, assuming that $L = 5$.

(b) Make a sketch of the output over the time interval $-5 \le n \le 10$, assuming that $L = 5$. Use MATLAB if necessary, but learn to do it by hand also.

(c) Finally, derive a general formula for $y[n]$ that will apply for any length L and for the range $n \ge 0$.

P-5.4 An LTI system is described by the difference equation

$$y[n] = 2x[n] - 3x[n-1] + 2x[n-2]$$

(a) When the input to this system is

$$x[n] = \begin{cases} 0 & n < 0 \\ n+1 & n = 0, 1, 2 \\ 5-n & n = 3, 4 \\ 1 & n \ge 5 \end{cases}$$

Compute the values of $y[n]$, over the index range $0 \le n \le 10$.

(b) For the previous part, plot both $x[n]$ and $y[n]$.

(c) Determine the response of this system to a unit impulse input, that is, find the output $y[n] = h[n]$ when the input is $x[n] = \delta[n]$. Plot $h[n]$ as a function of n.

P-5.5 An LTI system is described by the difference equation

$$y[n] = 4x[n] - 5x[n-1] + 1.2x[n-2]$$

(a) Draw the implementation of this system as a block diagram in direct form.

(b) Give the implementation as a block diagram in transposed direct form.

P-5.6 Consider a system defined by

$$y[n] = \sum_{k=0}^{M} b_k x[n-k]$$

(a) Suppose that the input $x[n]$ is nonzero only for $0 \le n \le N-1$. Show that $y[n]$ is nonzero at most over a finite interval of the form $0 \le n \le P-1$, which is the support of $y[n]$. Determine P in terms of the filter order M and the input signal length N.

(b) Suppose that the input $x[n]$ is nonzero only for $N_1 \le n \le N_2$. What is the length of $x[n]$? Show that $y[n]$ is nonzero at most over a finite interval of the form $N_3 \le n \le N_4$, which is the support of $y[n]$. Determine N_3 and N_4 and also the length of $y[n]$ in terms of N_1, N_2, and M. *Hint:* Draw a sketch similar to Fig. 5-4 on p. 173.

P-5.7 The unit-step $u[n]$ can be used to represent sequences that are zero for $n < 0$.

(a) Plot the sequence $x[n] = (-0.5)^n u[n]$ for $-2 \le n \le 8$.

(b) The L-point running sum is defined as

$$y[n] = \sum_{k=0}^{L-1} x[n-k]$$

For the input sequence $x[n] = (-0.5)^n u[n]$, compute the values of $y[n]$ over the index range $-2 \le n \le 8$, assuming that $L = 2$.

(c) For the input sequence $x[n] = a^n u[n]$, derive a general formula for the output of the running sum $y[n]$ that applies for any value a, for any length L, and for the index range $n \ge -2$. In doing so, you may have use for the formula:

$$\sum_{k=M}^{N} \alpha^k = \frac{\alpha^M - \alpha^{N+1}}{1-\alpha}$$

P-5.8 The impulse response $h[n]$ of an FIR filter is shown in Fig. P-5.8. Determine the filter coefficients $\{b_k\}$ of the difference equation for the FIR filter. If an input $x[n] = \{2, 1, -1\}$ is applied to it, obtain the output.

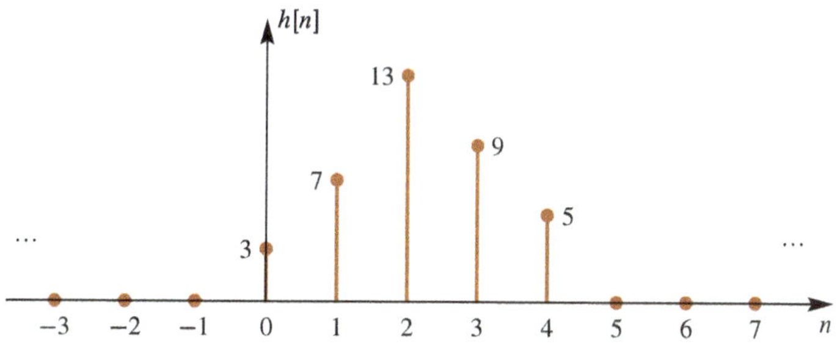

Figure P-5.8

P-5.9 If the filter coefficients of an FIR system are $\{b_k\} = \{-10,\ 10,\ 10\}$ and the input signal is

$$x[n] = \begin{cases} 0 & \text{for } n \text{ even} \\ 1 & \text{for } n \text{ odd} \end{cases}$$

determine the output signal $y[n]$ for all n. Give your answer as either a plot or a formula.

P-5.10 For each of the following systems, determine whether or not the system is (1) linear, (2) time-invariant, and (3) causal:

(a) $y[n] = x[n]\cos(0.2\pi n)$

(b) $y[n] = x[n] - x[n-1]$

(c) $y[n] = |x[n]|$

(d) $y[n] = Ax[n] + B$, where A and B are nonzero constants.

P-5.11 Suppose that $\mathcal{S}$ is an LTI system whose exact form is unknown. It is tested by running some inputs into the system, and then observing the output signals. Suppose that the following input–output pairs are the result of the tests:

Input: $x[n]$	**Output:** $y[n]$
$\delta[n] - \delta[n-1]$	$\delta[n] - \delta[n-1] + 2\delta[n-3]$
$\cos(\pi n/2)$	$2\cos(\pi n/2 - \pi/4)$

(a) Make a plot of the signal:

$$y[n] = \delta[n] - \delta[n-1] + 2\delta[n-3].$$

(b) Use linearity and time invariance to find the output of the system when the input is

$$x[n] = 7\delta[n] - 7\delta[n-2]$$

P-5.12 An LTI system has impulse response

$$h[n] = 2\delta[n] - 3\delta[n-1] + \delta[n-2] + 4\delta[n-5]$$

(a) Implement the above system in direct form and transposed direct form.

(b) If the difference equation was given instead, how would the implementations differ?

P-5.13 For a particular LTI system, when the input is the unit step, $x_1[n] = u[n]$, the corresponding output is

$$y_1[n] = \delta[n] + 2\delta[n-1] - \delta[n-2] = \begin{cases} 0 & n < 0 \\ 1 & n = 0 \\ 2 & n = 1 \\ -1 & n = 2 \\ 0 & n \geq 3 \end{cases}$$

(a) Determine the output when the input to the LTI system is $x_2[n] = 3u[n] - 6u[n-2]$. Give your answer as a formula expressing $y_2[n]$ in terms of known sequences, or give a list of values for $-\infty < n < \infty$.

(b) It is possible to write the unit impulse as the difference of two unit steps. Determine A and B so that the following is true:

$$\delta[n] = Au[n] + Bu[n-1]$$

(c) Use the result of part (b) to determine the impulse response $h[n]$ of the system.

P-5.14 For the FIR filter $y[n] = x[n] - 2x[n-1]$, determine the output signal $y[n]$ for two similar cases.

(a) The infinite-length input signal $x[n] = 2^n u[n]$.

(b) The length-7 input signal $x[n] = 2^n(u[n] - u[n-7])$.

(c) For part (b), plot $x[n]$ and $y[n]$ for $0 \leq n \leq 10$.

P-5.15 One form of the *deconvolution* process starts with the output signal and the filter's impulse response, from which it is sometimes possible to find the input signal.

(a) If the output of an FIR filter with $h[n] = \delta[n-4]$ is

$$y[n] = 3u[n-3] - 7u[n-5]$$

determine the input signal, $x[n]$.

(b) For the *first-difference* filter defined as

$$y[n] = x[n] - x[n-1]$$

determine the output signal when the input is $u[n-3]$.

(c) If the output of a first-difference FIR filter is

$$y[n] = \delta[n-3] - \delta[n-6],$$

determine the input signal, $x[n]$.

(d) If the output of a 2-point running sum filter is

$$y[n] = -6\delta[n] - 6\delta[n-3],$$

determine the input signal, $x[n]$.

P-5.16 Another form of the *deconvolution* process starts with the output signal and the input signal, from which it may be possible to find the impulse response.

(a) If the input signal is $x[n] = u[n]$, find the FIR filter that produces the output signal $y[n] = u[n-1]$.

(b) If the input signal is $x[n] = u[n]$, find the FIR filter that produces the output signal $y[n] = \delta[n]$.

(c) If the input signal is $x[n] = (\frac{1}{2})^n u[n]$, find the FIR filter that produces the output signal $y[n] = \delta[n-1]$.

P-5.17 It is not always possible to solve the *deconvolution* process for a given input–output pair. For example, show that there is no FIR filter that can process the input $x[n] = \delta[n] + \delta[n-1]$ to give the output $y[n] = \delta[n]$.

P-5.18 Suppose that three systems are connected in cascade. In other words, the output of $\mathcal{S}_1$ is the input to $\mathcal{S}_2$, and the output of $\mathcal{S}_2$ is the input to $\mathcal{S}_3$. The three systems are specified by the following difference equations:

$$\begin{aligned} \mathcal{S}_1: &\quad y_1[n] = x_1[n] + x_1[n-1] \\ \mathcal{S}_2: &\quad y_2[n] = x_2[n] - 2x_2[n-2] \\ \mathcal{S}_3: &\quad y_3[n] = x_3[n-1] - x_3[n-2] \end{aligned}$$

We wish to determine the equivalent system that is a single operation from the input $x[n]$ (into $\mathcal{S}_1$) to the output $y[n]$, which is the output of $\mathcal{S}_3$.

(a) Determine the impulse response $h[n]$ of the overall system, that is, find $h[n]$ so that $y[n] = x[n] * h[n]$.

(b) Write one difference equation that defines the overall system in terms of $x[n]$ and $y[n]$ only.

(c) If a new system is defined as $(\mathcal{S}_1 + \mathcal{S}_2) * \mathcal{S}_3$ determine its $h[n]$. Thus $x_1[n] = x_2[n] = x[n]$, $x_3[n] = y_2[n] + y_1[n]$, and $y[n] = y_3[n]$.

P-5.19 Consider a system implemented by the following MATLAB program:

```
% xx.mat is a binary file containing the
%    vector of input samples called "xx"
load xx
yy1 = conv(ones(1,4),xx);
yy2 = conv([1, -1, -1, 1],xx);
ww = yy1 + yy2;
yy = conv(ones(1,3),ww);
```

The overall system from input xx to output yy is an LTI system composed of three LTI systems.

(a) Draw a block diagram of the system that is implemented by the program given above. Be sure to indicate the impulse responses and difference equations of each of the component systems.

(b) The overall system is an LTI system. What is its impulse response, and what is the difference equation that is satisfied by the input $x[n]$ and the output $y[n]$?

CHAPTER 6

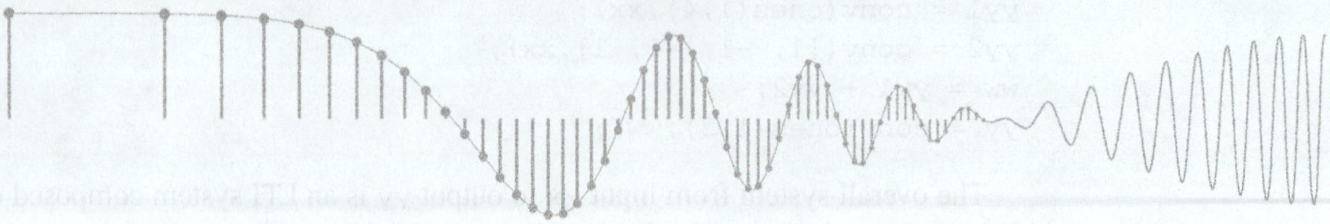

Frequency Response of FIR Filters

Chapter 5 introduced the class of FIR discrete-time systems. We showed that the weighted running average of a finite number of input sequence values defines a discrete-time system, and that such systems are linear and time-invariant (LTI). We also showed that the impulse response of an FIR system completely defines the system. In this chapter, we introduce the concept of the ***frequency response*** of an LTI FIR filter and show that the frequency response and impulse response are uniquely related.

It is remarkable that for LTI systems, when the input is a complex sinusoid, the corresponding output signal is another complex sinusoid of exactly the same frequency, but with different magnitude and phase. All LTI systems possess this ***sinusoid-in gives sinusoid-out*** property. The frequency-response function, when plotted over all frequencies, summarizes the response of an LTI system by giving the magnitude and phase change experienced by all possible sinusoids. Furthermore, since LTI systems obey the principle of superposition, the frequency-response function is a complete characterization of the behavior of the system for any input that can be represented as a sum of sinusoids. Since, as we will show in Chapter 7, almost any discrete-time signal can be represented by a superposition of sinusoids, the frequency response is sufficient therefore to represent the system for almost any input signal.

6-1 Sinusoidal Response of FIR Systems

Linear time-invariant systems behave in a particularly simple way when the input is a discrete-time complex exponential. To see this, consider an FIR system

$$y[n] = \sum_{k=0}^{M} b_k x[n-k] \tag{6.1}$$

and assume that the input is a complex exponential signal[1] with normalized radian frequency $\hat{\omega}$, amplitude A, and phase ϕ

$$x[n] = Ae^{j\varphi}e^{j\hat{\omega}n} \qquad -\infty < n < \infty$$

For such inputs, the corresponding output is

$$\begin{aligned} y[n] &= \sum_{k=0}^{M} b_k\, Ae^{j\varphi}e^{j\hat{\omega}(n-k)} \\ &= \left(\sum_{k=0}^{M} b_k e^{-j\hat{\omega}k}\right) Ae^{j\varphi}e^{j\hat{\omega}n} \\ &= \mathcal{H}(\hat{\omega})Ae^{j\varphi}e^{j\hat{\omega}n} \qquad -\infty < n < \infty \end{aligned} \tag{6.2}$$

where

$$\mathcal{H}(\hat{\omega}) = \sum_{k=0}^{M} b_k e^{-j\hat{\omega}k} \tag{6.3}$$

Because we have represented the frequency of the complex exponential signal as the general symbol $\hat{\omega}$, we have obtained an expression (6.3) that is a function of $\hat{\omega}$. In other words, (6.3) describes the response of the LTI system to a complex exponential signal of *any* frequency $\hat{\omega}$. The quantity $\mathcal{H}(\hat{\omega})$ defined by (6.3) is therefore called the ***frequency-response function*** for the system. Generally, we shorten this to ***frequency response***.

However, there is an issue of notation for the frequency response that is dictated by consistency with the z-transform (to be introduced in Chapter 9). In (6.3) the righthand side contains powers of the complex exponential $e^{j\hat{\omega}}$. This is true of many expressions

[1] Recall that this discrete-time signal could have been obtained by sampling the continuous-time signal $x(t) = Ae^{j\varphi}e^{j\omega t}$. If $x[n] = x(nT_s)$, then ω and $\hat{\omega}$ are related by $\hat{\omega} = \omega T_s$, where T_s is the sampling period.

for the frequency response. Therefore, we elect to use the notation $H(e^{j\hat{\omega}})$ instead of $\mathcal{H}(\hat{\omega})$ to emphasize the ubiquity of $e^{j\hat{\omega}}$. The function $H(e^{j\hat{\omega}})$ still depends on the variable $\hat{\omega}$. Furthermore, since the impulse response sequence of an FIR filter is the same as the sequence of filter coefficients, we can express the frequency response in terms of either the filter coefficients b_k or the impulse response $h[k]$, that is,

Frequency Response of an FIR System

$$H(e^{j\hat{\omega}}) = \sum_{k=0}^{M} b_k e^{-j\hat{\omega}k} = \sum_{k=0}^{M} h[k]e^{-j\hat{\omega}k} \tag{6.4}$$

Several important points can be made about (6.2) and (6.4). First of all, the precise interpretation of (6.2) is as follows: When the input is a discrete-time complex exponential signal, the output of an LTI FIR filter is also a discrete-time complex exponential signal with a different complex amplitude, but the same frequency $\hat{\omega}$. The frequency response multiplies the complex exponential input signal, thereby changing the complex amplitude. While it is tempting to express this fact by the mathematical statement $y[n] = H(e^{j\hat{\omega}})x[n]$, it is strongly recommended that this should never be done because it is too easy to forget that the mathematical statement is true *only* for complex exponential signals of frequency $\hat{\omega}$. The statement $y[n] = H(e^{j\hat{\omega}})x[n]$ is *meaningless* for any signal other than signals of precisely the form $x[n] = Ae^{j\varphi}e^{j\hat{\omega}n}$. *It is very important to understand this point.*

A second important point is that the frequency response $H(e^{j\hat{\omega}})$ is complex-valued so it can be expressed either as $H(e^{j\hat{\omega}}) = |H(e^{j\hat{\omega}})|e^{j\angle H(e^{j\hat{\omega}})}$ or as $H(e^{j\hat{\omega}}) = \Re\{H(e^{j\hat{\omega}})\} + j\Im\{H(e^{j\hat{\omega}})\}$. The magnitude and angle form of the frequency response is the most convenient form, since the complex multiplication needed in (6.2) is most conveniently accomplished in polar form. The effect of the LTI system on the magnitude (A) and phase (φ) of the input complex exponential signal is determined completely by the frequency response $H(e^{j\hat{\omega}})$. Specifically, if the input is $x[n] = Ae^{j\varphi}e^{j\hat{\omega}n}$, then using the polar form of $H(e^{j\hat{\omega}})$ gives the result

$$\begin{aligned} y[n] &= \left(|H(e^{j\hat{\omega}})|e^{j\angle H(e^{j\hat{\omega}})}\right) Ae^{j\varphi}e^{j\hat{\omega}n} \\ &= (|H(e^{j\hat{\omega}})| \cdot A)e^{j(\angle H(e^{j\hat{\omega}})+\varphi)}e^{j\hat{\omega}n} \end{aligned} \tag{6.5}$$

The magnitude of $H(e^{j\hat{\omega}})$ multiplies the signal amplitude (A) which changes the size of the signal. Thus, $|H(e^{j\hat{\omega}})|$ is usually referred to as the *gain* of the system.[2] The angle $\angle H(e^{j\hat{\omega}})$ simply adds to the phase (φ) of the input, thereby producing additional phase shift in the complex exponential signal.

[2]The gain can be less than or greater than 1, even though the word suggests greater than.

EXAMPLE 6-1 Formula for the Frequency Response

Consider an LTI system for which the difference equation coefficients are $\{b_k\} = \{1,\ 2,\ 1\}$. Substituting into (6.3) gives

$$H(e^{j\hat{\omega}}) = 1 + 2e^{-j\hat{\omega}} + e^{-j\hat{\omega}2}$$

To obtain formulas for the magnitude and phase of the frequency response of this FIR filter, we can manipulate the equation as follows:

$$\begin{aligned} H(e^{j\hat{\omega}}) &= 1 + 2e^{-j\hat{\omega}} + e^{-j\hat{\omega}2} \\ &= e^{-j\hat{\omega}}\left(e^{j\hat{\omega}} + 2 + e^{-j\hat{\omega}}\right) \\ &= e^{-j\hat{\omega}}\left(2 + 2\cos\hat{\omega}\right) \end{aligned}$$

Since $(2 + 2\cos\hat{\omega}) \geq 0$ for frequencies $-\pi < \hat{\omega} \leq \pi$, the magnitude is $|H(e^{j\hat{\omega}})| = (2 + 2\cos\hat{\omega})$ and the phase is $\angle H(e^{j\hat{\omega}}) = -\hat{\omega}$.

EXAMPLE 6-2 Complex Exponential Input

Consider the complex input $x[n] = 2e^{j\pi/4}e^{j(\pi/3)n}$ whose frequency is $\hat{\omega} = \pi/3$. If this signal is the input to the system of Example 6-1, then we must evaluate the frequency response at the input frequency in order to use the frequency-domain method defined in (6.5). The evaluation gives: $|H(e^{j\pi/3})| = 2 + 2\cos(\pi/3) = 3$ and $\angle H(e^{j\pi/3}) = -\pi/3$. Therefore, the output of the system for the given input is

$$\begin{aligned} y[n] &= 3e^{-j\pi/3} \cdot 2e^{j\pi/4}e^{j\pi n/3} \\ &= (3 \cdot 2) \cdot e^{(j\pi/4 - j\pi/3)}e^{j\pi n/3} \\ &= 6e^{-j\pi/12}e^{j\pi n/3} = 6e^{j\pi/4}e^{j\pi(n-1)/3} \end{aligned}$$

Thus, for this system and the given input $x[n]$, the output is equal to the input multiplied by 3, and the phase shift of $-\pi/3$ corresponds to a delay of one sample.

EXERCISE 6.1

When the sequence of coefficients is symmetrical ($b_0 = b_M$, $b_1 = b_{M-1}$, etc.), the frequency response can be manipulated as in Example 6-1. Following the style of that example, show that the frequency response of an FIR filter with coefficients $\{b_k\} = \{1,\ -2,\ 4,\ -2,\ 1\}$ can be expressed as

$$H(e^{j\hat{\omega}}) = [4 - 4\cos(\hat{\omega}) + 2\cos(2\hat{\omega})]e^{-j2\hat{\omega}}$$

6-2 Superposition and the Frequency Response

The principle of superposition makes it very easy to find the output of an LTI system if the input is a sum of complex exponential signals. This is why the frequency response is so important in the analysis and design of LTI systems. As an example, suppose that the input to an LTI system is a cosine wave with a specific normalized frequency $\hat{\omega}_1$ plus a DC level,

$$x[n] = A_0 + A_1 \cos(\hat{\omega}_1 n + \varphi_1)$$

If we represent the signal in terms of complex exponentials, the signal is composed of three complex exponential signals, with frequencies $\hat{\omega} = 0$, $\hat{\omega}_1$, and $-\hat{\omega}_1$, so

$$x[n] = A_0 e^{j0n} + \frac{A_1}{2} e^{j\varphi_1} e^{j\hat{\omega}_1 n} + \frac{A_1}{2} e^{-j\varphi_1} e^{-j\hat{\omega}_1 n}$$

By superposition, we can determine the output due to each term separately and then add them to obtain the output $y[n]$ corresponding to $x[n]$. Because the components of the input signal are all complex exponential signals, it is easy to find their respective outputs if we know the frequency response of the system; we just multiply each component by $H(e^{j\hat{\omega}})$ evaluated at the corresponding frequency to obtain

$$y[n] = H(e^{j0}) A_0 e^{j0n} + H(e^{j\hat{\omega}_1}) \frac{A_1}{2} e^{j\varphi_1} e^{j\hat{\omega}_1 n} + H(e^{-j\hat{\omega}_1}) \frac{A_1}{2} e^{-j\varphi_1} e^{-j\hat{\omega}_1 n}$$

Note that we have used the fact that a constant signal is a complex exponential with $\hat{\omega} = 0$. If we express $H(e^{j\hat{\omega}_1})$ as $H(e^{j\hat{\omega}_1}) = |H(e^{j\hat{\omega}_1})| e^{j\angle H(e^{j\hat{\omega}_1})}$, then the algebraic steps[3] in (6.6) show that $y[n]$ can finally be expressed as a modified cosine signal plus a constant.

$$\begin{aligned}
y[n] &= H(e^{j0}) A_0 e^{j0n} + H(e^{j\hat{\omega}_1}) \frac{A_1}{2} e^{j\varphi_1} e^{j\hat{\omega}_1 n} + H^*(e^{j\hat{\omega}_1}) \frac{A_1}{2} e^{-j\varphi_1} e^{-j\hat{\omega}_1 n} \\
&= H(e^{j0}) A_0 + |H(e^{j\hat{\omega}_1})| e^{j\angle H(e^{j\hat{\omega}_1})} \frac{A_1}{2} e^{j\varphi_1} e^{j\hat{\omega}_1 n} \\
&\quad + |H(e^{j\hat{\omega}_1})| e^{-j\angle H(e^{j\hat{\omega}_1})} \frac{A_1}{2} e^{-j\varphi_1} e^{-j\hat{\omega}_1 n} \\
&= H(e^{j0}) A_0 + |H(e^{j\hat{\omega}_1})| \frac{A_1}{2} e^{j(\hat{\omega}_1 n + \varphi_1 + \angle H(e^{j\hat{\omega}_1}))} \\
&\quad + |H(e^{j\hat{\omega}_1})| \frac{A_1}{2} e^{-j(\hat{\omega}_1 n + \varphi_1 + \angle H(e^{j\hat{\omega}_1}))} \\
&= H(e^{j0}) A_0 + |H(e^{j\hat{\omega}_1})| A_1 \cos\left(\hat{\omega}_1 n + \varphi_1 + \angle H(e^{j\hat{\omega}_1})\right)
\end{aligned} \tag{6.6}$$

Notice that the magnitude and phase change of the cosine input signal are taken from the positive frequency part of $H(e^{j\hat{\omega}})$, but also notice that it was crucial to express $x[n]$ as a

[3] It is assumed that $H(e^{j\hat{\omega}})$ has the ***conjugate-symmetry*** property $H(e^{-j\hat{\omega}}) = H^*(e^{j\hat{\omega}})$, which is always true when the filter coefficients are real. (See Section 6-4.3.)

sum of complex exponentials and then use the frequency response to find the output due to each complex component separately.

EXAMPLE 6-3 Cosine Input

For the FIR filter with coefficients $\{b_k\} = \{1,\ 2,\ 1\}$, find the output when the input is

$$x[n] = 3\cos\left(\frac{\pi}{3}n - \frac{\pi}{2}\right)$$

The frequency response of the system was determined in Example 6-1 to be

$$H(e^{j\hat{\omega}}) = (2 + 2\cos\hat{\omega})e^{-j\hat{\omega}}$$

Note that $H(e^{-j\hat{\omega}}) = H^*(e^{j\hat{\omega}})$, so $H(e^{j\hat{\omega}})$ has conjugate symmetry. Solution of this problem requires just one evaluation of $H(e^{j\hat{\omega}})$ at the frequency $\hat{\omega} = \pi/3$:

$$\begin{aligned} H(e^{j\pi/3}) &= e^{-j\pi/3}\,(2 + 2\cos(\pi/3)) \\ &= e^{-j\pi/3}\left(2 + 2(\tfrac{1}{2})\right) = 3e^{-j\pi/3} \end{aligned}$$

Therefore, the magnitude is $|H(e^{j\pi/3})| = 3$ and the phase is $\angle H(e^{j\pi/3}) = -\pi/3$, so the output is

$$\begin{aligned} y[n] &= (3)(3)\cos\left(\frac{\pi}{3}n - \frac{\pi}{3} - \frac{\pi}{2}\right) \\ &= 9\cos\left(\frac{\pi}{3}(n-1) - \frac{\pi}{2}\right) \end{aligned}$$

DEMO
DLTI Demo

Notice that the magnitude of the frequency response multiplies the amplitude of the cosine signal, and the phase angle of the frequency response adds to the phase of the cosine signal. This problem could also be studied and solved with the `dltidemo` MATLAB GUI.

LAB
Hearing Test via Frequency Response

If the input signal consists of the sum of many complex exponential signals, the frequency response can be applied to find the output due to each component separately, and the results added to determine the total output. This is the principle of superposition at work. If we can find a representation for a signal in terms of complex exponentials, the frequency response gives a simple and highly intuitive means for determining what an LTI system does to that input signal. In the general case, if the input to an LTI system is a real signal and can be represented as

$$\begin{aligned} x[n] &= X_0 + \sum_{k=1}^{N}\left(\frac{X_k}{2}e^{j\hat{\omega}_k n} + \frac{X_k^*}{2}e^{-j\hat{\omega}_k n}\right) \\ &= X_0 + \sum_{k=1}^{N}|X_k|\cos(\hat{\omega}_k n + \angle X_k) \end{aligned}$$

then the corresponding output is $y[n]$ in (6.7), if $H(e^{-j\hat{\omega}}) = H^*(e^{j\hat{\omega}})$, which is true when the filter coefficients are real.

$$\begin{aligned} y[n] &= H(e^{j0})X_0 + \sum_{k=1}^{N}\left(H(e^{j\hat{\omega}_k})\frac{X_k}{2}e^{j\hat{\omega}_k n} + H(e^{-j\hat{\omega}_k})\frac{X_k^*}{2}e^{-j\hat{\omega}_k n}\right) \\ &= H(e^{j0})X_0 + \sum_{k=1}^{N}|H(e^{j\hat{\omega}_k})||X_k|\cos\left(\hat{\omega}_k n + \angle X_k + \angle H(e^{j\hat{\omega}_k})\right) \end{aligned} \quad (6.7)$$

For $y[n]$ in (6.7), each individual complex exponential component in $x[n]$ has been modified by the frequency response evaluated at the frequency of that component.

EXAMPLE 6-4 Three Sinusoidal Inputs

For the FIR filter with coefficients $\{b_k\} = \{1, 2, 1\}$, find the output when the input is

$$x[n] = 4 + 3\cos\left(\frac{\pi}{3}n - \frac{\pi}{2}\right) + 3\cos\left(\frac{7\pi}{8}n\right) \quad (6.8)$$

There are three freqeuencies, 0, $\pi/3$, and $7\pi/8$, so the output has three sinusoids at those same frequencies. The frequency response of the system was determined in Example 6-1, and is the same as the frequency response in Example 6-3. The input in this example differs from that of Example 6-3 by the addition of a constant (DC) term and an additional cosine signal of frequency $7\pi/8$. The solution by superposition therefore requires that we evaluate $H(e^{j\hat{\omega}})$ at frequencies 0, $\pi/3$, and $7\pi/8$, giving

$$\begin{aligned} H(e^{j0}) &= 4 \\ H(e^{j\pi/3}) &= 3e^{-j\pi/3} \\ H(e^{j7\pi/8}) &= 0.1522e^{-j7\pi/8} \end{aligned}$$

Therefore, the output is

$$\begin{aligned} y[n] &= 4\cdot 4 + 3\cdot 3\cos\left(\frac{\pi}{3}n - \frac{\pi}{3} - \frac{\pi}{2}\right) + 0.1522\cdot 3\cos\left(\frac{7\pi}{8}n - \frac{7\pi}{8}\right) \\ &= 16 + 9\cos\left(\frac{\pi}{3}(n-1) - \frac{\pi}{2}\right) + 0.4567\cos\left(\frac{7\pi}{8}(n-1)\right) \end{aligned}$$

Notice that, in this case, the DC component is multiplied by 4, the component at frequency $\hat{\omega} = \pi/3$ is multiplied by 3, but the component at frequency $\hat{\omega} = 7\pi/8$ is multiplied by 0.1522. Because the frequency-response magnitude (gain) is so small at frequency $\hat{\omega} = 7\pi/8$, the component at this frequency is essentially *filtered out* of the input signal.

The examples of this section illustrate an approach to solving problems that is often called the ***frequency-domain*** approach. As these examples show, we do not need to deal with the ***time-domain*** description (i.e., the difference equation or impulse response) of the system when the input is a complex exponential signal. We can work exclusively with the frequency-domain description (i.e., the frequency response), if we think about how the spectrum of the signal is modified by the system rather than considering what happens to the individual samples of the input signal. We will have ample opportunity to visit both the time domain and the frequency domain in the remainder of this chapter.

DEMO
Introduction to FIR Filtering

6-3 Steady-State and Transient Response

In Section 6-1, we showed that if the input is

$$x[n] = Xe^{j\hat{\omega}n} \qquad -\infty < n < \infty \tag{6.9}$$

where $X = Ae^{j\varphi}$, then the corresponding output of an LTI FIR system is

$$y[n] = H(e^{j\hat{\omega}})Xe^{j\hat{\omega}n} \qquad -\infty < n < \infty \tag{6.10}$$

where

$$H(e^{j\hat{\omega}}) = \sum_{k=0}^{M} b_k e^{-j\hat{\omega}k} \tag{6.11}$$

In (6.9), the condition that $x[n]$ be a complex exponential signal existing over $-\infty < n < \infty$ is important. Without this condition, we cannot obtain the simple result of (6.10). However, this condition appears to be somewhat impractical. In any practical implementation, we surely would not have actual input signals that exist back to $-\infty$! Fortunately, we can relax the condition that the complex exponential input be defined over the doubly infinite interval and still take advantage of the convenience of (6.10). To see this, consider the following "suddenly applied" complex exponential signal that starts at $n = 0$ and is nonzero only for $0 \le n$:

$$x[n] = Xe^{j\hat{\omega}n}u[n] = \begin{cases} Xe^{j\hat{\omega}n} & 0 \le n \\ 0 & n < 0 \end{cases} \tag{6.12}$$

Note that multiplication by the unit-step signal is a convenient way to impose the suddenly applied condition. The output of an LTI FIR system for this input is obtained by using the definition of $x[n]$ in (6.12) in the FIR difference equation:

$$y[n] = \sum_{k=0}^{M} b_k \underbrace{Xe^{j\hat{\omega}(n-k)}u[n-k]}_{x[n-k]} \tag{6.13}$$

By considering different values of n and the fact that $u[n-k]=0$ for $k>n$, it follows that the sum in (6.13) can be expressed as three cases

$$y[n] = \begin{cases} 0 & n < 0 \\ \left(\sum_{k=0}^{n} b_k e^{-j\hat{\omega}k}\right) X e^{j\hat{\omega}n} & 0 \le n < M \\ \left(\sum_{k=0}^{M} b_k e^{-j\hat{\omega}k}\right) X e^{j\hat{\omega}n} & M \le n \end{cases} \tag{6.14}$$

That is, when the complex exponential signal is suddenly applied, the output can be considered to be defined over three distinct regions. In the first region, $n < 0$, the input is zero, and therefore the corresponding output is zero, too. The second region is a transition region whose length is M samples (i.e., the *order* of the FIR system). In this region, the complex multiplier of $e^{j\hat{\omega}n}$ depends upon n. This region is often called the ***transient*** part of the output. In the third region, $M \le n$, the output is identical to the output that would be obtained if the input were defined over the doubly infinite interval, that is,

$$y[n] = H(e^{j\hat{\omega}}) X e^{j\hat{\omega}n} \qquad M \le n \tag{6.15}$$

This part of the output is generally called the ***steady-state*** part. While we have specified that the steady-state part exists for all $n \ge M$, it should be clear that (6.15) holds only as long as the input remains equal to $Xe^{j\hat{\omega}n}$. If, at some time $n > M$, the input changes frequency or the amplitude becomes zero, another transient region occurs.

EXAMPLE 6-5 Steady-State Output

A simple example illustrates the above discussion. Consider the system of Exercise 6.1, whose filter coefficients are the sequence $\{b_k\} = \{1, -2, 4, -2, 1\}$. The frequency response of this system is

$$H(e^{j\hat{\omega}}) = [4 - 4\cos(\hat{\omega}) + 2\cos(2\hat{\omega})]e^{-j2\hat{\omega}}$$

If the input is the suddenly applied cosine signal

$$x[n] = \cos(0.2\pi n - \pi)u[n]$$

we can represent it as the sum of two suddenly applied complex exponential signals. Therefore, the frequency response can be used as discussed in Section 6-2 to determine the corresponding steady-state output. Since $H(e^{j\hat{\omega}})$ at $\hat{\omega} = 0.2\pi$ is

$$H(e^{j\hat{\omega}}) = [4 - 4\cos(0.2\pi) + 2\cos(0.4\pi)]e^{-j2(0.2\pi)} = 1.382e^{-j(0.2\pi)2}$$

and $M = 4$, the steady-state output is

$$y[n] = 1.382\cos(0.2\pi(n-2) - \pi) \qquad 4 \le n$$

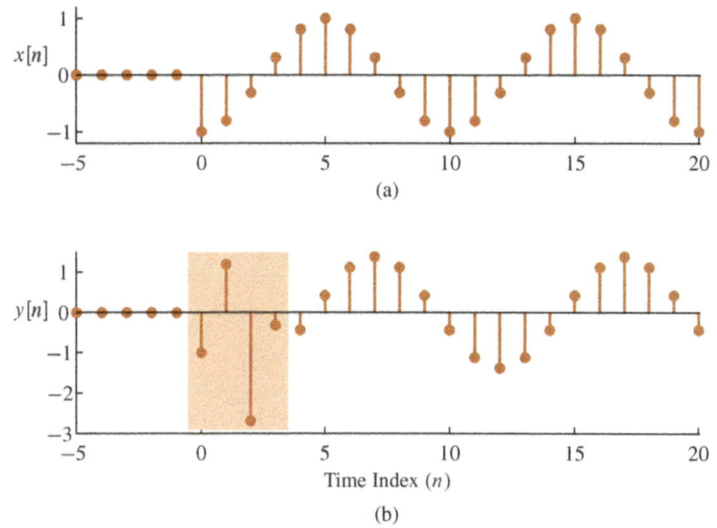

Figure 6-1 (a) Input $x[n] = \cos(0.2\pi n - \pi)u[n]$ and (b) corresponding output $y[n]$ for the FIR filter with coefficients $\{1, -2, 4, -2, 1\}$. The transient region is the shaded area in (b). (Note the different vertical scales.)

The frequency response has allowed us to find a simple expression for the output everywhere in the steady-state region. If we desire the values of the output in the transient region, we could compute them using the difference equation for the system.

The input and output signals for this example are shown in Fig. 6-1. Since $M = 4$ for this system, the transient region is $0 \le n \le 3$ (indicated by the shaded region), and the steady-state region is $n \ge 4$. Also note that, as predicted by the steady-state analysis above, the signal in the steady-state region is simply a scaled (by 1.382) and shifted (by two samples) version of the input.

6-4 Properties of the Frequency Response

The frequency response $H(e^{j\hat{\omega}})$ is a complex-valued function of the normalized frequency variable $\hat{\omega}$. This function has interesting properties that often can be used to simplify analysis.

6-4.1 Relation to Impulse Response and Difference Equation

$H(e^{j\hat{\omega}})$ can be calculated directly from the filter coefficients $\{b_k\}$. If (6.1) is compared to (6.4), we see that, given the difference equation, it is simple to write down an expression for $H(e^{j\hat{\omega}})$ by noting that each term $b_k x[n-k]$ in (6.1) corresponds to a term $b_k e^{-j\hat{\omega}k}$ or $h[k]e^{-j\hat{\omega}k}$ in (6.4), and vice versa. Likewise, $H(e^{j\hat{\omega}})$ can be determined directly from the impulse response since the impulse response of the FIR system consists of the sequence of filter coefficients (i.e., $h[k] = b_k$ for $k = 0, 1, \ldots, M$). To emphasize this point, we can write the correspondence

Time Domain	$\longleftrightarrow$	*Frequency Domain*
$h[n] = \sum_{k=0}^{M} h[k]\delta[n-k]$	$\longleftrightarrow$	$H(e^{j\hat{\omega}}) = \sum_{k=0}^{M} h[k]e^{-j\hat{\omega}k}$

The process of going from the FIR difference equation, or impulse response, to the frequency response is straightforward. It is also simple to go from the frequency response to the FIR difference equation or to the impulse response if we express $H(e^{j\hat{\omega}})$ in terms of powers of $e^{-j\hat{\omega}}$. These points are illustrated by the following two examples.

EXAMPLE 6-6 $h[n] \longleftrightarrow H(e^{j\hat{\omega}})$

Consider the FIR filter defined by the impulse response

$$h[n] = -\delta[n] + 3\delta[n-1] - \delta[n-2]$$

The values of $h[n]$ are the filter coefficients $\{b_k\} = \{-1, 3, -1\}$, so the difference equation corresponding to this impulse response is

$$y[n] = -x[n] + 3x[n-1] - x[n-2]$$

Thus, the frequency response of this system is

$$H(e^{j\hat{\omega}}) = -1 + 3e^{-j\hat{\omega}} - e^{-j2\hat{\omega}}$$

EXAMPLE 6-7 Difference Equation from $H(e^{j\hat{\omega}})$

Often the frequency response is given by an equation containing sines or cosines, for example,

$$H(e^{j\hat{\omega}}) = e^{-j\hat{\omega}}\left(3 - 2\cos\hat{\omega}\right)$$

To obtain the FIR difference equation, it is necessary to write $H(e^{j\hat{\omega}})$ using powers of $e^{-j\hat{\omega}}$. In this example, we can use the inverse Euler formula $\cos\hat{\omega} = \frac{1}{2}(e^{j\hat{\omega}} + e^{-j\hat{\omega}})$, and then obtain

$$\begin{aligned} H(e^{j\hat{\omega}}) &= e^{-j\hat{\omega}}\left[3 - 2\left(\frac{e^{j\hat{\omega}} + e^{-j\hat{\omega}}}{2}\right)\right] \\ &= -1 + 3e^{-j\hat{\omega}} - e^{-j\hat{\omega}2} \end{aligned}$$

Now, it should be easy to see that $H(e^{j\hat{\omega}})$ corresponds to the following FIR difference equation:

$$y[n] = -x[n] + 3x[n-1] - x[n-2]$$

The impulse response, likewise, is easy to determine directly from $H(e^{j\hat{\omega}})$, when expressed in terms of powers of $e^{-j\hat{\omega}}$.

EXERCISE 6.2 Use the inverse Euler formula for sines to find the impulse response and difference equation for $H(e^{j\hat{\omega}}) = 2j\sin(\hat{\omega}/2)e^{-j\hat{\omega}/2}$.

6-4.2 Periodicity of $H(e^{j\hat{\omega}})$

An important property of a discrete-time LTI system is that its frequency response $H(e^{j\hat{\omega}})$ is always a periodic function with period 2π. This can be seen by considering a frequency $\hat{\omega} + 2\pi$ where $\hat{\omega}$ is any frequency. Substituting into (6.3) gives

$$H(e^{j(\hat{\omega}+2\pi)}) = \sum_{k=0}^{M} b_k e^{-j(\hat{\omega}+2\pi)k}$$

$$= \sum_{k=0}^{M} b_k e^{-j\hat{\omega}k} \underbrace{e^{-j2\pi k}}_{=1} = H(e^{j\hat{\omega}})$$

since $e^{-j2\pi k} = 1$ when k is an integer. It is not surprising that $H(e^{j\hat{\omega}})$ should have this property, since, as we have seen in Chapter 4, a change in the frequency of a discrete-time complex exponential by 2π is not detectable, that is,

$$x[n] = Xe^{j(\hat{\omega}+2\pi)n} = Xe^{j\hat{\omega}n}\overset{1}{\cancel{e^{j2\pi n}}} = Xe^{j\hat{\omega}n}$$

In other words, two complex exponential signals with frequencies differing by 2π have identical time samples, so there is no reason to expect a discrete-time system to behave differently for two such frequencies. For this reason, it is always sufficient to specify the frequency response only over an interval of one period, usually, $-\pi < \hat{\omega} \leq \pi$.

6-4.3 Conjugate Symmetry

The frequency response $H(e^{j\hat{\omega}})$ is complex valued, but usually has a symmetry in its magnitude and phase that allows us to focus on just the interval $0 \leq \hat{\omega} \leq \pi$ when plotting. This is the property of ***conjugate symmetry***

$$H(e^{-j\hat{\omega}}) = H^*(e^{j\hat{\omega}}) \tag{6.16}$$

which is true whenever the filter coefficients are real (i.e., $b_k = b_k^*$, or, equivalently $h[k] = h^*[k]$). We can prove this property for the FIR case as follows:

$$H^*(e^{j\hat{\omega}}) = \left(\sum_{k=0}^{M} b_k e^{-j\hat{\omega}k}\right)^*$$

$$= \sum_{k=0}^{M} b_k^* e^{+j\hat{\omega}k}$$

$$= \sum_{k=0}^{M} b_k e^{-j(-\hat{\omega})k} = H(e^{-j\hat{\omega}})$$

The conjugate-symmetry property implies that the magnitude function is an even function of $\hat{\omega}$ and the phase is an odd function, so we write

$$|H(e^{-j\hat{\omega}})| = |H(e^{j\hat{\omega}})|$$

$$\angle H(e^{-j\hat{\omega}}) = -\angle H(e^{j\hat{\omega}})$$

Similarly, the real part is an even function of $\hat{\omega}$ and the imaginary part is an odd function, so

$$\Re\{H(e^{-j\hat{\omega}})\} = \Re\{H(e^{j\hat{\omega}})\}$$

$$\Im\{H(e^{-j\hat{\omega}})\} = -\Im\{H(e^{j\hat{\omega}})\}$$

As a result, plots of the frequency response are often shown only over half a period, $0 \leq \hat{\omega} \leq \pi$, because the negative frequency region can be constructed by symmetry. These symmetries are illustrated by the plots in Section 6-5.

EXERCISE 6.3 Prove that the magnitude is an even function of $\hat{\omega}$ and the phase is an odd function of $\hat{\omega}$ for a conjugate-symmetric frequency response. Start by proving that the magnitude-squared $H(e^{j\hat{\omega}})H^*(e^{j\hat{\omega}})$ is an even function.

6-5 Graphical Representation of the Frequency Response

DEMO

Introduction to FIR Filtering

Two important points should be emphasized about the frequency response of an LTI system. First, since the frequency response usually varies with frequency, sinusoids of different frequencies are treated differently by the system. Second, by appropriate choice of the coefficients, b_k, a wide variety of frequency response shapes can be realized. In order to visualize the variation of the frequency response with frequency, it is useful to plot the magnitude $|H(e^{j\hat{\omega}})|$ and phase $\angle H(e^{j\hat{\omega}})$ versus $\hat{\omega}$. Two plots are needed because $H(e^{j\hat{\omega}})$ is complex valued. We will see that the plots tell us at a glance what the system does to complex exponential signals and sinusoids of different frequencies. Several examples are provided in this section to illustrate the value of plotting the frequency response.

6-5.1 Delay System

The delay system is a simple FIR filter given by the difference equation

$$y[n] = x[n - n_0]$$

where the integer n_0 is the amount of delay. It has only one nonzero filter coefficient, $b_{n_0} = 1$, so its frequency response is

$$H(e^{j\hat{\omega}}) = e^{-j\hat{\omega}n_0} \tag{6.17}$$

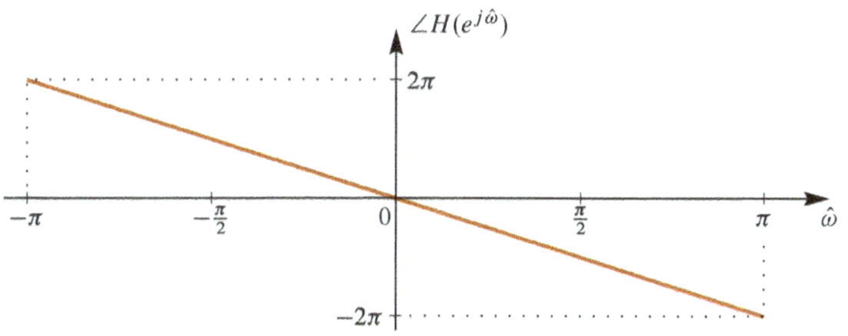

Figure 6-2 Phase response of pure delay ($n_0 = 2$) system, $H(e^{j\hat{\omega}}) = e^{-j2\hat{\omega}}$.

For this system, the two plots of the frequency response are easy to visualize: The magnitude response is one for all frequencies because $|e^{-j\hat{\omega}n_0}| = 1$. The phase plot is a straight line with a slope equal to $-n_0$, as in Fig. 6-2, because $\angle H(e^{j\hat{\omega}}) = -n_0\hat{\omega}$. As a result, we can associate the slope of a linear phase versus frequency with time delay as a general rule in all filters. Since time delay shifts a signal plot horizontally but does not change its shape, we often think of linear phase as an ideal phase response.

6-5.2 First-Difference System

As another simple example, consider the first-difference system

$$y[n] = x[n] - x[n-1]$$

which forms its output by subtracting adjacent values of the input sequence. The frequency response of this LTI system is

$$H(e^{j\hat{\omega}}) = 1 - e^{-j\hat{\omega}}$$

Since $H(e^{j\hat{\omega}}) = 1-(\cos(\hat{\omega})-j\sin(\hat{\omega})) = (1-\cos\hat{\omega})+j\sin\hat{\omega}$, the magnitude and phase can be written from the real and imaginary parts

$$\begin{aligned}|H(e^{j\hat{\omega}})| &= [(1-\cos\hat{\omega})^2 + \sin^2\hat{\omega}]^{1/2} \\ &= [2(1-\cos\hat{\omega})]^{1/2} = 2|\sin(\hat{\omega}/2)| \\ \angle H(e^{j\hat{\omega}}) &= \arctan\left(\frac{\sin\hat{\omega}}{1-\cos\hat{\omega}}\right)\end{aligned}$$

The magnitude and phase are plotted in Fig. 6-3 for $-3\pi < \hat{\omega} < 3\pi$. These extended plots allow us to verify that $H(e^{j\hat{\omega}})$ is periodic with period 2π, and to see the conjugate-symmetry properties discussed in Section 6-4.3. Normally, we would plot the frequency response of a discrete-time system only for $-\pi < \hat{\omega} \leq \pi$, or, when there is conjugate symmetry, $0 \leq \hat{\omega} \leq \pi$.

The utility of the magnitude and phase plots of $H(e^{j\hat{\omega}})$ can be seen even for this simple example. In Fig. 6-3 $H(e^{j0}) = 0$, so we conclude that the system completely removes components with $\hat{\omega} = 0$ (i.e., DC). Furthermore, in the magnitude plot (Fig. 6-3(a)) we can see that the system boosts components at higher frequencies (near $\hat{\omega} = \pi$) relative to the lower frequencies. This sort of filter is called a ***highpass filter***. This sort of behavior

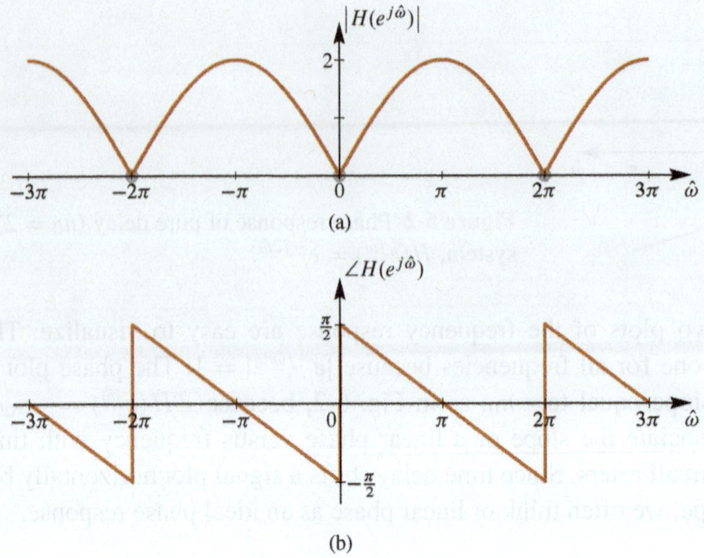

Figure 6-3 (a) Magnitude and (b) phase for $H(e^{j\hat{\omega}}) = 1 - e^{-j\hat{\omega}}$ over three periods showing periodicity and conjugate symmetry of $H(e^{j\hat{\omega}})$.

of a filter in low frequency and high frequency bands is another typical way of thinking about systems in the frequency domain.

The real and imaginary parts and the magnitude and phase can always be derived by standard manipulations of complex numbers. However, there is a simpler approach for getting the magnitude and phase when the sequence of coefficients is either symmetric or antisymmetric about a central point. The following algebraic manipulation of $H(e^{j\hat{\omega}})$ is possible because the $\{b_k\}$ coefficients satisfy the antisymmetry condition:

$$b_k = -b_{M-k} \tag{6.18}$$

where $M = 1$ for the first-difference filter. The trick, which we used previously in Example 6-1 on p. 217, is to factor out an exponential whose phase is half of the filter order $(M/2)$ times $\hat{\omega}$, and then use the inverse Euler formula to combine corresponding positive- and negative-frequency complex exponentials. For the first-difference filter, we exploit (6.18) to write

$$\begin{aligned}
H(e^{j\hat{\omega}}) &= 1 - e^{-j\hat{\omega}} \\
&= e^{-j\hat{\omega}/2}\left(e^{j\hat{\omega}/2} - e^{-j\hat{\omega}/2}\right) \\
&= 2je^{-j\hat{\omega}/2}\sin(\hat{\omega}/2) \\
&= \underbrace{2\sin(\hat{\omega}/2)}_{\text{Magnitude?}}\,\exp(j\underbrace{(\pi/2 - \hat{\omega}/2)}_{\text{Phase?}})
\end{aligned}$$

The form derived for $H(e^{j\hat{\omega}})$ is almost a valid polar form, but since $\sin(\hat{\omega}/2)$ is negative for $-\pi < \hat{\omega} < 0$, we must write the magnitude as $|H(e^{j\hat{\omega}})| = 2|\sin(\hat{\omega}/2)|$, and then absorb the negative sign[4] into the phase response for $-\pi < \hat{\omega} < 0$. Thus the correct phase in polar form is

$$\angle H(e^{j\hat{\omega}}) = \begin{cases} \pi/2 - \hat{\omega}/2 & 0 < \hat{\omega} \le \pi \\ -\pi + \pi/2 - \hat{\omega}/2 & -\pi \le \hat{\omega} < 0 \end{cases}$$

This formula for the phase is consistent with the phase plot in Fig. 6-3(b), which exhibits several straight line segments, but has discontinuities at $\hat{\omega} = 0$ and $\hat{\omega} = \pm 2\pi$. The size of these discontinuities is π, since they correspond to sign changes in $H(e^{j\hat{\omega}})$.

EXAMPLE 6-8 First-Difference Removes DC

Suppose that the input to a first-difference system is $x[n] = 4 + 2\cos(0.3\pi n - \pi/4)$. Since the output is related to the input by the difference equation $y[n] = x[n] - x[n-1]$, it follows that

$$\begin{aligned} y[n] &= 4 + 2\cos(0.3\pi n - \pi/4) - 4 - 2\cos(0.3\pi(n-1) - \pi/4) \\ &= 2\cos(0.3\pi n - \pi/4) - 2\cos(0.3\pi n - 0.55\pi) \end{aligned}$$

From this result, we see that the first-difference system removes the constant value and leaves two cosine signals of the same frequency, which could be combined by phasor addition. However, the solution using the frequency-response function is simpler. Since the first-difference system has frequency response

$$H(e^{j\hat{\omega}}) = 2\sin(\hat{\omega}/2)e^{j(\pi/2 - \hat{\omega}/2)}$$

the output of this system for the given input can be obtained more quickly via

$$y[n] = 4H(e^{j0}) + 2|H(e^{j0.3\pi})|\cos(0.3\pi n - \pi/4 + \angle H(e^{j0.3\pi}))$$

In other words, we need to evaluate the frequency response only at $\hat{\omega} = 0$ and $\hat{\omega} = 0.3\pi$. Since $H(e^{j0}) = 0$ and

$$\begin{aligned} H(e^{j0.3\pi}) &= 2\sin(0.3\pi/2)e^{j(0.5\pi - 0.15\pi)} \\ &= 0.908e^{j0.35\pi} \end{aligned}$$

the output is

$$\begin{aligned} y[n] &= (0.908)(2)\cos(0.3\pi n - \pi/4 + 0.35\pi) \\ &= 1.816\cos(0.3\pi n + 0.1\pi) \end{aligned}$$

6-5.3 A Simple Lowpass Filter

In Examples 6-1, 6-3, and 6-4, the system had frequency response

$$H(e^{j\hat{\omega}}) = 1 + 2e^{-j\hat{\omega}} + e^{-j2\hat{\omega}} = (2 + 2\cos\hat{\omega})e^{-j\hat{\omega}}$$

[4]Remember that $-1 = e^{\pm j\pi}$, so we can add either $+\pi$ or $-\pi$ to the phase for $-\pi < \hat{\omega} < 0$. In this case, we add $-\pi$ so that the resulting phase curve remains between $-\pi$ and $+\pi$ radians for all $\hat{\omega}$.

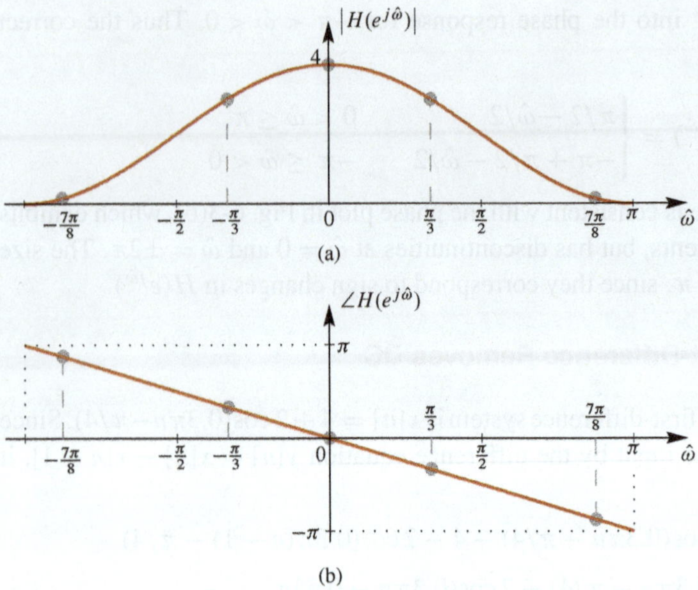

Figure 6-4 Magnitude (a) and phase (b) of system with frequency response $H(e^{j\hat{\omega}}) = (2 + 2\cos\hat{\omega})e^{-j\hat{\omega}}$. Gray dots indicate points where the frequency response is evaluated to calculate the sinusoidal response in Example 6-9.

Since the factor $(2 + 2\cos\hat{\omega}) \geq 0$ for all $\hat{\omega}$, it follows that

$$|H(e^{j\hat{\omega}})| = (2 + 2\cos\hat{\omega})$$

and

$$\angle H(e^{j\hat{\omega}}) = -\hat{\omega}$$

These functions are plotted in Fig. 6-4 for $-\pi < \hat{\omega} \leq \pi$. The phase response (Fig. 6-4(b)) is linear (a straight line) with a slope of -1, so we know that the system introduces a delay of one sample. The magnitude response (Fig. 6-4(a)) shows that the system tends to favor the low frequencies (close to $\hat{\omega} = 0$) with high gain, while it tends to suppress high frequencies (close to $\hat{\omega} = \pi$). In this case, there is a gradual decrease in gain from $\hat{\omega} = 0$ to $\hat{\omega} = \pi$, so that the midrange frequencies receive more gain than the high frequencies, but less than the low frequencies. Filters with magnitude responses that suppress the high frequencies of the input are called ***lowpass filters***.

EXAMPLE 6-9 Lowpass Filter

If we repeat Example 6-4, we can show how the plot of $H(e^{j\hat{\omega}})$ in Fig. 6-4 makes it easy to find the filter's output for sinusoidal inputs. In Example 6-4, the input was

$$x[n] = 4 + 3\cos\left(\frac{\pi}{3}n - \frac{\pi}{2}\right) + 3\cos\left(\frac{7\pi}{8}n\right) \tag{6.19}$$

as shown in Fig. 6-5(a), and the filter coefficients were $\{b_k\} = \{1,\ 2,\ 1\}$. In order to get the output signal, we must evaluate $H(e^{j\hat{\omega}})$ at frequencies 0, $\pi/3$, and $7\pi/8$ giving

$$H(e^{j0}) = 4$$
$$H(e^{j\pi/3}) = 3e^{-j\pi/3}$$
$$H(e^{j7\pi/8}) = 0.1522e^{-j7\pi/8}$$

These values are the points indicated with gray dots on the graphs of Fig. 6-4. As in Example 6-4, the output is

$$y[n] = 4 \cdot 4 + 3 \cdot 3\cos\left(\frac{\pi}{3}n - \frac{\pi}{3} - \frac{\pi}{2}\right) + 0.1522 \cdot 3\cos\left(\frac{7\pi}{8}n - \frac{7\pi}{8}\right)$$

$$= 16 + 9\cos\left(\frac{\pi}{3}(n-1) - \frac{\pi}{2}\right) + 0.4567\cos\left(\frac{7\pi}{8}(n-1)\right)$$

We can see two features in the output signal $y[n]$. The sinusoid with $\hat{\omega} = 7\pi/8$ has a very small magnitude because the magnitude response around $\hat{\omega} = \pi$ is relatively small. Also, the linear phase slope of -1 means that the filter introduces a time delay of one sample which is evident in the second and third terms of $y[n]$.

The output of the simple lowpass filter is the time waveform shown in Fig. 6-5(b). Note that the DC component is indicated in both parts of the figure as a gray horizontal line. The output appears to be the sum of a constant level of 16 plus a cosine that has amplitude 9 and seems to be periodic with period 6. Closer inspection reveals that this is not exactly true because there is a third output component at frequency $\hat{\omega} = 7\pi/8$, which is just barely visible in Fig. 6-5(b). Its size is about 5% of the size of the component with frequency $\hat{\omega} = \pi/3$.

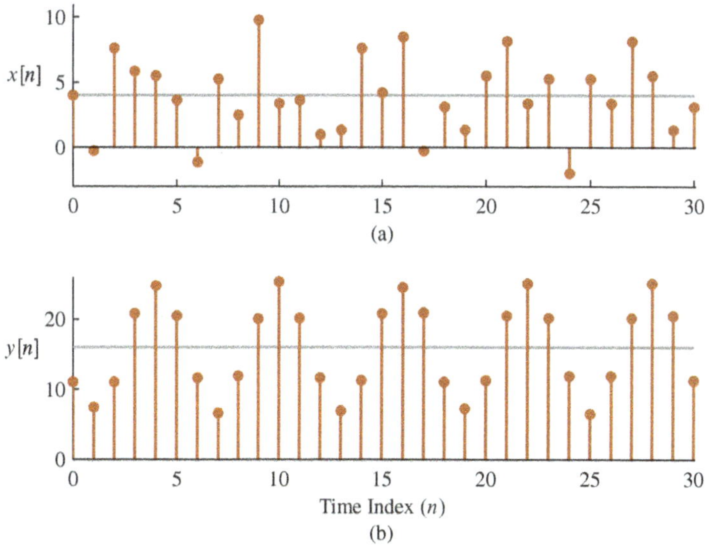

Figure 6-5 Input and output of a system with frequency response $H(e^{j\hat{\omega}}) = (2 + 2\cos\hat{\omega})e^{-j\hat{\omega}}$. (a) Segment of the input signal $x[n]$ given by (6.19), and (b) the corresponding segment of the output.

6-6 Cascaded LTI Systems

DEMO
Cascading FIR Filters

The cascade connection of systems allows us to treat small systems as building blocks in two ways: several low-order systems can be cascaded to create a more complicated high-order system, or a high-order system can be broken down into a few simple low-order systems that are easier to design and analyze. In Section 5-8, we showed that if two LTI systems are connected in cascade (output of the first is input to the second), then the overall impulse response is the convolution of the two individual impulse responses. In other words, the cascade system is equivalent to a single system whose impulse response is the convolution of the two individual impulse responses. In this section, we show that the frequency response of a cascade connection of two LTI systems is simply the product of the individual frequency responses.

Figure 6-6(a) shows the cascade connection of two LTI systems. To find the frequency response of the overall system (from input $x[n]$ to output $y_1[n]$), we let

$$x[n] = e^{j\hat{\omega}n}$$

where $\hat{\omega}$ is a fixed frequency. Then, the output of the first LTI system is

$$w_1[n] = H_1(e^{j\hat{\omega}})e^{j\hat{\omega}n}$$

The signal $w_1[n]$ is a complex exponential and the term $H_1(e^{j\hat{\omega}})$ is a *complex constant* that determines its magnitude and phase. Thus, the output of the second system is

$$y_1[n] = H_2(e^{j\hat{\omega}})\left(H_1(e^{j\hat{\omega}})e^{j\hat{\omega}n}\right) = H_2(e^{j\hat{\omega}})H_1(e^{j\hat{\omega}})e^{j\hat{\omega}n}$$

From a similar analysis of Fig. 6-6(b), it follows that

$$y_2[n] = H_1(e^{j\hat{\omega}})\left(H_2(e^{j\hat{\omega}})e^{j\hat{\omega}n}\right) = H_1(e^{j\hat{\omega}})H_2(e^{j\hat{\omega}})e^{j\hat{\omega}n}$$

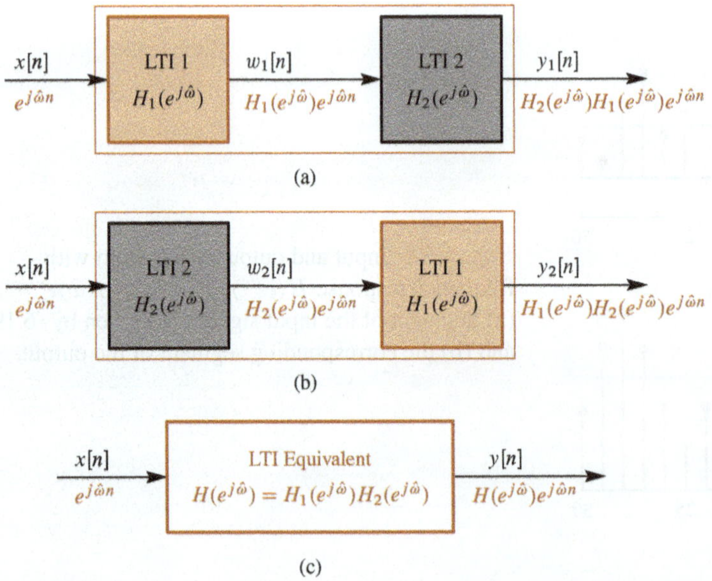

Figure 6-6 Frequency response of three equivalent cascaded LTI systems. All three systems have the same frequency response so that $y[n] = y_1[n] = y_2[n]$ for the same input $x[n]$.

Since $\hat{\omega}$ is a fixed frequency, the multiplication of frequency response *values*, $H_2(e^{j\hat{\omega}})H_1(e^{j\hat{\omega}})$ or $H_1(e^{j\hat{\omega}})H_2(e^{j\hat{\omega}})$, must give the same answer because complex multiplication is commutative, and it follows that $y_1[n] = y_2[n]$. Thus, the two cascade systems are equivalent for the same complex exponential input, and both of them are equivalent to a single LTI system with frequency response

$$H(e^{j\hat{\omega}}) = H_1(e^{j\hat{\omega}})H_2(e^{j\hat{\omega}}) \tag{6.20}$$

The output $y[n]$ of any system with frequency response $H(e^{j\hat{\omega}})$ given by (6.20) is the same as either $y_1[n]$ or $y_2[n]$. This is depicted in Fig. 6-6(c).

DEMO
Cascading FIR Filters

Recall from Section 5-8 on p. 202 that the overall impulse response is $h_1[n]*h_2[n]$. We can conclude therefore that "convolution in the time domain is equivalent to multiplication in the frequency domain." For cascaded systems, convolution of impulse responses corresponds to multiplication of the frequency responses.

Convolution $\longleftrightarrow$ ***Multiplication***

$$h_1[n] * h_2[n] \longleftrightarrow H_1(e^{j\hat{\omega}})H_2(e^{j\hat{\omega}}) \tag{6.21}$$

The correspondence shown in (6.21) is useful because it provides another way of representing and manipulating LTI systems. This is illustrated by the following example.

EXAMPLE 6-10 Frequency Responses in Cascade

Suppose that the first system in a cascade of two FIR LTI systems is defined by the set of coefficients $\{2, 1, 2\}$ and the second system is defined by the coefficients $\{0, 3, 0, -3\}$. Therefore, the frequency responses of the individual systems are

$$H_1(e^{j\hat{\omega}}) = 2 + e^{-j\hat{\omega}} + 2e^{-j\hat{\omega}2}$$

and

$$H_2(e^{j\hat{\omega}}) = 3e^{-j\hat{\omega}} - 3e^{-j\hat{\omega}3}$$

The overall frequency response is obtained by carrying out the multiplication of terms:

$$\begin{aligned} H(e^{j\hat{\omega}}) &= H_1(e^{j\hat{\omega}})H_2(e^{j\hat{\omega}}) \\ &= \left(2+e^{-j\hat{\omega}}+2e^{-j\hat{\omega}2}\right)\left(3e^{-j\hat{\omega}}-3e^{-j\hat{\omega}3}\right) \\ &= 6e^{-j\hat{\omega}}+3e^{-j\hat{\omega}2}+(6-6)e^{-j\hat{\omega}3}-3e^{-j\hat{\omega}4}-6e^{-j\hat{\omega}5} \end{aligned}$$

Thus, the overall equivalent impulse response is

$$h[n] = 6\delta[n-1] + 3\delta[n-2] - 3\delta[n-4] - 6\delta[n-5]$$

and the filter coefficients of the equivalent FIR system are $\{0, 6, 3, 0, -3, -6\}$.

This example illustrates that convolving two impulse responses is equivalent to multiplying their frequency responses. For FIR systems, the frequency response is a polynomial in the variable $e^{-j\hat{\omega}}$. Thus, multiplying two frequency responses requires polynomial multiplication. This result provides the theoretical basis for the "synthetic" polynomial multiplication algorithm discussed in Section 5-4.3.1 on p. 182.

EXERCISE 6.4 Suppose that two systems are cascaded. The first system is defined by the set of coefficients $\{1, 2, 3, 4\}$, and the second system is defined by the coefficients $\{-1, 1, -1\}$. Determine the frequency response and the impulse response of the overall cascade system.

6-7 Running-Sum Filtering

Another common LTI system is defined by the difference equation

$$\begin{aligned} y[n] &= \sum_{k=0}^{L-1} x[n-k] \\ &= x[n] + x[n-1] + \cdots + x[n-L+1] \end{aligned} \tag{6.22}$$

This system (6.22) is called an *L-point running sum* because the output at time n is computed as the sum of $x[n]$ and the $L-1$ previous samples of the input. The system defined by (6.22) can be implemented in MATLAB for $L = 11$ by the statements:

```
bb = ones(11,1);
yy = conv(bb, xx);
```

where `xx` is a vector containing the samples of the input. The vector `bb` contains the 11 filter coefficients, which are all equal to one in this case.

Using (6.4) on p. 216, the frequency response of the L-point running sum is

$$H(e^{j\hat{\omega}}) = \sum_{k=0}^{L-1} e^{-j\hat{\omega}k} \tag{6.23}$$

We can derive a simple formula for the magnitude and phase of the running sum by making use of the formula for the sum of the first L terms of a geometric series,

$$\sum_{k=0}^{L-1} \alpha^k = \frac{1-\alpha^L}{1-\alpha} \qquad (\text{for } \alpha \neq 1) \tag{6.24}$$

First of all, we identify $e^{-j\hat{\omega}}$ as α, and then do the following steps:

$$\begin{aligned}
H(e^{j\hat{\omega}}) = \sum_{k=0}^{L-1} e^{-j\hat{\omega}k} &= \left(\frac{1 - e^{-j\hat{\omega}L}}{1 - e^{-j\hat{\omega}}}\right) \\
&= \left(\frac{e^{-j\hat{\omega}L/2}(e^{j\hat{\omega}L/2} - e^{-j\hat{\omega}L/2})}{e^{-j\hat{\omega}/2}(e^{j\hat{\omega}/2} - e^{-j\hat{\omega}/2})}\right) \\
&= \left(\frac{\sin(\hat{\omega}L/2)}{\sin(\hat{\omega}/2)}\right) e^{-j\hat{\omega}(L-1)/2} \qquad (6.25)
\end{aligned}$$

The numerator and denominator in the second line are simplified by using the inverse Euler formula for sines. We find it convenient to express (6.25) in the form

$$H(e^{j\hat{\omega}}) = D_L(\hat{\omega}) e^{-j\hat{\omega}(L-1)/2} \qquad (6.26)$$

where

$$\boxed{D_L(\hat{\omega}) = \frac{\sin(\hat{\omega}L/2)}{\sin(\hat{\omega}/2)}} \qquad (6.27)$$

We refer to the function $D_L(\hat{\omega})$ as a ***Dirichlet form***,[5] and the subscript L indicates that it comes from an L-point running sum. In MATLAB, a scaled version of $D_L(\hat{\omega})$ can be evaluated with the `diric` function.

6-7.1 Plotting the Frequency Response

The frequency response of an 11-point running-sum filter is given by the equation

$$H(e^{j\hat{\omega}}) = D_{11}(\hat{\omega}) e^{-j\hat{\omega}5} \qquad (6.28)$$

where, in this case, $D_{11}(\hat{\omega})$ is the Dirichlet form defined by (6.27) with $L = 11$, that is,

$$D_{11}(\hat{\omega}) = \frac{\sin(11\hat{\omega}/2)}{\sin(\hat{\omega}/2)} \qquad (6.29)$$

As (6.28) makes clear, the frequency-response function, $H(e^{j\hat{\omega}})$, can be expressed as the product of the real amplitude function $D_{11}(\hat{\omega})$ and the complex exponential term $e^{-j5\hat{\omega}}$. The latter has a magnitude of 1 and a phase angle $-5\hat{\omega}$. Figure 6-7(a) shows a plot of the *amplitude* function, $D_{11}(\hat{\omega})$; the phase function $-5\hat{\omega}$ is shown in Fig. 6-7(b). We use the terminology *amplitude* rather than magnitude, because $D_{11}(\hat{\omega})$ can be negative. We can make the plot of the magnitude $|H(e^{j\hat{\omega}})|$ by taking the absolute value of $D_{11}(\hat{\omega})$.

[5] We cannot use the terminology "Dirichlet function" because it has a different well-defined meaning in mathematics. In Fourier analysis, $D_L(\hat{\omega})$ is called the Dirichlet kernel.

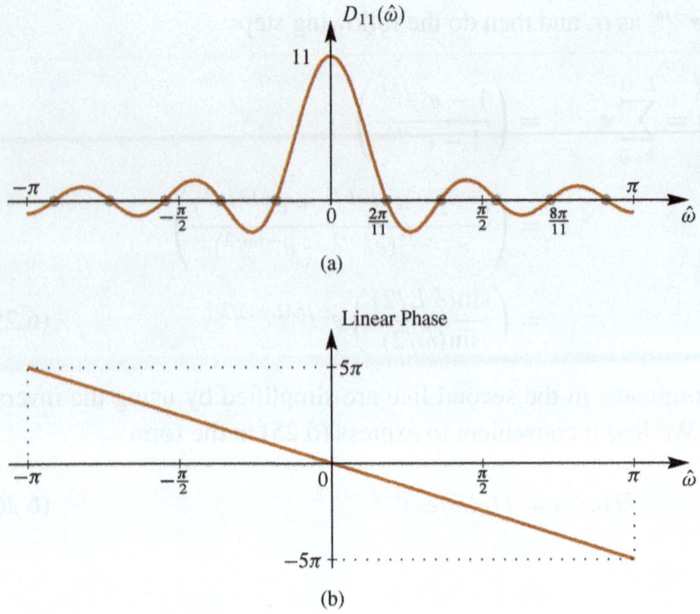

Figure 6-7 (a) *Amplitude* and (b) phase functions for frequency response of 11-point running-sum filter. The *amplitude* is a *Dirichlet form,* but it is not the magnitude since it has negative values.

Figure 6-7 shows only one period, $-\pi < \hat{\omega} \leq \pi$, but the frequency response is, of course, periodic with period 2π, so the plots in Fig. 6-7 would simply repeat with that period.

In the case of the 11-point running sum, the phase factor is easy to plot, since it is a straight line with slope of -5. The amplitude factor is somewhat more involved, but the following properties of the function $D_{11}(\hat{\omega})$ will aid in plotting:

(a) $D_{11}(\hat{\omega})$ is an even function of $\hat{\omega}$ that is periodic with period[6] 2π.

(b) $D_{11}(\hat{\omega})$ has a maximum value of 11 at $\hat{\omega} = 0$. In general, the max value is L.

(c) $D_{11}(\hat{\omega})$ has zeros at nonzero integer multiples of $2\pi/11$. (In general, the zeros of $D_L(\hat{\omega})$ are at nonzero multiples of $2\pi/L$.)

(d) $D_{11}(\hat{\omega})$ decays as $\hat{\omega}$ increases; the local extrema between the zeros (called *sidelobes*) decrease when going from $\hat{\omega} = 0$ to $\hat{\omega} = \pm\pi$.

To show evenness, note that $D_{11}(\hat{\omega})$ is the ratio of two sine functions. The ratio of two odd functions (i.e., sines) is an even function, so $D_{11}(-\hat{\omega}) = D_{11}(\hat{\omega})$. Since $D_{11}(\hat{\omega})$ is even and periodic with period 2π, we need only to consider its values in the interval $0 \leq \hat{\omega} \leq \pi$. All others can be inferred from symmetry and periodicity.

The behavior for $\hat{\omega} = 0$ is of particular interest because, at this frequency, $D_{11}(\hat{\omega})$ appears to be indeterminate, because the numerator and denominator are zero and,

$$D_{11}(0) = \frac{0}{0}$$

By L'Hôpital's rule, however, it is easily shown that $\lim_{\hat{\omega}\to 0} D_{11}(\hat{\omega}) = 11$.

[6]The Dirichlet form $D_L(\hat{\omega})$ has a period of 4π when L is even.

The numerator is $\sin(11\hat{\omega}/2)$, which oscillates between $+1$ and -1 and is zero whenever $11\hat{\omega}/2 = \pi k$, where k is an integer; solving for $\hat{\omega}$, $D_{11}(\hat{\omega})$ is zero at frequencies $\hat{\omega}_k = 2\pi k/11$ where k is a nonzero integer. The first zeros at $\pm 2\pi/11$ define the width of the *mainlobe*.

The denominator of $D_{11}(\hat{\omega})$ is $\sin(\hat{\omega}/2)$, which is zero at $\hat{\omega} = 0$ and increases monotonically to a maximum of one at $\hat{\omega} = \pi$. Therefore, $D_{11}(\hat{\omega})$ is large around $\hat{\omega} = 0$, where the denominator is small, and it oscillates with decreasing amplitude as $\hat{\omega}$ increases to π.

Together, the amplitude and phase plots of Fig. 6-7 completely define the frequency response of the 11-point running-sum filter. Normally, however, the frequency response is represented in polar form

$$H(e^{j\hat{\omega}}) = |H(e^{j\hat{\omega}})|e^{j\angle H(e^{j\hat{\omega}})}$$

This would require plotting $|H(e^{j\hat{\omega}})|$ and $\angle H(e^{j\hat{\omega}})$ as functions of $\hat{\omega}$. It is easy to see from (6.28) that $|H(e^{j\hat{\omega}})| = |D_{11}(\hat{\omega})|$. Figure 6-8(a) shows a plot of $|H(e^{j\hat{\omega}})| = |D_{11}(\hat{\omega})|$ for the 11-point running-sum filter, and we see that the negative sidelobes have been flipped to be positive. On the other hand, the phase response, $\angle H(e^{j\hat{\omega}})$, in Fig. 6-8(b) is more complicated to plot than the linear function shown in Fig. 6-7(b). There are two reasons for this:

(a) The algebraic sign of $D_{11}(\hat{\omega})$ must be represented in the phase function, since $|H(\hat{\omega})| = |D_{11}(\hat{\omega})|$ discards the sign of $D_{11}(\hat{\omega})$.

(b) The plot of the phase function is usually a plot of the *principal value* between $-\pi$ and $+\pi$.

The sign of $D_{11}(\hat{\omega})$ can be incorporated into the phase by noting that $D_{11}(\hat{\omega}) = |D_{11}(\hat{\omega})|e^{\pm j\pi}$ whenever $D_{11}(\hat{\omega}) < 0$. The principal value of the angle of a complex number is defined to be the angle between $-\pi$ and $+\pi$ radians. Using the result

$$e^{j(\theta \pm 2\pi k)} = e^{j\theta}e^{\pm j2\pi k} = e^{j\theta}$$

where k is any integer, we see that we can add or subtract integer multiples of 2π from the angle of a complex number without changing the value of the complex number. In fact, we can always find a multiple of 2π, which, when added to or subtracted from θ, produces a result in the range $-\pi < \theta \leq +\pi$.

This is called *reducing θ modulo 2π*. The principal value is generally what is computed when an inverse-tangent function is evaluated on a scientific calculator, in MATLAB, or in other computer languages. In Fig. 6-7(b), we were able to plot angle values outside the principal value range simply because we had an equation for the phase angle.

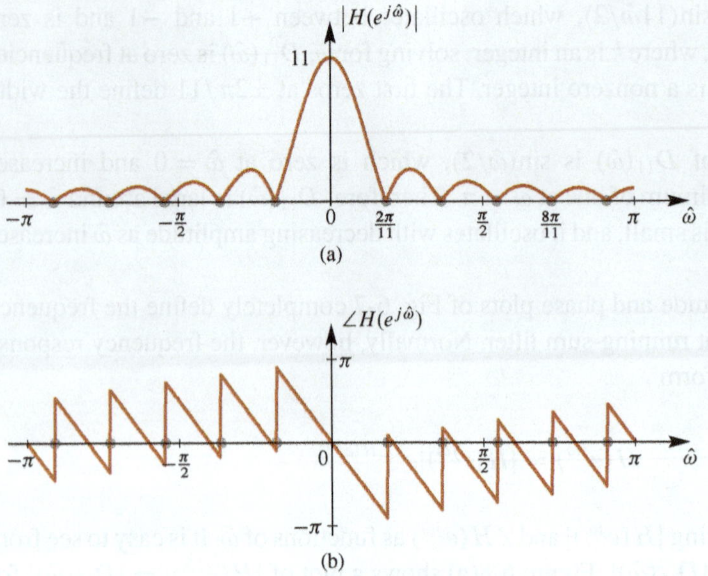

Figure 6-8 (a) Magnitude and (b) phase functions for frequency response of the 11-point running-sum filter. Magnitude is equal to the absolute value of the amplitude in Fig. 6-7(a). Phase has jumps of π whenever the magnitude is zero.

A plot like Figure 6-8 could be produced using the following MATLAB code:

```
omega = -pi:(pi/500):pi;
bb = ones(1,11)/11;
HH = freqz(bb,1,omega);
subplot(2,1,1),  plot(omega,abs(HH))
subplot(2,1,2),  plot(omega,angle(HH))
```

The MATLAB function `angle` uses the 4-quadrant arctangent function to return the principal value of the angle determined by the real and imaginary parts of the elements of the vector `HH`.

In Fig. 6-8, the phase curve is seen to have discontinuities that occur at the zeros of $D_{11}(\hat{\omega})$. These discontinuities arise from the multiples of π radians added to the phase due to the negative sign of $D_{11}(\hat{\omega})$ in the intervals $2\pi/11 < |\hat{\omega}| < 4\pi/11$, $6\pi/11 < |\hat{\omega}| < 8\pi/11$, and $10\pi/11 < |\hat{\omega}| < \pi$, as well as multiples of 2π that are added implicitly in the computation of the principal value. The equation for the phase curve plotted in Fig. 6-8 is dominated by the linear term $(-5\hat{\omega})$, but has the following special cases for frequencies $0 \le \hat{\omega} \le \pi$:

$$\angle H(e^{j\hat{\omega}}) = \begin{cases} -5\hat{\omega} & 0 \le \hat{\omega} < 2\pi/11 \\ -5\hat{\omega} + \pi & 2\pi/11 < \hat{\omega} < 4\pi/11 \\ -5\hat{\omega} + 2\pi & 4\pi/11 < \hat{\omega} < 6\pi/11 \\ -5\hat{\omega} + \pi + 2\pi & 6\pi/11 < \hat{\omega} < 8\pi/11 \\ -5\hat{\omega} + 4\pi & 8\pi/11 < \hat{\omega} < 10\pi/11 \\ -5\hat{\omega} + \pi + 4\pi & 10\pi/11 < \hat{\omega} \le \pi \end{cases}$$

The values for negative frequency, $-\pi < \hat{\omega} < 0$, can be filled in using the fact that the phase is an odd function (i.e., $\angle H(e^{-j\hat{\omega}}) = -\angle H(e^{j\hat{\omega}})$).

EXERCISE 6.5 Test yourself to see whether you understand the principal value of $\angle H(e^{j\hat{\omega}})$ by plotting (the principal value of) the phase of $H(e^{j\hat{\omega}}) = e^{-j4\hat{\omega}}$ for $-\pi < \hat{\omega} \leq \pi$.

6-7.2 Cascade of Magnitude and Phase

It can be seen from (6.25) that $H(e^{j\hat{\omega}})$ is the product of two functions, $H(e^{j\hat{\omega}}) = H_2(e^{j\hat{\omega}})H_1(e^{j\hat{\omega}})$, where

$$H_1(e^{j\hat{\omega}}) = e^{-j\hat{\omega}(L-1)/2} \tag{6.30}$$

and

$$H_2(e^{j\hat{\omega}}) = D_L(\hat{\omega}) = \frac{\sin(\hat{\omega}L/2)}{\sin(\hat{\omega}/2)} \tag{6.31}$$

which is the Dirichlet form defined earlier. The component $H_1(e^{j\hat{\omega}})$ contributes only to the phase of $H(e^{j\hat{\omega}})$, and this phase contribution is a linear function of $\hat{\omega}$. Earlier, we saw that a linear phase such as $\angle H_1(e^{j\hat{\omega}}) = -\hat{\omega}(L-1)/2$ corresponds to a time delay of $(L-1)/2$ samples. For $L = 11$ the linear-phase slope of $-(L-1)/2 = -5$ is clearly in evidence in Fig. 6-7(b). The frequency response of the second system $H_2(e^{j\hat{\omega}})$ is real. It contributes to the magnitude of $H(e^{j\hat{\omega}})$, and when it is negative, it also contributes $\pm\pi$ to the phase of $H(e^{j\hat{\omega}})$ causing the discontinuities at multiples of $2\pi/11$ in Fig. 6-8(b).

The product representation suggests the block diagram of Fig. 6-9, which shows that the running sum can be viewed as a cascade combination of a delay followed by a "lowpass filter" that accentuates low frequencies relative to high frequencies. This decomposition is useful for thinking about how the system affects the input. On the other hand, an implementation that conforms to the block diagram of Fig. 6-9 consists of a causal delay system followed by a noncausal filter $H_2(e^{j\hat{\omega}}) = D_L(\hat{\omega})$. Strictly speaking, we have not treated the implementation of noncausal filters in this chapter. Therefore, the overall *causal* moving average system would still be implemented by the difference equation (6.22) on p. 234.

Finally, there are issues with the implementation of the causal delay system in Fig. 6-9 as well. When $(L-1)/2$ is a integer, $w[n] = x[n-(L-1)/2]$ is a delay of the input samples. The case when $(L-1)/2$ is not an integer requires special interpretation, which will be provided in Section 6-8.2.

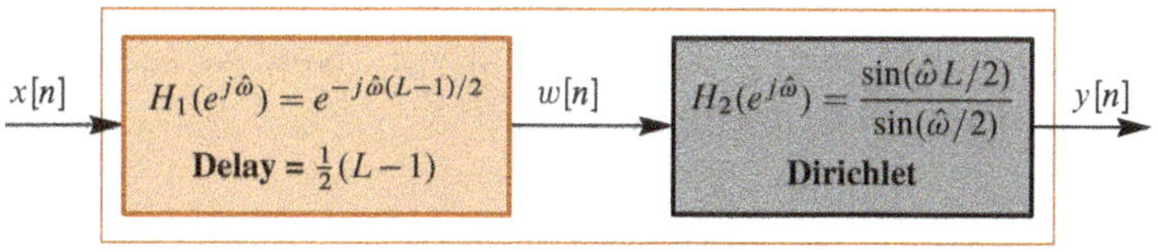

Figure 6-9 Representation of L-point running sum as the cascade of a delay and a real frequency response.

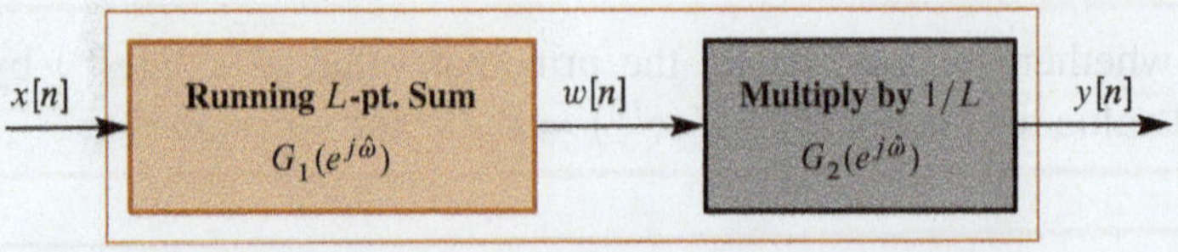

Figure 6-10 Representation of L-point running-average filter as the cascade of a running sum and a multiplier.

6-7.3 Frequency Response of Running Averager

The running sum and running-average filters are nearly the same because the running averager involves adding up L consecutive signal samples and dividing by L.

$$y[n] = \frac{1}{L}\underbrace{(x[n] + x[n-1] + \cdots + x[n-L+1])}_{\text{Running Sum}} \tag{6.32}$$

In hardware the divide-by-L operation would be done with a multiplier, so the running averager can be expressed as the cascade of two LTI systems as shown in Fig. 6-10. From this cascade representation, it is easy to write the frequency response of the L-pt. averager as

$$G_2(e^{j\hat{\omega}})G_1(e^{j\hat{\omega}}) = \frac{1}{L}D_L(\hat{\omega})e^{-j\hat{\omega}(L-1)/2} \tag{6.33}$$

6-7.4 Experiment: Smoothing an Image

As a simple experiment to show the filtering effect of the running-average system, consider the image in Fig. 6-11. The image is a two-dimensional discrete signal that can be represented as a two-dimensional array of samples $x_i[m, n]$. In an image, each sample value is called a *pixel*, which is shorthand for *picture element*. A single horizontal scan line (at $m = 40$) was extracted from the image yielding the one-dimensional signal $x[n] = x_i[40, n]$, plotted in Fig. 6-12. The position in the image from which $x[n]$ was extracted is shown by the orange line in the image. The values in the image signal are all positive integers in the range $0 \le x_i[m, n] \le 255$. These numbers can be represented by 8-bit binary numbers.[7] If you compare the one-dimensional plot to the gray levels in the region around the line, you can see that dark regions in the image have large values (near 255), and bright regions have low values (near zero). This is actually a "negative" image, but that is appropriate since it is a scan of black ink written on a white background.

An 11-point running averager was applied to $x[n]$, and the input and output were plotted on the same graph (Fig. 6-13(a)). Notice that the output $y[n]$ appears to be a smoother version of $x[n]$ with a slight shift to the right. The smoothness is a result of the lowpass Dirichlet filter that attenuates the higher frequencies in the signal, which correspond to the edges of the handwritten characters in the image. The shift is the 5-sample delay that we expect for an 11-point running averager. To verify the effect of this delay, Fig. 6-13(b) shows a plot of $w[n] = x[n-5]$ and $y[n]$. Now we see that the output appears to be aligned perfectly with the input.

[7] We often say that the total dynamic range of such image values is 8 bits.

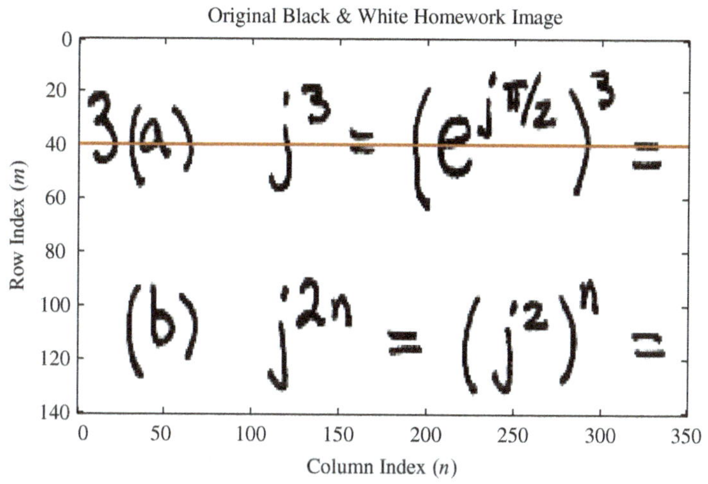

Figure 6-11 "Homework" image with row 40 indicated by the orange line.

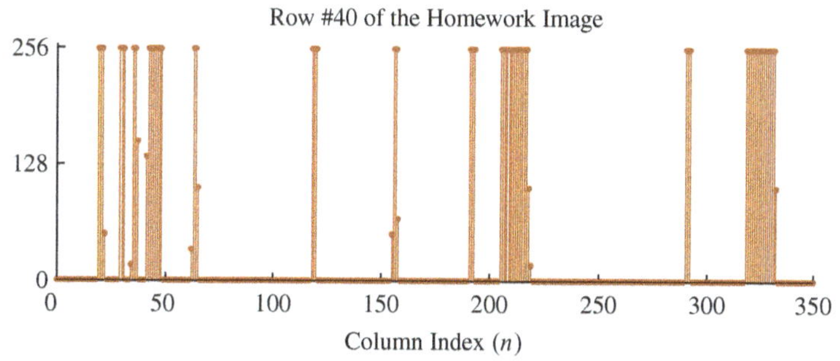

Figure 6-12 Signal values $x[n]$ along row 40 in the "homework" image.

The 11-point running averager is applied over all the rows of the image to get a visual assessment of the lowpass filtering operation.[8] The result is shown in Fig. 6-14, where it is obvious that the lowpass filter has blurred the image only in the horizontal dimension. There is no blurring in the vertical dimension as can be verified by examining the equals signs in Fig. 6-14. As we have seen, the filter attenuates the high-frequency components of the image. Thus, we can conclude that sharp edges in an image must be associated with high frequencies.

As another example of the effect of filtering, the "homework" image signal was corrupted by adding a cosine signal plus a constant to each row of the image. Along one row of the image the signal is now:

$$x_1[n] = x[n] + 128\cos(2\pi n/11) + 128$$

The cosine is the corrupting interference signal, while the constant (DC value) of 128 is needed to keep the image values positive. For line 40 of the image, this corrupted signal $x_1[n]$ was filtered with the 11-point running averager. The delayed input $x_1[n-5]$ and the corresponding output are shown in Fig. 6-15. By comparing Figs. 6-13(b) and 6-15, it is clear that the output is the same[9] for $x[n]$ and $x_1[n]$ except for the offset of 128. The

[8] In MATLAB, the function `filter2( )` does this two-dimensional filtering without using for loops.

[9] A careful comparison reveals that there is a small difference over the transient region $0 \le n \le 9$.

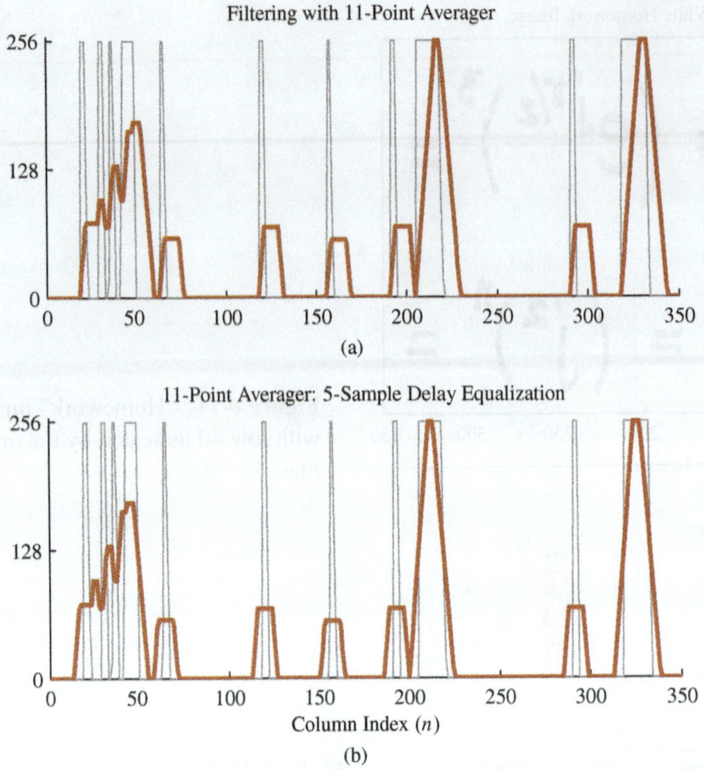

Figure 6-13 (a) Output $y[n]$ of 11-point running averager (solid orange line) and input (thin gray line). (b) Output of 11-point running averager (orange line) and input delayed by 5 samples (thin gray line).

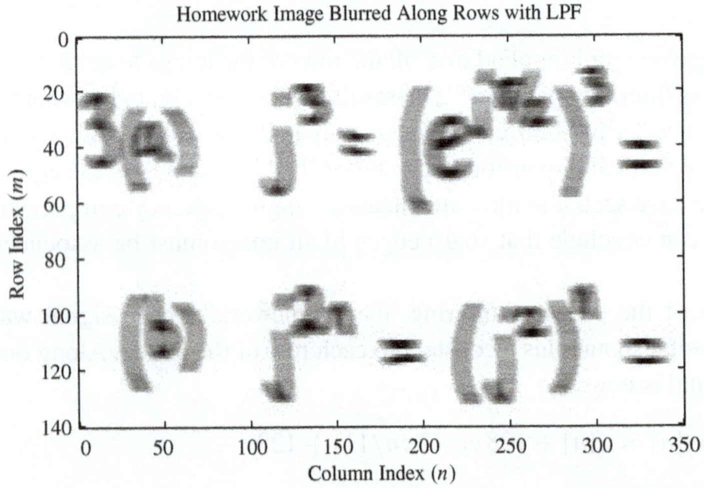

Figure 6-14 Result of filtering the rows of the "homework" image with an 11-point running averager. Note that the blurring is only in the horizontal dimension.

reason is clear: $\hat{\omega} = 2\pi/11$ is one of the frequencies that is completely removed by the averaging filter, because $H(e^{j2\pi/11}) = 0$. Since the system is LTI and obeys superposition, the output due to $x_1[n]$ must be the same as the output due to $x[n]$ alone. If the cosine is added to each row of an image, it appears visible as vertical stripes (Fig. 6-16(a)). When each row is processed with an 11-point running-average filter, the cosine is removed, but the image will still be blurred horizontally (Fig. 6-16(b)).

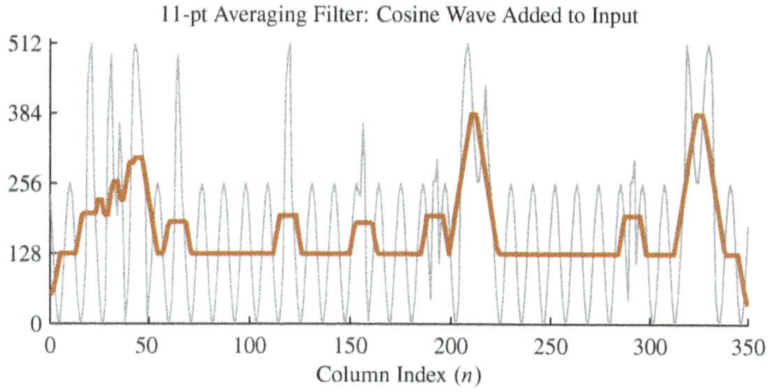

Figure 6-15 Output of an 11-point running averager with $128\cos(2\pi n/11)$ added to the image scan line. The solid orange line is the output; the thin gray line is the input delayed by five samples. Note that the vertical scale is twice that of Fig. 6-13(b).

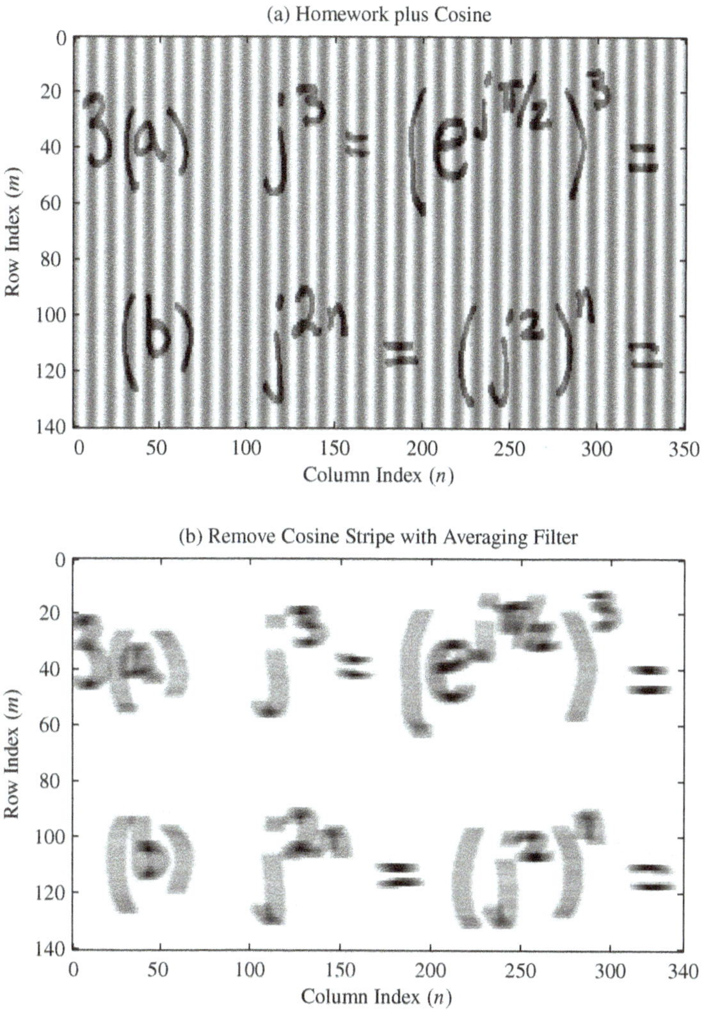

Figure 6-16 (a) "Homework plus cosine" input image. The periodic nature of the cosine across each row causes a vertical striping. (b) After filtering the rows of the "homework plus cosine" image with an 11-point running averager, the processed image is blurred, but has no traces of the cosine stripes. The filter transient on the left edge was removed , reducing the horizontal width by ten pixels. (Both input and output were rescaled for 8-bit image display.)

6-8 Filtering Sampled Continuous-Time Signals

Discrete-time filters can be used to filter bandlimited continuous-time signals that have been sampled. In this section, we study the relationship between the frequency response of the discrete-time filter and the ***effective frequency response*** applied to the continuous-time signal. When the input to a discrete-time system is a sequence derived by sampling a continuous-time signal, we can use our knowledge of sampling and reconstruction to relate the frequency response of the discrete-time filter to the effect of the filter on the spectrum of the original continuous-time signal.

Consider the system depicted in Fig. 6-17, and assume that the input is the complex sinusoid

$$x(t) = Xe^{j\omega t}$$

with $X = Ae^{j\varphi}$. After sampling, the input sequence to the discrete-time filter is

$$x[n] = x(nT_s) = Xe^{j\omega nT_s} = Xe^{j\hat{\omega}n}$$

The relationship between the discrete-time frequency $\hat{\omega}$ and continuous-time frequencies ω and f is

$$\boxed{\hat{\omega} = \omega T_s = \frac{2\pi f}{f_s}} \qquad (6.34)$$

If the frequency of the continuous-time complex exponential satisfies the condition of the sampling theorem (i.e., $|\omega| < \pi/T_s$), then there is no aliasing, and the normalized discrete-time frequency in (6.34) satisfies $|\hat{\omega}| < \pi$.

The frequency response of the discrete-time system gives us a quick way to calculate the output $y[n]$ in Fig 6-17.

$$y[n] = H(e^{j\hat{\omega}})Xe^{j\hat{\omega}n}$$

If we now make the substitution $\hat{\omega} = \omega T_s$, then we can write $y[n]$ in terms of the analog frequency ω as

$$y[n] = H(e^{j\omega T_s})Xe^{j\omega T_s n}$$

Finally, since no aliasing occurred in the original sampling, the *ideal* D-to-C converter reconstructs the original complex exponential, but with a new complex amplitude, giving

$$y(t) = H(e^{j\omega T_s})Xe^{j\omega t} \qquad (6.35)$$

This formula for $y(t)$ is correct only for frequencies such that $-\pi/T_s < \omega < \pi/T_s$, because the ideal D-to-C converter reconstructs all digital frequency components in the band

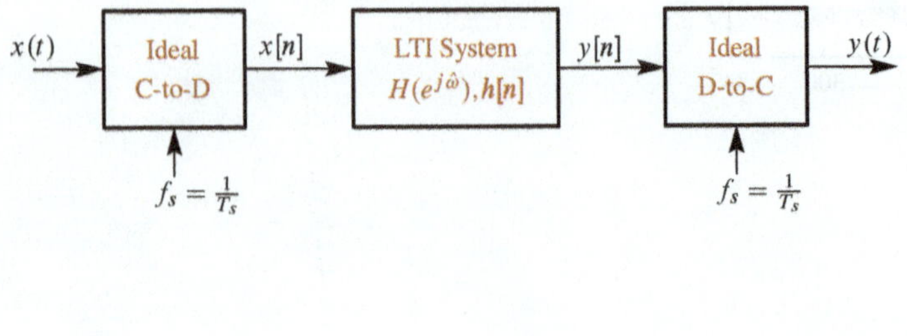

Figure 6-17 System for doing discrete-time filtering of bandlimited continuous-time signals.

$|\hat{\omega}| < \pi$ as analog frequencies in the band $|\omega| < \pi/T_s$. As long as there is no aliasing, the frequency band of the input signal $x(t)$ matches the frequency band of the output $y(t)$. Thus, the overall system of Fig. 6-17 behaves as though it is an LTI continuous-time system whose frequency response is $H(e^{j(\omega T_s)})$, but only for $|\omega| < \pi/T_s$.

When the system of Fig. 6-17 is used to implement LTI filtering operations on *bandlimited* continuous-time signals, there are two factors that control the effective frequency reponse. First, we can change the discrete-time system $H(e^{j\hat{\omega}})$ to vary the overall effective frequency response, for a fixed sampling period. Alternatively, for a fixed discrete-time system $H(e^{j\hat{\omega}})$, we can change the sampling period T_s (avoiding aliasing of the input) and obtain a new effective frequency response. Usually we select a sampling rate close to the Nyquist rate to avoid aliasing, and then design a discrete-time LTI filter whose effective frequency response $H(e^{j\omega T_s})$ has the desired frequency-selective properties.

6-8.1 Example: A Lowpass Averager for Continuous-Time Signals

As an example, we use the 11-point running averager

$$y[n] = \frac{1}{11}\sum_{k=0}^{10} x[n-k]$$

as the filter in Fig. 6-17. The frequency response of this discrete-time system is

$$H(e^{j\hat{\omega}}) = \frac{\sin(\hat{\omega}11/2)}{11\sin(\hat{\omega}/2)}e^{-j\hat{\omega}5}$$

The magnitude of this frequency response is shown in Fig. 6-18(a).

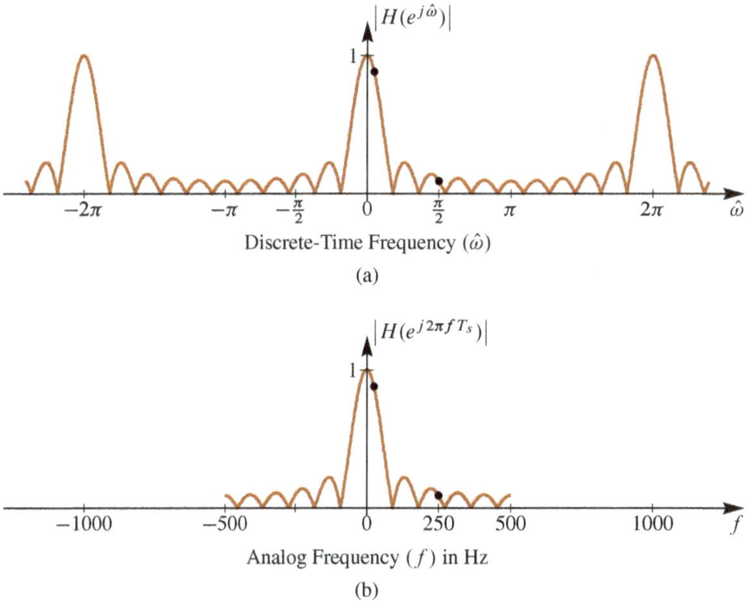

Figure 6-18 (a) Frequency response (magnitude) of an 11-point running averager and (b) effective analog-frequency response, when the system in Fig. 6-17 is used to filter bandlimited analog signals with sampling rate $f_s = 1000$ Hz. The black dots at $f = 25$ Hz and $f = 250$ Hz indicate two values of the frequency response needed for the input $x(t)$ in the example.

When the 11-point running averager is used as the discrete-time system in Fig. 6-17 with sampling frequency $f_s = 1000$, we want to answer two questions: What is the effective analog-frequency response, and how would the signal

$$x(t) = \cos(2\pi(25)t) + \sin(2\pi(250)t)$$

be processed by the system?

The frequency-response question is easy. The effective continuous-time frequency response is

$$H(e^{j\omega T_s}) = H(e^{j\omega/1000}) = H(e^{j2\pi f/1000})$$

for $|f| < 500$ Hz where f is in hertz. A plot of the effective continuous-time frequency response versus f is shown in Fig. 6-18(b). Note that the frequency response of the overall system *stops abruptly at* $|f| = f_s/2 = 500$ Hz, since the ideal D-to-C converter does not reconstruct frequencies above $f_s/2$.

The second question is also easy to answer if we track the two frequencies of the input signal through the three systems of Fig. 6-17 by using the effective continuous-time frequency response (6.35). The input $x(t)$ contains two sinusoids with frequencies at $\omega = 2\pi(25)$ and $2\pi(250)$. Since $f_s = 1000 > 2(250) > 2(25)$, there is no aliasing,[10] so the same frequency components appear in the output signal $y(t)$. The magnitude and phase changes are found by evaluating the effective frequency response at 25 and 250 Hz.

$$H(e^{j2\pi(25)/1000}) = \frac{\sin(\pi(25)(11)/1000)}{11\sin(\pi(25)/1000)} e^{-j2\pi(25)(5)/1000} = 0.8811e^{-j\pi/4}$$

$$H(e^{j2\pi(250)/1000}) = \frac{\sin(\pi(250)(11)/1000)}{11\sin(\pi(250)/1000)} e^{-j2\pi(250)(5)/1000} = 0.0909e^{-j\pi/2}$$

These values are marked on the plot in Fig. 6-18(b). Thus the final output is

$$y(t) = 0.8811\cos(2\pi(25)t - \pi/4) + 0.0909\sin(2\pi(250)t - \pi/2)$$

The lowpass nature of the running-average filter has greatly attenuated the 250-Hz component, while the 25-Hz component suffered only slight attenuation because it lies in the passband of the filter near 0 Hz.

EXERCISE 6.6 Assuming the same input signal and the same discrete-time system, work the example of this section again, but use a sampling rate of $f_s = 750$ Hz.

6-8.2 Interpretation of Delay

We have seen that a frequency response of the form $H(e^{j\hat{\omega}}) = e^{-j\hat{\omega}n_0}$ implies a time delay of n_0 samples. For n_0, an integer, the interpretation of this is straightforward. If the

[10]When there is aliasing, this sort of problem is less straightforward because the output signal $y(t)$ has different frequency components from the input.

input to the system is $x[n]$, the corresponding output is $y[n] = x[n - n_0]$. However, if n_0 is not an integer, the interpretation is less obvious. An example of where this can occur is the L-point running-average system whose frequency response is

$$H(e^{j\hat{\omega}}) = \tfrac{1}{L} D_L(\hat{\omega}) e^{-j\hat{\omega}(L-1)/2}$$

where $D_L(\hat{\omega})$ is the Dirichlet form defined in (6.27). The exponential term tells us that the L-point running averager includes a delay of $\frac{1}{2}(L-1)$ samples. If L is an odd integer, this delay causes the output to be shifted $\frac{1}{2}(L-1)$ samples with respect to the input. However, if L is an even integer, then $\frac{1}{2}(L-1)$ is not an integer. The analyis of this section provides an interpretation of this non-integer delay factor.

Suppose that the input to the ideal C-to-D converter is

$$x(t) = Xe^{j\omega t}$$

where $\omega < \pi/T_s$ so that there is no aliasing. Then the sampled input to the L-point running averager is

$$x[n] = Xe^{j\omega n T_s} = Xe^{j\hat{\omega} n}$$

where $\hat{\omega} = \omega T_s$, and the output is

$$y[n] = H(e^{j\hat{\omega}}) X e^{j\hat{\omega} n} = \tfrac{1}{L} D_L(\hat{\omega}) e^{-j\hat{\omega}(L-1)/2} X e^{j\hat{\omega} n}$$

Finally, since $\omega < \pi/T_s$, the ideal D-to-C converter reconstructs $y(t)$ as a complex exponential signal which can be written as

$$\begin{aligned} y(t) &= \tfrac{1}{L} D_L(\hat{\omega}) X e^{-j\hat{\omega}(L-1)/2} e^{j\omega t} \\ &= \tfrac{1}{L} D_L(\omega T_s) X e^{-j\omega T_s (L-1)/2} e^{j\omega t} \\ &= \tfrac{1}{L} D_L(\omega T_s) X e^{j\omega(t - T_s(L-1)/2)} \end{aligned}$$

The term $(t - T_s(L-1)/2)$ in the exponent indicates that the complex exponential has a time shift of $\frac{1}{2}(L-1)T_s$ seconds with respect to continuous-time signals sampled with sampling period T_s, and this is true regardless of whether or not $\frac{1}{2}(L-1)$ is an integer. Note that the exact $\frac{1}{2}(L-1)T_s$ second delay is a result of using an *ideal* D-to-C converter that performs perfect bandlimited reconstruction.

EXAMPLE 6-11 Time Delay of FIR Filter

To illustrate the effect of noninteger delay with the running-average filter, consider the cosine signal $x[n] = \cos(0.2\pi n)$, which could have resulted in Fig 6-17 from sampling the signal $x(t) = \cos(200\pi t)$ with sampling rate $f_s = 1000$ Hz. Figure 6-19(a) shows $x(t)$ and $x[n]$. If $x[n]$ is the input to a 5-point running-average filter, the steady-state part of the output is

$$y_5[n] = \frac{\sin(0.2\pi(5/2))}{5\sin(0.2\pi/2)} \cos(0.2\pi n - 0.2\pi(2)) = 0.6472\cos(0.2\pi(n-2))$$

For this filter output, the output of the *ideal* D-to-C converter in Fig. 6-17(b) would be

$$y_5(t) = y_5[n]\Big|_{n=1000t} = 0.6472\cos(200\pi(t - 0.002))$$

The delay is 2 samples. On the other hand, if the same signal $x[n]$ is the input to a 4-point running-average system, the steady-state part of the output (Fig. 6-19(c)) is

$$y_4[n] = \frac{\sin(0.2\pi(4/2))}{4\sin(0.2\pi/2)}\cos(0.2\pi n - 0.2\pi(3/2)) = 0.7694\cos(0.2\pi(n - 3/2))$$

Now the delay is $3/2$ samples, so we cannot write $y_4[n]$ as an integer shift with respect to the input sequence. In this case, the "$3/2$ samples" delay introduced by the filter can be interpreted in terms of the corresponding output of the *ideal* D-to-C converter in Fig. 6-17(c), which in this case would be

$$y_4(t) = y_4[n]\Big|_{n=1000t} = 0.7694\cos(200\pi(t - 0.0015))$$

Figure 6-19 shows the input and the corresponding outputs $y_5[n]$ and $y_5(t)$ and $y_4[n]$ and $y_4(t)$. In all cases, the solid gray curve is the continuous-time cosine signal that would be reconstructed by the ideal D-to-C converter for the given discrete-time signal.

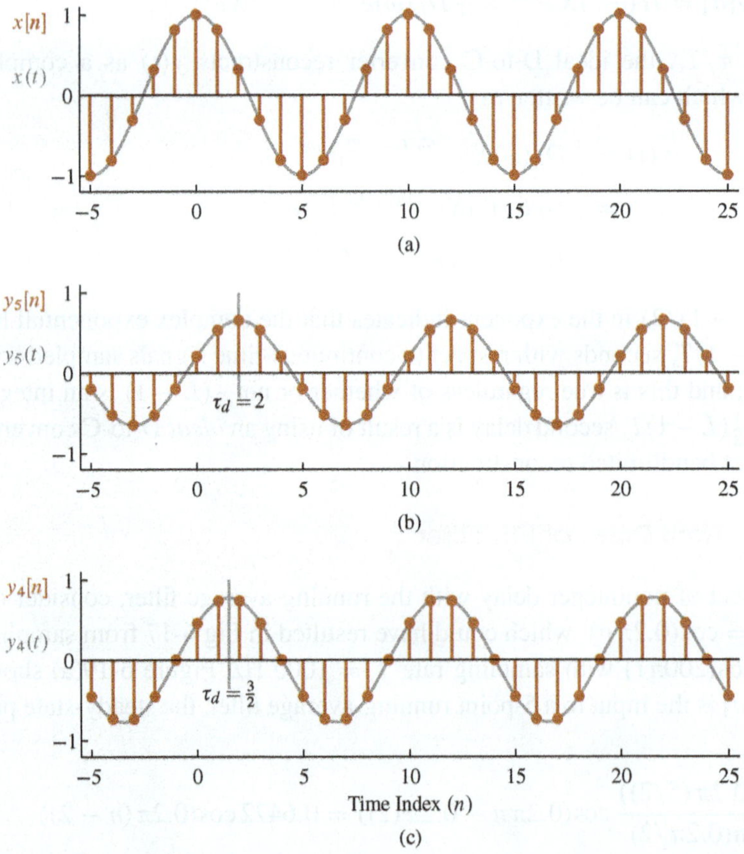

Figure 6-19 (a) Input signal sampled at $f_s = 1000$ Hz, (b) output of 5-point running averager, and (c) output of 4-point running averager. The solid gray curve is the corresponding continuous-time signal: (a) $x(t)$, (b) $y_5(t)$, and (c) $y_4(t)$.

The following specific points are illustrated by this example:

1. Both outputs are smaller than the input. This is because $D_L(0.2\pi) < 1$ for both cases.
2. The gray vertical lines in Fig. 6-19(b,c) show the peaks of the output cosine signals that correspond to the peak of the input at $n = 0$. Note that the delay is $(5-1)/2 = 2$ for the 5-point averager and $(4-1)/2 = 1.5$ for the 4-point running averager.
3. When the ideal D-to-C converter processes the fractional delay of 1.5, it has to implement an interpolation of the cosine signal at points halfway between the original samples.

6-9 Summary and Links

This chapter introduced the concept of the frequency response for the class of FIR filters. The frequency response applies to any LTI system, as we will see in upcoming chapters. The MATLAB GUI shown in Fig. 6-20 embodies the "sinusoid-in gives sinusoid-out" concept which is the primary result of this chapter. It shows that one evaluation of the frequency response is sufficient to predict how a sinusoid is processed by an LTI system. When the filter changes, the frequency response changes so sinusoidal

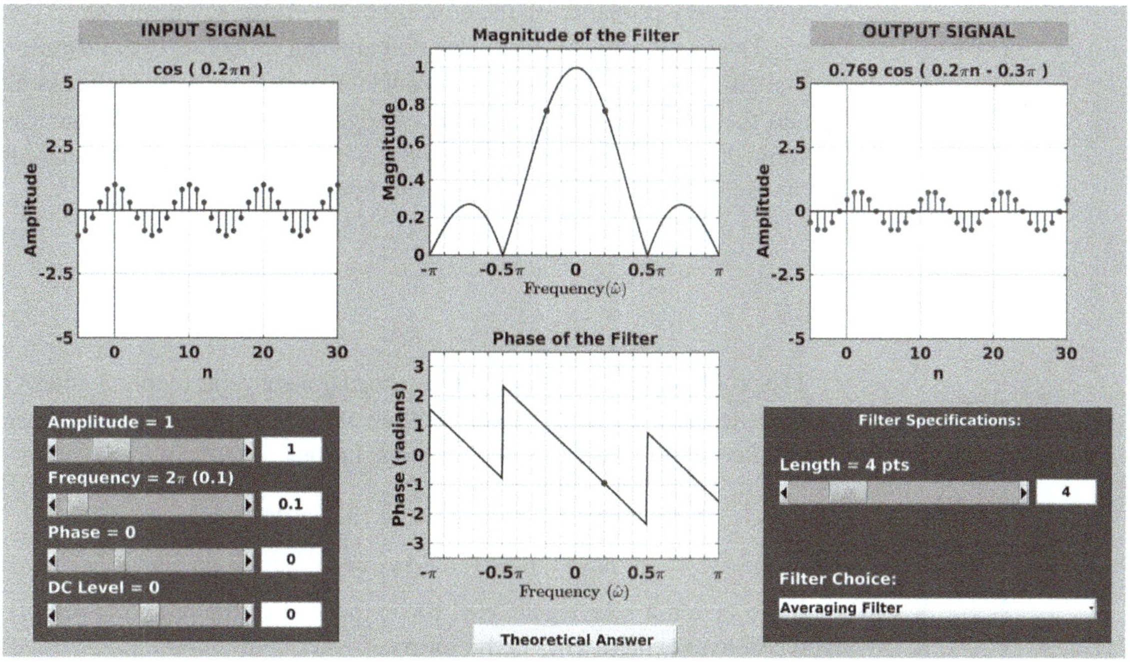

Figure 6-20 Screen shot of the `dltidemo` GUI which illustrates the *sinusoid-in gives sinusoid-out* concept. The user can change the LTI system as well as the input sinusoidal frequency, amplitude, and phase.

DEMO

DLTI: Sinusoid-in gives Sinusoid-out

frequency components are treated differently. The specific case shown in Fig. 6-20 is from Example 6-11 when $x[n] = \cos(0.2\pi n)$ is filtered with a 4-point running averager. The frequency response is shown in the middle with dots indicating the values of $H(e^{j\hat{\omega}})$ needed to determine the new amplitude and phase of the output sinusoid. The output signal on the right is exactly the same as $y_4[n]$ in Fig. 6-19(c).

EXERCISE 6.7 Use the `dltidemo` GUI in MATLAB to find a case where the input sinusoid is *nulled out* by the 4-point running-average filter. In other words, determine a frequency $\hat{\omega}_1$ for the input $x[n] = \cos(\hat{\omega}_1 n)$ such that the output signal is $y[n] = 0$.

LAB

Bandpass and Nullling Filters

This chapter extends the discussion of Chapter 5, which introduced the basics of FIR filtering. The labs, in particular, require the student to be familiar with both chapters. The lab entitled *Bandpass and Nulling Filters* is an experiment with the frequency response of two classes of useful FIR filters.

LAB

Encoding and Decoding Touch-Tones

Three other labs involving FIR filtering and the frequency response require the implementation of complete processing systems. These are: (1) a system to decode Touch-Tone signals, (2) an octave-band system that determines automatically which note is played on a piano, and (3) a filter bank system that simulates a cochlear implant device used to restore intelligible hearing for a patient with complete hearing loss. Each of these implementations involves many filters with a variety of frequency responses. Thus, these projects become easier to attack once the ideas of Chapters 7–10 are studied, but the projects rely primarily on the concept of the frequency response.

LAB

Octave Band Filtering

The Companion Website also contains the following demonstrations of lowpass and highpass filtering: filtering photographic images to show that lowpass filtering is blurring, and that highpass filtering enhances edges in images, cascade processing of images to show that a highpass filter can undo the blurring effects of a lowpass filter, and filtering of sound signals to illustrate bass and treble emphasis.

6-10 Problems

DEMO

Cascading FIR Filters

NOTE

Problems with Solutions

The problems in this section address the major concepts of this chapter. On the Companion Website there are many related homework problems which can be used for review and practice. Most of them have detailed solutions.

P-6.1 Suppose the input signal to an FIR system is

$$x[n] = e^{j(0.2\pi n - 0.7\pi)}$$

If we define a new signal $y[n]$ to be the first difference $y[n] = x[n] - x[n-1]$, it is possible to express $y[n]$ in the form

$$y[n] = Ae^{j(\hat{\omega}_0 n + \varphi)}$$

Determine the numerical values of A, φ, and $\hat{\omega}_0$.

P-6.2 Suppose that the input to a discrete-time system is the complex signal

$$x[n] = e^{j0.2\pi n} + e^{j0.5\pi n}$$

(a) Determine the output when the system is a 2-point running sum filter

(b) Explain why the output can be written as the sum of two complex exponential signals

$$y[n] = A_1 e^{j\varphi_1} e^{j\hat{\omega}_1 n} + A_2 e^{j\varphi_2} e^{j\hat{\omega}_2 n}$$

Determine the values of A_1, φ_1, A_2, φ_2 by using the frequency response of the system.

(c) Determine the output when the system is a 3-point running sum filter.

P-6.3 Suppose that a discrete-time system is described by the input–output relationship

$$y[n] = (x[n])^2$$

(a) Determine the output when the input is the sum of two complex exponential signals

$$x[n] = e^{j0.2\pi n} + e^{j0.6\pi n}$$

(b) Explain why the output cannot be written as the sum of two complex exponential signals.

$$y[n] \neq A_1 e^{j\varphi_1} e^{j\hat{\omega}_1 n} + A_2 e^{(j\varphi_2)} e^{j\hat{\omega}_2 n}$$

(c) Explain why this example is sufficient to conclude that this is not an LTI system.

P-6.4 An LTI system is described by the difference equation

$$y[n] = x[n] - \sqrt{2}x[n-1] + x[n-2]$$

(a) Determine an equation for the frequency response $H(e^{j\hat{\omega}})$ and express it in the form

$$H(e^{j\hat{\omega}}) = R(e^{j\hat{\omega}})e^{-j\hat{\omega}n_0}$$

where $R(e^{j\hat{\omega}})$ is a real function and n_0 is an integer.

(b) $H(e^{j\hat{\omega}})$ is a periodic function of $\hat{\omega}$; determine the period.

(c) Plot the magnitude of $H(e^{j\hat{\omega}})$ as a function of $\hat{\omega}$ for $\pi < \hat{\omega} < 3\pi$. Do this by hand, and then check with the MATLAB function `freqz`.

(d) Find all frequencies $\hat{\omega}$ in the interval $-\pi < \hat{\omega} \leq \pi$, for which the output response to the input $e^{-j\hat{\omega}n}$ is zero.

(e) When the input to the system is $x[n] = \cos(\pi n/3)$, determine the output signal and express it in the form $y[n] = A\cos(\hat{\omega}_0 n + \varphi)$.

P-6.5 An LTI filter is described by the difference equation

$$y[n] = 2x[n] + x[n-1] + 3x[n-2]$$

(a) Obtain an expression for the frequency response of this system.

(b) Determine the output when the input is $x[n] = 1 + 4\cos(0.6\pi n + 0.2\pi)$ for $-\infty < n < \infty$.

(c) Determine the step response, which is the output when the input is the unit-step sequence, $u[n]$.

(d) As $n \to \infty$ the step response is a constant. Explain why that constant is equal to the DC value of the frequency response.

P-6.6 A linear time-invariant filter is described by the difference equation

$$y[n] = x[n] - x[n-2]$$

(a) Obtain an expression for the frequency response of this system.

(b) Sketch the frequency response (magnitude and phase) as a function of frequency.

(c) Determine the output if the input signal is $x[n] = 4 + \cos(0.25\pi n - \pi/4)$.

(d) The signal $x_1[n] = (4 + \cos(0.25\pi n - \pi/4))\,u[n]$ is zero for $n < 0$, so it "starts" at $n = 0$. If the input signal is $x_1[n]$, for which values of n will the corresponding output be equal to the output obtained in (c)?

P-6.7 For each of the following frequency responses determine the corresponding impulse response

(a) $H(e^{j\hat{\omega}}) = 4 - e^{-j3\hat{\omega}}$

(b) $H(e^{j\hat{\omega}}) = 6e^{-j4\hat{\omega}}\cos(3\hat{\omega})$

(c) $H(e^{j\hat{\omega}}) = e^{-j3.5\hat{\omega}}\dfrac{\sin(4\hat{\omega})}{\sin(\hat{\omega}/2)}$

P-6.8 The frequency response of an LTI filter is given by the formula

$$H(e^{j\hat{\omega}}) = (1 - e^{-j\hat{\omega}})(1 - e^{j2\pi/3}e^{-j\hat{\omega}})(1 - e^{-j2\pi/3}e^{-j\hat{\omega}})$$

(a) Write the difference equation that gives the relation between the input $x[n]$ and the output $y[n]$.

(b) Determine the step response of this filter (i.e., the output when the input is $x[n] = u[n]$).

(c) If the input is of the form $x[n] = Ae^{j\varphi}e^{j\hat{\omega}n}$, for what values of $-\pi < \hat{\omega} \leq \pi$ is $y[n] = 0$ for all n?

P-6.9 The frequency response of an LTI filter is given by the formula

$$H(e^{j\hat{\omega}}) = (1 - e^{-j3\hat{\omega}})(2 - 4e^{-j\hat{\omega}} + 3e^{-j3\hat{\omega}})$$

(a) Write the difference equation and determine the impulse response of this LTI filter.

(b) If the input is of the form $x[n] = Ae^{j\varphi}e^{j\hat{\omega}n}$, for what values of $-\pi < \hat{\omega} \leq \pi$ is $y[n] = 0$ for all n?

(c) If $y[n] = 0.3$, what was the input $x[n]$?

P-6.10 Suppose that $\mathcal{S}$ is an LTI system whose exact form is unknown. It has been tested by observing the output signals corresponding to several different test inputs. Suppose that the following input–output pairs were the result of the tests:

Input: $x[n]$	**Output:** $y[n]$
$\delta[n]$	$\delta[n] - \delta[n-3]$
$\cos(2\pi n/3)$	0
$\cos(\pi n/3 + \pi/2)$	$2\cos(\pi n/3 + \pi/2)$

(a) Make a plot of the signal $x[n] = 3\delta[n] - 2\delta[n-2] + \delta[n-3]$.

(b) When the input is $x[n] = 3\delta[n] - 2\delta[n-2] + \delta[n-3]$, determine the output of the system and make a plot of it.

(c) Determine the output when the input is $x[n] = \cos(\pi(n-3)/3)$.

(d) Is the following statement true or false? Explain.

$$H(e^{j(\pi/2)}) = 0$$

P-6.11 The *Dirichlet* form is defined in (6.27) as

$$D_L(\hat{\omega}) = \frac{\sin(L\hat{\omega}/2)}{\sin(\hat{\omega})}$$

For the case $L = 6$:

(a) Make a plot of $D_6(\hat{\omega})$ over the range $-3\pi < \hat{\omega} \leq 3\pi$.

(b) Determine the period and maximum magnitude of $D_6(\hat{\omega})$.

(c) Determine the period of $H(e^{j\hat{\omega}}) = D_6(\hat{\omega})e^{-j2.5\hat{\omega}}$.

P-6.12 An LTI filter is described by the difference equation

$$y[n] = x[n] - 3x[n-1] + 3x[n-2] - x[n-3]$$

(a) Obtain an expression for the frequency response of this system, and, using the fact that $(1-a)^3 = 1 - 3a + 3a^2 - a^3$, show that $H(e^{j\hat{\omega}})$ can be expressed in the form

$$H(e^{j\hat{\omega}}) = 8\sin^3(\hat{\omega}/2)e^{j(-\pi/2-3\hat{\omega}/2)}$$

(b) Sketch the frequency response magnitude as a function of $\hat{\omega}$.

(c) What is the output if the input is $x[n] = 5 + 6\cos(0.5\pi(n-2))$?

(d) Determine the impulse response of this LTI filter.

(e) Use superposition to find the output when $x[n] = 5 + 6\cos(0.5\pi n) + 9\delta[n-3]$.

P-6.13 Suppose that three systems are connected in cascade, where the output of $\mathcal{S}_1$ is the input to $\mathcal{S}_2$, and the output of $\mathcal{S}_2$ is the input to $\mathcal{S}_3$. The three systems are specified as follows:

$$\mathcal{S}_1: \quad y_1[n] = x_1[n] + x_1[n-1]$$

$$\mathcal{S}_2: \quad y_2[n] = x_2[n] - x_2[n-2]$$

$$\mathcal{S}_3: \quad y_3[n] = x_3[n-2] - x_3[n-3]$$

where the output of $\mathcal{S}_i$ is $y_i[n]$ and its input is $x_i[n]$.

(a) Determine the frequency responses of the three individual systems.

(b) Determine the overall frequency response of the cascade system as a simple formula.

(c) Determine the difference equation of the equivalent system that is a single operation from the input $x[n]$ (into $\mathcal{S}_1$) to the overall output $y[n]$, which is the output of $\mathcal{S}_3$. Thus, $x[n]$ is $x_1[n]$ and $y[n]$ is $y_3[n]$.

P-6.14 An LTI filter is described by the difference equation

$$y[n] = \frac{1}{4}\Big(x[n] + x[n-1] + x[n-2] + x[n-3]\Big)$$

(a) What is the impulse response $h[n]$ of this system?

(b) Obtain an expression for the frequency response of this system.

(c) Suppose that the input is

$$x[n] = 5 + 4\cos(0.2\pi n) + 3\cos(0.5\pi n)$$

for $-\infty < n < \infty$. Obtain an expression for the output in the form $y[n] = A + B\cos(\hat{\omega}_0 n + \phi_0)$.

(d) Suppose that the input is

$$x_1[n] = \left[5 + 4\cos(0.2\pi n) + 3\cos(0.5\pi n)\right]u[n]$$

where $u[n]$ is the unit-step sequence. For what values of n will the output $y_1[n]$ be equal to the output $y[n]$ in (d)?

P-6.15 A system for filtering continuous-time signals is shown in Fig. P-6.15.

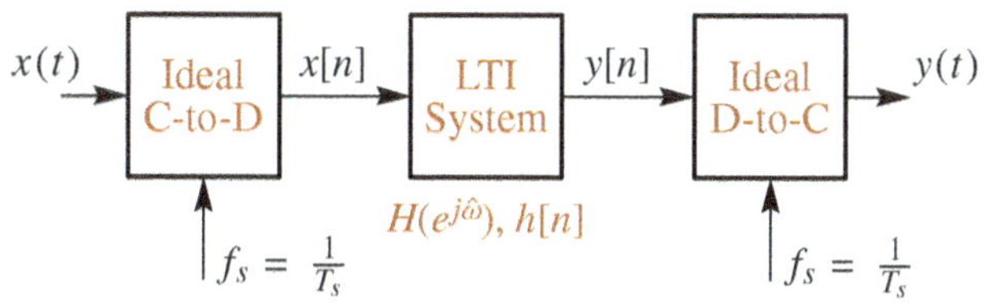

Figure P-6.15

The input to the C-to-D converter in this system is

$$x(t) = 10 + 8\cos(200\pi t) + 6\cos(500\pi t + \pi/4)$$

for $-\infty < t < \infty$. The impulse response of the LTI system is

$$h[n] = \frac{1}{4}\sum_{k=0}^{3}\delta[n-k]$$

If $f_s = 1000$ samples/sec, determine an expression for $y(t)$, the output of the D-to-C converter.

Hint: The results of Problem P-6.14 can be applied here.

P-6.16 The input to the C-to-D converter in Fig. P-6.15 is

$$x(t) = 10 + 20\cos(\omega_0 t + \pi/3) \qquad -\infty < t < \infty$$

(a) Suppose that the impulse response of the LTI system is $h[n] = \delta[n]$. If $\omega_0 = 2\pi(500)$, for what values of $f_s = 1/T_s$ is it true that $y(t) = x(t)$?

(b) Now suppose that the impulse response of the LTI system is changed to $h[n] = \delta[n-10]$. Determine the sampling rate $f_s = 1/T_s$ and a range of values for ω_0 so that the output of the overall system is

$$y(t) = x(t - 0.001) = 10 + 20\cos(\omega_0(t - 0.001) + \pi/3)$$

for $-\infty < t < \infty$.

(c) Suppose that the LTI system is a 5-point moving averager whose frequency response is

$$H(e^{j\hat{\omega}}) = \frac{\sin(5\hat{\omega}/2)}{5\sin(\hat{\omega}/2)} e^{-j\hat{\omega}2}$$

If the sampling rate is $f_s = 2000$ samples/sec, determine all values of ω_0 such that the output is equal to a constant (i.e., $y(t) = A$ for $-\infty < t < \infty$). Also, determine the constant A in this case.

P-6.17 The diagram in Fig. P-6.17 depicts a cascade connection of two linear time-invariant systems, where the first system is a 3-point moving averager and the second system is a first difference.

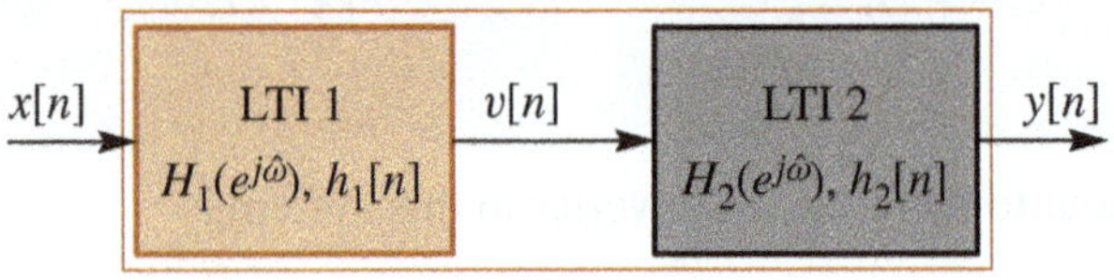

Figure P-6.17

(a) If the input is of the form $x[n] = 10 + x_1[n]$, the output, $y[n]$, of the overall system will be of the form $y[n] = y_1[n]$, where $y_1[n]$ is the output due only to $x_1[n]$. Explain why this is true.

(b) Determine the frequency-response function of the overall cascade system.

(c) Sketch the frequency-response (magnitude and phase) functions of the individual systems and the overall cascade system for $-\pi < \hat{\omega} \le \pi$.

(d) Obtain a single-difference equation that relates $y[n]$ to $x[n]$ for the overall cascade system.

P-6.18 A linear time-invariant system is described by the difference equation

$$y[n] = -x[n] + 2x[n-2] - x[n-4]$$

(a) Find the impulse response $h[n]$ and plot it.

(b) Determine an equation for the frequency response $H(e^{j\hat{\omega}})$ and express it in the form

$$H(e^{j\hat{\omega}}) = R(e^{j\hat{\omega}})e^{-j\hat{\omega}n_0}$$

where $R(e^{j\hat{\omega}})$ is a real function and n_0 is an integer.

(c) Carefully sketch and label a plot of $|H(e^{j\hat{\omega}})|$ for $-\pi < \hat{\omega} < \pi$.

(d) Carefully sketch and label a plot of the principal value of $\angle H(e^{j\hat{\omega}})$ for $-\pi < \hat{\omega} \leq \pi$.

P-6.19 For the cascade configuration in Fig. P-6.17, the two systems are defined by

$$H_1(e^{j\hat{\omega}}) = 1 + 2e^{-j\hat{\omega}} + e^{-j\hat{\omega}2} \qquad \text{and}$$
$$h_2[n] = \delta[n] - \delta[n-1] + \delta[n-2] - \delta[n-3]$$

(a) Determine the frequency response $H(e^{j\hat{\omega}})$ for the overall cascade system (i.e., from input $x[n]$ to output $y[n]$). Simplify your answer as much as possible.

(b) Determine and plot the impulse response $h[n]$ of the overall cascade system.

(c) Write down the difference equation that relates $y[n]$ to $x[n]$.

P-6.20 An LTI system whose frequency response has a first-order term such as

$$H(e^{j\hat{\omega}}) = (1 - e^{j\hat{\omega}_0}e^{-j\hat{\omega}})$$

completely nulls out the input signal of the form $e^{j\hat{\omega}_0 n}$. The concept of cascading systems can be used to null out additional complex exponential signals.

(a) Design an FIR filter that nulls out the signal $x[n] = 3 + 7\cos(0.2\pi n - 0.4\pi)$ by determining the filter coefficients of the difference equation for the nulling filter.

(b) Design the minimum-order FIR filter that nulls out the signal

$$x[n] = \sum_{k=0}^{5}(k^2 + 9)\cos(0.2\pi kn)$$

by determining the filter coefficients of the difference equation for the nulling filter.

P-6.21 The input to the C-to-D converter in Fig. P-6.15 is

$$x(t) = 10\cos(120\pi t) + 20\cos(240\pi t + \pi/3) \qquad -\infty < t < \infty$$

(a) If the sampling rate is $f_s = 1000$ Hz, determine the two normalized frequencies in the sampled signal $x[n] = x(n/f_s)$.

(b) Design the FIR filter $H(e^{j\hat{\omega}})$ so that the output $y(t)$ is zero. In other words, find the filter coefficients of the lowest possible order FIR system.

(c) Check your design by using MATLAB to plot the frequency response (magnitude) versus $\hat{\omega}$, and also versus the analog frequency f in hertz where the frequencies 60 Hz and 120 Hz should be nulled out.

P-6.22 The frequency response of a first-difference system is plotted in Fig. 6-3.

(a) Use these plots to find the output of the system if the input is $x[n] = 10 + 10\cos(0.5\pi n)$.

(b) Explain why the phase-response curve is discontinuous at frequencies around $\hat{\omega} = 0$ and at multiples of 2π.

P-6.23 The frequency response of an LTI system is plotted in Fig. P-6.23

(a) Use these plots to find the output of the system if the input is

$$x[n] = 10 + 10\cos(0.2\pi n) + 10\cos(0.5\pi n)$$

(b) Explain why the phase-response curve is discontinuous at frequencies around $\hat{\omega} = 2\pi(0.17)$ and also around $\hat{\omega} = 0.5\pi$.

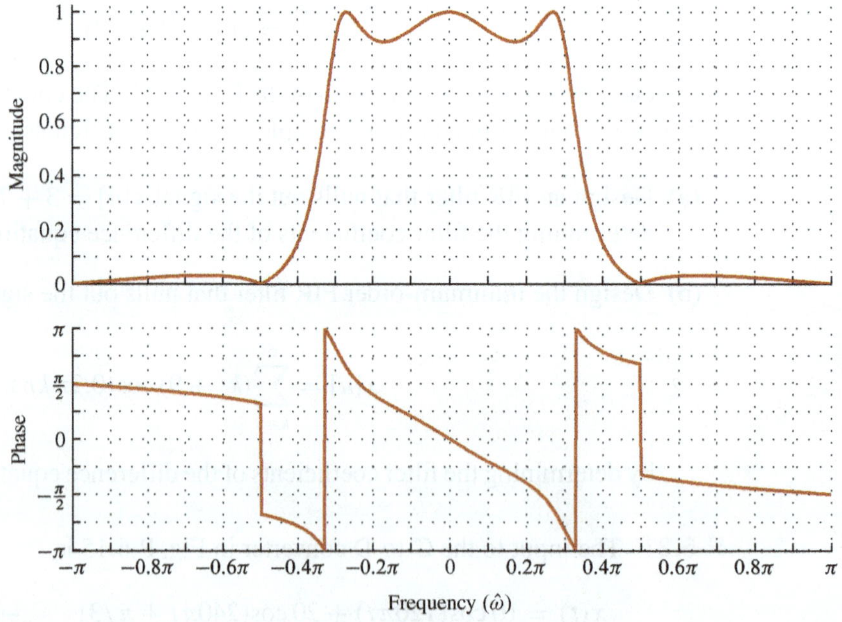

Figure P-6.23

CHAPTER

7

Discrete-Time Fourier Transform

In Chapter 3 and Appendix C, we showed that interesting continuous-time waveforms $x(t)$ can be synthesized by summing sinusoids, or complex exponential signals, having different frequencies f_k and complex amplitudes a_k. We also introduced the concept of the *spectrum* of a signal as the collection of information about the frequencies and corresponding complex amplitudes $\{f_k, a_k\}$ of the complex exponential signals, and found it convenient to display the spectrum as a plot of spectrum lines versus frequency, each labeled with amplitude and phase. This spectrum plot is a *frequency-domain representation* that tells us at a glance "how much of each frequency is present in the signal."

In Chapter 4, we extended the spectrum concept from continuous-time signals $x(t)$ to discrete-time signals $x[n]$ obtained by sampling $x(t)$. In the discrete-time case, the line spectrum is plotted as a function of normalized frequency $\hat{\omega}$. In Chapter 6, we developed the frequency response $H(e^{j\hat{\omega}})$ which is the frequency-domain representation of an FIR filter. Since an FIR filter can also be characterized in the time domain by its impulse response signal $h[n]$, it is not hard to imagine that the frequency response is the *frequency-domain representation*, or *spectrum*, of the sequence $h[n]$.

In this chapter we show that the frequency response is identical to the result of applying the more general concept of the discrete-time Fourier transform (DTFT) to the impulse response of the LTI system. The DTFT is a frequency-domain representation for a wide range of both finite- and infinite-length discrete-time signals $x[n]$. The DTFT is denoted as $X(e^{j\hat{\omega}})$, which shows that the frequency dependence always includes the complex exponential function $e^{j\hat{\omega}}$. The operation of taking the Fourier transform of a signal will become a common tool for analyzing signals and systems *in the frequency domain.*[1]

The application of the DTFT is usually called Fourier analysis, or spectrum analysis or "going into the Fourier domain or frequency domain." Thus, the words *spectrum, Fourier transform, and frequency-domain representation* become equivalent, even though each one retains its own distinct character.

7-1 DTFT: Fourier Transform for Discrete-Time Signals

The concept of frequency response discussed in Chapter 6 emerged from analysis showing that if an input to an LTI discrete-time system is of the form $x[n] = e^{j\hat{\omega}n}$, then the corresponding output has the form $y[n] = H(e^{j\hat{\omega}})e^{j\hat{\omega}n}$, where $H(e^{j\hat{\omega}})$ is called the frequency response of the LTI system. This fact, coupled with the principle of superposition for LTI systems leads to the fundamental result that the frequency response function $H(e^{j\hat{\omega}})$ is sufficient to determine the output due to any linear combination of signals of the form $e^{j\hat{\omega}n}$ or $\cos(\hat{\omega}n + \theta)$. For discrete-time filters such as the causal FIR filters discussed in Chapter 6, the frequency response function is obtained from the summation formula

$$H(e^{j\hat{\omega}}) = \sum_{n=0}^{M} h[n]e^{-j\hat{\omega}n} = h[0] + h[1]e^{-j\hat{\omega}} + \cdots + h[M]e^{-j\hat{\omega}M} \qquad (7.1)$$

where $h[n]$ is the impulse response. In a mathematical sense, the impulse response $h[n]$ is *transformed* into the frequency response by the operation of evaluating (7.1) for each value of $\hat{\omega}$ over the domain $-\pi < \hat{\omega} \leq \pi$. The operation of transformation (adding up the terms in (7.1) for each value $\hat{\omega}$) replaces a function of a discrete-time index n (a sequence) by a periodic function of the continuous frequency variable $\hat{\omega}$. By this transformation, the time-domain representation $h[n]$ is replaced by the frequency-domain representation $H(e^{j\hat{\omega}})$. For this concept to be complete and useful, we need to know that the result of the transformation is unique, and we need the ability to go back from the frequency-domain representation to the time-domain representation. That is, we need an *inverse transform* that recovers the original $h[n]$ from $H(e^{j\hat{\omega}})$. In Chapter 6, we showed that the sequence can be reconstructed from a frequency response represented in terms of powers of $e^{-j\hat{\omega}}$ as in (7.1) by simply picking off the coefficients of the polynomial since $h[n]$ is the coefficient of $e^{-j\hat{\omega}n}$. While this process can be effective if M is small, there is a much more powerful approach to inverting the transformation that holds even for infinite-length sequences.

[1] It is common in engineering to say that we "take the discrete-time Fourier transform" when we mean that we consider $X(e^{j\hat{\omega}})$ as our representation of a signal $x[n]$.

In this section, we present the general concept of the discrete-time Fourier transform which gives the frequency response when applied to the impulse response of an LTI system. We give an integral form for the inverse DTFT that can be used even when $H(e^{j\hat{\omega}})$ does not have a finite polynomial representation such as (7.1). Furthermore, we show that the DTFT can be used to represent a wide range of sequences, including sequences of infinite length, and that these sequences can be impulse responses, inputs to LTI systems, outputs of LTI systems, or indeed, any sequence that satisfies certain conditions to be discussed in this chapter.

7-1.1 Forward DTFT

The DTFT of a sequence $x[n]$ is defined as

Discrete-Time Fourier Transform

$$X(e^{j\hat{\omega}}) = \sum_{n=-\infty}^{\infty} x[n]e^{-j\hat{\omega}n} \tag{7.2}$$

The DTFT $X(e^{j\hat{\omega}})$ that results from this definition is a function of frequency $\hat{\omega}$. Going from the signal $x[n]$ to its DTFT is referred to as "taking the forward transform," and going from the DTFT back to the signal is referred to as "taking the inverse transform." The limits on the sum in (7.2) are shown as infinite so that the DTFT is defined for infinitely long signals as well as finite-length signals.[2] However, a comparison of (7.2) to (7.1) shows that if the sequence were a finite-length impulse response, then the DTFT of that sequence would be the same as the frequency response of the FIR system. More generally, if $h[n]$ is the impulse response of an LTI system, then the DTFT of $h[n]$ is the frequency response $H(e^{j\hat{\omega}})$ of that system. Examples of infinite-duration impulse response filters will be given in Chapter 10.

EXERCISE 7.1 Show that the DTFT function $X(e^{j\hat{\omega}})$ defined in (7.2) is always periodic in $\hat{\omega}$ with period 2π, that is,

$$X(e^{j(\hat{\omega}+2\pi)}) = X(e^{j\hat{\omega}}).$$

7-1.2 DTFT of a Shifted Impulse Sequence

Our first task is to develop examples of the DTFT for some common signals. The simplest case is the time-shifted unit-impulse sequence $x[n] = \delta[n - n_0]$. Its forward DTFT is by definition

$$X(e^{j\hat{\omega}}) = \sum_{n=-\infty}^{\infty} \delta[n - n_0]e^{-j\hat{\omega}n}$$

[2]The infinite limits are used to imply that the sum is over all n, where $x[n] \neq 0$. This often avoids unnecessarily awkward expressions when using the DTFT for analysis.

Since the impulse sequence is nonzero only at $n = n_0$ it follows that the sum has only one nonzero term, so

$$X(e^{j\hat{\omega}}) = e^{-j\hat{\omega}n_0}$$

To emphasize the importance of this and other DTFT relationships, we use the notation $\overset{\text{DTFT}}{\longleftrightarrow}$ to denote the forward and inverse transforms in one statement:

DTFT Representation of $\delta[n - n_0]$

$$x[n] = \delta[n - n_0] \overset{\text{DTFT}}{\longleftrightarrow} X(e^{j\hat{\omega}}) = e^{-j\hat{\omega}n_0} \tag{7.3}$$

7-1.3 Linearity of the DTFT

Before we proceed further in our discussion of the DTFT, it is useful to consider one of its basic properties. The DTFT is a linear operation, that is, the DTFT of a sum of two or more scaled signals results in the identical sum and scaling of their corresponding DTFTs. To verify this, assume that $x[n] = ax_1[n] + bx_2[n]$, where a and b are (possibly complex) constants. The DTFT of $x[n]$ is by definition

$$X(e^{j\hat{\omega}}) = \sum_{n=-\infty}^{\infty} (ax_1[n] + bx_2[n])e^{-j\hat{\omega}n}$$

If both $x_1[n]$ and $x_2[n]$ have DTFTs, then we can use the algebraic property that multiplication distributes over addition to write

$$X(e^{j\hat{\omega}}) = a \sum_{n=-\infty}^{\infty} x_1[n]e^{-j\hat{\omega}n} + b \sum_{n=-\infty}^{\infty} x_2[n]e^{-j\hat{\omega}n} = aX_1(e^{j\hat{\omega}}) + bX_2(e^{j\hat{\omega}})$$

That is, the frequency-domain representations are combined in exactly the same way as the signals are combined.

EXAMPLE 7-1 DTFT of an FIR Filter

The following FIR filter

$$y[n] = 5x[n-1] - 4x[n-3] + 3x[n-5]$$

has a finite-length impulse response:

$$h[n] = 5\delta[n-1] - 4\delta[n-3] + 3\delta[n-5]$$

Each impulse in $h[n]$ is transformed using (7.3), and then combined according to the linearity property of the DTFT which gives

$$H(e^{j\hat{\omega}}) = 5e^{-j\hat{\omega}} - 4e^{-j3\hat{\omega}} + 3e^{-j5\hat{\omega}}$$

7-1.4 Uniqueness of the DTFT

The DTFT is a *unique* relationship between $x[n]$ and $X(e^{j\hat{\omega}})$; in other words, two different signals cannot have the same DTFT. This is a consequence of the linearity property because if two different signals have the same DTFT, then we can form a third signal by subtraction and obtain

$$x_3[n] = x_1[n] - x_2[n] \overset{\text{DTFT}}{\longleftrightarrow} X_3(e^{j\hat{\omega}}) = \underbrace{X_1(e^{j\hat{\omega}}) - X_2(e^{j\hat{\omega}})}_{\text{identical DTFTs}} = 0$$

However, from the definition (7.2) it is easy to argue that $x_3[n]$ has to be zero if its DTFT is zero, which in turn implies that $x_1[n] = x_2[n]$.

The importance of uniqueness is that if we know a DTFT representation such as (7.3), we can start in either the time or frequency domain and easily write down the corresponding representation in the other domain. For example, if $X(e^{j\hat{\omega}}) = e^{-j\hat{\omega}3}$, then we know that $x[n] = \delta[n-3]$.

7-1.5 DTFT of a Pulse

Another common signal is the L-point rectangular pulse, which is a finite-length time signal consisting of all ones:

$$r_L[n] = u[n] - u[n-L] = \begin{cases} 1 & n = 0, 1, 2, \ldots, L-1 \\ 0 & \text{elsewhere} \end{cases}$$

Its forward DTFT is by definition

$$R_L(e^{j\hat{\omega}}) = \sum_{n=0}^{L-1} 1\, e^{-j\hat{\omega}n} = \frac{1 - e^{-j\hat{\omega}L}}{1 - e^{-j\hat{\omega}}} \tag{7.4}$$

where we have used the formula for the sum of L terms of a geometric series to "sum" the series and obtain a closed-form expression for $R_L(e^{j\hat{\omega}})$. This is a signal that we studied before in Chapter 6 as the impulse response of an L-point running-sum filter. In Section 6-7, the frequency response of the running-sum filter was shown to be the product of a Dirichlet form and a complex exponential. Referring to the earlier results in Section 6-7 or further manipulating (7.4), we obtain another DTFT pair:

DTFT Representation of L-Point Rectangular Pulse

$$r_L[n] = u[n] - u[n-L] \overset{\text{DTFT}}{\longleftrightarrow} R_L(e^{j\hat{\omega}}) = \frac{\sin(L\hat{\omega}/2)}{\sin(\hat{\omega}/2)} e^{-j\hat{\omega}(L-1)/2} \tag{7.5}$$

Since the filter coefficients of the running-sum filter are L times the filter coefficients of the running-average filter, there is no L in the denominator of (7.5).

7-1.6 DTFT of a Right-Sided Exponential Sequence

As an illustration of the DTFT of an infinite-duration sequence, consider a "right-sided" exponential signal of the form $x[n] = a^n u[n]$, where a can be real or complex. Such a signal is zero for $n < 0$ (on the *left-hand side* of a plot). It decays "exponentially" for $n \geq 0$ if $|a| < 1$; it remains constant at 1 if $|a| = 1$; and it grows exponentially if $|a| > 1$. Its DTFT is by definition

$$X(e^{j\hat{\omega}}) = \sum_{n=-\infty}^{\infty} a^n u[n] e^{-j\hat{\omega}n} = \sum_{n=0}^{\infty} a^n e^{-j\hat{\omega}n}$$

We can obtain a closed-form expression for $X(e^{j\hat{\omega}})$ by noting that

$$X(e^{j\hat{\omega}}) = \sum_{n=0}^{\infty} (ae^{-j\hat{\omega}})^n$$

which can now be recognized as the sum of all the terms of an infinite geometric series, where the ratio between successive terms is $(ae^{-j\hat{\omega}})$. For such a series there is a formula for the sum that we can apply to give the final result

$$X(e^{j\hat{\omega}}) = \sum_{n=0}^{\infty} (ae^{-j\hat{\omega}})^n = \frac{1}{1 - ae^{-j\hat{\omega}}}$$

There is one limitation, however. Going from the infinite sum to the closed-form result is only valid when $|ae^{-j\hat{\omega}}| < 1$ or $|a| < 1$. Otherwise, the terms in the geometric series either grow without bound or remain constant, and their sum is infinite.

This DTFT pair is another widely used result, worthy of highlighting as we have done with the shifted impulse and pulse sequences.

DTFT Representation of $a^n u[n]$

$$x[n] = a^n u[n] \stackrel{\text{DTFT}}{\longleftrightarrow} X(e^{j\hat{\omega}}) = \frac{1}{1 - ae^{-j\hat{\omega}}} \quad \text{if } |a| < 1 \tag{7.6}$$

EXERCISE 7.2 Use the uniqueness property of the DTFT along with (7.6) to find $x[n]$ whose DTFT is

$$X(e^{j\hat{\omega}}) = \frac{1}{1 - 0.5e^{-j\hat{\omega}}}$$

EXERCISE 7.3 Use the linearity of the DTFT and (7.6) to determine the DTFT of the following sum of two right-sided exponential signals: $x[n] = (0.8)^n u[n] + 2(-0.5)^n u[n]$.

7-1.7 Existence of the DTFT

In the case of finite-length sequences such as the impulse response of an FIR filter, the sum defining the DTFT has a finite number of terms. Thus, the DTFT of an FIR filter as in (7.1) always exists because $X(e^{j\hat{\omega}})$ is always finite. However, in the general case, where one or both of the limits on the sum in (7.2) are infinite, the DTFT sum may diverge (become infinite). This is illustrated by the right-sided exponential sequence in Section 7-1.6 when $|a| > 1$.

A sufficient condition for the existence of the DTFT of a sequence $x[n]$ emerges from the following manipulation that develops a bound on the size of $X(e^{j\hat{\omega}})$:

$$\begin{aligned}
|X(e^{j\hat{\omega}})| &= \left|\sum_{n=-\infty}^{\infty} x[n]e^{-j\hat{\omega}n}\right| \\
&\le \sum_{n=-\infty}^{\infty} \left|x[n]e^{-j\hat{\omega}n}\right| && \text{(magnitude of sum} \le \text{sum of magnitudes)} \\
&\le \sum_{n=-\infty}^{\infty} |x[n]| \overbrace{\left|e^{-j\hat{\omega}n}\right|}^{1} && \text{(magnitude of product = product of magnitudes)} \\
&\le \sum_{n=-\infty}^{\infty} |x[n]|
\end{aligned}$$

If the sum of $|x[n]|$ is finite, then the DTFT will exist.

Sufficient Condition for Existence of the DTFT

$$\left|X(e^{j\hat{\omega}})\right| \le \sum_{n=-\infty}^{\infty} |x[n]| < \infty \tag{7.7}$$

A sequence $x[n]$ satisfying (7.7) is said to be *absolutely summable*, and when (7.7) holds, the infinite sum defining the DTFT $X(e^{j\hat{\omega}})$ in (7.2) is said to *converge* to a finite result for all $\hat{\omega}$.

EXAMPLE 7-2 DTFT of a Complex Exponential?

Consider a right-sided complex exponential sequence, $x[n] = r^n e^{j\hat{\omega}_0 n} u[n]$ when $r = 1$. Applying the condition of (7.7) to this sequence leads to

$$\sum_{n=0}^{\infty} |e^{j\hat{\omega}_0 n}| = \sum_{n=0}^{\infty} 1 \rightarrow \infty$$

Thus, the DTFT of a right-sided complex exponential is not guaranteed to exist, and it is easy to verify that $|X(e^{j\hat{\omega}_0})| \rightarrow \infty$. On the other hand, if $r < 1$, the DTFT of $x[n] = r^n e^{j\hat{\omega}_0 n} u[n]$ exists and is given by the result of Section 7-1.6 with $a = re^{j\hat{\omega}_0}$. The non-existence of the DTFT is also true for the related case of a two-sided complex exponential, defined as $e^{j\hat{\omega}_0 n}$ for $-\infty < n < \infty$.

7-1.8 The Inverse DTFT

Now that we have a condition for the existence of the DTFT, we need to address the question of the inverse DTFT. The uniqueness property implies that if we have a table of known DTFT pairs such as (7.3), (7.5), and (7.6), we can always go back and forth between the time-domain and frequency-domain representations simply by table lookup as in Exercise 7.2. However, with this approach, we would always be limited by the size of our table of known DTFT pairs.

Instead, we want to continue the development of the DTFT by studying a general expression for performing the inverse DTFT. The DTFT $X(e^{j\hat{\omega}})$ is a function of the continuous variable $\hat{\omega}$, so an integral (7.8) with respect to normalized frequency $\hat{\omega}$ is needed to transform $X(e^{j\hat{\omega}})$ back to the sequence $x[n]$.

Inverse DTFT

$$x[n] = \frac{1}{2\pi}\int_{-\pi}^{\pi} X(e^{j\hat{\omega}})e^{j\hat{\omega}n}\,d\hat{\omega}. \tag{7.8}$$

Observe that n is an integer parameter in the integral, while $\hat{\omega}$ now is a dummy variable of integration that disappears when the definite integral is evaluated at its limits. The variable n can take on all integer values in the range $-\infty < n < \infty$, and hence, using (7.8) we can extract each sample of a sequence $x[n]$ whose DTFT is $X(e^{j\hat{\omega}})$. We could verify that (7.8) is the correct inverse DTFT relation by substituting the definition of the DTFT in (7.2) into (7.8) and rearranging terms.

Instead of carrying out a general proof, we present a simpler and more intuitive justification by working with the shifted impulse sequence $\delta[n - n_0]$, whose DTFT is known to be

$$X(e^{j\hat{\omega}}) = e^{-j\hat{\omega}n_0}$$

The objective is to show that (7.8) gives the correct time-domain result when operating on $X(e^{j\hat{\omega}})$. If we substitute this DTFT into (7.8), we obtain

$$\frac{1}{2\pi}\int_{-\pi}^{\pi} X(e^{j\hat{\omega}})e^{j\hat{\omega}n}\,d\hat{\omega} = \frac{1}{2\pi}\int_{-\pi}^{\pi} e^{-j\hat{\omega}n_0}e^{j\hat{\omega}n}\,d\hat{\omega} = \frac{1}{2\pi}\int_{-\pi}^{\pi} e^{j\hat{\omega}(n-n_0)}\,d\hat{\omega} \tag{7.9}$$

The definite integral of the exponential must be treated as two cases: first, when $n = n_0$,

$$\frac{1}{2\pi}\int_{-\pi}^{\pi} e^{j\hat{\omega}(n-n_0)}\,d\hat{\omega} = \frac{1}{2\pi}\int_{-\pi}^{\pi} d\hat{\omega} = 1 \tag{7.10a}$$

and then for $n \neq n_0$,

$$\frac{1}{2\pi}\int_{-\pi}^{\pi} e^{j\hat{\omega}(n-n_0)}\,d\hat{\omega} = \frac{1}{2\pi}\left.\frac{e^{j\hat{\omega}(n-n_0)}}{j(n-n_0)}\right|_{-\pi}^{\pi} = \frac{e^{j\pi(n-n_0)} - e^{-j\pi(n-n_0)}}{j2\pi(n-n_0)} = 0 \tag{7.10b}$$

since the numerator is zero when $n - n_0$ is a nonzero integer.

Equations (7.10a) and (7.10b) also show that the complex exponentials $e^{j\hat{\omega}n}$ and $e^{-j\hat{\omega}n_0}$ (when viewed as periodic functions of $\hat{\omega}$) are *orthogonal* to each other.[3]

Putting these two cases together, we have

$$\frac{1}{2\pi}\int_{-\pi}^{\pi} e^{-j\hat{\omega}n_0}e^{j\hat{\omega}n}\,d\hat{\omega} = \begin{Bmatrix} 1 & n = n_0 \\ 0 & n \neq n_0 \end{Bmatrix} = \delta[n-n_0] \tag{7.11}$$

Thus, we have shown that (7.8) correctly returns the sequence $x[n] = \delta[n-n_0]$, when the DTFT is $X(e^{j\hat{\omega}}) = e^{-j\hat{\omega}n_0}$.

This example is actually strong enough to justify that the inverse DTFT integral (7.8) will always work for finite-length sequences, because the DTFT of a general finite-length sequence is always a finite linear combination of complex exponential terms like $e^{-j\hat{\omega}n_0}$. The linearity property of the DTFT, therefore, guarantees that the inverse DTFT integral will recover a finite-length sequence that is the same linear combination of shifted impulses, which is the correct sequence for a finite-length signal. If the signal is of infinite extent, it can be shown that if $x[n]$ is absolutely summable as in (7.7) so that the DTFT exists, then (7.8) recovers the original sequence from $X(e^{j\hat{\omega}})$.

EXERCISE 7.4

Recall that $X(e^{j\hat{\omega}})$ defined in (7.2) is always periodic in $\hat{\omega}$ with period 2π. Use this fact and a change of variables to argue that we can rewrite the inverse DTFT integral with limits that go from 0 to 2π, instead of $-\pi$ to $+\pi$; that is, show that

$$\frac{1}{2\pi}\int_{-\pi}^{\pi} X(e^{j\hat{\omega}})e^{j\hat{\omega}n}\,d\hat{\omega} = \frac{1}{2\pi}\int_{0}^{2\pi} X(e^{j\hat{\omega}})e^{j\hat{\omega}n}\,d\hat{\omega}$$

7-1.9 Bandlimited DTFT

Ordinarily we define a signal in the time domain, but the inverse DTFT integral enables us to define a signal in the frequency domain by specifying its DTFT as a function of frequency. Once we specify the magnitude and phase of $X(e^{j\hat{\omega}})$, we apply (7.8) and carry out the integral to get the signal $x[n]$. An important example of this process is to define an *ideal bandlimited signal*, which is a function that is nonzero in the low frequency band $|\hat{\omega}| \le \hat{\omega}_b$ and zero in the high frequency band $\hat{\omega}_b < |\hat{\omega}| \le \pi$. If the nonzero portion of the DTFT is a constant value of one with a phase of zero, then we have

$$X(e^{j\hat{\omega}}) = \begin{cases} 1 & |\hat{\omega}| \le \hat{\omega}_b \\ 0 & \hat{\omega}_b < |\hat{\omega}| \le \pi \end{cases}$$

which is plotted in Fig. 7-1(a).

[3]This same property is used in Section C-1 of Appendix C to derive the Fourier series integral for periodic continuous-time signals. Note, for example, the similarity between equations (C.1) and (7.8).

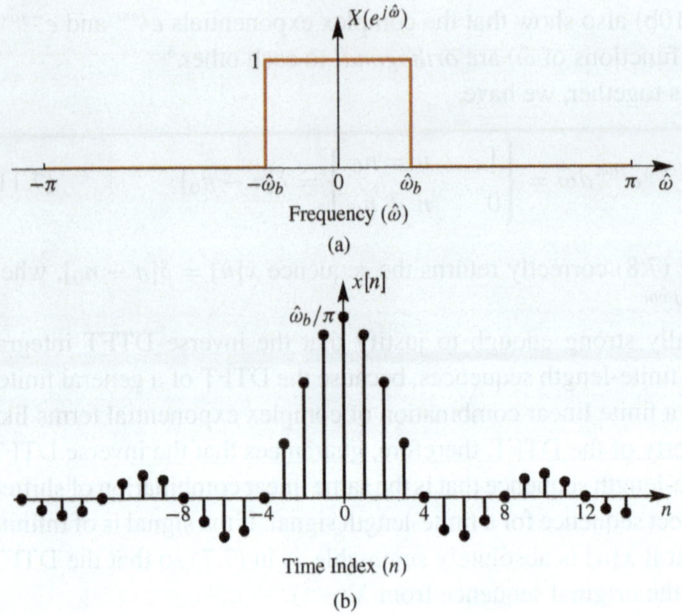

Figure 7-1 Bandlimited DTFT. (a) DTFT is a rectangle bandlimited to $|\hat{\omega}| < \hat{\omega}_b$. (b) Inverse DTFT is a sampled sinc function with $\hat{\omega}_b = 0.25\pi$.

For this simple DTFT function, the integrand of (7.8) is a piecewise constant function that is relatively easy to integrate after we substitute the definition of $X(e^{j\hat{\omega}})$ into the inverse DTFT integral (7.8)

$$x[n] = \frac{1}{2\pi}\int_{-\pi}^{\pi} X(e^{j\hat{\omega}})e^{j\hat{\omega}n}\,d\hat{\omega}$$

$$= \frac{1}{2\pi}\int_{-\pi}^{-\hat{\omega}_b} 0\,e^{j\hat{\omega}n}\,d\hat{\omega} + \frac{1}{2\pi}\int_{-\hat{\omega}_b}^{\hat{\omega}_b} 1\,e^{j\hat{\omega}n}\,d\hat{\omega} + \frac{1}{2\pi}\int_{\hat{\omega}_b}^{\pi} 0\,e^{j\hat{\omega}n}\,d\hat{\omega}$$

The integral has been broken into three cases for the three intervals, where $X(e^{j\hat{\omega}})$ is either zero or one. Only the middle integral is nonzero, and the integration yields

$$x[n] = \frac{1}{2\pi}\int_{-\hat{\omega}_b}^{\hat{\omega}_b} 1\,e^{j\hat{\omega}n}\,d\hat{\omega}$$

$$= \left.\frac{e^{j\hat{\omega}n}}{2\pi jn}\right|_{-\hat{\omega}_b}^{\hat{\omega}_b} = \frac{e^{j\hat{\omega}_b n} - e^{-j\hat{\omega}_b n}}{(2j)\pi n} = \frac{\sin(\hat{\omega}_b n)}{\pi n}$$

The last step uses the inverse Euler formula for sine.

The result of the inverse DTFT is the discrete-time signal

$$x[n] = \frac{\sin(\hat{\omega}_b n)}{\pi n} \qquad -\infty < n < \infty \tag{7.12}$$

where $0 < \hat{\omega}_b < \pi$. This mathematical form, which is called a "sinc function," is plotted in Fig. 7-1(b) for $\hat{\omega}_b = 0.25\pi$. Although the "sinc function" appears to be undefined at $n = 0$, a careful application of L'Hôpital's rule, or the small angle approximation to the sine function, shows that the value is actually $x[0] = \hat{\omega}_b/\pi$. Since a DTFT pair is unique, we have obtained another DTFT pair that can be added to our growing inventory.

DTFT Representation of a Sinc Function

$$x[n] = \frac{\sin(\hat{\omega}_b n)}{\pi n} \stackrel{\text{DTFT}}{\longleftrightarrow} X(e^{j\hat{\omega}}) = \begin{cases} 1 & |\hat{\omega}| \leq \hat{\omega}_b \\ 0 & \text{otherwise} \end{cases} \tag{7.13}$$

We will revisit this transform as a frequency response in Section 7-3 when discussing ideal filters.

> Our usage of the term sinc function refers to a form, rather than a specific function definition. The form of the "sinc function" has a sine function in the numerator and the argument of the sine in the denominator. In signal processing, the normalized sinc function is defined as
>
> $$\text{sinc}(\theta) = \frac{\sin \pi\theta}{\pi\theta}$$
>
> and this is the definition used in the MATLAB M-file `sinc`. If we expressed $x[n]$ in (7.12) in terms of this definition of the sinc function, we would write
>
> $$x[n] = \frac{\hat{\omega}_b}{\pi}\text{sinc}\left(\frac{\hat{\omega}_b}{\pi}n\right)$$
>
> While it is nice to have the name sinc for this function, which turns up often in Fourier transform expressions, it can be cumbersome to figure out the proper argument and scaling. On the other hand, the term "sinc function" is widely used as a convenient shorthand for any function of the general form of (7.12), so we use the term henceforth in that sense.

The sinc signal is important in discrete-time signal and system theory, but it is impossible to determine its DTFT by directly applying the forward transform summation (7.2). This can be seen by writing out the forward DTFT of a sinc function, which is the infinite summation below.

$$X(e^{j\hat{\omega}}) = \sum_{n=-\infty}^{\infty} \frac{\sin(\hat{\omega}_b n)}{\pi n} e^{-j\hat{\omega}n} = \begin{cases} 1 & |\hat{\omega}| \leq \hat{\omega}_b \\ 0 & \text{otherwise} \end{cases} \tag{7.14}$$

However, because of the uniqueness of the DTFT, we have already obtained the desired DTFT transform pair by starting in the frequency domain with the correct transform $X(e^{j\hat{\omega}})$ and taking the inverse DTFT of $X(e^{j\hat{\omega}})$ to get the sinc function sequence.

Another property of the sinc function sequence is that it is not absolutely summable. If we recall from (7.7) that absolute summability is a *sufficient condition* for the DTFT to exist, then the sinc function must be an exception. We know that its DTFT exists because the right-hand side of (7.14) is finite and well defined. Therefore, the transform pair (7.13) shows that the condition of absolute summability is a sufficient, but not a necessary condition, for the existence of the DTFT.

7-1.10 Inverse DTFT for the Right-Sided Exponential

Another infinite-length sequence is the right-sided exponential signal $x[n] = a^n u[n]$ discussed in Section 7-1.6. In this case, we were able to use a familiar result for geometric series to "sum" the expression for the DTFT and obtain a closed-form representation

$$X(e^{j\hat{\omega}}) = \frac{1}{1 - ae^{-j\hat{\omega}}} \qquad |a| < 1 \tag{7.15}$$

On the other hand, suppose that we want to determine $x[n]$ given $X(e^{j\hat{\omega}})$ in (7.15). Substituting this into the inverse DTFT expression (7.8) gives

$$x[n] = \frac{1}{2\pi} \int_{-\pi}^{\pi} \frac{e^{j\hat{\omega}n}}{1 - ae^{-j\hat{\omega}}} \, d\hat{\omega} \tag{7.16}$$

Although techniques exist for evaluating such integrals using the theory of complex variables, we do not assume knowledge of these techniques. Instead, we invoke the uniqueness property of the DTFT which tells us that we can always rely on the tabulated result in (7.6). In other words, we can write the inverse transform by "table lookup." The important point of this example and the sinc function example is that once a transform pair has been determined, by whatever means, we can use that DTFT relationship to move back and forth between the time and frequency domains without integrals or sums. Furthermore, in Section 7-2 we will introduce a number of general properties of the DTFT that can be employed to simplify forward and inverse DTFT manipulations even more.

7-1.11 The DTFT Spectrum

So far, we have not used the term "spectrum" when discussing the DTFT, but it should be clear at this point that it is appropriate to refer to the DTFT as a spectrum representation of a discrete-time signal. Recall that we introduced the term spectrum in Chapter 3 to mean the collection of frequency and complex amplitude information required to synthesize a signal using the Fourier synthesis equation (3.26) in Section 3-5. In the case of the DTFT, the synthesis equation is the inverse transform integral (7.8), and the analysis equation (7.2) provides a means for determining the complex amplitudes of the complex exponentials $e^{j\hat{\omega}n}$ in the synthesis equation. To make this a little more concrete, we can

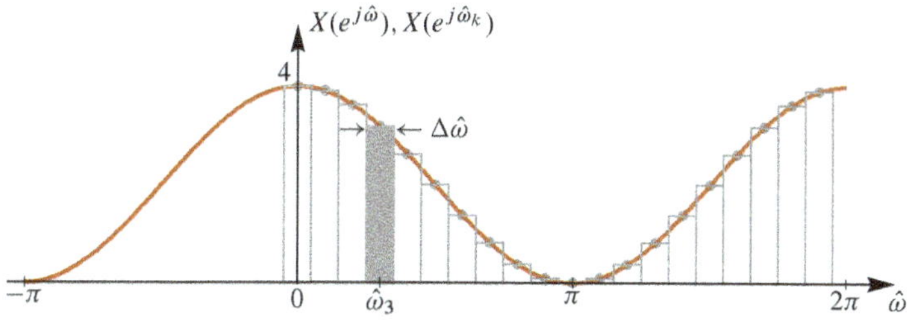

Figure 7-2 Riemann sum approximation to the integral of the inverse DTFT. Here, $X(e^{j\hat{\omega}}) = 2(1+\cos\hat{\omega})$, which is the DTFT of $x[n] = \delta[n+1] + 2\delta[n] + \delta[n-1]$.

view the inverse DTFT integral as the limit of a finite sum by writing (7.8) in terms of the Riemann sum definition[4] of the integral

$$x[n] = \frac{1}{2\pi}\int_{0}^{2\pi} X(e^{j\hat{\omega}})e^{j\hat{\omega}n}d\hat{\omega} = \lim_{\Delta\hat{\omega}\to 0}\sum_{k=0}^{N-1}\left(\frac{1}{2\pi}X(e^{j\hat{\omega}_k})\Delta\hat{\omega}\right)e^{j\hat{\omega}_k n} \qquad (7.17)$$

where $\Delta\hat{\omega} = 2\pi/N$ is the spacing between the frequencies $\hat{\omega}_k = 2\pi k/N$, with the range of integration[5] $0 \le \hat{\omega} < 2\pi$ being covered by choosing $k = 0, 1, \ldots, N-1$. The expression on the right in (7.17) contains a *sum of complex exponential signals* whose spectrum representation is the set of frequencies $\hat{\omega}_k$ together with the corresponding complex amplitudes $X(e^{j\hat{\omega}_k})\Delta\hat{\omega}/(2\pi)$. This is illustrated in Fig. 7-2 which shows the values $X(e^{j\hat{\omega}_k})$ as gray dots and the rectangles have area equal to $X(e^{j\hat{\omega}_k})\Delta\hat{\omega}$. Each one of the rectangles can be viewed as a spectrum line, especially when $\Delta\hat{\omega} \to 0$. In the limit as $\Delta\hat{\omega} \to 0$, the magnitudes of the spectral components become infinitesimally small, as does the spacing between frequencies. Therefore, (7.17) suggests that the inverse DTFT integral synthesizes the signal $x[n]$ as a sum of infinitely small complex exponentials with all frequencies $0 \le \hat{\omega} < 2\pi$ being used in the sum. The changing magnitude of $X(e^{j\hat{\omega}})$ specifies the relative amount of each frequency component that is required to synthesize $x[n]$. This is entirely consistent with the way we originally defined and subsequently used the concept of spectrum in Chapters 3–6, so we henceforth feel free to apply the term *spectrum* also to the DTFT representation.

7-2 Properties of the DTFT

We have motivated our study of the DTFT primarily by considering the problem of determining the frequency response of a filter, or more generally the Fourier representation of a signal. While these are important applications of the DTFT, it is also important to note that the DTFT also plays an important role as an "operator" in the theory of discrete-time signals and systems. This is best illustrated by highlighting some of the important properties of the DTFT operator.

[4]The Riemann sum approximation to an integral is $\int_0^u q(t)dt \approx \sum_{n=0}^{N_u-1} q(t_n)\Delta t$, where the integrand $q(t)$ is sampled at N_u equally spaced times $t_n = n\Delta t$, and $\Delta t = u/N_u$.

[5]For convenience in indexing the sum, we have used the result of Exercise 7.4 to change the limits to 0 to 2π.

7-2.1 Linearity Property

As we showed in Section 7-1.3, the DTFT operation obeys the principle of superposition (i.e., it is a linear operation). This is summarized in (7.18)

Linearity Property of the DTFT

$$x[n] = ax_1[n] + bx_2[n] \overset{\text{DTFT}}{\longleftrightarrow} X(e^{j\hat{\omega}}) = aX_1(e^{j\hat{\omega}}) + bX_2(e^{j\hat{\omega}}) \tag{7.18}$$

7-2.2 Time-Delay Property

When we first studied sinusoids, the phase was shown to depend on the time-shift of the signal. The simple relationship is "phase equals the negative of frequency times time-shift." This concept carries over to the general case of the Fourier transform. The time-delay property of the DTFT states that time-shifting results in a phase change in the frequency domain:

Time-Delay Property of the DTFT

$$y[n] = x[n - n_d] \overset{\text{DTFT}}{\longleftrightarrow} Y(e^{j\hat{\omega}}) = X(e^{j\hat{\omega}})e^{-j\hat{\omega}n_d} \tag{7.19}$$

The reason that the delay property is so important and useful is that (7.19) shows that multiplicative factors of the form $e^{-j\hat{\omega}n_d}$ in frequency-domain expressions always signify time delay.

EXAMPLE 7-3 Delayed Sinc Function

Let $y[n] = x[n - 10]$, where $x[n]$ is the sinc function of (7.13):

$$y[n] = \frac{\sin \hat{\omega}_b(n - 10)}{\pi(n - 10)}$$

Using the time-delay property and the result for $X(e^{j\hat{\omega}})$ in (7.13), we can write down the following expression for the DTFT of $y[n]$ with virtually no further analysis:

$$Y(e^{j\hat{\omega}}) = X(e^{j\hat{\omega}})e^{-j\hat{\omega}10} = \begin{cases} e^{-j\hat{\omega}10} & 0 \le |\hat{\omega}| \le \hat{\omega}_b \\ 0 & \hat{\omega}_b < |\hat{\omega}| \le \pi \end{cases}$$

Notice that the magnitude plot of $|Y(e^{j\hat{\omega}})|$ is still a rectangle as in Fig. 7-1(a); time delay only changes the phase in the frequency domain.

To prove the time-delay property, consider a sequence $y[n] = x[n - n_d]$, which is simply a time-shifted version of another sequence $x[n]$. We need to compare the DTFT of $y[n]$ to the DTFT of $x[n]$. By definition, the DTFT of $y[n]$ is

$$Y(e^{j\hat{\omega}}) = \sum_{n=-\infty}^{\infty} \underbrace{x[n - n_d]}_{y[n]} e^{-j\hat{\omega}n} \tag{7.20}$$

If we make the substitution $m = n - n_d$ for the index of summation in (7.20), we obtain

$$Y(e^{j\hat{\omega}}) = \sum_{m=-\infty}^{\infty} x[m]e^{-j\hat{\omega}(m+n_d)} = \sum_{m=-\infty}^{\infty} x[m]e^{-j\hat{\omega}m}e^{-j\hat{\omega}n_d} \tag{7.21}$$

Since the factor $e^{-j\hat{\omega}n_d}$ does not depend on m and is common to all the terms in the sum on the right in (7.21), we can write $Y(e^{j\hat{\omega}})$ as

$$Y(e^{j\hat{\omega}}) = \left(\sum_{m=-\infty}^{\infty} x[m]e^{-j\hat{\omega}m}\right) e^{-j\hat{\omega}n_d} = X(e^{j\hat{\omega}})e^{-j\hat{\omega}n_d} \tag{7.22}$$

Therefore, we have proved that time-shifting results in a phase change in the frequency domain.

7-2.3 Frequency-Shift Property

Consider a sequence $y[n] = e^{j\hat{\omega}_0 n}x[n]$ where the DTFT of $x[n]$ is $X(e^{j\hat{\omega}})$. The multiplication by a complex exponential causes a frequency shift in the DTFT of $y[n]$ compared to the DTFT of $x[n]$. By definition, the DTFT of $y[n]$ is

$$Y(e^{j\hat{\omega}}) = \sum_{n=-\infty}^{\infty} \underbrace{e^{j\hat{\omega}_0 n}x[n]}_{y[n]}\, e^{-j\hat{\omega}n} \tag{7.23}$$

If we combine the exponentials in the summation on the right side of (7.23), we obtain

$$Y(e^{j\hat{\omega}}) = \sum_{n=-\infty}^{\infty} x[n]e^{-j(\hat{\omega}-\hat{\omega}_0)n} = X(e^{j(\hat{\omega}-\hat{\omega}_0)}) \tag{7.24}$$

Therefore, we have proved the following general property of the DTFT:

Frequency-Shift Property of the DTFT

$$y[n] = e^{j\hat{\omega}_0 n}x[n] \stackrel{\text{DTFT}}{\longleftrightarrow} Y(e^{j\hat{\omega}}) = X(e^{j(\hat{\omega}-\hat{\omega}_0)}) \tag{7.25}$$

7-2.3.1 DTFT of a Finite-Length Complex Exponential

An excellent illustration of the frequency-shifting property comes from studying the DTFT of a finite-duration complex exponential. This is an important case for bandpass filter design and for spectrum analysis. In spectrum analysis, we would expect the DTFT to have a large value at the frequency of the finite-duration complex exponential signal, and the frequency-shifting property makes it easy to see that fact.

Consider a length-L complex exponential signal

$$x_L[n] = Ae^{j(\hat{\omega}_0 n+\varphi)} \qquad \text{for } n = 0, 1, 2, \ldots, L-1$$

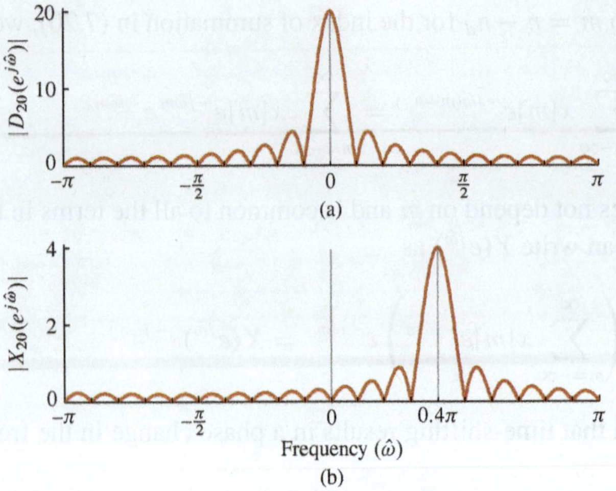

Figure 7-3 Illustration of DTFT frequency-shifting property. (a) DTFT magnitude for the length-20 unit amplitude rectangular pulse. (b) DTFT of a length-20 complex exponential with amplitude $A = 0.2$, and whose frequency is $\hat{\omega}_0 = 0.4\pi$.

which is zero for $n < 0$ and $n \geq L$. An alternative representation for $x_L[n]$ is the product of a length-L rectangular pulse times the complex exponential.

$$x_L[n] = Ae^{j\varphi} \underbrace{e^{j(\hat{\omega}_0 n)}\, r_L[n]}_{\text{frequency shift}}$$

where the rectangular pulse $r_L[n]$ is equal to one for $n = 0, 1, 2, \ldots, L-1$, and zero elsewhere.

To determine an expression for the DTFT of $x_L[n]$, we use the DTFT of $r_L[n]$ which is known to contain a Dirichlet form[6] as in (7.5) (also see Table 7-1 on p. 293). Figure 7-3(a) shows the DTFT magnitude for a length-20 rectangular pulse (i.e., $|R_{20}(e^{j\hat{\omega}})| = |D_{20}(\hat{\omega})|$). The peak value of the Dirichlet magnitude is 20 at $\hat{\omega} = 0$, and there are zeros at integer multiples of $2\pi/20 = 0.1\pi$, when $L = 20$.

Then the DTFT of $x_L[n]$ is obtained with the frequency-shifting property:

$$X_L(e^{j\hat{\omega}}) = Ae^{j\varphi}\, R_L(e^{j(\hat{\omega}-\hat{\omega}_0)}) \tag{7.26a}$$

$$= Ae^{j\varphi}\, D_L(\hat{\omega}-\hat{\omega}_0)\, e^{-j(\hat{\omega}-\hat{\omega}_0)(L-1)/2} \tag{7.26b}$$

where $D_L(\hat{\omega}-\hat{\omega}_0)$ is a frequency-shifted version of the Dirichlet form

$$D_L(\hat{\omega}-\hat{\omega}_0) = \frac{\sin((\hat{\omega}-\hat{\omega}_0)L/2)}{\sin((\hat{\omega}-\hat{\omega}_0)/2)} \tag{7.27}$$

Since the exponential terms in (7.26b) only contribute to the phase, this result says that the magnitude $|X_L(e^{j\hat{\omega}})| = A|D_L(\hat{\omega}-\hat{\omega}_0)|$ is a frequency-shifted Dirichlet that is scaled by A. For $\hat{\omega}_0 = 0.4\pi$, $A = 0.2$, and $L = 20$, the length-20 complex exponential, $x_{20}[n] = 0.2\exp(j0.4\pi n)r_{20}[n]$ has the DTFT magnitude shown in Fig. 7-3(b). Notice that the peak of the shifted Dirichlet envelope is at the frequency of the complex exponential, $\hat{\omega}_0 = 0.4\pi$. The peak height is the product of the Dirichlet peak height and the amplitude of the complex exponential, $AL = (0.2)(20) = 4$.

[6]The Dirichlet form was first defined in Section 6-7 on p. 235.

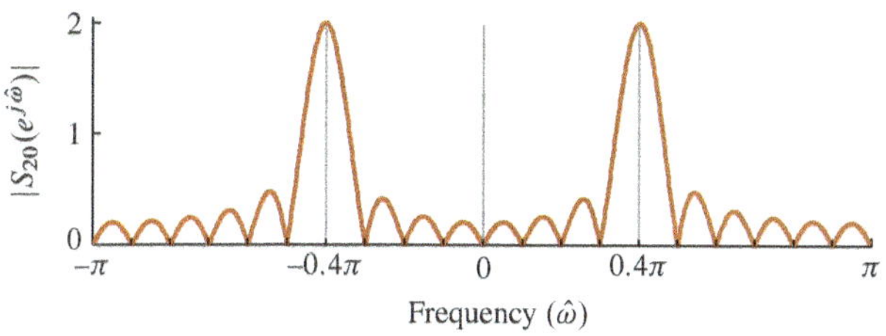

Figure 7-4 DTFT of a length-20 sinusoid with amplitude $A = 0.2$, and frequency $\hat{\omega}_0 = 0.4\pi$.

7-2.3.2 DTFT of a Finite-Length Real Cosine Signal

A sinusoid is composed of two complex exponentials, so the frequency-shifting property has to be applied twice to obtain the DTFT. Consider a length-L sinusoid

$$s_L[n] = A\cos(\hat{\omega}_0 n + \varphi) \qquad \text{for } n = 0, 1, \ldots, L-1 \tag{7.28a}$$

which we can write as the sum of complex exponentials at frequencies $+\hat{\omega}_0$ and $-\hat{\omega}_0$ as follows:

$$s_L[n] = \tfrac{1}{2}Ae^{j\varphi}e^{j\hat{\omega}_0 n} + \tfrac{1}{2}Ae^{-j\varphi}e^{-j\hat{\omega}_0 n} \qquad \text{for } n = 0, 1, \ldots, L-1 \tag{7.28b}$$

Using the linearity of the DTFT and (7.26b) for the two frequencies $\pm\hat{\omega}_0$ leads to the expression

$$\begin{aligned} S_L(e^{j\hat{\omega}}) = {} & \tfrac{1}{2}Ae^{j\varphi}D_L(\hat{\omega} - \hat{\omega}_0)e^{-j(\hat{\omega}-\hat{\omega}_0)(L-1)/2} \\ & + \tfrac{1}{2}Ae^{-j\varphi}D_L(\hat{\omega} + \hat{\omega}_0)e^{-j(\hat{\omega}+\hat{\omega}_0)(L-1)/2} \end{aligned} \tag{7.29}$$

where the function $D_L(\hat{\omega})$ is the Dirichlet form in (7.27). In words, the DTFT is the sum of two Dirichlets: one shifted up to $+\hat{\omega}_0$ and the other down to $-\hat{\omega}_0$.

Figure 7-4 shows $|S_{20}(e^{j\hat{\omega}})|$ as a function of $\hat{\omega}$ for the case $\hat{\omega}_0 = 0.4\pi$ with $A = 0.2$ and $L = 20$. The DTFT magnitude exhibits its characteristic even symmetry, and the peaks of the DTFT occur near $\hat{\omega}_0 = \pm 0.4\pi$. Furthermore, the peak heights are equal to approximately $\frac{1}{2}AL$. This approximation comes from evaluating (7.29) for $\hat{\omega}_0 = 0.4\pi$, and assuming that the value of $|S_{20}(e^{j\hat{\omega}_0})|$ is determined entirely by the first term in (7.29).

In Section 8-7, we will revisit the fact that isolated spectral peaks are often indicative of sinusoidal signal components. Knowledge that the peak height depends on both the amplitude A and the duration L is useful in interpreting spectrum analysis results for signals involving multiple frequencies.

7-2.4 Convolution and the DTFT

Perhaps the most important property of the DTFT concerns the DTFT of a sequence that is the discrete-time convolution of two sequences. The following property says that the DTFT *transforms convolution into multiplication.*

Convolution Property of the DTFT

$$y[n] = x[n] * h[n] \quad \overset{\text{DTFT}}{\longleftrightarrow} \quad Y(e^{j\hat{\omega}}) = X(e^{j\hat{\omega}})H(e^{j\hat{\omega}}) \tag{7.30}$$

To illustrate the convolution property, consider the convolution of two signals $h[n] * x[n]$, where $x[n]$ is a finite-length signal with three nonzero values

$$x[n] = 3\delta[n] + 4\delta[n-1] + 5\delta[n-2] \tag{7.31}$$

and $h[n]$ is a signal whose length may be finite or infinite. When the signal $x[n]$ is convolved with $h[n]$, we can write

$$\begin{aligned} y[n] &= \sum_{k=0}^{M} x[k]h[n-k] \\ &= x[0]h[n] + x[1]h[n-1] + x[2]h[n-2] + \underbrace{x[3]h[n-3] + \cdots}_{\text{these terms are zero}} \end{aligned} \tag{7.32}$$

If we then take the DTFT of (7.32), we obtain

$$Y(e^{j\hat{\omega}}) = x[0]H(e^{j\hat{\omega}}) + x[1]e^{-j\hat{\omega}}H(e^{j\hat{\omega}}) + x[2]\underbrace{e^{-j2\hat{\omega}}H(e^{j\hat{\omega}})} \tag{7.33}$$

where the delay property applied to a term like $h[n-2]$ creates the term $e^{-j2\hat{\omega}}H(e^{j\hat{\omega}})$. Then we observe that the term $H(e^{j\hat{\omega}})$ on the right-hand side of (7.33) can be factored out to write

$$Y(e^{j\hat{\omega}}) = \underbrace{\left(x[0] + x[1]e^{-j\hat{\omega}} + x[2]e^{-j2\hat{\omega}}\right)}_{\text{DTFT of } x[n]} H(e^{j\hat{\omega}}) \tag{7.34a}$$

and, therefore, we have the desired result which is multiplication of the DTFTs as asserted in (7.30).

$$Y(e^{j\hat{\omega}}) = X(e^{j\hat{\omega}})H(e^{j\hat{\omega}}) \tag{7.34b}$$

Substituting the signal values for $x[n]$ in this example we obtain

$$Y(e^{j\hat{\omega}}) = X(e^{j\hat{\omega}})H(e^{j\hat{\omega}}) = \underbrace{\left(3 + 4e^{-j\hat{\omega}} + 5e^{-j2\hat{\omega}}\right)}_{\text{coefficients are signal values}} H(e^{j\hat{\omega}}) \tag{7.34c}$$

The steps above do not depend on the numerical values of the signal $x[n]$, so a general proof could be constructed along these lines. In fact, the proof would be valid for signals of infinite length, where the limits on the sum in (7.32) would be infinite. The only additional concern for the infinite-length case would be that the DTFTs $X(e^{j\hat{\omega}})$ and $H(e^{j\hat{\omega}})$ must exist as we have discussed before.

EXAMPLE 7-4 Frequency Response of Delay

The delay property is a special case of the convolution property of the DTFT. To see this, recall that we can represent delay as the convolution with a shifted impulse

$$y[n] = x[n] * \delta[n - n_d] = x[n - n_d]$$

so the impulse response of a delay system is $h[n] = \delta[n - n_d]$. The corresponding frequency response (i.e., DTFT) of the delay system is

$$H(e^{j\hat{\omega}}) = \sum_{n=-\infty}^{\infty} \delta[n - n_d]e^{-j\hat{\omega}n} = e^{-j\hat{\omega}n_d}$$

Therefore, using the convolution property, the DTFT of the output of the delay system is

$$Y(e^{j\hat{\omega}}) = X(e^{j\hat{\omega}})H(e^{j\hat{\omega}}) = X(e^{j\hat{\omega}})e^{-j\hat{\omega}n_d}$$

which is identical to the delay property of (7.19).

7-2.4.1 Filtering is Convolution

The convolution property of LTI systems provides an effective way to think about LTI systems. In particular, when we think of LTI systems as "filters" we are thinking of their frequency responses, which can be chosen so that some frequencies of the input are blocked while others pass through with little modification. As we discussed in Section 7-1.11, the DTFT $X(e^{j\hat{\omega}})$ plays the role of spectrum for both finite-length signals and infinite-length signals. The convolution property (7.30) reinforces this view. If we use (7.30) to write the expression for the output of an LTI system using the DTFT synthesis integral, we have

$$y[n] = \frac{1}{2\pi}\int_{0}^{2\pi} H(e^{j\hat{\omega}})X(e^{j\hat{\omega}})e^{j\hat{\omega}n}d\hat{\omega} \tag{7.35a}$$

which can be approximated with a Riemann sum as in Section 7-1.11

$$y[n] = \lim_{\Delta\hat{\omega}\to 0} \sum_{k=0}^{N-1} H(e^{j\hat{\omega}_k})\left(\frac{\Delta\hat{\omega}}{2\pi}X(e^{j\hat{\omega}_k})\right)e^{j\hat{\omega}_k n} \tag{7.35b}$$

Now we see that the complex amplitude $\frac{\Delta\hat{\omega}}{2\pi}X(e^{j\hat{\omega}_k})$ at each frequency is modified by the frequency response $H(e^{j\hat{\omega}_k})$ of the system evaluated at the given frequency $\hat{\omega}_k$. This is exactly the same result as the *sinusoid-in gives sinusoid-out* property that we saw in Chapter 6 for case where the input is a discrete sum of complex exponentials. In Section 7-3, we will expand on this idea and define several ideal frequency-selective filters whose frequency responses are ideal versions of filters that we might want to implement in a signal processing application.

7-2.5 Energy Spectrum and the Autocorrelation Function

An important result in Fourier transform theory is *Parseval's Theorem*:

Parseval's Theorem for the Energy of a Signal

$$\sum_{n=-\infty}^{\infty} |x[n]|^2 = \frac{1}{2\pi}\int_{-\pi}^{\pi} |X(e^{j\hat{\omega}})|^2 d\hat{\omega} \tag{7.36}$$

The left-hand side of (7.36) is a scalar value which is the energy in the signal. Thus, the right-hand side of (7.36) is also the energy, but the DTFT $|X(e^{j\hat{\omega}})|^2$ shows how the energy is distributed versus frequency. Therefore, the DTFT $|X(e^{j\hat{\omega}})|^2$ is called the *magnitude-squared spectrum* or the *energy spectrum* of $x[n]$.[7] If we first define the energy of a signal as the sum of the squares

$$E = \sum_{n=-\infty}^{\infty} |x[n]|^2 \tag{7.37}$$

then the energy E is a single number that is often a convenient measure of the size of the signal.

EXAMPLE 7-5 Energy of the Sinc Signal

The energy of the sinc signal (expressed in the time domain) is

$$E = \sum_{n=-\infty}^{\infty} \left(\frac{\sin \hat{\omega}_b n}{\pi n}\right)^2$$

While it is impossible to evaluate this sum directly, application of Parseval's theorem yields

$$E = \sum_{n=-\infty}^{\infty} \left(\frac{\sin \hat{\omega}_b n}{\pi n}\right)^2 = \frac{1}{2\pi} \int_{-\hat{\omega}_b}^{\hat{\omega}_b} |1|^2 d\hat{\omega} = \frac{\hat{\omega}_b}{\pi}$$

because the DTFT of the sinc signal is one for $-\hat{\omega}_b \leq \hat{\omega} \leq \hat{\omega}_b$. A simple interpretation of this result is that the energy is proportional to the bandwidth of the sinc signal and evenly distributed in $\hat{\omega}$ across the band $|\hat{\omega}| \leq \hat{\omega}_b$.

7-2.5.1 Autocorrelation Function

The energy spectrum is the Fourier transform of a time-domain signal which turns out to be the autocorrelation of $x[n]$. The ***autocorrelation function*** is widely used in signal detection applications. Its usual definition (7.39) is equivalent to the following convolution operation:

$$c_{xx}[n] = x[-n] * x[n] = \sum_{k=-\infty}^{\infty} x[-k]x[n-k] \tag{7.38}$$

By making the substitution $m = -k$ for the dummy index of summation, we can write

$$c_{xx}[n] = \sum_{m=-\infty}^{\infty} x[m]x[n+m] \qquad -\infty < n < \infty \tag{7.39}$$

[7] In many physical systems, energy is calculated by taking the square of a physical quantity (like voltage) and integrating, or summing, over time.

which is the basic definition of the autocorrelation function when $x[n]$ is real. Observe that in (7.39), the index n serves to shift $x[n+m]$ with respect to $x[m]$ when both sequences are thought of as functions of m. It can be shown that $c_{xx}[n]$ is maximum at $n=0$, because then, $x[n+m]$ is perfectly aligned with $x[m]$. The independent variable n in $c_{xx}[n]$ is often called the "lag" by virtue of its meaning as a shift between two copies of the same sequence $x[n]$. When the lag is zero ($n=0$) in (7.39), the value of the autocorrelation is equal to the energy in the signal (i.e., $E = c_{xx}[0]$).

EXERCISE 7.5 In (7.38), we use the time-reversed signal $x[-n]$. Show that the DTFT of $x[-n]$ is $X(e^{-j\hat{\omega}})$.

EXERCISE 7.6 In Chapter 6, we saw that the frequency response for a real impulse response must be conjugate symmetric. Since the frequency response function is a DTFT, it must also be true that the DTFT of a real signal $x[n]$ is conjugate symmetric. Show that if $x[n]$ is real, $X(e^{-j\hat{\omega}}) = X^*(e^{j\hat{\omega}})$.

Using the results of Exercises 7.5 and 7.6 for real signals, the DTFT of the autocorrelation function $c_{xx}[n] = x[-n] * x[n]$ is

$$C_{xx}(e^{j\hat{\omega}}) = X(e^{-j\hat{\omega}})X(e^{j\hat{\omega}}) = X^*(e^{j\hat{\omega}})X(e^{j\hat{\omega}}) = |X(e^{j\hat{\omega}})|^2 \tag{7.40}$$

Since $C_{xx}(e^{j\hat{\omega}})$ can be written as a magnitude squared, it is purely real and $C_{xx}(e^{j\hat{\omega}}) \geq 0$. A useful relationship results if we represent $c_{xx}[n]$ in terms of its inverse DTFT, that is,

$$c_{xx}[n] = \frac{1}{2\pi}\int_{-\pi}^{\pi} |X(e^{j\hat{\omega}})|^2 e^{j\hat{\omega}n}\, d\hat{\omega} \tag{7.41}$$

If we evaluate both (7.39) and (7.41) at $n=0$, we see that the energy of the sequence can also be computed from $C_{xx}(e^{j\hat{\omega}}) = |X(e^{j\hat{\omega}})|^2$ as follows:

$$\sum_{m=-\infty}^{\infty} |x[n]|^2 = c_{xx}[0] = \frac{1}{2\pi}\int_{-\pi}^{\pi} |X(e^{j\hat{\omega}})|^2 d\hat{\omega} \tag{7.42}$$

Finally, equating (7.37) and (7.42) we obtain *Parseval's Theorem* (7.36).

7-3 Ideal Filters

In any practical application of LTI discrete-time systems, the frequency response function $H(e^{j\hat{\omega}})$ would be derived by a filter design procedure that would yield an LTI system that could be implemented with finite computation. However, in the early phases of the system design process it is common practice to start with *ideal filters* that have simple frequency responses that provide ideal frequency selectivity.

7-3.1 Ideal Lowpass Filter

An ideal lowpass filter (LPF) has a frequency response that consists of two regions: the passband near $\hat{\omega} = 0$ (DC), where the frequency response is one, and the stopband away from $\hat{\omega} = 0$, where it is zero. An ideal LPF is therefore defined as

$$H_{\text{LP}}(e^{j\hat{\omega}}) = \begin{cases} 1 & |\hat{\omega}| \leq \hat{\omega}_{\text{co}} \\ 0 & \hat{\omega}_{\text{co}} < |\hat{\omega}| \leq \pi \end{cases} \tag{7.43}$$

The frequency $\hat{\omega}_{\text{co}}$ is called the *cutoff frequency* of the LPF passband. Figure 7-5 shows a plot of $H_{\text{LP}}(e^{j\hat{\omega}})$ for the ideal LPF. The shape is rectangular and $H_{\text{LP}}(e^{j\hat{\omega}})$ is even symmetric about $\hat{\omega} = 0$. As discussed in Section 6-4.3, this property is needed because real-valued impulse responses lead to filters with a conjugate symmetric frequency responses. Since $H_{\text{LP}}(e^{j\hat{\omega}})$ has zero phase, it is real-valued, so being conjugate symmetric is equivalent to being an even function.

EXERCISE 7.7

In Chapter 6, we saw that the frequency response for a real impulse response must be conjugate symmetric. Show that a lowpass frequency response defined with linear phase

$$H(e^{j\hat{\omega}}) = e^{-j7\hat{\omega}} H_{\text{LP}}(e^{j\hat{\omega}})$$

is conjugate symmetric, which would imply that its inverse DTFT is real.

The impulse response of the ideal LPF, found by applying the DTFT pair in (7.13), is a sinc function form

$$h_{\text{LP}}[n] = \frac{\sin \hat{\omega}_{\text{co}} n}{\pi n} \qquad -\infty < n < \infty \tag{7.44}$$

The ideal LPF is impossible to implement because the impulse response $h_{\text{LP}}[n]$ is noncausal and, in fact, has nonzero values for large negative indices as well as large positive indices. However, that does not invalidate the ideal LPF concept—the idea of selecting the low-frequency band and rejecting all other frequency components. Even the running average filter discussed in Section 6-7.3 might be a satisfactory LPF in some applications. Figure 7-6 shows the frequency response of an 11-point running average filter. Note

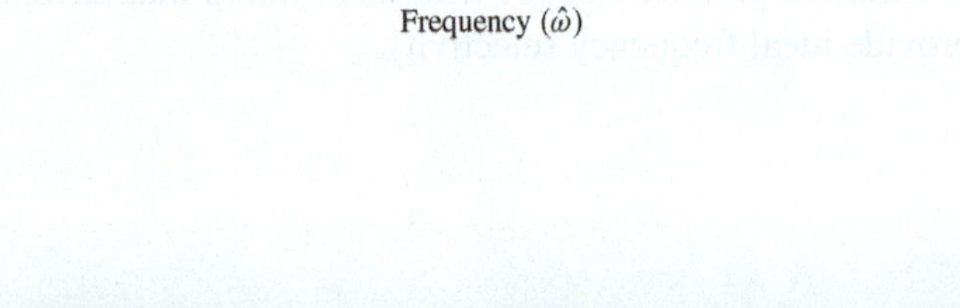

Figure 7-5 Frequency response of an ideal LPF with its cutoff at $\hat{\omega}_{\text{co}}$ rad. Recall that $H_{\text{LP}}(e^{j\hat{\omega}})$ must also be periodic with period 2π.

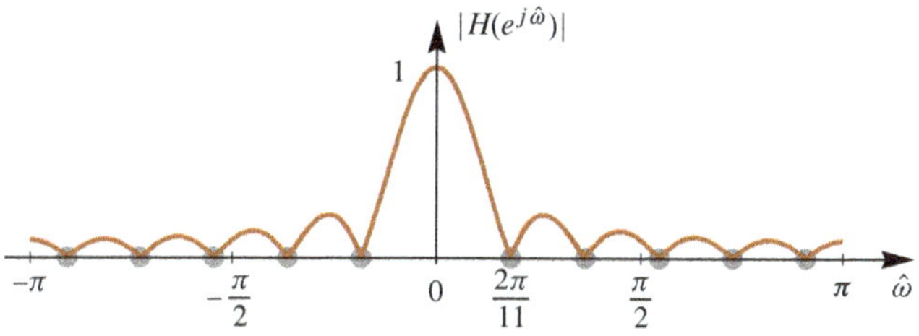

Figure 7-6 Magnitude response of an 11-point running average filter. (Also shown in full detail in Fig. 6-8.)

that this causal filter has a lowpass-like frequency response magnitude, but it is far away from zero in what might be considered the stopband. In more stringent filtering applications, we need better approximations to the ideal characteristic. In a practical application, the process of filter design involves mathematical approximation of the ideal filter with a frequency response that is close enough to the ideal frequency response while corresponding to an implementable filter.

The following example shows the power of the transform approach when dealing with filtering problems.

EXAMPLE 7-6 Ideal Lowpass Filtering

Consider an ideal LPF with frequency response given by (7.43) and impulse response in (7.44). Now suppose that the input signal $x[n]$ to the ideal LPF is a bandlimited sinc signal

$$x[n] = \frac{\sin \hat\omega_b n}{\pi n} \tag{7.45}$$

Working in the time domain, the corresponding output of the ideal LPF would be given by the convolution expression

$$y[n] = x[n] * h_{\text{LP}}[n] = \sum_{m=-\infty}^{\infty} \left(\frac{\sin \hat\omega_b m}{\pi m}\right)\left(\frac{\sin \hat\omega_{\text{co}}(n-m)}{\pi(n-m)}\right) \tag{7.46}$$

Evaluating this convolution directly in the time domain is impossible both analytically and via numerical computation. However, it is straightforward to obtain the filter output if we use the DTFT because in the frequency domain the transforms are both rectangles and they are multiplied. From (7.13), the DTFT of the input is

$$X(e^{j\hat\omega}) = \begin{cases} 1 & |\hat\omega| \le \hat\omega_b \\ 0 & \hat\omega_b < |\hat\omega| \le \pi \end{cases} \tag{7.47}$$

Therefore, the DTFT of the ideal filter's output is $Y(e^{j\hat\omega}) = X(e^{j\hat\omega})H(e^{j\hat\omega})$, which would be of the form

$$Y(e^{j\hat\omega}) = X(e^{j\hat\omega})H_{\text{LP}}(e^{j\hat\omega}) = \begin{cases} 1 & |\hat\omega| \le \hat\omega_a \\ 0 & \hat\omega_a < |\hat\omega| \le \pi \end{cases} \tag{7.48a}$$

When multiplying the DTFTs to get the right-hand side of (7.48a), the product of the two rectangles is another rectangle whose width is the smaller of $\hat{\omega}_b$ and $\hat{\omega}_{co}$, so the bandlimit frequency $\hat{\omega}_a$ is

$$\hat{\omega}_a = \min(\hat{\omega}_b, \hat{\omega}_{co}) \tag{7.48b}$$

Since we want to determine the output signal $y[n]$, we must take the inverse DTFT of $Y(e^{j\hat{\omega}})$. Thus, using (7.13) to do the inverse transformation, the convolution in (7.46) must evaluate to another sinc signal

$$y[n] = \sum_{m=-\infty}^{\infty} \left(\frac{\sin \hat{\omega}_b m}{\pi m}\right)\left(\frac{\sin \hat{\omega}_{co}(n-m)}{\pi(n-m)}\right) = \frac{\sin \hat{\omega}_a n}{\pi n} \tag{7.49}$$

EXERCISE 7.8 The result given (7.48a) and (7.48b) is easily seen from a graphical solution that shows the rectangular shapes of the DTFTs. With $\hat{\omega}_b = 0.4\pi$ and $\hat{\omega}_{co} = 0.25\pi$, sketch plots of $X(e^{j\hat{\omega}})$ from (7.47) and $H_{LP}(e^{j\hat{\omega}})$ in (7.43) on the same set of axes and then verify the result in (7.48b).

We can generalize the result of Example 7-6 in several interesting ways. First, when $\hat{\omega}_b > \hat{\omega}_{co}$ we can see that the output is the impulse response of the ideal LPF, so the input (7.45) in a sense acts like an impulse to the ideal LPF. Furthermore, for ideal filters where a band of frequencies is completely removed by the filter, many different inputs could produce the same output. Also, we can see that if the bandlimit $\hat{\omega}_b$ of the input to an ideal LPF is less than the cutoff frequency (i.e., $\hat{\omega}_b \le \hat{\omega}_{co}$), then the input signal passes through the filter unchanged. Finally, if the input consists of a desired bandlimited signal plus some sort of competing signal such as noise whose spectrum extends over the entire range $|\hat{\omega}| \le \pi$, then if the signal spectrum is concentrated in a band $|\hat{\omega}| \le \hat{\omega}_b$, it follows by the principle of superposition that an ideal LPF with cutoff frequency $\hat{\omega}_{co} = \hat{\omega}_b$ passes the desired signal without modification while removing all frequencies in the spectrum of the competing signal above the cutoff frequency. This is often the motivation for using a LPF.

7-3.2 Ideal Highpass Filter

The ideal highpass filter (HPF) has its stopband centered on low frequencies, and its passband extends from $|\hat{\omega}| = \hat{\omega}_{co}$ out to $|\hat{\omega}| = \pi$. (The highest normalized frequency in a sampled signal is, of course, π.)

$$H_{HP}(e^{j\hat{\omega}}) = \begin{cases} 0 & |\hat{\omega}| \le \hat{\omega}_{co} \\ 1 & \hat{\omega}_{co} < |\hat{\omega}| \le \pi \end{cases} \tag{7.50}$$

Figure 7-7 shows an ideal HPF with its cutoff frequency at $\hat{\omega}_{co}$ rad/s. In this case, the high frequency components of a signal pass through the filter unchanged while the

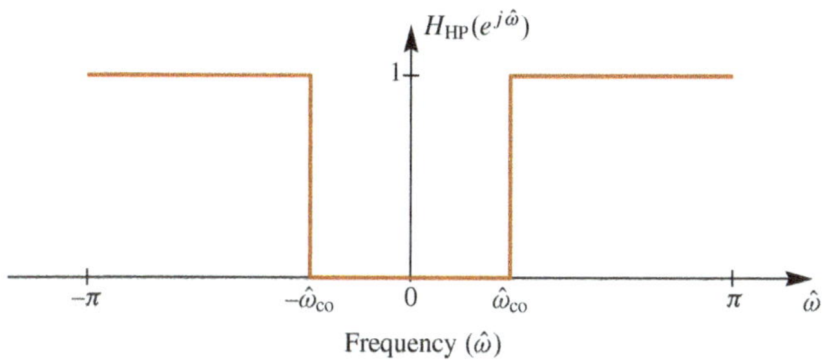

Figure 7-7 Frequency response of an ideal HPF with cutoff at $\hat{\omega}_{co}$ rad.

low frequency components are completely eliminated. Highpass filters are often used to remove constant levels (DC) from sampled signals. Like the ideal LPF, we should define the ideal highpass filter with conjugate symmetry $H_{HP}(e^{-j\hat{\omega}}) = H^*_{HP}(e^{j\hat{\omega}})$ so that the corresponding impulse response is a real function of time.

EXERCISE 7.9 If $H_{LP}(e^{j\hat{\omega}})$ is an ideal LPF with its cutoff frequency at $\hat{\omega}_{co}$ as plotted in Fig. 7-5, show that the frequency response of an ideal HPF with cutoff frequency $\hat{\omega}_{co}$ can be represented by

$$H_{HP}(e^{j\hat{\omega}}) = 1 - H_{LP}(e^{j\hat{\omega}})$$

Hint: Make a plot of the function $1 - H_{LP}(e^{j\hat{\omega}})$.

EXERCISE 7.10 Using the results of Exercise 7.9, show that the impulse response of the ideal HPF is

$$h_{HP}[n] = \delta[n] - \frac{\sin(\hat{\omega}_{co}n)}{\pi n}$$

7-3.3 Ideal Bandpass Filter

The ideal bandpass filter (BPF) has a passband centered away from the low-frequency band, so it has two stopbands, one near $\hat{\omega} = 0$ and the other at high frequencies. Two cutoff frequencies must be given to specify the ideal BPF, $\hat{\omega}_{co_1}$ for the lower cutoff, and $\hat{\omega}_{co_2}$ for the upper cutoff. Thus, the ideal BPF has frequency response

$$H_{BP}(e^{j\hat{\omega}}) = \begin{cases} 0 & |\hat{\omega}| < \hat{\omega}_{co_1} \\ 1 & \hat{\omega}_{co_1} \le |\hat{\omega}| \le \hat{\omega}_{co_2} \\ 0 & \hat{\omega}_{co_2} < |\hat{\omega}| \le \pi \end{cases} \tag{7.51}$$

Figure 7-8 shows an ideal BPF with its cutoff frequencies at $\hat{\omega}_{co_1}$ and $\hat{\omega}_{co_2}$. Once again we use a symmetrical definition of the passbands and stopbands which is required to make the corresponding impulse response real. In this case, all complex exponential components

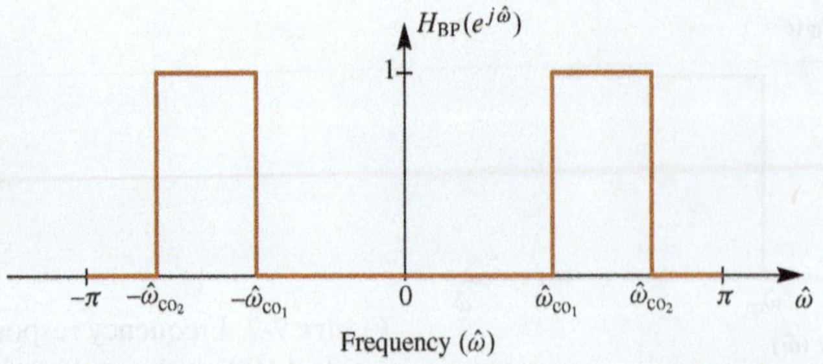

Figure 7-8 Frequency response of an ideal BPF with its passband from $\hat{\omega} = \hat{\omega}_{co_1}$ to $\hat{\omega} = \hat{\omega}_{co_2}$ rad.

of a signal that lie in the band $\hat{\omega}_{co_1} \le |\hat{\omega}| \le \hat{\omega}_{co_2}$ are passed unchanged through the filter, while those at other frequencies are completely removed.

EXERCISE 7.11 If $H_{BP}(e^{j\hat{\omega}})$ is an ideal BPF with its cutoffs at $\hat{\omega}_{co_1}$ and $\hat{\omega}_{co_2}$, show by plotting that the filter defined by $H_{BR}(e^{j\hat{\omega}}) = 1 - H_{BP}(e^{j\hat{\omega}})$ could be called an ideal *band-reject* filter. Determine the edges of the stopband of the band-reject filter.

7-4 Practical FIR Filters

Ideal filters are useful concepts, but not practical since they cannot be implemented with a finite amount of computation. Therefore, we perform *filter design* to approximate an ideal frequency response to get a practical filter. For FIR filters, a filter design method must produce filter coefficients $\{b_k\}$ for the time-domain implementation of the FIR filter as a difference equation

$$y[n] = b_0x[n] + b_1x[n-1] + b_2x[n-2] + \cdots + b_Mx[n-M]$$

The filter coefficients are the values of the impulse response, and the DTFT of the impulse response determines the actual magnitude and phase of the designed frequency response, which can then be assessed to determine how closely it matches the desired ideal response. There are many ways to approximate the ideal frequency response, but we concentrate on the method of *windowing* which can be analyzed via the DTFT.

7-4.1 Windowing

The concept of windowing is widely used in signal processing. The basic idea is to *extract a finite section* of a very long signal $x[n]$ *via multiplication* $w[n]x[n+n_0]$. This approach works if the window function $w[n]$ is zero outside of a finite-length interval. In filter design the window *truncates* the infinitely long ideal impulse response $h_i[n]$,[8] and then modifies the truncated impulse response. The simplest window function is the L-point *rectangular window* which is the same as the rectangular pulse studied in Section 7-1.5.

$$w_r[n] = r_L[n] = \begin{cases} 1 & 0 \le n \le L-1 \\ 0 & \text{elsewhere} \end{cases} \quad (7.52)$$

[8]The subscript i denotes an ideal filter of the type discussed in Section 7-3.

Multiplying by a rectangular window only truncates a signal.

The important idea of windowing is that the product $w_r[n]h_i[n+n_0]$ extracts L values from the signal $h_i[n]$ starting at $n = n_0$. For the rectangular window, we obtain

$$w_r[n]h_i[n+n_0] = \begin{cases} 0 & n < 0 \\ w_r[n]h_i[n+n_0] & 0 \le n \le L-1 \\ 0 & n \ge L \end{cases} \tag{7.53}$$

The name *window* comes from the idea that we can only "see" L values of the signal $h_i[n+n_0]$ within the window interval when we "look" through the window. Multiplying by $w[n]$ is looking through the window. When we change n_0, the signal shifts, and we see a different length-L section of the signal.

The nonzero values of the window function do not have to be all ones, but they should be positive. For example, the symmetric L-point Hamming window[9] is defined as

$$w_m[n] = \begin{cases} 0.54 - 0.46\cos(2\pi n/(L-1)) & 0 \le n \le L-1 \\ 0 & \text{elsewhere} \end{cases} \tag{7.54}$$

The MATLAB function `hamming(L)` computes a vector with values given by (7.54). The stem plot of the Hamming window in Fig. 7-9 shows that the values are larger in the middle and taper off near the ends. The window length can be even or odd, but an odd-length Hamming window is easier to characterize. Its maximum value is 1.0 which occurs at the midpoint index location $n = (L-1)/2$, and the window is symmetric about the midpoint, with *even symmetry* because $w_m[n] = w_m[L-1-n]$.

7-4.2 Filter Design

Ideal Filters are defined by their frequency responses, consisting of *perfect* passbands and stopbands. An ideal filter cannot be an FIR filter because there is no finite set of filter coefficients whose DTFT is equal to the ideal frequency response. Recall from Fig. 7-1 that the impulse response of the ideal LPF is an infinitely long sinc function as shown by the following DTFT pair:

$$h_i[n] = \frac{\sin(\hat{\omega}_{co} n)}{\pi n} \overset{\text{DTFT}}{\longleftrightarrow} H_i(e^{j\hat{\omega}}) = \begin{cases} 1 & |\hat{\omega}| \le \hat{\omega}_{co} \\ 0 & \hat{\omega}_{co} < |\hat{\omega}| \le \pi \end{cases} \tag{7.55}$$

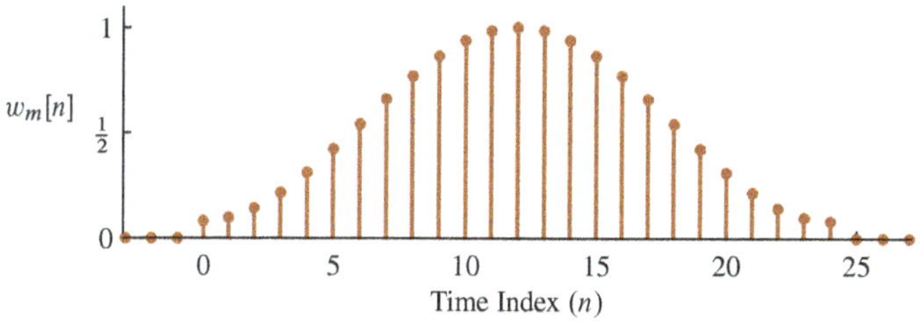

Figure 7-9 Length-25 Hamming window is nonzero only for $n = 0, 1, \ldots, 24$.

[9]This window is named for Richard Hamming who found that improved frequency-domain characteristics result from slight adjustments in the 0.5 coefficients of the Hann window (8.49) discussed in Chapter 8.

where $\hat{\omega}_{co}$ is the *cutoff frequency* of the ideal LPF, which separates the passband from the stopband. The sinc function is infinitely long.

7-4.2.1 Window the Ideal Impulse Response

In order to make a practical FIR filter, we can multiply the sinc function by a window to produce a length-L impulse response. However, we must also shift the sinc function so that its main lobe is in the center of the window, because intuitively we should use the largest values from the ideal impulse response. From the shifting property of the DTFT, the time shift of $(L-1)/2$ introduces a linear phase in the DTFT. Thus, the impulse response obtained from windowing is

$$h[n] = w[n]h_i[n-(L-1)/2]$$
$$= \begin{cases} w[n]\dfrac{\sin(\hat{\omega}_{co}(n-(L-1)/2))}{\pi(n-(L-1)/2)} & n = 0, 1, \ldots, L-1 \\ 0 & \text{elsewhere} \end{cases} \tag{7.56}$$

where $w[n]$ is the window, either rectangular or Hamming.[10] Since the nonzero domain of the window starts at $n = 0$, the resulting FIR filter is causal. We usually say that the practical FIR filter has an impulse response that is a *windowed* version of the ideal impulse response.

The windowing operation is shown in Fig. 7-10(a) for the rectangular window which truncates the ideal impulse response to length $L = 25$. In Fig. 7-10(b), the Hamming-windowed impulse response $h_m[n]$ results from truncating the ideal impulse response, and also weighting the values to reduce the ends more than the middle. The midpoint value is preserved (i.e., $h_m[12] = \hat{\omega}_{co}/\pi$) while the first and last points are 8% of their original values (e.g., $h_m[0] = 0.08h_i[0]$). A continuous outline of the Hamming window is drawn in (a) to show the weighting that is applied to the truncated ideal impulse response. The benefit of using the Hamming window comes from the fact that it smoothly tapers the ends of the truncated ideal lowpass impulse response.

7-4.2.2 Frequency Response of Practical Filters

A *practical filter* is a causal length-L filter whose frequency response closely approximates the desired frequency response of an ideal filter. Although it is possible to take the DTFT of a windowed sinc, the resulting formula is quite complicated and does not offer much insight into the quality of the frequency-domain approximation. Instead, we can evaluate the frequency reponse directly using MATLAB's `freqz` function because we have a simple formula for the windowed filter coefficients.

[10]We only consider rectangular and Hamming windows here, but there are many other window functions, and each of them results in filters with different frequency response characteristics.

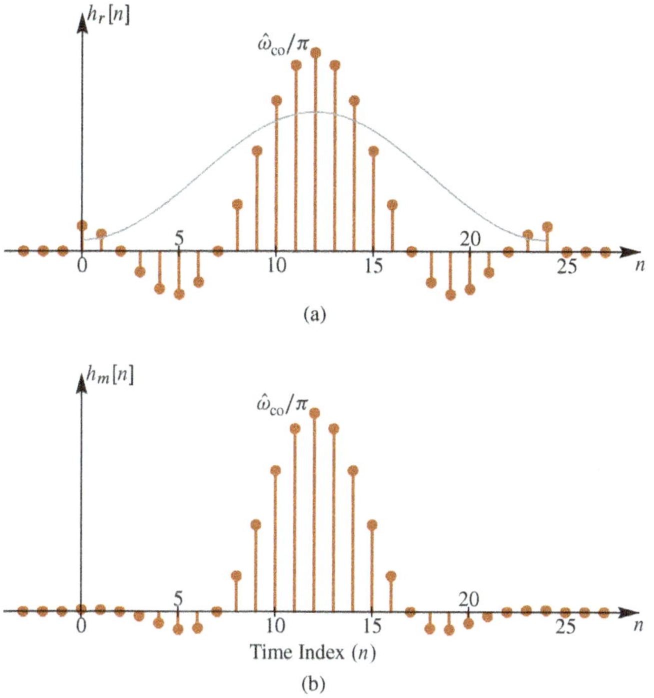

Figure 7-10 Impulse responses for two LPFs. (a) Length-25 LPF with rectangular window (i.e., truncated sinc function). (b) Length-25 LPF using 25-point Hamming window multiplying a sinc function. The shape of the Hamming window is shown in (a) because it truncates and multiplies the sinc function shown in (a) to give the plot in (b).

Figure 7-11(a) shows the magnitude response for an FIR filter whose impulse response is a length-25 truncated ideal impulse response with $\hat{\omega}_{co} = 0.4\pi$. The passband of the actual filter is not flat but it oscillates above and below the desired passband value of one; the same behavior is exhibited in the stopband. These *passband and stopband ripples* are usually observed for practical FIR filters that approximate ideal LPFs, and they are particularly noticeable with the rectangular window.

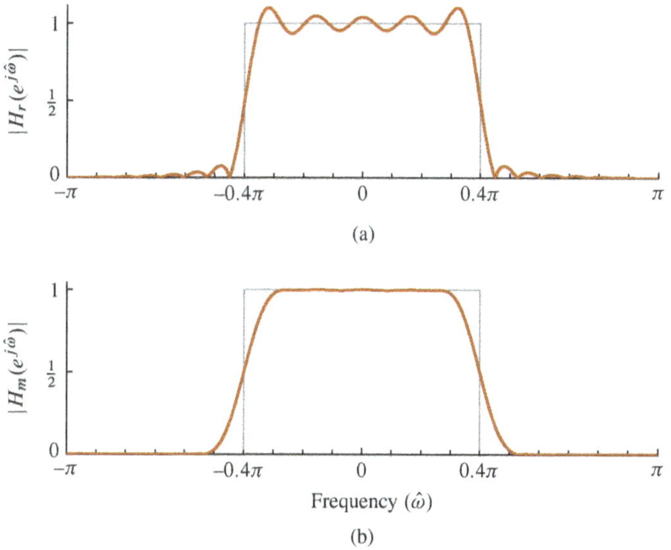

Figure 7-11 Frequency response magnitudes for LPFs whose impulse responses are shown in Fig. 7-10. (a) Length-25 LPF whose impulse response is a truncated sinc function obtained with a rectangular window. (b) Length-25 LPF whose impulse response is the product of a 25-point Hamming window and a sinc function. The ideal LPF with $\hat{\omega}_{co} = 0.4\pi$ is shown in gray. The phase response of both of these filters is a linear phase with slope $-(L-1)/2 = -12$.

Figure 7-11(b) shows the magnitude response for an FIR filter whose impulse response is a length-25 Hamming-windowed sinc. In this case, the ripples are not visible on the magnitude plot because the scale is linear and the ripples are tiny, less than 0.0033. In terms of approximating the value of one in the passband and zero in the stopband, the Hamming-windowed ideal LPF is much better. However, this improved approximation comes at a cost—the edge of the passband near the cutoff frequency has a much lower slope. In filter design, we usually say it "falls off more slowly" from the passband to the stopband. Before we can answer the question of which filter is better, we must decide whether ripples are more important than the fall off rate from passband to stopband, or vice versa.

7-4.2.3 Passband Defined for the Frequency Response

Frequency-selective digital filters (e.g., LPFs, BPFs, and HPFs) have a magnitude response that is close to one in some frequency regions, and close to zero in others. For example, the plot in Fig. 7-12(a) is an LPF whose magnitude response is within (approximately) $\pm 10\%$ of one when $0 \le \hat{\omega} < 0.364\pi$. This region where the magnitude is close to one is called the ***passband*** of the filter. It is useful to have a precise definition of the passband edges, so that the passband width can be measured and we can compare different filters.

From a plot of the magnitude response (e.g., via `freqz` in MATLAB) it is possible to determine the set of frequencies where the magnitude is very close to one, as defined by $\left||H(e^{j\hat{\omega}})| - 1\right|$ being less than δ_p, which is called the *passband ripple*. A common design choice for the desired passband ripple is a value between 0.01 and 0.1 (i.e., 1% to 10%). For a LPF, the passband region extends from $\hat{\omega} = 0$ to $\hat{\omega}_p$, where the parameter $\hat{\omega}_p$ is called the *passband edge*.

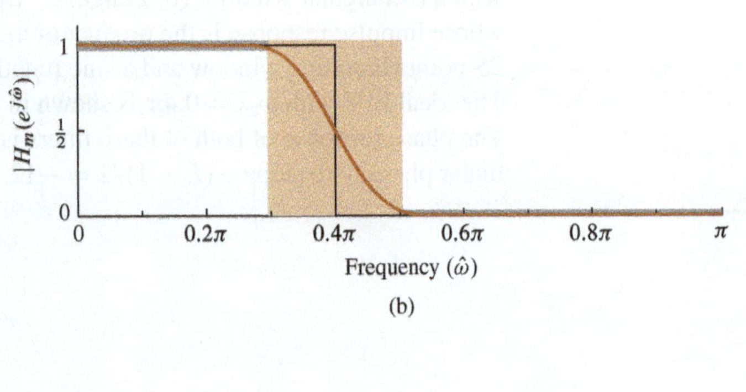

Figure 7-12 Magnitude response within an LPF template showing passband and stopband ripple tolerances along with the transition zone. (a) Length-25 LPF $H_r(e^{j\hat{\omega}})$ with rectangular window (i.e., truncated sinc). (b) Length-25 LPF $H_m(e^{j\hat{\omega}})$ using 25-point Hamming window. Only the positive half of the frequency axis is shown, because the magnitude response is an even function whenever the impulse response is real.

For the two LPFs shown in Fig. 7-12, we can make an accurate measurement of δ_p and $\hat{\omega}_p$ from the zoomed plots in Fig. 7-13. For the rectangular window case, a careful measurement gives a maximum passband ripple size of $\delta_p = 0.104$, with the passband edge at $\hat{\omega}_p = 0.364\pi$. For the Hamming window case, we need the zoomed plot of the passband region to see the ripples, as in Fig. 7-13(b). Then we can measure the passband ripple to be $\delta_p = 0.003$ for the Hamming window case—more than 30 times smaller. Once we settle on the passband ripple height, we can measure the passband edge—in Fig. 7-12(b) it is $\hat{\omega}_p = 0.2596\pi$. Notice that the *actual passband edges* are not equal to the design parameter $\hat{\omega}_{co}$ which is called the *cutoff frequency*. There is sometimes a confusion when the terminology "passband cutoff frequency" is used to mean passband edge, which then implies that $\hat{\omega}_{co}$ and $\hat{\omega}_p$ might be the same, but after doing a few examples it should become clear that this is never the case.

7-4.2.4 Stopband Defined for the Frequency Response

When the frequency response (magnitude) of the digital filter is close to zero, we have the ***stopband*** region of the filter. The stopband is a region of the form $\hat{\omega}_s \le \hat{\omega} \le \pi$, if the magnitude response of a LPF is plotted only for nonnegative frequencies. The parameter $\hat{\omega}_s$ is called the *stopband edge.* In the rectangular window LPF example of Figs. 7-12(a) and 7-13(a), the magnitude response is close to zero when $0.438\pi \le \hat{\omega} \le \pi$ (i.e., high frequencies). The stopband ripple for this region is expected to be less than 0.1, and is measured to be $\delta_s = 0.077$. For the rectangular windowed sinc, the stopband edge is $\hat{\omega}_s = 0.4383\pi$.

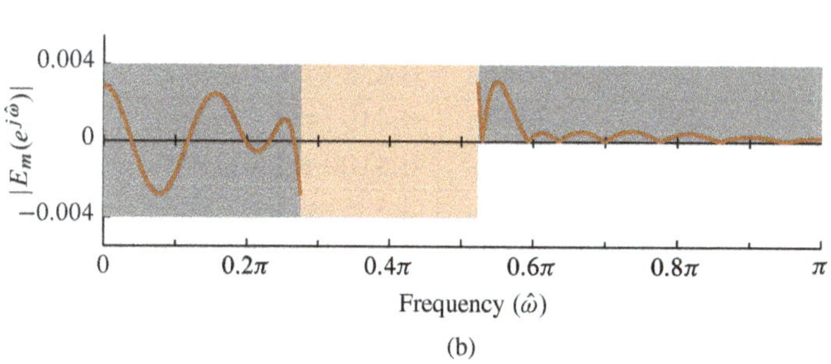

Figure 7-13 Blowup of the error between the actual magnitude response and the ideal, $E(e^{j\hat{\omega}}) = |H(e^{j\hat{\omega}})| - |H_i(e^{j\hat{\omega}})|$, which shows the passband and stopband ripples. (a) Length-25 LPF $H_r(e^{j\hat{\omega}})$ with rectangular window (i.e., truncated ideal lowpass impulse response). (b) Length-25 LPF $H_m(e^{j\hat{\omega}})$ using 25-point Hamming window. The ripples for the Hamming case are more than 30 times smaller.

We can repeat this process for the Hamming window LPF using Figs. 7-12(b) and 7-13(b) to determine the stopband ripple δ_s, and then the set of frequencies where the magnitude is less than δ_s. The result is a stopband ripple measurement of 0.0033, and a corresponding stopband edge of $\hat{\omega}_s = 0.5361\pi$.

7-4.2.5 Transition Zone of the LPF

Unlike an ideal LPF where the stopband begins at the same frequency where the passband ends, in a practical filter there is always a nonzero difference between the passband edge and the stopband edge. This difference is called the *transition width* of the filter: $\Delta\hat{\omega} = \hat{\omega}_s - \hat{\omega}_p$. The smaller the transition width, the better the filter because it is closer to the *ideal filter* which has a transition width of zero.

For the LPFs in Fig. 7-12, the measured transition width of the rectangular windowed sinc is

$$\Delta\hat{\omega}_r = \hat{\omega}_s - \hat{\omega}_p = 0.4383\pi - 0.3646\pi = 0.0737\pi \tag{7.57a}$$

while the Hamming windowed sinc has a much wider transition width—almost four times wider

$$\Delta\hat{\omega}_m = \hat{\omega}_s - \hat{\omega}_p = 0.5361\pi - 0.2596\pi = 0.2765\pi \tag{7.57b}$$

EXAMPLE 7-7 Transition Width versus Filter Order

One property of the transition width is that it can be controlled by changing the filter order. There is an approximate inverse relationship, so doubling the order reduces the transition width by roughly one half. We can test this idea on the length-25 rectangular window LPF in Fig. 7-12(b) which has an order equal to 24. If we design a new LPF that has the same cutoff frequency, $\hat{\omega}_{co} = 0.4\pi$, but twice the order (i.e., $M = 48$), then we can repeat the measurement of the bandedges $\hat{\omega}_p$, $\hat{\omega}_s$, and the transition width $\Delta\hat{\omega}$.

Filter	Order: M	$\hat{\omega}_p$	$\hat{\omega}_s$	$\Delta\hat{\omega}$
Rect	24	0.3646π	0.4383π	0.0737π
Rect	48	0.3824π	0.4192π	0.0368π
Rect	96	0.3909π	0.4100π	0.0191π
Hamming	24	0.2596π	0.5361π	0.2765π
Hamming	48	0.3308π	0.4687π	0.1379π
Hamming	96	0.3660π	0.4340π	0.0680π

Comparing the values of $\Delta\hat{\omega}$, the ratio is $(0.0737\pi)/(0.0368\pi) = 2.003$. Doubling the order once more to $M = 96$ gives a transition width of 0.0191π, so the ratio is $(0.0737\pi)/(0.0191\pi) = 3.86 \approx 4$. For the Hamming window case, the measured transition width for $M = 48$ is $\Delta\hat{\omega} = 0.1379\pi$, and for $M = 96$, $\Delta\hat{\omega} = 0.0680\pi$. The ratios of 0.2765π to 0.1379π and 0.0680π are 2.005 and 4.066 which matches the approximate inverse relationship expected.

LAB
FIR Filter Design

When comparing the transition widths in (7.57a) and (7.57b), we see that $\Delta\hat{\omega}_m = 3.75\Delta\hat{\omega}_r$, so the transition width of the Hamming window LPF is almost four times larger for $L = 25$. This empirical observation confirms the statement, "when comparing equal-order FIR filters that approximate a LPF, the one with larger transition width has smaller ripples." However, this statement does not mean that the ripples can be reduced merely by widening the transition width. Within one window type, such as Hamming window filters, changing the transition width does not change the ripples by more than a few percent.

7-4.2.6 Summary of Filter Specifications

The foregoing discussion of ripples, bandedges, and transition width can be summarized with the tolerance scheme shown in Fig. 7-12. The filter design process is to approximate the ideal frequency response very closely. Once we specify the desired ripples and bandedges, we can draw a template around the ideal frequency response. The template should, in effect, give the trade-off between ripple size and transition width. Then an acceptable filter design would be any FIR filter whose magnitude response lies entirely within the template. The Hamming window method is just one possible design method among many that have been developed.

7-4.3 GUI for Filter Design

DEMO
Filter Design GUI

The *DSP-First* GUI called `filterdesign` illustrates several filter design methods for LPF, BPF, and HPF filters. The interface is shown in Fig. 7-14. Both FIR and IIR filters can be designed, but we are presently interested in the FIR case which is selected with the `FIR` button in the upper right. The default design method is the *Window Method* using a Hamming window. The window type can be changed by selecting another window type from the drop-down list in the lower right. To specify the design it is necessary to set the order of the FIR filter and choose one or more cutoff frequencies; these parameters can be entered in the edit boxes.

The plot initially shows the frequency response magnitude on a linear scale, with a frequency axis in Hz. Clicking on the word [Magnitude] toggles the magnitude scale to a log scale in dB. Clicking on the word [Frequency] toggles the frequency axis to normalized frequency $\hat{\omega}$, and also lets you enter the cutoff frequency using $\hat{\omega}_{co}$ in rad.

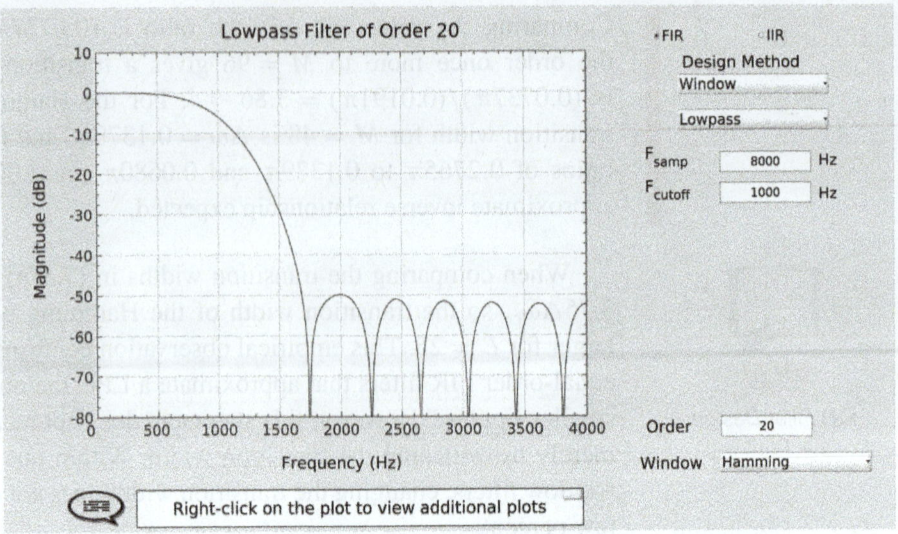

Figure 7-14 Interface for the `filterdesign` GUI. When the filter choice is set to `FIR`, many different window types can be selected, including the Hamming window and the rectangular window (i.e., only truncation to a finite length). The specification of one or more cutoff frequencies (f_{co}) must be entered using continuous-time frequency (in Hz), along with a sampling rate (f_s, also in Hz). In normalized frequency, the cutoff frequency is $\hat{\omega}_{co} = 2\pi(f_{co}/f_s)$.

Recall that $\hat{\omega}_{co} = 2\pi(f_{co}/f_s)$. The plotting region can also show the phase response of $H(e^{j\hat{\omega}})$, or the impulse response of the filter $h[n]$. Right click on the plot region to get a menu. The Options menu provides zooming and a grid via `Options->Zoom` and `Options->Grid`.

The filter coefficients can be "exported" from the GUI by using the menu `File->Export Coeffs`. To make some filters for comparison, redo the designs in Fig. 7-12 and export the filter coefficients to the workspace under unique names. Then you can make your own plot of the frequency response in MATLAB using the `freekz` function (or `freqz`) followed by a `plot` command. For example, an interesting activity would be to zoom in on the magnitude response plots as in Fig. 7-13 to check the measurements of ripples and transition width.

7-5 Table of Fourier Transform Properties and Pairs

Table 7-1 on p. 293 includes all the Fourier transform pairs that we have derived in this chapter as well as one pair (the left-sided exponential) that we did not derive. In addition, the basic properties of the Fourier transform, which make the DTFT convenient to use in designing and analyzing systems, are given in Table 7-2 on p. 294 for easy reference.

7-6 Summary and Links

In this chapter, we introduced the DTFT, and developed some of its basic properties for understanding the behavior of linear systems. The DTFT provides a frequency-domain representation for signals as well as systems, and like other Fourier transforms it generalizes the idea of a spectrum for a discrete-time signal. We obtained the DTFT by generalizing the concept of the frequency response, and showed how the inverse transform could be used to obtain the impulse response of various ideal filters. Therefore, it is not surprising that the DTFT plays an important role in practical filter design for methods based on rectangular and Hamming windowing.

DEMO
Visual DTFT Pairs

Table 7-1 Basic discrete-time Fourier transform pairs.

Table of DTFT Pairs	
Time-Domain: $x[n]$	***Frequency-Domain:*** $X(e^{j\hat{\omega}})$
$\delta[n]$	1
$\delta[n-n_d]$	$e^{-j\hat{\omega}n_d}$
$r_L[n]=u[n]-u[n-L]$	$\dfrac{\sin(\frac{1}{2}L\hat{\omega})}{\sin(\frac{1}{2}\hat{\omega})}e^{-j\hat{\omega}(L-1)/2}$
$r_L[n]\,e^{j\hat{\omega}_0 n}$	$\dfrac{\sin(\frac{1}{2}L(\hat{\omega}-\hat{\omega}_o))}{\sin(\frac{1}{2}(\hat{\omega}-\hat{\omega}_o))}e^{-j(\hat{\omega}-\hat{\omega}_o)(L-1)/2}$
$\dfrac{\sin(\hat{\omega}_b n)}{\pi n}$	$\begin{cases}1 & \lvert\hat{\omega}\rvert\le\hat{\omega}_b\\ 0 & \hat{\omega}_b<\lvert\hat{\omega}\rvert\le\pi\end{cases}$
$a^n u[n] \quad (\lvert a\rvert<1)$	$\dfrac{1}{1-ae^{-j\hat{\omega}}}$
$-b^n u[-n-1] \quad (\lvert b\rvert>1)$	$\dfrac{1}{1-be^{-j\hat{\omega}}}$

Table 7-2 Basic discrete-time Fourier transform properties.

Table of DTFT Properties		
Property Name	***Time-Domain:*** $x[n]$	***Frequency-Domain:*** $X(e^{j\hat{\omega}})$
Periodic in $\hat{\omega}$		$X(e^{j(\hat{\omega}+2\pi)}) = X(e^{j\hat{\omega}})$
Linearity	$ax_1[n] + bx_2[n]$	$aX_1(e^{j\hat{\omega}}) + bX_2(e^{j\hat{\omega}})$
Conjugate Symmetry	$x[n]$ is real	$X(e^{-j\hat{\omega}}) = X^*(e^{j\hat{\omega}})$
Conjugation	$x^*[n]$	$X^*(e^{-j\hat{\omega}})$
Time-Reversal	$x[-n]$	$X(e^{-j\hat{\omega}})$
Delay	$x[n - n_d]$	$e^{-j\hat{\omega}n_d}X(e^{j\hat{\omega}})$
Frequency Shift	$x[n]e^{j\hat{\omega}_0 n}$	$X(e^{j(\hat{\omega}-\hat{\omega}_0)})$
Modulation	$x[n]\cos(\hat{\omega}_0 n)$	$\frac{1}{2}X(e^{j(\hat{\omega}-\hat{\omega}_0)}) + \frac{1}{2}X(e^{j(\hat{\omega}+\hat{\omega}_0)})$
Convolution	$x[n] * h[n]$	$X(e^{j\hat{\omega}})H(e^{j\hat{\omega}})$
Autocorrelation	$x[-n] * x[n]$	$\lvert X(e^{j\hat{\omega}})\rvert^2$
Parseval's Theorem	$\sum_{n=-\infty}^{\infty} \lvert x[n]\rvert^2 = \frac{1}{2\pi}\int_{-\pi}^{\pi} \lvert X(e^{j\hat{\omega}})\rvert^2 d\hat{\omega}$	

7-7 Problems

NOTE

Problems with Solutions

The problems in this section address the major concepts of this chapter. On the Companion Website there are many related homework problems which can be used for review and practice. Most of them have detailed solutions.

P-7.1 Determine the DTFT of each of the following sequences:

(a) $x_1[n] = 3\delta[n-4] \overset{DTFT}{\longleftrightarrow} X_1(e^{j\hat{\omega}}) = 3e^{-j\hat{\omega}4}$

(b) $x_2[n] = 5\delta[n-7] - \delta[n-2] + 5\delta[n+7] \overset{DTFT}{\longleftrightarrow} X_2(e^{j\hat{\omega}}) = 5e^{j\hat{\omega}7} - e^{j\hat{\omega}2} + 5e^{j\hat{\omega}7}$

(c) $x_3[n] = 3u[n-2] - 3u[n-7] = \begin{cases} 3 & 2 \le n \le 6 \\ 0 & \text{otherwise} \end{cases}.$

(d) $x_4[n] = \frac{\sin(0.6\pi n)}{7\pi n} \overset{DTFT}{\longleftrightarrow} X_4(e^{j\hat{\omega}}) = \begin{cases} \frac{1}{7} & \hat{\omega} < 0.6\pi \\ 0 & 0.6\pi < \hat{\omega} < \pi \end{cases}$

P-7.2 Determine the inverse DTFT of each of the following transforms:

(a) $Y_1(e^{j\hat{\omega}}) = \pi \overset{DTFT}{\longleftrightarrow} y_1[n] = \pi\delta[n]$

(b) $Y_2(e^{j\hat{\omega}}) = 3e^{-j4\hat{\omega}} \overset{DTFT}{\longleftrightarrow} y_2[n] = 3\delta[n-4]$

(c) $Y_3(e^{j\hat{\omega}}) = 5j\sin(4\hat{\omega}) = \frac{5}{2}e^{j4\hat{\omega}} - \frac{5}{2}e^{-j4\hat{\omega}} \overset{DTFT}{\longleftrightarrow} y_3[n] = \frac{5}{2}\delta[n+4] - \frac{5}{2}\delta[n-4]$

(d) $Y_4(e^{j\hat{\omega}}) = 3\cos(5\hat{\omega}) = \frac{3}{2}e^{j5\hat{\omega}} + \frac{3}{2}e^{-j5\hat{\omega}} \overset{DTFT}{\longleftrightarrow} \frac{3}{2}\delta[n+5] + \frac{3}{2}\delta[n-5]$

P-7.3 Determine the inverse or forward DTFT as appropriate:

(a) $v_1[n] = \begin{cases} 0 & n < 0 \\ 1 & n = 0, 1, \ldots, 5 \\ 0 & n > 5 \end{cases}$

(b) $V_2(e^{j\hat{\omega}}) = \begin{cases} 1 & |\hat{\omega}| \le 0.4\pi \\ 0 & 0.4\pi < |\hat{\omega}| \le \pi \end{cases}$

(c) $v_3[n] = \begin{cases} 0 & n < 0 \\ (-1)^n & n = 0, 1, \ldots, 5 \\ 0 & n > 5 \end{cases}$

(d) $V_4\left(e^{j\hat{\omega}}\right) = \begin{cases} 0 & |\hat{\omega}| \le 0.4\pi \\ 1 & 0.4\pi < |\hat{\omega}| \le \pi \end{cases}$

P-7.4 Determine the DTFT of each of the following sequences:

(a) $h_1[n] = 4\delta[n] - \dfrac{\sin(0.16\pi n)}{0.25\pi n}$

(b) $h_2[n] = \dfrac{\sin(0.5\pi n)}{0.1\pi n} - \dfrac{\sin(0.2\pi n)}{0.1\pi n}$

(c) $h_3[n] = \dfrac{\sin(0.5\pi(n-6))}{\pi(n-6)} - \dfrac{\sin(0.2\pi(n-6))}{\pi(n-6)}$

P-7.5 Consider a linear time-invariant system defined by the difference equation

$$y[n] = 5x[n-1] + 5x[n-2] + 5x[n-3] + 5x[n-4] + 5x[n-5] = \sum_{k=1}^{5} 5x[n-k]$$

The frequency response of this system can be expressed in the following "Dirichlet-like" equation:

$$H(e^{j\hat{\omega}}) = \alpha \left(\frac{\sin(L\hat{\omega}/2)}{\sin(\hat{\omega}/2)} \right) e^{-j\beta\hat{\omega}}$$

Determine the parameters L, α, β.

P-7.6 The DTFT must be periodic with a period equal to 2π, so it is sufficient to define the DTFT for $-\pi \le \hat{\omega} < \pi$. For each of the following ideal filters, determine the inverse transform of the given DTFT:

(a) $G_1(e^{j\hat{\omega}}) = \begin{cases} 5e^{-j0.3\hat{\omega}} & 0 \le |\hat{\omega}| \le 0.4\pi \\ 0 & 0.4\pi < |\hat{\omega}| \le \pi \end{cases}$

(b) $G_2(e^{j\hat{\omega}}) = \begin{cases} 0 & 0 \le |\hat{\omega}| \le 0.7\pi \\ 4e^{-j0.1\hat{\omega}} & 0.7\pi < |\hat{\omega}| \le \pi \end{cases}$

(c) $G_3(e^{j\hat{\omega}}) = \begin{cases} 0 & 0 \le |\hat{\omega}| \le 0.4\pi \\ 9e^{-j0.2\hat{\omega}} & 0.4\pi < |\hat{\omega}| \le 0.7\pi \\ 0 & 0.7\pi < |\hat{\omega}| \le \pi \end{cases}$

P-7.7 The convolution property of the DTFT can simplify difficult operations that arise in LTI filtering. Simplify the following:

(a) $x_1[n] = \dfrac{\sin(0.25\pi n)}{9\pi n} * \dfrac{\sin(0.44\pi n)}{4\pi n}$

(b) $x_2[n] = \dfrac{\sin(0.25\pi n)}{9\pi n} * \dfrac{\sin(0.25\pi n)}{8\pi n} * \dfrac{\sin(0.25\pi n)}{7\pi n}$

(c) $x_3[n] = (0.3)^n u[n] * (10\delta[n-2] - 3\delta[n-3])$

P-7.8 The frequency response of an LTI system is $H(e^{j\hat{\omega}}) = \cos(3\hat{\omega})$. Determine the output of the system for each of the following input signals:

(a) $x_1[n] = \delta[n-3]$

(b) $x_2[n] = 3\cos(0.25\pi n)$

(c) $x_3[n] = u[n] - u[n-9]$

P-7.9 An example of the frequency-shifting property comes from modeling the response of a physical system as an oscillating signal whose amplitude is dropping. The precise form is an exponential times a sinusoid, $x[n] = a^n \cos(\hat{\omega}_0 n) u[n]$.

(a) Use the inverse Euler formula to show that we can express $x[n]$ as

$$x[n] = \tfrac{1}{2}e^{j\hat{\omega}_0 n}a^n u[n] + \tfrac{1}{2}e^{-j\hat{\omega}_0 n}a^n u[n]$$

(b) By using the DTFT pair in (7.6), applying the frequency-shift property (7.25) twice, and finding a common denominator for the resulting two fractions, show that we can write $X(e^{j\hat{\omega}})$ as

$$X(e^{j\hat{\omega}}) = \frac{1 - a\cos(\hat{\omega}_0)e^{-j\hat{\omega}}}{(1 - ae^{j\hat{\omega}_0}e^{-j\hat{\omega}})(1 - ae^{-j\hat{\omega}_0}e^{-j\hat{\omega}})} = \frac{1 - a\cos(\hat{\omega}_0)e^{-j\hat{\omega}}}{1 - 2a\cos(\hat{\omega}_0)e^{-j\hat{\omega}} + a^2 e^{-j\hat{\omega}2}}$$

P-7.10 For the signal $v[n] = a^n u[n]$, the DTFT $V(e^{j\hat{\omega}})$ can be found in Table 7-1. For the questions below, assume $a = 0.95$.

(a) Make a plot of $|V(e^{j\hat{\omega}})|^2$ for $-\pi < \hat{\omega} \le \pi$.

(b) If $V(e^{j\hat{\omega}})$ were a frequency response, is the filter lowpass, highpass, or bandpass?

(c) Define $w[n] = v[n]\cos(0.3\pi n)$, and then make a plot of $|W(e^{j\hat{\omega}})|^2$ for $-\pi < \hat{\omega} \le \pi$.

(d) If $W(e^{j\hat{\omega}})$ were a frequency response, is the filter lowpass, highpass, or bandpass?

P-7.11 Suppose that $H_{\text{LP}}(e^{j\hat{\omega}})$ is an ideal LPF with its bandedge at $\hat{\omega}_{\text{co}}$ as in Fig. 7-5.

(a) Show that the frequency response of an ideal HPF with cutoff frequency $\hat{\omega}_{\text{co}}$ can be represented by

$$H_{\text{HP}}(e^{j\hat{\omega}}) = 1 - H_{\text{LP}}(e^{j\hat{\omega}})$$

(b) Determine the cutoff frequency of the ideal HPF.

(c) Use the result of part (a) to determine an expression for the impulse response $h_{\text{HP}}[n]$ of the HPF.

P-7.12 Suppose that $H_{\text{BP}}(e^{j\hat{\omega}})$ is an ideal BPF with its bandedges at $\hat{\omega}_{\text{co}_1}$ and $\hat{\omega}_{\text{co}_2}$ as in (7.51) and as plotted in Fig. 7-8.

(a) Show that the frequency response of an ideal BPF with cutoff frequencies $\hat{\omega}_{\text{co}_1} = 0.3\pi$ and $\hat{\omega}_{\text{co}_2} = 0.6\pi$ can be represented as the difference of two ideal LPFs as in

$$H_{\text{BP}}(e^{j\hat{\omega}}) = H_{\text{LP}_1}(e^{j\hat{\omega}}) - H_{\text{LP}_2}(e^{j\hat{\omega}})$$

In particular, give the cutoff frequencies of the two LPFs, $H_{\text{LP}_1}(e^{j\hat{\omega}})$ and $H_{\text{LP}_2}(e^{j\hat{\omega}})$.

(b) Use the result of part (a) to determine an expression for the impulse response $h_{\text{BP}}[n]$ of the BPF.

P-7.13 Consider the left-sided exponential sequence defined by

$$x[n] = -b^n u[-n-1]$$

(a) Sketch $x[n]$ for the case $b = 2$.

(b) For what values of b does the DTFT of $x[n]$ exist?

(c) Show that for the values of b found in part (b), the DTFT of $x[n]$ is

$$X(e^{j\hat{\omega}}) = \frac{1}{1 - be^{-j\hat{\omega}}}$$

P-7.14 Use Parseval's Theorem and the DTFT to evaluate:

(a) $\displaystyle\sum_{n=-\infty}^{\infty} \left| \frac{\sin(0.25\pi n)}{3\pi n} \right|^2$

(b) $\displaystyle 4 \int_{-\pi}^{\pi} \left| \frac{\sin(5\hat{\omega})}{\sin(\hat{\omega}/2)} e^{-j(9/2)\hat{\omega}} \right|^2 d\hat{\omega}$

P-7.15 Use the DTFT or discrete-time convolution to determine the autocorrelation function for each of the following signals:

(a) $x_1[n] = \delta[n-4]$

(b) $x_2[n] = \delta[n-4] - \delta[n-6]$

(c) $x_3[n] = u[n-6] - u[n-16]$

(d) $x_4[n] = \dfrac{\sin(0.25\pi n)}{9\pi n}$

P-7.16 Consider the signal $x[n] = u[n] - u[n-5]$

(a) Determine $c_{xx}[n]$, the autocorrelation function (acf) of this signal.

(b) Determine the DTFT of the acf; verify that it is purely real and nonnegative.

(c) Determine the acf of the signal $y[n] = x[n-2]$.
Hints: Since the peak value of an acf must always be at zero lag, consider the possibility that $c_{yy}[n]$ might be the same as $c_{xx}[n]$. Working with the DTFT of the acf might be easier.

DEMO
Filter Design GUI

P-7.17 Use the `filterdesign` GUI to create the impulse response of an FIR HPF with the following specifications: filter order $M = 32$, cutoff frequency $\hat{\omega}_{co} = 0.6\pi$, Hamming window design method.

(a) Make a plot of the frequency response magnitude; and also two "zoomed in" plots to show details of the passband and stopband magnitudes as in Fig. 7-13.

(b) From the plots in the previous part measure the passband and stopband edges, as well as the passband and stopband ripples (i.e., the parameters $\hat{\omega}_p$, $\hat{\omega}_s$, δ_p, and δ_s).

(c) If the filter order were changed to $M = 80$, estimate the change in the passband and stopband edges. Note: the ripples should not change much.

(d) Design the order-80 filter and measure the new passband and stopband edges. Comment on the accuracy of the estimate in the previous part.

P-7.18 Use the `filterdesign` GUI to create the impulse response of an FIR HPF with the Hamming window method. Use the following specifications: filter order $M = 24$, cutoff frequency $\hat{\omega}_{co} = 0.6\pi$. The questions below deal with filtering specific input signals and determining the corresponding output signals.

(a) When the input $x_1[n]$ is DC with a DC level of 10, determine the output level.

(b) When the input is the sinusoid $x_2[n] = 10\cos(0.5\pi n)$ with zero phase, determine the output in the standard form $y_2[n] = A_2\cos(\hat{\omega}n + \varphi_2)$.

(c) When the input is the sinusoid $x_3[n] = 10\cos(0.75\pi n)$ with zero phase, determine the output in the standard form $y_3[n] = A_3\cos(\hat{\omega}n + \varphi_3)$.

P-7.19 Use the `filterdesign` GUI to create the impulse response of an FIR LPF with the rectangular window method. Use the following specifications: filter order $M = 37$, cutoff frequency $\hat{\omega}_{co} = 0.3\pi$. The questions below deal with filtering specific input signals and determining the corresponding output signals.

(a) When the input $x_1[n]$ is DC with a DC level of 10, determine the output level $y_1[n]$.

(b) When the input is the sinusoid $x_2[n] = 10\cos(0.5\pi n)$ with zero phase, determine the output in the standard form $y_2[n] = A_2\cos(\hat{\omega}n + \varphi_2)$.

(c) Use MATLAB to verify the outputs in the previous parts. Generate 100 values of the signal $s[n] = x_1[n] + x_2[n] = 10u[n] + 10\cos(0.5\pi n)u[n]$ in a vector `sn` and filter the signal `sn` with MATLAB's `filter` command to obtain `snOutput`. Plot `snOutput` for `n=0:99`. For comparison plot $y_1[n] + y_2[n]$ from parts (a) and (b) over the same range.
Note: since the MATLAB-filtered signal starts at $n = 0$, there is a *transient* interval at beginning of the output where the two plots differ, but after $n = 37$ the match should be perfect. Thus, it might be helpful to subtract the two signals being compared, and plot the difference in order to accentuate the transient region.

P-7.20 Use the `filterdesign` GUI to create the impulse response of an FIR HPF with the Hamming window method. Use the following specifications: filter order $M = 32$, cutoff frequency $\hat{\omega}_{co} = 0.6\pi$. The following questions deal with filtering specific input signals and determining the corresponding output signals. You can solve the following using the `filterdesign` GUI. Note that by placing the cursor on the frequency response curve, you can read out the frequency and magnitude of the point on the curve.

(a) Suppose the input sinusoid has unknown frequency and zero phase $x[n] = 9\cos(\hat{\omega}n)$. Use the `filterdesign` GUI to determine the lowest frequency $\hat{\omega}$ where the output has amplitude equal to 9 (i.e., $y[n] = 9\cos(\hat{\omega}n + \varphi)$). For the chosen frequency, use the `filterdesign` GUI to determine the phase φ.

(b) Determine all possible frequencies where the output is $y[n] = 9\cos(\hat{\omega}n + \varphi)$ when the input is $x[n] = 9\cos(\hat{\omega}n)$.

(c) Determine all possible frequencies where the output is zero when the input is $x[n] = 9\cos(\hat{\omega}n)$. *Hint:* Look for jumps of π radians in the phase curve.

Note: recall that the filter coefficients can be exported to the MATLAB workspace from the `filterdesign` GUI. Then `freqz` could be used to plot the frequency response in a figure window where the zoom tool could be used to examine details of the magnitude response.

CHAPTER

8

Discrete Fourier Transform

This chapter builds on the definition and discussion of the DTFT in Chapter 7. The objective here is to define a *numerical Fourier transform* called the discrete Fourier transform (or DFT) that results from taking *frequency samples of the DTFT.* We show how the DFT can be used to compute a spectrum representation of any finite-length sampled signal very efficiently with the fast Fourier transform (FFT) algorithm. The DFT not only gives a spectrum representation of any finite-length sequence, but it can also be used as a representation of a periodic sequence formed by infinite repetition of one (finite-length) period. Finally, the DFT is the core computation needed in the spectrogram which provides a time-frequency *spectrum analysis* of a very long signal by doing DFTs of successive short sections extracted from the long signal.

It is safe to say that spectrum analysis is one of the most common operations used in the practice of signal processing. Sooner or later, most scientists and engineers encounter a situation where a "sampled data signal" has been obtained from an A-to-D converter and must be analyzed to determine its spectral properties. Often, spectrum analysis is used to discover whether or not the signal contains strong periodic components. A listing of natural phenomena that are cyclic, or nearly so, would contain hundreds of entries. Common examples include speech and music signals, as well as other natural observations such as tides and yearly sun spot cycles. In the case of sampled signals, the question of interest is how to derive the spectrum by doing numerical operations on a signal that is defined only by a finite set of numbers. This process of going from the sampled signal to its spectrum is called *discrete-time spectrum analysis* and the DFT is its workhorse operator.

8-1 Discrete Fourier Transform (DFT)

The DTFT of a discrete-time signal $x[n]$ can be viewed as a generalization of the spectrum concept introduced in Chapters 3 and 4, where *discrete* lines in the frequency domain represented sums of complex exponentials in the time domain. The spectrum concept obtained from the DTFT is a *continuous* function of frequency denoted $X(e^{j\hat{\omega}})$, and its value as a Fourier transform is easiest to appreciate when $x[n]$ and $X(e^{j\hat{\omega}})$ are defined as mathematical expressions. Because the spectrum is so useful, however, it is natural to want to be able to determine the spectrum of a signal even if $x[n]$ is not described by a simple mathematical formula. Indeed, it would be handy to have a computer program that could *calculate the spectrum from samples of a signal.*

Two steps are needed to change the DTFT sum (7.2) into a computable form: the continuous frequency variable $\hat{\omega}$ must be sampled, and the limits on the DTFT sum must be finite. First, even though $\hat{\omega}$ is a continuous variable, it does have a finite range $-\pi \le \hat{\omega} < \pi$ so we can evaluate (7.2) at a finite set of frequencies, denoted by $\hat{\omega}_k$. Second, the DTFT sum has a finite number of terms when the signal duration is finite. We cannot compute the transform of an infinite-duration signal, but it is common to operate on finite sections of a very long signal.

For a finite-duration signal, the DTFT sampled at frequencies $\hat{\omega}_k$ is

$$X(e^{j\hat{\omega}_k}) = \sum_{n=0}^{L-1} x[n]e^{-j\hat{\omega}_k n} \qquad k = 0, 1, \ldots N-1 \tag{8.1}$$

where L is the length of $x[n]$. We must choose a specific set of frequency samples $\{\hat{\omega}_k\}$, but which frequencies should be used to evaluate (8.1)? The usual domain for the spectrum is $-\pi \le \hat{\omega} < \pi$, but any interval of length 2π would suffice. For reasons that will become apparent, it is common to choose that interval to be

$$0 \le \hat{\omega}_k < 2\pi$$

and to evaluate (8.1) at the N equally spaced frequencies

$$\hat{\omega}_k = \frac{2\pi k}{N} \qquad k = 0, 1, \ldots, N-1 \tag{8.2}$$

Note that this set of N frequencies covers the range $0 \le \hat{\omega} < 2\pi$ since $k = N$ would give $\hat{\omega}_N = 2\pi$, which is an alias frequency of $\hat{\omega}_0 = 0$. Substituting (8.2) into (8.1) gives N frequency samples of the DTFT

$$X\left(e^{j\frac{2\pi k}{N}}\right) = \sum_{n=0}^{L-1} x[n]e^{-j(2\pi k/N)n} \tag{8.3}$$

for $k = 0, 1, \ldots, N-1$. Equation (8.3) is a "finite Fourier sum" which is computable. The sum in (8.3) must be computed for N different values of the discrete frequency index k. The index n in the sum on the right is the counting index of the sum, and thus it disappears when the sum is computed. Since the left-hand side of (8.3) depends only on the frequency index k, we define $X[k] = X\left(e^{j\frac{2\pi k}{N}}\right)$.

When the number of frequency samples N is equal to the signal length L, the summation in (8.3) with $N = L$ becomes:

The Discrete Fourier Transform

$$X[k] = \sum_{n=0}^{N-1} x[n]e^{-j(2\pi/N)kn} \qquad k = 0, 1, \ldots, N-1 \tag{8.4}$$

Equation (8.4) is called the ***discrete Fourier transform*** or ***DFT*** in recognition of the fact that it is a Fourier transformation, and it is discrete in both time and frequency.[1] The DFT takes N samples in the time-domain and *transforms* them into N values $X[k]$ in the frequency domain. Typically, the values of $X[k]$ are complex, while the values of $x[n]$ are often real, but $x[n]$ in (8.4) could be complex.

EXAMPLE 8-1 Short-Length DFT

In order to compute the 4-point DFT of the sequence $x[n] = \{1,\ 1,\ 0,\ 0\}$, we carry out the sum (8.4) four times, once for each value of $k = 0, 1, 2, 3$. When $N = 4$, all the exponents in (8.4) are integer multiples of $\pi/2$ because $2\pi/N = \pi/2$.

$$\begin{aligned}
X[0] &= \overset{1}{\cancel{x[0]}}e^{-j0} + \overset{1}{\cancel{x[1]}}e^{-j0} + \overset{0}{\cancel{x[2]}}e^{-j0} + \overset{0}{\cancel{x[3]}}e^{-j0} \\
&= 1 + 1 + 0 + 0 = 2 \\
X[1] &= 1e^{-j0} + 1e^{-j\pi/2} + 0e^{-j\pi} + 0e^{-j3\pi/2} \\
&= 1 + (-j) + 0 + 0 = 1 - j = \sqrt{2}e^{-j\pi/4} \\
X[2] &= 1e^{-j0} + 1e^{-j\pi} + 0e^{-j2\pi} + 0e^{-j3\pi} \\
&= 1 + (-1) + 0 + 0 = 0 \\
X[3] &= 1e^{-j0} + 1e^{-j3\pi/2} + 0e^{-j3\pi} + 0e^{-j9\pi/2} \\
&= 1 + (j) + 0 + 0 = 1 + j = \sqrt{2}e^{j\pi/4}
\end{aligned}$$

Thus, we obtain the four DFT coefficients[2] $X[k] = \{2,\ \sqrt{2}e^{-j\pi/4},\ 0,\ \sqrt{2}e^{j\pi/4}\}$.

---

[1] In contrast, the term discrete-time Fourier transform (DTFT) emphasizes that only time is discrete.

[2] The term "coefficient" is commonly applied to DFT values. This is appropriate because $X[k]$ is the (complex amplitude) coefficient of $e^{j(2\pi/N)kn}$ in the IDFT (8.5).

Example 8-1 illustrates an interesting fact about the limits of summation in (8.4) where we have chosen to sum over N samples of $x[n]$ and to evaluate the DFT at N frequencies. Sometimes, the sequence length of $x[n]$ is shorter than N (i.e., $L < N$), and $x[n]$ is nonzero only in the interval $0 \le n \le L-1$. In Example 8-1, $L = 2$ and $N = 4$. In such cases, we can simply append $N-L$ zero samples to the nonzero samples of $x[n]$ and then carry out the N-point DFT[3] computation. These zero samples extend a finite-length sequence, but do not change its nonzero portion.

8-1.1 The Inverse DFT

The DFT is a legitimate transform because it is possible to invert the transformation defined in (8.4). In other words, there exists an *inverse discrete Fourier transform* (or IDFT), which is a computation that converts $X[k]$ for $k = 0, 1, \ldots, N-1$ back into the sequence $x[n]$ for $n = 0, 1, \ldots, N-1$. The inverse DFT is

Inverse Discrete Fourier Transform

$$x[n] = \frac{1}{N}\sum_{k=0}^{N-1} X[k]e^{j(2\pi/N)kn} \qquad n = 0, 1, \ldots, N-1 \tag{8.5}$$

Equations (8.5) and (8.4) define the unique relationship between an N-point sequence $x[n]$ and its N-point DFT $X[k]$. Following our earlier terminology for Fourier representations, the DFT defined by (8.4) is the *analysis* equation and IDFT defined by (8.5) is the *synthesis* equation. To prove that these equations are a consistent invertible Fourier representation, we note that (8.5), because it is a finite, well-defined computation, would surely produce *some* sequence when evaluated for $n = 0, 1, \ldots, N-1$. So let us call that sequence $v[n]$ until we prove otherwise. Part of the proof is given by the following steps:

$$v[n] = \frac{1}{N}\sum_{k=0}^{N-1} X[k]e^{j(2\pi k/N)n} \tag{8.6a}$$

$$= \frac{1}{N}\sum_{k=0}^{N-1} \underbrace{\left(\sum_{m=0}^{N-1} x[m]e^{-j(2\pi k/N)m}\right) e^{j(2\pi k/N)n}}_{\text{Forward DFT, } X[k]} \tag{8.6b}$$

$$= \frac{1}{N}\sum_{m=0}^{N-1} x[m] \underbrace{\left(\sum_{k=0}^{N-1} e^{j(2\pi k/N)(n-m)}\right)}_{=0,\text{ except for } m=n} \tag{8.6c}$$

$$= x[n] \tag{8.6d}$$

[3] The terminology "N-point DFT" means that the sequence $x[n]$ is known for N time indices, and the DFT is computed at N frequencies in (8.2).

Several things happened in the manipulations leading up to the equality assertion of (8.6d). On the second line (8.6b), we substituted the right-hand side of (8.4) for $X[k]$ after changing the index of summation from n to m. We are allowed to make this change because m is a "dummy index," and we need to reserve n for the index of the sequence $v[n]$ that is synthesized by the IDFT. On the third line (8.6c), the summations on k and m were interchanged. This is permissible since these finite sums can be done in either order.

Now we need to consider the term in parenthesis in (8.6c). Exercise 8.1 (below) states the required *orthogonality* result, which can be easily verified. If we substitute (8.8) into the third line (8.6c), we see that the only term in the sum on m that is nonzero is the term corresponding to $m = n$. Thus, $v[n] = x[n]$ for $0 \le n \le N - 1$ as we wished to show.

EXERCISE 8.1

Orthogonality Property of Periodic Discrete-Time Complex Exponentials

Use the formula

$$\sum_{k=0}^{N-1} \alpha^k = \frac{1-\alpha^N}{1-\alpha} \tag{8.7}$$

to show that

$$\begin{aligned} d[n-m] &= \frac{1}{N}\sum_{k=0}^{N-1} e^{j(2\pi/N)nk} e^{j(2\pi/N)(-m)k} && \text{(definition)} \\ &= \frac{1}{N}\sum_{k=0}^{N-1} e^{j(2\pi/N)(n-m)k} && \text{(alternate form)} \\ &= \frac{1}{N}\left(\frac{1-e^{j(2\pi)(n-m)}}{1-e^{j(2\pi/N)(n-m)}}\right) && \text{(use (8.7))} \end{aligned}$$

$$d[n-m] = \begin{cases} 1 & n-m = rN \\ 0 & \text{otherwise} \end{cases} \tag{8.8}$$

where r is any positive or negative integer including $r = 0$.

EXAMPLE 8-2 Short-Length IDFT

The 4-point DFT in Example 8-1 is the sequence $X[k] = \{2, \sqrt{2}e^{-j\pi/4}, 0, \sqrt{2}e^{j\pi/4}\}$. If we compute the 4-point IDFT of this $X[k]$, we should recover $x[n]$ when we apply the IDFT summation (8.5) for each value of $n = 0, 1, 2, 3$. As before, the exponents in (8.5) are all integer multiples of $\pi/2$ when $N = 4$.

$$\begin{aligned} x[0] &= \tfrac{1}{4}\left(X[0]e^{j0} + X[1]e^{j0} + X[2]e^{j0} + X[3]e^{j0}\right) \\ &= \tfrac{1}{4}\left(2 + \sqrt{2}e^{-j\pi/4} + 0 + \sqrt{2}e^{j\pi/4}\right) = \tfrac{1}{4}(2 + (1-j) + 0 + (1+j)) = 1 \end{aligned}$$

$$x[1] = \tfrac{1}{4}\left(X[0]e^{j0} + X[1]e^{j\pi/2} + X[2]e^{j\pi} + X[3]e^{j3\pi/2}\right)$$
$$= \tfrac{1}{4}\left(2+\sqrt{2}e^{j(-\pi/4+\pi/2)}+0+\sqrt{2}e^{j(\pi/4+3\pi/2)}\right) = \tfrac{1}{4}(2+(1+j)+(1-j)) = 1$$
$$x[2] = \tfrac{1}{4}\left(X[0]e^{j0} + X[1]e^{j\pi} + X[2]e^{j2\pi} + X[3]e^{j3\pi}\right)$$
$$= \tfrac{1}{4}\left(2+\sqrt{2}e^{j(-\pi/4+\pi)}+0+\sqrt{2}e^{j(\pi/4+3\pi)}\right) = \tfrac{1}{4}(2+(-1+j)+(-1-j)) = 0$$
$$x[3] = \tfrac{1}{4}\left(X[0]e^{j0} + X[1]e^{j3\pi/2} + X[2]e^{j3\pi} + X[3]e^{j9\pi/2}\right)$$
$$= \tfrac{1}{4}\left(2+\sqrt{2}e^{j(-\pi/4+3\pi/2)}+0+\sqrt{2}e^{j(\pi/4+9\pi/2)}\right) = \tfrac{1}{4}(2+(-1-j)+(-1+j))=0$$

Thus, we have verified that the length-4 signal $x[n] = \{1,\ 1,\ 0,\ 0\}$ can be recovered from its 4-point DFT coefficients, $X[k] = \left\{2,\ \sqrt{2}e^{-j\pi/4},\ 0,\ \sqrt{2}e^{j\pi/4}\right\}$.

8-1.2 DFT Pairs from the DTFT

Since the DFT is a frequency-sampled version of the DTFT for a finite-length signal, it is possible to construct DFT pairs by making the substitution $\hat{\omega} \to (2\pi/N)k$ in a DTFT pair. In Table 7-1, there are only four finite-length signals, so each of these has a DFT obtained by sampling $X(e^{j\hat{\omega}})$.

8-1.2.1 DFT of Shifted Impulse

The impulse signal $\delta[n]$ and the shifted impulse signal are the easiest cases because they are very simple signals. If we take the DFT of $x_1[n] = \delta[n]$, the DFT summation simplifies to one term:

$$X_1[k] = \sum_{n=0}^{N-1} \delta[n]e^{-j(2\pi k/N)n} = \underbrace{\delta[0]e^{-j0}}_{=1} + \sum_{n=1}^{N-1} \overbrace{\delta[n]}^{0}\, e^{-j(2\pi k/N)n} = 1$$

It is tempting to develop all DFT pairs by working directly with the DFT summation, but it is much easier to get some pairs by frequency sampling known DTFTs. Recall that we already know the DTFT of the shifted impulse $\delta[n-n_d]$, so

$$x[n] = \delta[n-n_d] \overset{\text{DTFT}}{\longleftrightarrow} e^{-j\hat{\omega}n_d} \quad \Rightarrow \quad X[k] = e^{-j\hat{\omega}n_d}\Big|_{\hat{\omega}=(2\pi k/N)} = e^{-j(2\pi k/N)n_d} \tag{8.9}$$

When the shift (n_d) is zero in (8.9), the DFT of $\delta[n]$ is $X[k] = 1$, as before. We want to write (8.9) as the pair $\delta[n-n_d] \overset{\text{DFT}}{\longleftrightarrow} e^{-j(2\pi k/N)n_d}$, but since frequency sampling the DFT requires that the finite-length signal be defined over the interval $0 \le n \le N-1$, for now it appears that we must require that n_d lie in that interval. In fact, we can remove this restriction, as will be discussed in Section 8-3.2.

8-1.2.2 DFT of Complex Exponential

The third and fourth cases in Table 7-1 involve the finite-length (rectangular) pulse whose DTFT involves a Dirichlet form[4] which is (7.27) in Chapter 7. Sampling the known DTFT avoids reworking a messy summation. The third case is the finite-length (rectangular) pulse

$$r_L[n] = \begin{cases} 1 & 0 \le n \le L-1 \\ 0 & L \le n \le N-1 \end{cases}$$

whose DTFT is the third entry in Table 7-1, so its DFT has one term that is a sampled Dirichlet form

$$R_L[k] = R_L(e^{j\hat{\omega}})\Big|_{\hat{\omega}=(2\pi k/N)} = \underbrace{\frac{\sin(\frac{1}{2}L(2\pi k/N))}{\sin(\frac{1}{2}(2\pi k/N))}}_{\text{sampled Dirichlet form}} e^{-j(2\pi k/N)(L-1)/2} \tag{8.10}$$

The fourth case is a finite-length complex exponential which can be written as a product of the rectangular pulse $r_L[n]$ and a complex exponential at frequency $\hat{\omega}_0$

$$x_4[n] = r_L[n]\, e^{j\hat{\omega}_0 n} = \begin{cases} e^{j\hat{\omega}_0 n} & 0 \le n \le L-1 \\ 0 & L \le n \le N-1 \end{cases}$$

The DTFT of $x_4[n]$ is the fourth entry in Table 7-1 which comes from the frequency-shifting property of the DTFT, so after frequency sampling we get the following DFT for the finite-length complex exponential:

$$X_4[k] = R_L(e^{j(\hat{\omega}-\hat{\omega}_0)})\Big|_{\hat{\omega}=(2\pi k/N)} = \underbrace{\frac{\sin(\frac{1}{2}L(2\pi k/N - \hat{\omega}_0))}{\sin(\frac{1}{2}(2\pi k/N - \hat{\omega}_0))}}_{\text{shifted and sampled Dirichlet}} e^{-j(2\pi k/N - \hat{\omega}_0)(L-1)/2} \tag{8.11}$$

The expression on the right-hand side of (8.11) appears rather complicated but it has a recognizable structure as long as it is written in terms of the Dirichlet form which comes from the DTFT for a rectangular pulse.

$$X_4[k] = D_L(2\pi k/N - \hat{\omega}_0)\, e^{-j(2\pi k/N - \hat{\omega}_0)(L-1)/2} \tag{8.12}$$

Since the exponential term in (8.12) only contributes to the phase, this result says that the magnitude depends only on the Dirichlet form $|X_4[k]| = |D_L(2\pi k/N - \hat{\omega}_0)|$. A typical magnitude plot of $|X[k]|$ is shown in Fig. 8-1 for the case when $N = L = 20$, and $\hat{\omega}_0 = 5\pi/20 = 2\pi(2.5)/N$. The continuous magnitude of the Dirichlet envelope (for the DTFT) has been plotted in gray so that it is obvious where the frequency samples are being taken. Notice that the peak of the Dirichlet envelope lies along the horizontal axis at the non-integer value of 2.5, which corresponds to the DTFT frequency $\hat{\omega} = (2\pi/N)(2.5)$.

[4]The Dirichlet form was first defined in Section 6-7 on p. 235.

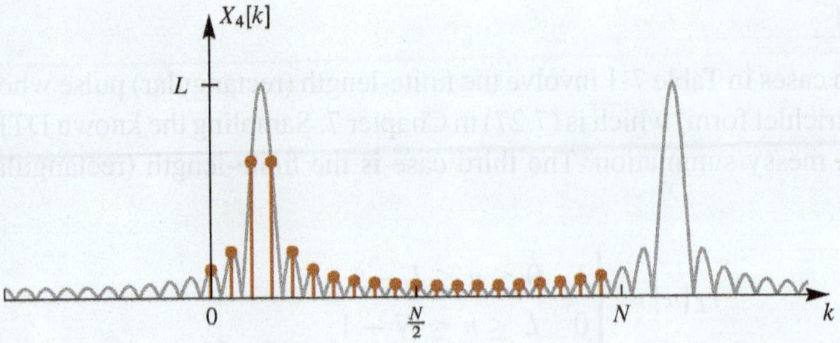

Figure 8-1 Magnitude of DFT coefficients for a 20-point DFT of a length-20 complex exponential whose frequency is $\hat{\omega}_0 = 0.25\pi$ which is *not an integer multiple of* $2\pi/N$ with $N = L = 20$. Gray line is the Dirichlet form magnitude being sampled.

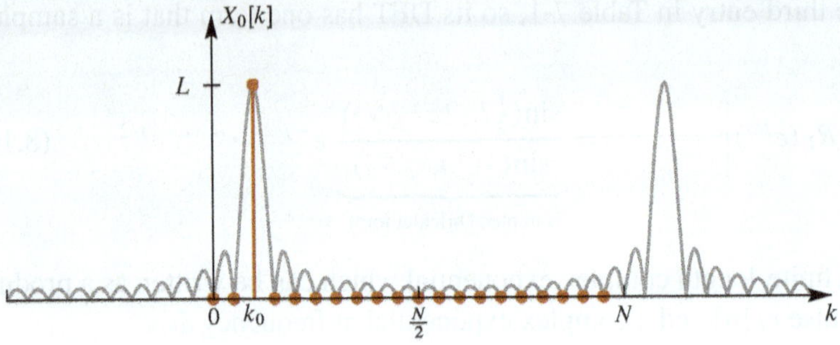

Figure 8-2 Magnitude of DFT coefficients for a 20-point DFT of a length-20 complex exponential whose frequency is $\hat{\omega}_0 = 0.2\pi$ which is *an integer multiple* of $2\pi/N$ (i.e., $N = L = 20$ and $k_0 = 2$).

The DFT is very simple when the frequency of the complex exponential signal is an exact integer multiple of $2\pi/N$ and the DFT length equals the signal length (i.e., $N = L$). For this very special case, we define $x_0[n] = e^{j(2\pi k_0/N)n}$ with $k_0 < N$, and then use $\hat{\omega}_0 = 2\pi k_0/N$ in (8.12) to obtain

$$X_0[k] = D_N(2\pi(k - k_0)/N)\, e^{-j(2\pi(k-k_0)/N))(N-1)/2}$$

The big simplification comes from the fact that the Dirichlet form $D_N(\hat{\omega})$ evaluated at integer multiples of $2\pi/N$ is zero, except for $D_N(0)$, so we get

$$X_0[k] = N\delta[k - k_0] \tag{8.13}$$

The scaled discrete impulse at $k = k_0$ means that $X_0[k_0] = N$ and all other DFT coefficients are zero. This result is confirmed in Fig. 8-2, where we can see that the DFT $X_0[k]$ is obtained by sampling the gray Dirichlet magnitude exactly at its peak and at its zero crossings. The peak value is $N = L$, and it is the only nonzero value in the DFT.

EXERCISE 8.2 Substitute (8.13) into the inverse N-point DFT relation (8.5) to show that the corresponding time-domain sequence is

$$x_2[n] = e^{j(2\pi/N)k_0 n} \qquad n = 0, 1, \ldots, N-1$$

8-1.3 Computing the DFT

The DFT representation in (8.4) and (8.5) is important in digital signal processing for two reasons: the expressions have finite limits, making it possible to compute numeric values of $x[n]$ and $X[k]$, and they are Fourier representations that have special properties like the DTFT that are useful in the analysis and design of DSP systems. Both the DFT and the IDFT summations can be regarded as computational methods for taking N numbers in one domain and creating N numbers in the other domain. The values in both domains might be complex. Equation (8.4), for example, is really N separate summations, one for each value of k. To evaluate one of the $X[k]$ terms, we need $N-1$ complex additions and $N-1$ complex multiplications when $k \neq 0$. If we count up all the arithmetic operations required to evaluate all of the $X[k]$ coefficients, the total is $(N-1)^2$ complex multiplications and $N^2 - N$ complex additions. For example, when $N = 4$ as in Example 8-2, the 9 complex multiplications and 12 complex additions are clearly shown. Note that terms involving $e^{j0} = 1$ do not require an actual multiplication. For the inverse transform, the multiplication by $\frac{1}{4}$ would require an additional 4 multiplications if done separately as shown in Example 8-2.

One of the most important discoveries[5] in the field of digital signal processing is the ***fast Fourier transform***, or ***FFT***, a set of algorithms that can evaluate (8.4) or (8.5) with a number of operations proportional to $N \log_2 N$ rather than $(N-1)^2$. When N is a power of two, the FFT algorithm computes the entire set of coefficients $X[k]$ with approximately $(N/2)\log_2 N$ complex operations. The $N \log_2 N$ behavior becomes increasingly significant for large N. For example, if $N = 1024$, the FFT computes the DFT coefficients $X[k]$ with $(N/2)\log_2 N = 5120$ complex multiplications, rather than $(N-1)^2 = 1{,}046{,}529$ as required by direct evaluation of (8.4). The algorithm is most often applied when the DFT length N is a power of two, but it also works efficiently if N has many small-integer factors. On the other hand, when N is a prime number, the standard FFT algorithm offers no savings over a direct evaluation of the DFT summation. FFT algorithms of many different variations are available in most computer languages, and for almost any computer hardware architecture. In MATLAB, the command is simply `fft`, and most other spectral analysis functions in MATLAB call `fft` to do the bulk of their work. The DFT of a vector `x` is computed using the statement `X = fft( x, N )`, where `X` is the DFT of `x`. MATLAB uses a variety of FFT algorithms for this computation depending on the value of N, with the best case being N equal to a power of 2. More details on the FFT and its derivation can be found in Section 8-8.

8-1.4 Matrix Form of the DFT and IDFT

Another easy way to gain insight into the computation is to write the DFT summation as a matrix-vector multiplication, where the N signal values and N DFT coefficients become N-element column vectors:

[5] J. W. Cooley and J. W. Tukey, "An Algorithm for the Machine Computation of Complex Fourier Series," *Mathematics of Computation,* vol. 19, pp. 297–301, April 1965. The basic idea of the FFT has been traced back as far as Gauss at the end of the 18th century.

$$\begin{bmatrix} X[0] \\ X[1] \\ X[2] \\ \vdots \\ X[N-1] \end{bmatrix} = \begin{bmatrix} 1 & 1 & 1 & \cdots & 1 \\ 1 & e^{-j2\pi/N} & e^{-j4\pi/N} & & e^{-j2(N-1)\pi/N} \\ 1 & e^{-j4\pi/N} & e^{-j8\pi/N} & \cdots & e^{-j4(N-1)\pi/N} \\ \vdots & \vdots & \vdots & & \vdots \\ 1 & e^{-j2(N-1)\pi/N} & e^{-j4(N-1)\pi/N} & \cdots & e^{-j2(N-1)(N-1)\pi/N} \end{bmatrix} \times \begin{bmatrix} x[0] \\ x[1] \\ x[2] \\ \vdots \\ x[N-1] \end{bmatrix} \tag{8.14}$$

In MATLAB, the DFT matrix can be obtained with the function `dftmtx(N)` for an $N \times N$ matrix. Then computing the DFT would be a matrix-vector product: `X = dftmtx(N)*x`, where `x` is the vector of signal samples and `X` the vector of DFT coefficients. However, it is much more efficient to compute the DFT of a vector using the statement `X = fft( x, N )`, where `X` is the DFT of `x`.

EXERCISE 8.3 The IDFT can also be expressed as a matrix-vector product. Write out the typical entry of the $N \times N$ IDFT matrix, and then use MATLAB to create a 6×6 IDFT matrix. Check your work by multiplying the IDFT matrix by the DFT matrix (in MATLAB). Explain why the expected result should be an identity matrix.

8-2 Properties of the DFT

In this section and Section 8-3, we examine many properties of the DFT. Some are tied to its interpretation as a frequency sampled version of the DTFT, some are inherited directly from DTFT properties, and others are needed to interpret the results of the DFT. One notable special property is the periodicity of $X[k]$ with respect to k which is inherited from the 2π-periodicity of the DTFT. This periodicity affects the placement of negative frequency components in $X[k]$, and also symmetries.

We show that both the IDFT and DFT summations define the signal $x[n]$ and its N-point DFT coefficients $X[k]$ to be periodic with period N.

$$X[k] = X[k+N] \quad -\infty < k < \infty \tag{8.15a}$$

$$x[n] = x[n+N] \quad -\infty < n < \infty \tag{8.15b}$$

These periodicity properties might be surprising because the DFT was defined initially for a finite-length N-point sequence, and the DFT yields N DFT coefficients. However, many transform properties implicitly involve operations that evaluate the indices n or k outside of the interval $[0, N-1]$, so periodicity is needed to explain properties such as the delay property and convolution.

8-2.1 DFT Periodicity for $X[k]$

We have shown that the DFT of a finite N-point sequence is a sampled version of the DTFT of that sequence, that is,

$$X[k] = \sum_{n=0}^{N-1} x[n]e^{-j(2\pi/N)kn} = X(e^{j\hat{\omega}})\Big|_{\hat{\omega}=(2\pi k/N)}$$
$$= X(e^{j(2\pi k/N)}) \quad k = 0, 1, \ldots, N-1 \qquad (8.16)$$

Since the DTFT $X(e^{j\hat{\omega}})$ is always periodic with a period of 2π, the DFT $X[k]$ must also be periodic. The definition of the DFT implies that the index k always remains between 0 and $N-1$, but there is nothing that prevents the DFT summation (8.4) from being evaluated for $k \geq N$, or $k < 0$. The frequency sampling relationship $\hat{\omega}_k = 2\pi k/N$ is still true for all integers k, so $\hat{\omega}_k + 2\pi = 2\pi k/N + 2\pi(N/N) = 2\pi(k+N)/N = \hat{\omega}_{k+N}$. In other words, the DFT coefficients $X[k]$ have a period equal to N, because

$$X[k] = \underbrace{X(e^{j2\pi(k)/N}) = X(e^{j(2\pi(k)/N+2\pi)})}_{\text{DTFT period is } 2\pi} = X(e^{j2\pi(k+N)/N}) = X[k+N]$$

When the DFT values $X[k]$ are needed outside the interval $0 \leq k < N$, they can be extended at no additional computational cost using the period of N. Thus, $X[N] = X[0]$, $X[N+1] = X[1]$, $X[N+2] = X[2]$, and so on; likewise, $X[-1] = X[N-1]$, $X[-2] = X[N-2], \ldots$. This periodicity is illustrated in Fig. 8-3.

8-2.2 Negative Frequencies and the DFT

The DFT and IDFT formulas use nonnegative indices, which is convenient for computation and mathematical expressions. As a result, the IDFT synthesis formula (8.5) appears to have positive frequencies only. However, when we make the spectrum plot of a discrete-time signal (as in Chapter 4, Figs. 4-8 through 4-11) we expect to see both positive and negative frequency lines, along with conjugate symmetry when the signal is real. Thus, we must reinterpret the DFT indexing to see a conjugate-symmetric spectrum plot.

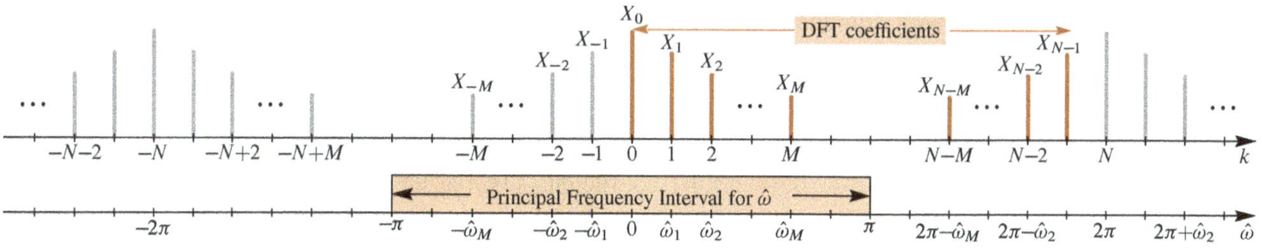

Figure 8-3 Periodicity of the DFT coefficients $X[k+N] = X[k]$, and the relationship of the frequency index k to samples of normalized frequency, $\hat{\omega}_k$. The DFT coefficients are denoted with subscripts (i.e., X_k instead of $X[k]$), in order to enhance the readability of the labels. The IDFT uses indexing that runs from $k=0$ to $k=N-1$, which is one period. The DTFT, on the other hand, typically uses the principal frequency interval $-\pi \leq \hat{\omega} < \pi$.

The signal defined by the IDFT in (8.5) has N equally spaced normalized frequencies $\hat{\omega}_k = (2\pi/N)k$ over the positive frequency range $0 \le \hat{\omega} < 2\pi$. These frequencies can be separated into two subsets

$$0 \le (2\pi/N)k < \pi \quad \text{for} \quad 0 \le k < N/2 \tag{8.17a}$$

$$\pi \le (2\pi/N)k < 2\pi \quad \text{for} \quad N/2 \le k \le N-1 \tag{8.17b}$$

Since the DTFT is periodic in $\hat{\omega}$ with a period of 2π, the frequency sampling index $k = N-1$ corresponding to the positive frequency $\hat{\omega}_{N-1} = 2\pi(N-1)/N$ aliases to the index $k = -1$ which corresponds to $\hat{\omega}_{-1} = -2\pi/N = \hat{\omega}_{N-1} - 2\pi$. In general, the sample of the DTFT at $\hat{\omega}_{-k} = (2\pi/N)(-k)$ has the same value as the sample at frequency $\hat{\omega}_{N-k} = (2\pi/N)(-k+N)$. This is the replication of frequency components discussed in Chapter 4. In fact, all the frequencies in the second subset above (8.17b) actually are aliases of the negative frequencies in the spectrum.[6] This reordering of the indices is important when we want to plot the spectrum for $-\pi \le \hat{\omega} < \pi$. When N is even, the indices in the second subset $\{\frac{N}{2}, \frac{N}{2}+1, \frac{N}{2}+2, \ldots, N-2, N-1\}$ would be aliases of $\{-\frac{N}{2}, -\frac{N}{2}+1, -\frac{N}{2}+2, \ldots, -2, -1\}$.

EXERCISE 8.4 Prove that the IDFT can be rewritten as a sum where half the indices are negative. Assume that the DFT length N is even.

$$x[n] = \frac{1}{N} \sum_{k=-N/2}^{N/2-1} X[k] e^{j(2\pi/N)kn}$$

8-2.3 Conjugate Symmetry of the DFT

When we have a real signal $x[n]$, there is conjugate symmetry in the DTFT, so the DFT coefficients must also satisfy the following property: $X[-1] = X^*[1]$, $X[-2] = X^*[2], \ldots, X[-k] = X^*[k], \ldots$. If we put the periodicity of $X[k]$ together with conjugate symmetry, we can make the general statement that the DFT of a real signal satisfies

$$X[N-k] = X^*[k] \quad \text{for } k = 0, 1, \ldots, N-1$$

Figure 8-4 shows a DFT coefficient $X[k_0]$ at the normalized frequency $\hat{\omega}_{k_0} = 2\pi k_0/N$. The corresponding negative-frequency component, which must be the conjugate, is shown in gray at $k = -k_0$. The spectrum component at $k = N - k_0$ is an alias of the negative-frequency component.

[6]The case where $k = N/2$ and N is even will be discussed further in Section 8-2.3.1. We follow the MATLAB convention (when using `fftshift`) which assigns this index to the second set.

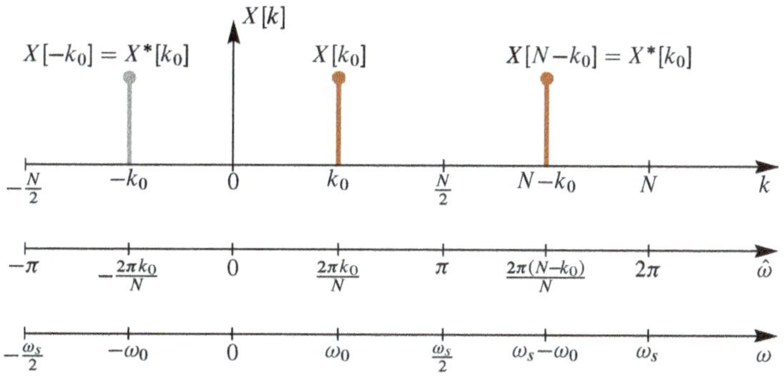

Figure 8-4 Illustration of the conjugate symmetry of the DFT coefficients showing that $X[N-k_0] = X^*[k_0]$. There are three "frequency scales." The top scale shows the DFT index k. The middle scale shows normalized frequency $\hat{\omega}$ for the discrete-time signal. The bottom scale shows the continuous-time frequency scale ($\omega = \hat{\omega}/T_s$) that would be appropriate if the sequence $x[n]$ had been obtained by sampling with a sampling frequency, $\omega_s = 2\pi/T_s$. Thus, the DFT index k_0 corresponds to the analog frequency $\omega_0 = 2\pi k_0/(NT_s)$ rad/s.

EXAMPLE 8-3 DFT Symmetry

In Example 8-1, the frequency indices of the 4-point DFT correspond to the four frequencies $\hat{\omega}_k = \{0, \pi/2, \pi, 3\pi/2\}$. The frequency $\hat{\omega}_3 = 3\pi/2$ is an alias of $\hat{\omega} = -\pi/2$. We can check that the DFT coefficients in Example 8-1 satisfy the conjugate-symmetric property, e.g., $X[1] = X^*[4-1] = X^*[3] = \sqrt{2}e^{-j\pi/4}$.

EXERCISE 8.5

It is easy to create a MATLAB example that demonstrates the conjugate-symmetry property by executing `Xk=fft(1:8)`, which computes the 8-point DFT of a real signal. List the values of the signal $x[n]$ for $n = 0, 1, 2, \ldots, 7$. Then tabulate the values of the MATLAB vector `Xk` in polar form from which you can verify that $X[N-k] = X^*[k]$ for $k = 0, 1, \ldots, 7$ and $N = 8$. Finally, list the value of $\hat{\omega}$ corresponding to each index k.

8-2.3.1 Ambiguity at $X[N/2]$

When the DFT length N is even, the transform $X[k]$ has a value at $k = N/2$; when N is odd, $k = N/2$ is not an integer so the following comments do not apply. The index $k = N/2$ corresponds to a normalized frequency of $\hat{\omega} = 2\pi(N/2)/N = \pi$. However, the spectrum is periodic in $\hat{\omega}$ with a period of 2π, so the spectrum value is the same at $\hat{\omega} = \pm\pi$. In other words, the frequency $\hat{\omega} = -\pi$ is an alias of $\hat{\omega} = \pi$.

When we plot a spectrum as in Chapter 3, the zero frequency point is placed in the middle, and the frequency axis runs from $-\pi$ to $+\pi$. Strictly speaking we need not include both end points because the values have to be equal, $X[N/2] = X[-N/2]$. Thus, we have to make a choice, either placing $X[N/2]$ on the positive frequency side, or $X[-N/2]$ on the negative frequency side of the spectrum. The following example for $N = 6$ shows the two reorderings that are possible:

$$\{\underbrace{X[0]}_{\hat{\omega}=0}, X[1], X[2], \underbrace{X[3]}_{\hat{\omega}=\pi}, X[4], X[5]\} \longrightarrow \{\underbrace{X[-3]}_{\hat{\omega}=-\pi}, X[-2], X[-1], \underbrace{X[0]}_{\hat{\omega}=0}, X[1], X[2]\}$$

$$\longrightarrow \{X[-2], X[-1], \underbrace{X[0]}_{\hat{\omega}=0}, X[1], X[2], \underbrace{X[3]}_{\hat{\omega}=\pi}\}$$

In this example, the periodicity of the DFT coefficients enabled the following replacements $X[5] = X[-1]$, $X[4] = X[-2]$, and $X[3] = X[-3]$. The DFT component $X[3]$ at $k = 3$ corresponding to the frequency $\hat{\omega} = 2\pi(3)/6 = \pi$ must have the same value as $X[-3]$ at $k = -3$ corresponding to $\hat{\omega} = 2\pi(-3)/6 = -\pi$. The choice of which reordering to use is arbitrary, but the MATLAB function `fftshift` does the first, putting $X[N/2]$ at the beginning so that it becomes $X[-N/2]$ and corresponds to $\hat{\omega} = -\pi$. Therefore, we adopt this convention in this chapter, keeping in mind this only applies when N is even.

EXERCISE 8.6 Prove that the DFT coefficient $X[N/2]$ must be real-valued when the vector $x[n]$ is real and the DFT length N is even.

8-2.4 Frequency-Domain Sampling and Interpolation

When we want to make a smooth plot of a DTFT such as the frequency response, we need frequency samples at a very fine spacing. Implicitly, the DFT assumes that the transform length is the same as the signal length L. Thus, the L-point DFT *computes samples of* the DTFT for $\hat{\omega} = (2\pi/L)k$, with $k = 0, 1, \ldots, L-1$. The L frequency samples are equally spaced samples in the interval $0 \le \hat{\omega} < 2\pi$, so we have

$$H[k] = H(e^{j(2\pi/L)k}) = H(e^{j\hat{\omega}})\Big|_{\hat{\omega}=(2\pi/L)k} \qquad k = 0, 1, \ldots, L-1 \tag{8.18}$$

If we want to have frequency samples at a finer spacing $(2\pi/N)$ where $N > L$, then a simple trick called zero-padding can be used so that we can compute an N-point DFT. In ***zero-padding***, we append zero-valued samples to the signal to make it longer prior to computing the N-point DFT. In other words, define a length-N signal $h_{zp}[n]$ as

$$h_{zp}[n] = \begin{cases} h[n] & n = 0, 1, \ldots, L-1 \\ 0 & n = L, L+1, \ldots, N-1 \end{cases} \tag{8.19}$$

Now take the N-point DFT of $h_{\text{zp}}[n]$ and then split the sum into two smaller summations:

$$H_{\text{zp}}[k] = \sum_{n=0}^{N-1} h_{\text{zp}}[n]e^{-j(2\pi/N)kn} \tag{8.20a}$$

$$= \sum_{n=0}^{L-1} h_{\text{zp}}[n]e^{-j(2\pi/N)kn} + \sum_{n=L}^{N-1} \overbrace{h_{\text{zp}}[n]}^{0} e^{-j(2\pi/N)kn} \tag{8.20b}$$

$$= \sum_{n=0}^{L-1} h[n]e^{-j(2\pi/N)kn} \tag{8.20c}$$

$$H_{\text{zp}}[k] = H(e^{j(2\pi/N)k}) \qquad \text{for } k = 0, 1, 2, \dots, N-1 \tag{8.20d}$$

Thus (8.20d) says the N-point DFT of an L-point (impulse response) signal augmented with zero padding gives frequency response samples at $\hat{\omega}_k = 2\pi k/N$.

In the special case where $N = 2L$, we get twice as many frequency samples as with $N = L$, but the even-indexed samples are identical to values from an L-point DFT, that is,

$$H_{\text{zp}}[2\ell] = H[\ell] \qquad \text{for } \ell = 0, 1, \dots, L-1 \tag{8.21}$$

To see that (8.21) is true when $N = 2L$, consider an even-indexed frequency such as $(2\pi/N)10$. The frequency $(2\pi/2L)10 = (2\pi/L)5$, i.e., the 10th value of $H_{\text{zp}}[k]$, is the same as the fifth value of $H[k]$ because these values come from sampling $H(e^{j\hat{\omega}})$ at exactly the same frequency. On the other hand, consider an odd-indexed frequency such as $(2\pi/N)7$ when $N = 2L$. Then the frequency $(2\pi/2L)7 = (2\pi/L)(3.5)$ lies between $(2\pi/L)3$ and $(2\pi/L)4$, so the odd-indexed frequencies correspond to evaluating the frequency response "in between." In effect, the $2L$-point DFT is *interpolating* the L-point DFT $H[\ell]$ which has L sample values of the frequency response. Once we recognize this interpolation behavior of the DFT, we can use even more zero-padding to produce very dense frequency grids for evaluating the frequency response prior to plotting. In fact, in the limit as $N \to \infty$, the DFT converges to the true frequency response function which is a continuous function of $\hat{\omega}$.

EXAMPLE 8-4 Frequency Response Plotting with DFT

Suppose that we want to plot the frequency response of an 8-point running sum FIR filter. Although MATLAB can do the job with its frequency response function called `freqz`, we can also do the evaluation directly with the DFT by calling `fft` in MATLAB. This provides insight into the internal structure of `freqz` which calls the FFT to do the actual work. To evaluate the frequency response at many closely spaced frequencies we need to use zero padding, so we pick $N = 80$ and do the following in MATLAB:

```
N = 80;
hpadded = [ ones(1,8), zeros(1,N-8) ];
Hk = fft(hpadded);
stem( 0:N-1, abs(Hk) )   % Plot H[k] vs. k
```

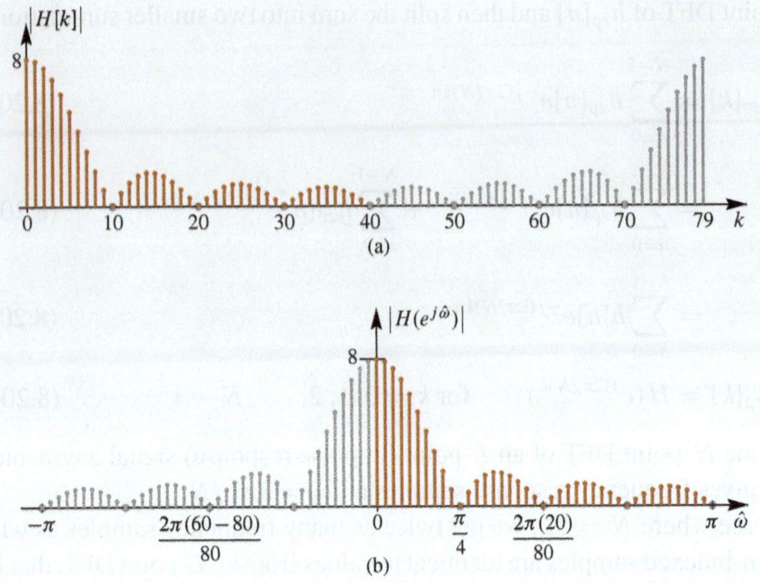

Figure 8-5 Frequency response of 8-point running sum filter. (a) Magnitude response obtained from length-80 FFT plotted versus DFT index (k). (b) Magnitude response plotted versus $\hat{\omega}$ after moving the negative frequency components shown in orange. The indices $0 \leq k < 40$ map to the frequencies $\hat{\omega} = 2\pi k/N$; while the indices $40 \leq k < 80$ map to the negative frequencies $\hat{\omega} = 2\pi(k - N)/N = 2\pi k/N - 2\pi$.

Zero-padding is a very common operation, so the `fft` function has a second argument that sets the FFT length and enables zero-padding. Thus, `fft(hn,N)` computes an N-point DFT of `hn`, with zero-padding if `N>length(hn)`. Figure 8-5(a) shows the resulting plot. If we compare the plot from this code to the magnitude plot of other running sum filters in Chapter 6, we see that the result is similar, but there is a big issue: the axis labels must be converted from "frequency index" k to $\hat{\omega}$. In addition, the frequency axis needs to have $\hat{\omega}$ running from $-\pi$ to π, so the negative frequency components must be moved as explained in Section 8-2.2. This is accomplished by treating the second half of the `Hk` vector differently from the first half as in the following code (which assumes `N` is even):

```
HkCentered = [ Hk(N/2+1:N), Hk(1:N/2) ]; %- fftshift.m does this
kCentered =   [-N+(N/2+1:N),      (1:N/2) ] - 1;
            %- make negative frequency indices
            % kCentered is [-N/2,-N/2+1,...,-2,-1,0,1,2,3,...,N/2-1]
stem( (2*pi/N)*kCentered, abs(HkCentered) )
```

The flipping operation above occurs so often when using the FFT in MATLAB that a special function called `fftshift` has been created to perform the reordering. The frequencies have to be reordered separately. When the frequency response has conjugate symmetry, it is customary to plot only the first half of `Hk` which corresponds to the positive frequency region.

8-2.5 DFT of a Real Cosine Signal

A built-in property of the DFT is linearity, which is true because the DTFT is linear and the DFT is obtained by frequency sampling the DTFT. Since a cosine signal is the sum of two complex exponentials by virtue of the inverse Euler formula, we get the DFT of a cosine by summing two terms obtained from (8.12).

The examples of the DFT of a complex exponential shown in Figs. 8-1 and 8-2 assume that the DFT length N is equal to L, the length of the complex exponential sequence. However, the general result in (8.12) holds for $L \leq N$, and it is useful to examine the case where $L < N$. Consider a length-L cosine signal

$$x[n] = A\cos(\hat{\omega}_0 n) \qquad \text{for } n = 0, 1, \ldots, L-1 \tag{8.22a}$$

which we can write as the sum of complex exponentials at frequencies $+\hat{\omega}_0$ and $2\pi - \hat{\omega}_0$ as follows:

$$x[n] = \frac{A}{2}e^{j\hat{\omega}_0 n} + \frac{A}{2}\underbrace{e^{j2\pi n}}_{=1}e^{-j\hat{\omega}_0 n} \qquad \text{for } n = 0, 1, \ldots, L-1 \tag{8.22b}$$

Using the linearity of the DFT and (8.12) with $\varphi = 0$ for these two complex exponentials leads to the expression

$$\begin{aligned} X[k] = &\frac{A}{2}D_L((2\pi k/N) - \hat{\omega}_0)\, e^{-j((2\pi k/N)-\hat{\omega}_0)(L-1)/2} \\ &+\frac{A}{2}D_L((2\pi k/N) - (2\pi - \hat{\omega}_0))\, e^{-j((2\pi k/N)+\hat{\omega}_0)(L-1)/2} \end{aligned} \tag{8.23}$$

where the function $D_L(\hat{\omega})$ is given by (7.27).

Figure 8-6 shows $|X[k]|$ as a function of k for the case $\hat{\omega}_0 = 0.4\pi$ with $N = 50$ and $L = 20$. An equivalent horizontal axis would be the normalized frequency axis $(\hat{\omega})$ where $\hat{\omega}_k = 2\pi k/N$. Note the DFT magnitude exhibits its characteristic symmetry where the maxima (peaks) of the DFT occur at $k = k_0$ and $k = N - k_0$, which correspond to the frequencies $(2\pi k_0/N) = \hat{\omega}_0$ and $(2\pi(N - k_0)/N) = 2\pi - \hat{\omega}_0$. Furthermore, the heights of the peaks are approximately $AL/2$. The latter can be shown by evaluating (8.23) under the assumption that $(2\pi k_0/N) = \hat{\omega}_0$ and that the value of $|X[k_0]|$ is determined mainly by the first term in (8.23).

In Section 8-7, we will revisit the fact that isolated spectral peaks are often indicative of sinusoidal signal components. Knowledge of the fact that the peak height depends on both the amplitude A and the duration L is useful in interpreting spectrum analysis results for signals involving multiple frequencies.

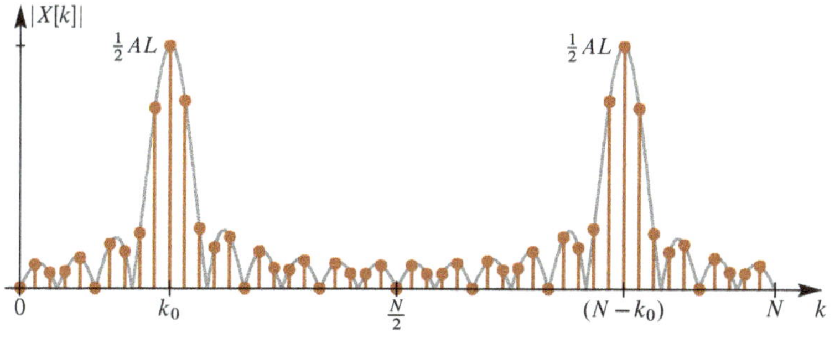

Figure 8-6 N-point DFT of a length-L real cosine signal whose frequency is an integer multiple of $2\pi/N$, where $N = 50$ and $L = 20$. Gray line is the magnitude of the DTFT being sampled. The integer multiple is $k_0 = 10$, so there are DFT samples at the peaks of the DTFT at $\hat{\omega}_0 = (2\pi/50)(10) = 0.4\pi$ and $2\pi - \hat{\omega}_0 = (2\pi/50)(50 - 10) = 1.6\pi$.

EXERCISE 8.7 Use MATLAB to compute samples of the signal $x_0[n] = 5\cos(0.211\pi n)$ for $n = 0, 1, \ldots, 499$. Compute and plot the magnitude of the $(N = 16384)$-point DFT of the length-500 sequence $x_0[n]$ using the MATLAB statements

```
x0=5*cos( 0.211*pi*(0:499) );
N=16384;
X0 = fft(x0,N);    %- fft takes care of zero padding
plot( (0:N/2)*2*pi/N , abs(X0(1:N/2+1)) );
```

Check to see if the peak height satisfies the relation $AL/2$ mentioned above. Repeat the above steps for the following length-5000 input signal:

$$x_3[n] = \begin{cases} 0.5\cos(0.4\pi n) & \text{for } n = 5000, 5001, \ldots, 9999 \\ 0 & \text{otherwise} \end{cases}$$

Overlay the two plots by using MATLAB's `hold on` command before plotting the second DFT. Explain why the spectral peak height is the same for both signals even though their amplitudes differ by a factor of ten.

EXERCISE 8.8 Use Euler's relation to represent the signal $x[n] = A\cos(2\pi k_0 n/N)$ as a sum of two complex exponentials. Assume k_0 is an integer. Use the fact that the DFT is a linear operation and the periodicity of the DFT coefficients to show that its DFT can be written as

$$X[k] = \frac{AN}{2}\delta[k - k_0] + \frac{AN}{2}\delta[k - (N - k_0)] \qquad k = 0, 1, \ldots N - 1$$

8-3 Inherent Time-Domain Periodicity of $x[n]$ in the DFT

In this section, we will study more properties of the DFT/IDFT representation. In the previous section, we examined properties of the DFT which are tied to its interpretation as a frequency sampled version of the DTFT, and the fact that $X[k]$ is periodic. This periodicity affects the placement of negative frequency components in $X[k]$, and also symmetries. In this section, we want to show that the IDFT summation requires that the time-domain signal $x[n]$ must also be periodic with period N.

$$x[n] = x[n + N] \qquad -\infty < n < \infty$$

This property might be surprising because the DFT has been defined for a finite-length N-point sequence, and the DFT yields N DFT coefficients. However, some transform properties implicitly involve operations that evaluate the indices n or k outside of $[0, N - 1]$, so periodicity is needed to explain properties such as the delay property and convolution.

8-3.1 DFT Periodicity for $x[n]$

It is less obvious that the IDFT summation implies periodicity. However, we can ask what happens when the IDFT sum (8.5) is evaluated for $n < 0$ or $n \geq N$. In particular, consider evaluating (8.5) at $n + N$ where n is in the interval $0 \leq n \leq N - 1$, that is,

$$\tilde{x}[n+N] = \frac{1}{N}\sum_{k=0}^{N-1} X[k]e^{j(2\pi/N)k(n+N)} \tag{8.24}$$

We have denoted the result in (8.24) as $\tilde{x}[n+N]$ because we are testing to see what value is computed at $n + N$.

$$\tilde{x}[(n+N)] = \frac{1}{N}\sum_{k=0}^{N-1} X[k]e^{j(2\pi/N)k(n+N)} \tag{8.25a}$$

$$= \frac{1}{N}\sum_{k=0}^{N-1} X[k]e^{j(2\pi/N)k(n)}\underbrace{e^{j(2\pi/N)k(N)}}_{1} \tag{8.25b}$$

$$= \frac{1}{N}\sum_{k=0}^{N-1} X[k]e^{j(2\pi/N)k(n)} = x[n] \tag{8.25c}$$

In (8.25b) observe that $e^{j(2\pi/N)kN} = e^{j2\pi k} = 1$ for all k.

The result in (8.25c) says that the IDFT does not give a value of zero at $n + N$, as it would if the IDFT returned the original finite-length sequence $x[n]$ with values $x[n] = 0$ outside the interval $0 \leq n \leq N - 1$. Instead, it repeats values of $x[n]$ from within the interval $0 \leq n \leq N - 1$. By the same process, it can be shown that $\tilde{x}[n + rN] = x[n]$ when we evaluate at any point $n + rN$ where r is an integer. In other words, the result from the IDFT sum is periodic in n with period N if (8.5) is evaluated outside the base interval $0 \leq n \leq N - 1$. We can express this succinctly as follows:

$$\frac{1}{N}\sum_{k=0}^{N-1} X[k]e^{j(2\pi/N)kn} = \tilde{x}[n] = \sum_{r=-\infty}^{\infty} x[n+rN] \quad \text{for } -\infty < n < \infty \tag{8.26}$$

where $x[n]$ is the original sequence whose DTFT was sampled as in (8.16). The infinite sum on the right-hand side, which involves shifted copies of the same signal, is illustrated in Fig. 8-7. Even though the discussion leading to (8.26) seems to assume that $x[n]$ is zero outside the interval $0 \leq n \leq N - 1$, it turns out that the relationship in (8.26) is also true for any signal that has a DTFT.

Equation (8.26) is an important observation about the IDFT because it provides the answer to the question: "when can a sequence $x[n]$ be reconstructed exactly from N samples of its DTFT $X(e^{j\hat{\omega}})$?" From (8.26) it is clear that if $x[n] = 0$ for n outside the base interval $0 \leq n \leq N - 1$, then $x[n] = \tilde{x}[n]$ for $0 \leq n \leq N - 1$. That is, the sequence $x[n]$ can be reconstructed exactly from N samples of its DTFT $X(e^{j\hat{\omega}})$ if it is a finite length-N sequence. This could be termed the *Sampling Theorem for the DTFT.*

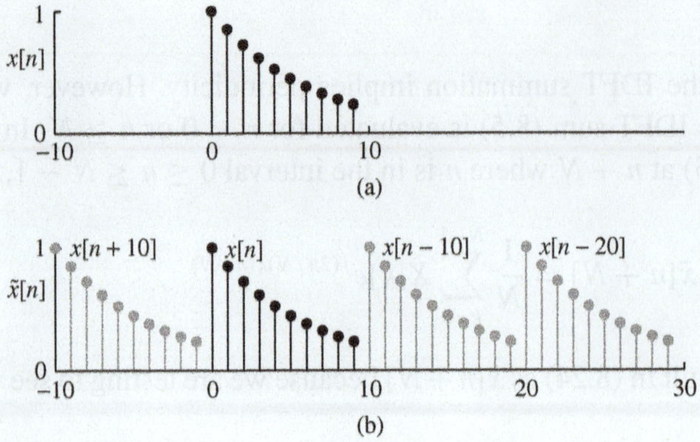

Figure 8-7 Illustration of the sum in (8.26). (a) A finite-length sequence $x[n]$ of length 10. (b) Shifted copies of $x[n]$ that are summed to make up the inherent periodic sequence $\tilde{x}[n]$ with period $N = 10$. The inherent periodic sequence $\tilde{x}[n]$ continues with period N outside the range $-10 \leq n < 30$.

The inherent periodicity of the DFT/IDFT representation in both n and k forces us to interpret some of the familiar properties of Fourier representations in a special way. Specifically, we must never lose sight of the periodic sequence $\tilde{x}[n]$ that is inherently represented by the DFT/IDFT representation. This is particularly important when considering signal operations such as delay and convolution. A good example of the time periodicity issue comes from reconsidering the DFT pair for a shifted impulse (8.9) where we want to write the pair as $\delta[n - n_d] \stackrel{\text{DFT}}{\longleftrightarrow} e^{-j(2\pi k/N)n_d}$. In this case, the IDFT periodicity issue arises when $n_d \geq N$, as the following example illustrates.

EXAMPLE 8-5 DFT of Shifted Impulse

Consider the 10-point DFT of $q[n] = \delta[n - 14]$ which should be $Q[k] = e^{-j0.2\pi(14)k}$ by virtue of the DFT pair given in (8.9). If we take the 10-point IDFT of $Q[k]$ we get a length-10 signal which is defined over the time index range $n = 0, 1, 2, \ldots 9$. Here is one way to determine the IDFT of $Q[k]$.

$$Q[k] = e^{-j0.2\pi(10+4)k} = e^{-j2\pi k} e^{-j0.2\pi(4)k} = e^{-j0.2\pi(4)k} \stackrel{\text{DFT}}{\longleftrightarrow} \delta[n - 4]$$

Thus, the result of the IDFT has a nonzero value at $n = 4$, and seems to be different from $q[n]$ which was nonzero at $n = 14$.

For the 10-point DFT of $q[n] = \delta[n - 14]$, we can only take the DFT if we extend $q[n]$ to a periodic signal $\tilde{q}[n]$ that has impulses at $n = \ldots, -16, -6, 4, 14, 24, 34, \ldots$. That is, we would write $\tilde{q}[n]$ as

$$\begin{aligned}\tilde{q}[n] &= \cdots + \underbrace{\delta[n+6]}_{q[n+20]} + \underbrace{\delta[n-4]}_{q[n+10]} + \underbrace{\delta[n-14]}_{q[n]} + \underbrace{\delta[n-24]}_{q[n-10]} + \cdots \\ &= \sum_{r=-\infty}^{\infty} \delta[n - 14 + 10r] = \sum_{r=-\infty}^{\infty} q[n + 10r]\end{aligned}$$

which expresses the periodic signal $\tilde{q}[n]$ in the form of (8.26).

8-3.2 The Time Delay Property for the DFT

As we showed in Chapter 7, the DTFT delay property is

$$y[n] = x[n - n_d] \stackrel{\text{DTFT}}{\longleftrightarrow} e^{-j\hat{\omega}n_d} X(e^{j\hat{\omega}})$$

when $y[n] = x[n - n_d]$ is a time-shifted version of $x[n]$. If we use frequency sampling, then we expect the DFT delay property would be

$$Y[k] = e^{-j\hat{\omega}n_d} X(e^{j\hat{\omega}})\Big|_{\hat{\omega}=(2\pi k/N)} = e^{-j(2\pi k/N)n_d} X[k] \tag{8.27}$$

However, when the IDFT is applied to $X[k]$ and $Y[k]$ it is the inherent periodic signal $\tilde{x}[n]$ that is shifted by n_d to give the inherent periodic signal $\tilde{y}[n]$. If we take the point of view that we only compute sequence values in the interval $0 \le n \le N - 1$, then we are led to an unexpected result. We can see that might be an issue by noting that if $x[n]$ is nonzero over the entire interval $0 \le n \le N - 1$, then there would not be room in that interval for $x[n - n_d]$. This is similar to the behavior already seen in Example 8-5 for a time-shifted impulse. For example, if $n_d = 4$, then $x[n - 4]$ is nonzero for $4 \le n \le N + 3$. This is illustrated in Fig. 8-8 for the case $N = 10$. Figure 8-8(a) shows an original 10-point sequence $x[n]$. If we represent this sequence by its 10-point DFT $X[k]$, then we

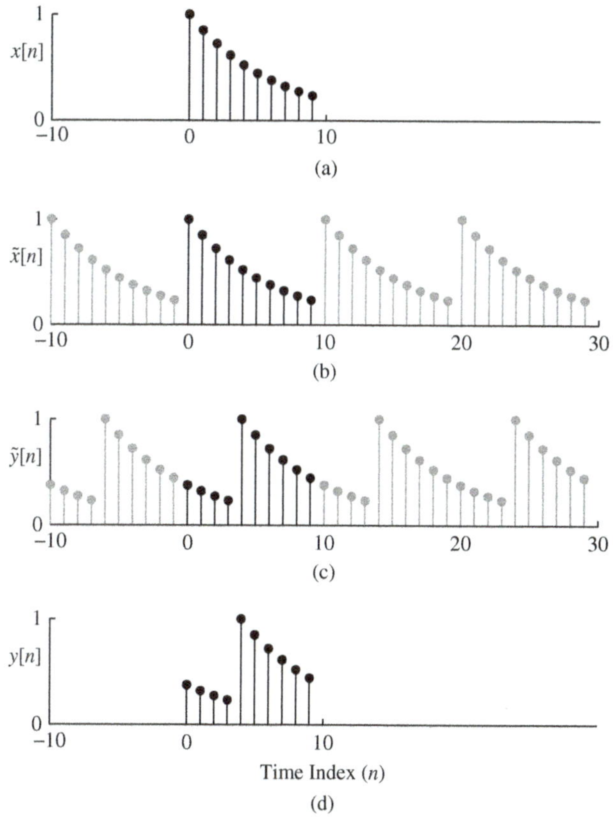

Figure 8-8 Illustration of the time-shift property of the DFT. (a) A finite-length sequence $x[n]$ of length 10. (b) The inherent periodic sequence $\tilde{x}[n]$ for a 10-point DFT representation. (c) Time-shifted periodic sequence $\tilde{y}[n] = \tilde{x}[n - 4]$ which is also equal to the IDFT of $Y[k] = e^{-j(2\pi k/10)(4)} X[k]$. (d) The sequence $y[n]$ obtained by evaluating the 10-point IDFT of $Y[k]$ only in the interval $0 \le n \le 9$.

are implicitly representing the periodic sequence $\tilde{x}[n]$ shown in Fig. 8-8(b), and if we form $Y[k] = e^{-j(2\pi k/10)(4)}X[k]$, then the corresponding periodic sequence is $\tilde{y}[n] = \tilde{x}[n-4]$ as shown in Fig. 8-8(c). The gray dots on gray lines depict the periodic extension of the sequence in the base interval $0 \le n \le 9$, which is shown with solid black dots and lines. Finally, if we form $Y[k] = e^{-j(2\pi k/10)(4)}X[k]$ and then use the 10-point IDFT to compute the sequence $y[n]$ for $0 \le n \le 9$, we obtain the result of Fig. 8-8(d).

Comparing Figs. 8-8(a) and (d), it is clear that $y[n] \neq x[n-4]$, however, the four samples on the left side of $y[n]$ were originally the four samples on the right side of $x[n]$. Thus, $y[n]$ is related to $x[n]$ by a shift that is sometimes called "circular" since the shifted samples appear to "rotate" or "wrap around" within the base interval. This effect is often called *time aliasing* in recognition of the close analogy with the frequency aliasing that occurs in sampling continuous-time signals. A convenient way to represent this ***circular shift*** is through the use of modular arithmetic.

EXAMPLE 8-6 Modulo-N Arithmetic

In number theory, a consistent algebraic system can be defined using remainders with respect to a fixed integer N, called the modulus. Recall that any integer n can be written uniquely as $n = qN + r$, where the quotient q is an integer and the remainder r is nonnegative and less than the modulus N. We write $r = n \bmod N$ to denote the remainder of n modulo-N. For example, $(-2) \bmod 10$ is equal to 8 because $-2 = (-1)(10) + 8$.

For signal delay and convolution, we only need addition and subtraction of integer indices. Suppose that $N = 10$, and we want to add 7 and 6. The result for modulo-10 arithmetic is 3 because we do normal addition $(7 + 6) = 13$, and then reduce modulo-10, taking the positive remainder which is 3. For mod-10 arithmetic the remainder must always be a positive integer in the range 0 to $N - 1 = 9$. If we subtract 4 from 2, the result $(2 - 4) \bmod 10 = -2 \bmod 10 = 8$. When we count up modulo-10, the sequence is $\{0, 1, 2, 3, 4, 5, 6, 7, 8, 9, 0, 1, 2, 3, \ldots\}$ because adding 1 to 9 gives 10, and $10 \bmod 10 = 0$. Furthermore, if we evaluate $(n - 4) \bmod 10$ for $n = 0, 1, \ldots 9$, we start at $(0 - 4) \bmod 10$ which is equal to 6 and get $\{6, 7, 8, 9, 0, 1, 2, 3, 4, 5\}$.

For a somewhat more compact general notation, we define index subtraction modulo-N via

$$((n - n_d))_N = (n - n_d) \bmod N \tag{8.28}$$

The modulo-N operation guarantees that $0 \le ((n - n_d))_N \le N - 1$. Thus, for $n = 0, 1, \ldots, N - 1$, the index $((n - n_d))_N$ cycles through the base interval starting at $n = (0 - n_d) \bmod N = N - n_d$. In terms of this notation (8.28) then, the correct DFT shift property is

$$y[n] = x[((n - n_d))_N] \stackrel{\text{DFT}}{\longleftrightarrow} Y[k] = e^{-j(2\pi k/N)n_d}X[k]$$

which emphasizes that the time shift is a circular shift with period N.

8-3.2.1 Zero Padding

While the signal in Fig. 8-8(d) is the correct result for the 10-point DFT, it is not what we would normally want for a time-shifted signal. Equation (8.26) and the example of Fig. 8-8 suggest a solution, however. First note that the signal $x[n-4]$ is not a 10-point sequence if we start counting at $n = 0$. With the delay, the first four values of $y[n]$ are zero and the shifted sequence goes from $n = 0$ to $n = 13$, so the length of the shifted sequence is 14 samples. Thus, if we augment the sequence $x[n]$ with 4 zero samples at the right-hand end, it is possible to shift the sequence to the right by 4 samples and still remain within the new base interval $0 \leq n \leq 13$. This augmentation is known as ***zero padding***. Therefore, we must compute a 14-point (or larger) DFT if we wish to use multiplication of the DFT $X[k]$ by $e^{-j(2\pi k/N)(4)}$ to effect a time shift of 4 samples that does not "wrap around." This is illustrated in Fig. 8-9.

Figure 8-9(a) shows the sequence $x[n]$ padded with 4 zero samples (which are, in fact, correct values for the finite-length sequence $x[n]$). Figure 8-9(b) shows the inherent periodic sequence $\tilde{x}[n]$ when using the DFT representation with $N = 14$, and Fig. 8-9(c) shows the shifted-by-4 periodic sequence $\tilde{y}[n]$ corresponding to $Y[k] = e^{-j(2\pi k/14)(4)}X[k]$. Finally, if we form $Y[k] = e^{-j(2\pi k/14)(4)}X[k]$ and then use the 14-point IDFT to compute the sequence $y[n]$ for $0 \leq n \leq 13$, we obtain the result of

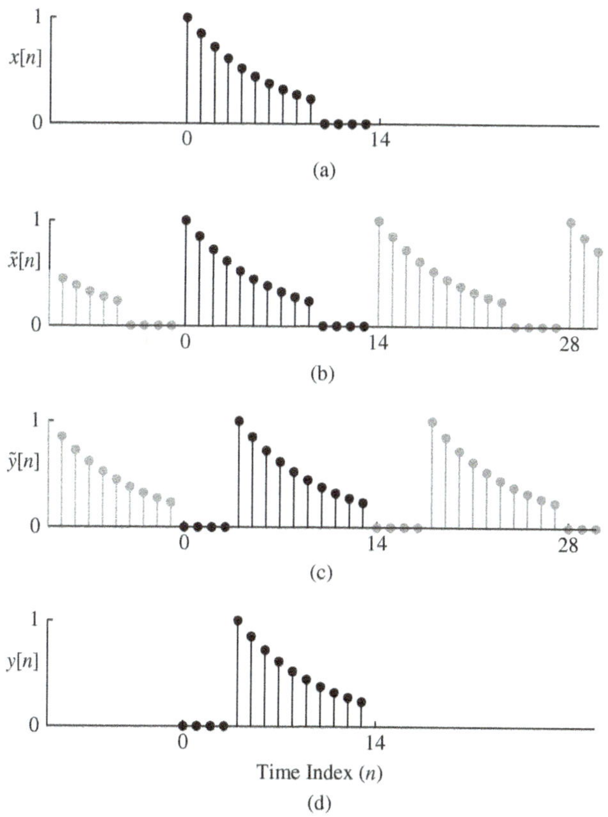

Figure 8-9 Illustration of the time-shift property of the DFT with zero padding. (a) A finite-length sequence $x[n]$ of length 10 padded with 4 zeros. (b) The inherent periodic sequence $\tilde{x}[n]$ for a 14-point DFT representation. (c) Time-shifted periodic sequence $\tilde{y}[n] = \tilde{x}[n-4]$ corresponding to $Y[k] = e^{-j(2\pi k/14)(4)}X[k]$. (d) The sequence $y[n]$ obtained by evaluating the 14-point IDFT of $Y[k]$ only in the interval $0 \leq n \leq 13$.

Fig. 8-9(d). Now, clearly, $y[n] = x[n-4]$ for $0 \le n \le 13$. A longer DFT with $N \ge 14$ would also get the correct shifted signal, but would have extra zeros for $n > 13$.

8-3.3 The Convolution Property for the DFT

One of the most important properties of the DTFT is that time-domain convolution becomes frequency-domain multiplication in the DTFT, that is,

$$y[n] = h[n] * x[n] \overset{\text{DTFT}}{\longleftrightarrow} Y(e^{j\hat{\omega}}) = H(e^{j\hat{\omega}})X(e^{j\hat{\omega}})$$

Continuing with our view that the DFT is a sampled version of the DTFT, we show that a similar property holds for the DFT, but keeping in mind what we have just learned about time delay and the DFT, it is not surprising to learn that the inherent periodicity of the DFT/IDFT representation leads to special considerations.

Since the DFT/IDFT representation deals with finite-length sequences, we want to study the convolution of two finite-length signals, each having a DFT representation. Therefore, the convolution expression can be written as a finite sum

$$y[n] = \sum_{m=0}^{M-1} h[m]x[n-m] \tag{8.29}$$

where we assume that $h[n] = 0$ except in the interval $n = 0, 1, \ldots, M-1$ and $x[n] = 0$ except in the interval $n = 0, 1, \ldots, L-1$; that is, $h[n]$ is an M-point sequence and $x[n]$ is an L-point sequence. In Chapter 5, we learned that $y[n]$ can be nonzero only in the interval $0 \le n \le L + M - 2$, so the length of the sequence $y[n]$, counting the sample at $n = 0$, is $L + M - 1$.

Now assume that we obtain a DFT $Y[k]$ by sampling $Y(e^{j\hat{\omega}})$ at the DFT frequencies $\hat{\omega}_k = (2\pi k/N)$, that is,

$$Y[k] = H(e^{j\hat{\omega}})X(e^{j\hat{\omega}})\Big|_{\hat{\omega}=(2\pi k/N)} = H(e^{j(2\pi k/N)})X(e^{j(2\pi k/N)}) = H[k]X[k] \tag{8.30}$$

where all three DFTs are N-point DFTs. If the sequences $x[n]$ and $h[n]$ are not of the same length, then at least the shorter one is assumed to be zero-padded to length N. Not surprisingly, convolution of two finite-length sequences is represented by the DFT as the product of their DFTs. However, we should not jump to conclusions because what the DFT $Y[k] = H[k]X[k]$ represents is not the ordinary convolution as in Chapter 5 and in (8.29), but rather, the periodic convolution of the inherent periodic sequences $\tilde{x}[n]$ and $\tilde{h}[n]$ as in

$$\tilde{y}[n] = \sum_{m=0}^{N-1} \tilde{h}[m]\tilde{x}[n-m] = \frac{1}{N}\sum_{k=0}^{N-1} Y[k]e^{j(2\pi/N)kn} \tag{8.31}$$

where $\tilde{y}[n]$ is the IDFT of $Y[k]$. Note that in periodic (or circular) convolution, both sequences have the same period (N) and the sum is over one period. The IDFT of $Y[k]$

is inherently periodic with period N as we have already seen. Furthermore, (8.26) tells us that the periodic sequence $\tilde{y}[n]$ is related to the sequence $y[n]$ in (8.29) as

$$\tilde{y}[n] = \sum_{r=-\infty}^{\infty} y[n + rN] \tag{8.32}$$

Now we can determine the length N that is needed to get $y[n]$. Since the desired convolution output $y[n]$ in (8.29) has length $L + M - 1$ samples, we would require $N \geq L + M - 1$ if we want $\tilde{y}[n] = y[n]$ over the base interval $0 \leq n \leq N - 1$. This can be achieved by computing the DFTs $X[k]$ and $H[k]$ with N satisfying this condition. Zero-padding of both the sequences $x[n]$ and $h[n]$ to length N is needed.

This convolution property is illustrated by the time-delay property of Section 8-3.2, which is a special case of the convolution property. First, recall that if we have a delay-by-n_d system then its impulse response is $h[n] = \delta[n - n_d]$, and

$$y[n] = h[n] * x[n] = \delta[n - n_d] * x[n] = x[n - n_d]$$

In this case, the length of the sequence $h[n]$ is $n_d + 1$ samples because we start counting at $n = 0$. If $x[n]$ is an L-point sequence, then the total length of $y[n]$ is $n_d + L$. We have shown that the DFT of $h[n] = \delta[n - n_d]$ is $H[k] = e^{-j(2\pi/N)kn_d}$. Thus, the DFT product, $Y[k] = H[k]X[k] = e^{-j(2\pi/N)kn_d}X[k]$, is identical to what we would obtain using the time-delay property for the DFT if $N \geq L + n_d$.

EXAMPLE 8-7 Convolution of Pulses

Suppose that $x[n]$ is a length-10 rectangular pulse, and $h[n]$ is a length-6 rectangular pulse. From Chapter 5, we know that the convolution result has a trapezoidal shape. If we use 10-point DFTs, it is possible to work out the answer analytically. The 10-point DFT of $x[n]$ is $X[k] = 10\delta[k]$, because $x[n]$ is all ones and we can use the DFT pair in (8.13) with $k_0 = 0$. Then we can carry out the multiplication of DFTs without knowing $H[k]$. Since $\delta[k]$ is nonzero only when $k = 0$, we obtain

$$Y[k] = X[k]H[k] = 10\delta[k]H[k] = 10H[0]\delta[k]$$

The value of $H[0]$ is the DFT coefficient at $k_0 = 0$, which is

$$H[0] = \sum_{n=0}^{9} h[n]\cancelto{1}{e^{-j0.2\pi(0)n}} = 6$$

Thus, we have $Y[k] = 60\delta[k]$ and we can take the length-10 IDFT to get $y[n]$

$$y[n] = \frac{1}{10}\sum_{k=0}^{9} 60\delta[k]e^{j0.2\pi kn} = \underbrace{6\delta[0]e^{j0}}_{k=0 \text{ term}} = 6 \quad \text{for } n = 0, 1, \ldots 9$$

The only nonzero term in the sum is the one for $k = 0$. The convolution result $y[n]$ is a constant which is not what we want.

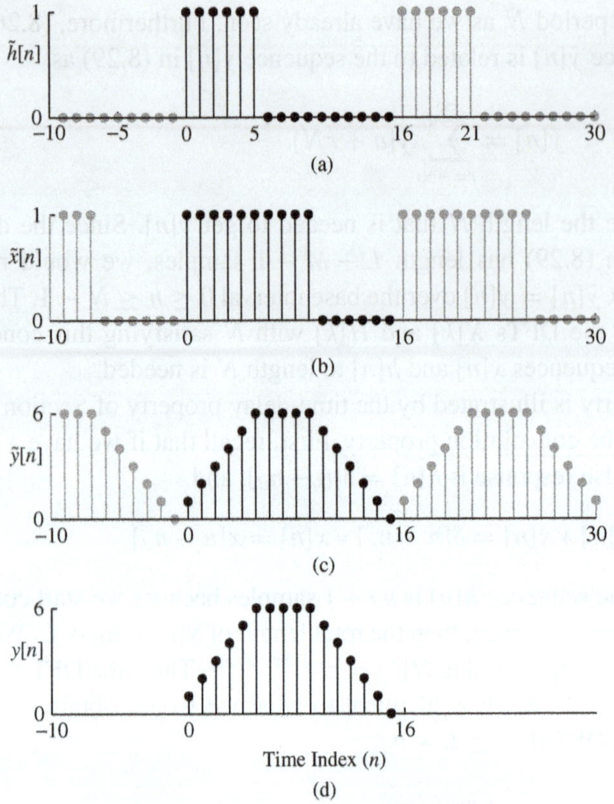

Figure 8-10 Illustration of convolving two rectangular pulses with length-16 DFTs. (a) The inherent periodic signal $\tilde{h}[n]$ corresponding to the 16-point DFT $H[k]$ of a length-6 pulse. (b) The inherent periodic signal $\tilde{x}[n]$ corresponding to the 16-point DFT $X[k]$ of a length-10 pulse. (c) The inherent periodic signal $\tilde{y}[n]$ corresponding to $Y[k] = H[k]X[k]$; $\tilde{y}[n]$ is also equal to the circular convolution defined in (8.31). (d) The 16-point sequence $y[n]$ obtained by evaluating the 16-point IDFT of $Y[k]$ only in the interval $0 \le n \le 15$.

If we change the length of the DFTs to $N = 16$, we can get the correct convolution of the two rectangular pulses where the length is $10+6-1 = 15$. This case cannot be done algebraically without a lot of tedious manipulations, so we would prefer to use MATLAB.

```
x = ones(1,10); h=[1,1,1,1,1,1];
y16 = ifft( fft( h, 16 ) .* fft( x, 16 ), 16);
y10 = ifft( fft( h, 10 ) .* fft( x, 10 ), 10);
```

The results for the $N = 16$ case are plotted in Fig. 8-10, but we omit the plots for $N = 10$ because the signal $\tilde{y}[n]$ is constant.

This example demonstrates the general result that periodic convolution as computed using the DFT and IDFT can be identical to the ordinary convolution of two finite-length sequences if the value of N in the DFT/IDFT representation is greater than or equal to the sum of the lengths of the two sequences minus one (i.e., $N \ge M + L - 1$).

8-4 Table of Discrete Fourier Transform Properties and Pairs

In this chapter, we have derived a number of useful transform pairs, and we have also derived several important properties of DFTs. Table 8-1 on p. 327 includes the DFT pairs that we have derived previously.

Table 8-1 Basic discrete Fourier transform pairs.

Table of DFT Pairs	
Time-Domain: $x[n]$	***Frequency-Domain:*** $X[k]$
$\delta[n]$	1
$\delta[n-n_d]$	$e^{-j(2\pi k/N)n_d}$
$r_L[n]=u[n]-u[n-L]$	$\underbrace{\dfrac{\sin(\frac{1}{2}L(2\pi k/N))}{\sin(\frac{1}{2}(2\pi k/N))}}_{=D_L(2\pi k/N)} e^{-j(2\pi k/N)(L-1)/2}$
$r_L[n]\,e^{j(2\pi k_0/N)n}$	$D_L(2\pi(k-k_0)/N)\,e^{-j(2\pi(k-k_0)/N)(L-1)/2}$

DEMO
Time-Frequency Visualization

The basic properties of the DFT are what make it convenient to use in designing and analyzing systems, so they are given in Table 8-2 on p. 328 for easy reference. It is important to emphasize that these properties must all be interpreted in terms of the inherent periodicity of the DFT/IDFT representation. For example, the time-delay property applies to the inherent periodic sequence that we have denoted $\tilde{x}[n]$ and the convolution property likewise concerns periodic convolution of the inherent periodic representations of $x[n]$ and $h[n]$. Some of the properties in Table 8-2 have not been derived or discussed in this chapter. They are nevertheless true and are included for completeness.

8-5 Spectrum Analysis of Discrete Periodic Signals

In Chapter 3, Section 3-6, the Fourier Series integral was presented as the operator that "Fourier analyzes" a continuous-time *periodic signal* to extract its spectrum. The resulting Fourier Series sum represents the periodic signal as a weighted sum of complex exponentials, and the frequencies of the complex exponentials are all integer multiples of a fundamental frequency. In this section, we consider the Fourier analyzer for a discrete-time periodic signal by developing the Discrete Fourier Series (DFS) representation. The key to the DFS is the IDFT sum which synthesizes a periodic signal when evaluated outside of the $0 \le n \le N-1$ interval, and which is also a weighted sum of complex exponentials whose frequencies are all integer multiples of $2\pi/N$. As a result, we can interpret the DFT as a method of Fourier analysis that extracts the spectrum of a discrete-time (sampled) periodic signal $\tilde{x}[n]$ by computing the DFT of one period of its samples. Once we have established the DFS, we are able to connect the spectrum analysis of a periodic discrete-time signal to the spectrum of a periodic continuous-time signal.[7]

[7] A similar relationship to the one that we will derive holds between the continuous-time Fourier transform and the discrete-time Fourier transform.

Table 8-2 Basic discrete Fourier transform properties.

Table of DFT Properties		
Property Name	***Time-Domain: $x[n]$***	***Frequency-Domain: $X[k]$***
Periodic	$x[n] = x[n+N]$	$X[k] = X[k+N]$
Linearity	$ax_1[n] + bx_2[n]$	$aX_1[k] + bX_2[k]$
Conjugate Symmetry	$x[n]$ is real	$X[N-k] = X^*[k]$
Conjugation	$x^*[n]$	$X^*[N-k]$
Time-Reversal	$x[((N-n))_N]$	$X[N-k]$
Delay	$x[((n-n_d))_N]$	$e^{-j(2\pi k/N)n_d} X[k]$
Frequency Shift	$x[n]e^{j(2\pi k_0/N)n}$	$X[k-k_0]$
Modulation	$x[n]\cos((2\pi k_0/N)n)$	$\frac{1}{2}X[k-k_0] + \frac{1}{2}X[k+k_0]$
Convolution	$\sum_{m=0}^{N-1} h[m]x[((n-m))_N]$	$H[k]X[k]$
Parseval's Theorem	$\sum_{n=0}^{N-1} \|x[n]\|^2 = \frac{1}{N}\sum_{k=0}^{N-1} \|X[k]\|^2$	

8-5.1 Periodic Discrete-Time Signal: Discrete Fourier Series

As we showed in Section 8-3.1, the IDFT representation of a finite-length sequence $x[n]$ is inherently periodic with period N. When we use the IDFT to represent a finite-length sequence we usually restrict the evaluation of the IDFT to the range $0 \le n \le N-1$. However, if we evaluate the IDFT outside that range we obtain the periodic sequence $\tilde{x}[n] = \tilde{x}[n+N]$,

$$\tilde{x}[n] = \frac{1}{N}\sum_{k=0}^{N-1} X[k]e^{j(2\pi k/N)n} = \sum_{r=\infty}^{\infty} x[n+rN] \tag{8.33}$$

which in (8.33) is the infinite repetition of the finite-length sequence $x[n]$ whose N-point DFT is $X[k]$.

Another notable fact about the IDFT summation (8.33) is that it is the sum of N complex exponentials with uniformly spaced frequencies, $\hat{\omega}_k = (2\pi/N)k$. In other words, all the frequencies are integer multiples of $\hat{\omega}_1 = (2\pi/N)$. If we alter our point of view away from the representation of a finite-length sequence and focus on the fact that the N-point IDFT gives a periodic result when evaluated outside the base interval, then it is reasonable also to call the representation in (8.33) the *Discrete Fourier Series* (DFS) for the periodic signal $\tilde{x}[n]$. Just as in the case of the Fourier series for continuous-time signals discussed in Section 3-5, we now have a representation of a discrete-time *periodic* signal as a sum of *harmonic* complex exponentials.[8]

We want to use the IDFT to synthesize a periodic signal but the usual Fourier Series synthesis involves negative frequencies, so we need to reassign some indices to be negative frequencies as in Section 8-2.2. When the periodic signal is real-valued (e.g., a sinusoid), then we would expect a synthesis formula with negative frequencies, as well as positive frequencies, that is,

$$\tilde{x}[n] = \sum_{m=-M}^{M} a_m e^{j(2\pi m/N)n} \qquad \text{for } -\infty < n < \infty \tag{8.34}$$

In addition, the coefficients $\{a_m\}$ have complex conjugate symmetry, $a_{-m} = a_m^*$ when $\tilde{x}[n]$ is real. In the discrete-time case, the summation limit (M) must be finite because the sum in (8.34) is equivalent to the IDFT (8.33) which has only N terms, and is a general representation for any periodic discrete-time signal whose period is N. The number of terms in (8.34) is $2M + 1$, so we must have $2M \leq N - 1$.

EXAMPLE 8-8 Synthesize a Periodic Signal from DFS

Suppose that a signal $\tilde{x}[n]$ is defined with a DFS summation like (8.34) with specific values for the coefficients $a_m = (m^2 - 1)$. If $M = 2$, then

$$\tilde{x}[n] = \sum_{m=-2}^{2} (m^2 - 1) e^{j(2\pi m/N)n} \qquad \text{for } -\infty < n < \infty$$

For the case $N = 5$, make a list of the values of $\tilde{x}[n]$ for $n = 0, 1, 2, \ldots, 10$ to show that $\tilde{x}[n]$ has a period equal to 5.

Solution: The summation formula for $\tilde{x}[n]$ can be written out

$$\begin{aligned}
\tilde{x}[n] &= ((-2)^2 - 1)e^{j(2\pi/5)(-2)n} + ((-1)^2 - 1)e^{j(2\pi/5)(-1)n} + ((0)^2 - 1)e^{j(2\pi/5)(0)n} \\
&\quad + ((1)^2 - 1)e^{j(2\pi/5)(1)n} + ((2)^2 - 1)e^{j(2\pi/5)(2)n} \\
&= (4 - 1)e^{j(2\pi/5)(-2)n} + (-1)e^{j(2\pi/5)(0)n} + (4 - 1)e^{j(2\pi/5)(2)n} \\
&= 3e^{j(2\pi/5)(-2)n} - e^{j(2\pi/5)(0)n} + 3e^{j(2\pi/5)(2)n} \\
&= 6\cos(4\pi n/5) - 1
\end{aligned}$$

---

[8] Recall that the Fourier Series synthesis formula for continuous-time signals (given in Chapter 3) represents a periodic signal as a (possibly infinite) sum of harmonic complex exponentials.

This expression can be evaluated by substituting integer values for n to obtain the following list of values for $0 \le n \le 10$:

$$\tilde{x}[n] = \{5, -5.045, 0.545, 0.545, -5.045, 5, -5.045, 0.545, 0.545, -5.045, 5\}$$

Thus, we see that $\tilde{x}[n]$ repeats with a period of 5.

The coefficients of the complex exponential terms $(1/N)X[k]$ in (8.33) become the Fourier Series coefficients. We already know that the DFT coefficients are periodic, that is, $X[k \pm N] = X[k]$ as shown in Fig. 8-3. Thus, we can identify the Fourier coefficients $\{a_m\}$ in (8.34) with respect to those in (8.33) to obtain

$$a_m = \begin{cases} \frac{1}{N}X[m] & m = 0, 1, 2, \ldots, M \\ \frac{1}{N}X[m+N] & m = -1, -2, \ldots, -M \end{cases} \tag{8.35}$$

The end result is that we can write a DFS for a periodic discrete-time signal by using the (scaled) DFT as an analysis summation to obtain the $\{a_m\}$ coefficients from one period of the signal.

$$a_k = \frac{X[k]}{N} = \frac{1}{N}\sum_{n=0}^{N-1} \tilde{x}[n]e^{-j(2\pi k/N)n} \tag{8.36a}$$

$$\tilde{x}[n] = \sum_{k=0}^{N-1} a_k e^{j(2\pi k/N)n} = \sum_{m=-M}^{M} a_m e^{j(2\pi m/N)n} \quad (\text{for } 2M+1 \le N) \tag{8.36b}$$

In the DFS, the factor of $(1/N)$ is associated with the analysis summation (8.36a).

EXAMPLE 8-9 Conjugate Symmetry of DFS Coefficients

When determining the DFS coefficients $\{a_m\}$ from the DFT $X[k]$, there are two cases to consider: N even and N odd. The notation is much easier when N is odd because we can write $N = 2M + 1$ where M is an integer. For example, when $N = 5$ we have $M = 2$, so the Fourier Series would be

$$\tilde{x}[n] = a_0 + a_1e^{j0.4\pi n} + a_2e^{j0.8\pi n} + a_{-1}e^{-j0.4\pi n} + a_{-2}e^{-j0.8\pi n}$$

The 5-point DFT of one period of $\tilde{x}[n]$ is $\{Na_0, Na_1, Na_2, Na_{-2}, Na_{-1}\}$. On the other hand, when N is even there is a complication. For example, when $N = 4$ the summation in (8.34) implies that $2M \le 3$, or $M = 1$, but using the 4-point IDFT we can write

$$\tilde{x}[n] = \tfrac{1}{N}X[0] + \tfrac{1}{N}X[1]e^{j0.5\pi n} + \tfrac{1}{N}X[2]e^{j\pi n} + \tfrac{1}{N}X[3]e^{j1.5\pi n}$$

so the DFS representation of $\tilde{x}[n]$ would be

$$\tilde{x}[n] = a_0 + a_1e^{j0.5\pi n} + a_2e^{j\pi n} + a_{-1}e^{-j0.5\pi n}$$

The case where $a_2 \neq 0$ is a special case that is similar to the ambiguity with $X[N/2]$ treated in Section 8-2.3.1. The value of a_2 (or $X[N/2]$) is real when the signal is real, so it does not require a complex conjugate term in negative frequency.

EXAMPLE 8-10 Period of a Discrete-Time Sinusoid

We might expect the fundamental period of a periodic signal to be the inverse of its fundamental frequency because this is true for continuous-time signals. However, for discrete-time signals this fact is often not true. The reason for this uncertainty is that *the period of the discrete-time signal must be an integer.*

Consider the signal $\tilde{x}_1[n] = \cos(0.125\pi n)$, whose frequency is $\pi/8$ rad. The period of this signal is $N = 16$; it is also the shortest period so we want to call 16 the fundamental period. If we take the 16-point DFT of one period of $\tilde{x}_1[n]$, we get $X_1[k] = 8\delta[k-1] + 8\delta[k-15]$. Then we can convert these DFT coefficients into a DFS representation with $a_1 = 8/16 = \frac{1}{2}$ and $a_{-1} = 8/16 = \frac{1}{2}$

$$\tilde{x}_1[n] = \tfrac{1}{2}e^{j(2\pi/16)n} + \tfrac{1}{2}e^{j(2\pi(-1)/16)n}$$

Now, consider the signal $\tilde{x}_2[n] = \cos(0.625\pi n)$, whose frequency is $5\pi/8$ rad; its period is not $2\pi/(5\pi/8) = 16/5$. Its period is also $N = 16$, and this is the shortest *integer* period. If we take the 16-point DFT of one period of $\tilde{x}_2[n]$, we get $X_2[k] = 8\delta[k-5] + 8\delta[k-11]$. Then we can convert these DFT coefficients into a DFS representation with $a_5 = 8/16 = \frac{1}{2}$ and $a_{-5} = 8/16 = \frac{1}{2}$

$$\tilde{x}_2[n] = \tfrac{1}{2}e^{j(2\pi(5)/16)n} + \tfrac{1}{2}e^{j(2\pi(-5)/16)n}$$

Our problem is that the period of $\tilde{x}_2[n]$ being 16 implies that the fundamental frequency is $2\pi/16$ so a_5 should be the fifth harmonic, but the definition of the cosine $\tilde{x}_2[n]$ has only one frequency $10\pi/16$ which has to be the fundamental frequency. In fact, this inconsistency happens whenever we take the DFT of a sinusoid with frequency $2\pi k_0/N$, and the integer k_0 is not a factor of N.

Therefore, this example illustrates that it is impossible to define a simple *consistent* relationship between the fundamental period and fundamental frequency of a periodic discrete-time signal.

8-5.2 Sampling Bandlimited Periodic Signals

The next task is to relate the DFS to the continuous-time Fourier Series. The connection is frequency scaling when sampling above the Nyquist rate (as done in Chapter 4). Consider a periodic bandlimited continuous-time signal represented by the following *finite* Fourier series

$$x(t) = \sum_{m=-M}^{M} a_m e^{j2\pi f_0 m t} \qquad -\infty < t < \infty, \tag{8.37}$$

where f_0 is the fundamental frequency (in Hz) and m denotes the integer index of summation.[9] This continuous-time signal is bandlimited because there is a maximum frequency, $2\pi M f_0$ rad/s, in the expression (8.37) for $x(t)$. When sampling $x(t)$, we must have $f_s > 2Mf_0$ Hz to satisfy the Nyquist rate criterion.

When $x(t)$ is sampled at a rate $f_s = 1/T_s$, the sampled signal $\tilde{x}[n]$ is a sum of discrete-time complex exponentials obtained from (8.37)

$$\tilde{x}[n] = x(nT_s) = \sum_{m=-M}^{M} a_m e^{j2\pi f_0 m n T_s} \qquad \text{for } -\infty < n < \infty. \tag{8.38}$$

The discrete-time signal defined in (8.38) might not be periodic, but if we restrict our attention to the special case where $f_s = Nf_0$, then $\tilde{x}[n]$ is guaranteed to be periodic with a period of N. The number of samples in each period (N) is equal to the duration of the fundamental period (T_0) times the sampling rate f_s, that is, $N = f_s/f_0 = (f_s)(T_0)$ is an integer. We can invert N to write $(1/N) = (1/f_s)(1/T_0) = (T_s)(f_0)$. When we make the substitution $f_0 T_s = 1/N$ in (8.38), the expression for $\tilde{x}[n]$ becomes

$$\tilde{x}[n] = \sum_{m=-M}^{M} a_m e^{j(2\pi/N)mn} \qquad \text{for } -\infty < n < \infty. \tag{8.39}$$

We recognize (8.39) as the *Discrete Fourier Series* of a periodic signal whose period is N. Comparing (8.37) and (8.39), we see that the Fourier Series coefficients a_m are identical for $x(t)$ and $\tilde{x}[n]$.

If we want to compute the DFS coefficients from the samples $x(nT_s)$ over one period, the number of samples in one period (N) must be large enough so that we can use the DFT of $\tilde{x}[n]$ over one period to extract a_m. Thus, we need to have at least $2M + 1$ DFT coefficients which can be used to get the $2M + 1$ DFS coefficients a_m from $X[k]$ using (8.35). The condition given in (8.36b) requires $N \geq 2M + 1$. In addition, the fact that $f_0 T_s = 1/N$ can be rewritten as $f_s = Nf_0$ means that the sampling rate is N times the fundamental frequency of $x(t)$. The Sampling Theorem requirement for no aliasing is $f_s > 2Mf_0$, that is, the sampling rate must be greater than the Nyquist rate, which is twice the highest frequency in the bandlimited signal $x(t)$. Since $f_s = Nf_0$, the Nyquist rate condition implies that $N > 2M$ which is the same requirement. Therefore, in the $f_s = Nf_0$ case, the number of samples per period must be greater than twice the number of (positive) frequency components in the continuous-time Fourier series representation of $x(t)$.

EXAMPLE 8-11 Fourier Series of a Sampled Signal

Suppose that we are given the following continuous-time signal

$$x(t) = 2\cos(42\pi t + 0.5\pi) + 4\cos(18\pi t)$$

[9]In this chapter, we use the index m for the Fourier Series to distinguish it from the IDFT where we want to use the index k. Previously, in Chapter 3, we used k for the continuous-time Fourier Series summation.

and we want to determine the Fourier Series representation of the resulting discrete-time signal. In particular, we would like to determine which Fourier Series coefficients are nonzero. We need a sampling rate that is an integer multiple of the fundamental frequency and is also greater than the Nyquist rate, which is 42 Hz. The two frequency components in $x(t)$ are at 9 Hz and 21 Hz, so the fundamental frequency is the least common divisor, which is $f_0 = 3$ Hz. We must pick $N > 14$ to satisfy the Nyquist rate condition, so for this example we use $f_s = 16 \times 3 = 48$ Hz.

The sum of sinusoids can be converted to a sum of complex exponentials,

$$x(t) = e^{j(42\pi t + 0.5\pi)} + e^{-j(42\pi t + 0.5\pi)} + 2e^{j(18\pi t)} + 2e^{-j(18\pi t)}$$

and then (8.38) can be employed to represent the sampled signal $\tilde{x}[n] = x(n/f_s)$ as

$$\tilde{x}[n] = e^{j(42\pi(n/48)+0.5\pi)} + e^{-j(42\pi(n/48)+0.5\pi)} + 2e^{j(18\pi(n/48))} + 2e^{-j(18\pi(n/48))} \quad (8.40)$$

The four discrete-time frequencies are $\hat{\omega} = \pm(42/48)\pi$ and $\pm(18/48)\pi$. In order to write (8.40) in the summation form of (8.39), we use $N = 16$. In (8.40), we want to emphasize the term $(2\pi/N) = (2\pi/16)$ in the exponents, so we write

$$\tilde{x}[n] = e^{j((2\pi/16)7n+0.5\pi)} + e^{-j((2\pi/16)7n+0.5\pi)} + 2e^{j((2\pi/16)3n)} + 2e^{-j((2\pi/16)3n)} \quad (8.41)$$

Now we can recognize this sum of four terms as a special case of (8.39) with $N = 16$ and $M = 7$ (i.e., the range of the sum is $-7 \le m \le 7$). The only nonzero Fourier coefficients in (8.39) are at those for $m = \pm 7, \pm 3$, and their values are $a_3 = 2$, $a_7 = e^{j0.5\pi} = j$, $a_{-3} = 2$, and $a_{-7} = e^{-j0.5\pi} = -j$.

These relationships between the DFS (8.39) and the continuous-time Fourier Series (8.37), and also between the DFS and the DFT, are illustrated in Fig. 8-11 where Fig. 8-11(a) shows a "typical" spectrum for a bandlimited continuous-time periodic signal (as a function of f), and Fig. 8-11(b) shows the spectrum for the corresponding periodic sampled signal (as a function of $\hat{\omega}$ and also k). Notice the alias images of the original spectrum on either side of the *base band* where all the frequencies lie in the interval $-\pi < \hat{\omega}_m \le \pi$. When $f_s > 2Mf_0$, the sampling rate f_s is greater than the Nyquist rate and the entire continuous-time spectrum is found in the base band because no aliasing distortion occurs.

We know from Chapter 4 that the spectrum of a sampled signal is periodic in $\hat{\omega}$ with a period of 2π. This attribute is shown in Fig. 8-11(b). Since the spectrum is periodic in $\hat{\omega}$ with a period of 2π with the spectral lines located at frequencies $\hat{\omega}_k = (2\pi/N)k$, an equivalent statement is that the spectrum is periodic in k with a period equal to N. To emphasize this point, Fig. 8-11(b) shows two horizontal plotting axes: one for normalized frequency $\hat{\omega}$ and the other for the indices k that appear in $\hat{\omega}_k$. The DFT shown in Fig. 8-11(c) is also subject to the same periodicity in k, so the DFT coefficients for $0 \le k \le N-1$ can be related to the DFS coefficients via $a_m = (1/N)X[m]$, for $-M \le m \le M$.

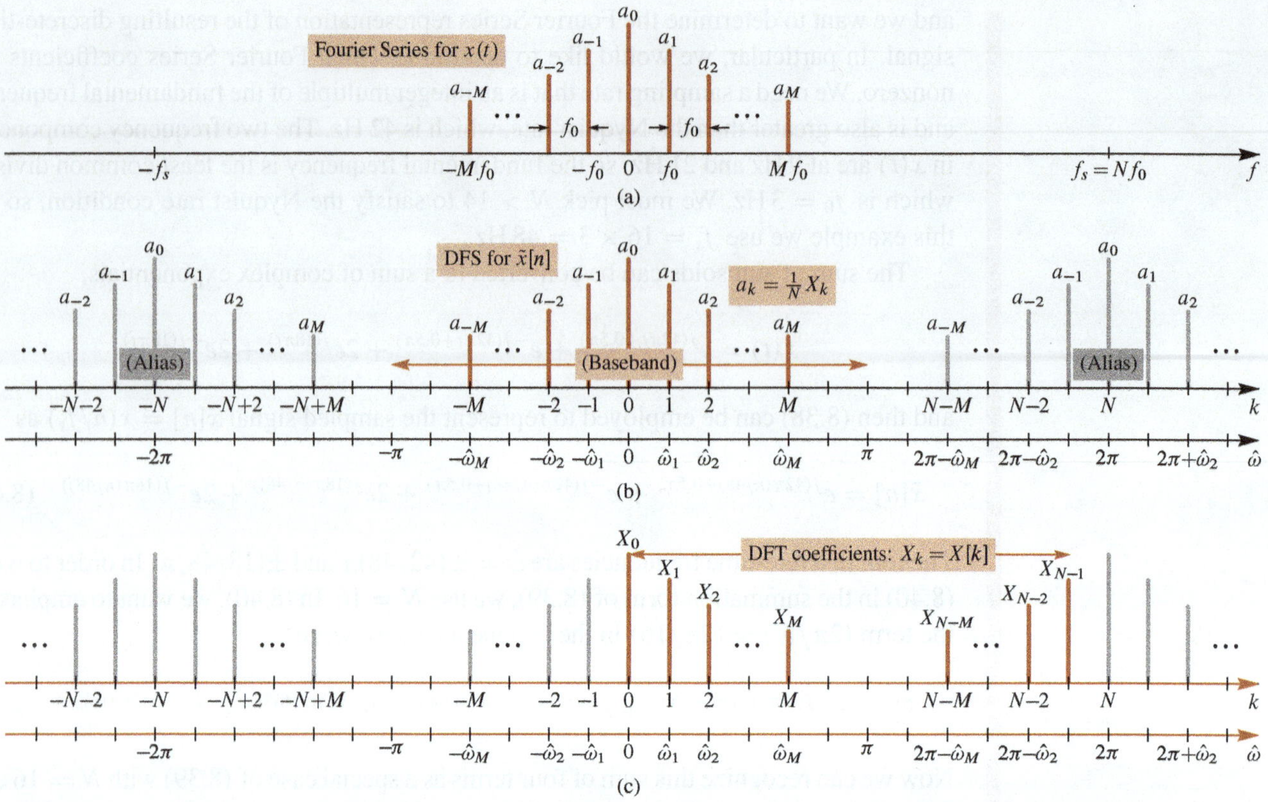

Figure 8-11 Frequency-domain view of sampling a periodic signal. (a) Line spectrum of a bandlimited continuous-time periodic signal whose fundamental frequency is f_0. (b) Line spectrum of the discrete-time periodic signal $\tilde{x}[n]$ obtained by sampling the signal in (a) above the Nyquist rate at $f_s = Nf_0$ where $N > 2M$, giving lines at $\hat{\omega}_k = (2\pi/N)k$. (c) DFT coefficients shown as a line spectrum, where the DFT is taken over one period of the periodic signal $\tilde{x}[n]$. The DFT coefficients are denoted with subscripts (i.e., X_k instead of $X[k]$), to enhance the readability of the labels.

8-5.3 Spectrum Analysis of Periodic Signals

In this section, we show how the DFT can be used to analyze continuous-time periodic signals that have been sampled. As a specific example, consider the periodic continuous-time signal

$$\begin{aligned} x_c(t) = 0.1223\cos(2\pi(200)t + 1.5077) \\ + 0.2942\cos(2\pi(400)t + 1.8769) \\ + 0.4884\cos(2\pi(500)t - 0.1852) \\ + 0.1362\cos(2\pi(1600)t - 1.4488) \\ + 0.0472\cos(2\pi(1700)t) \end{aligned} \quad (8.42)$$

The fundamental frequency of this signal is $\omega_0 = 2\pi(100)$ rad/s, and the signal consists of five harmonics at 2, 4, 5, 16, and 17 times the fundamental. This signal is the same as the synthetic vowel studied in Section 3-4.1. If we sample $x_c(t)$ at the rate

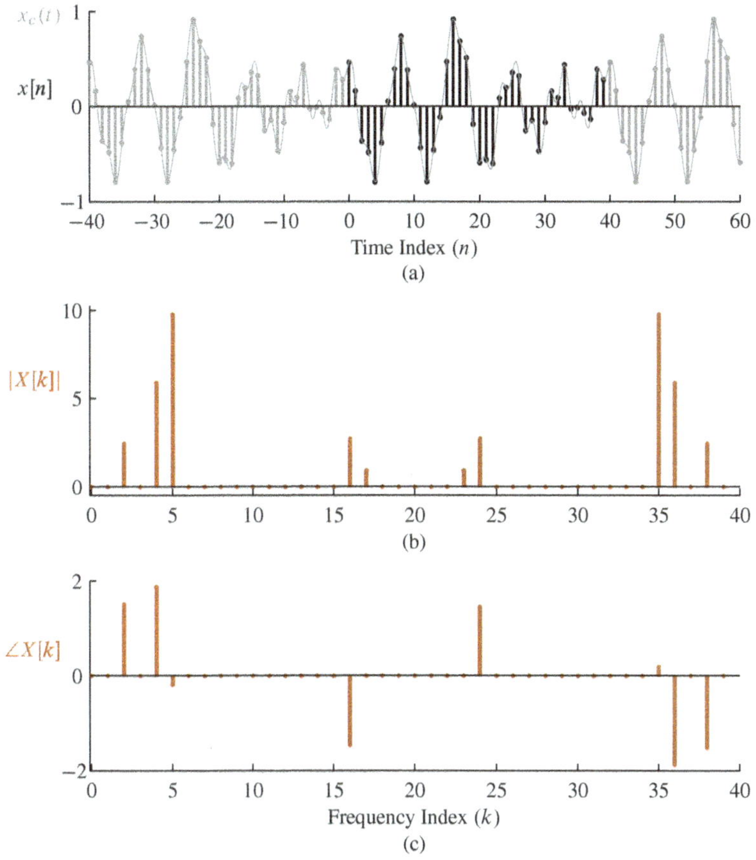

Figure 8-12 Periodic sequence $x[n]$ in (8.43) and the corresponding 40-point DFT spectrum $X[k]$. (a) Dark region indicates the 40-point interval (starting at $n = 0$) taken for analysis. (b) DFT magnitude spectrum. (c) DFT phase spectrum.

$f_s = 4000$ Hz, we obtain exactly 40 samples per period since $f_s = 4000$ Hz is exactly 40 times $f_0 = 100$ Hz. The sampled discrete-time signal $x[n] = x_c(n/4000)$ is

$$\begin{aligned}\tilde{x}[n] = {} & 0.1223\cos(0.1\pi n + 1.5077) \\ & + 0.2942\cos(0.2\pi n + 1.8769) \\ & + 0.4884\cos(0.25\pi n - 0.1852) \\ & + 0.1362\cos(0.8\pi n - 1.4488) \\ & + 0.0472\cos(0.85\pi n) \end{aligned} \tag{8.43}$$

The plot of $\tilde{x}[n]$ in Fig. 8-12(a) shows that it is periodic with a period of 40 (i.e., $\tilde{x}[n] = \tilde{x}[n+40]$). Another way to arrive at the same conclusion is to note that $\tilde{x}[n]$ has frequencies that are all multiples of $\hat{\omega}_0 = 0.05\pi = 2\pi/40$ which is therefore the fundamental frequency. Like the continuous-time case, $\tilde{x}[n]$ in (8.43) has harmonics at multiples of the normalized fundamental frequency numbered 2, 4, 5, 16, and 17.

Now we consider taking the DFT of $\tilde{x}[n]$ with a DFT length equal to the period. First, one length-40 period of $\tilde{x}[n]$ from Fig. 8-12(a) is extracted and used in a 40-point DFT (8.4). The 40 DFT coefficients $X[k]$ obtained are shown as magnitude and phase in Figs. 8-12(b) and (c), respectively. These 40 DFT coefficients represent one period of the sequence $\tilde{x}[n]$ *exactly* through the synthesis formula of the IDFT (8.5). The DFS is also a representation of $\tilde{x}[n]$ as a sum of complex exponentials at the harmonic frequencies, so the DFT coefficients $X[k]$ are related to the DFS coefficients as shown in (8.35).

EXERCISE 8.9 Use the results of Section 8-5.2, Exercise 8.8 to determine a formula for $X[k]$, the 40-point DFT of $x[n]$ in (8.43). Use 40 points of $x[n]$ taken over the period $0 \le n \le 39$. Verify that some of the nonzero DFT values are

$$X[k] = \begin{cases} 0.06115(40)e^{j1.5077} & k = 2 \\ 0.1471(40)e^{j1.8769} & k = 4 \\ 0.2442(40)e^{-j0.1852} & k = 5 \\ 0.0681(40)e^{-j1.4488} & k = 16 \\ 0.0236(40) & k = 17 \end{cases} \tag{8.44}$$

Determine the other nonzero values of $X[k]$, and also how many DFT coefficients $X[k]$ are equal to zero.

EXERCISE 8.10 Show that the IDFT of $X[k]$ defined in Exercise 8.9 gives $x[n]$ in (8.43). Furthermore, show that the signal $x[n]$ obtained via the IDFT is periodic with period N.

The plots of Figs. 8-12(b) and (c) are the spectrum for the sum of five sinusoids, or ten complex exponentials, so Fig. 8-12(b) consists of ten spectral lines, each one being just like the plot in Fig. 8-2 which shows the magnitude of the DFT for a single complex exponential signal of the form

$$x[n] = e^{j((2\pi/N)k_0 n+\varphi)} \qquad n = 0, 1, \ldots, N-1$$

where k_0 is an integer.

We have seen that the DFT can be used to obtain the DFS *when the DFT length exactly matches the period of* $x[n]$. Without an exact match, the DFT result might be a reasonable approximation, but it is not exact. To illustrate this point, consider a case where the sampling frequency is not an integer multiple of the fundamental frequency, that is, f_s/f_0 is not an integer and, therefore, cannot be equal to the DFT length. An example is sampling the signal in (8.42) at $f_s = 3950$ Hz (i.e., $f_s/f_0 = 39.5$), so the plot in Fig. 8-13(a) seems to have a period close to 40, but it is not exactly equal to 40.

If we proceed as though the period were 40, and compute the 40-point DFT of $N = 40$ samples of the sampled signal, we obtain the plot in Fig. 8-13(b). This DFT plot is somewhat similar to the DFT in Fig. 8-12(b), but none of the DFT coefficients are exactly zero. Instead, each cosine produces a contribution to the DFT that has many small values spread out over the entire range $0 \le k \le N-1$, as was seen in Fig. 8-1. The amplitude of the DFT values tend to peak around the DFT indices closest to the same peak frequencies as before, $(2\pi f_0/f_s)k$ where $k = 2, 4, 5, 16, 17$. However, the

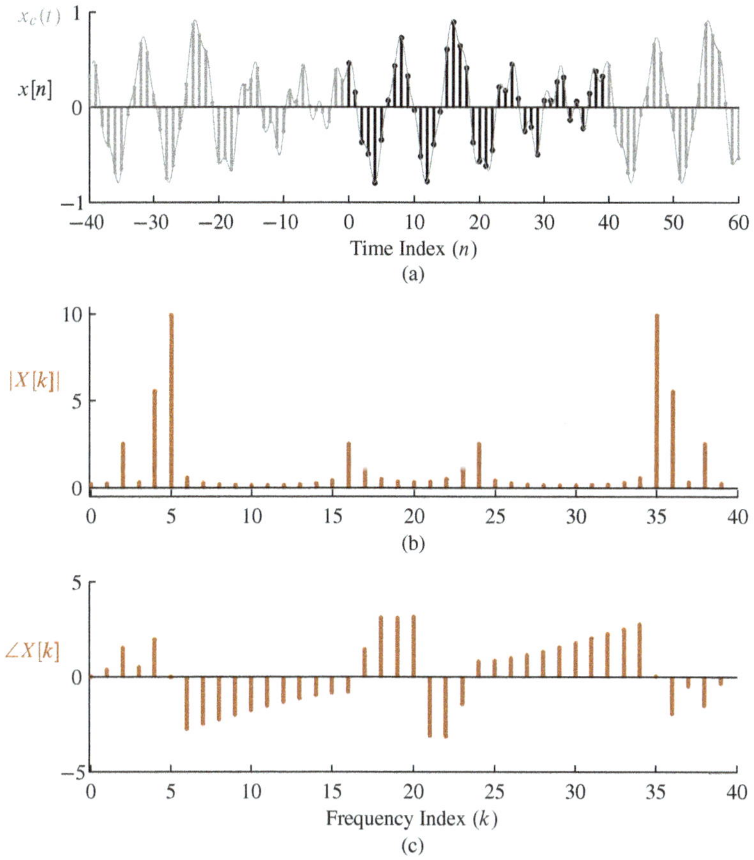

Figure 8-13 Discrete-time sequence $x[n]$ whose period is approximately equal to 40 obtained by sampling $x(t)$ in (8.42) at $f_s = 3950$ Hz, and the corresponding 40-point DFT spectrum $X[k]$. (a) Dark region indicates 40-point interval (starting at $n = 0$) taken for analysis. (b) DFT magnitude spectrum. (c) DFT phase spectrum.

presence of many additional spectral components gives a "blurred" (or imprecise) spectral representation of the original continuous-time signal.

EXERCISE 8.11 Use MATLAB to synthesize $N = 40$ samples of $x_c(t)$ in (8.42) with $f_s = 4100$ Hz instead of 4000 Hz; call the result `x4100`. Next, use the MATLAB statement `X41=fft(x4100,40)` to compute the 40-point DFT. Finally, make a plot of the DFT magnitude `stem(0:39,abs(X41))`, and compare the resulting magnitude spectrum to that of Fig. 8-12(b). Comment on expected zero regions of the DFT.

8-6 Windows

Since the DFT is a finite sum, it can only be used to analyze finite-length signals. Even infinite-length periodic signals that have a DFS are analyzed by taking the DFT of one period. We have examined the case where the finite-length signal is the impulse response of an FIR system, and have shown that samples of the frequency response, or equivalently the DTFT, can be computed by zero-padding and taking a long DFT. Remember from

Section 8-2.4 that the N-point DFT of a finite-length sequence is identical to the DTFT of the sequence evaluated at frequencies $\hat{\omega}_k = 2\pi k/N$ with $k = 0, 1, \ldots, N-1$, that is,

$$X[k] = X(e^{j(2\pi k/N)}) = \sum_{n=0}^{L-1} x[n]e^{-j(2\pi k/N)n} \tag{8.45}$$

We recognized that $X[k] = X(e^{j(2\pi k/N)})$ even if $L < N$ since $N-L$ zero samples are effectively appended to the end of the sequence $x[n]$. This operation was called ***zero-padding*** in Section 8-2.4. In MATLAB, the DFT is computed by the function `fft(x,N)`, which automatically zero-pads the sequence if the sequence length is smaller than the DFT length.

In this section, we will study another aspect of finite-length signals, which is their use as *windows* for local spectrum analysis of short sections of very long signals. The ***concept of windowing*** is widely used in signal processing. In fact, it has already been discussed in Section 7-4 for FIR filter design. The basic idea is to extract a finite section of a longer signal $x[n]$ via multiplication $w[n]x[n+n_0]$, where $w[n]$ is the window function and n_0 is the starting index of the extracted section. The product signal $w[n]x[n+n_0]$ is a finite-length sequence if the window function $w[n]$ is zero outside of an interval. The simplest window function is the L-point *rectangular window* defined as

$$w_r[n] = \begin{cases} 1 & 0 \le n \le L-1 \\ 0 & \text{elsewhere} \end{cases} \tag{8.46}$$

The essential idea of windowing is that the product $w_r[n]x[n+n_0]$ extracts L values from the signal $x[n]$ starting at $n = n_0$. Thus, the following expressions are equivalent

$$w_r[n]x[n+n_0] = \begin{cases} 0 & n < 0 \\ \overset{1}{\cancel{w}}_r[n]x[n+n_0] & 0 \le n \le L-1 \\ 0 & n \ge L \end{cases} \tag{8.47}$$

The name *window* comes from the idea that we can only "see" the L values of the signal $x[n+n_0]$ within the window interval when we "look" through the window. Multiplying by $w[n]$ is looking through the window. When we change n_0, the signal shifts, and we see a different length-L section of the signal.

The nonzero values of the window function do not have to be all ones, but they should be positive. For example, the symmetric L-point Hann, or von Hann, window[10] is defined as

$$w_h[n] = \begin{cases} 0.5 - 0.5\cos(2\pi(n+1)/(L+1)) & 0 \le n \le L-1 \\ 0 & \text{elsewhere} \end{cases} \tag{8.48}$$

[10]The Hann window is named for the Austrian scientist Julius von Hann. This window is often called the "hanning" window. Sometimes, the Hann window formula is given such that $w_h[0] = 0$, but our definition produces all positive values for $w_h[n]$.

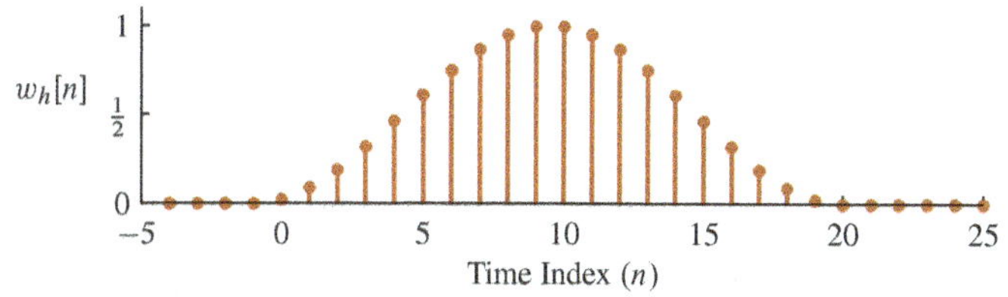

Figure 8-14 Time-domain plot of $L = 20$ Hann window. The window is zero for $n < 0$ and $n \geq 20$.

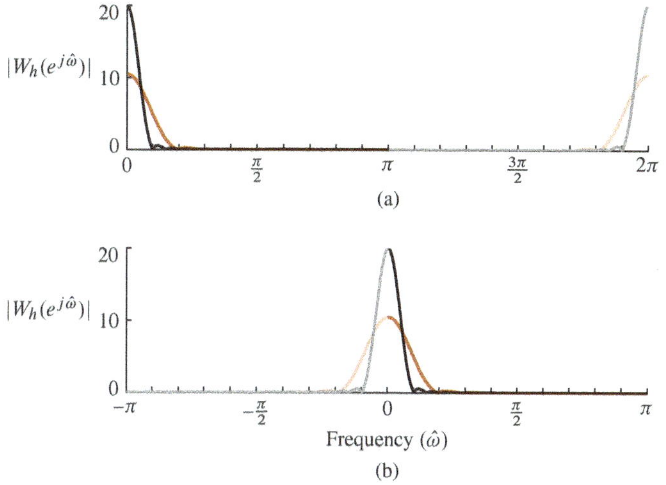

Figure 8-15 DTFT of two Hann windows; length $L = 20$ (orange lines), and length $L = 40$ (black/gray lines). (a) As computed by 1024-point FFT with zero padding. (b) As plotted with DFT values reordered to place $\hat{\omega} = 0$ in the middle. Gray and light orange indicate negative frequency components.

The stem plot of $w_h[n]$ for $L = 20$ in Fig. 8-14 shows that the window values are close to one in the middle (at $n = 9, 10$) and taper off near the ends. In MATLAB the function `hanning(L)` computes the window values as in (8.48).[11]

8-6.1 DTFT of Windows

Windows are used often in spectrum analysis because they have desirable properties in the frequency domain. Thus it is worth studying the DTFT of a few common windows. We already know the DTFT of the rectangular window—it involves the Dirichlet form.

$$W_r(e^{j\hat{\omega}}) = D_L(\hat{\omega})e^{-j(L-1)\hat{\omega}/2} = \frac{\sin(\frac{1}{2}L\hat{\omega})}{\sin(\frac{1}{2}\hat{\omega})}e^{-j\hat{\omega}(L-1)/2}$$

For the Hamming and von Hann windows, the analytic formulas for the DTFT do not provide much insight, so we prefer to obtain the DTFT numerically and make plots after computing a zero-padded DFT.

Figure 8-15 shows plots of the magnitude of the DTFT for two Hann windows with lengths $L = 20$ and $L = 40$. The DTFTs were obtained using a 1024-point DFT with zero padding. The 1024 frequency samples enable MATLAB's `plot` function to draw a smooth curve even though the sample values are connected by straight lines. Figure 8-15(b) shows

[11]A related window is the symmetric L-point Hamming window which was defined and discussed in Section 7-4; its values are generated by the MATLAB function `hamming(L)`.

that the DTFT of the Hann window is concentrated near $\hat{\omega} = 0$. The main lobe around $\hat{\omega} = 0$ covers the range $-4\pi/(L+1) < \hat{\omega} < 4\pi/(L+1)$; outside of this main lobe region the DTFT is very close to zero.

When we compare the DTFTs $|W_h(e^{j\hat{\omega}})|$ in Fig. 8-15(b) for the two different window lengths, we see that the DTFT magnitude becomes more concentrated around $\hat{\omega} = 0$ as we increase L. It can be shown that the first zero of $|W_h(e^{j\hat{\omega}})|$ occurs at $\hat{\omega} = 4\pi/(L+1)$ for the Hann window. Thus, increasing L from 20 to 40 cuts the width of the main lobe in half (approximately). The peak amplitude at $\hat{\omega} = 0$ is also approximately doubled, going from 10.5 to 20.5, which is consistent with the fact that, the general formula for the peak is $\frac{1}{2}(L+1)$. Plotting $|W_h(e^{j\hat{\omega}})|$ allows us to verify these results.

EXERCISE 8.12 A property of the DTFT is that the DC value ($\hat{\omega} = 0$) of the DTFT is equal to the sum of all the signal values in the time domain. Use the definition of the forward DTFT to prove this fact, which could then be used to determine $W_h(e^{j0})$.

Windows are used to extract sections of longer signals. An illustrative case is extracting a section of a sinusoid for spectrum analysis, which involves multiplying the sinusoid by the window and taking the DFT with zero padding to obtain samples of the DTFT. Within this context there are two questions that should be answered: first, how to choose a good window function, and second, how to choose the window length.

For the first question, consider an example that compares the rectangular window to the Hann window. Suppose that each window is applied to a sinusoid as $x_w[n] = w[n]\cos(0.4\pi n)$, with a window length of 40. Since the window is multiplied by a cosine, the frequency-shifting property of the DTFT predicts that the result will be the sum of two frequency-shifted copies of the DTFT of the window, that is,

$$X_w(e^{j\hat{\omega}}) = \tfrac{1}{2}W(e^{j(\hat{\omega}-0.4\pi)}) + \tfrac{1}{2}W(e^{j(\hat{\omega}+0.4\pi)}) \tag{8.49}$$

where $W(e^{j\hat{\omega}})$ is the DTFT of the window. Figure 8-16 shows the results, and it is easy to see the frequency shifting because the main lobes are centered at $\pm 0.4\pi$. For the rectangular window shown in Fig. 8-16(a), the main lobe is narrow, extending over the range $-2\pi/L \le \hat{\omega} \le 2\pi/L$ if measured between the first zero crossings on either side of the peak. However, the DTFT magnitude in Fig. 8-16(a) also exhibits many smaller lobes across the entire frequency range. These are called ***sidelobes***. For the rectangular window these sidelobes are rather high and are the main reason that this window is considered inferior to most others. On the other hand, the Hann window DTFT shown in Fig. 8-16(b) has a wider main lobe, but much smaller sidelobes. The main lobe of the Hann window DTFT extends over the range $-4\pi/(L+1) \le \hat{\omega} \le 4\pi/(L+1)$ when measured between the first zero crossings. Thus, the Hann window of length L has a frequency-domain width that is approximately twice that of a rectangular window of the same length. The sidelobes of the Hann DTFT magnitude in Fig. 8-16(b) are less than 1% of the peak heights.

If we pause to ask ourselves what the ideal DTFT should be for a windowed sinusoid of frequency $\hat{\omega}_0$, the answer is likely to be that we want to see only two spectrum lines

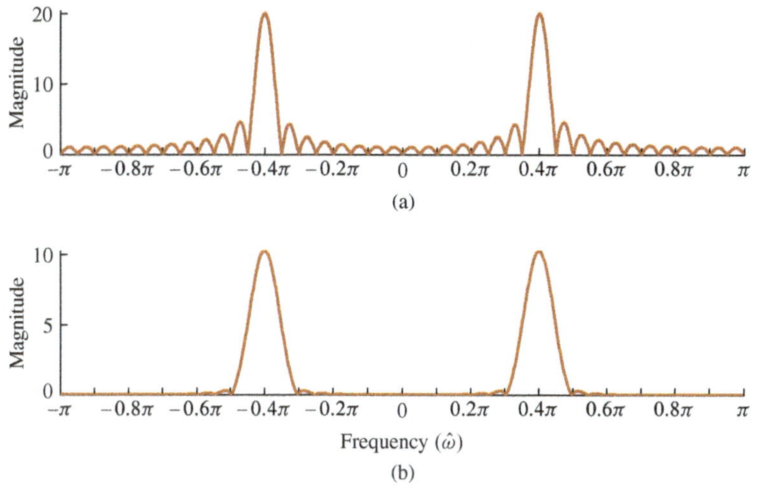

Figure 8-16 DTFT magnitudes of a windowed sinusoid via 1000-point DFT. (a) $L = 40$ rectangular window (peak heights are 20). (b) $L = 40$ Hann window (peak heights are 10.25). The sinusoid frequency is 0.4π rad.

at $\pm\hat{\omega}_0$. In other words, the ideal main lobe should be very narrow (concentrated at the correct frequencies $\pm\hat{\omega}_0$), and the ideal sidelobes should be zero (or nearly zero). It turns out that these two expectations are in conflict when using finite-length windows, because it is known that reducing the sidelobes can only be done at the expense of broadening the main lobe.

In addition, a comparison of the two windowed DTFTs in Fig. 8-16 shows that the peak heights are different. For the Hann window, we have already observed that the peak height is $\frac{1}{2}(L+1)$; for the rectangular window it is L. In either case, the window is contributing a known scale factor to the result. If the objective is spectrum analysis of a sinusoid, as in this example, we know that the result should be a spectrum line whose height is half the amplitude of the sinusoid, or $\frac{1}{2}$ in the unit-amplitude case. In Fig. 8-16(a) dividing by $L = 40$ gives $\frac{1}{2}$, and also in Fig. 8-16(b) dividing by $\frac{1}{2}(L+1) = 20.5$ gives $\frac{1}{2}$.

For the second question about length of the window, Fig. 8-15 demonstrates that the width of the main lobe in the DTFT is inversely proportional to the window length. We continue with the same example of spectrum analysis of a windowed unit-amplitude sinusoid, but this time use two different Hann windows with lengths 20 and 40. Figure 8-17 shows the resulting scaled DTFTs, where the known window scale factors have been divided out, so the peak values are $\frac{1}{2}$. If the main lobe width is measured between first zero crossings, then for Fig. 8-17(a) the main lobe extends from 0.2π to 0.6π, and in Fig. 8-17(b) from 0.3π to 0.5π. Thus for $L = 20$, the main lobe width $\Delta\hat{\omega}$ is $\Delta\hat{\omega} \approx 0.4\pi$, and for $L = 40$, $\Delta\hat{\omega} \approx 0.2\pi$. The dependence of $\Delta\hat{\omega}$ on L is an inverse relationship which can be approximated[12] as $\Delta\hat{\omega} \approx 8\pi/L$.

[12]A precise analysis would show that the first zero crossings are at $\pm 4\pi/(L+1)$, so the exact relationship is $8\pi/(L+1)$. For $L > 30$, the difference in the two formulas is less than 0.1%, and it is easier to remember *inversely proportional to L*.

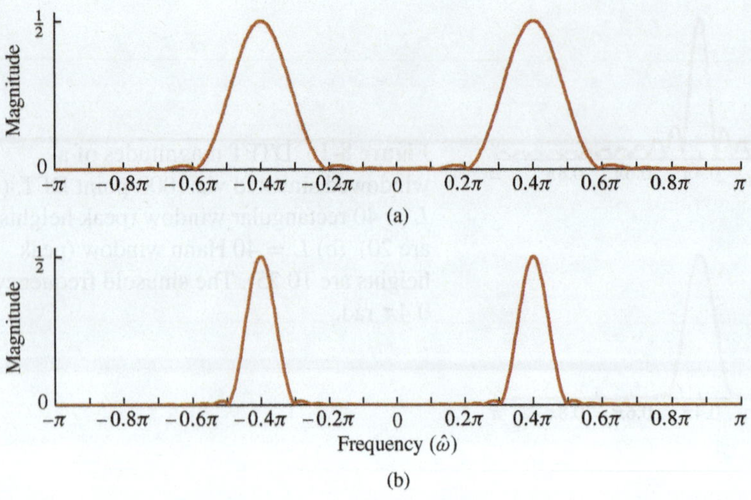

Figure 8-17 Scaled DTFT of windowed unit-amplitude sinusoid via 1000-point DFT. (a) $L = 20$ Hann window scaled by $1/5.25$. (b) $L = 40$ Hann window scaled by $1/10.25$. Sinusoid frequency is 0.4π rad.

EXAMPLE 8-12 Sum of Two Sinusoids

The importance of analyzing the frequency-domain window characteristics becomes clear when we consider signals that are composed of sums of sinusoids. If we have a signal with two frequencies

$$x[n] = A_1 \cos \hat{\omega}_1 n + A_2 \cos \hat{\omega}_2 n$$

where $\hat{\omega}_1 < \hat{\omega}_2$, then the windowed signal would be

$$x_w[n] = w[n]x[n] = A_1 w[n] \cos \hat{\omega}_1 n + A_2 w[n] \cos \hat{\omega}_2 n$$

and the corresponding DTFT would be

$$X_w(e^{j\hat{\omega}}) = \frac{A_1}{2} W(e^{j(\hat{\omega}-\hat{\omega}_1)}) + \frac{A_1}{2} W(e^{j(\hat{\omega}+\hat{\omega}_1)}) + \frac{A_2}{2} W(e^{j(\hat{\omega}-\hat{\omega}_2)}) + \frac{A_2}{2} W(e^{j(\hat{\omega}+\hat{\omega}_2)}) \quad (8.50)$$

If we want to compute samples of $X_w(e^{j\hat{\omega}})$ in order to estimate $\hat{\omega}_1$ and $\hat{\omega}_2$, the main lobes of the terms in (8.50) should not overlap. If the window is a Hann window of main lobe width $8\pi/L$, the main lobes do not overlap if $\hat{\omega}_2 - \hat{\omega}_1 > 8\pi/L$. In this case, we would obtain distinct peaks at $\pm\hat{\omega}_1$ and $\pm\hat{\omega}_2$ and the peak heights are nearly equal to $\frac{1}{2}A_1$ and $\frac{1}{2}A_2$. When there are two distinct peaks in the DTFT, we say that the two frequencies are *resolved*. The effect of window length on spectrum resolution is discussed further in Section 8-7.4.1.

8-7 The Spectrogram

We have seen that the DFT can *compute* exact frequency-domain representations of both periodic and finite-length discrete-time signals. An equally important case occurs when the sampled signal is indefinitely long, but not necessarily periodic. Then there are two different viewpoints to consider: the long signal has the same *global* spectral content for its entire duration, or the long signal is actually the concatenation of many short signals with changing *local* spectral content. In the first case, we might be able to analyze the signal mathematically with the DTFT, but we cannot *numerically calculate* its frequency domain representation unless the long signal is actually a finite-length signal. Then we could compute its DFT but we might have to wait an inordinately long time just to acquire all the samples, and even having done so, we would then be faced with a huge DFT computation. This would be unacceptable in a real-time system that needs to process signals on line. When the long signal is the concatenation of many short signals, the appropriate frequency-domain representation would be the collection of many short-length DFTs. Then the temporal variations of the *local* frequency content would stand out. The spectrogram arises from this viewpoint where a short-length DFT is used to repeatedly analyze different short sections of a longer signal.

Good examples of indefinitely long signals are audio signals such as speech or music. In both of these cases, it is the temporal variation of the frequency content that is of interest. The changing pitch and timbre of speech is how information is encoded in human speech. Likewise, we have already seen that music can be synthesized as a succession of tone combinations held nearly constant for short time intervals. For audio signals we may have digital recordings that give very long sequences obtained by sampling for many minutes or even hours. For example, one hour of stereophonic music sampled at 44.1 kHz would be represented by $44,100 \times 60 \times 60 = 158,760,000$ samples per channel. If we want to compute the (global) DFT of the entire one hour of audio, the closest power-of-two FFT needed would be $N = 2^{28} = 268,435,456$ per channel. On the other hand, the local spectrum approach above would use short-length FFTs to analyze short time segments within the long signal. This is reasonable (and generally preferable) because a long recording probably contains a succession of short passages where the spectral content does not vary. It might even be true that there is a natural segment length for the audio signal that will dictate the FFT length.

8-7.1 An Illustrative Example

To examine the points made above, we shall study a signal whose spectral properties vary with time. We begin by synthesizing a continuous-time signal that consists of four contiguous time intervals of cosine waves of different amplitudes and constant frequencies. Such a signal could be described by the equation

$$x(t) = \begin{cases} 5\cos(2\pi f_0 t) & 0 \le t < T_1 \\ 2\cos(2\pi f_1 t) & T_1 \le t < T_2 \\ 2\cos(2\pi f_2 t) & T_2 \le t < T_3 \\ 0.5\cos(2\pi f_3 t) & T_3 \le t < T_4 \\ \cdots & T_4 \le t \end{cases} \tag{8.51}$$

The signal $x(t)$ begins with a T_1-second interval of a cosine signal of amplitude 5 and frequency f_0 Hz. At $t = T_1$ the waveform switches to a cosine of amplitude 2 and frequency f_1, and retains these parameters throughout the interval $T_1 \le t < T_2$. The last two intervals switch to different amplitude and frequency parameters at T_2 and T_3 as described by (8.51). The signal might continue indefinitely with changing parameters after $t = T_4$, but it is sufficient to limit attention to the first four time intervals.

For a particular sampling frequency f_s, the resulting sampled signal would be

$$x[m] = x(m/f_s) = \begin{cases} 5\cos(\hat{\omega}_0 m) & 0 \le m \le 499 \\ 2\cos(\hat{\omega}_1 m) & 500 \le m \le 2999 \\ 2\cos(\hat{\omega}_2 m) & 3000 \le m \le 4999 \\ 0.5\cos(\hat{\omega}_3 m) & 5000 \le m \le 9999 \end{cases} \tag{8.52}$$

where $\hat{\omega}_0 = 2\pi f_0/f_s$, $\hat{\omega}_1 = 2\pi f_1/f_s$, $\hat{\omega}_2 = 2\pi f_2/f_s$, and $\hat{\omega}_3 = 2\pi f_3/f_s$ are dimensionless frequency quantities, and the frequencies change at $T_1 f_s = 500$, $T_2 f_s = 3000$, $T_3 f_s = 5000$, and $T_4 f_s = 10{,}000$ samples. The normalized frequencies $\hat{\omega}_0 = 0.211\pi$, $\hat{\omega}_1 = 0.111\pi$, $\hat{\omega}_2 = 0.8\pi$, and $\hat{\omega}_3 = 0.4\pi$ are obtained when $f_s = 2000$ Hz and $f_0 = 211$ Hz, $f_1 = 111$, $f_2 = 800$, and $f_3 = 400$ Hz. However, other values of f_0, f_1, f_2, f_3, and f_s could give the same normalized frequencies and, therefore, the same sequence of samples.

EXERCISE 8.13 Suppose that the sampling frequency is $f_s = 1000$ Hz. What values of f_0, f_1 f_2, and f_3 in (8.51) normalize to the values $\hat{\omega}_0 = 0.211\pi$, $\hat{\omega}_1 = 0.111\pi$, $\hat{\omega}_2 = 0.8\pi$, and $\hat{\omega}_3 = 0.4\pi$ in (8.52)?

EXERCISE 8.14 Use MATLAB to synthesize the sampled signal $x[m]$ in (8.52) with frequencies $\hat{\omega}_0 = 0.211\pi$, $\hat{\omega}_1 = 0.111\pi$, $\hat{\omega}_2 = 0.8\pi$, and $\hat{\omega}_3 = 0.4\pi$, and then use the D-to-A converter on your computer (`soundsc` in MATLAB) with $f_s = 2000$ Hz to listen to $x(t)$ in (8.51). During the listening observe the differences among the four segments of the signal. Is what you hear completely consistent with the specified signal parameters (duration, frequency, intensity) given above for (8.51)?

Now suppose that we compute the DFT of the entire 10,000-point sequence $x[m]$ in (8.52) using zero padding with $N = 16{,}384$. The resulting magnitude of the length-16,384 DFT is plotted in Fig. 8-18. The obvious features in this figure are four narrow peaks and these give some clues as to the nature of the signal $x[n]$, but the interpretation is far from complete and is not straightforward. The four strong peaks might suggest a signal comprised of a sum of four cosine waves of different frequencies and comparable amplitudes. With some assurance, we can state the following about the signal whose DFT is plotted in Fig. 8-18:

1. The four peaks in the DFT spectrum occur at frequencies corresponding to approximately $\hat{\omega} = 0.111\pi, 0.211\pi, 0.4\pi$, and 0.8π, so they probably represent sine or cosine wave components in $x[n]$ at those frequencies.

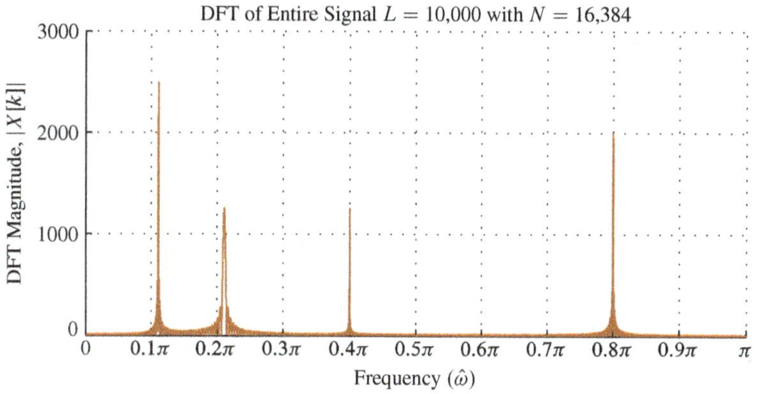

Figure 8-18 DFT of the entire sequence $x[n]$ in (8.52) with DFT length $N = 16{,}384$. Only the magnitude $|X[k]|$ is shown for $0 \le k \le 8192$. Horizontal axis is labeled with normalized frequency using $\hat{\omega}_k = 2\pi k/N$.

2. The peak heights of the DFT tell us something about the amplitudes of the sinusoidal components but, as demonstrated in Section 8-2.5, the peak height of the DFT of a cosine signal depends on both the amplitude of the cosine signal and its duration. The peak heights in Fig. 8-18 differ at most by a factor of two, but we know in this example that the actual amplitudes for frequencies $0.111\pi, 0.211\pi, 0.4\pi$, and 0.8π are 2, 5, 0.5, and 2, respectively. Although the height of the 0.211π peak is equal to the height of the peak at 0.4π, in fact, there is a 10:1 ratio between the amplitudes of those cosine components. Similarly, the DFT peak heights at 0.211π and 0.111π are in the ratio 1:2, while we know that the ratio of amplitudes of those two cosines is 5:2. The reason for this is that the DFT of a cosine of amplitude A and duration L achieves a maximum of approximately $AL/2$, as was demonstrated in Section 8-2.5. The amplitude and duration of the $\hat{\omega}_0 = 0.211\pi$ segment are $A_0 = 5$ and $L_0 = 500$, while the corresponding values for the $\hat{\omega}_3 = 0.4\pi$ segment are $A_3 = 0.5$ and $L_3 = 5000$; therefore, the peak heights are the same: $5 \times 500/2 = 0.5 \times 5000/2 = 1250$. Note that the different peak widths convey information about the duration of the sinusoids, as we saw during the discussion of windowing in Section 8-6.

3. Figure 8-18 tells us nothing about the time location or time variation of the signal properties. For example, we cannot tell whether the different components occur simultaneously (overlapping in time throughout the whole analysis interval) or sequentially (as they do in this example). This is because in either case, the signals are additively combined, and by the linearity of the DFT, the result is the sum of all the DFTs of the individual components. The change in starting time is a time shift of a signal which changes the phase of the DFT, but not the magnitude of the DFT.

8-7.2 Time-Dependent DFT

A long DFT gives useful information about the frequency content of a signal such as $x[n]$ in (8.52), but it would be much more useful if the DFT could be used also to track the time-varying properties of the signal. To formalize this concept for a signal $x[m]$ that is an indefinitely long sequence, we define the *time-dependent discrete Fourier transform* of this signal as

$$X[k, n_s] = \sum_{m=0}^{L-1} w[m]x[n_s + m]e^{-j(2\pi k/N)m} \qquad k = 0, 1, \ldots, N-1 \qquad (8.53)$$

where n_s is the ***analysis time index*** that specifies the starting index in $x[n]$, where a short-length DFT is taken. This equation involves two steps: windowing and short-length DFTs. First of all, $w[m]$ is called the ***analysis window***. It is a sequence such as the Hann window in (8.48) that is nonzero only in the interval $m = 0, 1, \ldots, L-1$, where L is assumed to be much smaller than the total length of the sequence $x[n]$. Therefore, the product $w[m]x[n_s + m]$ is also nonzero only for $m = 0, 1, \ldots, L-1$. In this way, the window *selects* a finite-length segment from the sequence $x[n]$ starting at the analysis time index n_s. Second, a short-length DFT of the finite-length windowed segment is computed to extract the spectrum for that local interval of the signal. By adjusting the analysis time index n_s, we can move any desired segment of $x[n]$ into the domain of the window, and thereby select any length-L segment of $x[n]$ for analysis. The length-L segments are often called *frames*. Because the window imposes time localization, $X[k, n_s]$ in (8.53) is also known as the ***short-time discrete Fourier transform*** or ***STDFT***.

The right-hand side of (8.53) is easily recognized as the N-point DFT of the finite-length sequence $w[m]x[n_s+m]$, so (8.53) can be evaluated efficiently for each choice of n_s by an FFT computation. The selection of analysis segments is illustrated by Fig. 8-19 for two values of the analysis starting time index, $n_s = 0$ and $n_s = 400$. Figure 8-19(a) shows the first 801 samples of the discrete-time signal $x[m]$ in (8.52) plotted versus m with dots for sample values. Also shown in Fig. 8-19(a) are two shaded regions of length 301 samples. One region starts at $m = 0$ and is shaded orange, and the other starts at $m = 400$ and is shaded in gray.[13] These are potential 301-sample analysis frames. Also shown in Fig. 8-19(a) is the outline of the (bell-shaped) Hann window $w[m]$ whose origin is fixed at $m = 0$.[14] The orange samples in Fig. 8-19(a) are the samples that are going to be multiplied by the window $w[m]$ when $n_s = 0$ and are thus selected for analysis with the DFT to obtain $X[k, 0]$, the DFT of $w[m]x[0+m]$.

Figure 8-19(b) shows the fixed window sequence (in orange) along with the sequence $x[400+m]$, which is the sequence $x[m]$ shifted to the left by 400 samples. The gray samples in Fig. 8-19(a) have become the orange samples in Fig. 8-19(b) where these samples are multiplied by the window and are thus selected for DFT analysis to compute $X[k, 400]$. To compute $X[k, n_s]$ for any other value of n_s, we simply form the "windowed" sequence $w[m]x[n_s + m]$ in the manner illustrated in Fig. 8-19, and then compute the DFT of the resulting finite-length sequence as in (8.53). The shift n_s is typically moved in jumps of R samples, where $1 \leq R < L$. As long as $R < L$ the windowed segments overlap and

[13]The shading of the samples is intended to emphasize two typical analysis intervals, with the implication that in general, all the samples would be plotted with the same color.

[14]The discrete-time window sequences are shown as continuous outlines to distinguish them from the samples of $x[m]$.

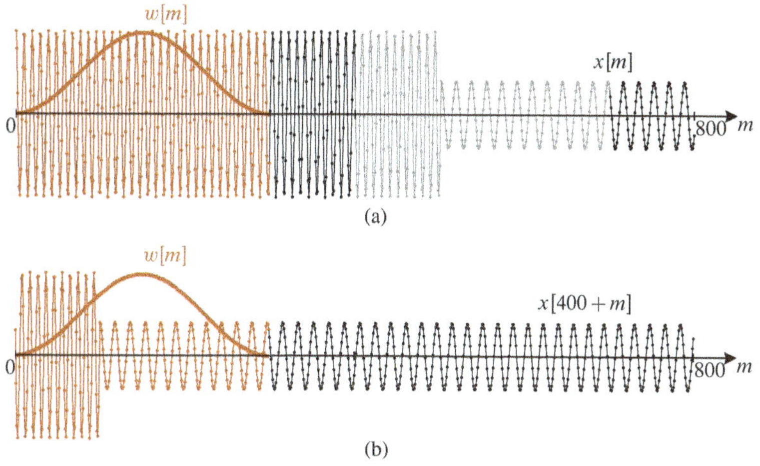

Figure 8-19 Time-dependent Hann windowing of a signal using a length-301 window. (a) The signal $x[0+m] = x[m]$ and fixed window $w[m]$. (b) Fixed window $w[m]$ with signal shifted by 400 samples (i.e., $x[400+m]$).

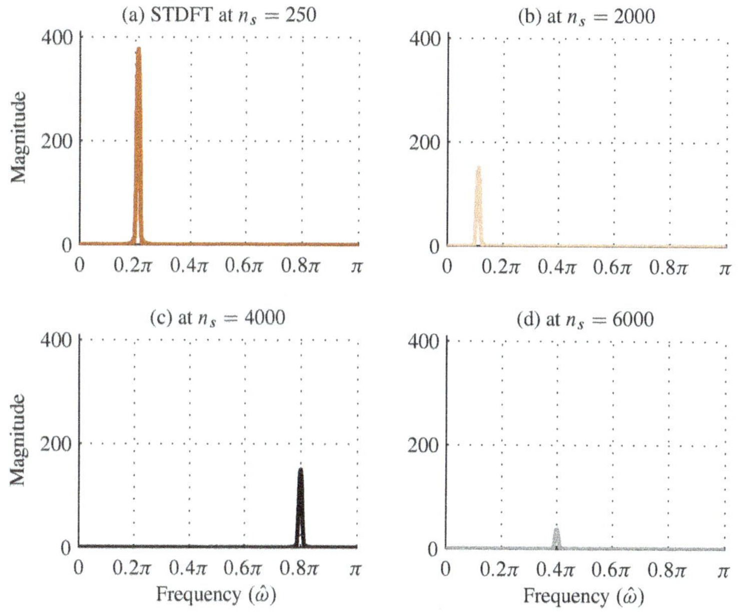

Figure 8-20 Short-time spectra for the signal defined in (8.52) with $L = 301$ and $N = 1024$. (a) STDFT at analysis time index $n_s = 250$. (b) STDFT at $n_s = 2000$. (c) STDFT at $n_s = 4000$. (d) STDFT at $n_s = 6000$. The frequency axis $\hat{\omega}$ is derived from k via $\hat{\omega} = 2\pi k/N$, and the DFT magnitudes for $0 \leq k \leq 512$ are connected as if plotting the corresponding DTFT at frequencies $0 \leq \hat{\omega} \leq \pi$.

all samples of the signal are included in at least one analysis window. A common default choice is $R = L/2$, but making R smaller provides a smoother spectrogram (discussed below) at the cost of more computation. The best value of R is often dictated by the need to track fast temporal changes in the signal being analyzed.

To illustrate the nature of the short-time DFT, Fig. 8-20 shows plots of $|X[k, 250]|$, $|X[k, 2000]|$, $|X[k, 4000]|$, and $|X[k, 6000]|$ for the signal of (8.52) with STDFT parameters $L = 301$ and $N = 1024$. Figure 8-20 confirms that $|X[k, 250]|$, $|X[k, 2000]|$, $|X[k, 4000]|$, and $|X[k, 6000]|$ each display a single peak centered at the frequency of the sinusoidal component at the corresponding *analysis* times. Furthermore, the amplitudes of the peaks are in correct proportion since for the analysis times chosen, there is only one frequency within the window and the window lengths are the same ($L = 301$).[15]

While plots like those in Fig. 8-20 are useful for displaying the frequency-domain properties in the neighborhood of a specific analysis time n_s, they do not give a complete overview of how the signal parameters vary with time. In this example, we selected the analysis time indices $n_s = 250, 2000, 4000$, and 6000, because we knew that these positions would show the different frequencies. In a practical setting, where we do not know the details of the signal in advance, the spectrogram display is much more effective and widely used.

8-7.3 The Spectrogram Display

The STDFT computation results in a two-dimensional function $X[k, n_s]$, where the k dimension represents frequency because $\hat{\omega}_k = (2\pi k/N)$ is the kth analysis frequency, and the n_s dimension represents time since the analysis window-position times are $t_s = n_s T_s$. Since the STDFT is a function of both frequency and time, there might be a different *local spectrum* for each analysis time. Although $X[k, n_s]$ is complex valued with a magnitude and phase, we usually plot only the magnitude versus (k, n_s) with a 3-D graphical display. A variety of display methods can be used to plot the magnitude $|X[k, n_s]|$ as a function of both k and n_s, including waterfall plots (a sequence of plots like those in Fig. 8-20 stacked vertically), perspective plots, contour plots, or grayscale images. In addition, plotting the log magnitude, $\log |X[k, n_s]|$ versus (k, n_s), is often used to show a wider amplitude range. The preferred form is the **spectrogram**, which is a grayscale (or pseudocolor) image where the gray level at point (k, n_s) is proportional to the magnitude $|X[k, n_s]|$ or the log magnitude, $\log |X[k, n_s]|$. Usually, large magnitudes are rendered as black, and small ones white, but pseudo-color images are also common.

DEMO
Spectrogram Practice

Spectrograms of the signal in (8.52) are shown in Fig. 8-21. The horizontal axis is analysis time and the vertical axis is frequency. Analysis time can be given as a sample index (n_s) as in Fig. 8-21, or converted to continuous-time $t_s = n_s/f_s$ via the sampling rate. Likewise, frequency can be converted from the index k to hertz, or to normalized frequency $\hat{\omega}$ as in Fig. 8-21. In the spectrogram image for a real signal, usually only the frequency range for $0 \le k \le N/2$, or $0 \le \hat{\omega} \le \pi$ is shown because the negative frequency region is conjugate symmetric. For the complex signal case, the frequency range is either $0 \le \hat{\omega} < 2\pi$, or $-\pi < \hat{\omega} \le \pi$, depending on whether or not the FFT values are reordered prior to being displayed.[16]

8-7.4 Interpretation of the Spectrogram

A glance at Fig. 8-21 confirms that the spectrogram provides a much clearer time-frequency picture of the characteristics of the signal defined in (8.52) than does either

[15] The absolute height of the peaks depends on the amplitude A of the sinusoid and the window—specifically, the peak height equals $0.5A\sum w[m]$.

[16] The *DSP-First* Toolbox functions `plotspec` and `spectgr`, which are similar to MATLAB's function `spectrogram`, use the ordering $-\pi < \hat{\omega} \le \pi$ for complex signals; MATLAB's `spectrogram` function uses $0 \le \hat{\omega} < 2\pi$.

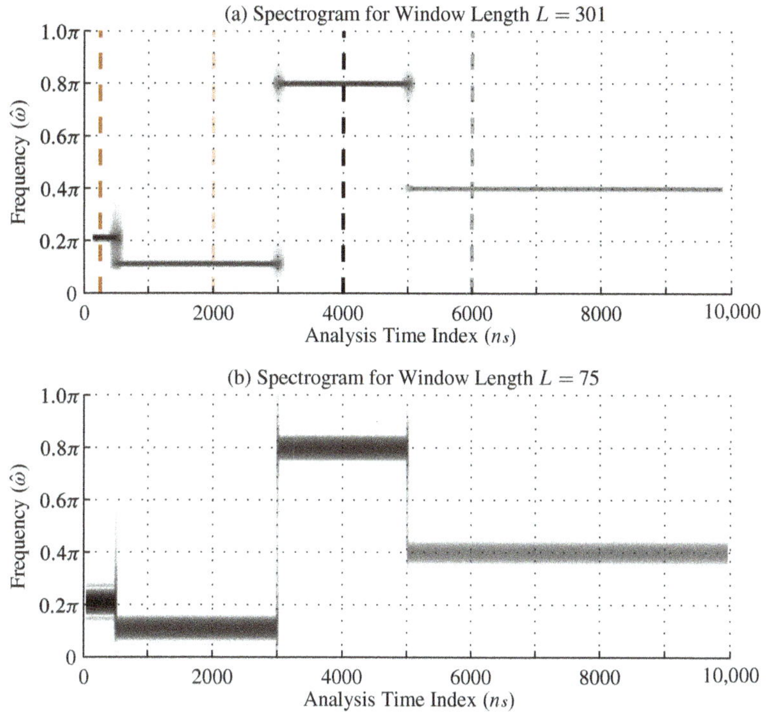

Figure 8-21 Spectrograms of test signal (8.52) calculated with `spectgr` (linear magnitude) for two different window lengths. (a) $L = 301$. (b) $L = 75$. For both cases, $N = 1024$ and the spacing between analysis times is $R \approx 0.1L$.

the single plot of a long DFT as in Fig. 8-18, or the four snapshots in Fig. 8-20. The four sinusoidal components and their starting/ending times are evident in Fig. 8-21(a). On the other hand, the dark horizontal bars in Fig. 8-21(b) are wider and less clearly defined, so it is hard to determine the exact frequency—however, it is clear in both spectrograms where the sinusoidal components are located in time. The dashed lines at $n_s = 250$, $n_s = 2000$, $n_s = 4000$, and $n_s = 6000$ show the locations of four individual DFTs that comprise Fig. 8-20; that is, the DFT plots in Figs. 8-20(a), (b), (c), and (d) are columns of the gray-scale spectrogram image shown in Fig. 8-21(a). The magnitude of the DFT is mapped to a gray-scale shading which is visible along the orange, black, and gray vertical marker lines, respectively. Figures 8-20(a–d) are, therefore, "vertical slices" of the spectrogram.

LAB

Spectrograms of AM and FM Signals

A comparison of the two spectrograms in Fig. 8-21 yields valuable insight into the effect of the window length in short-time Fourier analysis. The window length is $L = 301$ in Fig. 8-21(a) and $L = 75$ in Fig. 8-21(b). Along the time axis (n) at points where the frequency changes abruptly, there are "blurry" regions; these occur at $n_s = 500, 3000$, and 5000. The blurred regions occur whenever the analysis window straddles a region of rapid change since we have signal samples within the window from both sides of the change point. For example, when $n_s = 350$, the signal interval [350, 650] is shifted under the $L = 301$ point window prior to the DFT, so the first half of $x[350+m]$ has frequency 0.211π and the second half has frequency 0.111π. Thus, the STDFT $X[k, 350]$ should have lower peaks at both frequencies. The longer the window, the wider is this fuzzy region of ambiguity. This effect is seen in a comparison of Figs. 8-21(a) and (b) because the regions of fuzziness are much wider for the $L = 301$ window. This example illustrates the general principle that *precise location of temporal changes in the signal requires a short window*.

8-7.4.1 Frequency Resolution

Another important point is that the vertical width of the horizontal bars (for the sinusoids) is different in the two spectrograms. In fact, the width of the bars in Fig. 8-21(b) is approximately four times the width of the bars in Fig. 8-21(a), because the window lengths are 75 and 301, respectively. This suggests that for a signal having multiple frequency components overlapping in time, the spectrogram may not show two distinct peaks if the frequencies are too close together. In other words, the spectrogram may need a longer window to "resolve" two closely spaced frequency components into separate peaks in the frequency domain.

This ***resolution*** issue is illustrated by a test example where the signal is composed of multi-frequency sinusoids with closely spaced frequencies. The two spectrograms shown in Fig. 8-22 were computed for two different window lengths for the signal

$$x[m] = \begin{cases} \cos(0.2\pi m) + 1.5\cos(0.227\pi m) & 0 \le m < 5000 \\ \cos(0.2\pi m) & 5000 \le m < 7000 \\ 3\cos(0.6\pi m) + \cos(0.7\pi m) & 7000 \le m < 10{,}000 \end{cases} \tag{8.54}$$

By comparing the spectrograms we can see the effect of the window length on frequency resolution. During the first interval the two frequencies are very close together at 0.2π

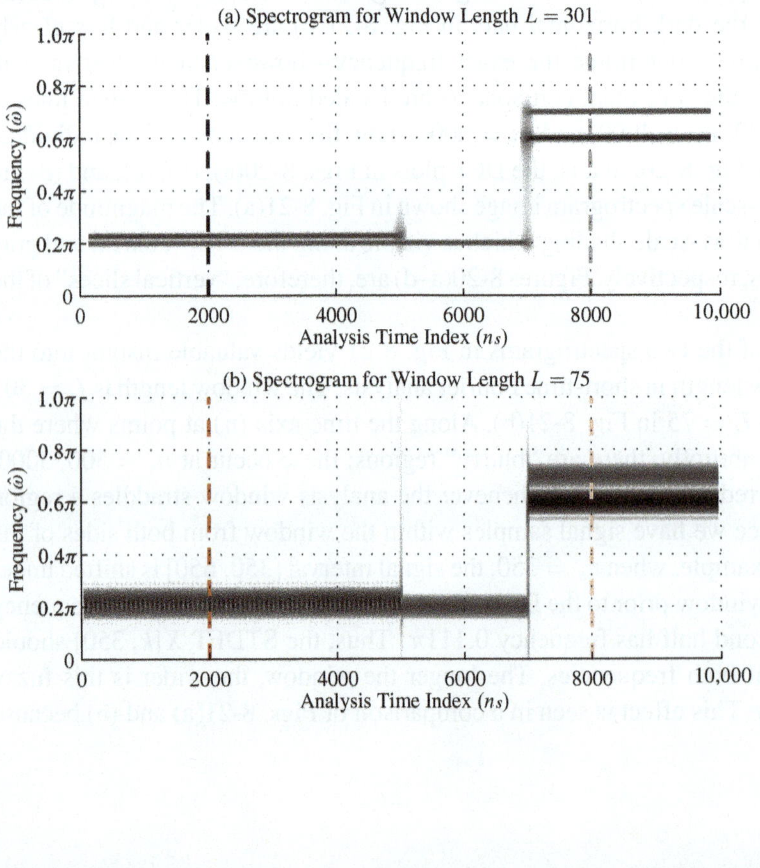

Figure 8-22 Spectrograms of test signal (8.54) for two different window lengths. (a) $L = 301$. (b) $L = 75$. FFT length is $N = 1024$ with zero padding.

and 0.227π (a spacing of 0.027π rad), while in the third interval, the two frequencies are farther apart at 0.6π and 0.7π (a spacing of 0.1π rad). The two frequencies in the first interval are both evident in Fig. 8-22(a), but not in Fig. 8-22(b) where they are merged into a single broad horizontal bar. On the other hand, during the third time interval, two clearly separated bars are evident in both spectrograms at the frequencies 0.6π and 0.7π, so, the two frequency components are *resolved.* This resolution is possible because, although the bars in Fig. 8-22(b) are 3–4 times wider than the corresponding features in Fig. 8-22(a), the spacing between the frequencies 0.7π and 0.6π is approximately four times the spacing between 0.227π and 0.2π. This example illustrates the general principle that *frequency resolution can be improved by lengthening the analysis window.*

LAB

Spectrogram: Harmonic Lines and Chirp Aliasing

To see the frequency resolution more clearly, Fig. 8-23 shows "vertical slices" taken along the dashed lines in Figs. 8-22(a) and (b) corresponding to analysis times $n_s = 2000$ and 8000, which are in the middle of the segments having two sinusoids with closely spaced frequencies. Figure 8-23(a) corresponds to the slice at $n_s = 2000$ with the window length $L = 301$, and it shows two distinct peaks around the frequencies 0.2π and 0.227π. However, Fig. 8-23(c) shows the slice at $n_s = 2000$ with the window length $L = 75$. In this case there is only a single broad peak, and we conclude that the two frequencies are "not resolved." On the other hand, for the slices at $n_s = 8000$ shown in Figs. 8-23(b) and (d), the two frequencies are more widely separated, and we see two distinct peaks in both of the spectrogram slices. However, for the slice at $n_s = 8000$, if the window length were reduced below 75, the two peaks in Fig. 8-23(d) would broaden and eventually as the window length decreases we would see only a single broad peak spanning the two frequencies at 0.6π and 0.7π. Finally, note that the vertical slice DFTs show clearly that the amplitude of the two sinusoids is different, but the amplitude difference is much harder to discern in the gray-scale spectrogram image.

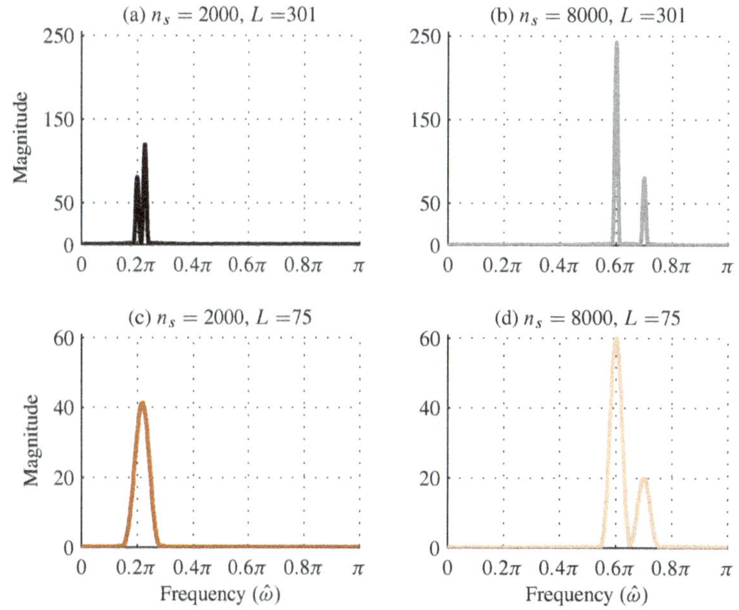

Figure 8-23 Short-time spectra for signal in (8.54) at two different time indices. These are the vertical slices shown in Fig. 8-22. (a) Hann window length $L = 301$, time index $n_s = 2000$. (b) $L = 301$, $n_s = 8000$. (c) $L = 75$, $n_s = 2000$. (d) $L = 75$, $n_s = 8000$. FFT length was 1024 with zero padding.

EXERCISE 8.15 Use MATLAB to generate the signal in the third time interval of (8.54)

$$x[m] = 3\cos(0.6\pi m) + \cos(0.7\pi m) \quad 7000 \le m < 10{,}000$$

Then using a Hann window, compute the spectrogram slice at $n_s = 8000$. Use window lengths of $L = 91$ and $L = 31$ and compute the DFTs to produce plots like Fig. 8-23. Determine whether or not the two sinusoidal components are "resolved."

As suggested by Fig. 8-23, the width of the peaks for the DFTs of the sinusoidal components depends inversely on the window length L. In the previous Section 8-6, we learned that what we are seeing in Fig. 8-23(a) is the DTFT of the window $w[n]$ shifted in frequency to the frequency of the sinusoid. Furthermore, it is a general principle of Fourier transforms that there is an inverse relation between window length and frequency width. Therefore, the frequency width of the DTFT of a window sequence such as the Hann window can be written as an inverse relationship

$$\Delta\hat{\omega} \approx \frac{C}{L}$$

where the constant C is a small number, usually $4\pi < C < 8\pi$ for "good" windows. For the Hann window, the zero-crossing width is $\Delta\hat{\omega} = 8\pi/L$, so $\Delta\hat{\omega} = 8\pi/75 = 0.107\pi$ for the length-75 Hann window, which is small enough to resolve the 0.6π and 0.7π frequency peaks at $n_s = 8000$. If two frequencies differ by less than $\Delta\hat{\omega}$, their STDFT peaks blend together. Thus, the ability to resolve two sinusoids (i.e., see two distinct peaks), depends inversely on the window length L. This makes sense intuitively because the longer the window, the longer the time to observe the periodicity (or non-periodicity) of a signal. On the other hand, if the window is very short, e.g., less than one period of the signal, then we have virtually no information about any periodicity of the signal, and it is reasonable to expect this to be reflected in the STDFT.

There is still one more point to make about the window length, and this is the issue of "time resolution." We observed in Fig. 8-21 that the window length is also important in tracking the temporal changes in a signal. The "fuzzy" transition regions are approximately four times as wide in Fig. 8-21(a) as in Fig. 8-21(b) because the window length goes from $L = 301$ to $L = 75$. Therefore, it is generally true that in order to track rapid spectrum changes, we need to keep the window length as short as possible.

We are forced, therefore, into a trade-off situation: the window should be short to track temporal changes, but long to resolve closely spaced frequency components. For this reason, when studying the properties of an unknown signal, it is common to compute spectrograms with differing window lengths. Features that are obvious in one case may be obscured in another, but with multiple spectrograms and a good understanding of how window length (and shape) affect the spectrogram image, it is often possible to determine much useful information about the signal. This time-frequency information can be combined with physical models for signal generation and/or detection to obtain very accurate models of the signal that can be useful in a variety of applications. A notable and ubiquitous application is audio coding for digital storage and transmission. Audio coders like MP3 are based upon the same type of frequency analysis computations employed to generate the spectrogram image. Because the human auditory system incorporates a type of frequency analysis, short-time frequency-domain analysis is a very natural way to incorporate auditory perception effects such as masking into the process of compressing the digital representation of an audio signal. By a careful frequency analysis "frame by frame," it is possible to mask coding errors by literally "hiding" them beneath prominent spectral components.

8-7.5 Spectrograms in MATLAB

Since the spectrogram can be computed by doing many FFTs of windowed signal segments, MATLAB is an ideal environment for doing the DFT calculation and displaying the image. Specifically, the MATLAB spectrogram computation evaluates (8.53) for length-L signal segments that are separated by R (i.e., $n_s = 0, R, 2R, 3R, \ldots$).

$$X[(2\pi k/N)f_s, rRT_s] = \sum_{m=0}^{L-1} w[m]x[rR+m]e^{-j(2\pi k/N)m} \quad k = 0, 1, \ldots, N/2 \tag{8.55}$$

The frequencies are scaled by $(2\pi/N)f_s$ and the times by T_s, because the default MATLAB display presents the spectrogram versus frequency in hertz and time in seconds. Quite often the default is $R = L/2$, so there is a 50% overlap of the signal segments analyzed by the DFT after windowing. The value of N is the DFT length, so zero padding would be done when $N > L$. Picking a large value for N gives many frequency samples along the k dimension because the frequency index k ranges from 0 to $N/2$. When the sampling rate $f_s = 1/T_s$ associated with $x[n]$ is available, the frequency locations for the DFT output can be scaled to hertz and the time axis can be converted from sample index to analysis time in seconds which is convenient for labeling the axes of the spectrogram display.[17]

The computation of the spectrogram can be expressed in a simple MATLAB program that involves one `for` loop. The core of the program is shown below for the special case of an even-length window with 50% overlap of signal segments.

```
% L = signal segment length, also window length (even integer)
% N = FFT (DFT) lengths
% win = window signal, L-point column vector
        such as the Hann window, or hanning.m
% xn = input signal (column vector)
% assume the overlap is 50%
Lx = length(xn);
NwinPos = 2*ceil(Lx/L);
X = zeros(N,NwinPos);
for ii=0:NwinPos-1
   X(:,ii+1) = fft(win.*xn(ii*L/2 + (1:L)),N);
end
```

[17]In other words, MATLAB's spectrogram function calibrates the frequency and time axes in continuous-time units assuming a given sampling rate f_s.

In recent versions of MATLAB the command that invokes the computation of (8.55) is

```
[S,F,T] = spectrogram(X,WINDOW,NOVERLAP,NFFT,Fs,'yaxis')
```

The final string `'yaxis'` is needed to override the default behavior that puts the frequency axis on the horizontal and the time axis vertical. The outputs from `spectrogram` are `S`, a two-dimensional array containing the *complex-valued* spectrogram values, `F`, a vector of all the analysis frequencies, and `T`, a vector containing the starting times of the signal segments being windowed. The inputs are the signal `X`, the window coefficients `WINDOW`, the overlap of signal segments `NOVERLAP`, the FFT length `NFFT`, and the sampling frequency `Fs`. In addition, the default `spectrogram` window is Hamming, so it should be replaced with a call to the Hann function, (e.g., `hanning(L)`). Note that the window skip parameter R in (8.55) is the window length minus the overlap, so the overlap should be less than the window length, but choosing `Noverlap` equal to `length(window)-1` would generate a lot of needless computation (and possibly a very large image!), because the window skip would be $R = 1$. It is common to pick the overlap to be somewhere between 50% and 90% of the window length (i.e., $0.1L \leq R \leq 0.5L$), depending on how smooth the final spectrogram image needs to be. See `help spectrogram` in MATLAB for more details.

The spectrogram image can be displayed by using any one of MATLAB's 3-D display functions. To get the inverted gray-scale images shown in this chapter, use the following:

```
imagesc( T, F, abs(S) )
axis  xy, colormap(1-gray)
```

The color map of `(1-gray)` gives a negative gray scale that is useful for printing, but on a computer screen it might preferable to use color (e.g., `colormap(jet)`). Finally, it may be advantageous to use a logarithmic amplitude scale in `imagesc` in order to see tiny amplitude components, as well as big ones.

8-8 The Fast Fourier Transform (FFT)

In Section 8-1.3, we discussed the FFT as an efficient algorithm for computing the DFT. The FFT is without a doubt the most famous algorithm in signal processing, and many of the systems that we use everyday rely on the FFT for an efficient implementation in hardware or software.

8-8.1 Derivation of the FFT

In this section, we give the basic divide-and-conquer method that leads to the FFT. From this derivation, it should be possible to write an FFT program that runs in time proportional to $(N/2)\log_2 N$, where N is the transform length. We need to assume that N is a power of two, so that the decomposition can be carried out recursively. Such algorithms are called *radix-2* algorithms.

The DFT summation (8.4) and the IDFT summation (8.5) are essentially the same, except for a minus sign in the exponent of the DFT and a factor of $1/N$ in the inverse DFT. Therefore, we concentrate on the DFT calculation, knowing that a program written for the DFT could be modified to do the IDFT by changing the sign of the complex exponentials and multiplying the final values by $1/N$. The DFT summation can be broken into two sets, one sum over the even-indexed points of $x[n]$ and another sum over the odd-indexed points.

$$X[k] = \text{DFT}_N\{x[n]\} \tag{8.56a}$$

$$= \sum_{n=0}^{N-1} x[n]e^{-j(2\pi/N)kn} \tag{8.56b}$$

$$= \left(x[0]e^{-j0} + x[2]e^{-j(2\pi/N)2k} + \cdots + x[N-2]e^{-j(2\pi/N)k(N-2)}\right) + \left(x[1]e^{-j(2\pi/N)k} + x[3]e^{-j(2\pi/N)3k} + \cdots + x[N-1]e^{-j(2\pi/N)k(N-1)}\right) \tag{8.56c}$$

$$X[k] = \sum_{\ell=0}^{N/2-1} x[2\ell]e^{-j(2\pi/N)k(2\ell)} + \sum_{\ell=0}^{N/2-1} x[2\ell+1]e^{-j(2\pi/N)k(2\ell+1)} \tag{8.56d}$$

At this point, two clever steps are needed: First, the exponent in the second sum must be broken into the product of two exponents, so we can factor out the one that does not depend on ℓ. Second, the factor of two in the exponents (2ℓ) can be associated with the N in the denominator of $2\pi/N$.

$$X[k] = \sum_{\ell=0}^{N/2-1} x[2\ell]e^{-j(2\pi k/N)(2\ell)} + e^{-j(2\pi k/N)} \sum_{\ell=0}^{N/2-1} x[2\ell+1]e^{-j(2\pi k/N)(2\ell)} \tag{8.56e}$$

$$X[k] = \sum_{\ell=0}^{N/2-1} x[2\ell]e^{-j(2\pi k/(N/2))\ell} + e^{-j(2\pi k/N)} \sum_{\ell=0}^{N/2-1} x[2\ell+1]e^{-j(2\pi k/(N/2))\ell} \tag{8.56f}$$

Now we have the correct form. Each of the summations in (8.56f) is a DFT of length $N/2$, so we can write

$$X[k] = \text{DFT}_{N/2}\{x[2\ell]\} + e^{-j(2\pi/N)k}\,\text{DFT}_{N/2}\{x[2\ell+1]\} \tag{8.57}$$

The formula (8.57) for reconstructing $X[k]$ from the two smaller DFTs has one hidden feature: It must be evaluated for $k = 0, 1, 2, \ldots, N-1$. The $N/2$-point DFTs produce output vectors that contain $N/2$ elements; for example, the DFT of the odd-indexed points would be

$$X^o_{N/2}[k] = \text{DFT}_{N/2}\{x[2\ell + 1]\}$$

for $k = 0, 1, 2, \ldots, N/2-1$. Thus, we need to recall that an $N/2$-point DFT is periodic with a period of $N/2$ in order to calculate $X[k]$, for $k \geq N/2$. It is easy to verify that

$$X^o_{N/2}[k + N/2] = X^o_{N/2}[k]$$

and likewise for the DFT of the even-indexed points, so we need merely to periodically extend the results of the $N/2$-point DFTs before doing the sum in (8.57). This requires no additional computation.

The decomposition in (8.57) is enough to specify the entire FFT algorithm: Compute two smaller DFTs and then multiply the outputs of the DFT over the odd indices by the exponential factor $e^{-j(2\pi/N)k}$. Refer to Fig. 8-24, where three levels of the recursive decomposition can be seen. If a recursive structure is adopted, the two $N/2$ DFTs can be decomposed into four $N/4$-point DFTs, and those into eight $N/8$-point DFTs, etc. If N is a power of two, this decomposition continues $(\log_2 N - 1)$ times and then eventually reaches the point where the DFT lengths are equal to two. For 2-point DFTs,

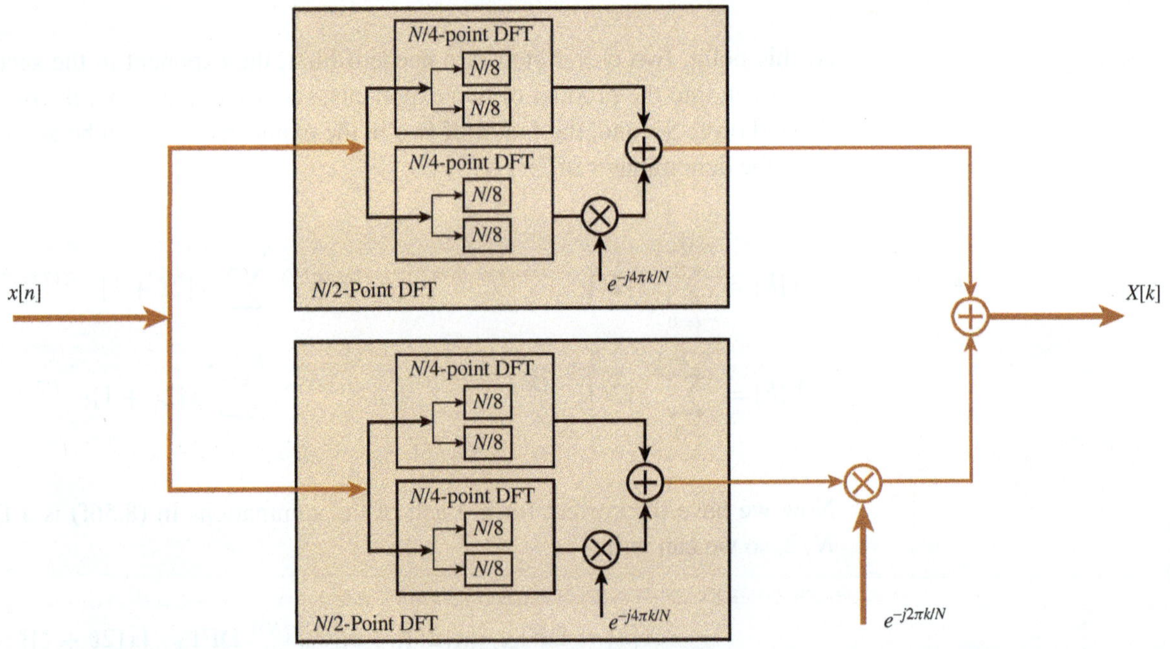

Figure 8-24 Block diagram of a radix-2 FFT algorithm for $N = 2^\nu$. The width of the lines is proportional to the amount of data being processed. For example, each $N/4$-point DFT must transform a data vector containing $N/4$ elements.

the computation is trivial:

$$X_2[0] = x_2[0] + x_2[1]$$
$$X_2[1] = x_2[0] + e^{-j2\pi/2}x_2[1] = x_2[0] - x_2[1]$$

The two outputs of the 2-point DFT are the sum and the difference of the inputs. Thus, the last stage of computation would require $N/2$ 2-point DFTs.

8-8.1.1 FFT Operation Count

The foregoing derivation is a brief sketch of the algorithm, but the basic idea for writing an FFT program using the two-point DFT and the complex exponential as basic operators has been presented. An important point about the FFT is not only writing the program, but also assessing the number of operations needed to complete the calculation. When it was first published, the FFT made a huge impact on how people thought about problems, because it made the frequency-domain accessible numerically. Spectrum analysis became a routine calculation, even for very long signals. Operations such as filtering, which seem to be natural for the time-domain, could be done more efficiently in the frequency-domain for very long FIR filters.

The number of operations needed to compute the FFT can be expressed in a simple formula. The structure of the algorithm given in (8.57) is sufficient to count the number of operations. The count goes as follows: the N point DFT can be done with two $N/2$ point DFTs followed by N complex multiplications and N complex additions, as we can see in (8.57).[18] Thus, we have

$$\mu_c(N) = 2\mu_c(N/2) + N$$
$$\alpha_c(N) = 2\alpha_c(N/2) + N$$

where $\mu_c(N)$ is the number of complex multiplications for a length-N DFT, and $\alpha_c(N)$ is the number of complex additions. This equation can be evaluated successively for $N = 2, 4, 8, \ldots$, because we know that $\mu_c(2) = 0$ and $\alpha_c(2) = 2$. Table 8-3 lists the number of operations for some transform lengths that are powers of two. The formula for each can be derived by matching the table:

$$\mu_c(N) = N(\log_2 N - 1)$$
$$\alpha_c(N) = N\log_2 N$$

Since complex number operations ultimately must be done as multiplies and adds between real numbers, it is useful to convert the number of operations to real adds and real multiplies. Each complex addition requires two real additions, but each complex multiplication is equivalent to four real multiplies and two real adds. Therefore, we can put two more columns in Table 8-3 with these counts. The last column in the table $4N^2$ is the number of real multiplications needed to compute an N-point DFT directly from its definition (8.4).

[18]Actually, the number of complex multiplications can be reduced to $N/2$, because $e^{-j2\pi(N/2)/N} = -1$.

Table 8-3 Number of operations for radix-2 FFT when N is a power of two. Notice how much smaller $\mu_r(N)$ is than $4N^2$.

N	$\mu_c(N)$	$\alpha_c(N)$	$\mu_r(N)$	$\alpha_r(N)$	$4N^2$
2	0	2	0	4	16
4	4	8	16	16	64
8	16	24	64	48	256
16	48	64	192	128	1024
32	128	160	512	320	4096
64	320	384	1280	768	16,384
128	768	896	3072	1792	65,536
256	1792	2048	7168	4096	262,144
⋮	⋮	⋮	⋮	⋮	⋮

In summary, the total operation count is proportional to $N \log_2 N$. The exact formulas from Table 8-3 are

$$\mu_r(N) = 4N(\log_2 N - 1)$$

$$\alpha_r(N) = 2N(\log_2 N - 1) + 2N = 2N \log_2 N$$

for the number of real multiplications and additions, respectively. Even these counts are a bit high because certain symmetries in the complex exponentials can be exploited to further reduce the computations.

8-9 Summary and Links

In this chapter, we introduced the DFT which is the finite form of the DTFT, and we have shown how the resulting concepts can be useful in computing spectrum representations of signals and for understanding the behavior of linear systems. We showed how the DFT can be derived from the DTFT by frequency sampling, which enables us to use the DTFT for analysis when explaining the result of a DFT/FFT computation. Finally, the spectrogram was introduced as an application of the DFT where a time-dependent spectrum is computed by taking the DFT of short windowed sections of a long signal.

DEMO
Spectrogram Practice

8-10 Problems

The problems in this section address the major concepts of this chapter. On the Companion Website there are many related homework problems which can be used for review and practice. Most of them have detailed solutions.

NOTE
Problems with Solutions

P-8.1 For each of the following length-8 signals, use frequency sampling of the DTFT to determine a formula for the 8-point DFT. Use MATLAB to check your work numerically.

(a) $x_a[n] = \begin{cases} 1 & n = 0 \\ 0 & n = 1, 2, \ldots, 7 \end{cases}$

(b) $x_b[n] = (-1)^n \quad \text{for } n = 0, 1, 2, \ldots, 7$

(c) $x_c[n] = \begin{cases} 1 & n = 3 \\ 0 & n \neq 3 \end{cases}$

(d) $x_d[n] = 0.5e^{j\pi n/2} \qquad \text{for } n = 0, 1, 2, \ldots, 7$

P-8.2 For each of the following 8-point DFTs, determine a formula for the 8-point IDFT. Use MATLAB to check your work numerically.

(a) $X_a[k] = 1$ for $k = 0, 1, 2, \ldots, 7$

(b) $X_b[k] = \begin{cases} 1 & k = 3, 5 \\ 0 & k = 0, 1, 2, 4, 6, 7 \end{cases}$

(c) $X_c[k] = \begin{bmatrix} 0 & -\sqrt{2}j & -2j & -\sqrt{2}j & 0 & \sqrt{2}j & 2j & \sqrt{2}j \end{bmatrix}$

(d) $X_d = 0.5\cos(\pi k/2) \qquad \text{for } k = 0, 1, 2, \ldots, 7$

P-8.3 For each of the following length-12 signals, determine a formula for the 12-point DFT. Use MATLAB to check your work numerically.

(a) $y_0[n] = 3(-1)^n$ for $n = 0, 1, \ldots, 11$.

(b) $y_1[n] = \begin{cases} 1 & n = 0, 1, 2, 3 \\ 0 & n = 4, 5, \ldots, 11 \end{cases}$

(c) $y_2[n] = \begin{cases} 1 & n = 0, 2, 4, 6, 8, 10 \\ 0 & n = 1, 3, 5, 7, 9, 11 \end{cases}$

P-8.4 For an unknown 4-point sequence $w[n]$, the corresponding 4-point DFT is

$$W[k] = \{0,\ j,\ 0,\ -j\}$$

(a) Define a new 4-point sequence $y[n] = 4 - 3w[n]$ and then determine the 4-point DFT $Y[k]$.
Hint: it is not necessary to find $w[n]$.

(b) Define a new 4-point sequence $v[n] = y[n] + 2(-1)^n w[n]$, and then determine the 4-point DFT $V[k]$.

P-8.5 An experienced engineer, who has taken many FFTs of *real-valued* sequences $x[n]$, has noticed that although many of the DFT coefficients are often complex-valued, $X[0]$ is always real-valued for any length N and $X[N/2]$ is real when N is even, but $X[N/2]$ is not available when N is odd.

(a) For the case $N = 255$ and $N = 256$, express $X[0]$ and $X[N/2]$ in such a way so as to prove this observation.

(b) Given the 6-point sequence $x[n] = [4, \ -1, \ 4, \ -1, \ 4, \ -1]$, determine its 6-point DFT sequence $X[k]$.

(c) If the 4-point DFT an unknown length-4 sequence $v[n]$ is $V[k] = \{1, \ 4 + j, \ -1, \ 4 - j\}$, determine $v[1]$.

(d) Given an 8-point sequence $z[n] = \sum_{m=0}^{7} 8(-1)^m \delta[n - m]$, the corresponding 8-point DFT is $Z[k]$. Determine $Z[0]$ and $Z[4]$.

P-8.6 Consider the real finite-length sequence $x[n]$ defined as

$$x[n] = \delta[n] + 2\delta[n - 1] + 3\delta[n - 2]$$

(a) Sketch $x[n]$ as a function of n.

(b) Sketch the finite-length sequence $y[n]$ whose 8-point DFT is $Y[k] = e^{-j0.5\pi k} Z[k]$ where $Z[k]$ is the 8-point DFT of $z[n] = 2x[n - 1]$.

P-8.7 Consider two finite-length sequences,

$$x[n] = \begin{cases} 0 & n < 0 \\ 1 & 0 \le n \le 3 \\ 0 & n > 3 \end{cases}$$

and $h[n] = 4\delta[n] - 2\delta[n - 1] - 2\delta[n - 2]$.

(a) Determine and plot the discrete convolution $y[n] = x[n] * h[n]$ for these two sequences.

(b) Note that the length of $x[n]$ is 4 and the length of $h[n]$ is 3 samples. Therefore, each sequence can be represented by an N-point DFT with $N \ge 4$. Zero padding would be used to extend the lengths when $N > 4$. Determine expressions for the N-point DFTs $X[k]$ and $H[k]$. Do not obtain numerical answers—instead express your answers in terms of $e^{-j2\pi(n)k/N}$.

(c) Now form the product $Y[k] = X[k]H[k]$ again expressed in terms of $e^{-j2\pi(n)k/N}$.

(d) From the result in part (c), determine the IDFT of $Y[k]$ when $N = 6$. Compare your answer to the result of part (a).

(e) Repeat when $N = 4$. In this case, the complex exponentials must be changed so that the exponents are less than 2π, i.e., $e^{-j(2\pi/N)(N+k)} = e^{-j(2\pi/N)k}$ if $0 \le k \le N-1$.

(f) In general, consider $x[n]$ to have length L and $h[n]$ to have length M. How should N be chosen so that the MATLAB statement `y = ifft( fft(x,N).*fft(h,N) )` produces the same result as the MATLAB statement `y = conv(x,h)`?

P-8.8 The inverse DTFT of $X(e^{j\hat{\omega}}) = 8e^{-j10\hat{\omega}} \dfrac{\sin(7.5\hat{\omega})}{\sin(0.5\hat{\omega})}$ is a length-15 sequence $x[n]$. If we evaluate a 60-point DFT of $x[n]$ (with zero padding), determine the DFT coefficient $X[8]$ and express it in polar form $Ae^{j\varphi}$ with $A \ge 0$ and $-\pi < \varphi < \pi$.

P-8.9 The DTFT of a length-32 real sinusoid signal sequence $q[n] = A\cos(\hat{\omega}_0 n)$ for $n = 0, 1, \ldots, 31$ consists of two "Dirichlet-form" peaks and can be written as (see Tables 8-2 and 8-3):

$$Q(e^{j\hat{\omega}}) = \tfrac{1}{2}A \frac{\sin(\frac{1}{2}(32)(\hat{\omega} - \hat{\omega}_0))}{\sin(\frac{1}{2}(\hat{\omega} - \hat{\omega}_0))} e^{-j\frac{31}{2}(\hat{\omega}-\hat{\omega}_0)} + \tfrac{1}{2}A \frac{\sin(\frac{1}{2}(32)(\hat{\omega} + \hat{\omega}_0))}{\sin(\frac{1}{2}(\hat{\omega} + \hat{\omega}_0))} e^{-j\frac{31}{2}(\hat{\omega}+\hat{\omega}_0)}$$

where $\hat{\omega}_0$ is the frequency of the sinusoid.

(a) Suppose that we define a length-32 signal as $s[n] = 0.1 + A\cos(\hat{\omega}_0 n)$, and take its length-32 DFT. Call the result $S[k]$, for $k = 0, 1, 2, \ldots, 31$. If the values of the DFT $S[k]$ are zero for 29 of the 32 DFT coefficients, and a maximum value of $|S[k]|$ is found at the index $k_1 = 10$, determine the frequency of the sinusoid, $\hat{\omega}_0$.

(b) For the 32-point DFT in part (a), determine the indices ***where the other two nonzero values*** of $S[k]$ are found (i.e., k_2 and k_3, where $S[k_2] \neq 0$ and $S[k_3] \neq 0$).

(c) If one of the two maximum values of $|S[k]|$ is found at the index $k_1 = 10$ and $|S[k_1]| = 50$, determine the amplitude A of the signal, $s[n]$, defined in (a).

P-8.10 Define a length-32 signal to be $r[n] = 1 + 7\cos(0.2\pi n)$ for $n = 0, 1, \ldots, 31$. Then form a length-N signal by zero-padding with 48 zeros, and take the N-point DFT of the zero-padded signal to obtain $R[k]$, for $k = 0, 1, 2, \ldots, N-1$.

(a) Determine the value of N for the N-point DFT taken after the zero padding.

(b) Determine the indices, k_{PEAKS}, where the two maximum values of $|R[k]|$ are found, and, give both indices where $|R[k]| = \max_k\{|R[k]|\}$.

(c) Determine the maximum value of $|R[k]|$ at the two indices found in part (b).

P-8.11 Suppose that a periodic signal $x[n]$ has a DFS given by

$$x[n] = \sum_{k=-10}^{10} (1 + k^2) e^{j(0.09\pi k)n}$$

Use the frequency response (or DTFT) to evaluate the convolution

$$y[n] = \frac{\sin(0.16\pi n)}{6\pi n} * x[n] = h[n] * x[n]$$

Express your answer in the simplest possible form.

P-8.12 Suppose that a discrete-time signal $x[n]$ is a sum of complex-exponential signals

$$x[n] = 2 + 2e^{j0.3\pi n} + 2e^{-j0.3\pi n} - 6je^{j0.8\pi n} + 6je^{-j0.8\pi n}$$

(a) Plot the spectrum of the DFS for $x[n]$ using only positive frequencies in the interval $0 \le \hat{\omega} \le 2\pi$.

(b) Suppose that $x_1[n] = (-1)^n x[n]$. Make a spectrum plot of the DFS for $x_1[n]$ using only positive frequencies.

(c) Suppose that $x_2[n] = x[n]e^{j0.5\pi n} - 6je^{j1.7\pi n}$. Make a spectrum plot of the DFS for $x_2[n]$ using only positive frequencies.

P-8.13 Suppose that the following continuous-time signal

$$x(t) = 4\cos(35\pi t) + 6\cos(15\pi t - 0.5\pi)$$

is sampled with rate $f_s = 50$ Hz to obtain the discrete-time signal $x[n]$ which is periodic with period N, and we want to determine the DFS representation of $x[n]$.

(a) Determine the period of $x[n]$ (in samples) and express $x[n]$ as a sum of exponential signals.

(b) Determine the values and indices k of the nonzero Fourier Series coefficients $\{a_k\}$ for the DFS summation. Recall that the range of the DFS summation is from $-M$ to M, where $M \le \frac{1}{2}N$. Express each nonzero a_k value in polar form.

(c) Make a spectrum plot of the DFS with the horizontal axis being $-\pi \le \hat{\omega} \le 3\pi$ to show the two periods of the spectrum as in Fig. 8-11(b).

(d) If $a_k = \sin\frac{\pi k}{10} - \cos\frac{\pi k}{5}$, determine the new $x[n]$.

P-8.14 Consider a 10-point running-sum FIR filter.

(a) Use the FFT to evaluate and plot its frequency response $H(e^{j\hat{\omega}})$. Write MATLAB code and show the plots of magnitude and phase versus frequency. Determine an appropriate value for the FFT length N to make a dense frequency grid for $\hat{\omega}$ so that the plots appear continuous.

(b) Suppose that the input signal to the 10-point running-sum filter is $x[n]$ whose discrete-time Fourier series has nonzero components only at $\hat{\omega}_1 = 0.2\pi$, $\hat{\omega}_2 = 0.4\pi$, and $\hat{\omega}_3 = 0.6\pi$, and a DC value of $X_0 = 4.5$. Explain why the output of the 10-point running-sum filter is then $y[n] = 45$ for all n. Use values from the plot of $|H(e^{j\hat{\omega}})|$ versus $\hat{\omega}$ to support your explanation.

P-8.15 A continuous-time signal $x_c(t)$ is bandlimited such that it has no spectral components for $|\omega| \geq 2\pi(1000)$. This signal is sampled with sampling rate $f_s = 1/T_s$ producing the sequence $x[n] = x_c(nT_s)$. Then length-L sections of the sampled signal are extracted via time windowing as in Fig. 8-19 and analyzed with an N-point FFT. The resulting N DFT coefficients are samples of the spectrum of the continuous-time signal $x_c(t)$ with a spacing of Δf Hz.

(a) For efficiency in computation, assume that N is a power of two. Both f_s and N can be chosen at will subject to the constraints that aliasing be avoided and $N = 2^{\nu}$. Determine the *minimum* value of N, and also f_s, so that the frequency spacing Δf between DFT coefficients is less than or equal to 5 Hz.

(b) Assume that the window is a Hann window whose length L is half the FFT length, N. For the value of N determined in (a), determine the spectral peak width (as measured between zero crossings). Give the answer in hertz. The result should be larger than 5 Hz, indicating that frequency resolution of narrow peaks is different from frequency sampling of the spectrum.

P-8.16 Suppose that a continuous-time signal $x(t)$ consists of several overlapping sinusoidal sections

$$x(t) = \begin{cases} \cos(2\pi(600)t) & 0 \leq t < 0.5 \\ \sin(2\pi(1100)t) & 0.3 \leq t < 0.7 \\ \cos(2\pi(500)t) & 0.4 \leq t < 1.2 \\ \cos(2\pi(700)t - \pi/4) & 0.4 \leq t < 0.45 \\ \sin(2\pi(800)t) & 0.35 \leq t < 1.0 \end{cases}$$

(a) If the signal is sampled with a sampling frequency of $f_s = 8000$ Hz, make a sketch of the *ideal* spectrogram that corresponds to the signal definition.

(b) Make a sketch of the *actual* spectrogram that would be obtained with an FFT length of $N = 500$ and a Hann window length of $L = 200$ with an overlap of 160 points. Make approximations to determine the frequency width in order to do the sketch without actually calculating the spectrogram in MATLAB.

P-8.17 In the spectrogram of the C-major scale in Fig. P-8.17, the analysis window was a Hann window of length $L = 256$, overlap was 200, and sampling rate $f_s = 4000$ Hz. Estimate the time duration of the notes, which is the same for all eight.

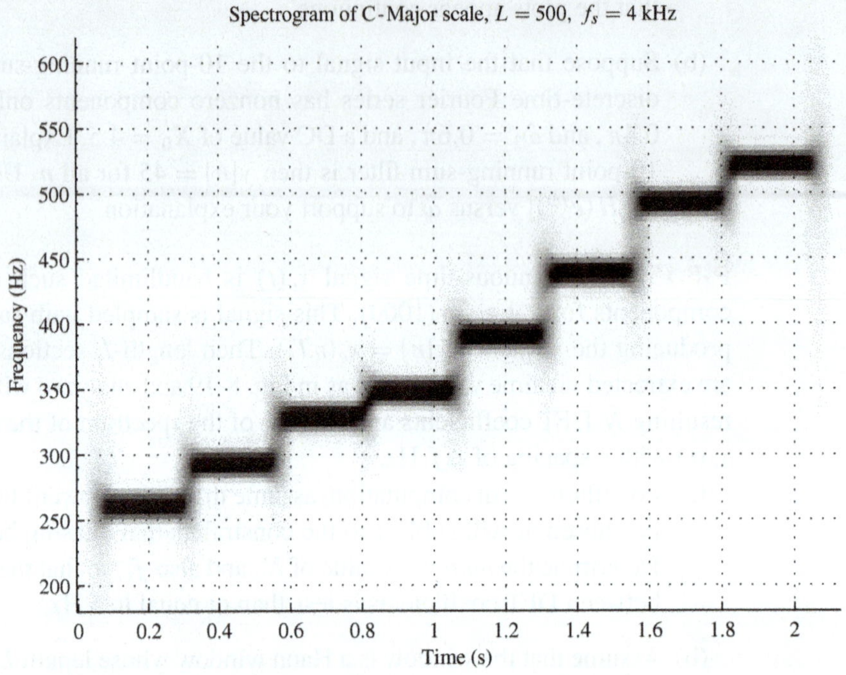

Figure P-8.17

P-8.18 Figure P-8.17 shows the spectrogram of a synthetic musical scale composed of sine waves. The analysis window was a Hann window of length $L = 256$, overlap was 200, and sampling rate $f_s = 4000$ Hz. Make a sketch of the spectrogram that would be obtained if the window length was $L = 100$. Explain how your sketch differs from Fig. P-8.17.

P-8.19 Assume that a speech signal has been sampled at 8000 Hz and then analyzed with MATLAB's `spectrogram` function using the following parameters: Hamming window with length $L = 100$, FFT length of $N = 256$, and overlap of 80 points. Determine the resolution of the resulting spectrogram image.

(a) Determine the frequency resolution (in Hz).

(b) Determine the time resolution (in sec).

P-8.20 The frequency resolution of a spectrum analysis system is determined by the discrete-time Fourier transform of the time window. Suppose that we sample a continuous-time signal at $f_s = 10{,}000$ Hz and we want to have frequency resolution of 250 Hz.

(a) If we use a Hann window for $w[n]$, what window length L would be needed? Estimate a minimum value for L.

(b) Use the specific value for L determined in part (a), and demonstrate that the resulting Hann window has sufficient frequency resolution by generating a signal with the sum of two equal-amplitude sinusoids whose frequencies are 3000 Hz and 3250 Hz and computing samples of the DTFT with a zero-padded FFT. The plot of the DTFT magnitude should exhibit two distinct peaks at the correct (known) frequencies.

P-8.21 The spectrogram of Fig. P-8.21 shows the period of time corresponding to the first nine notes of Beethoven's Für Elise in the treble clef of Fig. 3-22 on p. 101.

(a) Compute the theoretical frequencies (based on A440) from the sheet music for these nine notes.

(b) Measure the frequencies of the nine notes from the spectrogram of Fig. P-8.21, and compare your result to true frequencies from part (a).

(c) Measure the durations of each of the nine 16th notes.

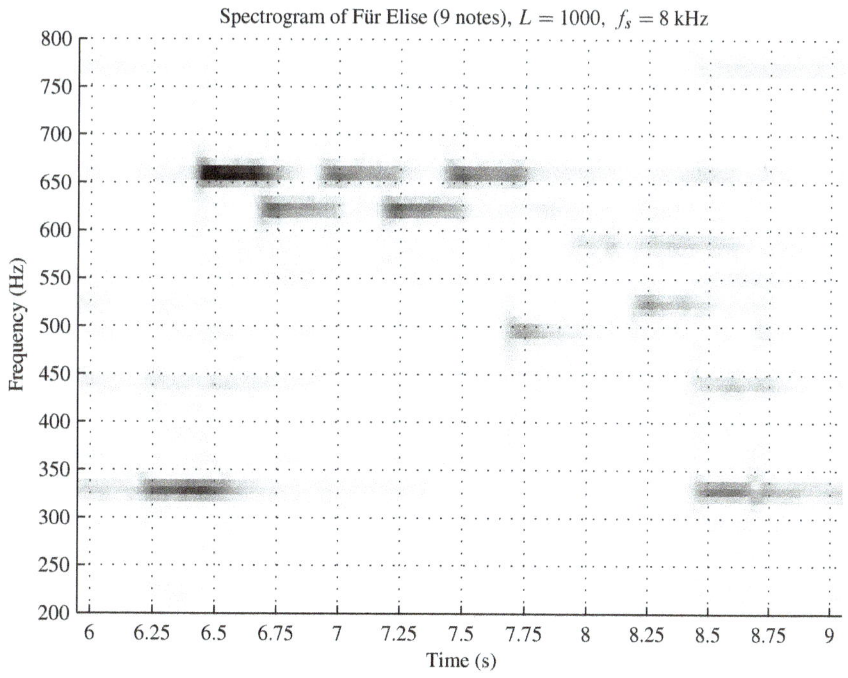

Figure P-8.21

P-8.22 In this problem, we consider a signal of the form

$$x[n] = \begin{cases} \cos(\hat{\omega}_0 n) + \cos(\alpha n^2) & 0 \le n \le 2000 \\ \cos(\hat{\omega}_0 n) + \cos(\hat{\omega}_1 n) & 2000 < n \le 4000 \end{cases} \tag{8.58}$$

Figure P-8.22 shows a spectrogram of this signal obtained with a DFT length of $N = 500$, a Hann window length of $L = 100$, and a window overlap of 90 (to obtain a very smooth image).

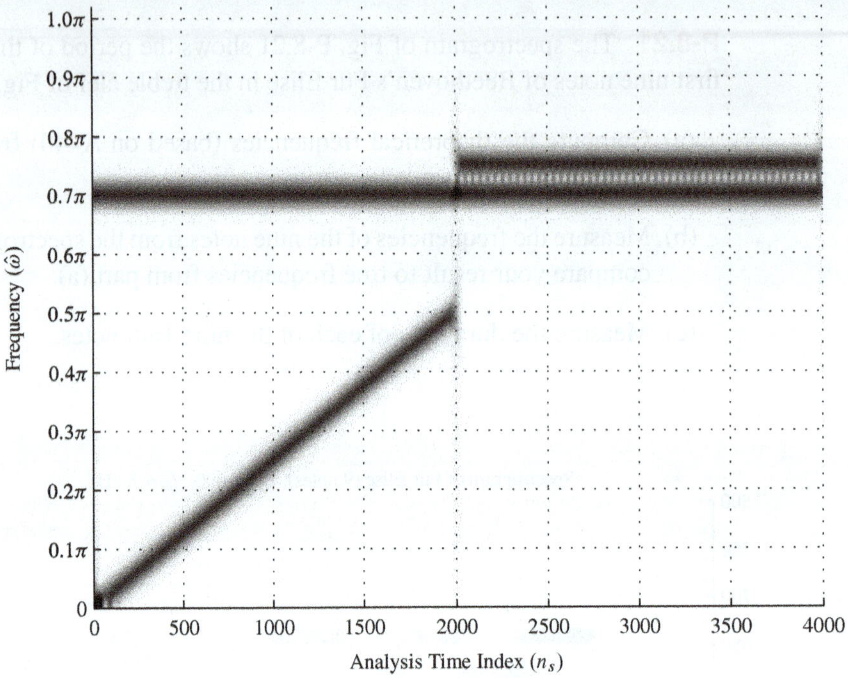

Figure P-8.22

(a) Using Fig. P-8.22, estimate the frequencies $\hat{\omega}_0$ and $\hat{\omega}_1$ and the parameter α.

(b) Describe how the spectrogram would change if the spectrum analysis window length is changed to $L = 400$.

CHAPTER

9

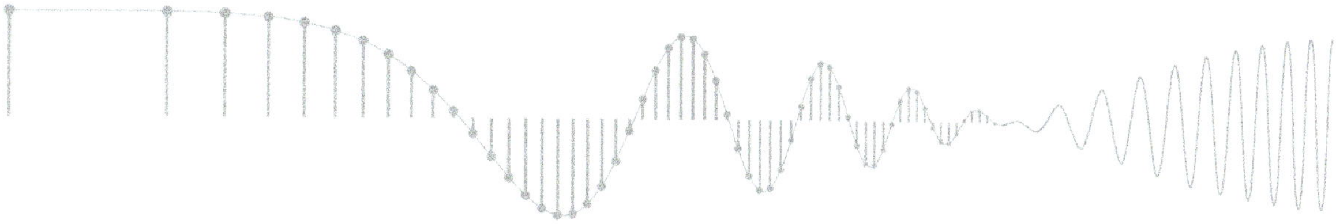

z-Transforms

In this chapter we introduce the z-transform, which brings polynomials and rational functions into the analysis of linear discrete-time systems. The key result is that FIR convolution is equivalent to polynomial multiplication. In fact, common algebraic operations, such as multiplying, dividing, and factoring polynomials, can be interpreted as combining or decomposing LTI systems. The most common z-transforms are rational functions (i.e., a numerator polynomial divided by a denominator polynomial). The roots of these polynomials are significant, because most properties of digital filters can be restated in terms of the locations of these roots.

The z-transform method is introduced in this chapter for FIR filters and finite-length sequences in general. We will use the FIR case to introduce the important concept of "domains of representation" for discrete-time signals and systems. Throughout this text, we consider three domains of representation of signals and systems: the ***n*-domain** or ***time domain*** (the domain of sequences, impulse responses, and difference equations), the ***$\hat{\omega}$*-domain** or ***frequency domain*** (the domain of frequency responses and spectrum representations), and the ***z*-domain** (the domain of z-transforms, operators, and poles and zeros).[1] The value of having three different domains of representation is that a difficult analysis in one domain is often much easier in one of the other domains. Therefore, increased understanding will result from developing skills for moving from one representation to another. For example, the cascade combination of LTI systems, which in the n-domain seems to require the new (less familiar) technique of convolution, is converted in the z-domain into the more familiar algebraic operation of polynomial multiplication.

All three domains can be used to analyze signals and systems. The z-domain exists primarily for its algebraic nature, which is convenient for mathematical analysis and synthesis. The frequency domain provides analysis techniques that also have physical significance when analyzing sound, and other tangible signals. It is important, however, to note that the n-domain is the primary *implementation domain* where signals are generated and processed, and where the implementation of filters often takes place. The frequency domain is sometimes used for processing because FFT algorithms make frequency-domain implementations practical and efficient.

9-1 Definition of the z-Transform

A finite-length signal $x[n]$ consists of a set of signal values $\{x[0], x[1], \ldots, x[L-1]\}$ that can be represented by the relation

$$x[n] = \sum_{k=0}^{L-1} x[k]\delta[n-k] \tag{9.1}$$

Each term in the summation, $x[k]\delta[n-k]$, is the value $x[k]$ at the time index $n = k$ which is the only index where $\delta[n-k]$ is nonzero. The ***z-transform*** of the signal $x[n]$ in (9.1) is defined by the formula

$$X(z) = \sum_{k=0}^{L-1} x[k]z^{-k} \tag{9.2}$$

where z, the independent variable of the z-transform $X(z)$, is a complex number. In (9.2) the signal values $\{x[0], x[1], \ldots, x[L-1]\}$ are used as coefficients of a polynomial in z^{-1}, and the exponent of z^{-k} denotes the fact that the polynomial coefficient $x[k]$ is the kth value of the signal. Although (9.2) is the conventional definition of the z-transform,[2] it is instructive to write $X(z)$ in the form

[1] Traditionally, signals and systems texts have identified just two domains: the time domain and the frequency domain. Many authors consider our $\hat{\omega}$-domain and z-domain together as the "frequency domain." This is mainly because, as we will see, the $\hat{\omega}$-domain can be viewed as a special case of the more general z-domain. However, we feel that because of the different character of the mathematical functions involved in the two domains, there is a distinct advantage in considering the $\hat{\omega}$- and z-domains as separate, but related, points of view.

[2] Some authors use positive powers of z in the definition of the z-transform, but this convention is not common in signal processing.

$$X(z) = \sum_{k=0}^{L-1} x[k](z^{-1})^k$$

which emphasizes that $X(z)$ is simply a polynomial of degree $L-1$ in the variable z^{-1}. The z-transform in (9.2) applies only to finite-length sequences because the upper limit of the sum is $L-1$ which implies that the z-transform polynomial has finite degree. In Section 9-2.3, we introduce the full-fledged definition of the z-transform, which has infinite limits.

When we use (9.2) to determine the z-transform of the signal $x[n]$, we say that we ***transform*** $x[n]$ into a new representation $X(z)$. Indeed, it is often said that we "take the z-transform of $x[n]$." This operation of ***taking the z-transform*** amounts to constructing a polynomial $X(z)$ whose coefficients are the values of the sequence $x[n]$. Specifically, the kth sequence value is the coefficient of the kth power of z^{-1} in the polynomial $X(z)$. It is just as easy to go from (9.2) back to (9.1), which is called ***taking the inverse z-transform***, because we can recover $x[n]$ from $X(z)$ simply by extracting the coefficient value of the kth power of z^{-1} and placing that value in the kth position of the sequence $x[n]$.

A ***z-transform pair*** is a sequence and its corresponding z-transform, which we denote as

$$x[n] \stackrel{z}{\longleftrightarrow} X(z) \tag{9.3}$$

We can emphasize the unique correspondence between a sequence $x[n]$ and its z-transform by writing

***n*-Domain**	$\stackrel{z}{\longleftrightarrow}$	***z*-Domain**
$x[n] = \sum_{k=0}^{L-1} x[k]\delta[n-k]$	$\stackrel{z}{\longleftrightarrow}$	$X(z) = \sum_{k=0}^{L-1} x[k]z^{-k}$

Notice that n is the independent variable of the sequence $x[n]$. Thus, we say that (9.1) represents the signal in the ***n-domain***. Since n is often an index that counts time in a sampled time waveform, we also refer to (9.1) as the ***time-domain*** representation of the signal. Similarly, observe that z is the independent variable of the z-transform $X(z)$. In other words, $X(z)$ in (9.2) represents the signal in the ***z-domain***, and when ***taking the z-transform*** of a signal, we move from the time domain to the z-domain.

As a simple, but very important, example of a z-transform pair, suppose that the signal is a shifted impulse (i.e., $x[n] = \delta[n-n_0]$). Then, applying the definition (9.2), it follows that $X(z) = z^{-n_0}$, so we can write

***n*-Domain**	$\stackrel{z}{\longleftrightarrow}$	***z*-Domain**
$x[n] = \delta[n-n_0]$	$\stackrel{z}{\longleftrightarrow}$	$X(z) = z^{-n_0}$

(9.4)

When the sequence is defined with a list of numerical values, we take the z-transform and get a general polynomial, as illustrated by the following example.

EXAMPLE 9-1 z-Transform of a Signal

Consider the sequence $x[n]$ given in the following table:

n	$n < -1$	-1	0	1	2	3	4	5	$n > 5$
$x[n]$	0	0	2	4	6	4	2	0	0

The z-transform of this sequence is

$$X(z) = 2 + 4z^{-1} + 6z^{-2} + 4z^{-3} + 2z^{-4}$$

The nonzero values of the sequence $\{2, 4, 6, 4, 2\}$ become the coefficients of the polynomial $X(z)$.

Example 9-1 shows how to determine the z-transform given the sequence values. The next example illustrates the *inverse* z-transform operation, that is, determining the sequence values when we are given its z-transform polynomial.

EXAMPLE 9-2 Inverse z-Transform

Consider the z-transform $X(z)$ given by the equation

$$X(z) = 1 - 2z^{-1} + 3z^{-3} - z^{-5}$$

In the z-transform definition (9.2), the signal values are the polynomial coefficients. Thus we can list $x[n]$ in tabular form as in Example 9-1, or we can give a case-by-case equation for the sequence values as a function of n in the form

$$x[n] = \begin{cases} 0 & n < 0 \\ 1 & n = 0 \\ -2 & n = 1 \\ 0 & n = 2 \\ 3 & n = 3 \\ 0 & n = 4 \\ -1 & n = 5 \\ 0 & n > 5 \end{cases}$$

Alternatively, we can go from the z-transform in (9.2) to the time-domain representation (9.1) in terms of impulse sequences, and then write the corresponding sequence $x[n]$ as

$$x[n] = \delta[n] - 2\delta[n-1] + 3\delta[n-3] - \delta[n-5]$$

9-2 Basic z-Transform Properties

In Section 9-1, we defined the z-transform, and showed for finite-length sequences that it is simple to go back and forth between the sequence $x[n]$ and its z-transform $X(z)$. In this section, we explore two universal transform properties that hold for the z-transform, and also show that when the z-transform is extended to the infinite-length case, these properties still hold.

9-2.1 Linearity Property of the z-Transform

The z-transform is a linear transformation because it satisfies the superposition property. To show this, we use superposition (addition and scaling) to form a new sequence $x[n] = ax_1[n] + bx_2[n]$ as the linear combination of $x_1[n]$ and $x_2[n]$, which are both assumed to have finite length less than or equal to L. Then we take the z-transform of the linear combination:

$$\begin{aligned} X(z) &= \sum_{n=0}^{L-1} (ax_1[n] + bx_2[n])\, z^{-n} \\ &= a\sum_{n=0}^{L-1} x_1[n]z^{-n} + b\sum_{n=0}^{L-1} x_2[n]z^{-n} \\ &= aX_1(z) + bX_2(z) \end{aligned}$$

These steps are true because multiplication *distributes* over addition for real and complex numbers. Thus, the z-transform $X(z)$ has the same linear combination (addition and scaling) as used in the time domain.

Linearity of the z-Transform

$$ax_1[n] + bx_2[n] \quad \overset{z}{\longleftrightarrow} \quad aX_1(z) + bX_2(z) \tag{9.5}$$

9-2.2 Time-Delay Property of the z-Transform

The second important property of the z-transform is that *multiplication* by the quantity z^{-1} in the z-domain corresponds to a *time shift* by one index in the n-domain. First, we illustrate this property with a numerical example. Consider the length-6 signal $x[n]$ defined by the following table of values:

n	$n < 0$	0	1	2	3	4	5	$n > 5$
$x[n]$	0	3	1	4	1	5	9	0

The z-transform of $x[n]$ is the following fifth-degree polynomial (in z^{-1})

$$X(z) = 3 + z^{-1} + 4z^{-2} + z^{-3} + 5z^{-4} + 9z^{-5}$$

Recall that the signal values $x[n]$ are the coefficients of the polynomial $X(z)$ and that the exponents correspond to the time locations of the values. For example, the term $4z^{-2}$ indicates that the signal value at $n = 2$ is 4 (i.e., $x[2] = 4$).

Now consider the effect of multiplying the polynomial $X(z)$ by z^{-1}:

$$\begin{aligned} Y(z) &= z^{-1}X(z) \\ &= z^{-1}(3 + z^{-1} + 4z^{-2} + z^{-3} + 5z^{-4} + 9z^{-5}) \\ &= 0z^{0} + 3z^{-1} + z^{-2} + 4z^{-3} + z^{-4} + 5z^{-5} + 9z^{-6} \end{aligned}$$

The resulting polynomial $Y(z)$ is the z-transform representation of a new signal $y[n]$, which is found by using the polynomial coefficients and exponents in $Y(z)$ to take the inverse z-transform and obtain the values of $y[n]$ at all time positions. Notice that there is now a term $4z^{-3}$ which indicates that the signal value at $n = 3$ is $y[3] = 4$. The result is the following table of values for $y[n]$:

n	$n < 0$	0	1	2	3	4	5	6	$n > 6$
$y[n]$	0	0	3	1	4	1	5	9	0

Each of the signal samples has moved over one position in the table, so $y[n] = x[n-1]$. In general, for any finite-length sequence, multiplication of the z-transform polynomial by z^{-1} simply subtracts one from each exponent in the polynomial, thereby changing all the time indices by one. Thus, we have the following fundamental relation:

Delay of One Sample

$$x[n-1] \quad \overset{z}{\longleftrightarrow} \quad z^{-1}X(z) \tag{9.6}$$

which we refer to as the ***unit-delay property*** of the z-transform.

The unit-delay property can be generalized for the case of shifting by more than one sample by simply applying (9.6) multiple times. The general result is

Delay of n_0 Samples

$$x[n-n_0] \quad \overset{z}{\longleftrightarrow} \quad z^{-n_0}X(z) \tag{9.7}$$

9-2.3 A General z-Transform Formula

So far, we have defined the z-transform only for finite-length signals using (9.2). Our definition assumes that the sequence is nonzero only in the interval $0 \le n \le N$, so the limits on the sum in (9.2) are finite. It is straightforward to extend the definition to signals of infinite length by simply extending the upper and lower limits to $+\infty$ and $-\infty$, respectively, that is,

$$X(z) = \sum_{n=-\infty}^{\infty} x[n]z^{-n} \tag{9.8}$$

It is not difficult to see that properties such as linearity and the delay property hold for any polynomial, even infinite-length polynomials, because the operations are performed on individual signal values which are individual polynomial coefficients.

However, infinite sums may cause subtle mathematical difficulties with convergence and thus require special attention. Summing an infinite number of complex terms as in (9.8) could result in an infinite result (i.e., the sum might not converge). A convenient example of the infinite-length case comes from taking the z-transform of the right-sided exponential:[3]

$$x[n] = (\tfrac{1}{2})^n u[n] \quad \overset{z}{\longleftrightarrow} \quad X(z) = \sum_{n=0}^{\infty} (\tfrac{1}{2})^n z^{-n} \tag{9.9}$$

The value of $\frac{1}{2}$ is chosen arbitrarily, but it affects the convergence of the sum. If we combine the $(\frac{1}{2})^n$ and z^{-n} terms as $(\frac{1}{2}z^{-1})^n$, then we can simplify the sum by recognizing that it is like the sum of a geometric sequence $\sum r^n = 1/(1-r)$. The result is

$$X(z) = \sum_{n=0}^{\infty} (\tfrac{1}{2}z^{-1})^n = \frac{1}{1-\frac{1}{2}z^{-1}} \qquad \text{if } |\tfrac{1}{2}z^{-1}| < 1 \tag{9.10}$$

which is very interesting because the sum with an infinite number of terms is reduced to a rational function with a first-degree polynomial in the denominator, and the only number that appears in the rational function is $\frac{1}{2}$. However, there is a condition on z that must be satisfied for the infinite sum to converge to the closed-form expression (9.10): namely, $|z| > \frac{1}{2}$. In other words, the z-transform $X(z)$ is defined for only part of the complex z-plane, called the ***region of convergence***.

The infinite-length case is considered in depth in Chapter 10 when we study recursive digital filters. Even then, a rigorous mathematical development of a complete z-transform theory for signal and system analysis (with regions of convergence) is better left to another, more advanced, course.

9-3 The z-Transform and Linear Systems

At this point, we have learned how to take the z-transform, and its inverse, but an obvious question to ask is what are the advantages of transforming from the n-domain to the z-domain? The key part of the answer is that the input–output (difference) equation of the FIR filter is equivalent to polynomial multiplication of z-transforms. As a result, the

[3]The right-sided exponential and its z-transform are given a detailed treatment in Chapter 10.

power of algebra for polynomials and rational functions can be employed to simplify much of the analysis of LTI systems. The remainder of this chapter develops this answer along with many other properties of the z-transform that are indispensable in the design and analysis of LTI systems.

9-3.1 Unit-Delay System

The unit-delay operator is one of the basic building blocks for the FIR difference equation, as shown in the block diagrams of Chapter 5, Section 5-5.2. In fact, the unit-delay operation can be interpreted as an FIR filter defined by

$$y[n] = x[n-1] \tag{9.11}$$

This is a first-order FIR filter whose two filter coefficients are $b_0 = 0$ and $b_1 = 1$. On the other hand, we have just learned the delay property of the z-transform in Section 9-2.2 which tells us that

$$\underbrace{Y(z)}_{\text{output}} = \underbrace{(z^{-1})}_{\text{delay}}\,\underbrace{X(z)}_{\text{input}} \tag{9.12}$$

Thus the output z-transform $Y(z)$ is the product of two z-transforms, (z^{-1}) and $X(z)$, and it seems reasonable to claim that the unit delay is represented by z^{-1}. This fact allows us to treat the block diagrams in Figs. 9-1(a) and (b) as equivalent.

9-3.2 z^{-1} Notation in Block Diagrams

The delay system representation in (9.12) and Fig. 9-1(b) can be used in block diagrams of general LTI systems as follows: All the unit delay operators become z^{-1} operators in the transform domain, and, owing to the linearity property of the z-transform, the scalar multipliers and adders remain the same as in the time-domain representation. Figure 9-2 shows a block diagram for a general length 2 FIR filter, $y[n] = b_0x[n] + b_1x[n-1]$, but uses the z-domain notation z^{-1} to represent the unit-delay operator. Since the block diagram uses z^{-1}, it is easy to write the z-transform of the output $Y(z)$ in terms of $X(z)$ and then simplify the algebraic expression to obtain

$$Y(z) = b_0X(z) + b_1z^{-1}X(z) = \underbrace{(b_0 + b_1z^{-1})}_{H(z)}\,X(z) \tag{9.13}$$

Once again, the output is the product of two z-transform functions: $X(z)$ and $H(z) = b_0 + b_1z^{-1}$ which depends on the filter coefficients. We use the notation $H(z)$ because we

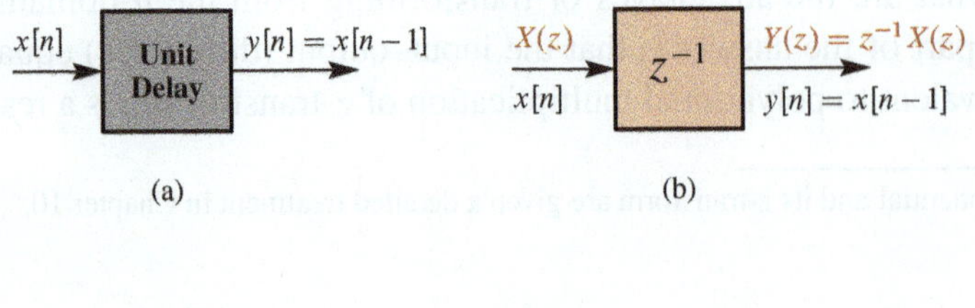

Figure 9-1 Representations of the unit delay system. (a) Explicit representation. (b) Representation using z^{-1}.

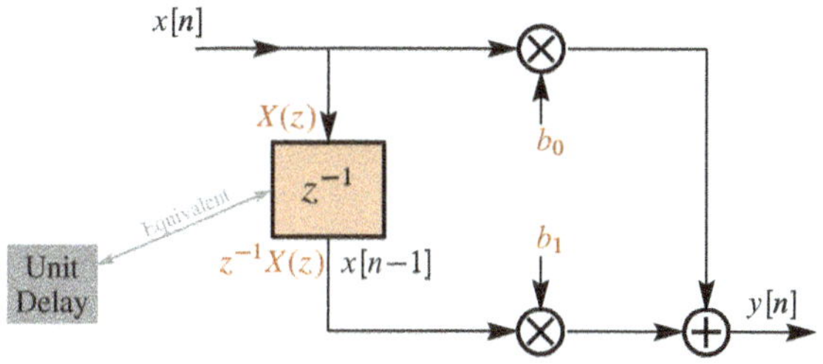

Figure 9-2 Computational structure for a first-order FIR filter $y[n] = b_0 x[n] + b_1 x[n-1]$. The block diagram uses z^{-1} to denote the unit delay, exploiting the equivalence shown in Figs. 9-1(a) and (b).

show in Section 9-3.4 that this $H(z)$ is actually the z-transform of the impulse response of the FIR filter.

EXERCISE 9.1 Draw a block diagram similar to Fig. 9-2 for the first difference system: $y[n] = x[n] - x[n-1]$

9-3.3 The z-Transform of an FIR Filter

There are two ways to obtain a z-domain representation of a general FIR filter. The first is to take the z-transform of the FIR difference equation, the second is to take the z-transform of the impulse response. The example in Fig. 9-2 suggests how to work with the FIR difference equation. Recall that the general difference equation of an FIR filter involves the sum of delayed copies of the input signal multiplied by scalars.

$$y[n] = \sum_{k=0}^{M} b_k\, x[n-k] \tag{9.14}$$

These operations are exactly the ones covered by the linearity property and delay property of the z-transform. The delay-by-n_0 property (9.4) tells us that the z-transform of a term like $x[n-k]$ is $z^{-k}X(z)$. The linearity property tells us that the same scaling coefficients b_k and sum will be used in the z-domain. As a result, we can take the z-transform of both sides of the difference equation (9.14) to obtain

$$Y(z) = \sum_{k=0}^{M} b_k \left(z^{-k} X(z)\right) = \underbrace{\left(\sum_{k=0}^{M} b_k\, z^{-k}\right)}_{H(z)} X(z) \tag{9.15}$$

Thus the output is again the product of $X(z)$ and a z-domain function $H(z)$ that depends solely on the filter coefficients. The z-transform function $H(z)$ is called the ***system function*** of the FIR filter.

9-3.4 z-Transform of the Impulse Response

The second way to obtain a z-domain representation of an FIR filter is to take the z-transform of the impulse response. In this approach, we recall that the FIR system is *completely characterized* by its impulse response $h[n]$, which is a finite-length signal. Therefore, when we take the z-transform of the impulse response $h[n]$

$$H(z) = \sum_{k=0}^{M} h[k]\, z^{-k} \tag{9.16}$$

we obtain a z-domain representation that must also be a *complete description* of the FIR system. The z-transform in (9.16) is called $H(z)$, but we need to recall one fact from Section 5-4.2 on p. 179 to claim that it is identical to the system function obtained in (9.15). That fact is

$$h[n] = \sum_{k=0}^{M} b_k\, \delta[n-k] \tag{9.17}$$

which says that the impulse response values for an FIR filter are equal to the filter coefficients.

To summarize, we have added the following important result for FIR systems:

System Function of FIR System

$$h[n] = \sum_{k=0}^{M} b_k \delta[n-k] \quad \stackrel{z}{\longleftrightarrow} \quad H(z) = \sum_{k=0}^{M} h[k] z^{-k} \tag{9.18}$$

Comparing (9.15) and (9.18), we see that the ***system function*** $H(z)$ can be written in two equivalent forms

$$H(z) = \sum_{k=0}^{M} b_k z^{-k} = \sum_{k=0}^{M} h[k] z^{-k} \tag{9.19}$$

because $h[k] = b_k$ for the FIR filter.

EXERCISE 9.2 Determine the system function $H(z)$ of an FIR filter whose impulse response is

$$h[n] = \delta[n] - 7\delta[n-2] - 3\delta[n-3]$$

9-3.5 Roots of a z-Transform Polynomial

To illustrate the power of algebra that comes with representing the FIR filter with a system function polynomial $H(z)$, we give two examples that rely on factoring polynomials. The z-transform of a finite-length signal, such as $H(z)$ for an FIR filter, is a function of the complex variable z, and it is also an Mth-degree polynomial (9.19) in the variable z^{-1}. Therefore, $H(z)$ has exactly M roots according to the fundamental theorem of algebra.

When $H(z)$ is expressed in terms of its roots, it can be written as the product of first-order factors that completely define the polynomial to within a multiplicative constant, that is,

$$H(z) = G\prod_{k=1}^{M}(1 - z_k z^{-1}) = G\prod_{k=1}^{M}\frac{(z - z_k)}{z} \tag{9.20}$$

where G is a constant. The first-degree term $(z - z_k)$ in the right-hand product makes it clear that these roots are also zeros of $H(z)$ (i.e., M values z_k, $k = 1, 2, \ldots, M$ such that $H(z_k) = 0$).

EXAMPLE 9-3 Zeros of System Function

Consider an FIR filter defined by the difference equation

$$y[n] = 6x[n] - 5x[n-1] + x[n-2]$$

The z-transform system function is $H(z) = 6 - 5z^{-1} + z^{-2}$ which factors as

$$H(z) = (3 - z^{-1})(2 - z^{-1}) = 6\,\frac{(z - \frac{1}{3})(z - \frac{1}{2})}{z^2}$$

Thus, the roots of $H(z)$ are $\frac{1}{3}$ and $\frac{1}{2}$. Note that a different filter

$$w[n] = x[n] - \tfrac{5}{6}x[n-1] + \tfrac{1}{6}x[n-2]$$

has a system function with the same roots, but each filter coefficient is 1/6 as big. This simply means that $w[n] = y[n]/6$, and the system function of the second system is $\frac{1}{6}H(z)$.

EXAMPLE 9-4 Difference Equation from Roots of $H(z)$

Suppose the system function $H(z)$ has two roots at $\{-1, \frac{1}{2}\}$. Then we can use (9.20) to form the $H(z)$ polynomial from the roots as

$$H(z) = G\,\frac{(z - (-1))}{z}\,\frac{(z - \frac{1}{2})}{z} = G(1 + z^{-1})(1 - \tfrac{1}{2}z^{-1})$$

where G is a scaling value. In order to find $H(z)$ we need to find G, which requires that we know an additional fact. In this example, we assume that the value of $H(z)$ at $z = 3$ is given as 4. Then $z^{-1} = \frac{1}{3}$, and

$$4 = H(z)\Big|_{z=3} = G(1 + (\tfrac{1}{3}))(1 - \tfrac{1}{2}(\tfrac{1}{3})) = (\tfrac{4}{3})(\tfrac{5}{6})G$$

Thus $G = 3.6$, and we can multiply the two factors to obtain $H(z) = 3.6 + 1.8z^{-1} - 1.8z^{-2}$. Finally, we can write the difference equation for this FIR filter as

$$y[n] = 3.6x[n] + 1.8x[n-1] - 1.8x[n-2]$$

EXERCISE 9.3 Determine the impulse response $h[n]$ of an FIR filter whose system function is

$$H(z) = 4(1 - e^{j\pi/2}z^{-1})(1 - e^{-j\pi/2}z^{-1})(1 + 0.8z^{-1})$$

Multiply the factors to get a polynomial and then determine the impulse response by "inverse z-transformation."
Note: The terms with complex conjugate roots can be simplified as follows:

$$(1 - re^{j\theta}z^{-1})(1 - re^{-j\theta}z^{-1}) = 1 - 2r\cos(\theta)z^{-1} + r^2z^{-2}$$

9-4 Convolution and the z-Transform

The power of the z-transform lies in the *convolution property* which states that polynomial multiplication of z-transforms is equivalent to time-domain convolution. In this section, we derive this important transform property by generalizing the result in (9.15). To show that the convolution of two signals can be "transformed" into a product of z-transforms, recall that discrete convolution of two sequences $x[n]$ and $h[n]$ is given by

$$y[n] = x[n] * h[n] = \sum_{k=0}^{M} h[k]x[n-k] \tag{9.21}$$

where $M+1$ is the length of the sequence $h[n]$. For the time being we assume that $x[n]$ also has finite length, which implies that the length of the result $y[n]$ is finite. To prove the desired general result, we take the z-transform of both sides of (9.21). On the right-hand side of (9.21), only the terms $x[n-k]$ depend on n, and their z-transforms are $z^{-k}X(z)$ which is obtained with the general delay property (9.7). The convolution sum is a superposition of these delayed signals (scaled by $h[k]$), so the linearity property (9.5) leads to

$$Y(z) = \sum_{k=0}^{M}\left(h[k]\underbrace{\left(z^{-k}X(z)\right)}_{z\text{-transform}}\right) \tag{9.22a}$$

$$= \left(\sum_{k=0}^{M} h[k]z^{-k}\right)X(z) = H(z)X(z). \tag{9.22b}$$

In (9.22a) $X(z)$ is common to all the terms in the sum so it can be factored out to obtain (9.22b). If $x[n]$ is a finite-length sequence, its z-transform $X(z)$ is a polynomial, so we have proven that the z-transform operator transforms convolution into polynomial multiplication.[4] The details of polynomial multiplication are illustrated in the following example.

[4]This derivation of the convolution property can be extended to infinite-length signals because only the linearity and delay properties are needed. However, with infinite-length signals there is always a question of convergence of the infinite sums, so we must say that the convolution property is true if the two z-transforms being multiplied have overlapping regions of convergence in the z-plane.

EXAMPLE 9-5 Convolution via $H(z)X(z)$

The z-transform method can be used to convolve the following signals:

$$x[n] = \delta[n-1] - \delta[n-2] + \delta[n-3] - \delta[n-4]$$
$$h[n] = \delta[n] + 2\delta[n-1] + 3\delta[n-2] + 4\delta[n-3]$$

The z-transforms of the sequences $x[n]$ and $h[n]$ are

$$X(z) = 0 + 1z^{-1} - 1z^{-2} + 1z^{-3} - 1z^{-4}$$
$$\text{and} \quad H(z) = 1 + 2z^{-1} + 3z^{-2} + 4z^{-3}$$

Both $X(z)$ and $H(z)$ are polynomials in z^{-1}, so we can compute the z-transform of the convolution by multiplying these two polynomials, that is,

$$\begin{aligned} Y(z) &= H(z)X(z) \\ &= (1 + 2z^{-1} + 3z^{-2} + 4z^{-3})(z^{-1} - z^{-2} + z^{-3} - z^{-4}) \\ &= z^{-1} + (-1+2)z^{-2} + (1-2+3)z^{-3} \\ &\quad + (-1+2-3+4)z^{-4} \\ &\quad + (-2+3-4)z^{-5} + (-3+4)z^{-6} + (-4)z^{-7} \\ &= z^{-1} + z^{-2} + 2z^{-3} + 2z^{-4} - 3z^{-5} + z^{-6} - 4z^{-7} \end{aligned}$$

Since the coefficients of any z-polynomial are the sequence values, with their position in time being indicated by the power of (z^{-1}), we can "inverse transform" $Y(z)$ to obtain

$$\begin{aligned} y[n] = \delta[n-1] + \delta[n-2] + 2\delta[n-3] + 2\delta[n-4] \\ - 3\delta[n-5] + \delta[n-6] - 4\delta[n-7] \end{aligned}$$

To verify that we have obtained the correct convolution result, we now perform the convolution sum (9.21) directly to compute the output. If we write out a few terms, we can detect a pattern that is similar to the z-transform polynomial multiplication.

$$\begin{aligned} y[0] &= h[0]x[0] = 1(0) = 0 \\ y[1] &= h[0]x[1] + h[1]x[0] = 1(1) + 2(0) = 1 \\ y[2] &= h[0]x[2] + h[1]x[1] + h[2]x[0] \\ &= 1(-1) + 2(1) + 3(0) = 1 \\ y[3] &= h[0]x[3] + h[1]x[2] + h[2]x[1] + h[3]x[0] \\ &= 1(1) + 2(-1) + 3(1) + 4(0) = 2 \\ y[4] &= h[0]x[4] + h[1]x[3] + h[2]x[2] + h[3]x[1] \\ &= 1(-1) + 2(1) + 3(-1) + 4(1) = 2 \\ \vdots &= \quad \vdots \end{aligned}$$

For each term above, notice how the index of $h[k]$ and the index of $x[n-k]$ sum to the same value (i.e., n) for all products that contribute to $y[n]$. The same thing happens in polynomial multiplication because exponents add.

In Section 5-4.3.1 on p. 182 we demonstrated a *synthetic multiplication* table for evaluating the convolution of $x[n]$ with $h[n]$. Now we have a justification for this procedure because we are, in fact, multiplying the polynomials $X(z)$ and $H(z)$. The procedure is repeated below for the numerical example of this section.

z	z^0	z^{-1}	z^{-2}	z^{-3}	z^{-4}	z^{-5}	z^{-6}	z^{-7}
$x[n]$, $X(z)$	0	+1	−1	+1	−1	0	0	0
$h[n]$, $H(z)$	1	2	3	4				
$X(z)$	0	+1	−1	+1	−1	0	0	0
$2z^{-1}X(z)$		0	+2	−2	+2	−2	0	0
$3z^{-2}X(z)$			0	+3	−3	+3	−3	0
$4z^{-3}X(z)$				0	+4	−4	+4	−4
$y[n]$, $Y(z)$	0	+1	+1	+2	+2	−3	+1	−4

In the z-transforms $X(z)$, $H(z)$, and $Y(z)$, the power of z^{-1} is implied by the horizontal position of the coefficient in the table. Each row is produced by multiplying the $x[n]$ row by one of the $h[n]$ values and shifting the result right by the implied power of z^{-1}. The final answer is obtained by summing down the columns, so the bottom row has the values of $y[n] = x[n] * h[n]$ or, equivalently, the coefficients of the polynomial $Y(z)$.

In this section we have established that convolution and polynomial multiplication are essentially the same thing.[5] Indeed, the most important result of z-transform theory is

Convolution Maps to Multiplication of z-Transforms

$$y[n] = h[n] * x[n] \quad \stackrel{z}{\longleftrightarrow} \quad Y(z) = H(z)X(z)$$

This result has many implications far beyond its use as a basis for understanding and implementing convolution.

[5] In MATLAB, there is no special function for multiplying polynomials. Instead, you simply use the convolution function `conv` to multiply polynomials since polynomial multiplication is identical to discrete convolution of the sequences of coefficients.

EXERCISE 9.4 Use the z-transform of

$$x[n] = \delta[n-1] - \delta[n-2] + \delta[n-3] - \delta[n-4]$$

and the system function $H(z) = 1 - z^{-1}$ to find the output of a first-difference filter when $x[n]$ is the input. Compute your answer by using polynomial multiplication and also by using the difference equation:

$$y[n] = x[n] - x[n-1]$$

What is the degree of the z-transform polynomial that represents the output $y[n]$?

9-4.1 Cascading Systems

One of the main applications of the z-transform in system design is its use in creating alternative filter structures that have exactly the same input–output behavior. An important example is the cascade connection of two or more LTI systems. In block diagram form, the cascade is drawn with the output of the first system connected to the input of the second, as in Fig. 9-3. The input signal is $x[n]$ and the overall output is $y[n]$. The sequence $w[n]$ is an intermediate signal that can be thought of as temporary storage.

As we have already seen in Section 5-8 on p. 202, if $h_1[n]$ and $h_2[n]$ are the respective impulse responses of the first and second systems, then the overall impulse response from input $x[n]$ to output $y[n]$ in Fig. 9-3 is $h[n] = h_1[n] * h_2[n]$ (i.e., convolution). Therefore, the z-transform of the overall impulse response $h[n]$ is the product of the individual z-transforms of the two impulse responses, that is,

Cascade of LTI Systems

$$h[n] = h_1[n] * h_2[n] \quad \stackrel{z}{\longleftrightarrow} \quad H(z) = H_1(z)H_2(z)$$

An important consequence of this result follows easily from the fact that multiplication is commutative; so that $H(z) = H_1(z)H_2(z) = H_2(z)H_1(z)$. This implies that convolution must also be a commutative operation and that the two systems can be cascaded in either order and we still obtain the same overall system response.

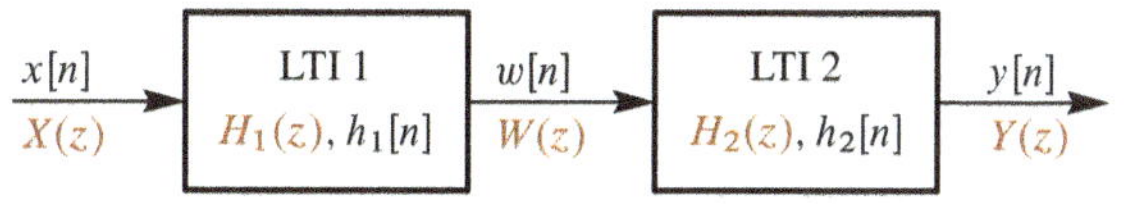

Figure 9-3 Cascade of two LTI systems.

EXAMPLE 9-6 $H(z)$ for Cascade

To give a simple example of this idea, consider a system described by the difference equations

$$w[n] = 3x[n] - x[n-1] \tag{9.23a}$$

$$y[n] = 2w[n] - w[n-1] \tag{9.23b}$$

which define a cascade of two first-order systems. As in Fig. 9-3, the output $w[n]$ of the first system is the input to the second system, and the overall output is the output of the second system. The intermediate signal $w[n]$ in (9.23a) must be computed prior to being used in (9.23b). We can combine the two filters into a single difference equation by substituting $w[n]$ from the first system into the second, which gives

$$\begin{aligned} y[n] &= 2w[n] - w[n-1] \\ &= 2(3x[n] - x[n-1]) - (3x[n-1] - x[n-2]) \\ &= 6x[n] - 5x[n-1] + x[n-2] \end{aligned} \tag{9.24}$$

Thus we have proved that the cascade of the two first-order systems is equivalent to a single second-order system. It is important to notice that the difference equation (9.24) defines an algorithm for computing $y[n]$ that is different from the algorithm specified by (9.23a) and (9.23b) together. However, the outputs of the two different implementations would be exactly the same, assuming perfectly accurate computation.

Working out the details of the overall difference equation as in (9.24) would be extremely tedious if the systems were higher order. The z-transform simplifies these operations into the multiplication of polynomials. The z-transform of the first-order systems gives the following system functions

$$H_1(z) = 3 - z^{-1} \qquad \text{and} \qquad H_2(z) = 2 - z^{-1}$$

Therefore, the overall system function is the product

$$H(z) = (3 - z^{-1})(2 - z^{-1}) = 6 - 5z^{-1} + z^{-2}$$

whose polynomial coefficients match the filter coefficients in the difference equation (9.24). Note that, even in this simple example, the z-domain solution is more straightforward than the n-domain solution.

EXERCISE 9.5 Use z-transforms to combine the following cascaded systems

$$w[n] = x[n] + x[n-1]$$

$$y[n] = w[n] - w[n-1] + w[n-2]$$

into a single difference equation for $y[n]$ in terms of $x[n]$.

9-4.2 Factoring z-Polynomials

If we can multiply z-transforms to get higher order systems, we can also factor z-transform polynomials to break down a large system into smaller modules. Since cascading systems is equivalent to multiplying their system functions, the factors of a high-order polynomial $H(z)$ represent component systems that make up $H(z)$ in a cascade connection.

EXAMPLE 9-7 Split $H(z)$ into a Cascade

Consider the following third-order system

$$H(z) = 1 - 2z^{-1} + 2z^{-2} - z^{-3}$$

which we want to split into the cascade of lower order systems with real coefficients. Since we want a product of system functions for the cascade, we need to find the roots and use the product form in (9.20). The MATLAB function `roots([1, -2, 2, -1])` will extract the roots giving

```
1.0,  0.5000 + 0.8660i, and 0.5000 - 0.8660i
```

One of the roots of $H(z)$ is $z = 1$, so $H_1(z) = (1 - z^{-1})$ is a factor of $H(z)$. The other two first-order factors have complex roots, but these roots are complex conjugates. Therefore, in order to obtain the other system function for the cascade we must form the product of these terms as

$$H_2(z) = (1 - e^{j\pi/3}z^{-1})(1 - e^{-j\pi/3}z^{-1}) = 1 - z^{-1} + z^{-2}$$

The resulting second-order factor has real coefficients. This is an instance of the general property that the complex-valued roots of a polynomial with real coefficients must occur in conjugate pairs. The factorization of $H(z)$ as

$$H(z) = (1 - z^{-1})(1 - z^{-1} + z^{-2})$$

gives the cascade shown in the block diagram of Fig. 9-4. The resulting difference equations for the cascade are obtained by inverse transforming the factors of $H(z)$

$$w[n] = x[n] - x[n-1]$$
$$y[n] = w[n] - w[n-1] + w[n-2]$$

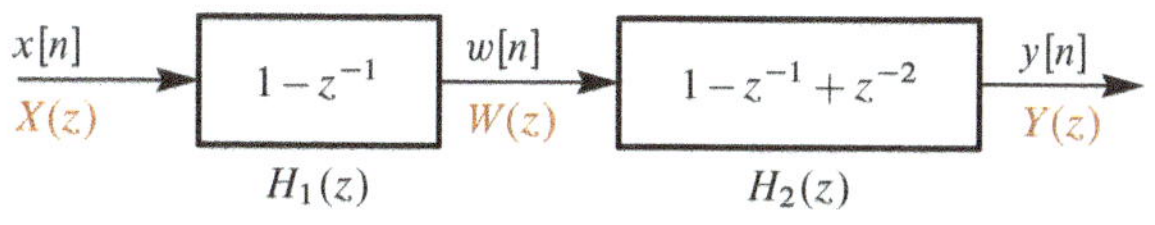

Figure 9-4 Factoring $H(z) = 1 - 2z^{-1} + 2z^{-2} - z^{-3}$ into the product of a first-order system and a second-order system.

9-4.3 Deconvolution

The cascading property leads to an interesting question that has practical application. Can we use the second filter in a cascade to undo the effect of the first filter? What we would like is for the output of the second filter to be equal to the input to the first. Stated more precisely, suppose that we have the cascade of two filters $H_1(z)$ and $H_2(z)$, and $H_1(z)$ is known. Is it possible to find $H_2(z)$ so that the overall system has its output equal to the input? If so, the z-transform analysis tells us that its system function would have to be $H(z) = 1$, so that

$$Y(z) = H_1(z)H_2(z)X(z) = H(z)X(z) = X(z)$$

Since the first system processes the input via convolution, the second filter tries to undo convolution, so the process is called ***deconvolution***. Another term for this is ***inverse filtering***, and if $H_1(z)H_2(z) = 1$, then $H_2(z)$ is said to be the *inverse* of $H_1(z)$ (and vice versa).

EXAMPLE 9-8 Deconvolution

If we take a specific example, we can generate a solution in terms of z-transforms. Suppose that $H_1(z) = 1 + 0.1z^{-1} - 0.72z^{-2}$. We want $H(z) = 1$, so we require that

$$H_1(z)H_2(z) = 1$$

Since $H_1(z)$ is known, we can solve for $H_2(z)$ to get

$$H_2(z) = \frac{1}{H_1(z)} = \frac{1}{1 + 0.1z^{-1} - 0.72z^{-2}}$$

What are we to make of this example? It seems that the deconvolver for an FIR filter must have a system function that is not a polynomial, but a rational function (ratio of two polynomials) with a denominator instead. This means that the inverse filter for an FIR filter cannot be another FIR filter, and deconvolution suddenly is not as simple as it appeared. Since we have not yet considered the possibility of anything but polynomial system functions, we cannot give the solution in the form of a difference equation. However, in Chapter 10, we will see that other types of LTI systems exist that do have rational system functions with denominators. We will explore aspects of the inverse filtering problem in Problem P-10.27.

9-5 Relationship Between the z-Domain and the $\hat{\omega}$-Domain

The FIR system function $H(z)$ has a functional form that is identical to the form of the frequency response formula $H(e^{j\hat{\omega}})$. The same is true for the z-transform of a length-L signal $x[n]$ where the functional form of $X(z)$ is identical to the form of the DTFT

$X(e^{j\hat{\omega}})$. This is quite easy to see if we display the frequency-domain formulas alongside the z-transform formulas:

***z*-Domain**	$\overset{z=e^{j\hat{\omega}}}{\longleftrightarrow}$	**$\hat{\omega}$-Domain**
$H(z) = \sum_{k=0}^{M} b_k z^{-k}$		$H(e^{j\hat{\omega}}) = \sum_{k=0}^{M} b_k e^{-j\hat{\omega}k}$
$X(z) = \sum_{n=0}^{L-1} x[n] z^{-n}$		$X(e^{j\hat{\omega}}) = \sum_{n=0}^{L-1} x[n] e^{-j\hat{\omega}n}$

After comparing the left- and right-hand sides, we see a clear correspondence between the z- and $\hat{\omega}$-domains if we make the following substitution in $H(z)$ or $X(z)$

$$\boxed{z = e^{j\hat{\omega}}} \tag{9.25}$$

because[6] then $z^{-n} = (e^{j\hat{\omega}})^{-n}$. The substitution (9.25) connects the frequency domain to the z-domain, and allows us to get the DTFT $X(e^{j\hat{\omega}})$ from the z-transform $X(z)$

$$\boxed{X(e^{j\hat{\omega}}) = X(z)\Big|_{z=e^{j\hat{\omega}}}} \tag{9.26}$$

The same is true for getting the frequency response $H(e^{j\hat{\omega}})$ from the z-transform system function $H(z)$. For example, when we analyze an FIR filter whose system function is $H(z) = 1 - z^{-2}$, its frequency response is $H(e^{j\hat{\omega}}) = 1 - e^{-j2\hat{\omega}}$, after changing z to $e^{j\hat{\omega}}$.

DEMO
Three Domains—FIR

We have implied by the notation $\overset{z=e^{j\hat{\omega}}}{\longleftrightarrow}$ that going from the z-transform system function to the DTFT can be reversed. This can be accomplished by expressing $H(e^{j\hat{\omega}})$ in terms of powers of $e^{-j\hat{\omega}}$ and then replacing each term $e^{-j\hat{\omega}k}$ by z^{-k}. The following example illustrates this process.

EXAMPLE 9-9 Moving from $\hat{\omega}$ to z

Consider a DTFT function expressed as

$$H(e^{j\hat{\omega}}) = (1 + \cos 2\hat{\omega})e^{-j3\hat{\omega}}$$

Using the inverse Euler formula for the cosine term gives

$$H(e^{j\hat{\omega}}) = \left(1 + \frac{e^{j2\hat{\omega}} + e^{-j2\hat{\omega}}}{2}\right) e^{-j3\hat{\omega}} = (e^{j\hat{\omega}})^{-3} + \tfrac{1}{2}(e^{j\hat{\omega}})^{-1} + \tfrac{1}{2}(e^{j\hat{\omega}})^{-5}$$

Making the substitution $e^{j\hat{\omega}} = z$ gives

$$H(z) = z^{-3} + \tfrac{1}{2}z^{-1} + \tfrac{1}{2}z^{-5} = \tfrac{1}{2}z^{-1} + z^{-3} + \tfrac{1}{2}z^{-5}$$

[6] The variables n and k in the summations are dummy indices for counting through the terms in the sum.

If the impulse response of the system is needed, the inverse z-transform gives

$$h[n] = \tfrac{1}{2}\delta[n-1] + \delta[n-3] + \tfrac{1}{2}\delta[n-5]$$

9-5.1 The z-Plane and the Unit Circle

The z-transform $X(z)$ is a complex-valued function of the complex variable $z = x + jy$, so a plot of the magnitude $|X(z)|$ would be a three-dimensional (3-D) plot of magnitude versus x and y, such as Fig. 9-5. The notation $X(e^{j\hat{\omega}})$ in (9.26) emphasizes the connection between the $\hat{\omega}$-domain and the z-domain because it indicates explicitly that the DTFT $X(e^{j\hat{\omega}})$ is obtained from $X(z)$ by *evaluating $X(z)$ for a specific set of values in the z-plane.* These values $z = e^{j\hat{\omega}}$ are a set of complex numbers whose magnitude is always one and whose angle is $\hat{\omega}$, which is also the frequency variable of the DTFT. In other words, the values of $z = e^{j\hat{\omega}}$ define a circle of radius one in the z-plane, which is called the ***unit circle***, and the DTFT is the z-transform evaluated on the unit circle. This is illustrated in Fig. 9-6, which shows the unit circle and a typical point $z = e^{j\hat{\omega}}$, which is at a distance of 1 from the origin and at an angle of $\hat{\omega}$ with respect to the real axis of the z-plane.

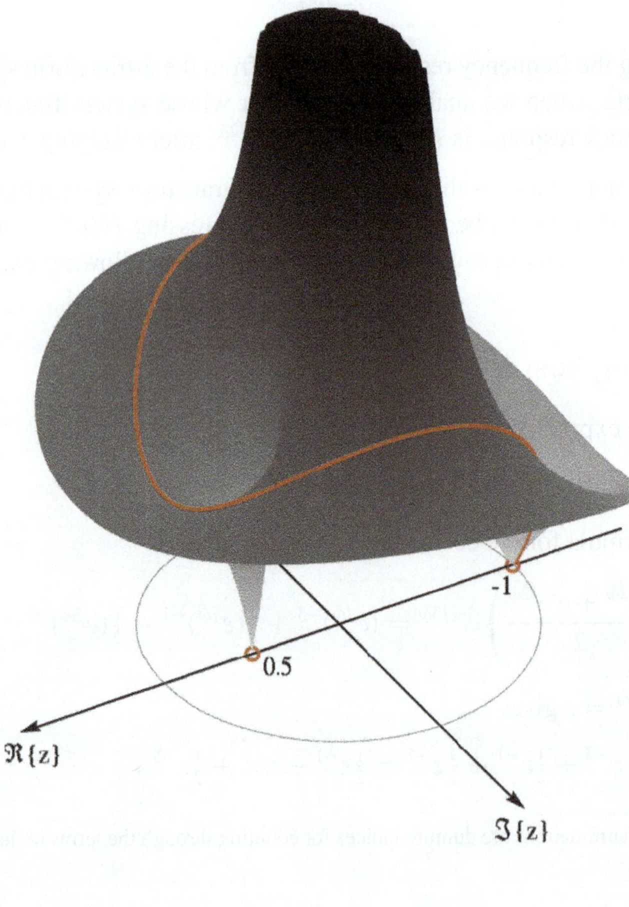

Figure 9-5 The z-transform $X(z) = 1 + \frac{1}{2}z^{-1} - \frac{1}{2}z^{-2} = (1 + z^{-1})(1 - \frac{1}{2}z^{-1})$ evaluated over the region $\{|z| \le 1.5\}$ of the z-plane that includes the unit circle. The evaluation of $H(z)$ on the unit circle is the orange line. The view is from the first quadrant, so the real axis runs right to left, and the point $z = 1$ is on the left. The two small orange circles indicate where $X(z) = 0$; these zeros are at $z = -1, \frac{1}{2}$. Since $|X(z)| \to \infty$ when $|z| \to 0$, the height of the plot is clipped to $|H(z)| \le 3$ to reduce the vertical range and make the zeros easier to see.

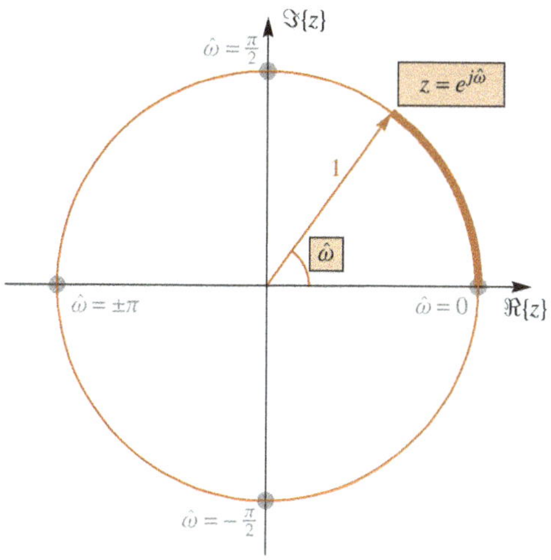

Figure 9-6 The complex z-plane including the unit circle, where $z=e^{j\hat{\omega}}$. The angle around the circle is determined by the frequency $\hat{\omega}$. The thick colored shading of the unit circle corresponds to a frequency interval from 0 to $\hat{\omega}$. The gray dots on the unit circle are the points $\{1,\ j,\ -1,\ -j\}$ which correspond to $\hat{\omega}=\{0,\ \frac{\pi}{2},\ \pm\pi,\ -\frac{\pi}{2}\}$.

The graphical representation of Fig. 9-6 gives us a convenient way of visualizing the relationship between the $\hat{\omega}$-domain and the z-domain. Since the $\hat{\omega}$-domain lies on a special part of the z-domain—the unit circle—many properties of the DTFT are evident from z-plane plots, as well as from root locations when $X(z)$ has numerator and denominator polynomials. For example, the periodicity of the DTFT is obvious from Fig. 9-6, which shows that evaluating the system function at all points on the unit circle requires moving through an angle of 2π rad. But, since the frequency $\hat{\omega}$ is also the angle of $e^{j\hat{\omega}}$ in the z-plane, a frequency range of $-\pi < \hat{\omega} \leq \pi$ covers all the angles with one trip around the unit circle. If the frequency starts at $\hat{\omega}=-\pi$ and increases to π, the circle is traversed in a counterclockwise manner starting at $z=e^{-j\pi}=-1$, then to $z=e^{-j\pi/2}=-j$, then to $z=e^{j0}=1$, and $z=+j$, and finally back to $z=e^{j\pi}=-1$. Continuing around the unit circle more times simply cycles through more periods of the DTFT. The gray dots in Fig. 9-6 highlight specific points in the z-plane and their corresponding frequencies.

In Sections 9-2 and 9-4, a number of useful properties of the z-transform were derived and their application was demonstrated. It should be clear from the discussion so far in this section that all these properties are inherited by the DTFT by virtue of the substitution $z=e^{j\hat{\omega}}$ in (9.26), see Table 9-1. To use the z-transform and the DTFT to solve signals and systems problems, it is essential to become so familiar with these properties that Table 9-1 no longer needs to be searched as a reference.

9-5.2 The z-Transform and the DFT

In Chapter 8, we defined the discrete Fourier transform (DFT) for finite-length sequences and showed that it is identical to frequency samples of the discrete-time Fourier transform (DTFT). Not surprisingly, given the relationship between the z-transform and the DTFT, there is also a direct relationship between the z-transform of a length-N sequence and its N-point DFT. This is easily seen by placing the relevant equations side by side as in

Table 9-1 Shared properties of z-transforms and DTFTs.

Property	n-Domain	z-Domain	$\hat{\omega}$-Domain
Definition	$x[n]$	$X(z) = \sum_{n=-\infty}^{\infty} x[n]z^{-n}$	$X(e^{j\hat{\omega}}) = \sum_{n=-\infty}^{\infty} x[n]e^{-j\hat{\omega}n}$
Linearity	$ax_1[n] + bx_2[n]$	$aX_1(z) + bX_2(z)$	$aX_1(e^{j\hat{\omega}}) + bX_2(e^{j\hat{\omega}})$
Time delay	$x[n - n_0]$	$z^{-n_0}X(z)$	$e^{-j\hat{\omega}n_0}X(e^{j\hat{\omega}})$
Convolution	$h[n] * x[n]$	$H(z)X(z)$	$H(e^{j\hat{\omega}})X(e^{j\hat{\omega}})$
Cascade LTI	$h_1[n] * h_2[n]$	$H_1(z)H_2(z)$	$H_1(e^{j\hat{\omega}})H_2(e^{j\hat{\omega}})$

***z*-Domain**		**DFT**
$X(z) = \sum_{n=0}^{N-1} x[n]z^{-n}$	$\xrightarrow{z=e^{j(2\pi/N)k}}$	$X[k] = \sum_{n=0}^{N-1} x[n]e^{-j(2\pi/N)kn}$

From (9.26), we have the general relation

$$X[k] = X(e^{j(2\pi/N)k}) = X(z)\Big|_{z=e^{j(2\pi/N)k}} \qquad k = 0, 1, \ldots, N-1 \tag{9.27}$$

which states that the DFT consists of samples of $X(z)$ evaluated at N points on the unit circle. Note that in representing the transition from the z-transform to the DFT, we have used the notation $\xrightarrow{z=e^{j(2\pi/N)k}}$, which has a unidirectional arrow to indicate that it may not be possible to go from the DFT values uniquely back to the z-transform polynomial $X(z)$ by reversing the substitution. This is because the DFT is a set of N complex numbers obtained directly by computation from a length-N sequence $x[n]$. If we know that the length of the underlying signal is N or less, then the N DFT coefficients would be sufficient to recover the signal and its z-transform.

Figure 9-7 shows the z-plane and the unit circle with gray dots corresponding to the values of z where the z-transform is sampled when we evaluate (9.27), assuming $N = 8$, and smaller colored dots for the sample locations when $N = 60$. Note that the points are all on the unit circle at a regular angular spacing of $(2\pi/N)$. This figure helps to make three things clear about the DFT. First, the figure shows that the set of DFT coefficients $X[k]$ is a periodic sequence having period N, since if we let k go outside the range $0 \le k \le N-1$, the z-plane points $z = e^{j2\pi k/N}$ repeat around the unit circle. Second, the DFT value for $k = 0$ corresponds to DC (i.e., $\hat{\omega} = 0$ and $z = 1$), while $k = N/2$ (assuming N is even)

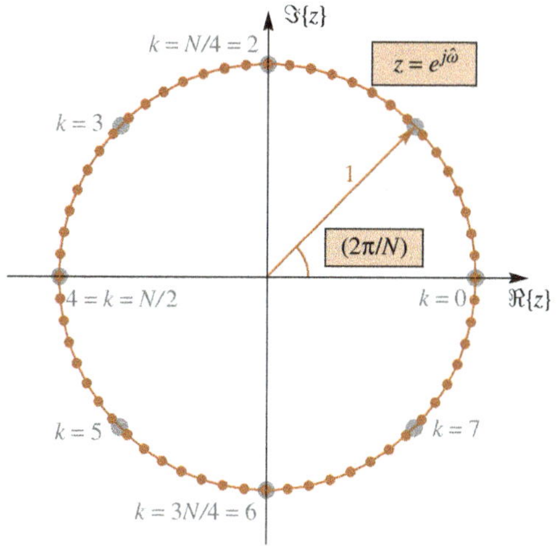

Figure 9-7 The complex z-plane including the unit circle, where $z = e^{j\hat{\omega}}$. The gray dots on the unit circle are locations where values of $X(z)$ are evaluated by the $N = 8$-point DFT, i.e., at the points $z_k = e^{j(2\pi/8)k}$ with $k = 0, 1, \ldots, 7$. The orange dots are locations where values of $X(z)$ are evaluated by a 60-point DFT.

corresponds to the frequency $\hat{\omega} = \pi$ and $z = -1$. Finally, if we compare the $N = 60$ case to the $N = 8$ case, we can say something about zero-padding. Suppose that $x[n]$ is a sequence of length 8, then its z-transform is a 7th-degree polynomial, and the 8-point DFT of $x[n]$ is the sampled z-transform

$$X_8[k] = \left(\sum_{n=0}^{7} x[n]z^{-n}\right)\Bigg|_{z=e^{j(2\pi/8)k}} \qquad k = 0, 1, \ldots, 7 \tag{9.28}$$

These are the samples at the eight gray dots in Fig. 9-7. If we want more frequency samples, say $N = 60$, then we can "zero-pad" the 8-point sequence with 52 zeros and take a 60-point DFT. The result is to sample the *same* $X(z)$, but at a finer angular spacing:

$$X_{60}[k] = \left(\sum_{n=0}^{7} x[n]z^{-n}\right)\Bigg|_{z=e^{j(2\pi/60)k}} \qquad k = 0, 1, \ldots, 59 \tag{9.29}$$

These are the samples of $X(z)$ at the 60 orange dots in Fig. 9-7. It is interesting to note that the eight gray dots are not a subset of the 60 colored dots, which means that zero padding will not always interpolate new frequency samples while keeping the existing ones. Finally, we note that choosing $N \gg 8$ will increase the density of the dots in Fig. 9-7, and in the limit as $N \to \infty$, for most practical purposes, we will have evaluated the z-transform everywhere on the unit circle, which means that we have computed the DTFT.

9-6 The Zeros and Poles of $H(z)$

The roots of the z-transform polynomials play a key role in simplifying the use of the system function $H(z)$ for problem solving. The product form of $H(z)$ given in (9.20) shows that $H(z)$ is a product of first-order factors $(1 - z_k z^{-1})$. For example, in Example 9-4 the system function was factored as

$$H(z) = 1 - 2z^{-1} + 2z^{-2} - z^{-3} = (1 - 1z^{-1})(1 - e^{j\pi/3}z^{-1})(1 - e^{-j\pi/3}z^{-1}) \quad (9.30a)$$

If we use the identity $(1 - z_k z^{-1}) = (z - z_k)/z$, then $H(z)$ also can be written as

$$H(z) = \frac{(z-1)(z - e^{j\pi/3})(z - e^{-j\pi/3})}{z^3} \quad (9.30b)$$

The roots of $H(z)$ are locations where $H(z) = 0$, so these locations, which come from the factors in the numerator of (9.30b), are called the ***zeros of the system function***. The factored form in (9.30b) shows clearly that the zeros are located at $z_1 = 1$, $z_2 = e^{j\pi/3}$, and $z_3 = e^{-j\pi/3}$ in the z-plane, but notice that the same zeros are evident in (9.30a) as coefficients of z^{-1}. The zeros of $H(z)$ can be complex as is the case in (9.30b).

Equation (9.30b) also shows that $H(z) \to \infty$ when $z \to 0$. Values of z for which $H(z)$ is infinite (and might be undefined) are called ***poles*** of $H(z)$. In (9.30b) we say that the denominator term z^3 represents three poles at $z = 0$, or equivalently that $H(z)$ has a third-order pole at $z = 0$.

When $H(z)$ contains only negative powers of z, it is usually most convenient to use (9.30a), since the negative powers of z have a direct correspondence to the difference equation and the impulse response. Furthermore, the factored form in (9.30a) does show explicitly that the zeros are located at $z_1 = 1$, $z_2 = e^{j\pi/3}$, and $z_3 = e^{-j\pi/3}$ in the z-plane, and less directly, that each factor has a pole at $z = \infty$. This factored form with negative powers of z is the most commonly used factorization in z-transform theory for signal processing. Initially, using the negative powers of z can be confusing, but there are long-term benefits to this notation.

9-6.1 Pole-Zero Plot

A graphical presentation of the zeros and poles makes it easy to visualize their locations relative to each other, and their positions relative to the unit circle, so we plot the zeros and poles of $H(z)$ as points in the complex z-plane. The plot in Fig. 9-8 shows the three zeros and the three poles for Example 9-7. Such a plot is called a ***pole-zero plot***. This plot was generated in MATLAB using the `zplane` function.[7] Each zero location is denoted

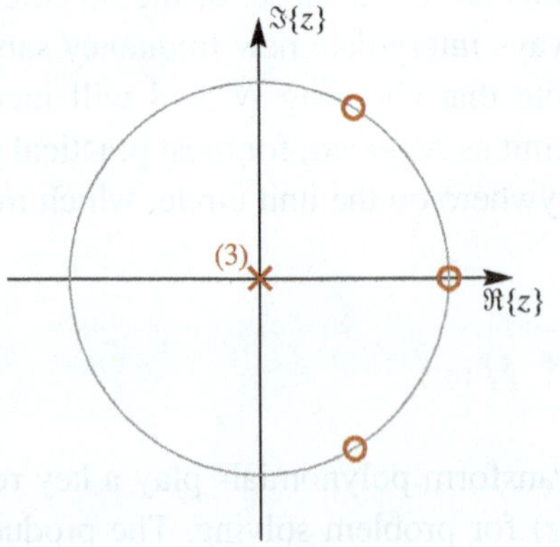

Figure 9-8 Zeros and poles of $H(z)$ marked in a z-plane that includes the unit circle.

[7] In the *DSP-First* Toolbox, there is a function called `zzplane` that makes pole-zero plots.

by a small circle, and the three repeated poles at $z = 0$ are indicated by a single **x** with a numeral 3 beside it. In general, for non-repeated poles the **x** symbol alone marks the location of each pole. Since the unit circle is where $H(z)$ is evaluated to obtain the DTFT of the LTI system whose system function is $H(z)$, it is also shown for reference as a gray circle in Fig. 9-8.

9-6.2 Significance of the Zeros of $H(z)$

The z-domain zeros can provide a link between the time-domain response of a system and the system function. In Section 9-6, we showed that the zeros of a polynomial system function are sufficient to determine $H(z)$ except for a constant multiplier. The system function determines the FIR difference equation of the filter because the polynomial coefficients of $H(z)$ are also the coefficients of the FIR difference equation. The difference equation (or convolution) is the computational method that turns an input $x[n]$ into the corresponding output $y[n]$. However, we can find certain inputs where knowledge of the zero locations is sufficient to make a precise statement about the output without actually using the difference equation to compute $y[n]$. Such signals are of the form $x[n] = z_0^n$ for all n, where the subscript signifies that z_0 is a fixed complex number. In this case, the output is

$$y[n] = \sum_{k=0}^{M} b_k\, x[n-k] \tag{9.31a}$$

$$y[n] = \sum_{k=0}^{M} b_k z_0^{n-k} \tag{9.31b}$$

$$y[n] = \sum_{k=0}^{M} b_k z_0^n z^{-k} = \overbrace{\left(\sum_{k=0}^{M} b_k z_0^{-k}\right)}^{=H(z_0)} z_0^n \tag{9.31c}$$

The quantity $H(z_0)$ is a complex constant obtained by evaluating $H(z)$ at $z = z_0$. The value of $H(z_0)$, through complex multiplication, causes a magnitude and phase change of the input signal z_0^n. A very interesting case occurs when z_0 is one of the zeros of $H(z)$. Then $H(z_0) = 0$ so the output from $x[n] = z_0^n$ is zero. This zero-output case is called ***nulling***.

EXAMPLE 9-10 Nulling Signals with Zeros

The system function studied in Example 9-7 is $H(z) = 1 - 2z^{-1} + 2z^{-2} - z^{-3}$, and its zeros are

$$
\begin{aligned}
z_1 &= 1 && \Rightarrow H(z_1) = 0 \\
z_2 &= \tfrac{1}{2} + j\tfrac{1}{2}\sqrt{3} = 1e^{j\pi/3} && \Rightarrow H(z_2) = 0 \\
z_3 &= \tfrac{1}{2} - j\tfrac{1}{2}\sqrt{3} = 1e^{-j\pi/3} && \Rightarrow H(z_3) = 0
\end{aligned}
$$

As shown in Fig. 9-8, these zeros all lie on the unit circle. Since the angle of a vector from the origin to a point on the unit circle corresponds to $\hat{\omega}$ in the frequency domain, complex sinusoids with frequencies 0, $\pi/3$, and $-\pi/3$ are "nulled out" by the system. That is, the output resulting from each of the following three signals will be zero:

$$x_1[n] = (z_1)^n = 1$$

$$x_2[n] = (z_2)^n = e^{j(\pi/3)n}$$

$$x_3[n] = (z_3)^n = e^{-j(\pi/3)n}$$

For example, when the input is $x_2[n]$, the output is $y_2[n] = H(z_2)(z_2)^n = 0$.

DEMO
PeZ GUI

As illustrated by this example, zeros of the system function that lie on the unit circle are very special. Their angles correspond to frequencies at which the gain of the system is zero (i.e., where the DTFT (or frequency response) is zero). Thus, complex sinusoids at those frequencies are blocked, or "nulled" by the system.

EXERCISE 9.6 Double check the fact that the input $x_2[n]$ determined in Example 9-10 produces an output that is zero everywhere by substituting this signal into the difference equation $y[n] = x[n] - 2x[n-1] + 2x[n-2] - x[n-3]$ to show that the complex phasors cancel out for all values of n.
Also show that the filter nulls out signals such as $2\cos(\pi n/3)$, which is the sum of $x_2[n]$ and $x_3[n]$.

9-6.3 Nulling Filters

We have just shown that if the zeros of $H(z)$ lie on the unit circle, then specific complex exponential or sinusoidal input signals are nulled (i.e., removed, by the filter). Therefore, it should be possible to use this result in designing a practical FIR filter that can null an interfering sinusoidal input. Such capability is often needed to eliminate jamming signals in a radar or communications system. Similarly, interference from a 60-Hz power line could be eliminated by placing a null at the correct $\hat{\omega}$ frequency.

Zeros in the z-plane can remove only signals that have the special form $x[n] = z_0^n$. If we want to eliminate a sinusoidal input signal, then we actually have to remove two signals, because

$$x[n] = \cos(\hat{\omega}_0 n) = \tfrac{1}{2}(e^{j\hat{\omega}_0})^n + \tfrac{1}{2}(e^{-j\hat{\omega}_0})^n$$

Each complex exponential can be removed with a first-order FIR filter, and then the two filters would be cascaded to form the second-order nulling filter that removes the cosine. The first complex exponential signal is nulled by a filter with system function

$$H_1(z) = 1 - z_1 z^{-1}$$

where $z_1 = e^{j\hat{\omega}_0}$ because $H_1(z_1) = 0$ at $z = z_1$, that is,

$$H_1(z_1) = 1 - z_1(z_1)^{-1} = 1 - 1 = 0$$

Similarly, $H_2(z) = 1 - z_2 z^{-1}$ removes the second complex exponential if $z_2 = e^{-j\hat{\omega}_0}$. Then, we cascade these two filters to create the second-order nulling filter whose system function is the product

$$\begin{aligned} H(z) &= H_1(z)H_2(z) \\ &= (1 - z_1 z^{-1})(1 - z_2 z^{-1}) \\ &= 1 - (z_1 + z_2)z^{-1} + (z_1 z_2)z^{-2} \\ &= 1 - (e^{j\hat{\omega}_0} + e^{-j\hat{\omega}_0})z^{-1} + (e^{j\hat{\omega}_0}e^{-j\hat{\omega}_0})z^{-2} \\ &= 1 - 2\cos(\hat{\omega}_0)z^{-1} + z^{-2} \end{aligned}$$

This second-order FIR filter has two zeros, at $z_1 = e^{j\hat{\omega}_0}$ and $z_2 = e^{-j\hat{\omega}_0}$, along with two poles at the origin. Figure 9-9 shows a special case where the two zeros remove components at $z = e^{\pm j\pi/4}$. From these zeros the numerical values for the coefficients of $H(z)$ are

$$H(z) = 1 - 2\cos(\pi/4)z^{-1} + z^{-2}$$

Thus the nulling filter that removes the signal $\cos(0.25\pi n)$ from its input is the FIR filter whose difference equation is

$$y[n] = x[n] - \sqrt{2}\,x[n-1] + x[n-2] \tag{9.32}$$

9-6.4 Graphical Relation Between z and $\hat{\omega}$

The equation $z = e^{j\hat{\omega}}$ provides the link between the z-domain and the $\hat{\omega}$-domain. As we have shown in (9.26), the DTFT is obtained by evaluating the system function on the unit circle of the z-plane. The graphical interpretation is that we visualize (or draw) the 3-D

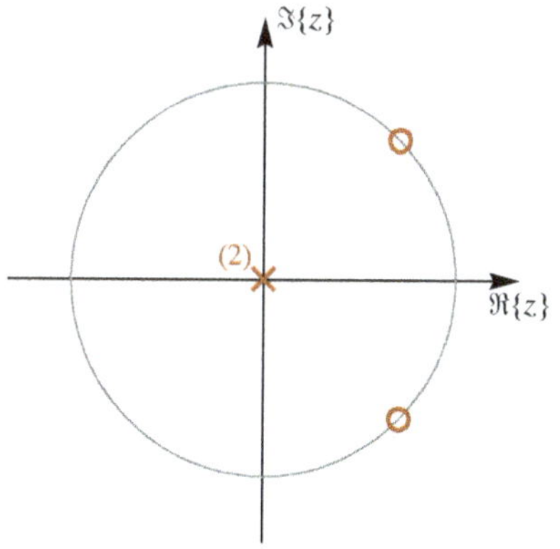

Figure 9-9 Zeros on unit circle for second-order nulling filter to remove complex exponential components at $\hat{\omega}_0 = \pm\pi/4$.

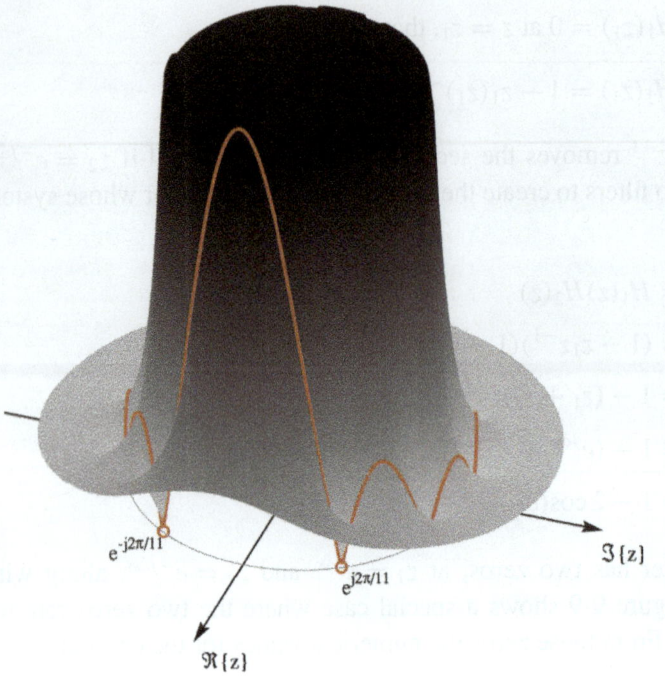

Figure 9-10 z-Transform system function for an FIR filter evaluated over the region $\{|z| \leq 1.5\}$ of the z-plane that includes the unit circle. Values along the unit circle are shown as the orange line where the DTFT (magnitude) is evaluated. The view is from the first quadrant, so the point $z = 1$ is in the foreground on the left. The height of the plot is clipped at $|H(z)| \leq 13$. Zeros on the unit circle at $z = e^{\pm j2\pi/11}$ are marked with small orange circles.

plot of $|H(z)|$, and then restrict our attention to the unit circle to get the DTFT plot of $|H(e^{j\hat{\omega}})|$. The poles and zeros play an important role since they tell us where the 3-D plot is either infinite or equal to zero. In a 3-D magnitude plot, the area around a zero forms a "valley," while the area around a pole is an extremely high "peak." When we restrict our attention to the unit circle, we are interested in the proximity of valleys and peaks to the path in 3-D that is traced out along the unit circle as in Figs. 9-5 and 9-10.

By considering only the pole-zero plot of the system function, it is generally possible to skip the 3-D plot and just imagine how a 3-D plot of $|H(z)|$ restricted to the unit circle would form the peaks and nulls of the plot of $|H(e^{j\hat{\omega}})|$ versus $\hat{\omega}$. As an example, we show in Fig. 9-10 a 3-D plot obtained by evaluating the z-transform magnitude $|H(z)|$ over a region of the z-plane that includes the unit circle as well as values both inside and outside the unit circle. The system in this case is an 11-point running sum FIR filter, where the coefficients are all equal to one. The frequency response of the 11-point running-sum filter was discussed in detail in Section 6-7 on p. 234, and the plot of its magnitude response is repeated in Fig. 9-11. Note that there are ten zeros at $\hat{\omega} = 2\pi k/11$,

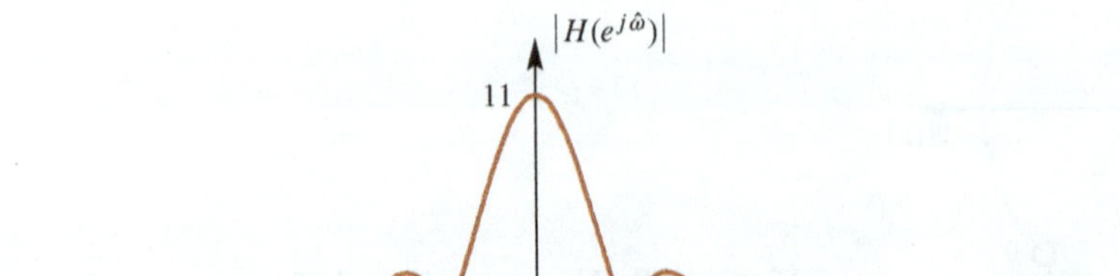

Figure 9-11 Frequency response (magnitude only) for the 11-point running sum. These are the values of $H(z)$ along the unit circle in the z-plane. There are 10 zeros spread out uniformly along the frequency axis except for $\hat{\omega} = 0$.

for $k = \pm1, \pm2, \pm3, \pm4, \pm5$, so we expect to have ten zeros of $|H(z)|$ around the unit circle in Fig. 9-10, where two of them are easy to see.

The system function for the 11-point running-sum filter is

$$H(z) = \sum_{k=0}^{10} z^{-k} \tag{9.33}$$

From the frequency response plot we know the angles of the zeros on the unit circle, so the polynomial in (9.33) can be represented as the product of ten factors

$$H(z) = (1 - e^{j2\pi/11}z^{-1})(1 - e^{j4\pi/11}z^{-1})\cdots(1 - e^{j18\pi/11}z^{-1})(1 - e^{j20\pi/11}z^{-1}) \tag{9.34}$$

Recall that each factor of the form $(1 - e^{j2\pi k/11}z^{-1})$ represents a zero at $z = e^{j2\pi k/11}$ and a pole at $z = 0$. Thus, (9.34) displays the 10 zeros of $H(z)$ at $z = e^{j2\pi k/11}$, for $k = 1, 2, \ldots, 10$ along with 10 poles at $z = 0$. In Section 9-7, we will use the roots of unity to factor $H(z)$ and confirm that the ten zeros of $H(z)$ in (9.33) are on the unit circle at angles $\hat{\omega} = 2\pi k/11$, for $k = 1, 2, \ldots, 10$.

In the 3-D magnitude plot of Fig. 9-10, the ten zeros pin down the 3-D plot around the unit circle. Inside the unit circle, the values of $H(z)$ become very large owing to the ten poles at $z = 0$. The frequency response magnitude $|H(e^{j\hat{\omega}})|$ is obtained by selecting the values of the z-transform along the unit circle in Fig. 9-10, which conforms to the plot of $|H(e^{j\hat{\omega}})|$ versus $\hat{\omega}$ given in Fig. 9-11. The shape of the frequency response can be explained in terms of the zero locations shown in Fig. 9-12 by recognizing that the ten poles at $z = 0$ push $|H(e^{j\hat{\omega}})|$ up, but the zeros along the unit circle make $H(e^{j\hat{\omega}}) = 0$ at regular intervals except for the region near $\hat{\omega} = 0$ where the regular pattern of zeros is broken by the missing zero at $z = 1$. The unit circle values follow the ups and downs of $|H(z)|$ as $\hat{\omega}$ goes from $-\pi$ to $+\pi$ with $|z| = 1$.

This example illustrates that an intuitive picture of the frequency response of an LTI system can be visualized from a plot of the poles and zeros of the system function $H(z)$. We simply need to remember that a pole "pushes up" the frequency response and a zero "pulls it down." Furthermore, a zero on the unit circle forces the frequency response to be exactly zero at the frequency corresponding to the angle of the zero.

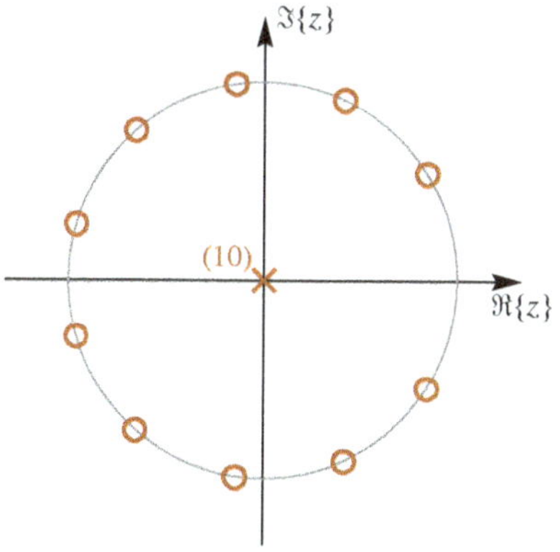

Figure 9-12 Zero and pole distribution for the 11-point running sum. There are ten zeros spread out uniformly around the unit circle, except for $z = 1$, and ten poles at the origin.

9-6.5 Three-Domain Movies

DEMO
3-Domain Movies

With the z-transform, DTFT and impulse response (or difference equation) we now have three ways to describe an FIR LTI system, and these three methods involve three different domains characterized with the variables z, $\hat{\omega}$, and n. One vital skill for working with LTI systems is the ability to visualize all three domains simultaneously for certain classes of filters. We developed such a visualization as a movie that presents plots for all three domains and then shows how the plots change as one parameter is varied. A typical example shows how the frequency response and impulse response change as one or two zeros of the FIR filter are varied. Figure 9-13 shows a frame from a 3-Domain Movie that illustrates the tie between the z, $\hat{\omega}$, and n domains. The screen layout has a pole-zero plot and $H(z)$ for the z domain, magnitude and phase plots of the frequency response for the $\hat{\omega}$ domain, and a stem plot of $h[n]$ for the n domain. The control for these movies is the z domain because it is easy to think about moving the zeros and watching what happens in the $\hat{\omega}$ and n domains. In the example of Fig. 9-13, the movie sequence moves the two complex-conjugate zeros around the unit circle by changing their angles. At the same time, the coefficients of $H(z)$ are updated, along with the plots of $h[n]$ and $H(e^{j\hat{\omega}})$.

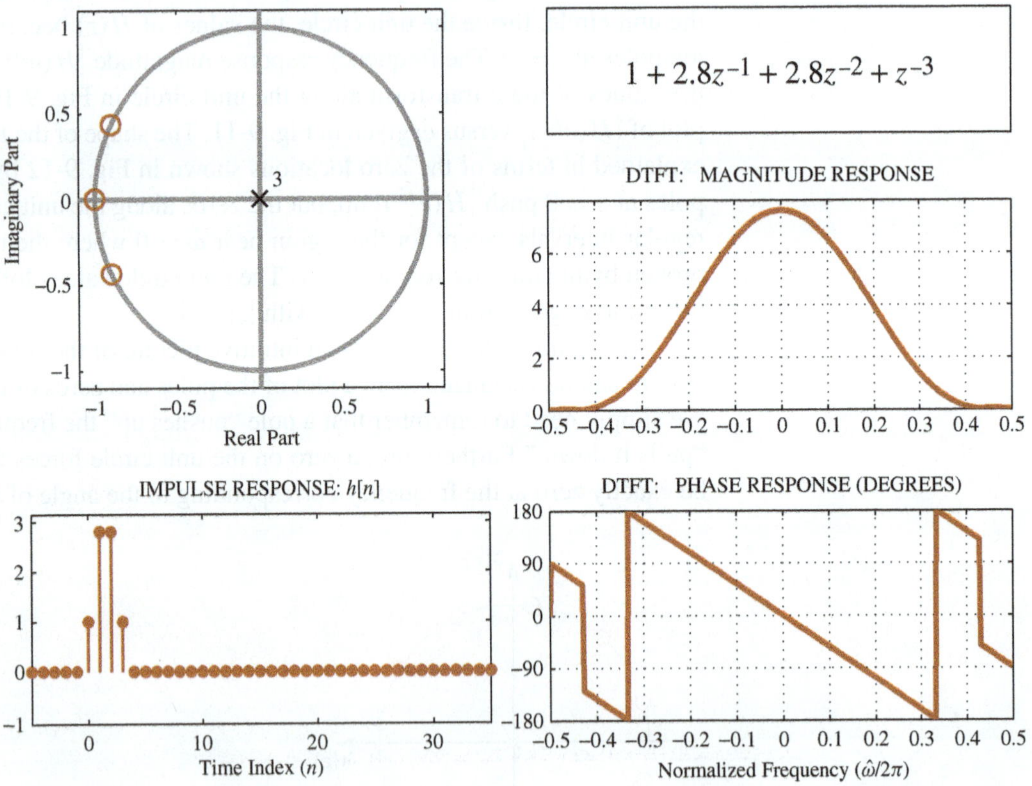

Figure 9-13 One frame of a *3-Domain Movie* showing the different domains: time domain (n), frequency domain $(\hat{\omega})$, the z-transform domain $H(z)$, and a pole-zero plot. In this movie, the conjugate zero pair moves around on the unit circle.

DEMO
PeZ GUI, Three Domains—FIR

Finally, as another way to show the same interrelationships, we mention that there is also a 3-Domain GUI (called `PeZ`) in which the user can move the zeros by clicking and dragging with a pointing tool, and simultaneously see changes in the frequency response and impulse response (see Fig. 10-28 on p. 473 for more details).

9-7 Simple Filters

Now that we have new knowledge of the tie between the z and $\hat{\omega}$ domains, we can exploit that knowledge to design FIR filters with desirable frequency response characteristics. In this section, we examine a special class of FIR filters that are all close relatives of the running-sum filter.

9-7.1 Generalize the L-Point Running-Sum Filter

A generalization of the L-point running-sum filter is

$$y[n] = \sum_{k=0}^{L-1} a^k x[n-k] \tag{9.35}$$

Note that the running sum is obtained when $a = 1$. It is still easy to determine that the system function for (9.35) is

$$H(z) = \sum_{k=0}^{L-1} a^k z^{-k} \tag{9.36}$$

The filter order is $M = L - 1$ which is also the degree of the $H(z)$ polynomial. Recalling the formula (6.24) from p. 234 for the sum of L terms of a geometric series, $H(z)$ can be rewritten in two different rational forms

$$H(z) = \sum_{k=0}^{L-1} a^k z^{-k} = \frac{1 - a^L z^{-L}}{1 - az^{-1}} = \frac{z^L - a^L}{z^{L-1}(z-a)} \tag{9.37}$$

The second form for $H(z)$ is a rational function where the numerator polynomial is $z^L - a^L$ and the denominator is $z^{L-1}(z-a)$. The zeros of $H(z)$ are determined by the roots of the numerator polynomial, which are the values of z such that

$$z^L - a^L = 0 \qquad \Longrightarrow z^L = a^L e^{j2\pi k} \tag{9.38}$$

because $e^{j2\pi k} = 1$ for k, an integer. Then we can take the Lth root of both sides to obtain the L roots

$$z = ae^{j2\pi k/L} \quad \text{for } k = 0, 1, 2, \ldots, L-1 \tag{9.39}$$

Notice that when $a = 1$ the values in (9.39) satisfy the equation $z^L = 1$, so they are *the Lth roots of unity*. In general, all the numerator roots (9.39) lie on a circle of radius $|a|$. The roots of the denominator in (9.37), which are either $z = 0$ (of order $L - 1$) or $z = a$, would normally be the poles of $H(z)$. However, since one of the numerator roots is $z = a$ (when $k = 0$ in (9.39)) the numerator has a term $(z - a)$ that cancels the corresponding term of the denominator, so that only the denominator term z^{L-1} actually causes the $L - 1$ poles of $H(z)$. Therefore, it follows that $H(z)$ has degree $L - 1$ and can be expressed in the factored form

$$H(z) = \sum_{k=0}^{L-1} a^k z^{-k} = \prod_{k=1}^{L-1}\left(1 - ae^{j2\pi k/L}z^{-1}\right) \tag{9.40}$$

Notice that the product (9.40) does not include the $k = 0$ term which would be the cancelled root at $z = a$.

EXAMPLE 9-11 $H(e^{j\hat{\omega}})$ from $H(z)$

The 11-point running-sum filter ($L = 11$) is the special case where $a = 1$ in (9.37), so one way to write its system function is

$$H(z) = \sum_{k=0}^{10} z^{-k} = \frac{1 - z^{-11}}{1 - z^{-1}} \tag{9.41}$$

Then the frequency response is obtained by substituting $e^{j\hat{\omega}}$ for z, followed by reducing the rational form to a Dirichlet form as was done in (6.25) and (6.26) of Chapter 6

$$H(e^{j\hat{\omega}}) = H(z)\Big|_{z=e^{j\hat{\omega}}} = \frac{1 - e^{-j11\hat{\omega}}}{1 - e^{-j\hat{\omega}}} = D_{11}(\hat{\omega})e^{-j5\hat{\omega}} \tag{9.42}$$

The pole-zero diagram for this case was shown previously in Fig. 9-12, along with the corresponding frequency response in Fig. 9-11. The 11 factors of the numerator polynomial $(z^{11} - 1)$ in (9.41) include a term $(z - 1)$ for the zero at $z = 1$ which is canceled by the corresponding term $(z - 1)$ in the denominator. This explains why we have only ten zeros around the unit circle with the gap at $z = 1$. The ten zeros around the unit circle in Fig. 9-12 show up as zeros of $H(e^{j\hat{\omega}})$ along the $\hat{\omega}$ axis in Fig. 9-11 at integer multiples of $2\pi/11$, except for $\hat{\omega} = 0$. Note that it is the gap at $z = 1$ that allows the frequency response to be larger at $\hat{\omega} = 0$. The other zeros around the unit circle keep $H(e^{j\hat{\omega}})$ small, thereby creating the "lowpass" filter frequency response shown in Fig. 9-11.

EXERCISE 9.7

For the Mth-order FIR filter whose impulse response is

$$h[n] = \sum_{k=0}^{M} a^k \delta[n - k] \tag{9.43}$$

the system function in (9.37) or (9.40) has zeros off the unit circle when $a \neq 1$. From the pole-zero plot of $H(z)$ shown in Fig. 9-14, determine the value of a and also the filter length L.

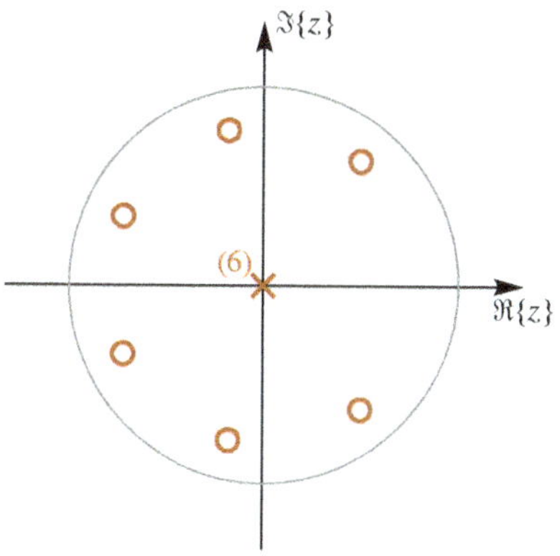

Figure 9-14 Pole-zero distribution for the a^k filter in (9.43) and (9.37) with $M = L - 1 = 6$. The six zeros lie on a circle of radius $a = 0.8$.

9-7.2 A Complex Bandpass Filter

Now we have a new insight that tells us how to control the frequency response of an FIR filter by placing equally spaced zeros on the unit circle to make stopbands, but leaving out one zero to make a passband. This viewpoint makes it easy to create other FIR filters where we control the location of the passband. If we move the passband to a center frequency away from $\hat{\omega} = 0$, and the filter passes a small band of frequency components, we call the filter a *bandpass* filter, or BPF.

One way to move the passband is to use the frequency-shifting property of the DTFT

$$h[n]e^{j\hat{\omega}_0 n} \quad \stackrel{\text{DTFT}}{\longleftrightarrow} \quad H(e^{j(\hat{\omega}-\hat{\omega}_0)})$$

For a simple BPF we can use the running-sum impulse response for $h[n]$, which is all ones, and pick $\hat{\omega}_0$ as the center frequency. The z-transform of $h[n]e^{j\hat{\omega}_0 n}$ is obtained by using (9.40) with $a = e^{j\hat{\omega}_0}$. The result is

$$H(z) = \sum_{n=0}^{L-1} 1 e^{j\hat{\omega}_0 n} z^{-n} = \prod_{k=1}^{L-1} (1 - e^{j\hat{\omega}_0} e^{j2\pi k/L} z^{-1}) \tag{9.44}$$

Therefore, the $L - 1$ zeros are at $z = e^{j(\hat{\omega}_0 + 2\pi k/L)}$ for $k = 1, 2, \ldots L - 1$, which are still on the unit circle, equally spaced but missing the zero at $z = e^{j\hat{\omega}_0}$. An example is shown in Fig. 9-15 for $\hat{\omega}_0 = 5\pi/L$ and $L = 11$. We expect that the passband of the filter will move from the frequency interval around DC ($z = 1$) to a like interval centered around $\hat{\omega}_0$. Figure 9-16 confirms our intuition, because the frequency location of the peak is $\hat{\omega} = 5\pi/11$. This filter is a *bandpass filter*, since $|H(e^{j\hat{\omega}})|$ is much smaller for frequencies outside the narrow passband around $\hat{\omega} = 5\pi/11$.

9-7.3 A Bandpass Filter with Real Coefficients

The previous strategy of frequency shifting gives complex-valued filter coefficients $b_k = a^k = e^{j\hat{\omega}_0 k}$ which might be undesirable if we want to filter real signals. However, there is

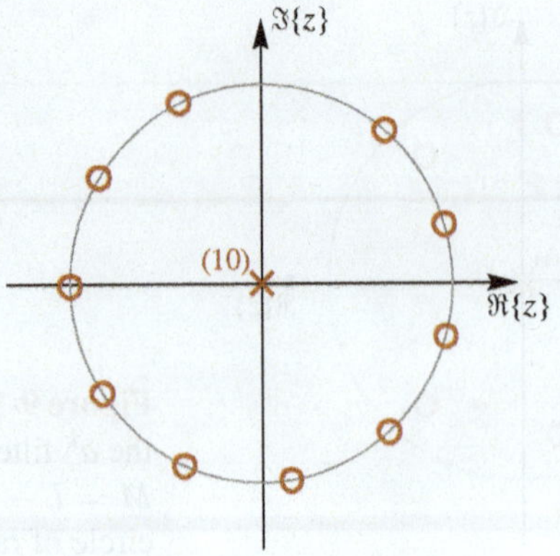

Figure 9-15 Zero and pole distribution for an 11-point complex BPF with $\hat{\omega}_0 = 5\pi/11$. The zero at angle $5\pi/11$ is missing, but there are ten zeros spread out at angles spaced by $2\pi/11$ around the unit circle.

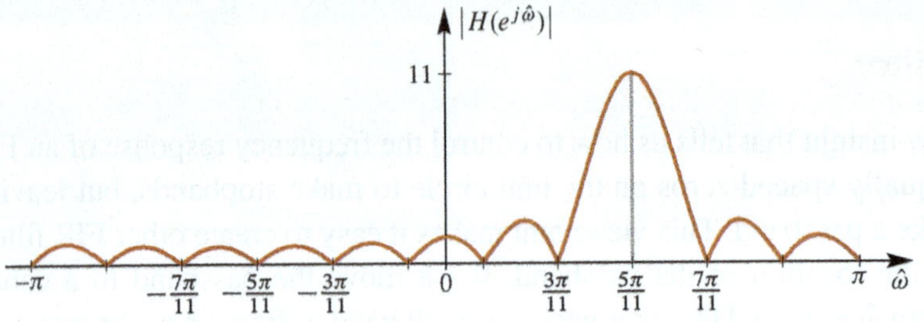

Figure 9-16 Frequency response (magnitude only) for the 11-point complex BPF with $\hat{\omega}_0 = 5\pi/11$. These are the values along the unit circle in the z-plane. There are ten zeros spaced by $2\pi/11$ along the frequency axis.

a second strategy that yields real filter coefficients—we can use a cosine to do frequency shifting (up and down). Thus the filter coefficients are now

$$b_k = \cos(\hat{\omega}_0(k - k_m)) \quad \text{for } k = 0, 1, \ldots, L-1$$

where k_m is the midpoint of the impulse response, $k_m = \frac{1}{2}(L-1)$. Any length-L section of the cosine could be used, but this choice guarantees that the linear phase condition (to be discussed in Section 9-9) is satisfied.

With these real filter coefficients, this new BPF can be written as the sum of two complex BPFs. By expanding the cosine coefficients of z^{-k} in terms of complex exponentials we obtain

$$H(z) = \sum_{k=0}^{L-1} \left(\cos(\hat{\omega}_0(k - k_m))\right) z^{-k} \tag{9.45a}$$

$$= \sum_{k=0}^{L-1} \left(\tfrac{1}{2}e^{j\hat{\omega}_0(k-k_m)} + \tfrac{1}{2}e^{-j\hat{\omega}_0(k-k_m)}\right) z^{-k}$$

$$= \tfrac{1}{2}e^{-j\hat{\omega}_0 k_m} \sum_{k=0}^{L-1} e^{j\hat{\omega}_0 k} z^{-k} + \tfrac{1}{2}e^{j\hat{\omega}_0 k_m} \sum_{k=0}^{L-1} e^{-j\hat{\omega}_0 k} z^{-k}$$

$$= H_1(z) + H_2(z) \tag{9.45b}$$

Thus we expect that the frequency response will have two "Dirichlet" peaks, which is the case in Fig. 9-17. Note that k_m in the exponent only contributes to the phases of $H_1(z)$ and

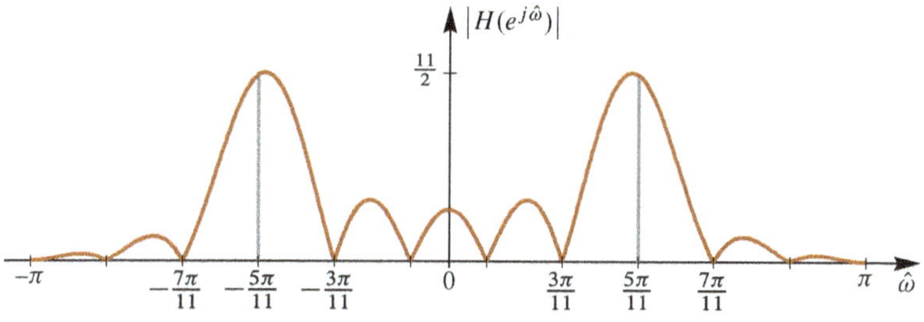

Figure 9-17 Frequency response (magnitude only) for the 11-point real BPF exhibits two passbands near $\hat{\omega} = \pm 5\pi/11 = \pm\hat{\omega}_0$. Notice that the peak magnitudes are not located exactly at $\hat{\omega} = \pm 5\pi/11 = 0.4545\pi$ since the sidelobes of one Dirichlet form add to the other peak to produce $H(e^{j\hat{\omega}})$. The peak value is 5.569 at $\hat{\omega} = \pm 0.439\pi$.

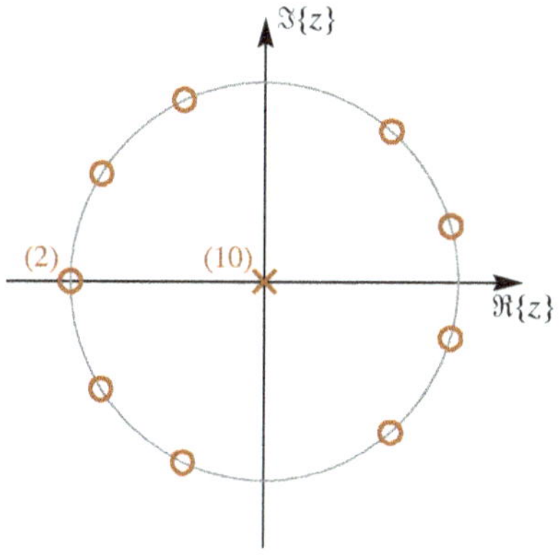

Figure 9-18 Pole-zero distribution for the 11-point real BPF shown in Fig 9-17. Of the original 11 roots of unity, two have been dropped off the unit circle at the angles $\pm 5\pi/11$, but there is now a double zero on the real axis at $z = -1$. There are ten poles at the origin.

$H_2(z)$. In fact, the sum in (9.45b) is the addition of complex quantities, not magnitudes, so the phase matters when doing the sum. There is no problem in constructing a frequency response plot in MATLAB which is shown in Fig. 9-17. In addition, the pole-zero diagram in Fig. 9-18 can be made with the `zplane` function in MATLAB.

9-8 Practical Bandpass Filter Design

Although much better filters can be designed by more sophisticated methods, the example of BPF design discussed in Section 9-7 is a useful illustration of the power of the z-transform to simplify the analysis of such problems. Its characterization of a filter by the zeros (and poles) of $H(z)$ is used frequently in filter design and implementation. The underlying reason is that the z-transform converts difficult problems involving convolution and frequency response into simple algebraic ideas based on multiplying and factoring polynomials. Thus, skills with basic algebra become essential everyday tools in design.

As a final example of FIR filters, we present a high-order FIR filter that has been designed by a computer-aided filter-design program. Most digital filter design is carried out by software tools that permit much more sophisticated control of the passband and stopband characteristics than we were able to achieve with the simple analysis of Section 9-7. Design programs such as `firpm` and `fir1` can be found in the MATLAB software augmented with the Signal Processing Toolbox. The discussion in Section 7-4 described the underlying methodology of the window method for FIR filter design which

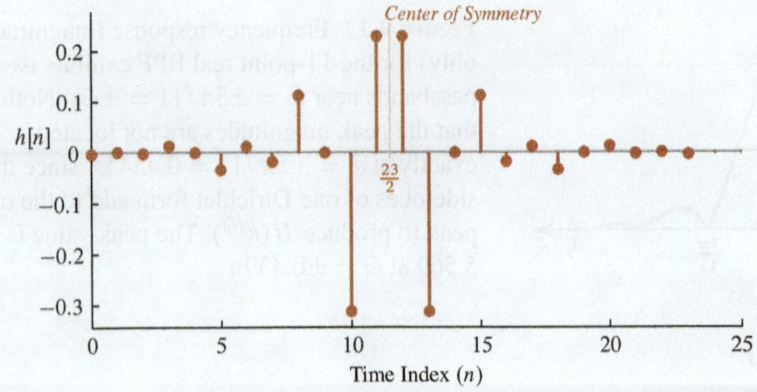

Figure 9-19 Impulse response for the 24-point BPF designed with MATLAB's `fir1` function. These are the FIR filter coefficients $\{b_k\}$, $k = 0, 1, 2, \ldots, 23$, which are needed in the difference equation implementation.

underlies `fir1`, so it is interesting to examine the output from the program `fir1` to get a notion of what can be achieved with a sophisticated design method. The software allows us to specify the frequency range for the passband, and then computes a good approximation to an ideal filter that has unity gain over the passband and zero gain in the stopband. An example is shown in Figs. 9-19, 9-20, and 9-21 for a length-24 FIR BPF.

The impulse response of the 24-point FIR BPF is shown in Fig. 9-19. Notice that the coefficients would be labeled $\{b_k\}$, from $k = 0, 1, 2, \ldots, 23$. The plot shows an obvious symmetry about a midpoint at $n = 23/2$. Table 9-2 lists the values for the impulse response and filter coefficients. These are the coefficients used in the difference equation

$$y[n] = \sum_{k=0}^{23} b_k x[n-k]$$

to implement the filter whose frequency response is shown in Fig. 9-20. Notice the wide passbands in the frequency ranges corresponding to $2\pi(0.24) < |\hat{\omega}| < 2\pi(0.36)$.

Table 9-2 Filter coefficients of 24-point FIR bandpass.

$b_0 = -0.0108 = b_{23}$
$b_1 = 0.0037 = b_{22}$
$b_2 = -0.0052 = b_{21}$
$b_3 = 0.0300 = b_{20}$
$b_4 = 0.0000 = b_{19}$
$b_5 = -0.0526 = b_{18}$
$b_6 = 0.0164 = b_{17}$
$b_7 = -0.0219 = b_{16}$
$b_8 = 0.1275 = b_{15}$
$b_9 = 0.0000 = b_{14}$
$b_{10} = -0.3236 = b_{13}$
$b_{11} = 0.2330 = b_{12}$

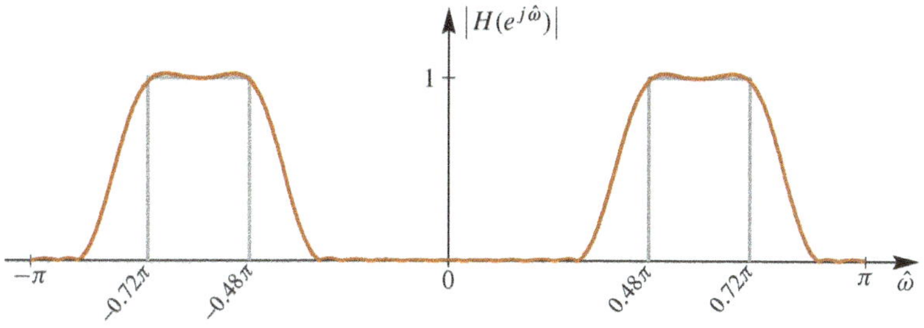

Figure 9-20 Frequency response (magnitude only) for the 24-point BPF. The passbands are the intervals $0.48\pi \leq |\hat{\omega}| \leq 0.72\pi$. The design was performed with `fir1(23,2*[0.2,0.4],kaw)`, where `kaw = kaiser(24,2.75)` is a Kaiser window.

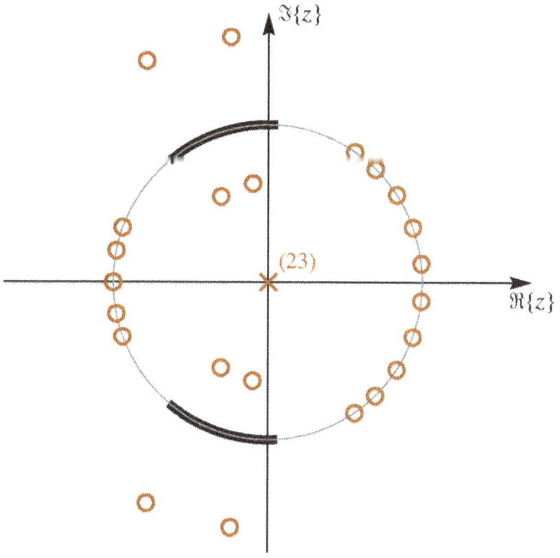

Figure 9-21 Zero and pole distribution for a 24-point BPF designed with MATLAB's `fir1` function. The section of unit circle corresponding to the passband of $H(e^{j\hat{\omega}})$ in Fig. 9-20 is outlined in black.

In these intervals, the gain of the filter deviates only slightly from one, so the amplitudes of sinusoidal input signals with these frequencies are not affected by the filter. We will see in Section 9-9 that this filter has linear phase, so these frequency components only experience delay.

Also note that in the regions $|\hat{\omega}| < 2\pi(0.16)$ and $2\pi(0.44) < |\hat{\omega}| < 2\pi(0.5)$ the gain is very nearly equal to zero. These regions are the "stopbands" of the filter, since the amplitudes of sinusoids with these frequencies are multiplied by a very small gain and thus blocked from appearing in the output.

Observe that the frequency response tapers smoothly from the passbands to the stopbands. In these transition regions, sinusoids are reduced in amplitude according to the gain shown in Fig. 9-20. In many applications, we would want such transition regions to be very narrow. Ideally, we might even want them to have zero width. While this is theoretically achievable, it comes with a high price, as discussed in Section 7-4.2.5. It turns out that for an FIR frequency selective (lowpass, bandpass, highpass) filter, the width of the transition region is inversely proportional to M, the order of the system function polynomial. The higher the order, the narrower the transition regions can be, and as $M \to \infty$, the transition regions shrink to zero. Unfortunately, increasing M also increases the amount of computation required to compute each sample of the output, so a trade-off is always required in any practical application.

Figure 9-21 shows the pole-zero plot of the FIR filter. Notice the distinctive pattern of locations of the zeros. In particular, note how the zeros off the unit circle seem to be grouped into sets of four zeros. Indeed, for each zero that is not on the unit circle, there are also zeros at the conjugate location, at the reciprocal location, and at the conjugate reciprocal location. These groups of four zeros are strategically placed automatically by the design process to form the passband of the filter. Similarly, the design process places zeros on (or close to) the unit circle to ensure that the gain of the filter is low in the stopband regions of the frequency axis. Also, note that all complex zeros appear in conjugate pairs, and since the system function is a 23rd order polynomial, there are 23 poles at $z = 0$. In Section 9-9, we will show that these properties of the pole-zero distribution are the direct result of the symmetry of the sequence of filter coefficients.

9-9 Properties of Linear-Phase Filters

The filter that was discussed in Section 9-8 is an example of a class of systems where the sequence of coefficients (impulse response) has symmetry of the form $b_k = b_{M-k}$, for $k = 0, 1, \ldots, M$. Such systems have a number of interesting properties that are easy to show in the z-domain representation.

9-9.1 The Linear-Phase Condition

FIR systems that have symmetric filter coefficients (and, therefore, symmetric impulse responses) have frequency responses with linear phase. An example that we have already studied is the L-point running sum, whose coefficients are all the same and therefore clearly satisfy the condition $b_k = b_{M-k}$, for $k = 0, 1, \ldots, M$. The example of Section 9-8 also has linear phase because it satisfies the same symmetry condition. To see why linear phase results from this symmetry, let us consider a simple example where the system function is of the form

$$H(z) = b_0 + b_1 z^{-1} + b_2 z^{-2} + b_1 z^{-3} + b_0 z^{-4} \tag{9.46}$$

Thus, $M = 4$ and the length of the sequence is $L = M + 1 = 5$ samples. After working out the frequency response for this special case, the generalization will be obvious. First, observe that we can write $H(z)$ as

$$H(z) = \left(b_0(z^2 + z^{-2}) + b_1(z^1 + z^{-1}) + b_2\right) z^{-2}$$

If M were greater, we would simply have more groups of factors of the form $(z^k + z^{-k})$. Now when we substitute $z = e^{j\hat{\omega}}$, each of these factors becomes a cosine term, that is,

$$H(e^{j\hat{\omega}}) = \left(2b_0 \cos(2\hat{\omega}) + 2b_1 \cos(\hat{\omega}) + b_2\right) e^{-j\hat{\omega}M/2}$$

In this specific example, we have shown that $H(e^{j\hat{\omega}})$ is of the form

$$H(e^{j\hat{\omega}}) = R(e^{j\hat{\omega}}) e^{-j\hat{\omega}M/2} \tag{9.47}$$

where, in this case, $M = 4$ and $R(e^{j\hat{\omega}})$ is the real function

$$R(e^{j\hat{\omega}}) = \big(2b_0 \cos(2\hat{\omega}) + 2b_1 \cos(\hat{\omega}) + b_2\big)$$

By following this style of analysis for the general case, it is easy to show that (9.47) holds whenever $b_k = b_{M-k}$, for $k = 0, 1, \ldots, M$. In the general case, it can be shown that the result depends on whether the integer M is even or odd

$$R(e^{j\hat{\omega}}) = \begin{cases} b_{\frac{M}{2}} + \sum\limits_{k=0}^{\frac{M-2}{2}} 2b_k \cos[(\frac{M}{2} - k)\hat{\omega}] & M \text{ even} \\ \sum\limits_{k=0}^{\frac{M-1}{2}} 2b_k \cos[(\frac{M}{2} - k)\hat{\omega}] & M \text{ odd} \end{cases} \tag{9.48}$$

Equation (9.47) shows that the frequency response of any symmetric filter has the form of a real amplitude function $R(e^{j\hat{\omega}})$ times a linear-phase factor $e^{-j\hat{\omega}M/2}$. The phase factor, as we have seen in Section 6-5.1 on p. 226, corresponds to a delay of $M/2$ samples. Thus, the analysis presented in Section 6-7 for the running-average filter is typical of what happens in the general symmetric FIR case. The main difference is that by choosing the filter coefficients carefully, as in Section 9-8, we can shape the function $R(e^{j\hat{\omega}})$ to have a much more selective frequency response.

9-9.2 Locations of the Zeros of FIR Linear-Phase Systems

If the filter coefficients satisfy the symmetry condition $b_k = b_{M-k}$, for $k = 0, 1, \ldots, M$, the z-transform polynomial obeys the following "reciprocal property"

$$H(1/z) = z^M H(z) \tag{9.49}$$

To demonstrate this reciprocal property of linear-phase filters, consider a 4-point system function $H(z)$ and replace z with $1/z$

$$\begin{aligned} H(1/z) &= b_0 + b_1(1/z)^{-1} + b_1(1/z)^{-2} + b_0(1/z)^{-3} \\ &= b_0 + b_1 z^1 + b_1 z^2 + b_0 z^3 \\ &= z^3(b_0 + b_1 z^{-1} + b_1 z^{-2} + b_0 z^{-3}) \\ &= z^3 H(z) \end{aligned}$$

Following the same style for a general M, (9.49) is easily proved for the general case.

The reciprocal property of linear-phase filters is responsible for the distinctive pattern of zeros in the pole-zero plot of Fig. 9-21. The zeros that are not on the unit circle occur as quadruples. Furthermore, these quadruples are responsible for creating the passband of the BPF. Zeros on the unit circle occur in pairs because the complex conjugate partner must be present, and these zeros are mainly responsible for creating the stopband of the filter.

These properties can be shown to be true in general for linear-phase filters. Specifically,

> When $b_k = b_{M-k}$, for $k = 0, 1, \ldots, M$, then if z_0 is a zero of $H(z)$, so are its conjugate, its inverse, and its conjugate inverse, that is, $\{z_0,\ z_0^*,\ 1/z_0,\ 1/z_0^*\}$ are all zeros of $H(z)$.

The conjugate zeros are included because the filter coefficients, which are also the coefficients of $H(z)$, are real. Therefore, all of the zeros of $H(z)$ must occur in complex-conjugate pairs. The inverse property of the roots is a consequence of the reciprocal property (9.49). Using (9.49), and assuming that z_0 is a zero of $H(z)$, we get

$$H(1/z_0) = z_0^M H(z_0) = 0$$

Since $H(z_0) = 0$, then we must have $H(1/z_0) = 0$ also. Most FIR filters are designed with a symmetry property, so the zero pattern of Fig. 9-21 is typical.

9-10 Summary and Links

LAB
Encoding and Decoding Touch-Tone Signals

The z-transform method was introduced in this chapter for FIR filters and finite-length sequences in general. The z-transform reduces the manipulation of LTI systems into simple operations on polynomials and rational functions. Roots of these z-transform polynomials are quite important because filter properties such as the frequency response can be inferred directly from the root locations.

We also introduced the important concept of "domains of representation" for discrete-time signals and systems. There are three domains: the n-domain or time domain, the $\hat{\omega}$-domain or frequency domain, and the z-domain. With three different domains at our disposal, even the most difficult problems generally can be simplified by switching to one of the other domains.

LAB
Octave Band Filtering

Among the laboratory projects on the website we have already provided two on the topic of FIR filtering in Chapters 5 and 6. The first one deals with FIR filtering of sinusoidal waveforms, convolution, and deconvolution; the second one explores the frequency responses of common filters such as the first difference and the L-point averager. In two other labs with projects, FIR filters are used in practical systems, such as a Touch-Tone decoding system, and an octave band filter bank for piano note detection. Each of these labs should be easier to understand and simpler to carry out with the newly acquired background on z-transforms.

DEMO
Three Domains—FIR

The Companion Website also contains some demonstrations of the relationship between the z-plane and the frequency domain and time domain.

(a) A set of 3-domain movies that show how the frequency response and the impulse response of an FIR filter change as a zero location is moved. Several different filters are demonstrated.

DEMO
Z to Freq

(b) A movie that animates the relationship between the z-plane and the unit circle where the frequency response lies.

LAB
PeZ 3 Domains

(c) The MATLAB program `PeZ` that facilitates exploring the three domains. There is also a lab based on `PeZ` which presents various useful filters (e.g., nulling filters) in the three domains. Please refer to Section 10-12.4 for a concise description of how to use the interactive controls in `PeZ`.

9-11 Problems

NOTE
Problems with Solutions

The problems in this section address the major concepts of the z-transform presented this chapter. In addition, there are many related homework problems available on the website, which can be used for review and practice because most of them have detailed solutions.

P-9.1 Use the superposition and time-delay properties of (9.5) and (9.6) to find the z-transforms of the following signals:

$$x_1[n] = 6\delta[n]$$
$$x_2[n] = 7\delta[n-3]$$
$$x_3[n] = 2\delta[n] + 9\delta[n-5]$$
$$x_4[n] = 6\delta[n] - 7\delta[n-3] - 2\delta[n] - 9\delta[n-5]$$

P-9.2 Use the superposition and time-delay properties of (9.5) and (9.6) to determine the z-transform $Y(z)$ in terms of $X(z)$ if

$$y_1[n] = x[n] - x[n-1] \quad \text{and}$$
$$y_2[n] = x[n] + x[n-1]$$

and in the process obtain the two *system functions*.

P-9.3 Suppose that an LTI system has a system function

$$H(z) = 1 - 3z^{-2} + 3z^{-4} + 4z^{-6} + 7z^{-7}$$

(a) Determine the difference equation that relates the output $y[n]$ of the system to the input $x[n]$.

(b) Determine and plot the output sequence $y[n]$ when the input is $x[n] = \delta[n]$.

P-9.4 An LTI system is described by the difference equation

$$y[n] = 0.2\,(x[n] - 1.5x[n-1] + 0.5x[n-2])$$

(a) Determine the system function $H(z)$ for this system.

(b) Plot the poles and zeros of $H(z)$ in the z-plane.

(c) From $H(z)$ obtain an expression for $H(e^{j\hat{\omega}})$, the frequency response of this system.

(d) Sketch the frequency response (magnitude and phase) as a function of frequency for $-\pi \le \hat{\omega} \le \pi$.

(e) Determine the output if the input is

$$x[n] = 7 - 6\cos[(\pi/3)(n-1)] + 5\cos[0.75\pi n].$$

P-9.5 Consider an LTI system whose system function is the product of five terms

$$H(z) = (1 - z^{-1})(1 - 0.7e^{-j\pi/2}z^{-1})(1 - 0.7e^{j\pi/2}z^{-1})(1 + e^{j2\pi/3}z^{-1})(1 + e^{-j2\pi/3}z^{-1})$$

(a) Write the difference equation that gives the relation between the input $x[n]$ and the output $y[n]$.

Hint: Multiply the factors of $H(z)$ to obtain a polynomial.

(b) Plot the poles and zeros of $H(z)$ in the complex z-plane.

(c) If the input is of the form $x[n] = Ae^{j\phi}e^{jn\hat{\omega}}$, for what values of $-\pi < \hat{\omega} \le \pi$ is it true that $y[n] = 0$?

P-9.6 The diagram in Fig. P-9.6(a) depicts a cascade connection of two LTI systems, where the output of the first system is the input to the second system and the overall output is the output of the second system.

(a) Suppose that LTI 1 is a 3-point averager described by the difference equation $y_1[n] = \frac{1}{3}(x[n] + x[n-1] + x[n-2])$ and LTI 2 is a 4-point averager described by the system function $H_2(z) = \frac{1}{4}(1 + z^{-1} + z^{-2} + z^{-3})$. Determine the system function of the overall cascade system in Fig. P-9.6(a).

(b) For the cascade system in Fig. P-9.6(b), assume that LTI 1 is a 4-point averager and LTI 2 is a 3-point averager. Determine the z-transforms, $Y_2(z)$ and $W(z)$ when $x[n] = \delta[n] - \delta[n-1]$.

(c) Obtain a single difference equation that relates $y[n]$ to $x[n]$ in Fig. P-9.6(a). Is the cascade of the two averagers the same as a 7-point averager?

(d) Plot the poles and zeros of $H(z) = H_1(z)H_2(z)$ in the complex z-plane.

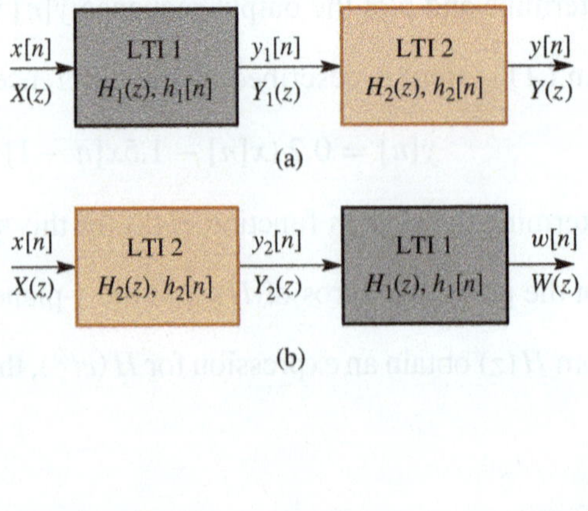

Figure P-9.6

P-9.7 Factor the following polynomial and plot its roots in the complex plane.

$$P(z) = 1 + z^{-1} + z^{-2} - z^{-3}$$

In MATLAB, use the functions called `roots` and `zplane`.

P-9.8 An LTI system is described by the difference equation

$$y[n] = \frac{1}{4}\{x[n] + x[n-1] + x[n-2] + x[n-3]\}$$
$$= \frac{1}{4}\sum_{k=0}^{3} x[n-k]$$

(a) What is $h[n]$, the impulse response of this system?

(b) Determine the system function $H(z)$ for this system.

(c) Plot the poles and zeros of $H(z)$ in the complex z-plane.
Hint: Remember the Lth roots of unity.

(d) From $H(z)$, obtain an expression for the frequency response $H(e^{j\hat{\omega}})$ of this system.

(e) Sketch the frequency response (magnitude and phase) as a function of frequency (or plot it using `freqz( )` in MATLAB, or `freekz( )` in the *SP-First Toolbox*).

(f) Suppose that the input is

$$x[n] = 5 + 4\cos(0.2\pi n) + 3\cos(0.5\pi n + \pi/4)$$

for $-\infty < n < \infty$. Obtain an expression for the output in the form $y[n] = A + B\cos(\hat{\omega}_0 n + \phi_0)$.

P-9.9 Suppose that a system $H(z)$ is the cascade of three first-difference and two 2-point running sum filters.

(a) Determine the system function $H(z)$ for the overall system.

(b) Plot the poles and zeros of $H(z)$ in the z-plane.

(c) From $H(z)$, obtain an expression for the frequency response $H(e^{j\hat{\omega}})$ of the overall cascade system.

(d) Use multiplication of z-transform polynomials to determine the impulse response $h[n]$ of the overall cascade system.

P-9.10 Suppose that an LTI system has system function equal to

$$H(z) = z^{-6} - 3z^{-7} + 2z^{-8} - z^{-9}$$

The input to the system is the sequence

$$x[n] = 3\delta[n] + 2\delta[n-3] - 7\delta[n-6] + 4\delta[n-9]$$

(a) Without actually computing the output, determine the values of N_1 and N_2 so that the following is true:

$$y[n] = 0 \quad \text{for } n < N_1 \text{ and } n > N_2$$

(b) Using the above N_1 and N_2, determine the (nonzero) values of $y[N_1]$, $y[(N_1 + N_2)/2]$ and $y[N_2]$.

(c) Use z-transforms and polynomial multiplication to find the entire sequence $y[n] = x[n] * h[n]$.

P-9.11 The intention of the following MATLAB program is to filter a sinusoid using the conv function.

```
omega = pi/6;
nn = [ 0:29 ];
xn = cos(omega*nn - pi/4);
bb = [ 1  0  0  1 ];
yn = conv( bb, xn );
```

(a) Determine $H(z)$ and the zeros of the FIR filter.

(b) Determine a formula for $y[n]$, the signal contained in the vector yn. Ignore the first few points, so your formula must be correct only for $n \geq 3$. This formula should give numerical values for the amplitude, phase and frequency of $y[n]$.

(c) Give a value of omega such that the output is guaranteed to be zero, for $n \geq 3$.

P-9.12 A system is defined by the system function

$$H(z) = (1 - z^{-1})(1 + z^{-1} + z^{-2})(1 + z^{-3})$$

(a) Write the time-domain description of this system in the form of a difference equation.

(b) Write a formula for the frequency response of the system.

(c) Derive simple formulas for the magnitude response versus $\hat{\omega}$ and the phase response versus $\hat{\omega}$. These formulas must contain no complex terms and no square roots.

(d) This system can "null" certain input signals. For which input frequencies $\hat{\omega}_0$ is the response to $x[n] = \cos(\hat{\omega}_0 n)$ equal to zero?

(e) When the input to the system is $x[n] = \cos(\pi n/4)$, determine the output signal $y[n]$ in the form:

$$A\cos(\hat{\omega}_0 n + \phi)$$

Give numerical values for the constants A, $\hat{\omega}_0$, and ϕ.

P-9.13 An LTI system has zeros at ± 1 and ± 3, and four poles at $z = 0$. The input to this system is

$$x[n] = 50 - 60\delta[n] + 20\cos(0.5\pi n + \pi/3)$$

for $-\infty < n < \infty$. If $h[0] = 1$, determine the output of the system $y[n]$ corresponding to the above input $x[n]$. Give an equation for $y[n]$ that is valid for all n.
Note: Superposition makes this problem easier.

P-9.14 The input to the C-to-D converter in Fig. P-9.14 is

$$x(t) = 3 + \cos(600\pi t - \pi/6) - 2\cos(1200\pi t)$$

The system function for the LTI system is

$$H(z) = 1 - z^{-1} + z^{-4} - z^{-5}$$

If $f_s = 2400$ samples/s, determine an expression for $y(t)$, the output of the D-to-C converter.

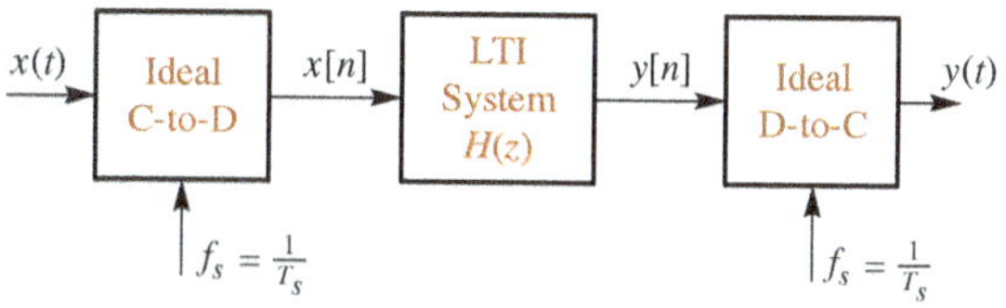

Figure P-9.14

P-9.15 Refer to the system for discrete-time filtering of a continuous-time signal shown in Fig. P-9.14. Assume that the discrete-time LTI system has a system function $H(z)$ defined as

$$H(z) = 5\left(1 + b_1 z^{-1} + z^{-2}\right)$$

(a) For the case where $b_1 = 0.8$, determine a formula for the frequency response of the discrete-time filter. Express your answer in the following form by finding α and β:

$$H(e^{j\hat{\omega}}) = e^{-j\hat{\omega}}\left(\alpha + \beta\cos(\hat{\omega})\right)$$

(b) For the case where $b_1 = 0.8$, plot the magnitude response of the digital filter versus frequency for $-\pi < \hat{\omega} \leq \pi$.

(c) The overall frequency response of this system from $x(t)$ to $y(t)$ is able to null out one continuous-time sinusoid (e.g., a system that removes a 60-Hz sinusoidal interference from an EKG signal). The value of the filter coefficient b_1 controls the frequency location of the null. If the sampling rate is $f_s = 8000$ Hz, determine the value of b_1 so that the overall frequency response has a null at 1600 Hz.

P-9.16 In the cascade system of Fig. P-9.6(a), it is known that the system function $H(z)$ of the overall system is the product of four terms

$$H(z) = (1 - z^{-2})(1 - 0.8e^{j\pi/4}z^{-1})(1 - 0.8e^{-j\pi/4}z^{-1})(1 + z^{-2})$$

(a) Determine the poles and zeros of $H(z)$ and plot them in the complex z-plane.

(b) It is possible to determine two system functions $H_1(z)$ and $H_2(z)$ so that (1) the overall cascade system has the given system function $H(z)$, and (2) the output of the first system is $y_1[n] = x[n] - x[n-4]$. Find $H_1(z)$ and $H_2(z)$.

(c) Determine the difference equation that relates $y[n]$ to $y_1[n]$ for your answer in (b).

P-9.17 A causal FIR filter is known to have four zeros: $\{0.8e^{\pm j\pi/3}, 1.25e^{\pm j\pi/3}\}$, but the zeros do not completely define the system.

(a) If the first sample of the impulse response is $h[0] = 7$, determine the impulse response of the filter.

(b) If the DC value of the frequency response is $H(e^{j0}) = 2$, determine the system function of the filter.

P-9.18 In Section 9-9, we showed that symmetric FIR filters have special properties. In this problem, we consider the antisymmetric case where $b_k = -b_{M-k}$, for $k = 0, 1, \ldots, M$. Consider the specific example

$$H(z) = b_0 + b_1 z^{-1} - b_1 z^{-3} - b_0 z^{-4}$$

where in this case of even order, $b_2 = -b_2 = 0$.

(a) Show that, for this example, $H(z)\big|_{z=1} = 0$.

(b) Show that, for this example,

$$H(e^{j\hat{\omega}}) = [2b_0 \sin(2\hat{\omega}) + 2b_1 \sin(\hat{\omega})]e^{j(\pi/2 - j\hat{\omega}2)}$$

(c) Show that, for this example,

$$H(1/z) = -z^4 H(z)$$

(d) Generalize these results for any M; the even-order and odd-order cases are slightly different.

P-9.19 Refer to Fig. P-9.6(a) which depicts a cascade connection of two LTI systems.

(a) Suppose that System 1 and System 2 both have a "square pulse" impulse response of the form

$$h_1[n] = h_2[n] = \delta[n] + \delta[n-1] + \delta[n-2] + \delta[n-3]$$

Determine the system functions $H_1(z)$ and $H_2(z)$ for the two systems.

(b) Use z-transforms to determine the system function $H(z)$ of the overall system.

(c) Determine the impulse response $h[n]$ for the overall system.

(d) Obtain a single difference equation that relates $y[n]$ to $x[n]$ in the cascade system.

(e) Show that $H_1(z) = H_2(z)$ can be expressed as

$$H_1(z) = H_2(z) = \frac{1 - z^{-4}}{1 - z^{-1}}$$

(f) For the overall system $H(z)$, plot the poles and zeros in the complex z-plane.

(g) Show that the frequency responses of the two systems are

$$H_1(e^{j\hat{\omega}}) = H_2(e^{j\hat{\omega}}) = \frac{\sin(2\hat{\omega})}{\sin(\hat{\omega}/2)} e^{-j3\hat{\omega}/2}$$

(h) From the frequency responses found in (g) obtain an expression for the frequency response $H(e^{j\hat{\omega}})$ and sketch the magnitude of the frequency response of the overall cascade system for $-\pi < \hat{\omega} \leq \pi$.

CHAPTER

10

IIR Filters

This chapter introduces a new class of linear time-invariant (LTI) systems that have infinite duration impulse responses. These systems are often called IIR filters, where IIR is short for infinite-impulse-response. In contrast to the FIR case, the output of an IIR filter is computed using previously computed values of the output signal as well as values of the input signal. Since the output is "fed back" to be combined with the input, these systems are examples of the general class of ***feedback systems***. From a computational point of view, the output samples are determined from previously computed output values, so the term ***recursive filter*** is also used.

The z-transform system functions for IIR filters are rational functions that have poles at the roots of the denominator polynomial and zeros at the roots of the numerator. Just as for the FIR case, we will show that many insights into the important properties of IIR filters can be obtained directly from the locations of the poles and zeros in the z-plane.

We begin this chapter with the first-order IIR system, which is the simplest case because it involves feedback of only the previous output sample. We construct the impulse response of this system and show that its duration is infinite. Then the frequency response and the z-transform are developed for the first-order filter. After the relationships among the three domains of representation for this simple case have been established, we consider second-order filters. These filters are particularly important because they can be used to model ***resonances*** such as would occur in a speech synthesizer, as well as many other natural phenomena that exhibit vibratory behavior. The frequency response for the second-order case can exhibit a narrowband character that leads to the concepts of bandwidth and center frequency, both of which can be controlled by appropriate choice of the feedback coefficients of the filter. The analysis and insights developed for the first- and second-order cases form the basis for understanding and implementing higher-order systems.

10-1 The General IIR Difference Equation

DEMO
IIR Filtering

FIR filters are extremely useful and have many nice properties, but the class of FIR filters is not the most general class of LTI systems. This is because the output $y[n]$ is formed solely from a finite set of samples of the input signal $x[n]$. The most general class of discrete-time LTI systems that can be implemented with a finite amount of computation is obtained when the output is formed not only from the input, but also from previously computed outputs.

The defining equation for this wider class of digital filters is the following general difference equation:

Difference Equation with Feedback

$$y[n] = \sum_{\ell=1}^{N} a_\ell y[n-\ell] + \sum_{k=0}^{M} b_k x[n-k] \tag{10.1}$$

The filter coefficients in the two sets $\{b_k\}$ and $\{a_\ell\}$ are constants. For reasons that will become obvious in the following simple example, the coefficients $\{a_\ell\}$ are called the ***feedback*** coefficients, and the $\{b_k\}$ are called the ***feed-forward*** coefficients. In all, $N + M + 1$ coefficients are needed to define the difference equation[1] (10.1).

Notice that if the feedback coefficients $\{a_\ell\}$ are all zero, the difference equation (10.1) reduces to the difference equation of an FIR system. Indeed, we have asserted that (10.1) defines the most general class of LTI systems that can be implemented with finite computation, so FIR systems must be a special case. When discussing FIR systems we have referred to M as the ***order*** of the system because M is the degree or order of the polynomial system function, as well as the number of delay terms in the FIR difference equation. For IIR systems, we have both M and N as measures of the number of delay terms, and we see that the system function of an IIR system is the ratio of an Mth-order polynomial to an Nth-order polynomial. Thus, there can be some ambiguity as to the order of an IIR system. In general, we define N, the number of feedback terms, to be the order of an IIR system.

EXAMPLE 10-1 IIR Block Diagram

As a first step toward understanding the general form given in (10.1), consider the first-order case where $M = N = 1$, that is,

[1] The difference equation of the IIR filter is also called a recursion formula for generating the output sequence. Thus, IIR systems are often called *recursive filters*.

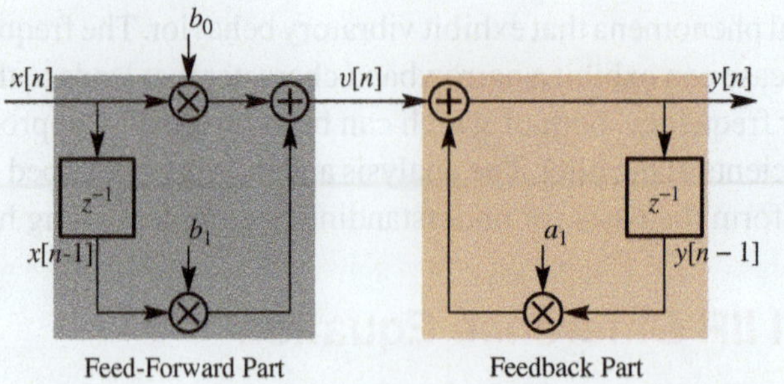

Figure 10-1 First-order IIR system showing one feedback coefficient a_1 and two feed-forward coefficients b_0 and b_1.

$$y[n] = a_1 y[n-1] + \underbrace{b_0 x[n] + b_1 x[n-1]}_{v[n]} \tag{10.2}$$

The block diagram representation of this difference equation, which is shown in Fig. 10-1, is constructed by noting that the signal $v[n] = b_0 x[n] + b_1 x[n-1]$ is computed by the left half of the diagram, and we "close the loop" by computing $a_1 y[n-1]$ from the delayed output and adding it to $v[n]$ to produce the output $y[n]$. In the diagram of Fig. 10-1, all the paths in the *feed-forward* section go left-to-right, which is forward from the input to the output; in the *feedback* section, the path through the delay goes right-to-left which is from the output back to the input.

We begin by studying a simplified version of the system defined by (10.2) and depicted in Fig. 10-1. Our objective is to characterize the filter in each of the three domains: time domain, frequency domain, and z-domain. Since the filter is defined by a time-domain difference equation (10.2), we first study how the difference equation is used to compute the output from the input, and then we illustrate how feedback results in an impulse response of infinite duration.

10-2 Time-Domain Response

To illustrate how the difference equation can be used to implement an IIR system, consider a numerical example where the filter coefficients in (10.2) are $a_1 = 0.8$, $b_0 = 5$, and $b_1 = 0$, so that

$$y[n] = 0.8y[n-1] + 5x[n] \tag{10.3a}$$

In addition, assume that the input signal is

$$x[n] = 2\delta[n] - 3\delta[n-1] + 2\delta[n-3] \tag{10.3b}$$

so that the total duration of the input is four samples, as shown in Fig. 10-2.

Although it is not a requirement, it is logical that the output sequence values should be computed in normal order (i.e., from left to right in a plot of the sequence). Furthermore, since the input is zero before $n = 0$, it would be natural to assume that $n = 0$ is the starting

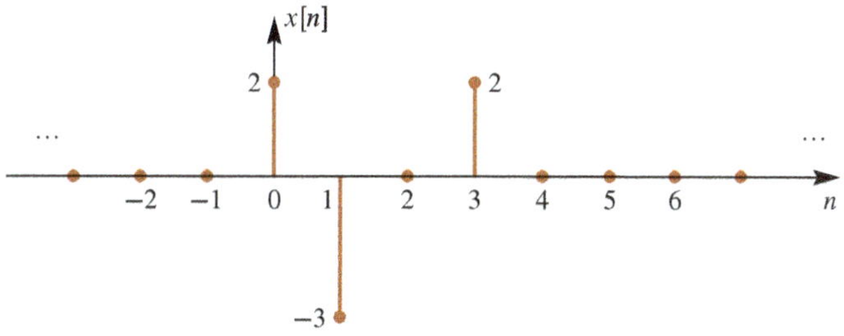

Figure 10-2 Input signal to recursive difference equation.

index of the output. Thus, we proceed to compute the output from the difference equation (10.3a) in the order $y[0], y[1], y[2], \ldots$. For example, the value of $x[0]$ is 2, so we can evaluate (10.3a) at $n = 0$ obtaining

$$y[0] = 0.8y[0-1] + 5x[0] = 0.8y[-1] + 5(2) \tag{10.4}$$

Immediately we run into a problem: The value of $y[n]$ at $n = -1$ is not known. We cannot solve this problem by changing where we start computing the output $y[n]$, because at any point along the n-axis, we still need to know the output at the previous time $n-1$. If we know the value $y[n-1]$, however, we can use the difference equation (10.3a) to compute the next value of the output signal at time n. Once the process is started, it can be continued indefinitely by iteration of the difference equation. The solution requires the following two assumptions, which taken together are called the ***initial rest conditions***:

Initial Rest Conditions

1. The input must be assumed to be zero prior to some starting time n_0 (i.e., $x[n] = 0$ for $n < n_0$). We say that such inputs are *suddenly applied*.

2. The output is likewise assumed to be zero prior to the starting time of the signal (i.e., $y[n] = 0$ for $n < n_0$). We say that the system is *initially at rest* if its output is zero prior to the application of a suddenly applied input.

These conditions are not particularly restrictive, especially in the case of a real-time system, where a new output must be computed as each new sample of the input is taken. Real-time systems must, of course, be ***causal*** in the sense that the computation of the present output sample must not involve future samples of the input or output. Furthermore, any practical device would have a time at which it first begins to operate. All that is needed is for the memory containing the delayed output samples to be set initially to zero.[2]

With the initial rest assumption, we let $y[n] = 0$ for $n < 0$, so now we can evaluate $y[0]$ as

$$y[0] = 0.8y[-1] + 5(2) = 0.8(0) + 5(2) = 10$$

[2]In the case of a digital filter applied to sampled data stored completely in computer memory, the causality condition is not required, but, generally, the output is computed in the same order as the natural order of the input samples. The difference equation could be *recursed* backwards through the sequence, but this would require a different definition of initial conditions.

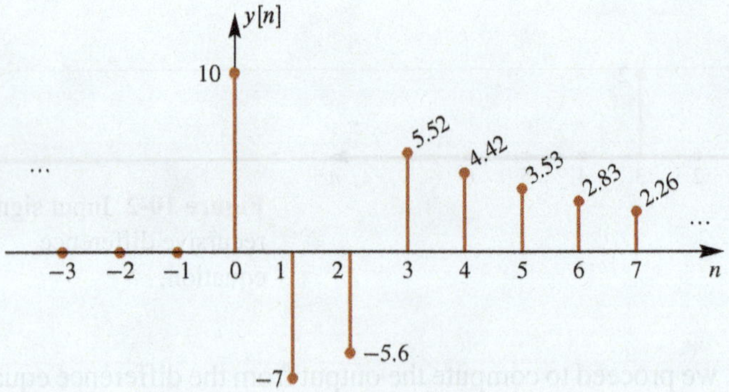

Figure 10-3 Output signal from recursive difference equation of (10.4) for the input of Fig. 10-2. For $n > 3$, the sequence is proportional to $(0.8)^n$, because the input signal ends at $n = 3$.

Once we have started the recursion, the rest of the values follow easily, since the input signal and previous outputs are known.

$$
\begin{aligned}
y[1] &= 0.8y[0] + 5x[1] = 0.8(10) + 5(-3) = -7 \\
y[2] &= 0.8y[1] + 5x[2] = 0.8(-7) + 5(0) = -5.6 \\
y[3] &= 0.8y[2] + 5x[3] = 0.8(-5.6) + 5(2) = 5.52 \\
y[4] &= 0.8y[3] + 5x[4] = 0.8(5.52) + 5(0) = 4.416 \\
y[5] &= 0.8y[4] + 5x[5] = 0.8(4.416) + 5(0) = 3.5328 \\
&\vdots \qquad\qquad \vdots
\end{aligned}
$$

This output sequence is plotted in Fig. 10-3 up to $n = 7$.

One key feature to notice in Fig. 10-3 is the structure of the output signal after the input becomes zero. When $x[n] = 0$ for $n > 3$, the difference equation for $n > 3$ becomes

$$y[n] = 0.8y[n-1] + \cancelto{0}{5x[n]} \qquad \text{for } n > 3$$

Then the ratio between successive terms is a constant, and the output signal decays exponentially with a rate determined by $a_1 = 0.8$. Therefore, we can write the following expression

$$y[n] = y[3](0.8)^{n-3} \qquad \text{for } n \geq 3$$

for the rest of the sequence $y[n]$ once the value for $y[3]$ is known.

10-2.1 Linearity and Time Invariance of IIR Filters

When applied to the general IIR difference equation of (10.1), the condition of initial rest is sufficient to guarantee that the system implemented by iterating the difference equation is both linear and time-invariant. Although the feedback terms make the proof more complicated than the FIR case (see Section 5-6.3 on p. 197), we can show that, for suddenly applied inputs and initial rest conditions, the principle of superposition holds because the difference equation involves only linear combinations of the input and output samples. Furthermore, since the initial rest condition must be true just before the beginning of a suddenly applied input, time invariance also holds.

EXERCISE 10.1 Assume that the input to the difference equation (10.3a) is $x_1[n] = 10x[n-4]$, where $x[n]$ is given by (10.3b) and Fig. 10-2. Thus the input is 10 times bigger and delayed by 4. Use iteration of (10.3a) to compute the corresponding output $y_1[n]$ for $n = 0, 1, \ldots, 11$ assuming the initial rest condition. Compare your result to the output plotted in Fig. 10-3, and verify that the system behaves as if it is both linear and time-invariant, that is, the output should be 10 times bigger and delayed by 4.

10-2.2 Impulse Response of a First-Order IIR System

In Chapter 5, we showed that the response to a unit impulse sequence characterizes an LTI system completely. Recall that when $x[n] = \delta[n]$, the resulting output signal, denoted by $h[n]$, is by definition the ***impulse response***. Since any other input signal can be written as a superposition of weighted and delayed impulses, the corresponding output for that signal can be constructed from weighted and shifted versions of the impulse response, $h[n]$. In other words, since the recursive difference equation with initial rest conditions is an LTI system, its output can always be represented via the weighting and shifting of $h[n]$ in the convolution sum

$$y[n] = \sum_{k=-\infty}^{\infty} x[k]h[n-k] \tag{10.5}$$

Therefore, it is of interest to characterize the recursive difference equation by its impulse response.

To illustrate the nature of the impulse response of an IIR system, consider the first-order recursive difference equation with $b_1 = 0$,

$$y[n] = a_1 y[n-1] + b_0 x[n] \tag{10.6}$$

By definition, the impulse response $h[n]$ is the output when the input is $x[n] = \delta[n]$, so the difference equation

$$h[n] = a_1 h[n-1] + b_0 \delta[n] \tag{10.7}$$

must be satisfied by $h[n]$ for all values of n. However, (10.7) is *not an explicit formula* for $h[n]$. If we want an explicit general formula for the impulse response in terms of the parameters a_1 and b_0, we could use (10.7) to construct a table of the first few values, and then, in this simple case, write down the general formula by inspection. The following table shows the sequences involved in the computation of $h[n]$:

n	$n<0$	0	1	2	3	4	$\cdots$
$\delta[n]$	0	1	0	0	0	0	$\cdots$
$h[n-1]$	0	0	b_0	$b_0(a_1)$	$b_0(a_1)^2$	$b_0(a_1)^3$	$\cdots$
$h[n]$	0	b_0	$b_0(a_1)$	$b_0(a_1)^2$	$b_0(a_1)^3$	$b_0(a_1)^4$	$\cdots$

From this table, we see that the explicit formula for the impulse response is

$$h[n] = \begin{cases} b_0(a_1)^n & \text{for } n \geq 0 \\ 0 & \text{for } n < 0 \end{cases} \tag{10.8a}$$

If we recall the definition of the unit step sequence from (5.17),

$$u[n] = \begin{cases} 1 & \text{for } n \geq 0 \\ 0 & \text{for } n < 0 \end{cases} \tag{10.8b}$$

(10.8a) can be expressed in a more compact form

$$h[n] = b_0(a_1)^n u[n] \tag{10.8c}$$

where multiplication of $b_0(a_1)^n$ by $u[n]$ enforces the conditions in (10.8a) for $n \geq 0$ and $n < 0$.

EXAMPLE 10-2 Impulse Response

For the system in (10.3a) with $a_1 = 0.8$ and $b_0 = 5$, the impulse response is

$$h[n] = 5(0.8)^n u[n] = \begin{cases} 5(0.8)^n & \text{for } n \geq 0 \\ 0 & \text{for } n < 0 \end{cases} \tag{10.9}$$

This impulse response is nonzero for $0 \leq n < \infty$ (i.e., it has infinite duration), which is what we mean by the designation IIR.

A slightly more general problem is to determine the impulse response of the first-order system when a shifted version of the input signal is also included in the difference equation, that is,

$$y[n] = a_1 y[n-1] + \underbrace{b_0 x[n] + b_1 x[n-1]}_{\text{input = superposition of two terms}} \tag{10.10a}$$

Because this system is linear and time-invariant, and the input consists of two terms, it follows that its impulse response can be written as a superposition of two terms as in

$$h[n] = b_0(a_1)^n u[n] + b_1(a_1)^{n-1} u[n-1] \tag{10.10b}$$

When plotting $h[n]$, this compact form should be rewritten as three cases

$$h[n] = \begin{cases} 0 & n < 0 \\ b_0 & n = 0 \\ (b_0 + b_1 a_1^{-1})(a_1)^n & n \geq 1 \end{cases} \tag{10.10c}$$

from which it is easy to see that the impulse response still decays exponentially (for $n \geq 1$) at a rate dependent only on a_1.

EXERCISE 10.2 Determine the impulse response $h[n]$ of the following first-order system:

$$y[n] = -0.5y[n-1] - 4x[n] + 5x[n-1].$$

Plot the resulting impulse response as a function of n.

10-2.3 Response to Finite-Length Inputs

For finite-length inputs, the output of an IIR (or FIR) system can be written as a *weighted sum of shifted copies of the impulse response.* The finite-length signal can be written as

$$x[n] = \sum_{k=N_1}^{N_2} x[k]\delta[n-k]$$

where $x[n] = 0$ for $n < N_1$ and $n > N_2$. Then the corresponding output is obtained via a convolution sum (10.5), but with finite limits as in

$$y[n] = \sum_{k=N_1}^{N_2} x[k]h[n-k]$$

Each term $x[k]h[n-k]$ is a shifted copy of the impulse response weighted by $x[k]$.

EXAMPLE 10-3 IIR Response to General Input

Consider again the LTI system defined by the difference equation (10.3a), whose impulse response was shown in Example 10-2 to be $h[n] = 5(0.8)^n u[n]$. For the input of (10.3b) and Fig. 10-2,

$$x[n] = 2\delta[n] - 3\delta[n-1] + 2\delta[n-3]$$

it is easily seen that

$$\begin{aligned} y[n] &= 2h[n] - 3h[n-1] + 2h[n-3] \\ &= 10(0.8)^n u[n] - 15(0.8)^{n-1} u[n-1] + 10(0.8)^{n-3} u[n-3] \end{aligned}$$

To evaluate this expression for a specific time index, we need to take into account the different regions over which $h[n]$, $h[n-1]$, and $h[n-3]$ are nonzero. Once we do, we obtain

$$y[n] = \begin{cases} 0 & n < 0 \\ 10 & n = 0 \\ 10(0.8) - 15 = -7 & n = 1 \\ 10(0.8)^2 - 15(0.8) = -5.6 & n = 2 \\ 10(0.8)^3 - 15(0.8)^2 + 10 = 5.52 & n = 3 \\ 5.52(0.8)^{n-3} & n > 3 \end{cases}$$

A comparison to the output obtained by iterating the difference equation (see Fig. 10-3 on p. 418) shows that we have obtained the same values of the output sequence by superposition of scaled and shifted impulse responses.

Example 10-3 illustrates two important points about IIR systems.

1. The initial rest condition guarantees that the output sequence does not begin until the input sequence begins (or later).

2. Because of the feedback, the impulse response can be infinite in extent, and the output due to a finite-length input sequence, being a superposition of scaled and shifted impulse responses, is generally (but not always) infinite in extent. This is in contrast to the FIR case, where a finite-length input always produces a finite-length output sequence.

10-2.4 Step Response of a First-Order Recursive System

When the input signal is infinitely long, the computation of the output of an IIR system using the difference equation is no different than for an FIR system; we simply continue to iterate the difference equation as long as samples of the output are desired. In the FIR case, the difference equation and the convolution sum are the same thing. This is not true in the IIR case, and computing the output using convolution is practical only in cases where simple formulas exist for both the input and the impulse response. Thus, IIR filters are generally implemented by iterating the difference equation with feedback.

The computation of the response of a first-order IIR system to a unit step input provides a relatively simple illustration of the issues involved. We use the same system as in (10.6)

$$y[n] = a_1 y[n-1] + b_0 x[n]$$

and assume that the input is the unit step sequence (10.8b) (i.e., $x[n] = u[n]$). For $n \geq 0$, the input term $b_0 u[n]$ is equal to b_0, so iterating the difference equation is equivalent to "multiplying the previous output by a_1 and adding b_0" to produce the output sequence one sample at a time. The first few values are tabulated here.

n	$x[n]$	$y[n]$
$n < 0$	0	0
0	1	b_0
1	1	$a_1(b_0) + b_0$
2	1	$a_1(a_1(b_0) + b_0) + b_0$
3	1	$a_1(a_1^2 + a_1 + 1)(b_0) + b_0$
4	1	$(a_1^4 + a_1^3 + a_1^2 + a_1 + 1)b_0$
$\vdots$	1	$\vdots$

Work through the table to be sure that you understand the computation. From the tabulated values, it can be seen that a general formula for the ***step response*** $y[n]$ is

$$\begin{aligned} y[n] &= b_0(1 + a_1 + a_1^2 + \cdots + a_1^n) \\ &= b_0 \sum_{k=0}^{n} a_1^k \end{aligned} \tag{10.11}$$

With a bit of manipulation, we can get a simple closed-form expression for the general term in the sequence $y[n]$. For this we need to recall the formula

$$\sum_{k=0}^{n} r^k = \begin{cases} \dfrac{1 - r^{n+1}}{1 - r} & r \neq 1 \\ n + 1 & r = 1 \end{cases} \tag{10.12}$$

which is the formula for summing the first $n + 1$ terms of a geometric series. Armed with this fact, the expression (10.11) for $y[n]$ (when $a_1 \neq 1$) can be manipulated into the form

$$y[n] = b_0 \frac{1 - a_1^{n+1}}{1 - a_1} \qquad \text{for } n \geq 0, \tag{10.13}$$

Three cases can be identified: $|a_1| > 1$, $|a_1| < 1$, and $|a_1| = 1$. Further investigation of these cases reveals two types of behavior.

1. When $|a_1| > 1$, the term a_1^{n+1} in the numerator dominates as $n \to \infty$, and the values for $y[n]$ grow without bound. This is called an ***unstable*** condition and is usually a situation to avoid. We will say more about the issue of stability later in Sections 10-5.3 and 10-10.

2. When $|a_1| < 1$, the term a_1^{n+1} decays to zero as $n \to \infty$. In this case, the system is ***stable***. Therefore, we can determine a limiting value for step response $y[n]$ as $n \to \infty$

$$\lim_{n\to\infty} y[n] = \lim_{n\to\infty} b_0 \frac{1 - a_1^{n+1}}{1 - a_1} = \frac{b_0}{1 - a_1}$$

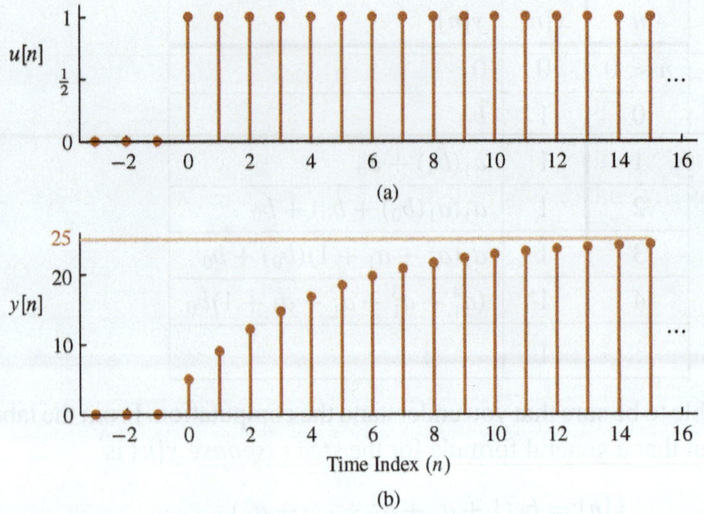

Figure 10-4 (a) Unit-step signal $u[n]$. (b) Step response of the first-order IIR filter in (10.3a) where $a_1 = 0.8$.

3. When $|a_1| = 1$, we might have an unbounded output, but not always. For example, when $a_1 = 1$, the step response (10.11) gives $y[n] = (n+1)b_0$ for $n \geq 0$, and the output $y[n]$ grows as $n \to \infty$. On the other hand, for $a_1 = -1$, the step response alternates; it is $y[n] = b_0$ for n even, but $y[n] = 0$ for n odd.[3]

The MATLAB plot in Fig. 10-4 shows the unit-step signal and the step response for the first-order system

$$y[n] = 0.8y[n-1] + 5x[n]$$

Since $a_1 = 0.8 < 1$, this is the second case above and the limiting value as $n \to \infty$ is 25, which can be calculated from the filter coefficients

$$\lim_{n\to\infty} y[n] = \frac{b_0}{1-a_1} = \frac{5}{1-0.8} = 25$$

It is also possible to compute the step response by the convolution sum

$$y[n] = x[n] * h[n] = \sum_{k=-\infty}^{\infty} x[k]h[n-k] \tag{10.14}$$

Since the input and output are given by the formulas $x[n] = u[n]$ and $h[n] = b_0(a_1)^n u[n]$, the result of the convolution is a compact formula that can be obtained with the z-transform. We will revisit this convolution in Section 10-9.1 once we have the partial fraction method for z-transforms.

---

[3]Even though the output may sometimes be bounded when $a_1 = 1$, the system is still considered to be unstable since *not every* bounded input produces a bounded output.

10-3 System Function of an IIR Filter

We saw in Chapter 9 for the FIR case that the system function $H(z)$ is the z-transform of the impulse response of the system. The system function for an FIR filter is always a polynomial, and one of its key properties is the *convolution property*:

> ***Convolution in the n-domain corresponds to multiplication in the z-domain.***
>
> $$y[n] = x[n] * h[n] \quad \stackrel{z}{\longleftrightarrow} \quad Y(z) = X(z)H(z)$$

This is also true for the IIR case. However, when the difference equation has feedback, it turns out that the system function $H(z)$ is the ratio of two polynomials. Ratios of polynomials are called ***rational*** functions. In this section, we study the system function for the case of a first-order IIR system, and show how the system function, impulse response, and difference equation are related.

10-3.1 The General First-Order Case

The general form of the first-order difference equation with feedback (10.10a) is repeated here:

$$y[n] = a_1 y[n-1] + b_0 x[n] + b_1 x[n-1] \tag{10.15a}$$

Since this equation must be satisfied for all values of n, we can use the delay property (9.7) on p. 372 to determine the z-transform of both sides of the equation and obtain the following equation relating the output z-transform $Y(z)$ to the input $X(z)$:

$$Y(z) = a_1 z^{-1} Y(z) + b_0 X(z) + b_1 z^{-1} X(z) \tag{10.15b}$$

Subtracting the term $a_1 z^{-1} Y(z)$ from both sides of the equation leads to the following manipulations:

$$Y(z) - a_1 z^{-1} Y(z) = b_0 X(z) + b_1 z^{-1} X(z) \tag{10.15c}$$

$$(1 - a_1 z^{-1}) Y(z) = (b_0 + b_1 z^{-1}) X(z) \tag{10.15d}$$

Since the system is an LTI system, it should be true that $Y(z) = H(z)X(z)$, where $H(z)$ is the system function of the LTI system. Solving (10.15d) for $H(z) = Y(z)/X(z)$, we obtain

$$H(z) = \frac{Y(z)}{X(z)} = \frac{b_0 + b_1 z^{-1}}{1 - a_1 z^{-1}} = \frac{B(z)}{A(z)} \tag{10.15e}$$

Thus, we have shown that $H(z)$ for the first-order IIR system is a ratio of two polynomials. The numerator polynomial $B(z) = b_0 + b_1 z^{-1}$ is defined by the weighting coefficients b_0 and b_1 that multiply the input signal $x[n]$ and its delayed versions; the denominator polynomial $A(z) = 1 - a_1 z^{-1}$ is defined by the feedback coefficient a_1. That this correspondence is true in general should be clear from the analysis that leads to the formula for $H(z)$. Indeed, the following is true for IIR systems of *any* order:

> *The coefficients of the* ***numerator*** *polynomial of the system function of an IIR system are the coefficients of the* ***feed-forward*** *terms of the difference equation. For the* ***denominator*** *polynomial, the constant term is one, and the remaining coefficients are the negatives of the* ***feedback*** *coefficients.*

In other words, there is one-to-one correspondence between $H(z)$ and the difference equation. If the system function contains higher degree polynomials

$$H(z) = \frac{\sum_{k=0}^{M} b_k z^{-k}}{1 - \sum_{\ell=1}^{N} a_\ell z^{-\ell}} \tag{10.16a}$$

then the corresponding difference equation is

$$y[n] = \sum_{\ell=1}^{N} a_\ell y[n-\ell] + \sum_{k=0}^{M} b_k x[n-k] \tag{10.16b}$$

In MATLAB, the `filter` function follows this same format. The statement

```
yy = filter(bb,aa,xx)
```

implements an IIR filter, where the vectors `bb` and `aa` hold the filter coefficients for the numerator and denominator polynomials, respectively.

EXAMPLE 10-4 MATLAB for IIR Filter

The following feedback filter:

$$y[n] = 0.5y[n-1] - 3x[n] + 2x[n-1]$$

would be implemented in MATLAB by

```
yy = filter([-3,2], [1,-0.5], xx)
```

where `xx` and `yy` are the input and output signal vectors, respectively. Notice that the `aa` vector has 1 for its first element and $-a_1$ for its second element, just like in the denominator polynomial $A(z)$ in (10.15e).

In general, we arrange the difference equation so that the coefficient multiplying $y[n]$ is 1 (e.g., by rescaling coefficients if necessary).

EXERCISE 10.3 Find the z-transform system function $H(z)$ of the following feedback filter:

$$y[n] = 0.5y[n-1] - 3x[n] + 2x[n-1]$$

EXERCISE 10.4 Determine $H(z)$ for the system implemented by the following MATLAB statement:

```
yy = filter(5, [1,0.8], xx).
```

10-3.2 $H(z)$ from the Impulse Response

In the analysis of Section 10-3.1, we assumed implicitly that the system function is the z-transform of the impulse response of an IIR system. While this is true, we have demonstrated only that it is true for the FIR case. In the first-order IIR case, we need to be able to take the z-transform of an infinitely long sequence. As an example of such a sequence, consider the impulse response $h[n] = a_1^n u[n]$.

Applying the definition of the z-transform from (9.8) on p. 373, we would write

$$H(z) = \sum_{n=0}^{\infty} h[n]z^{-n} = \sum_{n=0}^{\infty} a_1^n z^{-n} = \sum_{n=0}^{\infty} (a_1 z^{-1})^n \tag{10.17a}$$

which defines $H(z)$ as an infinite power series where the ratio between successive terms is $a_1 z^{-1}$. In order for $H(z)$ to be a legitimate function, the sum of all the terms in the geometric series (10.17a) must be finite, which is true if $|a_1 z^{-1}| < 1$. When the sum is finite, $H(z)$ is given by the closed-form expression

$$H(z) = \sum_{n=0}^{\infty} a_1^n z^{-n} = \frac{1}{1 - a_1 z^{-1}} \tag{10.17b}$$

The condition for the infinite sum to equal the closed-form expression can be expressed as $|a_1| < |z|$. The set of values of z in the complex plane satisfying this condition is called the ***region of convergence (ROC)***. From the preceding analysis, we can state the following z-transform pair:

$$\boxed{a_1^n u[n] \quad \stackrel{z}{\longleftrightarrow} \quad \frac{1}{1 - a_1 z^{-1}}} \tag{10.18}$$

which is the building block for all higher-order IIR z-transforms.

EXAMPLE 10-5 $H(z)$ from Impulse Response

As an example of using (10.18), we can find the difference equation that generates an impulse response such as

$$h[n] = 2(0.9)^n u[n] + 5(0.9)^{n-1} u[n-1] \tag{10.19}$$

Using the linearity and delay properties of the z-transform, as well as the z-transform pair in (10.18), the system function for the system having the given impulse response is

$$H(z) = 2\left(\frac{1}{1-0.9z^{-1}}\right) + 5z^{-1}\left(\frac{1}{1-0.9z^{-1}}\right) = \frac{2+5z^{-1}}{1-0.9z^{-1}} \tag{10.20}$$

Note that this is identical in form to $H(z)$ in (10.15e) which we obtained previously in Section 10-3.1 by taking the z-transform of a difference equation (10.15a). In other words, the one-to-one correspondence between the system function and difference equation given in (10.16a) and (10.16b) can be used to write the following difference equation that generates $h[n]$ in (10.19):

$$y[n] = 0.9y[n-1] + 2x[n] + 5x[n-1] \tag{10.21}$$

In this section, focusing on the first-order case, we have presented the three notable concepts of the ***z-transform method*** which can be linked to one another: the difference equation (10.15a), the system function (10.15e), and the impulse response $h[n] = a_1^n u[n]$. The coefficients of the numerator and denominator polynomials of the system function are the same as the coefficients of the difference equation. The system function $H(z)$ can be obtained by taking the z-transform of the impulse response, and for the first-order case, it is possible by taking the inverse z-transform to go directly from the system function to the impulse response of the system without the tedious process of iterating the difference equation. These methods also work for higher-order systems but the process of inverse z-transformation must be based on breaking the z-transform into a sum of first-order terms via the method of partial fractions, which is discussed in Section 10-9.2.

10-4 The System Function and Block Diagram Structures

The system function displays the coefficients of the difference equation in a convenient way that makes it easy to move back and forth between the difference equation and the system function. In this section, we exploit this correspondence to derive other difference equations, which are alternative implementation equations of exactly the same system function.

10-4.1 Direct Form I Structure

To illustrate the connection between the system function and the block diagram, recall the block diagram of Fig. 10-1, which is repeated in Fig. 10-5 for convenience. Block

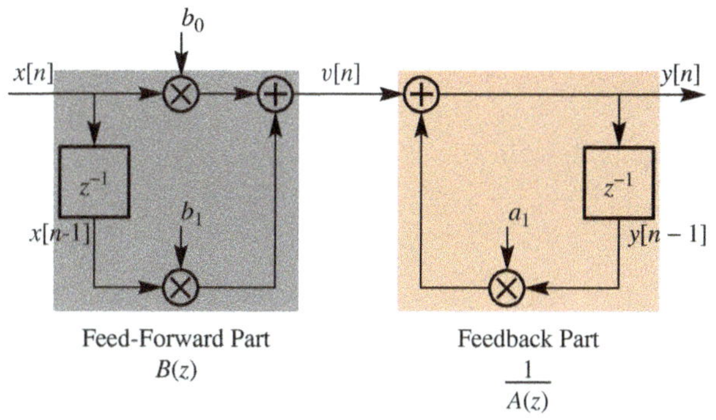

Figure 10-5 First-order IIR system showing one feedback coefficient a_1 and two feed-forward coefficients b_0 and b_1 in a Direct Form I structure, which is the cascade of a feed-forward filter and a feedback filter.

diagrams such as Fig. 10-5 are called *implementation structures*, or, more commonly, simply ***structures***, because they give a pictorial representation of the computations in the difference equations that can be used to implement the system.

Recall that the product of two z-transform system functions corresponds to the cascade of two systems. Therefore, the system function for the first-order feedback filter can be factored into an FIR piece and an IIR piece, as in

$$H(z) = \frac{b_0 + b_1 z^{-1}}{1 - a_1 z^{-1}} = \left(\frac{1}{1 - a_1 z^{-1}}\right)\left(b_0 + b_1 z^{-1}\right) = \left(\frac{1}{A(z)}\right) B(z) \tag{10.22}$$

The conclusion to be drawn from the algebraic factors in (10.22) is that a valid implementation for $H(z)$ is the pair of difference equations

$$v[n] = b_0 x[n] + b_1 x[n-1] \tag{10.23a}$$

$$y[n] = a_1 y[n-1] + v[n] \tag{10.23b}$$

We see in Fig. 10-5 that the polynomial $B(z)$ is the system function of the feed-forward part (10.23a) of the block diagram, and that $1/A(z)$ is the system function of the feedback part (10.23b) that completes the system. The system implemented in this way is called the ***Direct Form I*** implementation because it is possible to go *directly* from the system function to the filter coefficients of this block diagram (or the difference equations that it represents) by using the coefficients of the numerator and denominator polynomials in $H(z)$.

10-4.2 Direct Form II Structure

We know that for an LTI cascade system, we can change the order of the systems without changing the overall system response. This is obvious from the system function since

$$H(z) = \left(\frac{1}{A(z)}\right) B(z) = B(z)\left(\frac{1}{A(z)}\right) \tag{10.24}$$

Using the correspondences that we have established, the right-hand side of (10.24) leads to the block diagram shown in Fig. 10-6 where the feedback block is first in the

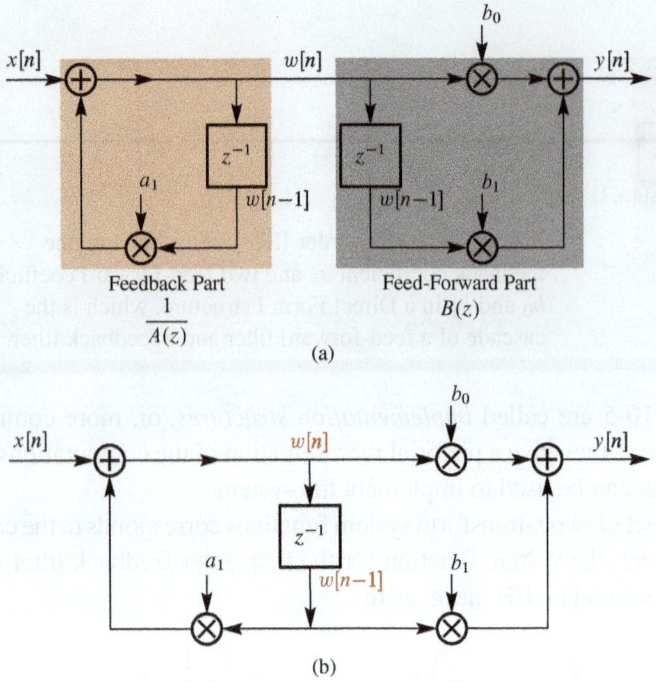

Figure 10-6 (a) First-order IIR system showing one feedback coefficient a_1 and two feed-forward coefficients b_0 and b_1 in a cascade structure with the feedback part first. (b) First-order IIR system in Direct Form II structure. The two delays in (a) have been merged into one.

cascade. Note that we have defined a different intermediate variable $w[n]$ as the output of the feedback part and input to the feed-forward part. Thus, the block diagram shows that an equivalent implementation of the system defined by (10.25) is

$$w[n] = a_1 w[n-1] + x[n] \tag{10.25a}$$

$$y[n] = b_0 w[n] + b_1 w[n-1] \tag{10.25b}$$

The block diagram representation of Fig. 10-6(a) can be simplified according to the following observation. Notice that the input to each of the unit delay operators is the same signal $w[n]$. Thus, there is no need for two distinct delay operations; they can be combined into a single delay, as in Fig. 10-6(b). Since delay operations are implemented with memory in a computer, the implementation of Fig. 10-6(b) would require less memory than the implementation of Fig. 10-6(a). Note, however, that both block diagrams represent the same difference equations (10.25a) and (10.25b). Again, because there is such a direct and simple correspondence between the filter coefficients in Fig. 10-6(b) and the polynomial coefficients in $H(z)$, this implementation is called the ***Direct Form II*** implementation of the first-order IIR system with system function

$$H(z) = \frac{b_0 + b_1 z^{-1}}{1 - a_1 z^{-1}}$$

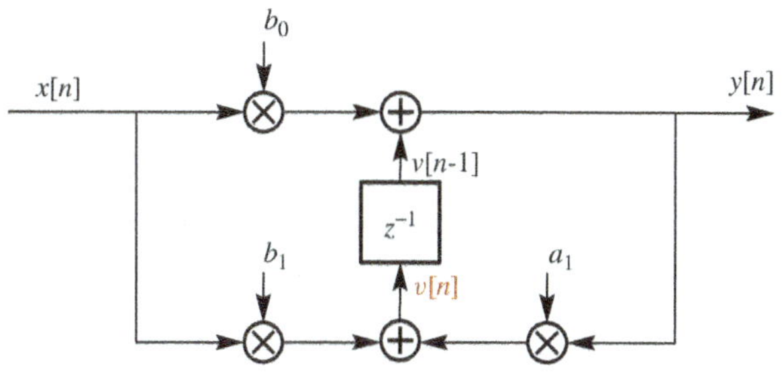

Figure 10-7 Computational structure from (10.26) for a general first-order IIR filter as a Transposed Direct Form II structure.

EXERCISE 10.5 Find the z-transform system function of the following set of cascaded difference equations:

$$w[n] = -0.5w[n-1] + 7x[n]$$

$$y[n] = 2w[n] - 4w[n-1]$$

Draw the block diagrams of this system in both Direct Form I and Direct Form II.

10-4.3 The Transposed Form Structure

In the direct form structures of Figs. 10-5 and 10-6, the sequence of operations on the signals is delay, multiply, and then add. If the z-transform of the general first-order difference equation (10.15a) is rewritten as

$$Y(z) = b_0 X(z) + \underbrace{z^{-1}}_{\text{delay}} \underbrace{(b_1 X(z) + a_1 Y(z))}_{V(z) \text{ is mult. \& add}} \tag{10.26}$$

a different computational structure is obtained, where the order of operations is multiply, add, and then delay, as shown in the block diagram of Fig. 10-7. The corresponding difference equations are

$$y[n] = b_0 x[n] + v[n-1] \tag{10.27a}$$

$$v[n] = b_1 x[n] + a_1 y[n] \tag{10.27b}$$

Specifically, the signals $x[n]$ and $y[n]$ are multiplied by b_1 and a_1, respectively, and then added together. The result is the signal $v[n]$ in (10.27b) and in Fig. 10-7. In (10.27a), the previous version of $v[n]$ is added to $b_0 x[n]$ to get the output $y[n]$. The feedback structure given in the block diagram of Fig. 10-7 is called the ***transposed*** form[4] of the Direct Form II structure shown in Fig. 10-7.

[4]For directed graphs, transposing is a well-defined mathematical operation that reverses all the branches. For block diagrams, nodes with multiple inputs must become adders while nodes with multiple outputs become branch points; also we must relabel the input as the output, and vice versa, to get the same system function.

In order to verify the actual difference equation for the transposed form, we need to write the equations that are defined by the block diagram. There is an orderly procedure for doing this if we follow two rules:

1. Assign variable names to the inputs of all delay elements. For example, $v[n]$ is used in Fig. 10-7, so the output of the delay is $v[n-1]$.
2. Write equations at all of the summing nodes; there are two in this case. The result is the two difference equations in (10.27a) and (10.27b).

The block diagram specifies an actual computation, so (10.27a) and (10.27b) require three multiplications and two additions at each time step n. Equation (10.27a) must be done first, because $y[n]$ is needed in (10.27b). Owing to the feedback, it is impossible to manipulate these n-domain equations into one of the cascade forms by eliminating variables. However, we can recombine these two equations in the z-transform domain to verify that we have the correct system function. First, we take the z-transform of each difference equation obtaining

$$Y(z) = b_0 X(z) + z^{-1} V(z) \tag{10.28a}$$

$$V(z) = b_1 X(z) + a_1 Y(z) \tag{10.28b}$$

Now we eliminate $V(z)$ by substituting the second equation into the first as follows:

$$Y(z) = b_0 X(z) + z^{-1}(b_1 X(z) + a_1 Y(z)) \tag{10.28c}$$

The result is identical to (10.26) which is one form of the z-transform of the general first-order difference equation (10.15a). Since they have the same system function and implement the same difference equation, (10.27a) and (10.27b) are equivalent to the Direct Form I (10.23a) and (10.23b), and to the Direct Form II (10.25a) and (10.25b).

Why are these different implementations of the same system function of interest to us? They all use the same number of multiplications and additions to compute exactly the same output from a given input. However, this is true only when the arithmetic is perfect. On a computer with finite precision (e.g., 16-bit words), each calculation involves round-off errors, which means that each implementation behaves slightly differently. In practice, the implementation of high-quality digital filters in hardware demands careful engineering to control round-off errors and overflows.

10-5 Poles and Zeros

Since the z-transform system function for an IIR filter is a rational function, its numerator and denominator polynomials have roots. These root locations in the complex z-plane provide an efficient way to characterize the system. Although we prefer to write the system function in terms of z^{-1} to facilitate the correspondence with the difference equation, when finding roots it might be better to rewrite the polynomials with powers of z. Thus we multiply the numerator and denominator of the first-order system function (10.15e) by z to obtain

$$H(z) = \frac{b_0 + b_1 z^{-1}}{1 - a_1 z^{-1}} = \frac{b_0 z + b_1}{z - a_1} \qquad (10.29)$$

The roots are zeros of the numerator and denominator polynomials, so the numerator has one root which is obtained by solving

$$b_0 z + b_1 = 0 \quad \Longrightarrow \quad \text{Root at } z = -\frac{b_1}{b_0}$$

Likewise, the denominator root comes from solving

$$z - a_1 = 0 \quad \Longrightarrow \quad \text{Root at } z = a_1$$

If we consider $H(z)$ as a function of z over the entire complex z-plane, the root of the numerator is a zero of the numerator polynomial and thus a ***zero*** of the system function $H(z)$, that is,

$$H(z)\Big|_{z=-(b_1/b_0)} = 0$$

Furthermore, the root of the denominator is a location in the z-plane where the denominator polynomial is zero, and thus where the function $H(z)$ *blows up*

$$H(z)\Big|_{z=a_1} \to \infty$$

so this location ($z = a_1$) is called a ***pole*** of the system function $H(z)$.

EXERCISE 10.6 Verify that the poles and zeros of the following z-transform system function are located at $z = -0.5$ and $z = -4/3$:

$$H(z) = \frac{3 + 4z^{-1}}{1 + 0.5z^{-1}}$$

EXERCISE 10.7 For the system function $H(z)$ of the following IIR system

$$y[n] = 0.5y[n-1] - x[n] + 3x[n-1]$$

determine the locations of the poles and zeros.

10-5.1 Roots in MATLAB

Since MATLAB is used extensively in the labs and demonstrations that accompany this text, it is worthwhile to point out how MATLAB handles polynomials and their roots. Not surprisingly, the MATLAB format for polynomials is a vector of coefficients. The ordering in the vector is compatible with the order of the coefficients in the functions `filter` and `freqz`. For example, the system function in (10.29) would be represented in MATLAB by the vectors `b=[b0,b1]` and `a=[1,-a1]` where the coefficients are thought of in terms of powers of z^{-1} moving left-to-right in the vector as in the numerator and denominator of the first rational form in (10.29). This is the format assumed by the MATLAB function `filter(b,a,xx)` for implementing the difference equation of the system, and also for `freqz(b,a,omega)` which computes samples of its frequency response. When using the MATLAB function `roots`, we can use exactly the same vectors `b` and `a`, because the `roots` function assumes that the first coefficient in a vector of polynomial coefficients is the coefficient of the *highest* power of z as in the numerator and denominator polynomials of the right-hand form of (10.29). Thus, the MATLAB statements `zeros=roots(b)` and `poles=roots(a)` compute the zeros and poles for a system function represented as a ratio of polynomials in the variable z.

DEMO
PeZ GUI

The MATLAB function `zplane`, or the *DSP-First* function `zzplane`, makes a pole-zero plot in the z-plane. Several other MATLAB-based resources are available to illustrate and demonstrate the pole-zero plot and its relation to the impulse response and frequency response of an LTI system. These relationships are illustrated by the MATLAB `PeZ` GUI.

10-5.2 Poles or Zeros at $z = 0$ or ∞

When the numerator and denominator polynomials have a different number of coefficients, we can have either zeros or poles at $z = 0$, or at $z = \infty$. We saw this behavior in Chapter 9, where causal FIR systems, whose system functions have only a numerator polynomial, had a number of poles at $z = 0$ equal to the number of zeros of the polynomial. In the general IIR case, if we include all the poles and zeros at $z = \infty$, as well as all at $z = 0$, then we can assert that ***the number of poles of $H(z)$ equals the number of zeros***. The following examples illustrate this.

EXAMPLE 10-6 Find Poles and Zeros

The system function of the IIR system

$$y[n] = 0.5y[n-1] + 3x[n]$$

can be written down directly from the filter coefficients as

$$H(z) = \frac{3}{1 - 0.5z^{-1}}$$

When we express $H(z)$ in positive powers of z

$$H(z) = \frac{3z}{z - 0.5}$$

we see that there is one pole at $z = 0.5$ and one zero at $z = 0$.

In the next example, the poles at $z = \infty$ are found by taking the limit as $z \to \infty$.

EXAMPLE 10-7 Zeros at $z = \infty$

The system function of the IIR system

$$y[n] = 0.5y[n-1] + 3x[n-2]$$

can be written down from the filter coefficients as

$$H(z) = \frac{3z^{-2}}{1 - 0.5z^{-1}} = \frac{3}{z(z - 0.5)}$$

The system has poles at $z = 0$ and $z = 0.5$, and taking the limit $\lim_{z\to\infty} H(z)$, we get $H(z) \to 3/z^2 \to 0$. Thus, it also has two zeros at $z = \infty$.

10-5.3 Output Response from Pole Location

The pole location of a first-order filter determines the shape of the impulse response. In Example 10-5 on p. 428, we showed that a first-order system with a numerator $(b_0 + b_1 z^{-1})$ has a system function that can be written as the sum of two first-order terms

$$H(z) = b_0\left(\frac{1}{1 - a_1 z^{-1}}\right) + b_1 z^{-1}\left(\frac{1}{1 - a_1 z^{-1}}\right)$$

Thus the corresponding impulse response has two terms:

$$h[n] = b_0(a_1)^n u[n] + b_1(a_1)^{n-1}u[n-1] \tag{10.30a}$$

$$= \begin{cases} 0 & \text{for } n < 0 \\ b_0 & \text{for } n = 0 \\ (b_0 + b_1 a_1^{-1})a_1^n & \text{for } n \geq 1 \end{cases} \tag{10.30b}$$

This IIR system has a single pole at $z = a_1$ and its impulse response (10.30b) is proportional to a_1^n for $n \geq 1$. If we change the numerator polynomial to include a term with z^{-n_0}, this statement is modified to be true for $n \geq n_0$, although the constant of proportionality would be different. Similar behavior was observed in Example 10-3 for a first-order IIR filter with a finite-length input. For all of these cases, eventually the response of the IIR system behaves as $(\text{pole})^n$ once $n \geq n_0$, and this $(\text{pole})^n$ behavior occurs when the input signal is zero.

Since the pole of the system function is at $z = a_1$, we see that the location of the pole can tell us whether the impulse response decays or grows as $n \to \infty$. If $|a_1| < 1$, the response dies out as $n \to \infty$, but if $|a_1| \geq 1$, the response does not die out. In fact, if $|a_1| > 1$, the output signal grows exponentially without bound, which is undesirable because a system that produces unbounded outputs is generally useless. A system that exhibits this property is called an unstable system. As we will see in Section 10-6, systems that produce bounded outputs when the input is bounded are called ***stable systems***.

For the first-order case, stability can be determined by checking the pole location with respect to the unit circle. If $|a_1| < 1$, the pole of the system function is inside the unit circle of the z-plane, and the system is stable. In Section 10-6, we will show that this ***poles inside the unit-circle condition*** is true even for causal higher-order IIR systems. Thus, stability of a system can be determined at a glance from a z-plane pole-zero plot of the system function.

EXAMPLE 10-8 Stability from Pole Location

The causal system whose system function is

$$H(z) = \frac{1 - 2z^{-1}}{1 - 0.8z^{-1}} = \frac{z - 2}{z - 0.8}$$

has a zero at $z = 2$ and a pole at $z = 0.8$. Therefore, the system is stable. The location of the zero, which is outside the unit circle, has nothing to do with stability of the system.

EXERCISE 10.8 Assume that the system function of an IIR filter is

$$H(z) = \frac{2 + 2z^{-1}}{1 - 1.25z^{-1}}$$

Plot the pole and zero in the z-plane, and determine whether or not the system is stable.

10-6 Stability of IIR Systems

We have just seen that the pole location of a first-order IIR system is sufficient to determine whether or not the system is stable. In this section, a general discussion of stability is presented which applies to all causal LTI systems. The general condition for stability is shown to be the absolute summability of the impulse response, which, in the case of a causal IIR system with multiple poles, leads to the statement that ***all the poles must be inside the unit circle***.

Because IIR systems have impulse responses of infinite duration due to the feedback inherent in their difference equations, it is important that the response does not "blow up" as the difference equation is iterated indefinitely. This requires that the system be *stable*. One common definition of stability is that the output of a stable system can always be bounded, $|y[n]| < M_y < \infty$, for any input that is bounded as $|x[n]| < M_x < \infty$. The finite constants M_x and M_y can be different. This definition for stability is called ***bounded-input, bounded-output*** or BIBO stability; it is a mathematically precise way of saying that "the output doesn't blow up."

For an LTI system, the BIBO definition can be reduced to one simple test on the impulse response of the LTI system, because the input and output are related by convolution. In the case of a *causal* LTI system, the lower limit on the convolution sum is zero, so

$$y[n] = \sum_{m=0}^{\infty} h[m]x[n-m]$$

Therefore, a bound on the size of $|y[n]|$ can be obtained as follows:

$$\left|y[n]\right| = \left|\sum_{m=0}^{\infty} h[m]x[n-m]\right| \le \sum_{m=0}^{\infty} |h[m]| \underbrace{|x[n-m]|}_{\le M_x} \tag{10.31a}$$

where we have used the fact that the magnitude of a sum of terms is less than or equal to the sum of the magnitudes. We can strengthen the inequality (10.31a) by replacing $|x[n-m]|$ for all m by its bound M_x.

$$|y[n]| < \underbrace{M_x \sum_{m=0}^{\infty} |h[m]|}_{=M_y\ ?} \tag{10.31b}$$

Therefore, a sufficient condition[5] for bounding $|y[n]|$ with $M_y < \infty$, whenever $|x[n]| < M_x < \infty$, is

Condition for Stability of a Causal LTI System

$$\sum_{m=0}^{\infty} \left|h[m]\right| < \infty \tag{10.32}$$

When (10.32) is satisfied, we say that the impulse response is *absolutely summable*, and this property guarantees that the LTI system is stable.

In the case of FIR systems, the impulse response has finite duration, and (10.32) is satisfied as long as the values of $h[n]$ are finite. Therefore, all practical FIR systems are stable. However, in the case of IIR systems, the impulse response is an infinitely long sequence, so some IIR systems might not be stable.

[5] It can be shown that (10.32) is also a necessary condition—any system that violates (10.32) is an *unstable* system.

EXAMPLE 10-9 Unstable System

Consider the first-order difference equation where the lone feedback coefficient is $a_1 = 1$,

$$y[n] = y[n-1] + x[n] \tag{10.33}$$

This system is often called an *accumulator system* because it simply adds the current sample of the input, $x[n]$, to the total of previous samples, $y[n-1]$. The impulse response of this system can be shown by iteration (10.8c) to be the unit-step signal $h[n] = u[n]$. From the z-transform pair in (10.18) with $a = 1$, it follows that the system function is

$$H(z) = \sum_{n=0}^{\infty} z^{-n} = \frac{1}{1 - z^{-1}} \tag{10.34}$$

where the associated ROC for the infinite sum is $1 < |z|$.

Applying the condition for stability in (10.32) to $h[n] = u[n]$, we conclude that this is NOT a stable system, because the absolutely summability test is

$$\sum_{n=0}^{\infty} |u[n]| = \sum_{n=0}^{\infty} 1 \to \infty$$

Thus, it must be true that there is some bounded input that produces an unbounded output. One such example is a shifted step input $x[n] = u[n-1]$, for which the bound is $M_x = 1$. The result of Exercise 10.9 is that the output $y[n] = nu[n-1]$ grows linearly with n, so we cannot find a constant M_y such that $|y[n]| < M_y$ for all n.

EXERCISE 10.9 Show by iteration of (10.33) that the output due to the shifted step input, $x[n] = u[n-1]$, is

$$y[n] = nu[n-1] \tag{10.35}$$

Note that having this one example (10.35) of an unbounded output is enough to prove that the accumulator system (10.33) is *unstable*. On the other hand, observing that one or even many bounded inputs produce corresponding bounded outputs is *not sufficient* to prove that the accumulator system is stable. For this particular system, there are many bounded inputs that do produce a bounded output. One such input is the signal $(-1)^n u[n]$, which is bounded by one and produces as output the response, $y[n] = (0.5+0.5(-1)^n)u[n]$, which is bounded by one. Fortunately, (10.32) is a necessary and sufficient condition for stability of an LTI system with impulse response $h[n]$, and we do not have to test the system for all inputs. We only need to apply the absolute summability test (10.32) once to the impulse response.

10-6.1 The Region of Convergence and Stability

In this section, we show that stability can be tested in the z-transform domain. Specifically, the ROC of $H(z)$ can be related to stability of the IIR system because both concepts are based on absolute summability. The definition of the system function is the z-transform of the impulse response, and, if the system is causal, the lower limit on the z-transform sum is $n = 0$, so

$$H(z) = \sum_{n=0}^{\infty} h[n]z^{-n} \tag{10.36}$$

The infinite sum (10.36) might not be finite everywhere in the z-plane, but within the ROC the sum converges to finite (complex) values. Finiteness of $|H(z)|$ in (10.36) can be guaranteed by requiring *absolute convergence of the infinite sum*

$$\underbrace{\left|\sum_{n=0}^{\infty} h[n]z^{-n}\right|}_{|H(z)|} \le \underbrace{\sum_{n=0}^{\infty} |h[n]|\,|z^{-n}| < \infty}_{\text{absolute convergence}} \tag{10.37}$$

The inequality in (10.37) relies on the fact that the magnitude of a sum is less than the sum of the magnitudes of the individual terms. Also, on the right-hand side, we have used the fact that the magnitude of a product of two terms is equal to the product of the magnitudes. Thus, the ROC is defined to be the set of z values where the sum (10.36) converges absolutely.

The unit circle in the z-plane is the case where $z = e^{j\hat{\omega}}$, or $|z| = 1$, and if we evaluate the right-hand sum in (10.37) on the unit circle we get

$$\sum_{n=0}^{\infty} |h[n]|\,\overset{1}{\cancel{|e^{-j\hat{\omega}n}|}} = \sum_{n=0}^{\infty} |h[n]| < \infty \tag{10.38}$$

which implies that if the unit circle lies within the ROC, then the impulse response is absolutely summable. Thus, we can state a z-domain condition for stability of an LTI system based on the ROC:

An IIR system is stable, if and only if the region of convergence for its system function $H(z)$ includes the unit circle of the z-plane.

An easy case to remember is the first-order case (10.18), $h[n] = a_1^n u[n]$, where the ROC is $\{z : |a_1| < |z|\}$ which is a subset of the z-plane that is the exterior of a disk of radius $|a_1|$. When $|a_1| < 1$, the exterior of the disk contains the unit circle. Note that it is also true that the first-order impulse response is absolutely summable when $|a_1| < 1$.

For a general causal LTI IIR system, we can prove that the ROC must be the exterior of a disk. In the following sections, we will show that the impulse response of a higher-order causal LTI system is the weighted sum of exponential terms

$$h[n] = \sum_{k=1}^{N} A_k p_k^n u[n] \tag{10.39}$$

where the p_k are the poles of the system and A_k are constant weights.[6] To have a valid system function $H(z)$, each of the exponential terms $p_k^n u[n]$ must have a z-transform that is absolutely convergent, so the ROC for each term is the disk exterior $\{z : |p_k| < |z|\}$. Then the ROC of $H(z)$ is the intersection of all the disk exteriors which is also a disk exterior of the form $\{z : R < |z|\}$, where R is the magnitude of the pole that is farthest from the origin of the z-plane. For stability, we need to have the unit circle inside the ROC which is true only if all the poles are inside the unit circle.

10-7 Frequency Response of an IIR Filter

In Chapter 6, we introduced the concept of frequency response $H(e^{j\hat{\omega}})$ as the complex function that determines the amplitude and phase change experienced by a complex exponential input to an LTI system. That is, if $x[n] = e^{j\hat{\omega}n}$, then the output is

$$y[n] = H(e^{j\hat{\omega}})e^{j\hat{\omega}n} \tag{10.40}$$

In Chapter 7, we generalized the frequency response to be the same as the DTFT of the impulse response. In Section 9-5 on p. 384, we showed that the frequency response of an FIR system is related to the system function by

$$H(e^{j\hat{\omega}}) = H(z)\Big|_{z=e^{j\hat{\omega}}} \tag{10.41}$$

That is, the frequency response is identical to the system function evaluated on the unit circle of the z-plane ($z = e^{j\hat{\omega}}$). The relation (10.41) also holds for IIR systems, but we need one extra provision: The IIR system must be stable in order for the frequency response to exist and to be given by (10.41). Recall that the condition for stability in (10.32) is the absolute summability of the impulse response, which in turn guarantees that the unit circle lies in the ROC, so the substitution in (10.41) is then legitimate. In Chapter 7, the condition for existence of the DTFT (7.7) was also shown to be absolute summability of the sequence.

Recall that the system function for the general first-order IIR system has the form

$$H(z) = \frac{b_0 + b_1 z^{-1}}{1 - a_1 z^{-1}}$$

where the ROC is $|a_1| < |z|$. This system is stable if $|a_1| < 1$, so we can obtain a formula for the frequency response by evaluating $H(z)$ on the unit circle:

$$H(e^{j\hat{\omega}}) = H(z)\Big|_{z=e^{j\hat{\omega}}} = \frac{b_0 + b_1 e^{-j\hat{\omega}}}{1 - a_1 e^{-j\hat{\omega}}} \tag{10.42}$$

This rational function (10.42) contains terms with $e^{-j\hat{\omega}}$, which are periodic in $\hat{\omega}$ with period 2π, so $H(e^{j\hat{\omega}})$ is periodic with a period equal to 2π, as expected for the frequency response of a discrete-time system.

[6] Equation (10.39) is true only when all the poles $\{p_k\}$ are distinct.

The frequency response $H(e^{j\hat{\omega}})$ in (10.42) is a complex-valued function of frequency $\hat{\omega}$, so we can reduce (10.42) to two separate real formulas for the magnitude and the phase as functions of frequency. For the magnitude response, it is expedient to compute the magnitude squared first, and then take the positive square root. The magnitude squared can be formed by multiplying $H(e^{j\hat{\omega}})$ in (10.42) by its complex conjugate (denoted by H^*). For our first-order example,

$$\begin{aligned}
\left|H(e^{j\hat{\omega}})\right|^2 &= H(e^{j\hat{\omega}})H^*(e^{j\hat{\omega}}) \\
&= \frac{b_0 + b_1 e^{-j\hat{\omega}}}{1 - a_1 e^{-j\hat{\omega}}} \cdot \frac{b_0^* + b_1^* e^{+j\hat{\omega}}}{1 - a_1^* e^{+j\hat{\omega}}} \\
&= \frac{|b_0|^2 + |b_1|^2 + b_0 b_1^* e^{+j\hat{\omega}} + b_0^* b_1 e^{-j\hat{\omega}}}{1 + |a_1|^2 - a_1^* e^{+j\hat{\omega}} - a_1 e^{-j\hat{\omega}}} \\
&= \frac{|b_0|^2 + |b_1|^2 + 2\Re\{b_0^* b_1 e^{-j\hat{\omega}}\}}{1 + |a_1|^2 - 2\Re\{a_1 e^{-j\hat{\omega}}\}}
\end{aligned}$$

This derivation does not assume that the filter coefficients are real. If the coefficients are real, we get the further simplification

$$\left|H(e^{j\hat{\omega}})\right|^2 = \frac{|b_0|^2 + |b_1|^2 + 2b_0 b_1 \cos(\hat{\omega})}{1 + |a_1|^2 - 2a_1 \cos(\hat{\omega})} \tag{10.43}$$

This formula is not particularly informative, because it is difficult to use it to visualize the shape of $|H(e^{j\hat{\omega}})|$. However, it could be used to write a program for evaluating and plotting the frequency response. The phase response is even messier. Arctangents are needed to compute the angles of the numerator and denominator, and then the two phases must be subtracted. When the filter coefficients are real, the phase is

$$\angle H(e^{j\hat{\omega}}) = \tan^{-1}\left(\frac{-b_1 \sin\hat{\omega}}{b_0 + b_1 \cos\hat{\omega}}\right) - \tan^{-1}\left(\frac{a_1 \sin\hat{\omega}}{1 - a_1 \cos\hat{\omega}}\right)$$

Again, the formula is so complicated that we cannot gain much insight from it directly. In a later section, we will use the poles and zeros of the system function together with the relationship (10.41) to construct an approximate plot of the frequency response without recourse to such complicated formulas.

10-7.1 Frequency Response Using MATLAB

Frequency responses can be computed and plotted easily by many signal processing software packages. In MATLAB, for example, the function `freqz` is provided for just that purpose.[7] The frequency response is evaluated over an equally spaced grid in the $\hat{\omega}$ domain, and then its magnitude and phase can be plotted. In MATLAB, the functions

[7]The function `freqz` is part of MATLAB's Signal Processing Toolbox. In case a substitute is needed, there is a similar function called `freekz` that is part of the *DSP-First* Toolbox.

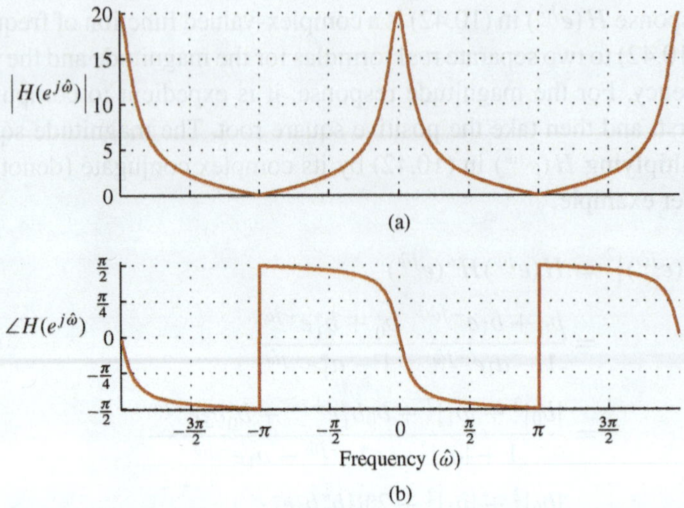

Figure 10-8 Frequency response (magnitude and phase) for a first-order feedback filter. The pole is at $z = 0.8$, and the numerator has a zero at $z = -1$.

`abs` and `angle` compute the magnitude and the angle of each element in a complex vector.

EXAMPLE 10-10 Plot $H(e^{j\hat{\omega}})$ via MATLAB

Consider the example

$$y[n] = 0.8y[n-1] + 2x[n] + 2x[n-1]$$

In order to define the filter coefficients in MATLAB, we put all the terms with $y[n]$ on one side of the equation, and the terms with $x[n]$ on the other.

$$y[n] - 0.8y[n-1] = 2x[n] + 2x[n-1]$$

Then we read off the filter coefficients and define the vectors `aa` and `bb` as

```
aa = [ 1, −0.8 ]        bb = [ 2, 2 ]
```

Thus, the vectors `aa` and `bb` are in the same form as for the `filter` function. The following call to `freqz` generates a 401-point vector `HH` containing the values of the frequency response at the vector of frequencies specified by the third argument, `2*pi*[-1:.005:1]`:

```
HH = freqz( bb, aa, 2*pi*[-1:.005:1] );
```

Plots of the resulting magnitude and phase are shown in Fig. 10-8. An extended frequency interval $-2\pi \leq \hat{\omega} \leq +2\pi$ is shown so that the 2π-periodicity of $H(e^{j\hat{\omega}})$ is evident.[8]

In this example, we can look for a connection between the poles and zeros and the shape of the frequency response. From the coefficients of the difference equation, we obtain the system function

[8]The labels on the graph were created using special plotting functions.

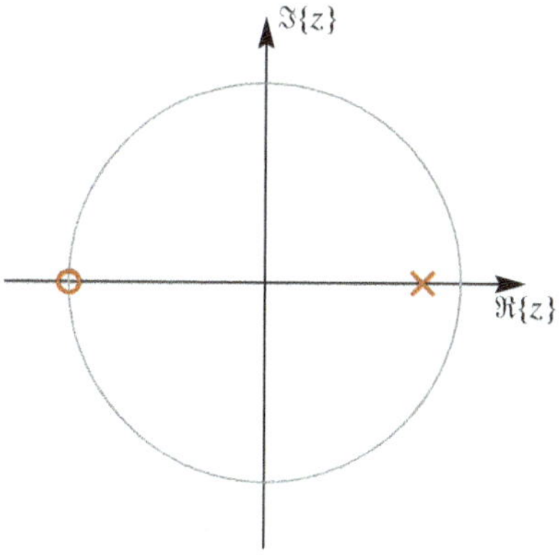

Figure 10-9 Pole-zero plot for a first-order IIR filter. The pole is at $z = 0.8$, and the numerator has a zero at $z = -1$.

$$H(z) = \frac{2 + 2z^{-1}}{1 - 0.8z^{-1}}$$

which, as shown in Fig. 10-9, has a pole at $z = 0.8$ and a zero at $z = -1$.

The point $z = -1$ is the same as $\hat{\omega} = \pi$ because $z = -1 = e^{j\pi} = e^{j\hat{\omega}}\big|_{\hat{\omega}=\pi}$. Thus, $H(e^{j\hat{\omega}})$ has the value zero at $\hat{\omega} = \pi$, since $H(z)$ is zero at $z = -1$. In a similar manner, the pole at $z = 0.8$ has an effect on the frequency response near $\hat{\omega} = 0$. Since $H(z)$ blows up at $z = 0.8$, the nearby points on the unit circle must have large values. The closest point on the unit circle is at $z = e^{j0} = 1$. In this case, we can evaluate the frequency response directly from the $H(z)$ formula to get

$$H(e^{j\hat{\omega}})\Big|_{\hat{\omega}=0} = H(z)\Big|_{z=1} = \frac{2 + 2z^{-1}}{1 - 0.8z^{-1}}\Big|_{z=1} = \frac{2 + 2(1)}{1 - 0.8(1)} = \frac{4}{0.2} = 20$$

10-7.2 Three-Dimensional Plot of a System Function

The relationship between $H(e^{j\hat{\omega}})$ and the pole-zero locations of $H(z)$ can be illustrated by making a 3-D plot of $H(z)$ over the complex z-plane. The frequency response $H(e^{j\hat{\omega}})$ is obtained by selecting the values of $H(z)$ only along the unit circle, because $z = e^{j\hat{\omega}}$ defines the unit circle for $-\pi < \hat{\omega} \leq \pi$.

In this section, we use as an example the system function

$$H(z) = \frac{1}{1 - 0.8z^{-1}} \tag{10.44}$$

to illustrate the relationship between $H(z)$ and the frequency response. Figures 10-10 and 10-11 are plots of the magnitude and phase of $H(z)$ over the region $|z| \leq 1.4$ of the z-plane. In the magnitude plot of Fig. 10-10, we observe that the pole (at $z = 0.8$) creates a large peak that makes all nearby values very large. At the precise location of the pole, $H(z) \to \infty$, but the 3-D surface has been truncated vertically, so the plot stays within a finite vertical scale. The phase response in Fig. 10-11 also exhibits its most rapid transition at $\hat{\omega} = 0$ which is the point $z = 1$ (the closest point on the unit circle to the pole at $z = 0.8$).

DEMO
Z to Freq

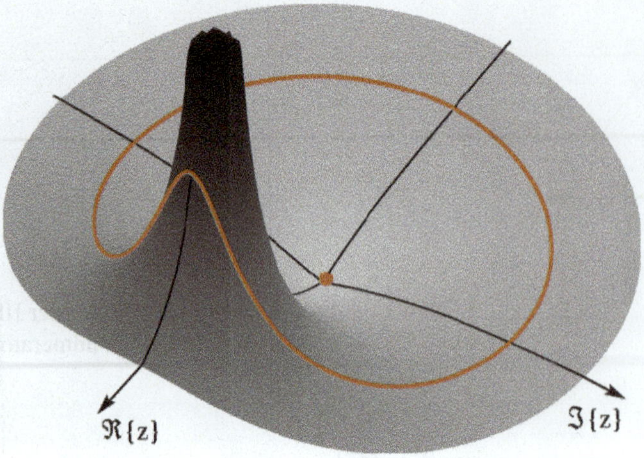

Figure 10-10 Magnitude of $H(z)$ in (10.44) evaluated over a region of the z-plane including the unit circle. The first-order filter has a pole at $z = 0.8$ and a zero at $z = 0$ (shown with an orange dot). Values of $|H(z)|$ along the unit circle are shown as a solid orange line where the frequency response (magnitude) is evaluated. The view is from the first quadrant, so the point $z = 1$ is in the foreground on the left.

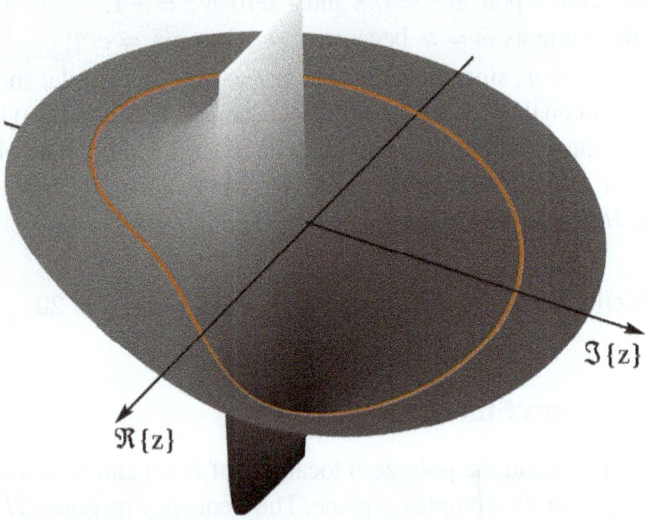

Figure 10-11 Phase of $H(z)$ evaluated over a region of the z-plane that includes the unit circle. Values of the phase along the unit circle are shown as a solid orange line. View is from the first quadrant, so the point $z = 1$ lies to the left in the foreground.

DEMO
Three Domains - IIR

The frequency response $H(e^{j\hat{\omega}})$ is obtained by selecting the values of the z-transform along the unit circle in Figs. 10-10 and 10-11. Two-dimensional plots of $H(e^{j\hat{\omega}})$ versus $\hat{\omega}$ are given in Fig. 10-12. The shape of the frequency response can be explained in terms of the pole location by recognizing that in Fig. 10-10 the pole at $z = 0.8$ pushes $H(e^{j\hat{\omega}})$ up in the region near $\hat{\omega} = 0$, which is the same as the point $z = 1$. The unit circle values follow the ups and downs of $H(z)$ as $\hat{\omega}$ goes from $-\pi$ to $+\pi$.

10-8 Three Domains

As depicted in Fig. 10-13, the coefficients $\{a_k, b_k\}$ play a key role in linking the n, $\hat{\omega}$, and z domain representations of signals and LTI systems.

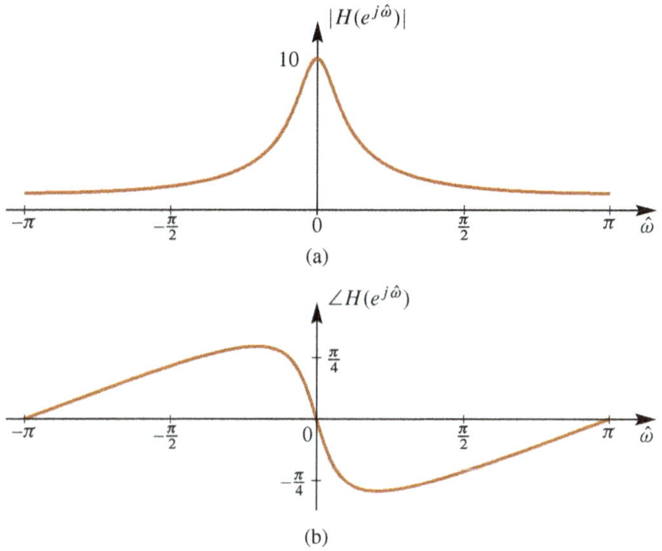

Figure 10-12 Frequency response for the first-order IIR filter (10.44). (a) Magnitude. (b) Phase. The pole is at $z = 0.8$ and the numerator has a zero at $z = 0$. These are the values of $H(z)$ along the unit circle in the z-plane of Figs. 10-10 and 10-11.

To illustrate the use of the analysis tools that we have developed in these three domains, we consider the general second-order IIR digital filter which is the following feedback difference equation

$$y[n] = a_1 y[n-1] + a_2 y[n-2] + b_0 x[n] + b_1 x[n-1] + b_2 x[n-2]$$

with constant filter coefficients $\{a_1, a_2, b_0, b_1, b_2\}$. This equation provides the algorithm for computing the output signal from the input signal by iteration. For example, the output is the impulse response $h[n]$ when the input is $x[n] = \delta[n]$.

PeZ - The z, n, and $\hat{\omega}$ Domains

Following the procedures illustrated for the first-order case, we can obtain the z-transform system function directly from the filter coefficients as

$$H(z) = \frac{b_0 + b_1 z^{-1} + b_2 z^{-2}}{1 - a_1 z^{-1} - a_2 z^{-2}}$$

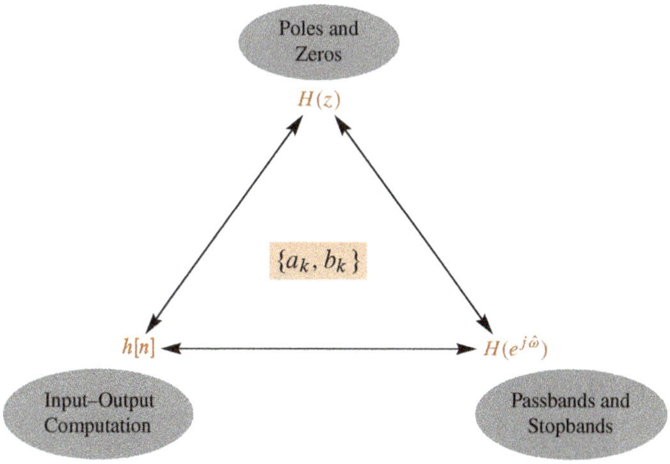

Figure 10-13 Relationship among the n-, z-, and $\hat{\omega}$-domains for an LTI system. The filter coefficients $\{a_k, b_k\}$ play a central role in connecting the three domains for an IIR system. For an FIR system the set of filter coefficients is only $\{b_k\}$.

and, when the poles are inside the unit circle, we can obtain the frequency response, or the DTFT, by evaluating on the unit circle ($z = e^{j\hat{\omega}}$)

$$H(e^{j\hat{\omega}}) = \frac{b_0 + b_1 e^{-j\hat{\omega}} + b_2 e^{-j2\hat{\omega}}}{1 - a_1 e^{-j\hat{\omega}} - a_2 e^{-j2\hat{\omega}}}$$

Since $H(z)$ is a ratio of polynomials, either the coefficients $\{a_k, b_k\}$ or the poles and zeros of $H(z)$ comprise a small set of parameters that completely define the filter.

Finally, the shapes of the passbands and stopbands of the frequency response are highly dependent on the pole and zero locations with respect to the unit circle, and the character of the impulse response can be related to the pole locations. The PeZ GUI discussed in Section 10-12.4 is useful for visualizing these relationships among the three domains for a system. To make this last point for the general case, we require one more tool—a technique for getting $h[n]$ directly from $H(z)$. This process, which applies to any z-transform and its corresponding sequence, is called the inverse z-transform and is presented in the next section.

DEMO
PeZ GUI

10-9 The Inverse z-Transform and Applications

We have seen how the three domains are connected for the first-order IIR system. Many of the concepts that we have introduced for the first-order system can be extended to higher-order systems in a straightforward manner. However, finding the impulse response from the system function is not an obvious extension of what we have done for the first-order case. We need to develop a process for inverting the z-transform that can be applied to systems with more than one pole. This process is called the ***inverse z-transform***. In this section, we show how to find the inverse for a general rational z-transform and then illustrate the process with some examples of determining the impulse responses of second-order and higher-order systems.

10-9.1 Revisiting the Step Response of a First-Order System

In Section 10-2.4, we computed the step response of a first-order system by iteration. Now we show how the inverse z-transform can be used to obtain the same result with some simple algebra. Consider a system whose system function is

$$H(z) = \frac{b_0 + b_1 z^{-1}}{1 - a_1 z^{-1}}$$

Recall that the z-transform of the output of this system is $Y(z) = H(z)X(z)$, so the z-transform approach to finding the output for a given input $x[n]$ is as follows:

1. Determine the z-transform $X(z)$ of the input signal $x[n]$.
2. Multiply $X(z)$ by $H(z)$ to get $Y(z)$.
3. Determine the inverse z-transform of $Y(z)$ to get the output $y[n]$.

This procedure will work and will avoid iteration of the difference equation (or convolution) if we can determine $X(z)$ and if we can perform the necessary inverse transformation. Our focus in this section is on deriving a general procedure for step 3 which is actually two steps.

In the case of the step response, the input $x[n] = u[n]$ is a special case of the more general sequence $a^n u[n]$ (i.e., $a = 1$). Therefore, from (10.18) it follows that the z-transform of $x[n] = u[n]$ is

$$X(z) = \frac{1}{1 - z^{-1}}$$

so $Y(z) = H(z)X(z)$ is

$$Y(z) = \frac{b_0 + b_1 z^{-1}}{(1 - a_1 z^{-1})(1 - z^{-1})} \tag{10.45a}$$

$$= \frac{b_0 + b_1 z^{-1}}{1 - (1 + a_1)z^{-1} + a_1 z^{-2}} \tag{10.45b}$$

Now we need to go back to the n-domain by inverse z-transformation. A standard approach is to use a table of z-transform pairs and simply look up the answer in the table. Our previous discussions in Chapter 9 and earlier in this chapter have developed the basis for a simple version of such a table. A summary of the z-transform knowledge that we have developed so far is given in Table 10-1. Although more extensive tables can be constructed, the results assembled in Table 10-1 are adequate for most purposes in this text.

The problem of finding $y[n]$ from $Y(z)$ in (10.45a) is that none of the entries in Table 10-1 are exactly the same as the given $Y(z)$, so we have to rearrange $Y(z)$ into a form where entries 1, 5, and 6 in the table can be used. The algebraic technique that we

Table 10-1 Summary of important z-transform properties and pairs.

Short Table of z-Transforms			
	$x[n]$	$\stackrel{z}{\longleftrightarrow}$	$X(z)$
1.	$ax_1[n] + bx_2[n]$	$\stackrel{z}{\longleftrightarrow}$	$aX_1(z) + bX_2(z)$
2.	$x[n - n_0]$	$\stackrel{z}{\longleftrightarrow}$	$z^{-n_0}X(z)$
3.	$y[n] = x[n] * h[n]$	$\stackrel{z}{\longleftrightarrow}$	$Y(z) = H(z)X(z)$
4.	$\delta[n]$	$\stackrel{z}{\longleftrightarrow}$	1
5.	$\delta[n - n_0]$	$\stackrel{z}{\longleftrightarrow}$	z^{-n_0}
6.	$a^n u[n]$	$\stackrel{z}{\longleftrightarrow}$	$\dfrac{1}{1 - az^{-1}}$

use is based on the ***partial fraction expansion***[9] of $Y(z)$, which is a general method for expressing a rational function $Y(z)$ as a *sum of first-order* rational functions, that is,

$$Y(z) = H(z)X(z) \tag{10.46a}$$

$$= \frac{b_0 + b_1 z^{-1}}{(1 - a_1 z^{-1})(1 - z^{-1})} \tag{10.46b}$$

$$= \frac{A}{1 - a_1 z^{-1}} + \frac{B}{1 - z^{-1}} \tag{10.46c}$$

If the expression in (10.46c) is pulled together over a common denominator, it should be possible to find A and B so that the numerator of the resulting rational function is equal to $b_0 + b_1 z^{-1}$ as in (10.46b). Equating the two numerators would lead to solving the following two equations in the two unknowns A and B:

$$A + B = b_0$$

$$-A - a_1 B = b_1$$

However, there is a much quicker way that also generalizes to higher-order z-transforms. A systematic procedure for finding A and B is based on the observation that, for this example,

$$Y(z)(1 - a_1 z^{-1}) = A + \frac{B(1 - a_1 z^{-1})}{1 - z^{-1}}$$

Then we can evaluate at the pole $z = a_1$ to isolate A as follows:

$$Y(z)(1 - a_1 z^{-1})\Big|_{z=a_1} = A + \frac{B(1 - a_1 z^{-1})}{1 - z^{-1}}\Bigg|_{z=a_1} = A$$

Since multiplying $Y(z)$ by $(1 - a_1 z^{-1})$ cancels the same factor in the denominator, we obtain

$$A = Y(z)(1 - a_1 z^{-1})\big|_{z=a_1} = \frac{b_0 + b_1 z^{-1}}{1 - z^{-1}}\Bigg|_{z=a_1} = \frac{b_0 + b_1 a_1^{-1}}{1 - a_1^{-1}}$$

Similarly, we can find B by evaluating at the pole $z = 1$ as

$$B = Y(z)(1 - z^{-1})\big|_{z=1} = \frac{b_0 + b_1 z^{-1}}{1 - a_1 z^{-1}}\Bigg|_{z=1} = \frac{b_0 + b_1}{1 - a_1}$$

Now using the superposition property of the z-transform (entry 1 in Table 10-1), and the exponential z-transform pair (entry 6 in Table 10-1), we can use (10.46c) to write down the desired answer as $y[n] = Aa_1^n u[n] + Bu[n]$ or

$$y[n] = \left(\frac{b_0 + b_1 a_1^{-1}}{1 - a_1^{-1}}\right) a_1^n u[n] + \left(\frac{b_0 + b_1}{1 - a_1}\right) u[n]$$

[9]The partial fraction expansion is an algebraic decomposition often presented in calculus for evaluating integrals that contain rational forms.

The two terms above can be put over a common denominator to express $y[n]$ as

$$y[n] = \left(\frac{(b_0 + b_1) - (b_0 a_1 + b_1) a_1^n}{1 - a_1} \right) u[n] \tag{10.47}$$

If we substitute the value $b_1 = 0$ into (10.47), we get

$$y[n] = b_0 \left(\frac{1 - a_1^{n+1}}{1 - a_1} \right) u[n]$$

which is the same result obtained in Section 10-2.4 by iteration of the difference equation (10.13).

With this two-pole example, we have established the framework for using the basic properties of z-transforms together with the partial fraction expansion[10] and a few basic z-transform pairs to perform inverse z-transformation for any rational z-transform. We summarize this procedure in the following subsection.

10-9.2 A General Procedure for Inverse z-Transformation

Let $X(z)$ be any rational z-transform of degree N in the denominator and M in the numerator. Assuming that $M < N$, we can find the sequence $x[n]$ that corresponds to $X(z)$ by the following procedure:

Procedure for
Inverse z-Transformation ($M < N$)

1. Factor the denominator polynomial of $X(z)$ and express the pole factors as $(1 - p_k z^{-1})$ for $k = 1, 2, \ldots, N$.

2. Make a partial fraction expansion of $X(z)$ to obtain a sum of terms of the form

$$X(z) = \sum_{k=1}^{N} \frac{A_k}{1 - p_k z^{-1}}$$

where $A_k = X(z)(1 - p_k z^{-1})\big|_{z=p_k}$

3. Write down the inverse transform as

$$x[n] = \sum_{k=1}^{N} A_k (p_k)^n u[n]$$

[10] In MATLAB the function `residuez` computes the parameters of a partial fraction expansion assuming pole factors of the form $(1 - p_k z^{-1})$.

This procedure will always work for any rational z-transform if the poles, p_k, are distinct. Repeated poles complicate the process, but can be handled systematically. In this text, we restrict our attention to the case of non-repeated poles. Next we illustrate the use of this procedure with two examples that have real poles.

EXAMPLE 10-11 Inverse z-Transform

We want to determine the signal $x[n]$ when its z-transform $X(z)$ is

$$X(z) = \frac{1 - 2.1z^{-1}}{1 - 0.3z^{-1} - 0.4z^{-2}} = \frac{1 - 2.1z^{-1}}{(1 + 0.5z^{-1})(1 - 0.8z^{-1})}$$

After factoring the denominator, the next step is to write $X(z)$ in the additive form

$$X(z) = \frac{A}{1 + 0.5z^{-1}} + \frac{B}{1 - 0.8z^{-1}}$$

Invoking the procedure for partial fraction expansion, we obtain

$$A = X(z)(1 + 0.5z^{-1})\big|_{z=-0.5} = \left.\frac{1 - 2.1z^{-1}}{1 - 0.8z^{-1}}\right|_{z=-0.5} = \frac{1 + 4.2}{1 + 1.6} = 2$$

and $$B = X(z)(1 - 0.8z^{-1})\big|_{z=0.8} = \left.\frac{1 - 2.1z^{-1}}{1 + 0.5z^{-1}}\right|_{z=0.8} = \frac{1 - 2.1/0.8}{1 + 0.5/0.8} = -1$$

Therefore,

$$X(z) = \frac{2}{1 + 0.5z^{-1}} - \frac{1}{1 - 0.8z^{-1}} \tag{10.48}$$

and the two first-order terms can be inverted using entry 6 in Table 10-1 to obtain

$$x[n] = 2(-0.5)^n u[n] - (0.8)^n u[n]$$

Note that the poles at $z = p_1 = -0.5$ and $z = p_2 = 0.8$ give rise to terms in $x[n]$ of the form $p_1^n u[n]$ and $p_2^n u[n]$.

In Example 10-11, the degree of the numerator is $M = 1$, which is less than the degree of the denominator, which is $N = 2$. This is important because the partial fraction expansion works only for rational functions such that $M < N$. The next example shows why this is so, and illustrates a method of dealing with this complication.

EXAMPLE 10-12 Long Division and Partial Fractions

In this case, define $Y(z)$ with a second-order numerator as

$$Y(z) = \frac{2 - 2.4z^{-1} - 0.4z^{-2}}{1 - 0.3z^{-1} - 0.4z^{-2}} = \frac{2 - 2.4z^{-1} - 0.4z^{-2}}{(1 + 0.5z^{-1})(1 - 0.8z^{-1})}$$

Now we must add a constant term to the partial fraction expansion, otherwise, we cannot generate the term $-0.4z^{-2}$ in the numerator when we combine the partial fractions over a common denominator—that is, we must assume the following form for $Y(z)$:

$$Y(z) = \frac{A}{1+0.5z^{-1}} + \frac{B}{1-0.8z^{-1}} + C$$

How can we determine the constant C? One way is to perform long division of the denominator polynomial into the numerator polynomial until we get a remainder whose degree is lower than that of the denominator. In this case, the polynomial long division is carried out as follows:

$$\begin{array}{r|l} & \quad 1 \\ -0.4z^{-2} - 0.3z^{-1} + 1 & \overline{-0.4z^{-2} - 2.4z^{-1} + 2} \\ & \underline{-0.4z^{-2} - 0.3z^{-1} + 1} \\ & \qquad \underbrace{-2.1z^{-1} + 1}_{\text{remainder}} \end{array}$$

Thus, if we place the remainder $(1 - 2.1z^{-1})$ over the denominator (in factored form), we can write $Y(z)$ as a rational part (fraction) plus the constant 1, that is,

$$Y(z) = \frac{1 - 2.1z^{-1}}{(1+0.5z^{-1})(1-0.8z^{-1})} + 1$$

The next step is to apply the partial fraction expansion technique to the rational part of $Y(z)$. Since the rational part turns out to be identical to $X(z)$ in (10.48) from Example 10-11, the results would be the same as in that example, so we can write $Y(z)$ as

$$Y(z) = \frac{2}{1+0.5z^{-1}} - \frac{1}{1-0.8z^{-1}} + 1$$

Therefore, from entries 4 and 6 in Table 10-1,

$$y[n] = 2(-0.5)^n u[n] - (0.8)^n u[n] + \delta[n]$$

Notice again that the time-domain sequence has terms of the form $p_k^n u[n]$. The constant term in the system function is the z-transform of an impulse, which is nonzero only at $n = 0$ (entry 4 in Table 10-1).

10-10 Steady-State Response and Stability

We can use the inverse z-transform method developed in Section 10-9 to demonstrate an important point about stability, the frequency response, and the sinusoidal steady-state response. To illustrate this point, consider the first-order LTI system defined by

$$y[n] = a_1 y[n-1] + b_0 x[n]$$

From our presentation of the three domains, we can write down the impulse response, system function, and frequency response directly from the difference equation. The impulse response is

$$h[n] = b_0 a_1^n u[n]$$

and the system function of this system is

$$H(z) = \frac{b_0}{1 - a_1 z^{-1}}$$

We can then obtain the frequency response by evaluating $H(z)$ on the unit circle

$$H(e^{j\hat{\omega}}) = H(z)\Big|_{z=e^{j\hat{\omega}}} = \frac{b_0}{1 - a_1 e^{-j\hat{\omega}}}$$

but this is true only if the system is stable ($|a_1| < 1$). The objective of this section is to refine the concept of stability and demonstrate its impact on the response to a sinusoid applied at $n = 0$.

Recall from Section 10-7 and (10.40) and (10.42) that the output of this system for a complex exponential input $x[n] = e^{j\hat{\omega}_0 n}$ for $-\infty < n < \infty$ is

$$y[n] = H(e^{j\hat{\omega}_0})e^{j\hat{\omega}_0 n} = \left(\frac{b_0}{1 - a_1 e^{-j\hat{\omega}_0}}\right) e^{j\hat{\omega}_0 n} \tag{10.49}$$

for *all time* (i.e., $-\infty < n < \infty$). What happens if the complex exponential input sequence is suddenly applied instead of existing for all n? The z-transform tools that we have developed make it easy to solve this problem. Indeed, the z-transform is ideally suited for situations where the sequences are suddenly applied exponentials. For the suddenly applied complex exponential sequence with frequency $\hat{\omega}_0$

$$x[n] = e^{j\hat{\omega}_0 n} u[n]$$

the z-transform is found from entry 6 of Table 10-1 to be

$$X(z) = \frac{1}{1 - e^{j\hat{\omega}_0} z^{-1}}$$

and the z-transform of the output of the LTI system is

$$\begin{aligned} Y(z) = H(z)X(z) &= \left(\frac{b_0}{1 - a_1 z^{-1}}\right)\left(\frac{1}{1 - e^{j\hat{\omega}_0} z^{-1}}\right) \\ &= \frac{b_0}{(1 - a_1 z^{-1})(1 - e^{j\hat{\omega}_0} z^{-1})} \end{aligned}$$

Using the technique of partial fraction expansion, we can show that

$$Y(z) = \frac{\left(\dfrac{b_0 a_1}{a_1 - e^{j\hat{\omega}_0}}\right)}{1 - a_1 z^{-1}} + \frac{\left(\dfrac{b_0}{1 - a_1 e^{-j\hat{\omega}_0}}\right)}{1 - e^{j\hat{\omega}_0} z^{-1}}$$

Therefore, the output due to the suddenly applied complex exponential sequence is

$$y[n] = \underbrace{\left(\frac{b_0 a_1}{a_1 - e^{j\hat{\omega}_0}}\right)(a_1)^n u[n]}_{\text{transient component}} + \underbrace{\left(\frac{b_0}{1 - a_1 e^{-j\hat{\omega}_0}}\right) e^{j\hat{\omega}_0 n} u[n]}_{\text{steady-state component}} \tag{10.50}$$

Equation (10.50) shows that the output consists of two terms that start at $n = 0$. One term is proportional to an exponential sequence a_1^n that is determined solely by the pole at $z = a_1$. If $|a_1| < 1$, this term dies out with increasing n, in which case it would be called appropriately the ***transient component***. The second term is a complex amplitude times the input complex exponential signal, and the result is identical to (10.49). The complex amplitude is $H(e^{j\hat{\omega}_0})$, the frequency response of the system evaluated at $\hat{\omega}_0$, the frequency of the suddenly applied complex sinusoid. This complex exponential component is the ***sinusoidal steady state component*** of the output, because this component remains after the exponentially decaying transient component becomes negligibly small.

Now we see that the location of the pole of $H(z)$ is crucial if we want the output to reach the sinusoidal steady state, and the transient to die out. When $|a_1| < 1$, the system is stable and the pole is inside the unit circle. Under this condition, the exponential a_1^n dies out and we find that the limiting value for large n is

$$y[n] \rightarrow \left(\frac{b_0}{1 - a_1 e^{-j\hat{\omega}_0}}\right) e^{j\hat{\omega}_0 n} = H(e^{j\hat{\omega}_0}) e^{j\hat{\omega}_0 n}$$

Otherwise, if $|a_1| \geq 1$, the term in (10.50) that is proportional to a_1^n grows with increasing n and soon dominates the output. The following example gives a specific numerical illustration.

EXAMPLE 10-13 Transient and Steady-State Responses

If $b_0 = 5$, $a_1 = -0.8$, and $\hat{\omega}_0 = 2\pi/10$, the transient component in (10.50) is

$$\begin{aligned} y_t[n] &= \left(\frac{-4}{-0.8 - e^{j0.2\pi}}\right)(-0.8)^n u[n] \\ &= 2.3351 e^{-j0.3502}(-0.8)^n u[n] \\ &= 2.1933(-0.8)^n u[n] - j0.8012(-0.8)^n u[n] \end{aligned}$$

Similarly, the steady-state component in (10.50) is

$$\begin{aligned} y_{ss}[n] &= \left(\frac{5}{1 + 0.8 e^{-j0.2\pi}}\right) e^{j0.2\pi n} u[n] \\ &= 2.9188 e^{j0.2781} e^{j0.2\pi n} u[n] \\ &= 2.9188 \cos(0.2\pi n + 0.2781)\, u[n] + j\, 2.9188 \sin(0.2\pi n + 0.2781)\, u[n] \end{aligned}$$

Figure 10-14 shows the real parts of the total output (a), the transient component (b), and the steady-state component (c). The signals all start at $n = 0$ which is when the complex exponential is applied at the input. Note how the transient component $(-0.8)^n$ oscillates, but dies away, which explains why the steady-state component eventually equals the total output. In Fig. 10-14, $y[n]$ in (a) and $y_{ss}[n]$ in (c) look identical for $n > 15$.

On the other hand, if the pole were at $z = 1.1$, the system would be unstable and the output would "blow up" as shown in Fig. 10-15. In this case, the output contains a term $(1.1)^n$ that eventually dominates and grows without bound.

The result of Example 10-13 can be generalized by observing that the only difference between this example and a system with a higher-order system function is that the total

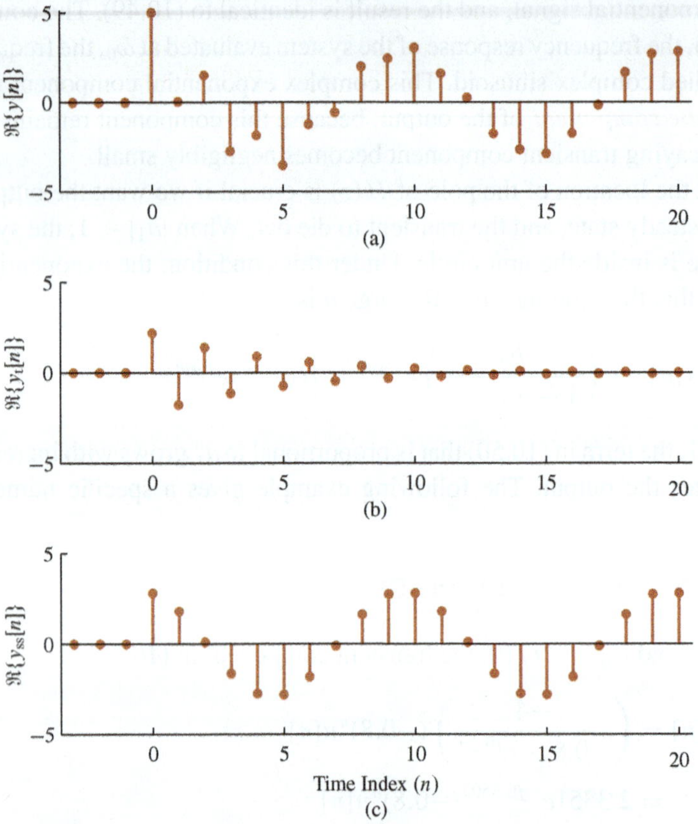

Figure 10-14 Illustration of transient and steady-state responses of a stable IIR system: (a) Total output $\Re\{y[n]\}$, (b) decaying transient component $\Re\{y_t[n]\}$, and (c) steady-state response $\Re\{y_{ss}[n]\}$.

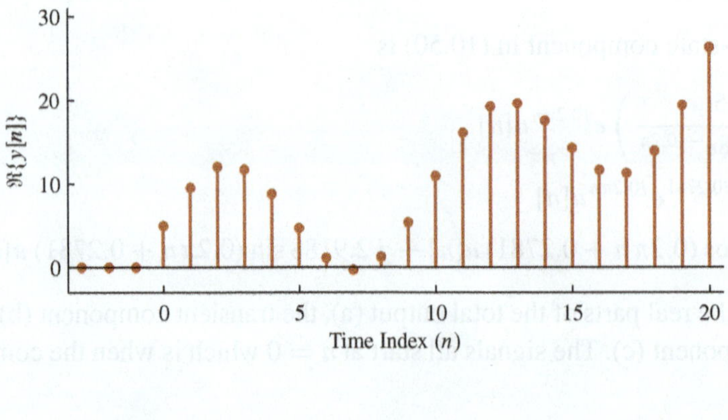

Figure 10-15 Illustration of an unstable IIR system where the output is dominated by the growing exponential component. Pole is at $z = 1.1$.

output would include N exponential terms for the N poles of the system function as well as the term $H(e^{j\hat{\omega}_0})e^{j\hat{\omega}_0 n}u[n]$. Specifically, it can be shown that for a suddenly applied exponential input sequence $x[n] = e^{j\hat{\omega}_0 n}u[n]$, the output of an Nth-order IIR $(N > M)$ system is always of the form

$$y[n] = \underbrace{\sum_{k=1}^{N} A_k (p_k)^n u[n]}_{\text{transient response}} + \underbrace{H(e^{j\hat{\omega}_0})e^{j\hat{\omega}_0 n}u[n]}_{\text{steady-state response}}$$

where p_k, $k = 1, 2, \ldots, N$ are the distinct poles of the system function. Therefore, the sinusoidal steady state exists and will dominate in the total response as $n \to \infty$ if the poles of the system function all lie strictly inside the unit circle. This makes the concept of frequency response useful in a practical setting where the system is stable and all signals must have a starting point at some finite time.

10-11 Second-Order Filters

We now focus our attention on filters with two feedback and three feed-forward coefficients. The general difference equation (10.1) becomes the second-order difference equation

$$y[n] = a_1 y[n-1] + a_2 y[n-2] + b_0 x[n] + b_1 x[n-1] + b_2 x[n-2] \qquad (10.51)$$

DEMO
PeZ GUI

As in Section 10-8, we can characterize the second-order filter (10.51) in each of the three domains: time domain, frequency domain, and z-domain. We start with the z-transform domain because by now we have demonstrated that the poles and zeros of the system function give a great deal of insight into most aspects of both the time and frequency responses.

10-11.1 z-Transform of Second-Order Filters

Using the approach followed in Section 10-3.1 for the first-order case, we can take the z-transform of the second-order difference equation (10.51) by replacing each delay with z^{-1} (second entry in Table 10-1). Then we can solve for $H(z)$ by moving all the $Y(z)$ terms to the left-hand side and then dividing to find $H(z)$ as $Y(z)/X(z)$. For the second-order filter we obtain

$$H(z) = \frac{Y(z)}{X(z)} = \frac{b_0 + b_1 z^{-1} + b_2 z^{-2}}{1 - a_1 z^{-1} - a_2 z^{-2}} \qquad (10.52)$$

Thus, the system function $H(z)$ for an IIR filter is a ratio of two second-degree polynomials, where the numerator polynomial depends on the feed-forward coefficients $\{b_k\}$ and the denominator depends on the feedback coefficients $\{a_\ell\}$. In order to work problems such as Exercise 10.10, it is only necessary to read the filter coefficients from the difference equation and then substitute them directly into the z-transform expression (10.52) for $H(z)$. Conversely, given the system function $H(z)$, it is a simple matter to write down the difference equation.

EXERCISE 10.10 Find the system function $H(z)$ of the IIR filter defined by the difference equation:

$$y[n] = 0.5y[n-1] + 0.3y[n-2] - x[n] + 3x[n-1] - 2x[n-2]$$

EXERCISE 10.11 For the system function

$$H(z) = \frac{1 + 2z^{-1} + z^{-2}}{1 - 0.8z^{-1} + 0.64z^{-2}}$$

write down the difference equation that relates the input $x[n]$ to the output $y[n]$.

10-11.2 Structures for Second-Order IIR Systems

The difference equation (10.51) with two feedback terms can be interpreted as an algorithm for computing the output sequence from the input. As we showed in Section 10-4, other computational orderings are possible, and the z-transform has the power to derive alternative structures through polynomial manipulations. Two alternative computational orderings that implement the system defined by $H(z)$ in (10.52) are given in Figs. 10-16 and 10-17.

In order to verify that the block diagram in Fig. 10-16 has the correct system function, we need to write the equations of the structure at the adders, and then eliminate the internal variable(s). For the case of the Direct Form II in Fig. 10-16, the equations at the output of the summing nodes on the output and input sides are

$$\begin{aligned} y[n] &= b_0 w[n] + b_1 w[n-1] + b_2 w[n-2] \\ w[n] &= x[n] + a_1 w[n-1] + a_2 w[n-2] \end{aligned} \tag{10.53}$$

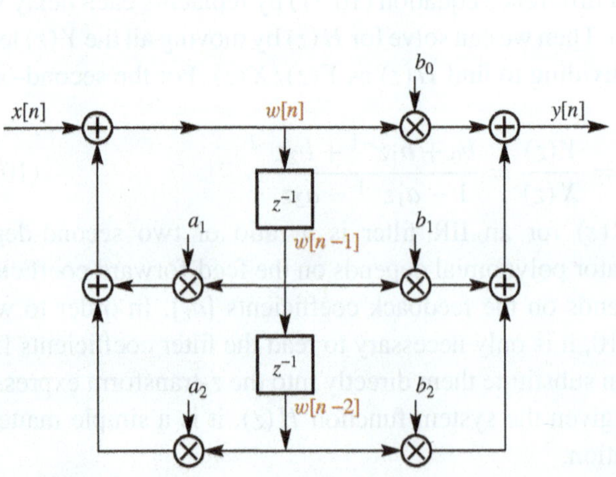

Figure 10-16 Direct Form II (DF-II) computational structure for the second-order recursive filter. The signal $w[n]$ is delayed twice, then multiplied by filter coefficients and followed by adding to form the output signal $y[n]$ and to update $w[n]$.

It is impossible to eliminate $w[n]$ in these two time-domain equations, but easy if we work with algebra in the z-transform domain. The corresponding z-transform equations are

$$Y(z) = b_0 W(z) + b_1 z^{-1} W(z) + b_2 z^{-2} W(z)$$
$$W(z) = X(z) + a_1 z^{-1} W(z) + a_2 z^{-2} W(z)$$

which can be rearranged into the form

$$Y(z) = (b_0 + b_1 z^{-1} + b_2 z^{-2}) W(z)$$
$$X(z) = (1 - a_1 z^{-1} - a_2 z^{-2}) W(z)$$

Since the system function $H(z)$ is the ratio of $Y(z)$ to $X(z)$, we divide the two equations to obtain

$$H(z) = \frac{Y(z)}{X(z)} = \frac{b_0 + b_1 z^{-1} + b_2 z^{-2}}{1 - a_1 z^{-1} - a_2 z^{-2}}$$

Thus we have proved that the Direct Form II structure in Fig. 10-16 which implements the pair of difference equations (10.53) is identical to the system defined by the single difference equation (10.51).

The Transposed Direct Form II in Fig. 10-17 can be worked out similarly. The difference equations represented by the block diagram are

$$\begin{aligned} y[n] &= b_0 x[n] + v_1[n-1] \\ v_1[n] &= b_1 x[n] + a_1 y[n] + v_2[n-1] \\ v_2[n] &= b_2 x[n] + a_2 y[n] \end{aligned} \tag{10.54}$$

Taking the z-transform of each of the three equations gives

$$\begin{aligned} Y(z) &= b_0 X(z) + z^{-1} V_1(z) \\ V_1(z) &= b_1 X(z) + a_1 Y(z) + z^{-1} V_2(z) \\ V_2(z) &= b_2 X(z) + a_2 Y(z) \end{aligned} \tag{10.55}$$

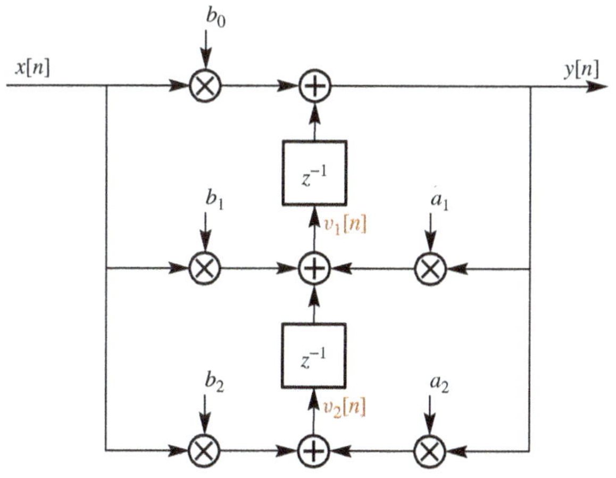

Figure 10-17 Transposed Direct Form II (TDF-II) computational structure for the second-order recursive filter. The signals $x[n]$ and $y[n]$ are first multiplied by filter coefficients, followed by adding, delaying, and adding once more to form the output.

Using these equations, we can eliminate $V_1(z)$ and $V_2(z)$ as follows:

$$Y(z) = b_0X(z) + z^{-1}(b_1X(z) + a_1Y(z) + z^{-1}V_2(z))$$
$$Y(z) = b_0X(z) + z^{-1}(b_1X(z) + a_1Y(z) + z^{-1}(b_2X(z) + a_2Y(z)))$$

Moving all the $X(z)$ terms to the right-hand side and the $Y(z)$ terms to the left-hand side gives

$$(1 - a_1z^{-1} - a_2z^{-2})Y(z) = (b_0 + b_1z^{-1} + b_2z^{-2})X(z)$$

After division, we obtain the same result again

$$H(z) = \frac{Y(z)}{X(z)} = \frac{b_0 + b_1z^{-1} + b_2z^{-2}}{1 - a_1z^{-1} - a_2z^{-2}}$$

Thus we have shown that the Transposed Direct Form II is equivalent to the system function for the basic Direct Form I difference equation in (10.51). Both cases illustrate the power of the z-transform approach in manipulating the difference equation sets that correspond to different structures.

In theory, the system with system function given by (10.52) can be implemented by iterating any of the difference equation sets (10.51), (10.53), or (10.54). For example, the MATLAB function `filter` uses the Transposed Direct Form II structure. However, as mentioned before, the reason for having different block diagram structures is that the order of calculation defined by the difference equations in (10.51), (10.53), or (10.54) is different. In a hardware implementation, the different structures behave differently, especially when using fixed-point arithmetic where rounding error is fed back into the structure. With double-precision floating-point arithmetic as in MATLAB, there is little difference.

10-11.3 Poles and Zeros

Finding the poles and zeros of $H(z)$ is less confusing if we rewrite the polynomials as functions of z rather than z^{-1}. Thus, the general second-order rational z-transform would become

$$H(z) = \frac{b_0 + b_1z^{-1} + b_2z^{-2}}{1 - a_1z^{-1} - a_2z^{-2}} = \frac{b_0z^2 + b_1z + b_2}{z^2 - a_1z - a_2}$$

after multiplying the numerator and denominator by z^2. Recall from algebra the following important property of polynomials:

Property of Real Polynomials

A polynomial of degree N has N roots.
If all the coefficients of the polynomial are real, the roots either must be real or must occur in complex conjugate pairs.

Therefore, in the second-order case, the numerator and denominator polynomials each have two roots, and there are two possibilities: Either the roots are complex conjugates of each other, or they are both real. We will now concentrate on the roots of the denominator, but exactly the same results hold for the numerator. From the quadratic formula, we get two poles at

$$\frac{a_1 + \sqrt{a_1^2 + 4a_2}}{2} \quad \text{and} \quad \frac{a_1 - \sqrt{a_1^2 + 4a_2}}{2}$$

When $a_1^2 + 4a_2 \geq 0$, both poles are real, and when $a_1^2 + 4a_2 = 0$, they are real and equal. However, when $a_1^2 + 4a_2 < 0$, the square root produces an imaginary result, and we have complex conjugate poles with values

$$p_1 = \tfrac{1}{2}a_1 + j\tfrac{1}{2}\sqrt{-\left(a_1^2 + 4a_2\right)} \quad \text{and} \quad p_2 = \tfrac{1}{2}a_1 - j\tfrac{1}{2}\sqrt{-\left(a_1^2 + 4a_2\right)} \tag{10.56}$$

The complex poles can be expressed in polar form as $p_1 = re^{j\theta}$ and $p_2 = re^{-j\theta}$, where r is the radius and $\pm\theta$ the angles. A simple way to express the relationship between complex roots and filter coefficients is to expand the complex conjugate factors

$$(1 - re^{j\theta}z^{-1})(1 - re^{-j\theta}z^{-1}) = 1 - 2r\cos(\theta)z^{-1} + r^2z^{-2} \tag{10.57}$$

To summarize we write

Complex Poles

$$a_2 = -r^2 \qquad a_1 = 2r\cos\theta$$
$$r = \sqrt{-a_2} \qquad \theta = \cos^{-1}\left(\frac{a_1}{2\sqrt{-a_2}}\right) \tag{10.58}$$

EXAMPLE 10-14 Complex Poles

The following $H(z)$ has two poles and two zeros.

$$H(z) = \frac{2 + 2z^{-1}}{1 - z^{-1} + z^{-2}} = 2\frac{z^2 + z}{z^2 - z + 1}$$

The system function can be written in factored form as either of the two forms

$$H(z) = \frac{2z(z+1)}{(z - e^{j\pi/3})(z - e^{-j\pi/3})} \tag{10.59a}$$

$$= \frac{2(1 + z^{-1})}{(1 - e^{j\pi/3}z^{-1})(1 - e^{-j\pi/3}z^{-1})} \tag{10.59b}$$

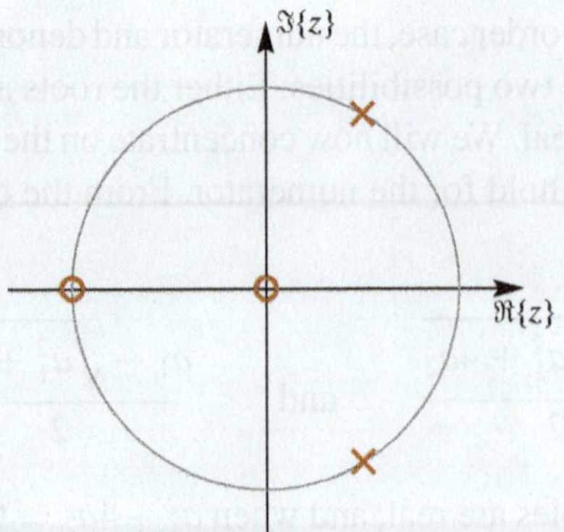

Figure 10-18 Pole-zero plot for system with $H(z) = \dfrac{2+2z^{-1}}{1-z^{-1}+z^{-2}}$. The two poles are on the unit circle at angles $\theta = \pm\pi/3$. There is one zero at $z=-1$ and another zero at $z=0$.

In the form (10.59a), it is easy to see that the poles $\{p_1,\ p_2\}$ and zeros $\{z_1,\ z_2\}$ are

$$p_1 = \tfrac{1}{2} + j\tfrac{1}{2}\sqrt{3} = e^{j\pi/3}$$
$$p_2 = \tfrac{1}{2} - j\tfrac{1}{2}\sqrt{3} = e^{-j\pi/3}$$
$$z_1 = 0$$
$$z_2 = -1$$

The poles and zeros are plotted in Fig. 10-18.

10-11.4 Impulse Response of a Second-Order IIR System

We have derived the general z-transform system function for the second-order filter

$$H(z) = \frac{B(z)}{A(z)} = \frac{b_0 + b_1 z^{-1} + b_2 z^{-2}}{1 - a_1 z^{-1} - a_2 z^{-2}} \tag{10.60}$$

and we have seen that the denominator polynomial $A(z)$ has two roots that define the poles of the second-order filter. Expressing $H(z)$ in terms of the poles gives

$$H(z) = \frac{b_0 + b_1 z^{-1} + b_2 z^{-2}}{(1-p_1 z^{-1})(1-p_2 z^{-1})} \tag{10.61}$$

Using the partial fraction expansion technique developed in Section 10-9, we can express the system function (10.60) as

$$H(z) = (-b_2/a_2) + \frac{A_1}{1-p_1 z^{-1}} + \frac{A_2}{1-p_2 z^{-1}}$$

where A_1 and A_2 can be evaluated by $A_k = H(z)(1-p_k z^{-1})\big|_{z=p_k}$ (as long as $p_1 \neq p_2$). Therefore, the impulse response has the form

$$h[n] = (-b_2/a_2)\delta[n] + A_1(p_1)^n u[n] + A_2(p_2)^n u[n] \tag{10.62}$$

In (10.62), the poles may both be real or they may be a pair of complex conjugates. In the following examples, we examine these two cases separately.

10-11.4.1 Distinct Real Poles

If p_1 and p_2 are real, the impulse response is composed of two real exponentials of the form $p_1^n u[n]$ and $p_2^n u[n]$. This case is illustrated by the following example.

EXAMPLE 10-15 Second-Order System Real Poles

Assume that the system function is

$$H(z) = \frac{1}{1 - \frac{5}{6}z^{-1} + \frac{1}{6}z^{-2}} = \frac{1}{(1 - \frac{1}{2}z^{-1})(1 - \frac{1}{3}z^{-1})} \qquad (10.63)$$

After factoring the denominator, we see that the poles are at $z = \frac{1}{2}$ and $z = \frac{1}{3}$. In addition, there are two zeros at $z = 0$. The poles and zeros of $H(z)$ are plotted in Fig. 10-19. We can extract the filter coefficients directly from the denominator of $H(z)$ and write the following difference equation

$$y[n] = \tfrac{5}{6}y[n-1] - \tfrac{1}{6}y[n-2] + x[n] \qquad (10.64)$$

which must be satisfied for any input and its corresponding output. Specifically, the impulse response would satisfy the difference equation

$$h[n] = \tfrac{5}{6}h[n-1] - \tfrac{1}{6}h[n-2] + \delta[n] \qquad (10.65)$$

which can be iterated to compute $h[n]$ if we know the values of $h[-1]$ and $h[-2]$. Specifically, to start the iteration, we need to know the values of the impulse response sequence just prior to $n = 0$ where the impulse first becomes nonzero. These values are supplied by the initial rest condition, which means that $h[-1] = 0$ and

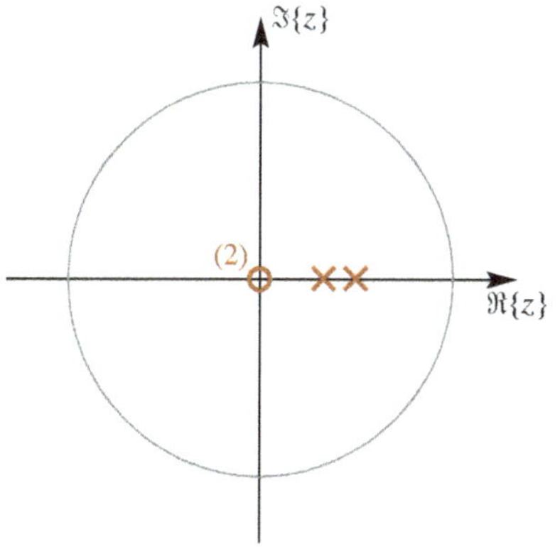

Figure 10-19 Pole-zero plot for system of Example 10-15. The poles are at $z = \frac{1}{2}$ and $z = \frac{1}{3}$; there are two zeros at $z = 0$.

$h[-2] = 0$. The following table shows the computation of a few values of the impulse response:

n	$n < 0$	0	1	2	3	4	…
$x[n]$	0	1	0	0	0	0	…
$h[n-2]$	0	0	0	1	$\frac{5}{6}$	$\frac{19}{36}$	…
$h[n-1]$	0	0	1	$\frac{5}{6}$	$\frac{19}{36}$	$\frac{65}{216}$	…
$h[n]$	0	1	$\frac{5}{6}$	$\frac{19}{36}$	$\frac{65}{216}$	$\frac{211}{1296}$	…

In contrast to the simpler first-order case, iteration is not too helpful because it is very difficult to guess the general n^{th} term for the impulse response sequence. Fortunately, we can rely on the inverse z-transform technique to give us the general formula. Applying the partial fraction expansion to (10.63), we obtain

$$H(z) = \frac{3}{1 - \frac{1}{2}z^{-1}} - \frac{2}{1 - \frac{1}{3}z^{-1}}$$

Then the inverse z-transform yields $h[n] = 3(\frac{1}{2})^n\, u[n] - 2(\frac{1}{3})^n\, u[n]$. Since both poles are inside the unit circle, the impulse response dies out for n large (i.e., the system is stable).

EXERCISE 10.12 Find the impulse response of the following second-order system:

$$y[n] = \tfrac{1}{4}y[n-2] + 5x[n] - 4x[n-1]$$

Hint: It will help to show that the poles are at $z = \pm\frac{1}{2}$.

10-11.5 Complex Poles

In this section, we study the case where the coefficients a_1 and a_2 in the second-order difference equation are such that the poles of $H(z)$ are complex. If we express the poles in polar form

$$p_1 = re^{j\theta} \qquad \text{and} \qquad p_2 = re^{-j\theta} = p_1^*$$

it is convenient to rewrite the denominator polynomial in terms of the parameters r and θ. From (10.58), we can write

$$\begin{aligned} A(z) &= \left(1 - p_1 z^{-1}\right)\left(1 - p_2 z^{-1}\right) \\ &= 1 - 2r\cos\theta z^{-1} + r^2 z^{-2} \end{aligned} \tag{10.66}$$

The system function is therefore

$$H(z) = \frac{b_0 + b_1 z^{-1} + b_2 z^{-2}}{1 - 2r\cos\theta z^{-1} + r^2 z^{-2}} \tag{10.67}$$

We can also identify the two feedback filter coefficients as $a_1 = 2r\cos\theta$ and $a_2 = -r^2$, so the corresponding difference equation is

$$y[n] = (2r\cos\theta)y[n-1] - r^2 y[n-2] + b_0 x[n] + b_1 x[n-1] + b_2 x[n-2] \tag{10.68}$$

This parameterization is significant because it allows us to see directly how the poles define the feedback terms of the difference equation (10.68). For example, if we want to change the angles of the poles, $\pm\theta$, then we vary only the coefficient a_1.

EXAMPLE 10-16 Second-Order System: Complex Poles

Consider the following system

$$y[n] = y[n-1] - y[n-2] + 2x[n] + 2x[n-1]$$

whose system function is

$$H(z) = \frac{2+2z^{-1}}{1-z^{-1}+z^{-2}} = \frac{2(1+z^{-1})}{(1-e^{j\pi/3}z^{-1})(1-e^{-j\pi/3}z^{-1})} \tag{10.69}$$

A pole-zero plot for this $H(z)$ was already given in Fig. 10-18. Using the partial fraction expansion technique, we can write $H(z)$ in the form

$$H(z) = \frac{\left(\dfrac{2+2e^{-j\pi/3}}{1-e^{-j2\pi/3}}\right)}{1-e^{j\pi/3}z^{-1}} + \frac{\left(\dfrac{2+2e^{j\pi/3}}{1-e^{j2\pi/3}}\right)}{1-e^{-j\pi/3}z^{-1}} = \frac{2e^{-j\pi/3}}{1-e^{j\pi/3}z^{-1}} + \frac{2e^{j\pi/3}}{1-e^{-j\pi/3}z^{-1}}$$

The right-hand side is the sum of two first-order terms for which entry 6 in Table 10-1 applies with $a = e^{j\pi/3}$ and $a = e^{-j\pi/3}$. The individual terms are complex conjugates, so their sum is purely real:

$$h[n] = 2e^{-j\pi/3}e^{j(\pi/3)n}u[n] + 2e^{j\pi/3}e^{-j(\pi/3)n}u[n] = 4\cos\left((2\pi/6)(n-1)\right)u[n]$$

The two complex exponentials with frequencies $\pm\pi/3$ combine to form the cosine. The impulse response is plotted in Fig. 10-20.

An important observation about the system in Example 10-16 is that it produces a pure sinusoid when stimulated by an impulse. Such a system is an example of a ***sine wave***

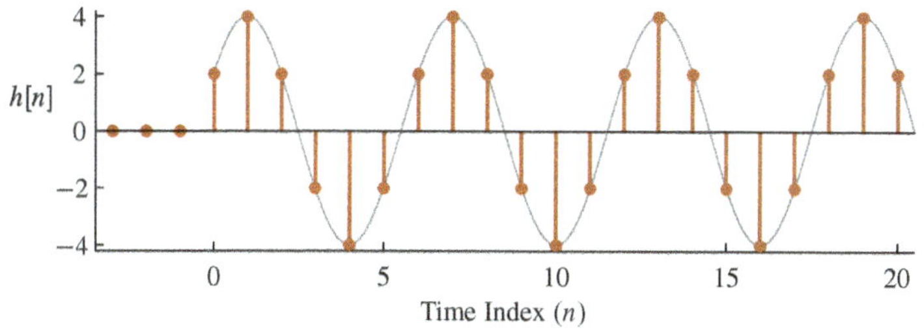

Figure 10-20 Impulse response for system with $H(z) = \dfrac{2+2z^{-1}}{1-z^{-1}+z^{-2}}$.

oscillator. After being stimulated by the single input sample from the impulse, the system continues indefinitely to produce a sinusoid of frequency $\hat{\omega}_0 = (2\pi/6)$. *This frequency is equal to the angle of the poles.* A first-order filter (or a filter with all real poles) can only decay (or grow) as $(p)^n$ or oscillate up and down as $(-1)^n$, but a second-order system can oscillate with different periods. This is important when modeling physical signals such as speech, music, or other sounds.

Note that in order to produce the continuing sinusoidal output, the system must have its poles on the unit circle[11] of the z-plane (i.e., $r = 1$). Also note that the angle of the poles is exactly equal to the radian frequency of the sinusoidal output. Thus, we can control the frequency of the sinusoidal oscillator by adjusting the a_1 coefficient of the difference equation (10.68) while leaving a_2 fixed at $a_2 = -1$.

EXAMPLE 10-17 Poles on Unit Circle

As an example of an oscillator with a different frequency, we can use (10.68) to define a difference equation with prescribed pole locations. If we take $r = 1$ and $\theta = \pi/2$, as shown in Fig. 10-21, we get $a_1 = 2r\cos\theta = 0$ and $a_2 = -r^2 = -1$, and the difference equation (10.68) becomes

$$y[n] = -y[n-2] + x[n] \qquad (10.70)$$

The system function of this system is

$$H(z) = \frac{1}{1+z^{-2}} = \frac{1}{(1-e^{j\pi/2}z^{-1})(1-e^{-j\pi/2}z^{-1})}$$
$$= \frac{\frac{1}{2}}{1-e^{j\pi/2}z^{-1}} + \frac{\frac{1}{2}}{1-e^{-j\pi/2}z^{-1}}$$

The inverse z-transform gives a general formula for $h[n]$:

$$h[n] = \tfrac{1}{2}e^{j(\pi/2)n}u[n] + \tfrac{1}{2}e^{-j(\pi/2)n}u[n] = \cos((2\pi/4)n)\,u[n] \qquad (10.71)$$

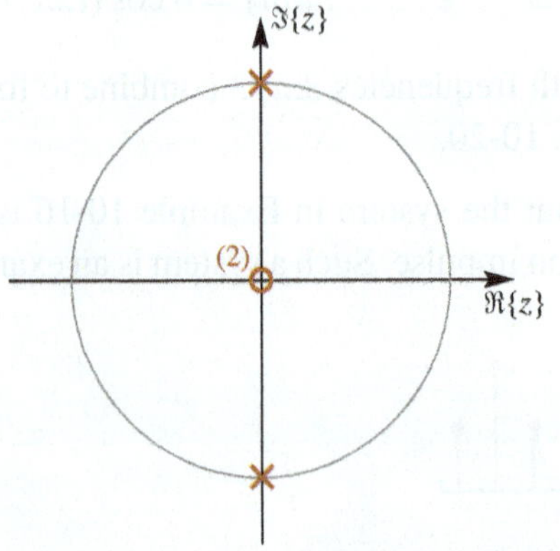

Figure 10-21 Pole-zero plot for system with $H(z) = \dfrac{1}{1+z^{-2}}$. The poles are on the unit circle at $z = \pm j$, and there are two zeros at $z = 0$.

[11] Strictly speaking, a system with poles on the unit circle is unstable, so for some inputs the output may increase without bound. As in the case of the accumulator system in Section 10-9, the impulse response of this second-order system does not blow up.

Once again, the frequency of the cosine term in the impulse response is equal to the angle of the pole, $\pi/2 = (2\pi/4)$.

The preceding two examples have shown that if the complex conjugate poles of the second-order system lie on the unit circle, the output oscillates sinusoidally and does not decay to zero. If the poles lie outside the unit circle, the output grows exponentially, whereas if they are inside the unit circle, the output decays exponentially to zero.

EXAMPLE 10-18 Stable Complex Poles

As an example of a stable system, we make the radius of the poles less than one with $r = \frac{1}{2}$ and the angles $\pm\pi/3$, as shown in Fig. 10-22. Then the filter coefficients are $a_1 = 2r\cos\theta = 2(\frac{1}{2})(\frac{1}{2}) = \frac{1}{2}$ and $a_2 = -r^2 = -(\frac{1}{2})^2 = -\frac{1}{4}$, so the difference equation (10.68) becomes

$$y[n] = \tfrac{1}{2}y[n-1] - \tfrac{1}{4}y[n-2] + x[n] \qquad (10.72)$$

The system function of this system is

$$H(z) = \frac{1}{1 - \frac{1}{2}z^{-1} + \frac{1}{4}z^{-2}} = \frac{\frac{1}{\sqrt{3}}e^{-j\pi/6}}{1 - \frac{1}{2}e^{j\pi/3}z^{-1}} + \frac{\frac{1}{\sqrt{3}}e^{j\pi/6}}{1 - \frac{1}{2}e^{-j\pi/3}z^{-1}}$$

and the inverse z-transform of the right-hand side gives the general formula for $h[n]$ after combining two complex exponential terms into a real sinusoid as

$$h[n] = \tfrac{2}{\sqrt{3}}(\tfrac{1}{2})^n \cos\left((2\pi/6)n - \pi/6\right) u[n] \qquad (10.73)$$

In this case, the general formula for $h[n]$ has an exponential decay of $(\frac{1}{2})^n$ multiplying a periodic cosine with period 6. The frequency of the cosine term in the impulse response (10.73) is again the angle of the pole, $\pi/3 = 2\pi/6$; while the decaying term $r^n = (\frac{1}{2})^n$ is controlled by the radius of the pole r. Figure 10-23 shows $h[n]$ for this example. Note the rapid decay of the impulse response. This rapid decay can be predicted from the pole-zero plot in Fig. 10-22, since, in general, poles that are not close to the unit circle correspond to rapidly decaying exponential components of the impulse response.

DEMO
IIR Filtering

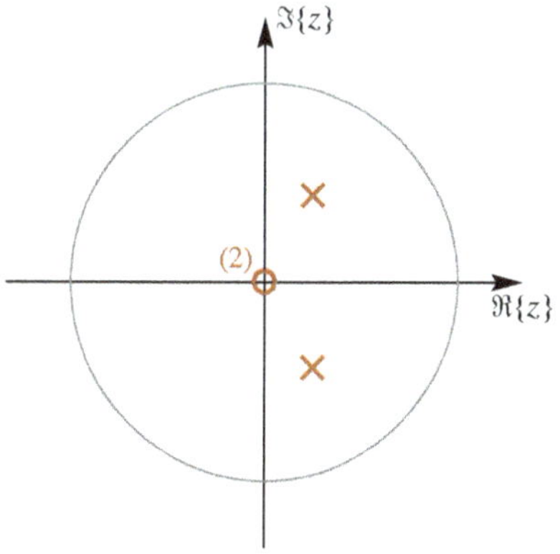

Figure 10-22 Pole-zero plot for system with $H(z) = \dfrac{1}{1 - \frac{1}{2}z^{-1} + \frac{1}{4}z^{-2}}$. The two poles are inside the unit circle at a radius of $r = 1/2$ and at angles $\theta = \pm\pi/3$. There are two zeros at $z = 0$.

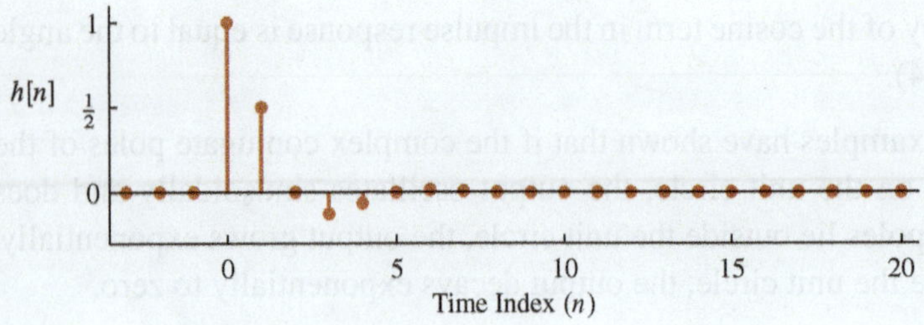

Figure 10-23 Impulse response for a two-pole system with system function $H(z) = \frac{1}{1-\frac{1}{2}z^{-1}+\frac{1}{4}z^{-2}}$.

10-12 Frequency Response of Second-Order IIR Filter

The general formula for the frequency response of a second-order system is

$$H(e^{j\hat{\omega}}) = \frac{b_0 + b_1 e^{-j\hat{\omega}} + b_2 e^{-j2\hat{\omega}}}{1 - a_1 e^{-j\hat{\omega}} - a_2 e^{-j2\hat{\omega}}} \tag{10.74}$$

The numerator and denominator of (10.74) contain terms like $e^{-j\hat{\omega}}$ and $e^{-j2\hat{\omega}}$, so $H(e^{j\hat{\omega}})$ is a periodic function with a period of 2π. Formulas for the magnitude and phase of $H(e^{j\hat{\omega}})$ can be obtained but they do not provide much insight. Rather than work out a general formula, we take a specific numerical example to show the kind of formulas that result.

EXAMPLE 10-19 Frequency Response of a Second-Order System

Consider the case where the system function is

$$H(z) = \frac{1 - z^{-2}}{1 - 0.9z^{-1} + 0.81z^{-2}} \tag{10.75}$$

DEMO
Three Domains–IIR

In order to find the magnitude of the frequency response, we work with the magnitude squared and then take the positive square root at the end. The magnitude squared is derived by multiplying out all the terms in the numerator and denominator of $H(e^{j\hat{\omega}})H^*(e^{j\hat{\omega}})$, and then collecting terms where the inverse Euler formula applies.

$$\begin{aligned}|H(e^{j\hat{\omega}})|^2 &= H(e^{j\hat{\omega}})H^*(e^{j\hat{\omega}})\\ &= \frac{1 - e^{-j2\hat{\omega}}}{1 - 0.9e^{-j\hat{\omega}} + 0.81e^{-j2\hat{\omega}}} \cdot \frac{1 - e^{j2\hat{\omega}}}{1 - 0.9e^{j\hat{\omega}} + 0.81e^{j2\hat{\omega}}}\\ &= \frac{2 - 2\cos(2\hat{\omega})}{2.4661 - 3.258\cos\hat{\omega} + 1.62\cos(2\hat{\omega})}\end{aligned}$$

This formula is useful for computations because it is expressed completely in terms of cosine functions. The procedure is general, so a similar formula could be derived for any IIR filter. Since the cosine is an even function, $|H(e^{j\hat{\omega}})|^2$ is an even function of $\hat{\omega}$, that is,

$$|H(e^{-j\hat{\omega}})|^2 = |H(e^{j\hat{\omega}})|^2$$

This is true for any system function when the difference equation has real coefficients. The magnitude $|H(e^{j\hat{\omega}})|$, which requires a square root, is also an even function.

The phase response is a bit messier. Recall that the angle of a complex number z can be found via $\tan^{-1}(\Im\{z\}/\Re\{z\})$. If arctangents are used to extract the angles of the numerator and denominator, then the two angles must be subtracted to get the phase. The filter coefficients in this example are real, so the phase is

$$\angle H(e^{j\hat{\omega}}) = \tan^{-1}\left(\frac{\sin(2\hat{\omega})}{1-\cos(2\hat{\omega})}\right) - \tan^{-1}\left(\frac{0.9\sin\hat{\omega} - 0.81\sin(2\hat{\omega})}{1-0.9\cos\hat{\omega}+0.81\cos(2\hat{\omega})}\right)$$

which can be shown to be an odd function of $\hat{\omega}$ (i.e., $\angle H(e^{-j\hat{\omega}}) = -\angle H(e^{j\hat{\omega}})$).

The even magnitude and odd phase are general characteristics of the frequency response (or any DTFT) of a real impulse response (or any real sequence). As discussed in Section 6-4.3 for the FIR case and now demonstrated for an IIR filter, the frequency response has conjugate symmetry such that

$$H(e^{-j\hat{\omega}}) = H^*(e^{j\hat{\omega}})$$

whenever the impulse response is real.

The formulas obtained in this example are too complicated to provide much insight directly. Subsequent discussions will show how to use the poles and zeros of $H(z)$ to construct an approximate plot of the frequency response without recourse to such formulas.

10-12.1 Frequency Response via MATLAB

As discussed in Sect. 10-7.1, the MATLAB function `freqz` can be used to evaluate the frequency response over a grid in the $\hat{\omega}$ domain for $-\pi < \hat{\omega} \leq \pi$. Then the magnitude and phase of $H(e^{j\hat{\omega}})$ can be extracted with the functions `abs` and `angle` prior to plotting the frequency response.

EXAMPLE 10-20 MATLAB for $H(e^{j\hat{\omega}})$

Consider the system introduced in Example 10-19:

$$y[n] = 0.9y[n-1] - 0.81y[n-2] + x[n] - x[n-2] \tag{10.76}$$

In order to define the filter coefficients in MATLAB, we put all the terms with $y[n]$ on one side of the equation, and the terms with $x[n]$ on the other.

$$y[n] - 0.9y[n-1] + 0.81y[n-2] = x[n] - x[n-2]$$

Then we read off the filter coefficients and define the vectors `aa` and `bb` as follows:

```
aa = [ 1, -0.9, 0.81 ]
bb = [ 1, 0, -1 ]
```

The following call to `freqz` generates a vector `HH` containing the values of the frequency response evaluated at the vector of frequencies specified by the third argument, `[-pi:(pi/100):pi]`:

```
HH = freqz(bb, aa, [-pi:(pi/100):pi])
```

A plot of the resulting magnitude and phase is shown in Fig. 10-24. Since $H(e^{j\hat{\omega}})$ is always periodic with a period of 2π, it is sufficient to make the frequency response plot over the range $-\pi < \hat{\omega} \leq \pi$.

In this example, we can show a connection between the poles and zeros and the shape of the frequency response. For this system, we have the system function

$$H(z) = \frac{1 - z^{-2}}{1 - 0.9z^{-1} + 0.81z^{-2}}$$

which has its poles at $z = 0.9e^{\pm j\pi/3}$ and zeros at $z = 1$ and $z = -1$. Since $z = -1$ is the same as $z = e^{j\pi}$, we conclude that $H(e^{j\hat{\omega}})$ is zero at $\hat{\omega} = \pi$, because $H(z) = 0$ at $z = -1$; likewise, the zero of $H(z)$ at $z = +1$ is a zero of $H(e^{j\hat{\omega}})$ at $\hat{\omega} = 0$. The poles have angles of $\pm\pi/3$ rad, so the poles have an effect on the frequency response near $\hat{\omega} = \pm\pi/3$. Since $H(z)$ is infinite at $z = 0.9e^{\pm j\pi/3}$, the nearby points on the unit circle (at $z = e^{\pm j\pi/3}$) must have large values. In this case, we can evaluate the frequency response directly from the formula to get

$$\begin{aligned}
H(e^{j\hat{\omega}})\Big|_{\hat{\omega}=\pi/3} &= H(z)\Big|_{z=e^{j\pi/3}} \\
&= \left.\frac{1 - z^{-2}}{1 - 0.9z^{-1} + 0.81z^{-2}}\right|_{z=e^{j\pi/3}} \\
&= \frac{1 - (-\frac{1}{2} - j\frac{1}{2}\sqrt{3})}{1 - 0.9(\frac{1}{2} - j\frac{1}{2}\sqrt{3}) + 0.81(-\frac{1}{2} - j\frac{1}{2}\sqrt{3})} \\
\Rightarrow \left|H(e^{j\pi/3})\right| &= \frac{\left|1.5 + j0.5(\sqrt{3})\right|}{\left|0.145 + j0.045(\sqrt{3})\right|} = 10.522
\end{aligned}$$

This value of the frequency response magnitude is a good approximation to the true maximum value of 10.5257, which occurs at $\hat{\omega} = 0.334\pi$.

10-12.2 3-dB Bandwidth

The width of the peak of the frequency response in Fig. 10-24 is called the ***bandwidth***. For consistency, bandwidth must be measured at some standard point on the plot of $|H(e^{j\hat{\omega}})|$. The most common practice is to plot $|H(e^{j\hat{\omega}})|$ on a dB scale and use the 3-dB width, which is calculated as follows:

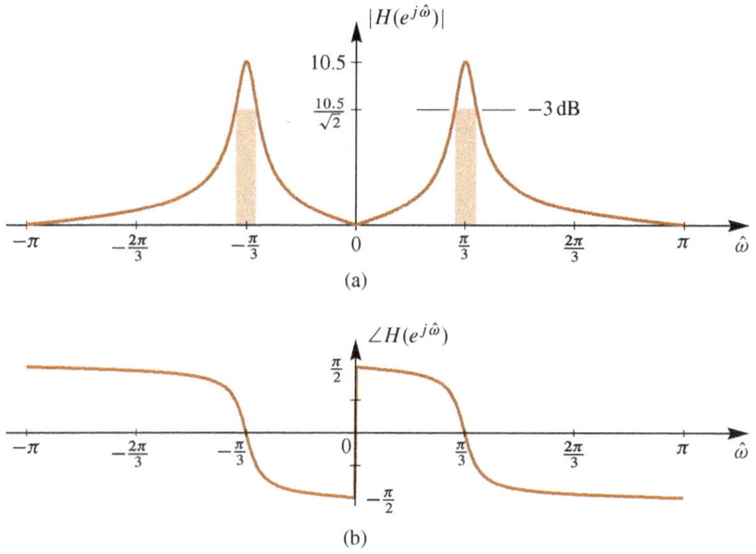

Figure 10-24 Frequency response (magnitude and phase) for a second-order IIR filter. The poles are at $z = 0.9e^{\pm j\pi/3}$ and the numerator has zeros at $z = 1$ and $z = -1$. (a) Magnitude response with shaded regions showing the 3-dB bandwidth around the peak at $\hat{\omega} = \pi/3$. (b) Phase response which is nonlinear.

Determine the peak value for $|H(e^{j\hat{\omega}})|$ and then find the nearest frequency on each side of the peak where the value of the frequency response is $\frac{1}{\sqrt{2}}H_{\text{peak}}$. The ***3-dB width*** is the difference $\Delta\hat{\omega}$ between these two frequencies. In decibels, the term $\frac{1}{\sqrt{2}}$ is approximately -3 dB, so the 3-dB bandwidth $\Delta\hat{\omega}$ is measured 3 dB below the peak.

In Fig. 10-24, the true peak value is 10.5257 at $\hat{\omega} = 0.334\pi$, so we look for points where $|H(e^{j\hat{\omega}})| = \frac{1}{\sqrt{2}}H_{\text{peak}} = (0.707)(10.5257) = 7.443$. These occur at $\hat{\omega} = 0.302\pi$ and $\hat{\omega} = 0.369\pi$, so the bandwidth is $\Delta\hat{\omega} = 0.067\pi = 2\pi(0.0335) = 0.2105$ rad. Alternatively, if we plot $20\log_{10}|H(e^{j\hat{\omega}})|$ instead of $|H(e^{j\hat{\omega}})|$ as in Fig. 10-25, the maximum value is 20.42 dB so the 3-dB bandwidth points are the frequencies where $20\log_{10}|H(e^{j\hat{\omega}})| = 20.42 - 3 = 17.42$ dB, which is 3 dB down from the peak. Note how $20\log_{10}|H(e^{j\hat{\omega}})|$ approaches $-\infty$ at the frequencies corresponding to the zeros of $H(z)$ at $z = \pm 1$ (where $|H(e^{j0})| = 0$ and $|H(e^{j\pi})| = 0$).

The 3-dB bandwidth calculation can be carried out efficiently with a computer program, but it is also helpful to have an approximate formula that can give quick

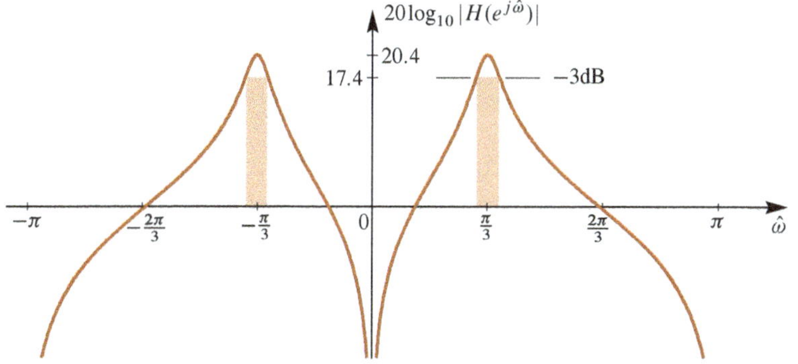

Figure 10-25 Log magnitude of frequency response (in dB) corresponding to Fig. 10-24(a). The shaded region shows the 3-dB bandwidth around the peak at $\hat{\omega} = \pi/3$.

"back-of-the-envelope" calculations. For poles at $re^{\pm j\theta}$, an excellent approximation for the second-order case with narrow peaks is given by the formula

$$\Delta\hat{\omega} \approx 2\,\frac{|1-r|}{\sqrt{r}} \tag{10.77}$$

which shows that the distance of the pole from the unit circle $|1-r|$ controls the bandwidth.[12] In Fig. 10-24 where $r = 0.9$, the bandwidth estimate (10.77) evaluates to

$$\Delta\hat{\omega} \approx 2\,\frac{(1-0.9)}{\sqrt{0.9}} \approx \frac{0.2}{0.9487} \approx 0.2108 \text{ rad}$$

The measured value of $\Delta\hat{\omega}$ was 0.2105 rad, so we see that the approximation is quite good in this case, where the pole is rather close to the unit circle (radius = 0.9).

10-12.3 Three-Dimensional Plot of a System Function

Since the frequency response $H(e^{j\hat{\omega}})$ is the system function evaluated on the unit circle, we can illustrate the connection between the z and $\hat{\omega}$ domains with a 3-D plot such as the one shown in Fig. 10-26. The poles and zeros of $H(z)$ provide the connection because they have a significant impact on the 3-D plot, and thus have a predictable influence on the frequency response, which is $H(z)$ evaluated on the unit circle.

DEMO

Z to Freq

Figure 10-26 shows a plot of the system function $H(z)$ at points inside, outside, and on the unit circle. The peaks located at the poles, $0.85e^{\pm j\pi/2}$, determine the frequency response behavior near $\hat{\omega} = \pm\pi/2$. If the poles were moved closer to the unit circle, the frequency response would have higher and narrower peaks. The zeros at $z = \pm 1$ create valleys that, in this case, lie on the unit circle at $\hat{\omega} = 0, \pi$.

Given the intuition derived from the 3-D example in Fig. 10-26, we can estimate any value of $|H(e^{j\hat{\omega}})|$ directly from the poles and zeros. This can be done systematically by writing $H(z)$ in the following form:

$$H(z) = G\,\frac{(z-z_1)(z-z_2)}{(z-p_1)(z-p_2)}$$

where z_1 and z_2 are the zeros and p_1 and p_2 are the poles of the second-order filter. The parameter G is a gain term that has been factored out. Then the magnitude of the frequency response is

$$\left|H(e^{j\hat{\omega}})\right| = G\,\frac{|e^{j\hat{\omega}}-z_1|\,|e^{j\hat{\omega}}-z_2|}{|e^{j\hat{\omega}}-p_1|\,|e^{j\hat{\omega}}-p_2|} \tag{10.78}$$

Equation (10.78) has a simple geometric interpretation. Each term $|e^{j\hat{\omega}}-z_i|$ or $|e^{j\hat{\omega}}-p_i|$ is the vector length from the zero z_i or the pole p_i to the position $e^{j\hat{\omega}}$, as shown in Fig. 10-27. The frequency response at a fixed value of $\hat{\omega}$ is the product of G times the

[12] This approximate formula for bandwidth is good only when the poles are isolated from one another. The approximation breaks down, for example, when a second-order system has two poles with small angles.

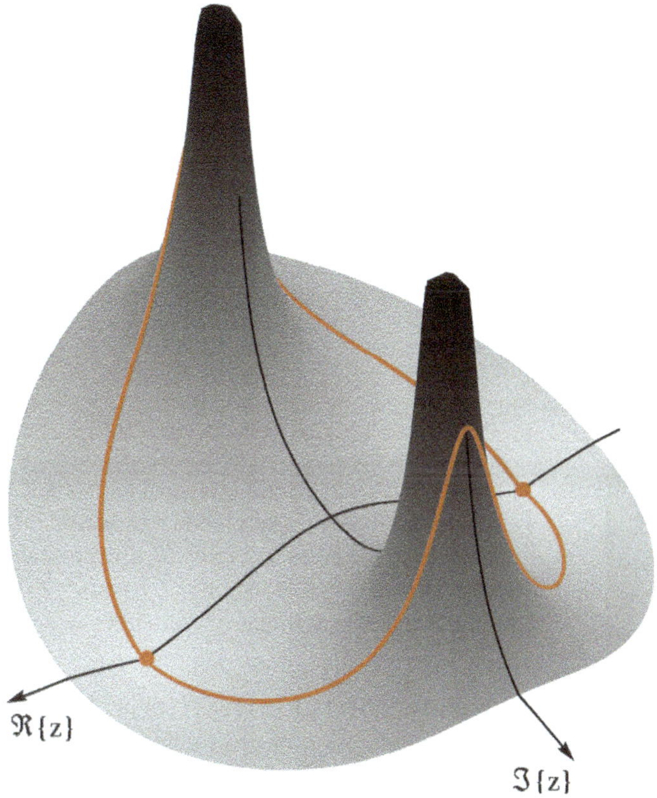

Figure 10-26 z-transform magnitude evaluated over a region of the z-plane including the unit circle. The view is from the first quadrant, so the $z = 1$ point is in the foreground at the left. Values along the unit circle are the frequency response (magnitude) for the second-order IIR filter defined in (10.75). The poles are at $z = 0.85e^{\pm j\pi/2}$ and the numerator has zeros at $z = \pm 1$. The zeros are denoted by orange dots that can be seen along the real axis.

product of all the lengths of the vectors from the zeros to the point $z = e^{j\hat{\omega}}$ divided by the product of all the lengths of the vectors from the poles to $e^{j\hat{\omega}}$.

$$\left|H(e^{j\hat{\omega}})\right| = G\,\frac{\overline{Z_1 Z}\cdot\overline{Z_2 Z}}{\overline{P_1 Z}\cdot\overline{P_2 Z}}$$

As we go around the unit circle, these vector lengths change. When we are on top of a zero, one of the numerator lengths is zero, so $|H(e^{j\hat{\omega}})| = 0$ at that frequency. When we are close to a pole, one of the denominator lengths is very small, so $\left|H(e^{j\hat{\omega}})\right|$ is large at that frequency.

We can apply this geometric reasoning to estimate the magnitude of $H(e^{j\hat{\omega}})$ at $\hat{\omega} = \pi/4$ which is the configuration shown in Fig. 10-27. The gain G is assumed to be 1. Rough estimates for the vector lengths are $\overline{Z_1 Z} \approx 1.8$ and $\overline{Z_2 Z} \approx 0.7$ for the zeros, and $\overline{P_1 Z} \approx 0.7$ and $\overline{P_2 Z} \approx 1.6$ for the poles. Thus we obtain the following estimate for the magnitude of the frequency response at $\hat{\omega} = \pi/4$:

$$\left|H(e^{j\pi/4})\right| = G\,\frac{\overline{Z_1 Z}\cdot\overline{Z_2 Z}}{\overline{P_1 Z}\cdot\overline{P_2 Z}} = (1)\frac{(1.8)(0.7)}{(0.7)(1.6)} = \frac{1.8}{1.6} = 1.125$$

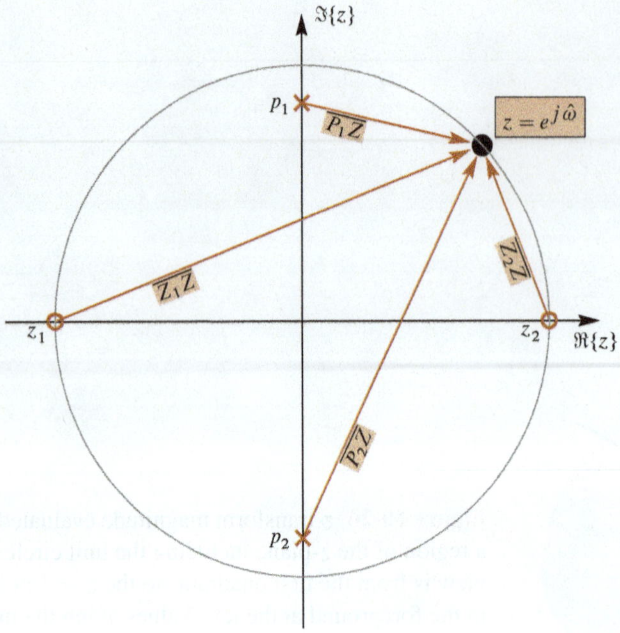

Figure 10-27 z-transform evaluated on the unit circle ($z = e^{j\hat{\omega}}$) by using a product of vector lengths from the poles and zeros. $\overline{P_iZ}$ denotes the vector from the ith pole to ($z = e^{j\hat{\omega}}$), and $\overline{Z_jZ}$ denotes the vector from the jth zero to ($z = e^{j\hat{\omega}}$).

The actual value obtained from evaluating the formula (10.75) for $H(z)$ is $|H(e^{j\pi/4})| = 1.132$. See Problems P-10.19 and P-10.23 for application of this method to notch filters and all-pass filters.

EXERCISE 10.13 For the pole-zero plot in Fig. 10-27, apply this geometric reasoning to estimate the magnitude of $H(e^{j\hat{\omega}})$ in Fig. 10-26 at $\hat{\omega} = \pi/2$, which is close to the pole at $z = j0.85$. Assume that the gain G is equal to 1. In this case, the estimate can be exact because the vector lengths do not have to be approximated.

10-12.4 Pole-Zero Placing with the `PeZ` GUI

In order to build an intuitive understanding of the relationship between the location of poles and zeros in the z-domain, the impulse response $h[n]$ in the n-domain, and the frequency response $H(e^{j\hat{\omega}})$ in the $\hat{\omega}$-domain, a GUI called `PeZ` was written in MATLAB for doing interactive explorations of the three domains. The `PeZ` GUI is contained in the *DSP-First* Toolbox. To run `PeZ`, type `pezdemo` at the command prompt and you will see the GUI shown in Fig. 10-28. `PeZ` is based primarily on the factored form of the system function, which represents the rational z-transform in terms of the poles and zeros:

$$H(z) = \frac{B(z)}{A(z)} = G\,\frac{\displaystyle\prod_{k=1}^{M}(1 - z_k z^{-1})}{\displaystyle\prod_{\ell=1}^{N}(1 - p_\ell z^{-1})} \tag{10.79}$$

where M is the number of zeros and N is the number of poles.

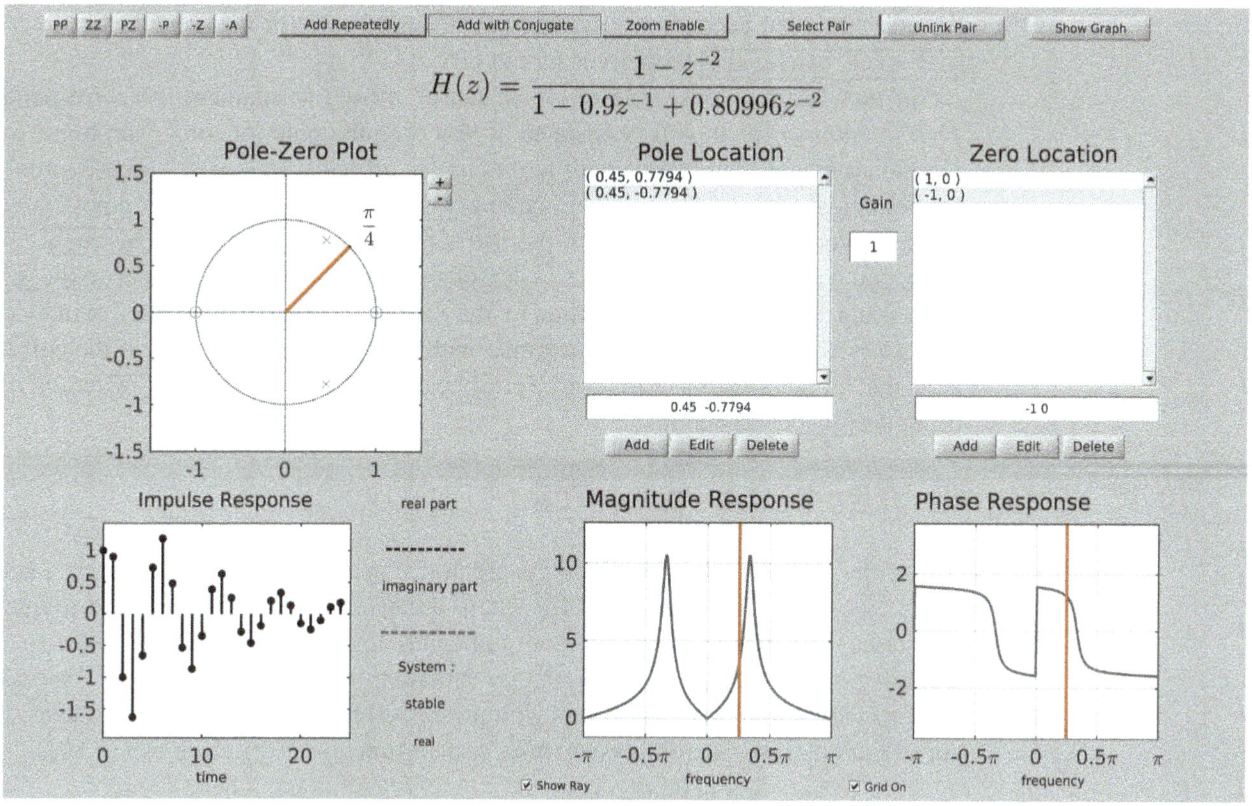

Figure 10-28 Graphical user interface for PeZ which illustrates the *three-domains* concept. The user can enter poles and zeros and see the resulting impulse response and frequency response. In addition, the poles and zeros can be moved with a mouse pointer and the other plots will be updated in real time. The example shown here is the same as $H(z)$ in (10.75) and the plots in Fig. 10-24.

DEMO
PeZ GUI

The PeZ GUI is controlled by the `Pole-Zero Plot` panel in the upper left where the user can add (or delete) poles and zeros, as well as move them around with the pointing device. The button denoted `ZZ` is used to add zeros; the button `PP` is used to add poles. By default, the `Add with Conjugate` property is turned on, so poles and zeros are typically added in pairs to satisfy the complex conjugate property, giving real coefficients. For more detailed information about the PeZ controls and menus, access the menu item called "Help."

Here are some of the features of PeZ. In the `Pole-Zero Plot` panel, you can selectively place pole pairs or zero pairs in the z-plane, and then observe (in the other plotting panels) how their placement affects the impulse and frequency responses. The orange ray in the z-domain window is tied to the orange vertical lines on the frequency responses, and they move together[13]. This helps identify frequency domain features that are caused by pole locations or zero locations, because the angle around

[13]The orange lines highlight an important feature in PeZ. In MATLAB the Pez display is rendered in color and these lines are actually red.

the unit circle corresponds to frequency $\hat{\omega}$. A single pole (or zero) can be added if the `Add with Conjugate` property is turned off.

In `PeZ`, an individual pole/zero pair can be moved around and the corresponding $H(e^{j\hat{\omega}})$ and $h[n]$ plots will be updated as you drag the pole (or zero) pair. Since exact placement of poles and zeros with the mouse is difficult, an `Edit` button is provided for numerical entry of the real and imaginary parts. Before you can edit a pole or zero, however, you must first select it in the list of `Pole Locations` or `Zero Locations`. Next, you will have to make the necessary change in the edit field (box above the `Edit` button) and then press the `Edit` button for the change to take effect. Removal of individual pole pairs or zero pairs can be performed with the `Delete` button below the edit box. Note that all poles and/or zeros can be easily cleared by clicking on the `-A` button or by using the `-P` or `-Z` buttons.

10-13 Example of an IIR Lowpass Filter

First-order and second-order IIR filters are useful and provide simple examples, but, in many cases, we use higher-order IIR filters because they can realize better frequency responses with flatter passbands and stopbands and sharper transition regions. The `butter`, `cheby1`, `cheby2`, and `ellip` functions in MATLAB's *Signal Processing Toolbox* can be used to design filters with prescribed frequency-selective characteristics. As an example, consider the system with system function $H(z)$ given by (10.80a).

$$H(z) = \frac{0.0798(1+z^{-1}+z^{-2}+z^{-3})}{1-1.556z^{-1}+1.272z^{-2}-0.398z^{-3}} \qquad (10.80a)$$

$$= \frac{0.0798(1+z^{-1})(1-e^{j\pi/2}z^{-1})(1-e^{-j\pi/2}z^{-1})}{(1-0.556z^{-1})(1-0.846e^{j0.3\pi}z^{-1})(1-0.846e^{-j0.3\pi}z^{-1})} \qquad (10.80b)$$

$$= -\frac{1}{5} + \frac{0.62}{1-.556z^{-1}} + \frac{0.17e^{j0.96\pi}}{1-.846e^{j0.3\pi}z^{-1}} + \frac{0.17e^{-j0.96\pi}}{1-.846e^{-j0.3\pi}z^{-1}} \qquad (10.80c)$$

This system is an example of a lowpass ***elliptic filter*** whose numerator and denominator coefficients were obtained using the MATLAB function `ellip`. The exact call was `ellip(3,1,30,0.3)`. Each of the three different forms above is useful: (10.80a) for identifying the coefficients used in the difference equation implementation of the filter, (10.80b) for sketching the pole-zero plot and the frequency response, and (10.80c) for finding the impulse response.[14] Figure 10-29 shows the poles and zeros of this filter. Note that all the zeros are on the unit circle and that all the poles are strictly inside the unit circle, as they must be for a stable system.

EXERCISE 10.14 From (10.80a), determine the difference equation (using Direct Form I) for implementing this system.

[14] Factoring polynomials and obtaining the partial fraction expansion was done in MATLAB using the functions `roots` and `residuez`, respectively.

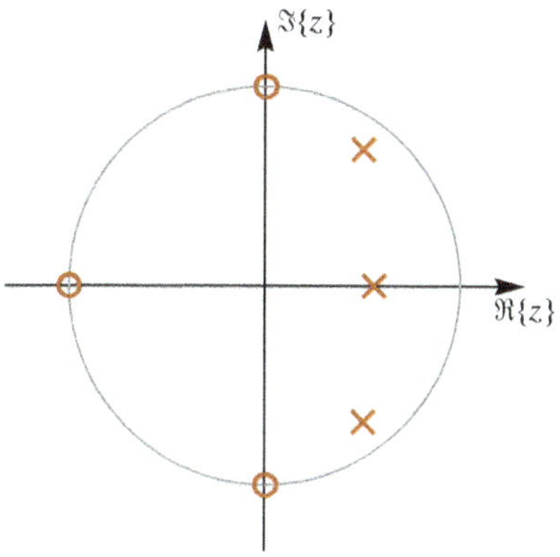

Figure 10-29 Pole-zero plot for a third-order IIR filter (10.80b). Three zeros lie on the unit circle at $z = \pm j$ and $z = -1$.

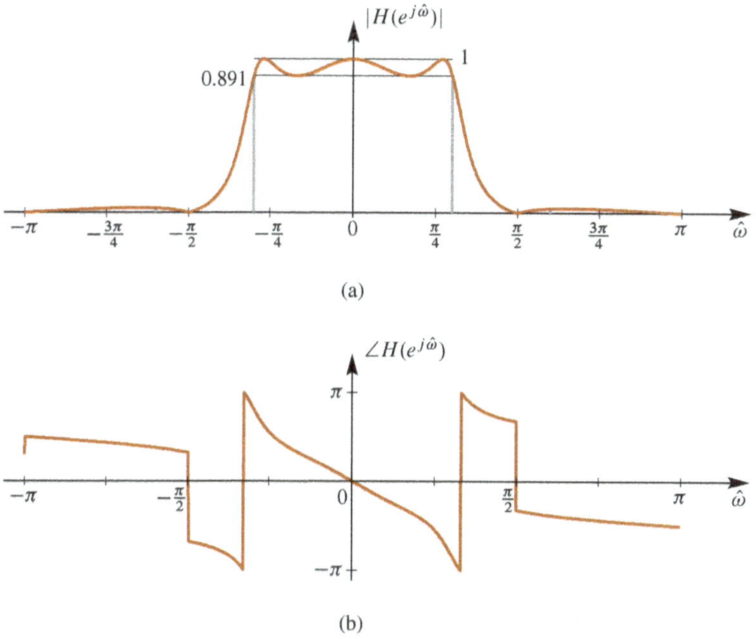

Figure 10-30 Frequency response for a third-order IIR filter; (a) magnitude, (b) phase.

The system function was evaluated on the unit circle using the MATLAB function `freqz`. A plot of this result is shown in Fig. 10-30. Note that the frequency response is large in the vicinity of the poles and small around the zeros. In particular, the passband of the frequency response is $|\hat{\omega}| \leq 2\pi(0.15)$, which corresponds to the poles with angles at $\pm 0.3\pi$. Also, the frequency response is exactly zero at $\hat{\omega} = \pm 0.5\pi$ and $\hat{\omega} = \pi$ since the zeros of $H(z)$ are at these angles and lie on the unit circle.

EXERCISE 10.15 From (10.80a) or (10.80b), determine the value of the frequency response at $\hat{\omega} = 0$.

Finally, Fig. 10-31 shows the impulse response of the system. It oscillates and dies out with increasing n because the dominant terms come from the two complex conjugate poles at angles $\pm 0.3\pi$ and radius 0.846. The decaying envelope is $(0.846)^n$.

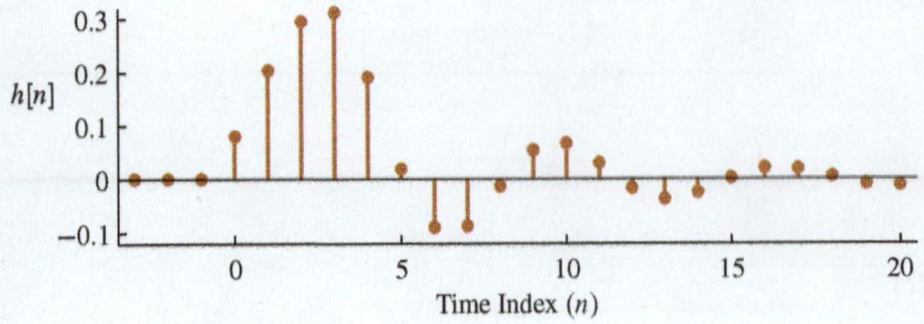

Figure 10-31 Impulse response for a third-order IIR filter.

EXERCISE 10.16 Use the partial fraction form (10.80c) to determine an equation for the impulse response of the filter.

The elliptic filter example described in this section is a simple example of a practical IIR lowpass filter. Higher-order filters can exhibit much better frequency-selective filter characteristics.

10-14 Summary and Links

The class of IIR filters was introduced in this chapter, along with new methods for using the z-transform to deal with poles. The z-transform changes problems about impulse responses, frequency responses, and system structures into algebraic manipulations of polynomials and rational functions. Poles of the system function $H(z)$ turn out to be the most important elements for IIR filters because properties such as the shape of the frequency response or the form of the impulse response can be inferred quickly from the pole locations.

IIR filters can be used for many tasks; frequency-selective filtering is the most common, but nulling or notch filters are also very useful for removing narrowband interference (e.g., 60 Hz power line signals). The MATLAB GUI called `PeZ` (see Fig. 10-28) is an ideal tool for studying the poles and zeros of IIR filters, and the connection between the time domain (n), the frequency domain ($\hat{\omega}$), and the z-transform domain. Some of the things that can be done with `PeZ` are the following:

DEMO

PeZ GUI

1. *Placing Zeros:* When placed on the unit circle, zeros of $B(z)$ force the frequency response to be zero which can then be used to null out sinusoids at one frequency.
2. *Placing Poles:* When placed near the unit circle (but inside), roots of $A(z)$ create peaks in the frequency response which can be used to form BPFs.
3. *Designing IIR Notch Filters:* By placing conjugate zeros on the unit circle with poles at the same angle, but just inside the unit circle, a notch or narrow region with small gain can be formed. The frequency response of the notch is much sharper than a nulling filter which is an FIR filter formed only from the conjugate zeros on the unit circle (see Problem P-10.19).

4. *Designing Allpass Filters:* By placing poles inside the unit circle with zeros at corresponding conjugate reciprocal locations, the frequency response has constant magnitude for all frequencies. Such systems are called all-pass filters (see Problem P-10.23).

LAB
PeZ—the z, n, and $\hat{\omega}$ Domains

We also continued to stress the important concept of "domains of representation." The n-domain or time domain, the $\hat{\omega}$-domain or frequency domain, and the z-domain give us three domains for thinking about the characteristics of an LTI system. We completed the ties between domains by introducing the inverse z-transform for constructing a signal from its z-transform. As a result, even difficult problems such as convolution can be simplified by working in the most convenient domain (z) and then transforming back to the original domain (n).

DEMO
Three Domains—IIR

In a lab devoted to studying the three domains, the `PeZ` GUI is used to explore the properties of IIR filters. The `PeZ` tool is useful for studying IIR systems because it presents the user with multiple views of an LTI system: pole-zero domain, frequency response, and impulse response. (See Fig. 10-28.)

The Companion Website also contains the following demonstrations of the relationship between the z-plane and the frequency domain and time domain:

(a) A set of "three-domain" movies that show how the frequency response and impulse response of an IIR filter change as a pole location is varied. Several different filters are demonstrated.

DEMO
Z to Freq

(b) A movie that animates the relationship between the z-plane and the unit circle where the frequency response lies.

(c) The `PeZ` GUI can be used to construct different IIR and FIR filters. There is also a tutorial on how to use `PeZ`.

10-15 Problems

NOTE
Problems with Solutions

The problems in this section address the major concepts of this chapter. On the Companion Website there are many related homework problems which can be used to review and practice. Most of them have detailed solutions.

P-10.1 The Fibonacci sequence is generated by a recursion.

(a) Determine a general formula for the Fibonacci sequence by finding the impulse response of the following second-order system:

$$y[n] = y[n-1] + y[n-2] + x[n]$$

(b) Determine the two poles of this system. Is this a stable system?

P-10.2 Find the impulse response of the second-order system

$$y[n] = 0.8(y[n-1] - y[n-2]) + x[n-1]$$

Express your answer as separate formulas for the cases where $n < 1$ and $n \geq 1$, thus covering the entire range of n.

P-10.3 Use time invariance to determine the impulse response of the following first order system:

$$y[n] = 0.4y[n-1] + 6x[n-8]$$

Assume that the system is at rest for $n < 0$. Plot the resulting signal $h[n]$ as a function of n. Pay careful attention to where the nonzero portion of the impulse response begins.

P-10.4 Find the impulse response of a system having a pole at 0.4 and a zero at 0 and $h[0] = 4$. Use this $h[n]$ to determine the output due to the input signal

$$x[n] = \begin{cases} 0 & n < 1, n \geq 7 \\ 2 & n = 1 \\ -3 & n = 2 \\ 0 & n = 3 \\ 5 & n = 4 \\ 0 & n = 5 \\ -1.5 & n = 6 \end{cases}$$

Write the formula as the sum of four terms, each of which is a shifted impulse response. Assume the initial rest condition. Plot the resulting signal $y[n]$ as a function of n for $0 \leq n \leq 12$.

P-10.5 The impulse response of an LTI system is $h[n] = 8\left(\sqrt{3}\right)^{-n} u[n]$. When the input signal is $x[n] = \delta[n] + \delta[n-2] + \beta\delta[n-6]$, determine the value of β so that the output $y[n] = x[n] * h[n]$ is zero for $n \geq 6$.

P-10.6 For the following feedback filters, determine the locations of the poles and zeros and plot the positions in the z-plane:

(a) $y[n] = \frac{1}{2}y[n-1] + \frac{1}{3}y[n-2] - x[n] - 2x[n-1] + 3x[n-2]$

(b) $y[n] = \frac{1}{2}y[n-2] - \frac{1}{3}y[n=2] - x[n] + 2x[n-1] + 3x[n-2]$

In part (b), only the signs on $y[n-2]$ and $x[n-1]$ were changed from those in part (a). Compare the two results.

P-10.7 In this problem, the degrees of the numerator and denominator polynomials are different, so there should be zeros (or poles) at $z = 0$ or $z = \infty$. Determine the poles and zeros of the following filters:

(a) $y[n] = \frac{1}{2}y[n-1] - \frac{1}{3}y[n-2] - x[n]$

(b) $y[n] = \frac{1}{2}y[n-1] - \frac{1}{3}y[n-2] - x[n-2]$

(c) $y[n] = \frac{1}{2}y[n-1] - \frac{1}{3}y[n-2] - x[n-5]$

Plot the positions of the poles and zeros in the z-plane. Make sure that the number of poles and zeros is the same, by allowing zeros (or poles) at $z = \infty$ or $z = 0$.

P-10.8 Given an IIR filter defined by the difference equation

$$y[n] = -\tfrac{3}{4}y[n-1] + x[n]$$

(a) Determine the system function $H(z)$. What are its poles and zeros?

(b) When the input to the system is three shifted impulses

$$x[n] = \begin{cases} +1 & n = 0, 3, 5 \\ -2 & n = 6 \\ 0 & \text{otherwise} \end{cases}$$

determine the functional form for the output signal $y[n]$. Assume that the output signal $y[n]$ is zero for $n < 0$.

P-10.9 An LTI allpass filter is described by the difference equation

$$y[n] = \tfrac{2}{3}y[n-1] - 12x[n] + 18x[n-1]$$

(a) Determine the system function $H(z)$ for this system. Express $H(z)$ as a ratio of polynomials in z^{-1} (negative powers of z) and also as a ratio of polynomials in positive powers of z.

(b) Plot the poles and zeros of $H(z)$ in the z-plane.

(c) From $H(z)$, obtain an expression for $H(e^{j\hat{\omega}})$, the frequency response of this system.

(d) Show that $|H(e^{j\hat{\omega}})|^2$ is constant for all $\hat{\omega}$, and determine the constant value.

P-10.10 Given an IIR filter defined by the difference equation

$$y[n] = -0.6y[n-2] + x[n]$$

(a) Determine the system function $H(z)$, and then find its poles and zeros.

(b) When the input to the system is

$$x[n] = 6\delta[n] + 2\delta[n-1] + 4.5\delta[n-2] + 3\delta[n-4]$$

determine the output signal $y[n]$. Assume that $y[n]$ is zero for $n < 0$.

P-10.11 An IIR filter has 6 poles at the 6th root of 0.8 and 6 zeros at $z = 0$

(a) Find the z-transform system function of the system. Plot its poles in the z-plane.

(b) Find the difference equation of the system.

(c) When the input signal is $x[n] = \delta[n] + \beta\delta[n-p]$, determine values of p for which the output signal is zero for $n \geq 6$.

P-10.12 Determine the inverse z-transform of the following:

(a) $H_a(z) = \dfrac{1 - 0.8z^{-1}}{1 + 0.7z^{-1}}$

(b) $H_b(z) = \dfrac{z^{-3}}{1 - 0.6z^{-1}}$

(c) $H_c(z) = 2 - 0.5z^{-1} - 3z^{-3} + 4z^{-4}$

(d) $H_d(z) = \dfrac{2z^{-5}}{z + 0.68} - z^{-7} + \dfrac{z^{-2}}{1 - 0.4z^{-1}}$

P-10.13 Determine the inverse z-transform of the following:

(a) $X_a(z) = \dfrac{1 - z^{-1}}{1 - \frac{1}{12}z^{-1} - \frac{1}{12}z^{-2}}$

(b) $X_b(z) = \dfrac{1 + z^{-2}}{1 + 0.8z^{-1} + 0.64z^{-2}}$

(c) $X_c(z) = \dfrac{z^2}{(z-2)(z-0.25)^2}$

P-10.14 For each of the pole-zero plots (1–6) in Fig. P-10.14, determine which of the following systems (specified by either an $H(z)$ or a difference equation) matches the pole-zero plot:

S_1: $y[n] = 0.9y[n-1] + \frac{1}{2}x[n] + \frac{1}{2}x[n-1]$

S_2: $y[n] = -0.9y[n-1] + 9x[n] + 10x[n-1]$

S_3: $H(z) = \dfrac{\frac{1}{2}(1 - z^{-1})}{1 + 0.9z^{-1}}$

S_4: $y[n] = \frac{1}{4}x[n] + x[n-1] + \frac{3}{2}x[n-2] + x[n-3] + \frac{1}{4}x[n-4]$

S_5: $H(z) = 1 - z^{-1} + z^{-2} - z^{-3} + z^{-4}$

S_6: $y[n] = \sum_{k=0}^{2} x[n-k]$

S_7: $y[n] = x[n] + x[n-1] + x[n-2] + x[n-3] + x[n-4] + x[n-5]$

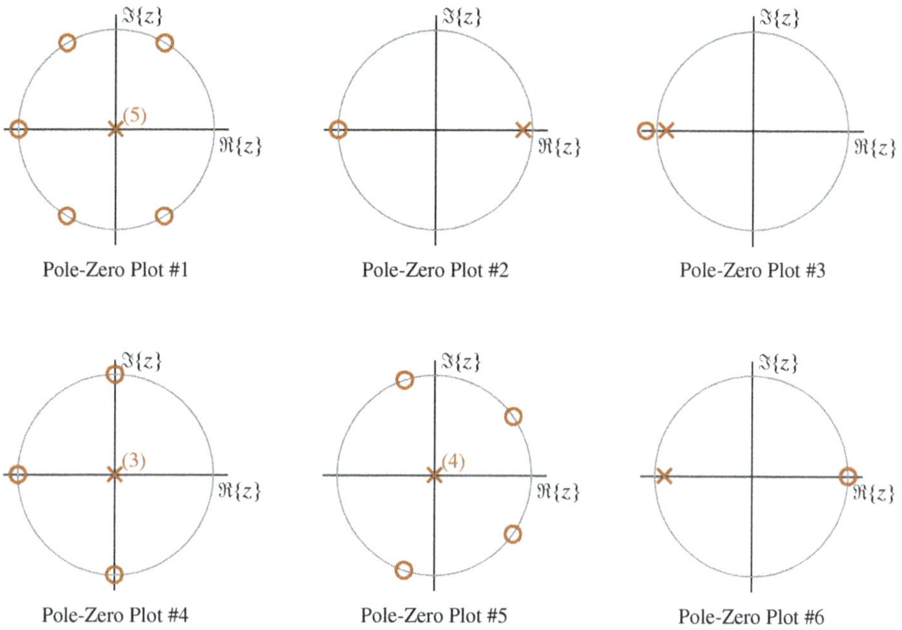

Figure P-10.14

P-10.15 Given an IIR filter defined by the difference equation

$$y[n] = \tfrac{1}{4}y[n-1] + x[n]$$

(a) When the input to the system is a unit-step sequence, $u[n]$, determine the functional form for the output signal $y[n]$. Use the inverse z-transform method. Assume that the output signal $y[n]$ is zero for $n < 0$.

(b) Find the output when $x[n]$ is a complex exponential that starts at $n = 0$:

$$x[n] = e^{j(\pi/6)n}u[n]$$

(c) From (b), identify the steady-state component of the response, and compare its magnitude and phase to the frequency response at $\hat{\omega} = \pi/6$.

P-10.16 The system function of an LTI system is given by the formula

$$H(z) = \frac{(1+z^{-1})(1-e^{j0.4\pi}z^{-1})(1-e^{-j0.4\pi}z^{-1})}{(1-0.7e^{j0.8\pi}z^{-1})(1-0.7e^{-j0.8\pi}z^{-1})}$$

(a) Write the difference equation that gives the relation between the input $x[n]$ and the output $y[n]$.

(b) Plot the poles and zeros of $H(z)$ in the complex z-plane.
Hint: Express $H(z)$ as a ratio of factored polynomials in z instead of z^{-1}.

(c) If the input is of the form $x[n] = Ae^{j\varphi}e^{j\hat{\omega}_0 n}$, for what values of $-\pi \le \hat{\omega}_0 \le \pi$ is it true that $y[n] = 0$?

P-10.17 For each of the frequency response plots (1–6) in Fig. P-10.17, determine which of the following systems (specified by either an $H(z)$ or a difference equation) matches the frequency response:

$$\mathcal{S}_1: \quad y[n] = 0.9y[n-1] + \tfrac{1}{2}x[n] + \tfrac{1}{2}x[n-1]$$

$$\mathcal{S}_2: \quad y[n] = -0.9y[n-1] + 9x[n] + 10x[n-1]$$

$$\mathcal{S}_3: \quad H(z) = \frac{\frac{1}{2}(1 - z^{-1})}{1 + 0.9z^{-1}}$$

$$\mathcal{S}_4: \quad y[n] = \tfrac{1}{4}x[n] + x[n-1] + \tfrac{3}{2}x[n-2] + x[n-3] + \tfrac{1}{4}x[n-4]$$

$$\mathcal{S}_5: \quad H(z) = 1 - z^{-1} + z^{-2} - z^{-3} + z^{-4}$$

$$\mathcal{S}_6: \quad y[n] = \sum_{k=0}^{2} x[n-k]$$

$$\mathcal{S}_7: \quad y[n] = x[n] + x[n-1] + x[n-2] + x[n-3] + x[n-4] + x[n-5]$$

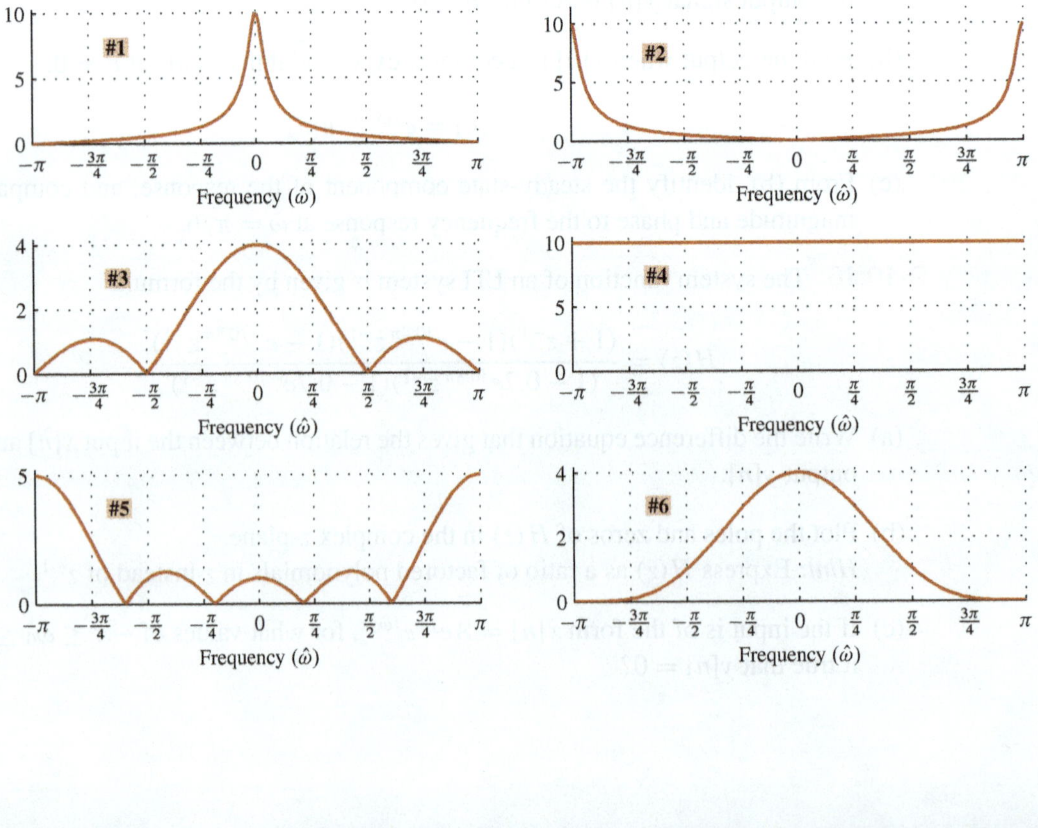

Figure P-10.17

P-10.18 In Fig. P-10.18, match five of the six pole-zero plots (PZ 1–6) with the correct one of five possible frequency responses (J–N). In the pole-zero plots, poles are denoted by **x** and zeros by **o**.

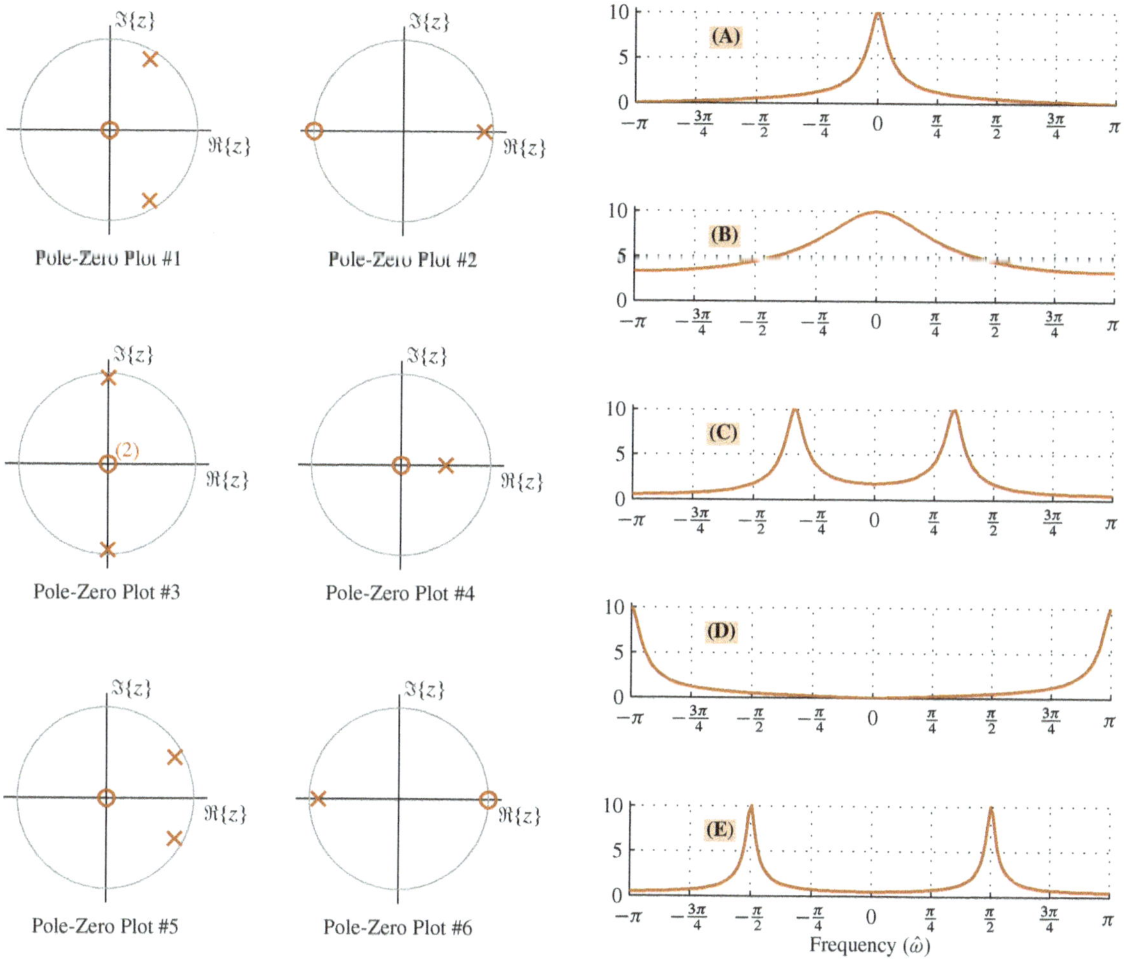

Figure P-10.18

P-10.19 An IIR notch filter is defined by the system function

$$H(z) = \frac{(1 - e^{j\pi/3}z^{-1})(1 - e^{-j\pi/3}z^{-1})}{(1 - re^{j\pi/3}z^{-1})(1 - re^{-j\pi/3}z^{-1})}$$

(a) Determine the difference equation that relates the input $x[n]$ to the output $y[n]$.

(b) Plot the poles and zeros of $H(z)$ in the z-plane for the case $r = 0.9$.

(c) Use the techniques of Section 10-12.3 to sketch the magnitude of the frequency response of the system for the case $r = 0.9$.

(d) How does the frequency response change as $r \to 1$?

(e) How would you change the frequency location of the "notch"?

P-10.20 In Fig. P-10.20, match five of the six pole-zero plots (PZ 1–6) with the correct one of five possible impulse responses (J–N). In the pole-zero plots, poles are denoted by **x** and zeros by **o**.

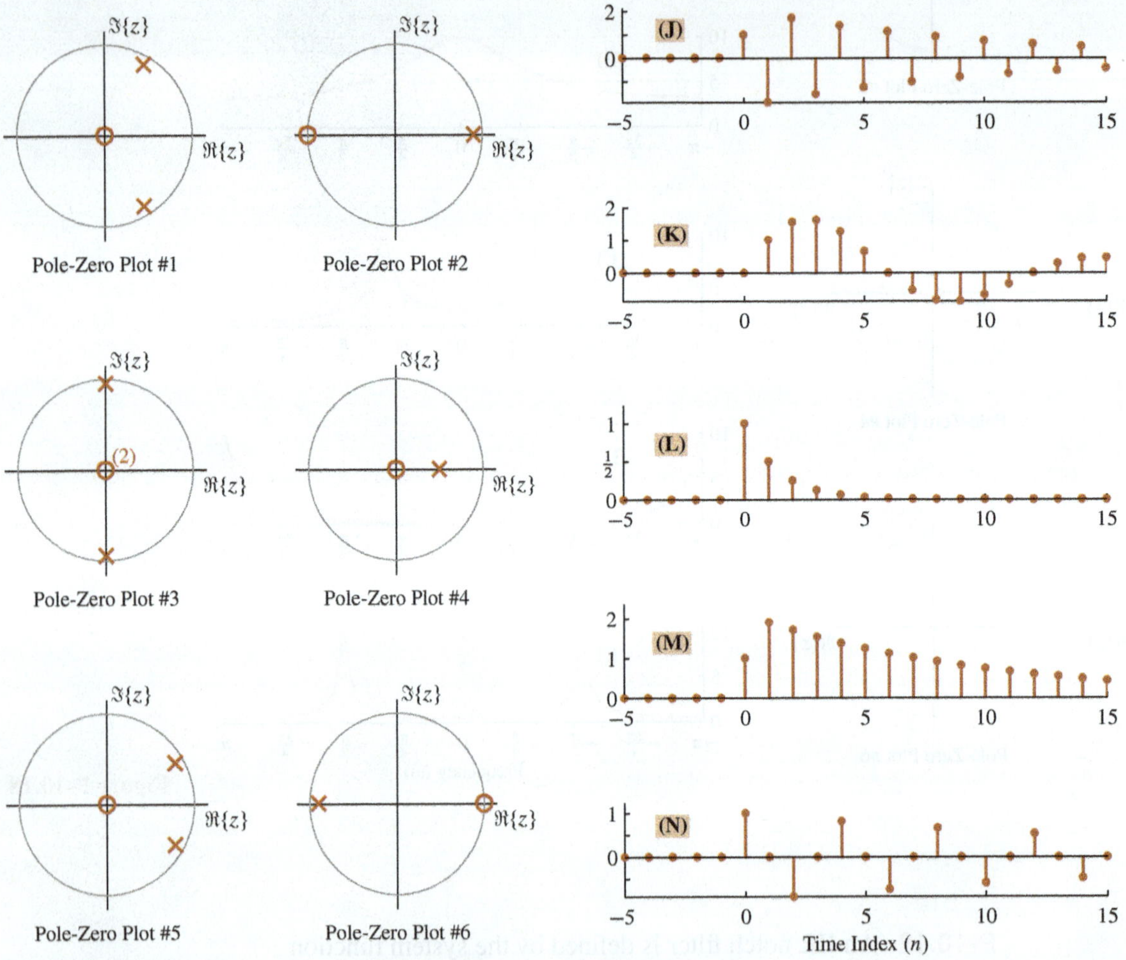

Figure P-10.20

P-10.21 The IIR filter described by the difference equation

$$y[n] = 0.9y[n-1] - 0.9x[n] + x[n-1]$$

is an all-pass filter.

(a) Determine the system function $H(z)$ for this system. Express $H(z)$ as a ratio of polynomials in z^{-1} and as a ratio of polynomials in z.

(b) For this first-order system, determine the pole and zero locations. Verify that the zero and the pole are reciprocals.

(c) For $H(z)$ in part (a), show that $|H(e^{j\hat{\omega}})|^2 = 1$.

(d) If the input to the system is

$$x[n] = 6 + 7\cos(0.3\pi n) + 8\cos(0.8\pi n)$$

Explain why the output as

$$x[n] = 6 + 7\cos(0.3\pi n + \varphi_1) + 8\cos(0.8\pi n + \varphi_2)$$

(e) Determine the value of the two phase terms φ_1 and φ_2.

P-10.22 The input to the C-to-D converter in Fig. P-10.22 is

$$x(t) = 4 + \cos(500\pi t) - 3\cos[(2000\pi/3)t]$$

The system function for the LTI system is

$$H(z) = \frac{(1 - z^{-1})(1 - e^{j\pi/2}z^{-1})(1 - e^{-j\pi/2}z^{-1})}{(1 - 0.9e^{j2\pi/3}z^{-1})(1 - 0.9e^{-j2\pi/3}z^{-1})}$$

If $f_s = 1000$ samples/sec, determine an expression for $y(t)$, the output of the D-to-C converter.

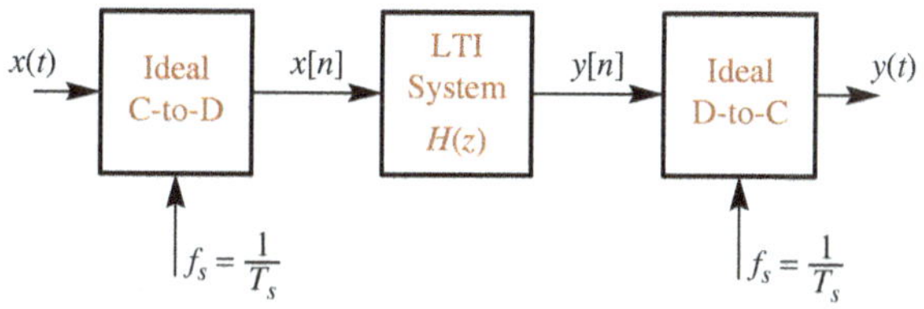

Figure P-10.22

P-10.23 An IIR system is defined by the system function

$$H(z) = \frac{(1 - 1.25e^{j\pi/3}z^{-1})(1 - 1.25e^{-j\pi/3}z^{-1})}{(1 - 0.8e^{j\pi/3}z^{-1})(1 - 0.8e^{-j\pi/3}z^{-1})}$$

(a) Plot the poles and zeros of $H(z)$ in the z-plane.

(b) Use the geometric estimation technique of Fig. 10-27 in Section 10-12.3 to sketch the magnitude of the frequency response of the system.

(c) This system is called an *all-pass* filter. If the poles were moved to a radius of $r = 0.9$, how would the zeros have to be moved to maintain the all-pass character of $H(e^{j\hat{\omega}})$?

P-10.24 The system function of an IIR system is

$$H(z) = \frac{1 + z^{-2}}{(1 - 0.8z^{-1})(1 + 0.3z^{-1})}$$

(a) Determine the difference equation that relates the input $x[n]$ to the output $y[n]$.

(b) Plot the poles and zeros of $H(z)$ in the z-plane. Is the system stable?

(c) At what frequency does $|H(e^{j\hat{\omega}})|$ have its maximum value? Estimate that value.

(d) At what frequencies does $|H(e^{j\hat{\omega}})| = 0$?

P-10.25 Answer the following questions about the system whose z-transform system function is

$$H(z) = \frac{1 + 0.8z^{-1}}{1 - 0.7z^{-1}}$$

(a) Determine the poles and zeros of $H(z)$.

(b) Derive a simple expression (purely real) for the magnitude squared of the frequency response $|H(e^{j\hat{\omega}})|^2$.

(c) Is this filter a lowpass filter or a highpass filter? Explain your answer in terms of the poles and zeros of $H(z)$.

P-10.26 The diagram in Fig. P-10.26 depicts a *cascade connection* of two LTI systems; where the output of the first system is the input to the second system, and the overall output is the output of the second system.

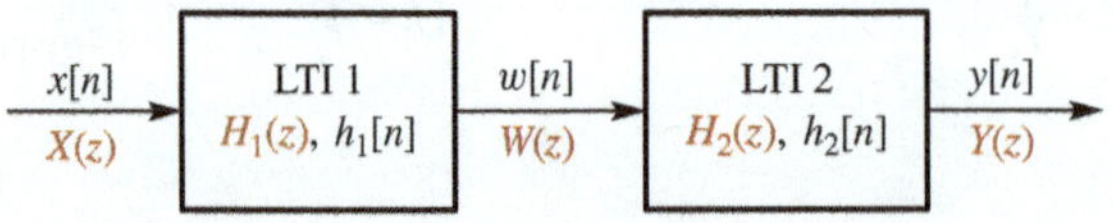

Figure P-10.26

(a) If the two systems are given as

$$H_1(z) = \frac{3}{2 + z^{-1}} \quad \text{and} \quad H_2(z) = \frac{10 + 5z^{-1}}{1 - 0.7z^{-1}}$$

determine $H(z)$, the system function of the overall cascade system. In addition, verify that $H(z)$ is a first-order system by finding the pole and zero of $H(z)$.

(b) Suppose that the two systems are IIR filters described below:

$$y_1[n] = \tfrac{1}{2}y_1[n-1] + x[n] + \tfrac{1}{2}x[n-1] \qquad \text{and} \qquad H_2(z) = \frac{3z^{-1}}{1 - \frac{5}{6}z^{-1} + \frac{1}{6}z^{-2}}$$

Determine the system function of the overall cascade system.

(c) For the systems in (b), obtain a single difference equation that relates $y[n]$ to $x[n]$ in Fig. P-10.26.

(d) For the systems in (b), plot the poles and zeros of $H(z)$ in the complex z-plane.

P-10.27 The diagram in Fig. P-10.26 depicts a *cascade connection* of two LTI systems.

(a) Use z-transforms to show that the system function for the overall system (from $x[n]$ to $y[n]$) is $H(z) = H_2(z)H_1(z)$, where $Y(z) = H(z)X(z)$.

(b) Derive a condition on $H(z)$ that guarantees that the output signal will always be equal to the input signal.

(c) If System 1 is the difference equation $y_1[n] = x[n] + \frac{5}{6}x[n-1]$, find a system function $H_2(z)$ so that output of the cascaded system will always be equal to its input. In other words, find $H_2(z)$, which will undo the filtering action of $H_1(z)$. This is called *deconvolution* and $H_2(z)$ is the *inverse* of $H_1(z)$.

(d) Suppose that $H_1(z)$ represents a general FIR filter. What conditions on $H_1(z)$ must hold if $H_2(z)$ is to be a stable and causal inverse filter for $H_1(z)$?

P-10.28 Define a discrete-time signal using the formula:

$$y[n] = (0.99)^n \cos(2\pi(0.123)n + \phi)$$

(a) Make a sketch of $y[n]$ versus n, as a "stem" plot. Take the range of n to be $0 \le n \le 20$.

(b) Design an IIR filter that will synthesize $y[n]$. Give your answer in the form of a difference equation with numerical values for the coefficients. Assume that the synthesis will be accomplished by using an impulse input to "start" the difference equation (which is at rest; i.e., has zero initial conditions).

P-10.29 In the following cascade of systems, all of the individual system functions, $H_i(z)$, are known.

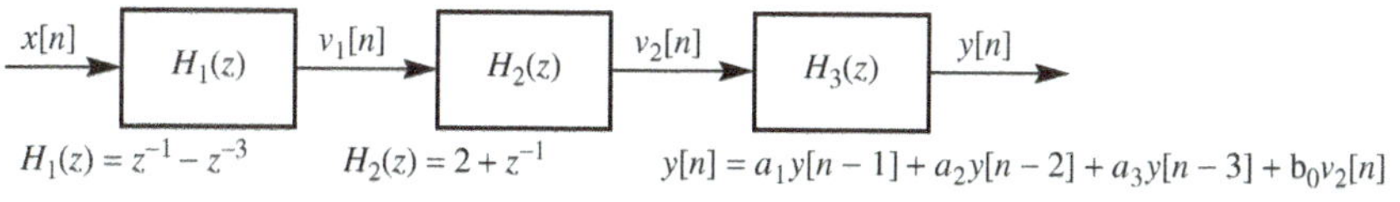

Figure P-10.29

(a) Determine $H_3(z)$, the z-transform of the last system.

(b) When $a_1 = a_2 = \frac{1}{2}$, $a_3 = 0$ and $b_0 = 7$, determine the system function, $H(z)$ for the cascaded system.

(c) Consider the impulse response $h[n]$ of the cascaded system (i.e., the output when the input is $x[n] = \delta[n]$). Determine values for $\{a_1, a_2, a_3, b_0\}$ so that the impulse response is $h[n] = 8\delta[n - n_d]$. Also determine the value of the time shift n_d.

P-10.30 We have shown several equivalent ways to describe an LTI system. Below is a list of time-domain descriptions (e.g., impulse response, difference equation, MATLAB code, or filter coefficients) of LTI systems. For each one, choose a corresponding system function or frequency response from the list on the right. Note that there are more entries on the right than needed.

Time-Domain Description	System Function/Frequency Response
(a) `yy=filter([0,0,1,0,1],1,xx);`	1 $H(e^{j\hat{\omega}}) = 2\cos(\hat{\omega})\, e^{-j3\hat{\omega}}$
(b) $y[n] = 0.9y[n-1] + 0.8y[n-2] + x[n] - x[n-3]$	2 $H(z) = \dfrac{z^{-2}(1 - z^{-1})}{1 - 0.8z^{-1}}$
(c) $h[n] = u[n-2] - u[n-5]$	3 $H(e^{j\hat{\omega}}) = \dfrac{1 - e^{-j3\hat{\omega}}}{1 - 0.9e^{-j\hat{\omega}} - 0.8e^{-j2\hat{\omega}}}$
(d) $y[n] = \sum_{k=1}^{3} \left(\sqrt{k} - 1\right) x[n-k]$	4 $H(e^{j\hat{\omega}}) = (1 + 2\cos(\hat{\omega}))e^{-j3\hat{\omega}}$
(e) $h[n] = 0.8^{n-2}u[n-2] - 0.8^{n-3}u[n-3]$	5 $H(z) = z^{-2}\left(1 + 0.707z^{-1}\right)$
	6 $H(e^{j\hat{\omega}}) = \dfrac{1 - e^{-j3\hat{\omega}}}{1 + 0.9e^{-j\hat{\omega}} + 0.8e^{-j2\hat{\omega}}}$

APPENDIX A

Complex Numbers

The basic manipulations of complex numbers are reviewed in this appendix. The algebraic rules for combining complex numbers are reviewed, and then a geometric viewpoint is taken to visualize these operations by drawing vector diagrams. This geometric view is a key to understanding how complex numbers can be used to represent signals. Specifically, the following three significant ideas about complex numbers are treated in this appendix:

- ***Simple Algebraic Rules:*** Operations on complex numbers ($z = x + jy$) follow exactly the same rules as real numbers, with j^2 replaced everywhere by -1.[1]
- ***Elimination of Trigonometry:*** Euler's formula, $z = re^{j\theta} = r\cos\theta + jr\sin\theta$, for the complex exponential provides a connection between trigonometric identities and simple algebraic operations on complex numbers.
- ***Representation by Vectors:*** A vector drawn from the origin to a point (x, y) in a two-dimensional plane is equivalent to $z = x + jy$. The algebraic rules for z are, in effect, the basic rules of vector operations. More important, however, is the *visualization* gained from the vector diagrams.

[1] Mathematicians and physicists use the symbol i for $\sqrt{-1}$; electrical engineers prefer to reserve the symbol i for current in electric circuits.

The first two ideas concern the algebraic nature of $z = x + jy$, the third its role as a ***representer*** of signals. Skill in algebraic manipulations is necessary, but the use of complex numbers in representation is more important in the long run. Complex numbers are used as a convenience in signal processing because when they stand for sinusoidal signals they can simplify manipulations of the signals. Thus, a sinusoidal problem (such as the solution to a differential equation) is converted into a complex number problem that can be (1) solved by the simple rules of algebra and (2) visualized through vector geometry. The key to all this is the higher order thinking that permits abstraction of the problem into the world of complex numbers. Ultimately, we are led to the notion of a "transform" such as the discrete Fourier transform or z-transform to reduce many other sophisticated problems to algebra. Because complex numbers are so crucial to almost everything we do in the study of signals and systems, a mastery of the concepts reviewed in this appendix is advised.

Once such insight is gained, it still is necessary to return occasionally to the lower level reality of calculations. When you have to manipulate complex numbers, software such as MATLAB or a calculator with built-in complex arithmetic functionality is a necessity. It is worthwhile to invest some time to learn how to use this feature on a calculator, or in software. However, it is also important to do some calculations by hand, so that you can *understand* and check what your calculator is doing.

Finally, it is too bad that complex numbers are called "complex." Most students, therefore, think of them as complicated. However, their elegant mathematical properties usually simplify mathematical expressions and calculations quite a bit.

A-1 Introduction

A complex number system is an extension of the real number system. Complex numbers are necessary to solve equations such as

$$z^2 = -1 \tag{A.1}$$

The symbol j is introduced to stand for $\sqrt{-1}$, so this equation (A.1) has the two solutions $z = \pm j$. More generally, complex numbers are needed to solve for the two roots of a quadratic equation

$$az^2 + bz + c = 0$$

which, according to the quadratic formula, has the two solutions:

$$z = \frac{-b \pm \sqrt{b^2 - 4ac}}{2a}$$

Whenever the discriminant $(b^2 - 4ac)$ is negative, the solution must be expressed as a complex number. For example, the roots of

$$z^2 + 6z + 25 = 0$$

are $z = -3 \pm j4$, because

$$\frac{\sqrt{b^2 - 4ac}}{2a} = \frac{\sqrt{36 - 4(25)}}{2} = \frac{\sqrt{-64}}{2} = \pm j4$$

A-2 Notation for Complex Numbers

Several different mathematical notations can be used to represent complex numbers. The two basic types are polar form and rectangular form. Converting between them quickly and easily is an important skill.

A-2.1 Rectangular Form

In ***rectangular form***, all of the following notations define the same complex number.

$$\begin{aligned} z &= (x, y) \\ &= x + jy \\ &= \Re\{z\} + j\Im\{z\} \end{aligned}$$

The ordered pair (x, y) can be interpreted as a point in the two-dimensional plane.[2]

A complex number can also be drawn as a vector whose tail is at the origin of the complex plane and whose head is at the point (x, y), in which case x is the horizontal coordinate of the vector and y the vertical coordinate (see Fig. A-1).

Figure A-2 shows some numerical examples. The complex number $z = 2 + j5$ is represented by the point (2, 5), which lies in the first quadrant of the two-dimensional plane; likewise, $z = 4 - j3$ lies in the fourth quadrant at the location $(4, -3)$. Since the complex number notation $z = x + jy$ represents the point (x, y) in the two-dimensional plane, the number j represents (0, 1), which is drawn as a vertical vector from the origin up to (0, 1), as in Fig. A-2. Thus multiplying a real number, such as 5, by j changes it from pointing along the horizontal axis to pointing vertically (i.e., $j(5 + j0) = 0 + j5$).

Rectangular form is also referred to as ***Cartesian form***. The horizontal coordinate x is called the ***real part***, and the vertical coordinate y the ***imaginary part***. The

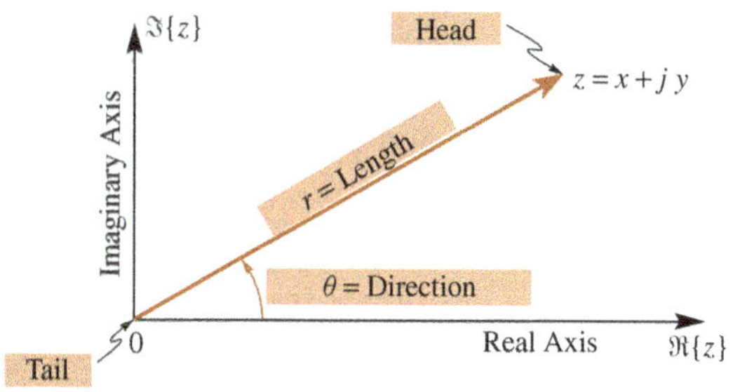

Figure A-1 Complex number represented as a vector from the origin to (x, y).

[2]This is also the notation used on some calculators when entering complex numbers.

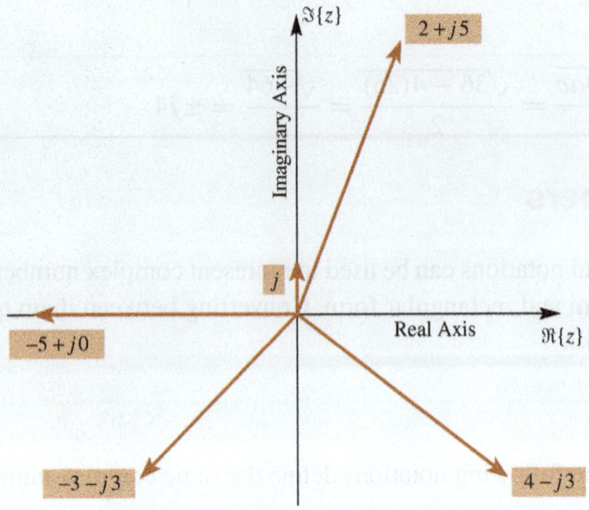

Figure A-2 Complex numbers plotted as vectors in the two-dimensional "complex plane." Each $z = x + jy$ is represented by a vector from the origin to the point with coordinates (x, y) in the complex plane.

operators $\Re\{z\}$ and $\Im\{z\}$ are provided to extract the real and imaginary parts of $z = x + jy$:

$$x = \Re\{z\} \tag{A.2a}$$

$$y = \Im\{z\} \tag{A.2b}$$

A-2.2 Polar Form

In ***polar form***, the vector is defined by its length (r) and its direction (θ) as in Fig. A-1. Therefore, we use the following descriptive notation sometimes: $r\angle\,\theta$. Three examples are shown in Fig. A-3, where the direction θ is given in degrees. The angle is always measured from the positive x-axis and may be either positive (counterclockwise) or negative (clockwise). However, we generally specify the ***principal value*** of the angle so that $-180^\circ < \theta \le 180^\circ$. This requires that integer multiples of 360° be subtracted from or added to the angle until the result is between -180° and $+180^\circ$. Thus, the vector $3\angle{-80^\circ}$ is the principal value of $3\angle 280^\circ$.

A-2.3 Conversion: Rectangular and Polar

DEMO
Zdrill

Both the polar and rectangular forms are commonly used to represent complex numbers. The prevalence of the polar form, for sinusoidal signal representation, makes it necessary to convert quickly and accurately between the two representations.

From Fig. A-4, we see that the x and y coordinates of the vector are given by

$$x = r\cos\theta \tag{A.3a}$$

$$y = r\sin\theta \tag{A.3b}$$

Therefore, a valid formula for z is

$$z = r\cos\theta + j\,r\sin\theta \tag{A.4}$$

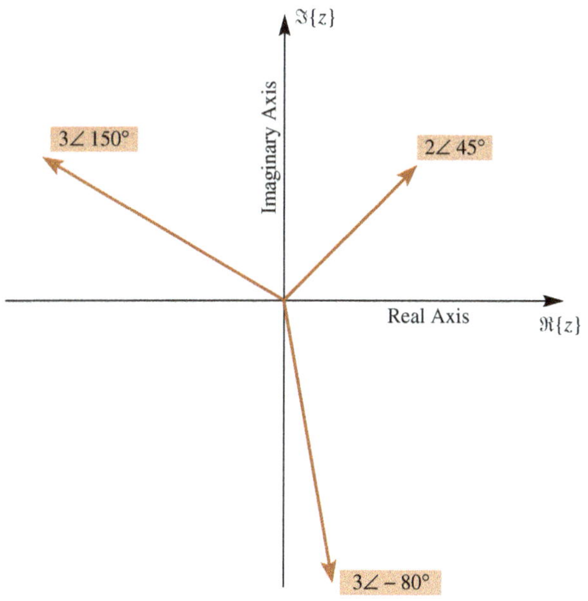

Figure A-3 Three complex numbers plotted in terms of length (r) and direction (θ) of their vector representation. The angle is always measured with respect to the positive real axis; its units are usually radians, but are shown as degrees in this case.

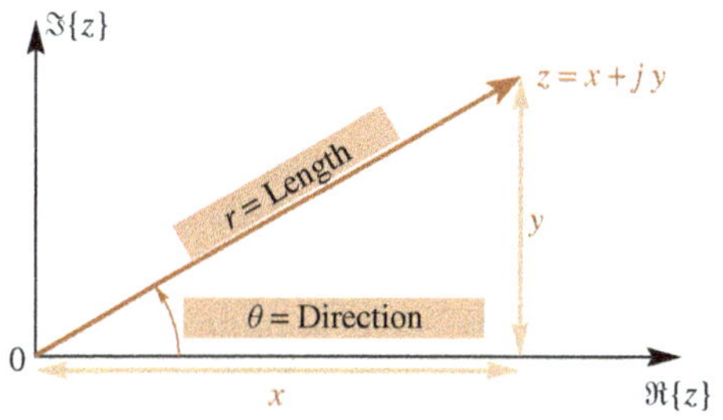

Figure A-4 Basic layout for relating Cartesian coordinates (x, y) to polar coordinates r and θ.

EXAMPLE A-1 Polar to Rectangular

In Fig. A-3, the three complex numbers are

$r\angle\theta$	Cartesian Form
$2\angle 45°$	$z = \sqrt{2} + j\sqrt{2}$
$3\angle 150°$	$z = -\dfrac{3\sqrt{3}}{2} + j\dfrac{3}{2}$
$3\angle -80°$	$z = 0.521 - j2.954$

The conversion from polar to rectangular was done via (A.4).

The conversion from (x, y) to $r\angle\theta$ is a bit trickier. From Fig. A-4, the formulas are

$$r = \sqrt{x^2 + y^2} \qquad \text{(Length)} \qquad \text{(A.5a)}$$

$$\theta = \arctan(y/x) \qquad \text{(Direction)} \qquad \text{(A.5b)}$$

The arctangent must give a four-quadrant answer, and the direction θ is usually given in radians rather than degrees.

EXERCISE A.1 At this point, the reader should convert to polar form the five complex numbers shown in Fig. A-2. The answers, in a random order, are $1\angle 90^\circ$, $5\angle -36.87^\circ$, $4.243\angle -135^\circ$, $5.385\angle 68.2^\circ$, and $5\angle 180^\circ$.

In Section A-3, we will introduce two other polar notations:

$$z = re^{j\theta}$$
$$= |z|e^{j\arg z}$$

where $|z| = r = \sqrt{x^2 + y^2}$ is called the ***magnitude*** of the vector and $\arg z = \theta$ is its ***phase*** in radians (not degrees). This exponential notation, which relies on Euler's formula, has the advantage that when it is used in algebraic expressions, the standard laws of exponents apply.

A-2.4 Difficulty in Second or Third Quadrant

The formula for the angle, $\theta = \arg z$, is the $\arctan(y/x)$ as in (A.5b). It must be used with care, especially when the real part is negative (see Fig. A-5). For example, the complex number $z = -3 + j2$ would require that we evaluate $\arctan(-2/3)$ to get the angle; the same calculation would be needed if $z = 3 - j2$. The arctangent of $-2/3$ is -0.588 rad, or about -33.7°, which is the correct angle for $z = 3 - j2$. However, for $z = -3 + j2$, the vector lies in the second quadrant and the angle must satisfy $90^\circ \le \theta \le 180^\circ$. We get the correct angle by adding 180°. In this case, the correct angle is $-33.7^\circ + 180^\circ = 146.3^\circ$, and if we prefer radians, the angle is $\pi - 0.588 = 2.55$ rad. The general method for getting the angle of z in polar form (in radians) is

$$\theta = \begin{cases} \arctan(y/x) & \text{if } x \ge 0 \\ \pm\pi + \arctan(y/x) & \text{if } x < 0 \end{cases} \tag{A.6}$$

Since π and $-\pi$ are effectively the same direction, when $x < 0$ we should pick the appropriate one to make the final answer satisfy $|\theta| \le \pi$.

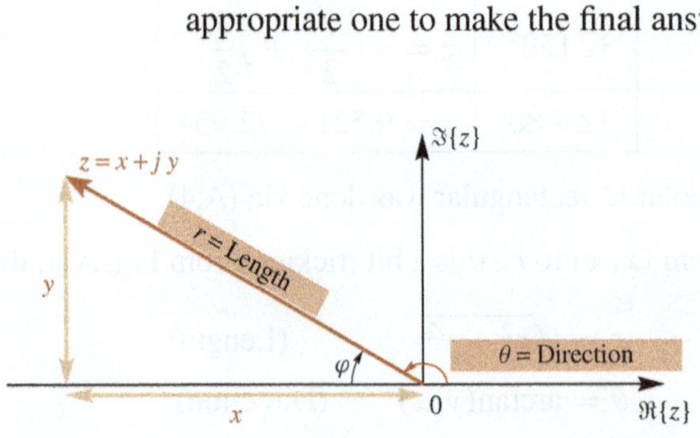

Figure A-5 In the second quadrant, the interior angle φ is easily calculated from x and y, but is not the correct angle for the polar form, which requires the exterior angle θ with respect to the positive real axis.

A-3 Euler's Formula

The conversion from polar to rectangular form (A.3) suggests the following formula:

Euler's Formula

$$e^{j\theta} = \cos\theta + j\sin\theta \tag{A.7}$$

Equation (A.7) defines the ***complex exponential*** $e^{j\theta}$, which is equivalent to $1\angle\theta$ (i.e., a vector of length 1 at angle θ). A proof of Euler's formula based on power series is outlined in Problem P-2.4 of Chapter 2.

The amazing property of (A.7) is that the laws of exponents apply to $e^{j\theta}$. Euler's formula (A.7) is so important that it must be instantly recalled; likewise, the inverse Euler formulas (A.8) and (A.9) should also be committed to memory.

EXAMPLE A-2 Euler's Formula

Here are some examples:

$$(\mathbf{90^\circ}): 1\angle\pi/2 \qquad e^{j\pi/2} = \cos(\pi/2) + j\sin(\pi/2) = 0 + j1 = j$$

$$(\mathbf{180^\circ}): 1\angle\pi \qquad e^{j\pi} = \cos(\pi) + j\sin(\pi) = -1 + j0 = -1$$

$$(\mathbf{45^\circ}): 1\angle\pi/4 \qquad e^{j\pi/4} = \cos(\pi/4) + j\sin(\pi/4) = \frac{\sqrt{2}}{2} + j\frac{\sqrt{2}}{2}$$

$$(\mathbf{60^\circ}): 1\angle\pi/3 \qquad e^{j\pi/3} = \cos(\pi/3) + j\sin(\pi/3) = \frac{1}{2} + j\frac{\sqrt{3}}{2}$$

EXAMPLE A-3 Degrees to Radians

Referring back to Fig. A-3, the three complex numbers can be rewritten as follows:

$$2\angle 45^\circ \quad \text{converts to} \quad z = 2e^{j\pi/4}$$

$$3\angle 150^\circ \quad \text{converts to} \quad z = 3e^{j5\pi/6}$$

$$3\angle -80^\circ \quad \text{converts to} \quad z = 3e^{-j4\pi/9} = 3e^{-j1.396}$$

Numbers like -1.396 rad are difficult to visualize, because we are more used to thinking of angles in degrees. It may be helpful to express the angles used in the exponents as a fraction of π rad (i.e., $-1.396 = -(1.396/\pi)\pi = -0.444\pi$ rad). This is a good habit to adopt, because it simplifies the conversion between degrees and radians. If θ is given in radians, the conversion is

$$\theta\left(\frac{180}{\pi}\right) = \text{direction in degrees}$$

A-3.1 Inverse Euler Formulas

Euler's formula (A.7) can be solved separately for the cosine and sine parts. The results are called the ***inverse Euler relations***

$$\cos\theta = \frac{e^{j\theta} + e^{-j\theta}}{2} \tag{A.8}$$

$$\sin\theta = \frac{e^{j\theta} - e^{-j\theta}}{2j} \tag{A.9}$$

Recalling that $\cos(-\theta) = \cos\theta$ and $\sin(-\theta) = -\sin\theta$, the proof of the $\sin\theta$ version (A.9) is as follows:

$$\begin{aligned} e^{-j\theta} &= \cos(-\theta) + j\sin(-\theta) \\ &= \cos\theta - j\sin\theta \\ e^{+j\theta} &= \cos\theta + j\sin\theta \\ \Longrightarrow e^{j\theta} - e^{-j\theta} &= 2j\sin\theta \\ \Longrightarrow \sin\theta &= \frac{e^{j\theta} - e^{-j\theta}}{2j} \end{aligned}$$

A-4 Algebraic Rules for Complex Numbers

The basic arithmetic operators for complex numbers follow the usual rules of algebra as long as the symbol j is treated as a special token that satisfies $j^2 = -1$. In rectangular form, all of these rules are relatively straightforward. The five fundamental rules are the following:

Addition:

$$\begin{aligned} z_1 + z_2 &= (x_1 + jy_1) + (x_2 + jy_2) \\ &= (x_1 + x_2) + j(y_1 + y_2) \end{aligned}$$

Subtraction:

$$\begin{aligned} z_1 - z_2 &= (x_1 + jy_1) - (x_2 + jy_2) \\ &= (x_1 - x_2) + j(y_1 - y_2) \end{aligned}$$

Multiplication:

$$\begin{aligned} z_1 \times z_2 &= (x_1 + jy_1) \times (x_2 + jy_2) \\ z_1 z_2 &= x_1x_2 + j^2 y_1y_2 + jx_1y_2 + jx_2y_1 \\ &= (x_1x_2 - y_1y_2) + j(x_1y_2 + x_2y_1) \end{aligned}$$

Division:

$$z_1 \div z_2 = (x_1 + jy_1)/(x_2 + jy_2)$$

$$\frac{z_1}{z_2} = \frac{z_1 z_2^*}{z_2 z_2^*} = \frac{z_1 z_2^*}{|z_2|^2}$$

$$= \frac{(x_1 + jy_1)(x_2 - jy_2)}{x_2^2 + y_2^2}$$

$$= \frac{(x_1 x_2 + y_1 y_2) + j(x_2 y_1 - x_1 y_2)}{x_2^2 + y_2^2}$$

Conjugate:

$$z_1^* = (x_1 + jy_1)^*$$

$$= x_1 - jy_1$$

Addition and subtraction are straightforward because we need only add or subtract the real and imaginary parts. On the other hand, addition (or subtraction) in polar form cannot be carried out directly on r and θ; instead, an intermediate conversion to rectangular form is required. In contrast, the operations of multiplication and division, which are rather messy in rectangular form, reduce to simple manipulations in polar form. For multiplication, multiply the magnitudes and add the angles; for division, divide the magnitudes and subtract the angles. The conjugate in polar form requires only a change of sign of the angle.

Multiplication:

$$(z_1)(z_2) = \left(r_1 e^{j\theta_1}\right)\left(r_2 e^{j\theta_2}\right)$$

$$= (r_1 r_2)\, e^{j(\theta_1 + \theta_2)}$$

Division:

$$z_1 \div z_2 = \frac{r_1 e^{j\theta_1}}{r_2 e^{j\theta_2}}$$

$$= \frac{r_1}{r_2}\, e^{j(\theta_1 - \theta_2)}$$

Conjugate:

$$z_1^* = (r_1 e^{j\theta_1})^*$$

$$= r_1 e^{-j\theta_1}$$

EXERCISE A.2

The inverse or reciprocal of a complex number z is the number z^{-1} such that

$$z^{-1} z = 1.$$

A common mistake with the inverse is to invert $z = x + jy$ by taking the inverse of x and y separately. To show that this is wrong, take the specific case where $z = 4 + j3$ and $w = \frac{1}{4} + j\frac{1}{3}$. Show that w is not the inverse of z, because $wz \neq 1$. Determine the correct inverse of z.

Polar form presents difficulties when adding (or subtracting) two complex numbers and expressing the final answer in polar form. An intermediate conversion to rectangular form must be done. The recipe for adding or subtracting complex numbers in polar form is:

1. Starting in polar form, we have

$$z_3 = z_1 \pm z_2 = r_1 e^{j\theta_1} \pm r_2 e^{j\theta_2}$$

2. Convert both z_1 and z_2 to Cartesian form:

$$z_3 = (r_1 \cos\theta_1 + j\, r_1 \sin\theta_1) \pm (r_2 \cos\theta_2 + j\, r_2 \sin\theta_2)$$

3. Perform the addition in Cartesian form:

$$z_3 = (r_1 \cos\theta_1 \pm r_2 \cos\theta_2) + j\,(r_1 \sin\theta_1 \pm r_2 \sin\theta_2)$$

4. Identify the real and imaginary parts of z_3:

$$x_3 = \Re\{z_3\} = r_1 \cos\theta_1 \pm r_2 \cos\theta_2$$
$$y_3 = \Im\{z_3\} = r_1 \sin\theta_1 \pm r_2 \sin\theta_2$$

5. Convert back to polar form using (A.5a) and (A.5b):

$$z_3 = x_3 + j y_3 \quad \Longleftrightarrow \quad z_3 = r_3 e^{j\theta_3}$$

If you have a calculator that converts between polar and rectangular form, learning how to use it will save much hand calculation and also be more accurate. Most scientific calculators even have the capability to use both notations, so the conversion into and out of Cartesian form is transparent to the user.

EXAMPLE A-4 Adding Polar Forms

Here is a numerical example of adding two complex numbers given in polar form:

$$z_3 = 7e^{j4\pi/7} + 5e^{-j5\pi/11}$$
$$z_3 = (-1.558 + j6.824) + (0.712 - j4.949)$$
$$z_3 = -0.846 + j1.875$$
$$z_3 = 2.057e^{j1.995} = 2.057e^{j0.635\pi} = 2.057\angle 114.3^\circ$$

Remember: When the angle appears in the exponent, its units must be in radians.

A-4.1 Complex Number Exercises

To practice computations for complex numbers, try the following exercises.

EXERCISE A.3

Add and multiply the following, then plot the results in the complex plane:

$$z_4 = 5e^{j4\pi/5} + 7e^{-j5\pi/7}$$

$$z_5 = \left(5e^{j4\pi/5}\right)\left(7e^{-j5\pi/7}\right)$$

The answers are $z_4 = -8.41 - j2.534 = 8.783e^{-j0.907\pi}$ and $z_5 = 35e^{j3\pi/35}$.

EXERCISE A.4

For the conjugate the simple rule is to change the sign of all the j terms. Work the following:

$$(3 - j4)^* = \ ?$$

$$(j(1 - j))^* = \ ?$$

$$(e^{j\pi/2})^* = \ ?$$

DEMO
zdrill

More drill problems can be generated by using the MATLAB program `zdrill`, which presents a GUI (graphical user interface) that asks questions for each of the complex operations, and also plots the vectors that represent the solutions. The `zdrill` GUI has both a novice and an advanced level. A screen shot of the GUI is shown in Fig. A-6.

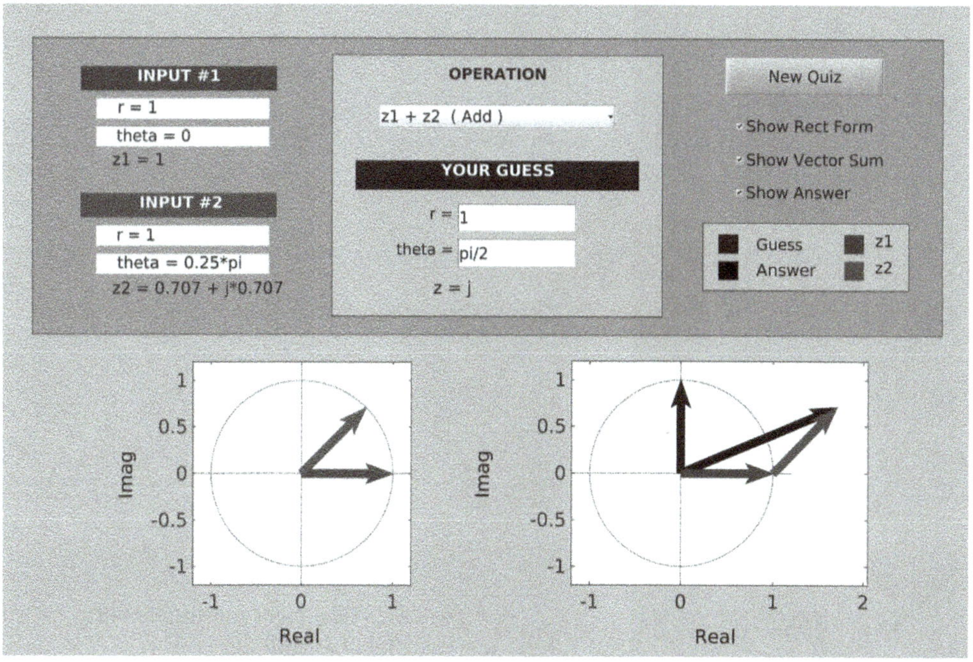

Figure A-6 The MATLAB GUI `zdrill` for practicing complex number operations. Addition of the complex numbers $z_1 = 1$ and $z_2 = e^{j0.25\pi}$ is illustrated with vectors in the panel at the lower right. The user's guess of $z = j$ is the vertical vector in that same panel.

EXERCISE A.5 Prove that the following three identities are true:

$$\Re\{z\} = (z + z^*)/2$$
$$\Im\{z\} = (z - z^*)/(2j)$$
$$|z|^2 = zz^*$$

A-5 Geometric Views of Complex Operations

It is important to develop an ability to visualize complex number operations. This is done by plotting the vectors that represent the numbers in the (x, y) plane, where $x = \Re\{z\}$ and $y = \Im\{z\}$. The key is to recall that, as shown in Fig. A-1, the complex number $z = x + jy$ is a vector whose tail is at the origin and whose head is at (x, y).

A-5.1 Geometric View of Addition

For complex addition, $z_3 = z_1 + z_2$, both z_1 and z_2 are viewed as vectors with their tails at the origin. The sum z_3 is the result of vector addition, and is constructed as follows (see Fig. A-7):

1. Draw a copy of z_1 with its tail at the head of z_2. Call this displaced vector $\hat{z}_1$.
2. Draw the vector from the origin to the head of $\hat{z}_1$. This vector is the sum z_3.

This method of addition can be generalized to many vectors. Figure A-8 shows the result of adding the four complex numbers

$$(1 + j) + (-1 + j) + (-1 - j) + (1 - j)$$

where the answer happens to be zero.

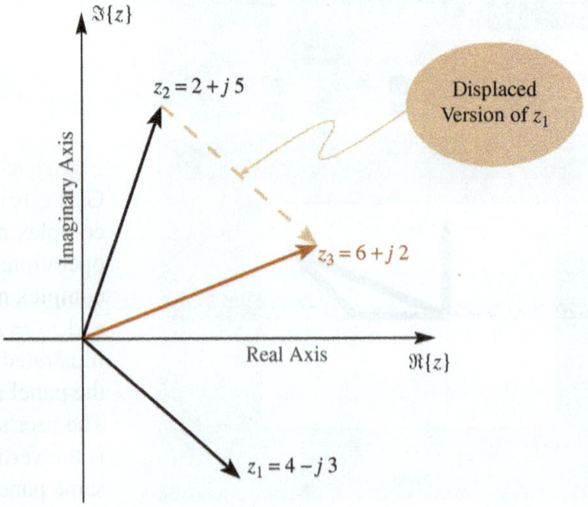

Figure A-7 Graphical construction of complex number addition $z_3 = z_1 + z_2$.

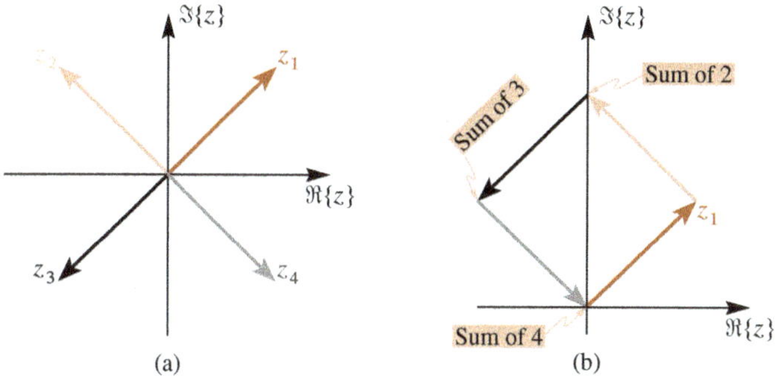

Figure A-8 Result of adding the four vectors $\{1+j,\ -1+j, -1-j,\ 1-j\}$ shown in (a) is zero. The "head-to-tail" graphical method is illustrated in (b). Note that (a) also shows that $z_1 + z_3 = 0$ and $z_2 + z_4 = 0$.

A-5.2 Geometric View of Subtraction

The visualization of subtraction requires that we interpret the operation as addition:

$$z_3 = z_2 - z_1 = z_2 + (-z_1)$$

Thus we must flip z_1 to $-z_1$ (by adding π radians to its angle) and then add the result to z_2. This method is shown in Fig. A-9 and is equivalent to the visualization of addition, as in Fig. A-7. There are two additional comments to be made about subtraction:

1. Since $z_2 = z_1 + z_3$, a displaced version of the difference vector could be drawn with its tail at the head of z_1 and its head at the head of z_2.

2. The triangle with vertices at the three points 0, z_1, and z_2 has sides equal to z_1, z_2, and $z_2 - z_1$.

EXERCISE A.6 Define a vector w to have its tail at z_1 and its head at z_2. Prove that w has the same length and direction as $z_2 - z_1$.

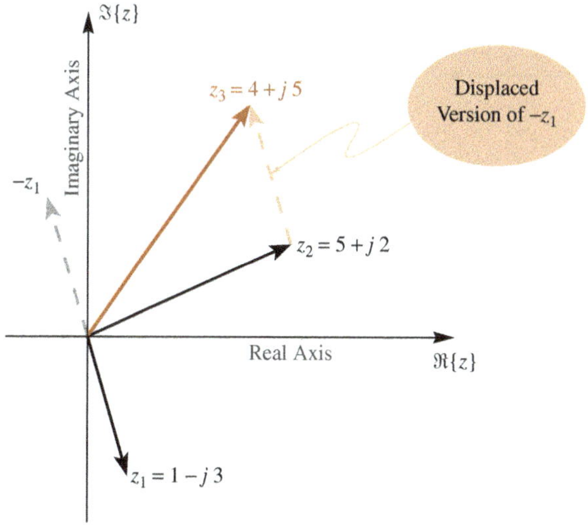

Figure A-9 Geometric view of subtraction. The vector z_1 is flipped to make $(-z_1)$ and then a displaced version is added to z_2 to obtain $z_3 = z_2 + (-z_1)$.

EXERCISE A.7 Prove the triangle inequality:

$$|z_2 - z_1| \le |z_1| + |z_2|$$

Use either an algebraic method by squaring both sides, or a geometric argument based on the intuitive idea that "the shortest distance between two points is a straight line."

A-5.3 Geometric View of Multiplication

It is best to view multiplication in terms of polar form where we multiply the magnitudes, and add the angles. In order to draw the product vector z_3, we must decide whether or not $|z_1|$ and/or $|z_2|$ are greater than 1. In Fig. A-10, it is assumed that only $|z_1|$ is larger than 1.

A special case occurs when $|z_2| = 1$, because then there is no scaling of the vector length, and the multiplication by $z_2 = e^{j\theta_2}$ becomes a rotation. Figure A-11 shows the case where $z_2 = j$, which gives a rotation by $\pi/2$ or 90°, because $j = e^{j\pi/2}$.

A-5.4 Geometric View of Division

Division is very similar to the visualization of multiplication, except that we must now subtract the angles and divide the magnitudes (Fig. A-12).

$$z_3 = \frac{z_1}{z_2} = \frac{r_1}{r_2} e^{j(\theta_1 - \theta_2)}$$

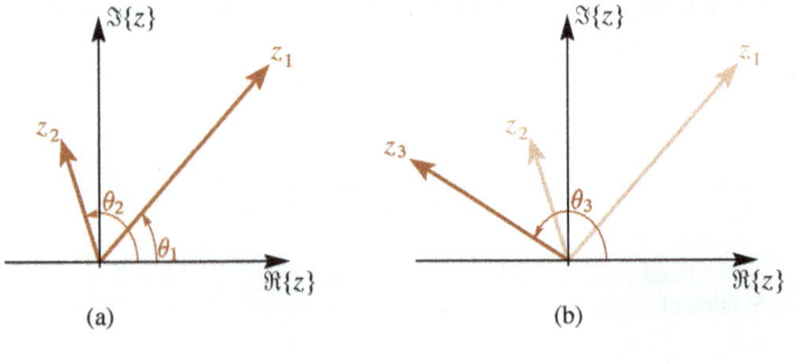

Figure A-10 (a) Two complex numbers to be multiplied shown as vectors. (b) Geometric view of complex multiplication $z_3 = z_1 z_2 = (r_1 r_2)e^{j(\theta_1+\theta_2)}$. Since $|z_2| < 1$, the length of z_3 is less than the length of z_1.

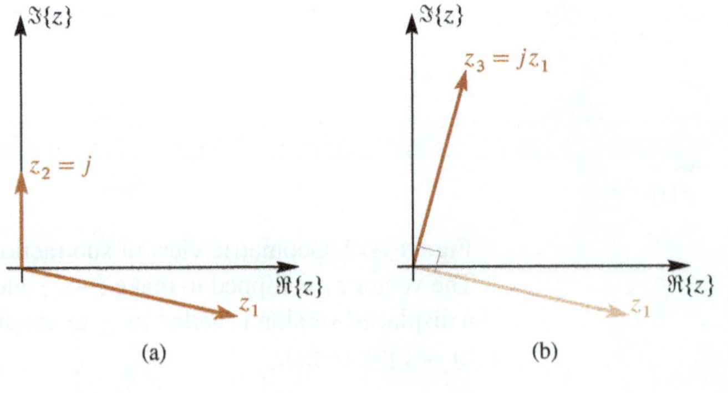

Figure A-11 (a) Two complex numbers to be multiplied shown as vectors. (b) Complex multiplication becomes a rotation when $|z_2| = 1$. The case where $z_2 = j$ gives a rotation by 90°, and the length of z_3 equals the length of z_1.

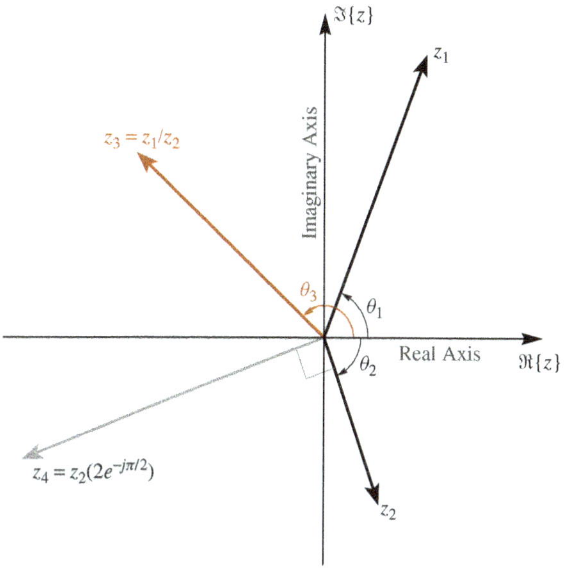

Figure A-12 Graphical visualization of complex number division $z_3 = z_1/z_2$. Notice that the angles subtract (i.e., $\theta_3 = \theta_1 - \theta_2$). Since $|z_2| > 1$, the vector z_3 is shorter than z_1.

EXERCISE A.8 Given two complex numbers z_4 and z_2, as in Fig. A-12, where the angle between them is 90° and the magnitude of z_4 is twice that of z_2. Evaluate z_4/z_2.

A-5.5 Geometric View of the Inverse, z^{-1}

This is a special case of division where $z_1 = 1$, so we simply negate the angle and take the reciprocal of the magnitude.

$$z^{-1} = \frac{1}{z} = \frac{1}{r}e^{-j\theta}$$

Refer to Fig. A-13 for examples of the inverse.

A-5.6 Geometric View of the Conjugate, z^*

In this case, we negate the angle, which has the effect of flipping the vector about the horizontal axis. The length of the vector remains the same.

$$z^* = x - jy = re^{-j\theta}$$

Figure A-14 shows two examples of the geometric interpretation of complex conjugation.

EXERCISE A.9 Prove the following fact:

$$\frac{1}{z^*} = \frac{1}{r}e^{j\theta}$$

Plot an example for $z = 1 + j$; also plot $1/z$ and z^*.

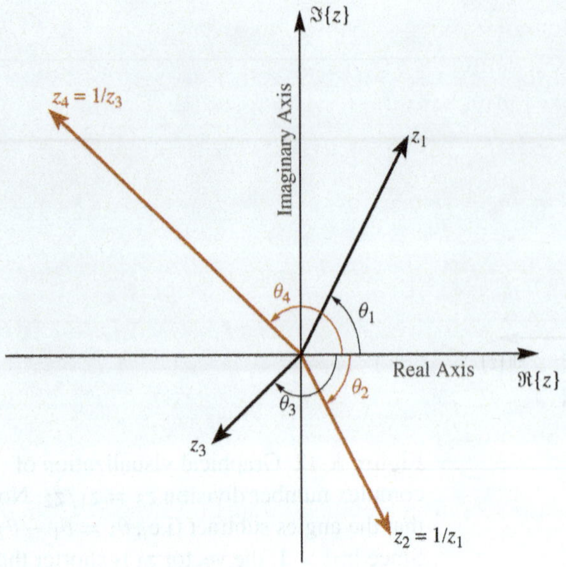

Figure A-13 Graphical construction of complex number inverse $1/z$. For the vectors shown, $|z_1| > 1$ and $|z_3| < 1$. Since the angle is negated when taking the inverse, $\theta_2 = -\theta_1$ and $\theta_4 = -\theta_3$.

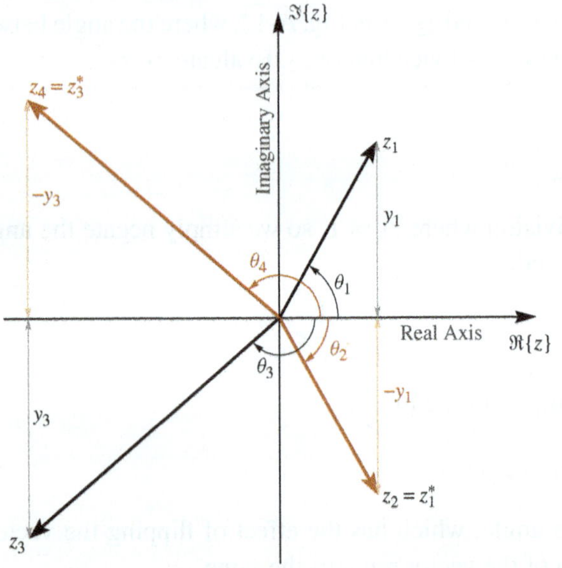

Figure A-14 Graphical construction of the complex conjugate z^*; only the imaginary part changes sign. The vectors flip about the real axis: z_1 flips down, and z_3 flips up. Since the angle is negated when doing the conjugate, $\theta_2 = -\theta_1$ and $\theta_4 = -\theta_3$; the magnitudes are unchanged, $|z_2| = |z_1|$ and $|z_4| = |z_3|$.

A-6 Powers and Roots

Integer powers of a complex number can be defined in the following manner:

$$z^N = \left(re^{j\theta}\right)^N = r^N e^{jN\theta}$$

In other words, the rules of exponents still apply, so the angle θ is multiplied by N and the magnitude is raised to the Nth power. Figure A-15 shows a sequence of these:

$$\{z^\ell\} = \{z^0,\ z^1,\ z^2,\ z^3, \ldots\}$$

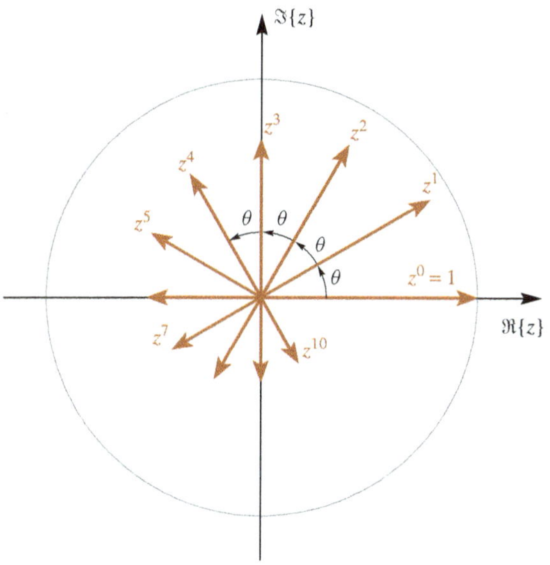

Figure A-15 Vectors representing a sequence of powers z^ℓ for $\ell = 0, 1, 2, \ldots, 10$. Since $|z| = 0.9 < 1$, the vectors spiral in toward the origin. The angular change between successive powers is constant: $\theta = \arg z = \pi/6$.

where the angle steps by a constant amount; in this case, exactly $\pi/6$ rad. The magnitude of z is less than one, so the successive powers spiral in toward the origin. If $|z| > 1$, the points would spiral outward; if $|z| = 1$, all the powers z^N would lie on the ***unit circle*** (a circle of radius 1). One famous identity is DeMoivre's formula:

$$(\cos\theta + j\sin\theta)^N = \cos N\theta + j\sin N\theta$$

The proof of this seemingly difficult trigonometric identity is actually trivial if we invoke Euler's formula (A.7) for $e^{jN\theta}$.

EXERCISE A.10 Let z^{N-1}, z^N, and z^{N+1} be three consecutive members of a sequence such as shown in Fig. A-15. If $z = 0.98e^{-j\pi/6}$ and $N = 11$, plot the three numbers.

A-6.1 Roots of Unity

In a surprising number of cases, the following equation must be solved:

$$z^N = 1 \tag{A.10}$$

In this equation, N is an integer. One solution is $z = 1$, but there are many others, because (A.10) is equivalent to finding all the roots of the Nth-degree polynomial $z^N - 1$, which must have N roots. It turns out that all the solutions are given by

$$z = e^{j2\pi\ell/N} \qquad \text{for } \ell = 0, 1, 2, \ldots, N-1$$

which are called the ***N^{th} roots of unity***. As shown in Fig. A-16, these N solutions are complex numbers equally spaced around the unit circle. The angular spacing between them is $2\pi/N$.

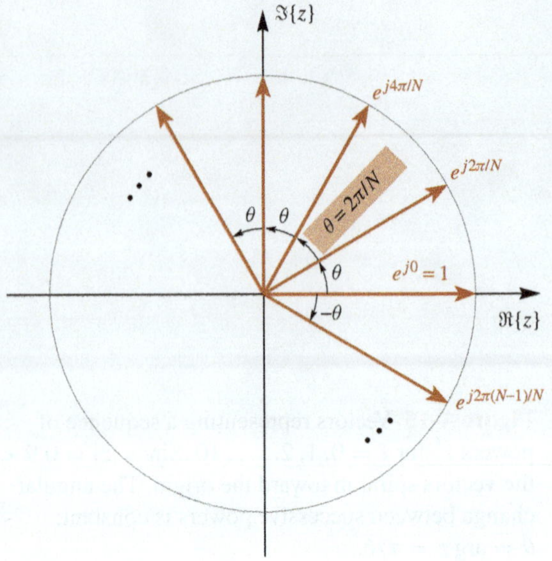

Figure A-16 Graphical display of the Nth roots of unity ($N = 12$). These are the solutions of $z^N = 1$. Notice that there are only N distinct roots.

A-6.1.1 Procedure for Finding Multiple Roots

Once we know that the solutions to (A.10) are the Nth roots of unity, we can describe a structured approach to solving equations with multiple roots. In order to get a general result, we consider a slightly more general situation

$$z^N = c$$

where c is a complex constant, $c = |c|e^{j\varphi}$. The procedure involves the following six steps:

1. Write z^N as $r^N e^{jN\theta}$.

2. Write c as $|c|e^{j\varphi}e^{j2\pi\ell}$, where ℓ is an integer. Note that when $c = 1$, we would write the number 1 as

$$1 = e^{j2\pi\ell} \qquad \text{for } \ell = 0, \pm 1, \pm 2, \ldots$$

3. Equate the two sides and solve for the magnitudes and angles separately.

$$r^N e^{jN\theta} = |c|e^{j\varphi}e^{j2\pi\ell}$$

4. The magnitude is the positive N^{th} root of a positive number $|c|$:

$$r = |c|^{1/N}$$

5. The angle contains the interesting information, because there are N different solutions:

$$N\theta = \varphi + 2\pi\ell \qquad \ell = 0, 1, \ldots, N-1$$
$$\Longrightarrow \quad \theta = \frac{\varphi + 2\pi\ell}{N}$$
$$\theta = \frac{\varphi}{N} + \frac{2\pi\ell}{N}$$

6. Thus, the N different solutions all have the same magnitude, but their angles are equally spaced with a difference of $2\pi/N$ between each adjacent pair.

EXAMPLE A-5 7th Roots of Unity

Solve $z^7 = 1$, using the procedure above.

$$r^7 e^{j7\theta} = e^{j2\pi\ell}$$
$$\Longrightarrow r = 1$$
$$\Longrightarrow 7\theta = 2\pi\ell$$
$$\theta = \frac{2\pi}{7}\ell \qquad \ell = 0, 1, 2, 3, 4, 5, 6$$

Therefore, these solutions are equally spaced around the unit circle, as shown in Fig. A-17. In this case, the solutions are called the ***7th roots of unity***.

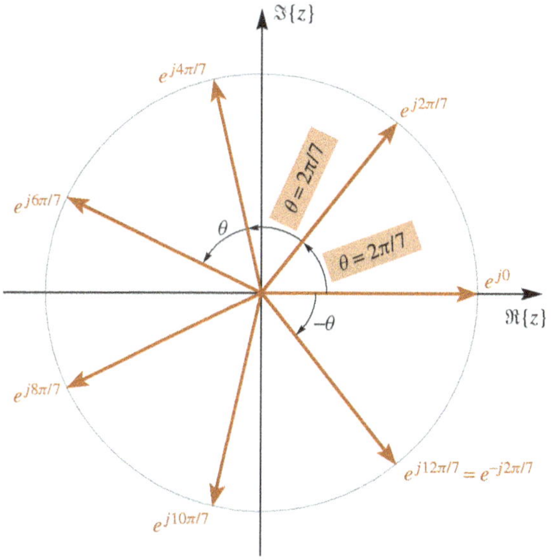

Figure A-17 Graphical display of the 7th roots of unity. Notice that the sequence $e^{j2\pi\ell/7}$ repeats with a period equal to 7.

EXERCISE A.11 Solve the following equation:

$$z^5 = -1$$

Use the fact that $-1 = e^{j\pi}$. Plot all the solutions as vectors. In this case, the resulting values would be called the fifth roots of minus one.

A-7 Summary and Links

LAB
Introduction to Complex Exponentials

DEMO
Complex Numbers via MATLAB

This appendix has presented a concise review of complex numbers and their visualization as vectors in the two-dimensional complex plane. Although this material should have been seen previously by most students during high school algebra, our intense use of complex notation demands much greater familiarity. The labs in Chapter 2 deal with various aspects of complex numbers, and also introduce MATLAB. These labs also introduce some new MATLAB functions for plotting vectors from complex numbers (`zvect, zcat`) and for changing between Cartesian and polar forms (`zprint`). The complex numbers via MATLAB demo is a quick reference to these routines.

In addition to the labs, we have written a MATLAB GUI that will generate drill problems for each of the complex operations studied here: addition, subtraction, multiplication, division, inverse, and conjugate. A screen shot of the `zdrill` GUI is shown in Fig. A-6.

A-8 Problems

DEMO
zdrill

The problems in this section provide a review of the major concepts of complex numbers and complex exponentials. The MATLAB GUI `zdrill` on the Companion Website can be used to generate additional problems for practice.

P-A.1 Convert the following to polar form:

(a) $z = 0 + j2$ (c) $z = (-1, 1)$

(b) $z = -3 - j4$ (d) $z = (0, -1)$

P-A.2 Convert the following to rectangular form:

(a) $\sqrt{2}\, e^{j(3\pi/4)}$ (c) $1.6\angle(\pi/6)$

(b) $3e^{-j(\pi/2)}$ (d) $7\angle(7\pi)$

P-A.3 Evaluate the following by reducing the answer to rectangular form:

(a) j^3 (c) $e^{j(\pi+2\pi m)}$ (m an integer)

(b) j^{2n} (n an integer) (d) $j^{1/2}$ (find two answers)

P-A.4 Simplify the following complex-valued expressions into both Cartesian and polar form:

(a) $(\sqrt{2} - j2)^8$

(b) $(\sqrt{2} - j2)^{-1}$

(c) $(\sqrt{2} - j2)^{1/2}$

(d) $3e^{j2\pi/3} - 4e^{-j\pi/6}$

(e) $\Im\{je^{-j\pi/3}\}$

P-A.5 Evaluate each expression and give the answer in both rectangular and polar form. In all cases, assume that $z_1 = -4 + j3$ and $z_2 = 1 - j$.

(a) z_1^*

(b) z_2^2

(c) $z_1 + z_2^*$

(d) jz_2

(e) $z_1^{-1} = 1/z_1$

(f) z_1/z_2

(g) e^{z_2}

(h) $z_1 z_1^*$

(i) $z_1 z_2$

P-A.6 Simplify the following complex-valued sum:

$$z = e^{j9\pi/3} + e^{-j5\pi/8} + e^{j13\pi/8}$$

Give the numerical answer for z in polar form. Draw a vector diagram for the three vectors and their sum (z).

P-A.7 Simplify the following complex-valued expressions. Give your answers in polar form. Reduce the answers to a simple numerical form.

(a) For $z = -3 + j4$, evaluate $1/z$.

(b) For $z = -2 + j2$, evaluate z^5.

(c) For $z = -5 + j13$, evaluate $|z|^2$.

(d) For $z = -2 + j5$, evaluate $\Re\{ze^{-j\pi/2}\}$

P-A.8 Solve the following equation for z:

$$z^4 = j$$

Be sure to find all possible answers, and express your answer(s) in polar form.

P-A.9 Let $z_0 = e^{j2\pi/N}$. Prove that $z_0^{N-1} = 1/z_0$.

P-A.10 Evaluate $(-j)^{1/2}$ and plot the result(s).

P-A.11 For each of the following complex numbers, simplify the number into the standard polar form (i.e., $z = re^{j\theta}$ with $r \geq 0$ and $-\pi < \theta \leq \pi$):

$$z_1 = \left(\frac{3+5j}{8-2j}\right)^* \qquad z_2 = \frac{-1}{4j} \qquad z_3 = je^{j\pi/3}$$

P-A.12 One of the special properties of the roots of unity is the following fact:

$$\sum_{k=0}^{N-1} e^{j2\pi k/N} = 0$$

Determine the value of z when $z = -\sum_{k=0}^{N-2} e^{j2\pi k/N}$, and $N = 56$. Express the answer in polar form.

P-A.13 Find the *real* numbers p and q such that the following equation is true:

$$\frac{p}{200 + jq} = 0.08e^{-j\pi/3}$$

P-A.14 Evaluate the following definite integral, and express the answer in polar form as indicated:

$$\int_{1}^{1.5} e^{j\pi t}\,dt = re^{j\theta}$$

APPENDIX

B

Programming in MATLAB

MATLAB is used extensively in the laboratory exercises provided on the Companion Website and listed in Appendix D. This appendix provides an overview of MATLAB and some of its capabilities. We focus on programming issues since a wealth of information is already available on syntax and basic commands.[1] MATLAB has an extensive online help system, which can be used to answer any questions not addressed in this brief presentation. In fact, an ideal way to read this appendix would be to have MATLAB running so that `helpwin` can be used whenever necessary, and the examples can be run and modified.

MATLAB (short for Matrix Laboratory) is an environment for numerical analysis and computing. It originated as an interface to the collections of numerical routines from the LINPACK and EISPACK projects, and it is now a widely used product of The Mathworks, Inc. MATLAB has evolved into a powerful programming environment containing many built-in functions for doing signal processing, linear algebra, and other mathematical

[1] One very useful reference is *MATLAB Tutorials and Learning Resources* at the Mathworks website: `http://www.mathworks.com/academia/student_center/tutorials/launchpad.html`.

calculations. The language has also been extended by means of toolboxes containing additional functions for MATLAB. For example, the Companion Website contains a toolbox of functions needed for the laboratory exercises and demo GUIs. The toolboxes are installed as separate directories within the MATLAB directory. Follow the instructions on the Companion Website to install the ***DSP First toolbox*** before doing the laboratory exercises.

NOTE
DSP First Toolbox

Since MATLAB is extensible, users find it convenient to write new functions whenever the built-in functions fail to do a vital task. The programming necessary to create new functions and scripts is not too difficult if the user has some experience with a high-level computer language. This appendix gives a brief overview of MATLAB for the purpose of programming in the context of signal processing.

B-1 MATLAB Help

MATLAB provides an online help system accessible by using the `helpwin` command. For example, to get information about the function `filter`, enter the following at the command prompt:

```
>> helpwin filter
```

The command prompt is indicated by » in the command window. The `helpwin` command opens a new window and displays help information. Help is also available for categories; for example, `helpwin punct` summarizes punctuation as used in MATLAB's syntax. The help system can also show the full documentation of any function in the help window via the command `doc`.

Users new to MATLAB should get started with the command `intro`, which covers the basic concepts in the MATLAB language. There are many demonstration programs that illustrate the various capabilities of MATLAB; these can be started with the command `demo`. Finally, if you are searching for other tutorials, some are freely available on the Web.

When unsure about a command, use `helpwin`.

B-2 Matrix Operations and Variables

The basic variable type in MATLAB is a matrix.[2] To declare a variable, simply assign it a value at the MATLAB prompt. For example,

```
>> A = [1 2 6 8 3 5; 5 2 7 7 7 0; 4 3 7 7 7 9]
   A =
       1   2   6   8   3   5
       5   2   7   7   7   0
       4   3   7   7   7   9
```

[2]Recent versions of MATLAB offer many other data types such as structures found in conventional programming languages. This appendix will only discuss matrices which are two-dimensional arrays.

When the definition of a matrix involves a long formula or many entries, then a very long MATLAB command can be broken onto two (or more) lines by placing an ellipsis (...) at the end of the line to be continued. For example,

```
P = [ 1, 2, 4, 6, 8 ] + [ pi, 4, exp(1), 0, -1 ] + ...
      [ cos(0.1*pi), sin(pi/3), tan(3), atan(2), sqrt(pi) ];
```

If an expression is followed by a semicolon (`;`), then the result is not echoed to the command window. This is very useful when dealing with large matrices. Also note that `;` is used to separate rows in a matrix definition. Likewise, spaces or commas (`,`) are used to separate column entries.

The dimensions of a matrix can always be extracted with the `size` operator:

```
>> Asize = size(A)
   Asize =
           3   6
```

Therefore, it becomes unnecessary to assign separate variables to track the number of rows and number of columns. Two special types of matrix variables are worthy of mention: *scalars* and *vectors*. A scalar is a matrix with only one element; its size is 1×1. A vector is a matrix that has only one row or one column. In the laboratory exercises on the Companion Website, signals are often stored as vectors.

Individual elements of a matrix variable may be accessed by giving the row index and the column index, for example:

```
>> A26 = A(2,6)
   A26 =
         0
```

Submatrices may be accessed in a similar manner by using the colon operator as explained in Section B-2.1.

B-2.1 The Colon Operator

The colon operator `(:)` is useful for creating index arrays, creating vectors of evenly spaced values, and accessing submatrices. Use `helpwin colon` for a detailed description of its capabilities.

The colon notation is based on the idea that an index range can be generated by giving a start, a skip, and then the end. Therefore, a regularly spaced vector of numbers is obtained by means of

```
iii = start:skip:end
```

Without the `skip` parameter, the default increment is 1. MATLAB exploits this sort of counting when doing matrix indexing. For example, a 2×3 submatrix can be extracted from the larger matrix `A` with `A(2:3,3:5)`, and the result would be a matrix of all sevens for `A` defined above. This is a generalization of `A(2,6)` which is the scalar element located at the 2nd row and 6th column of `A`. The colon can also serve as a wild card—that is, `A(2,:)`

is a row vector consisting of the entire 2nd row of the matrix `A`. Indexing backward flips a vector. For example, `A(3,6:-1:1)` reverses the order of the elements in a length-6 vector, which is the third row of `A`. A more convenient form is `A(3,end:-1:1)` in which `end` determines the last index in the vector. Finally, it is sometimes necessary to work with a list of all the values in a matrix, so `A(:)` gives an 18×1 column vector that is merely the columns of `A` concatenated together into one long column. This is an example of ***reshaping*** the matrix. More general reshaping of the matrix `A` can be accomplished with the `reshape(A,M,N)` function. For example, the 3×6 matrix `A` can be reshaped into a 9×2 matrix with: `Anew = reshape(A,9,2)`.

B-2.2 Matrix and Array Operations

The default operations in MATLAB are matrix operations. Thus `A*B` means matrix multiplication, which is defined and reviewed next.

B-2.2.1 A Review of Matrix Multiplication

The operation of matrix multiplication AB can be carried out only if the two matrices have compatible dimensions (i.e., the number of columns in A must equal the number of rows in B). For example, a 5×8 matrix can multiply an 8×3 matrix to give a result AB that is 5×3. In general, if A is $m \times n$, then B must be $n \times p$, and the product AB would be $m \times p$. Usually matrix multiplication is *not* commutative (i.e., $AB \neq BA$). If $p \neq m$, then the product BA cannot be defined, but even when BA is defined, we find that the commutative property applies only in special cases.

Each element in the product matrix is calculated with an inner product. To generate the first element in the product matrix, $C = AB$, simply take the first row of A and multiply *element-by-element* with the first column of B, and then sum. For example, if

$$A = \begin{bmatrix} a_{1,1} & a_{1,2} \\ a_{2,1} & a_{2,2} \\ a_{3,1} & a_{3,2} \end{bmatrix} \quad \text{and} \quad B = \begin{bmatrix} b_{1,1} & b_{1,2} \\ b_{2,1} & b_{2,2} \end{bmatrix}$$

then the first element of $C = AB$ is

$$c_{1,1} = a_{1,1}b_{1,1} + a_{1,2}b_{2,1}$$

which is, in fact, the inner product between the first row of A and the first column of B. Likewise, $c_{2,1}$ is found by taking the inner product between the *second* row of A and the *first* column of B, and so on for $c_{1,2}$ and $c_{2,2}$. The final result would be

$$C = \begin{bmatrix} c_{1,1} & c_{1,2} \\ c_{2,1} & c_{2,2} \\ c_{3,1} & c_{3,2} \end{bmatrix} = \begin{bmatrix} a_{1,1}b_{1,1} + a_{1,2}b_{2,1} & a_{1,1}b_{1,2} + a_{1,2}b_{2,2} \\ a_{2,1}b_{1,1} + a_{2,2}b_{2,1} & a_{2,1}b_{1,2} + a_{2,2}b_{2,2} \\ a_{3,1}b_{1,1} + a_{3,2}b_{2,1} & a_{3,1}b_{1,2} + a_{3,2}b_{2,2} \end{bmatrix} \tag{B.1}$$

Some special cases of matrix multiplication are the *outer product* and the *inner product*. In the **outer product**, a column vector multiplies a row vector to give a matrix. If we let one of the vectors be all 1s, then we can get a repeating result:

$$\begin{bmatrix} a_1 \\ a_2 \\ a_3 \end{bmatrix} \begin{bmatrix} 1 & 1 & 1 & 1 \end{bmatrix} = \begin{bmatrix} a_1 & a_1 & a_1 & a_1 \\ a_2 & a_2 & a_2 & a_2 \\ a_3 & a_3 & a_3 & a_3 \end{bmatrix}$$

With all 1s in the row vector, we end up repeating the column vector four times.

For the **inner product**, a row vector multiplies a column vector, so the result is a scalar. If we let one of the vectors be all 1s, then we will sum the elements in the other vector:

$$\begin{bmatrix} a_1 & a_2 & a_3 & a_4 \end{bmatrix} \begin{bmatrix} 1 \\ 1 \\ 1 \\ 1 \end{bmatrix} = a_1 + a_2 + a_3 + a_4$$

If the vectors are identical, then the inner product sums the squares of all elements in the vector:

$$\begin{bmatrix} a_1 & a_2 & a_3 \end{bmatrix} \begin{bmatrix} a_1 \\ a_2 \\ a_3 \end{bmatrix} = a_1^2 + a_2^2 + a_3^2$$

All these operations are done in MATLAB with `A*B` as long as `A` and `B` have compatible dimensions.

B-2.2.2 Pointwise Array Operations

If we want to do a pointwise multiplication between two arrays, some confusion can arise. In the pointwise case, we want to multiply the matrices together element-by-element, so they must have exactly the same size in both dimensions. For example, two 5×8 matrices can be multiplied pointwise, although we cannot do matrix multiplication between two 5×8 matrices. To obtain pointwise multiplication in MATLAB, we use the "point-star" operator `A.*B`. For example, if A and B are both 3×2, then

$$\texttt{D} = \texttt{A.*B} = \begin{bmatrix} d_{1,1} & d_{1,2} \\ d_{2,1} & d_{2,2} \\ d_{3,1} & d_{3,2} \end{bmatrix} = \begin{bmatrix} a_{1,1}\,b_{1,1} & a_{1,2}\,b_{1,2} \\ a_{2,1}\,b_{2,1} & a_{2,2}\,b_{2,2} \\ a_{3,1}\,b_{3,1} & a_{3,2}\,b_{3,2} \end{bmatrix}$$

where $d_{i,j} = a_{i,j}\,b_{i,j}$. We will refer to this type of multiplication as **array multiplication**. Notice that array multiplication is commutative because we would get the same result if we computed `D = B.*A`.

A general rule in MATLAB is that when "point" is used with another arithmetic operator, it modifies that operator's usual matrix definition to a pointwise one. Thus we have `./` for pointwise division and `.^` for pointwise exponentiation. For example,

```
xx = (0.9) .^ (0:49)
```

generates a vector whose values are equal to $(0.9)^n$, for $n = 0, 1, 2, \ldots, 49$.

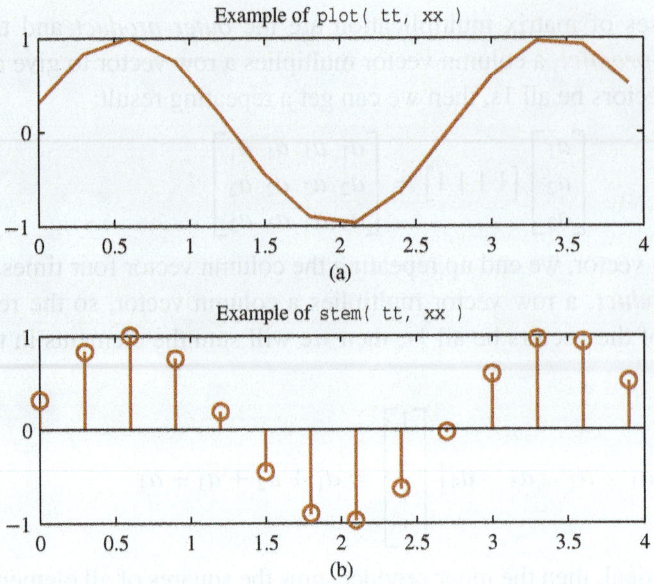

Figure B-1 Two MATLAB plotting formats, (a) `plot` and (b) `stem`. In both cases, the horizontal axis is defined by the vector `tt=0:0.4:4`, and the signal values are `xx = sin(0.7*pi*tt + 0.1*pi)`.

B-3 Plots and Graphics

MATLAB is capable of producing two-dimensional x-y plots and three-dimensional plots, displaying images, as well as creating and playing movies. The two most common plotting functions that are used in signal processing applications are `plot` and `stem`. The calling syntax for both `plot` and `stem` takes two vectors, one for the x-axis points, and the other for the y-axis.[3] The statement `plot(x,y)` produces a connected plot with straight lines between the data points

```
{(x(1),y(1)),(x(2),y(2)),...,(x(N),y(N))}
```

as shown in Fig. B-1(a). The same call with `stem(x,y)` produces the "lollipop" presentation of the same data in Fig. B-1(b). MATLAB has numerous plotting options that can be studied by using `helpwin stem`, `helpwin plotedit`, or by using `helpwin zoom`, `helpwin plot3d`, or `helpwin spectrogram`.

B-3.1 Figure Windows

Whenever MATLAB makes a plot, it displays the graphics in a ***figure window***. You can have multiple figure windows open, but only one of them is considered the *active* window. Any plot command executed in the command window will direct its graphical output to the active window. The command `figure(n)` will pop up a new figure window that can be referred to by the number `n`, or makes it active if it already exists. Control over many of the window attributes (size, location, color, etc.) is also possible with the `figure` command, which does initialization of the plot window.

[3] If only one argument is given, `plot(y)` uses the single argument as the y-axis, and uses `[1:length(y)]` for the x-axis. These functions also have optional arguments. To learn about them type `helpwin plot`.

B-3.2 Multiple Plots

Multiple plots per window can be done with the **`subplot`** function. This function does not do the actual plotting; it merely divides the figure window into tiles. To set up a 3×2 tiling of the figure window, use `subplot(3,2,tile_number)`. For example, `subplot(3,2,3)` will direct the next `plot` to the third tile, which is in the second row, left side. The graphs in Fig. B-1 were done with `subplot(2,1,1)` and `subplot(2,1,2)`. The program that creates Fig. B-1 is given in Section B-5.

B-3.3 Printing and Saving Graphics

Plots and graphics may be printed to a printer or saved to a file using the **`print`** command. To send the current figure window to the default printer, simply type `print` without arguments. To save the plot to a file, a device format and filename must be specified. The device format specifies which language will be used to store the graphics commands. For example, a useful format for including the file in a document is to make a PNG (Portable Network Graphics) image, which can be produced as follows:

```
>> print -dpng myplot
```

Postscript and PDF formats are also convenient when the plots are kept for printing at a later time. For a complete list of available file formats, supported printers, and other options, see `helpwin print`.

B-4 Programming Constructs

MATLAB supports the paradigm of "functional programming" in which it is possible to nest a sequence of function calls. Consider the following equation, which can be implemented with one line of MATLAB code.

$$\sum_{n=1}^{L} \log\left(|x_n|\right)$$

Here is the MATLAB equivalent:

```
sum( log( abs(x) ) )
```

where `x` is a vector containing the elements x_n. This example illustrates MATLAB in its most efficient form, where individual functions are combined to get the output. Writing efficient MATLAB code requires a programming style that generates small functions that are vectorized. Loops should be avoided. The primary way to avoid loops is to use calls to built-in functions and toolbox functions as much as possible.

B-4.1 MATLAB Built-In Functions

Many MATLAB functions operate on arrays just as easily as they operate on scalars. For example, if `x` is an array, then `cos(x)` returns an array of the same size as `x` containing the cosine of each element of `x`.

$$\texttt{cos(x)} = \begin{bmatrix} \cos(x_{1,1}) & \cos(x_{1,2}) & \cdots & \cos(x_{1,n}) \\ \cos(x_{2,1}) & \cos(x_{2,2}) & \cdots & \cos(x_{2,n}) \\ \vdots & \vdots & \vdots & \vdots \\ \cos(x_{m,1}) & \cos(x_{m,2}) & \cdots & \cos(x_{m,n}) \end{bmatrix}$$

Notice that no loop is needed, even though `cos(x)` does apply the cosine function to every array element. Most transcendental functions follow this pointwise rule. In some cases, it is crucial to make this distinction, such as the matrix exponential `(expm)` versus the pointwise exponential `(exp)`:

$$\texttt{exp(A)} = \begin{bmatrix} \exp(a_{1,1}) & \exp(a_{1,2}) & \cdots & \exp(a_{1,n}) \\ \exp(a_{2,1}) & \exp(a_{2,2}) & \cdots & \exp(a_{2,n}) \\ \vdots & \vdots & \vdots & \vdots \\ \exp(a_{m,1}) & \exp(a_{m,2}) & \cdots & \exp(a_{m,n}) \end{bmatrix}$$

B-4.2 Program Flow

Program flow can be controlled in MATLAB using `if` statements, `while` loops, and `for` loops. There is also a `switch` statement for cases. These are similar to any high-level language. Descriptions and examples for each of these program constructs can be viewed by using the MATLAB `helpwin` command.

B-5 MATLAB Scripts

Any expression that can be entered at the MATLAB prompt can also be stored in a text file and executed as a script. The text file is usually created with the built-in MATLAB editor. The file extension must be `.m`, and the script is executed in MATLAB simply by typing the filename (without the extension). These programs are usually called ***M-files***. Here is an example:

```
tt = 0:0.4:4;
xx = sin(0.7*pi*tt);
subplot(2,1,1)
plot( tt, xx )
title('Example of plot( tt, xx)')
subplot(2,1,2)
stem( tt, xx )
title('Example of stem( tt, xx)')
```

If these commands are saved in a file named `plotstem.m` that is on MATLAB's path or in the current directory,[4] then typing `plotstem` at the command prompt will run the file, and all eight commands will be executed as if they had been typed in at the command prompt. The result is the two plots that were shown in Fig. B-1.

B-6 Writing a MATLAB Function

You can write your own functions and add them to the MATLAB environment. These functions are another type of M-file, and are created as an ASCII file with a text editor. The first word in the M-file must be the keyword `function` to tell MATLAB that this file is to be treated as a function with arguments. On the same line as the word `function` is the calling template that specifies the input and output arguments of the function. The filename for the M-file must end in `.m`, and the filename will become the name of the new command for MATLAB. For example, consider the following file, which extracts the last L elements from a vector:

```
function  y = foo( x, L )
%FOO    get last L points of x
%  usage:
%         y = foo( x, L )
%  where:
%
%        x = input vector
%        L = number of points to get
%        y = output vector
N = length(x);
if( L > N )
  error('input vector too short')
end
y = x((N-L+1):N);
```

If this file is called `foo.m`, the operation may be invoked from the MATLAB command line by typing

```
aa = foo( (1:2:37), 7 );
```

The output will be the last seven elements of the vector `(1:2:37)`, i.e.,

```
aa = [ 25 27 29 31 33 35 37 ]
```

B-6.1 Creating a Clip Function

Most functions can be written according to a standard format. Consider a clip ***function M-file*** that takes two input arguments (a signal vector and a scalar threshold) and returns an output signal vector. You can use an editor to create an ASCII file `clipper.m` that contains the statements in Fig. B-2. We can break down the M-file `clipper.m` into four elements:

[4]Type `helpwin path` to learn how to set the MATLAB path and current directory.

```
1  function  y = clipper( x, Limit )
2  %CLIPPER saturate mag of x[n] at Limit
3  %    when |x[n]| > Limit, make |x[n]| = Limit
4  %       keeping the sign of x[n]
5  %  usage:   y = clipper( x, Limit )
6  %
7  %        x = input signal vector
8  %  Limit = limiting value
9  %        y = output vector after clipping
10
11 [nrows, ncols] = size(x);
12
13 if( ncols ~= 1 & nrows ~= 1 )   %-- NEITHER
14    error('CLIPPER: input not a vector')
15 end
16 Lx = max([nrows ncols]);
17
18 for n=1:Lx                    %-- Loop over entire vector
19    if( abs(x(n)) > Limit )
20       x(n) = sign(x(n))*Limit;   %-- saturate
21    end
22 end
23
24 y = x;    %-- copy to output vector
```

Lines 2–9 are comments that start with percent (%)

Determine matrix dimensions of x

Check that x is a row or a column vector

Length of x, which will also be the length of y

Since x is local, we can change it value without affecting the value in the calling workspace

y is the name of the vector passed back to the caller

Figure B-2 Illustration of a MATLAB function called `clipper.m`.

Definition of Input–Output: Function M-files must have the word `function` as the very first item in the file. The information that follows `function` on the same line is a declaration of how the function is to be called and what arguments are to be passed. The name of the function should match the name of the M-file; if there is a conflict, it is the name of the M-file stored on the disk that is known to the MATLAB command environment.

Input arguments are listed inside the parentheses following the function name. Each input is a matrix. The output argument (also a matrix) is on the left side of the equals sign. Multiple output arguments are also possible if square brackets surround the list of output arguments. For example, notice how the `size(x)` function in line 11 returns the number of rows and number of columns into separate output variables. Finally, observe that there is no explicit command for returning the outputs; instead, MATLAB returns whatever value is contained in the output matrix `y` when the function completes. For `clipper` the last line of the function assigns the clipped vector to `y`, so that the clipped vector is returned. MATLAB does have a command called `return`, but it just exits the function, it does not take an argument.

The essential difference between the function M-file and the script M-file is dummy variables versus permanent variables. MATLAB uses ***call by value*** so that the function makes local copies of its arguments. These local variables disappear after the function completes. For example, the following statement creates a clipped vector `wwclipped` from the input vector `ww`.

```
>> wwclipped = clipper(ww, 0.9999);
```

The arrays `ww` and `wwclipped` are permanent variables in the MATLAB workspace. The temporary arrays created inside `clipper` (i.e., `y`, `nrows`, `ncols`, `Lx`, and `n`) exist only while `clipper` runs; then they are deleted. Furthermore, these variable names are local to `clipper.m`, so the name `x` may also be used in the workspace as a permanent name. These ideas should be familiar to anyone experienced with a high-level computer language.

Self-Documentation: A line beginning with the `%` sign is a comment line. The first group of these in a function is used by MATLAB's help facility to make M-files automatically self-documenting. That is, you can now type `helpwin clipper` and the comment lines immediately following the function declaration line in *your* M-file will appear on the screen as help information. The format suggested in `clipper.m` (lines 2–9) follows the convention of giving the function name, its calling sequence, a brief explanation, and then the definitions of the input and output arguments.

Size and Error Checking: The function should determine the size of each vector or matrix that it will operate on. This information does not have to be passed as a separate input argument, but can be extracted with the `size` function (line 11). In the case of the `clipper` function, we want to restrict the function to operating on vectors, but we would like to permit either a row ($1 \times L$) or a column ($L \times 1$). Therefore, one of the variables `nrows` or `ncols` must be equal to 1; if not, we terminate the function with the bail-out function `error`, which prints a message to the command line and quits the function (lines 13–15).

Actual Function Operations: In the case of the `clipper` function, the actual clipping is done by a `for` loop (lines 18–22), which examines each element of the `x` vector for its value compared to the threshold `Limit`. In the case of negative numbers, the clipped value must be set to `-Limit`, hence the multiplication by `sign(x(n))`. This assumes that `Limit` is passed in as a positive number, a fact that might also be tested in the error-checking phase.

This particular implementation of `clipper` is very inefficient owing to the `for` loop. In Section B-7.1, we will show how to vectorize this program for speed.

B-6.2 Debugging a MATLAB M-file

Since MATLAB is an interactive environment, debugging can be done by examining variables in the workspace. MATLAB contains a symbolic debugger that is integrated with the text editor. Since different functions can use the same variable names, it is important to keep track of the local context when setting breakpoints and examining variables. Several useful debugging commands are listed here, others can be found with `helpwin debug`. Within the visual editor, these commands are accessed with menu items or function keys.[5]

dbstop is used to set a breakpoint in an M-file. It can also be used to give you a prompt when an error occurs by typing `dbstop if error` before executing the M-file. This allows you to examine variables within functions and also the calling workspace (by typing `dbup`).

dbstep incrementally steps through your M-file, returning you to a prompt after each line is executed.

dbcont causes normal program execution to resume, or, if there was an error, returns you to the MATLAB command prompt.

dbquit quits the debug mode and returns you to the MATLAB command prompt.

keyboard can be inserted into the M-file to cause program execution to pause, giving you a MATLAB prompt of the form `K>` to indicate that it is not the command-line prompt. When stopped at a breakpoint, the values of variables can be displayed by using the `K>` prompt, or by hovering over the variable name in the visual editor.

B-7 Programming Tips

This section presents a few programming tips that should improve the speed of your MATLAB programs. For more ideas and tips, examine some of the function M-files in the toolboxes of MATLAB by using the `edit` command. For example,

```
edit angle
edit conv
edit trapz
```

Copying the style of other (good) programmers is always an efficient way to improve your own knowledge of a computer language. In the following hints, we discuss some of the most important points involved in writing good MATLAB code. These comments will become increasingly useful as you develop more experience in MATLAB.

B-7.1 Avoiding Loops

Since MATLAB is an interpreted language, certain common programming habits are intrinsically inefficient. The primary one is the use of `for` loops to perform simple operations over an entire matrix or vector. *Whenever possible*, you should try to find a vector function (or the nested composition of a few vector functions) that will accomplish

[5]More details about the MATLAB debugger can be found at:
`http://www.mathworks.com/help/matlab/matlab_prog/debugging-process-and-features.html`

```
function ss = mysum1(xx)
%
[Nrows,Ncols] = size(xx);
ss = 0;
for jj=1:Nrows
  for kk=1:Ncols
    ss = ss + xx(jj,kk);
  end
end
```

Determine matrix dimensions of xx

Two for loops needed to index all rows and columns when xx is a matrix

(a)

```
function ss = mysum2(xx)
%
ss = sum( sum(xx) );
```

Sum down the columns to produce a vector

Sum elements of the vector to produce the scalar sum ss

(b)

```
function ss = mysum3(xx)
%
ss = sum(xx(:));
```

Colon turns the matrix xx into a column vector

Sum elements of the vector to produce the scalar sum ss

(c)

Figure B-3 Three ways to add all of the elements of a matrix. (a) Index all the array elements individually and add them one at a time, (b) call the sum function twice which will be the fastest method, and (c) call the sum function once which is the easiest code to understand and almost as fast as (b).

the desired result, rather than writing a loop. For example, if the operation is summing all the elements in a matrix, the difference between calling `sum` and writing a loop can be dramatic; the loop is slow owing to the interpreted nature of MATLAB[6]. Consider the three methods for summing all the elements in a matrix, shown in Fig. B-3. The first method in Fig. B-3(a) is the MATLAB equivalent of conventional programming. The last two methods rely on the built-in function `sum`, which has different characteristics depending on whether its argument is a matrix or a vector (called "operator overloading"). When acting on a matrix as in Fig. B-3(b), the inner `sum` returns a row vector containing the column sums; when acting on a row (or column) vector as in Fig. B-3(c), the sum is a scalar. To get the third method, the matrix `xx` is converted to a column vector with the colon operator. Then one call to `sum` will suffice.

B-7.2 Repeating Rows or Columns

Often it is necessary to form a matrix from a vector by replicating the vector in the rows or columns of the matrix.[7] If the matrix is to have all the same values, then functions such as `ones(M,N)` and `zeros(M,N)` can be used. But to replicate a column vector `x`

[6]In recent versions of MATLAB, optimizations have been introduced which improve the running time of for loops significantly.

[7]The function `repmat` is a very general operator for repeating matrices and vectors in MATLAB.

to create a matrix that has identical columns, a loop can be avoided by using the outer-product matrix multiply operation discussed in Section B-2.2. The following MATLAB code fragment will do the job for 11 columns:

```
x = (12:-2:0).';
% dot prime indicates transpose; prime alone is conjugate transpose
X = x * ones(1,11)
```

If x is a length L column vector, then the matrix X formed by the outer product is $L \times 11$. In this example, $L = 7$. Note that MATLAB is case-sensitive, so the variables x and X are different. We have used capital X to indicate a matrix, as would be done in mathematics.

B-7.3 Vectorizing Logical Operations

It is also possible to vectorize programs that contain if, else conditionals. The clipper function (Fig. B-2) offers an excellent opportunity to demonstrate this type of vectorization. The for loop in that function contains a logical test and might not seem like a candidate for vector operations. However, the relational and logical operators in MATLAB, such as greater than, apply to matrices. For example, a greater than test applied to a 3×3 matrix returns a 3×3 matrix of ones and zeros.

```
>> %-- create a test matrix
>> X = [ 1  2 -3; 3 -2  1; 4  0  -1]
   X = [ 1  2 -3
         3 -2  1
         4  0 -1 ]
>> %-- check the greater than condition to create a mask
>> Mask = X > 0
   Mask = [ 1  1  0
            1  0  1
            1  0  0 ]
>> %-- pointwise multiply by mask
>> Y = Mask .* X
   Y = [1  2  0
        3  0  1
        4  0  0 ]
```

The zeros mark where the condition was false; the ones denote where the condition was true. Thus, when we do the pointwise multiply of X by the masking matrix Mask we get a result that has all negative elements set to zero. Note that these last two statements process the entire matrix without ever using a for loop.

Since the saturation done in clipper.m requires that we change the large values in X, we can implement the entire for loop with three array multiplications. This leads to a vectorized saturation operator that works for matrices as well as vectors:

```
y = Limit*(X > Limit) - Limit*(X < -Limit) + X.*(abs(X) <= Limit);
```

Three different masking matrices are needed to represent the three cases of positive saturation, negative saturation, and no action. The additions correspond to the logical OR of these cases. The number of arithmetic operations needed to carry out this statement is $3N$ multiplications and $2N$ additions, where N is the total number of elements in `X`. This is actually more work than the loop in `clipper.m` if we counted only arithmetic operations. However, the cost of code interpretation is high. This vectorized statement is interpreted only once, whereas the three statements inside the `for` loop must be reinterpreted N times. If the two implementations are timed with `etime`, the vectorized version will be faster for long vectors.

B-7.4 Creating an Impulse

Another simple example of use of logical operators is given by the following trick for creating an impulse signal vector:

```
nn = [-10:25];
impulse = (nn==0);
```

This result may be plotted with `stem(nn, impulse)`. In a sense, this code fragment is perfect because it captures the essence of the mathematical formula that defines the impulse as existing only when $n = 0$.

$$\delta[n] = \begin{cases} 1 & n = 0 \\ 0 & n \neq 0 \end{cases}$$

B-7.5 The Find Function

An alternative to masking is to use the `find` function. This is not necessarily more efficient; it just gives a different approach. The `find` function will determine the list of indices in a vector where a condition is true. For example, `find( y > Limit )` will return the list of indices where the vector `y` is greater than the `Limit` value.[8] Thus we can do saturation as follows:

```
y = x;
yAbove = find(y > Limit);
y( yAbove ) =  Limit;
yBelow = find(y < -Limit);
y( yBelow ) = -Limit;
```

[8]When applied to matrix, the statement `find( y > Limit )` treats `y` as the vector `y(:)`, so the example code works for matrices as well as vectors.

B-7.6 Seek to Vectorize

The dictum to "avoid `for` loops" is not always easy to obey, because it means the algorithm must be cast in a vector form. If matrix-vector notation is incorporated into MATLAB programs, the resulting code will run much faster. Thus, a reasonable goal is the following:

Eliminate all `for` *loops.*

Even loops with logical tests can be ***vectorized*** if masks are created for all possible conditions. The MATLAB function `profile` measures various statistics about the execution time of a program which provides useful information for modifying the code to make it run faster.

B-7.7 Programming Style

If there were a proverb to summarize good programming style, it would probably read something like

May your functions be short
and your variable names long.
–Anon

This is certainly true for MATLAB. Each function should have a single purpose. This will lead to short, simple modules that can be linked together by functional composition to produce more complex operations. Avoid the temptation to build super functions with many options and a plethora of outputs.

MATLAB supports long variable names, so you can take advantage of this feature to give variables descriptive names. In this way, the number of comments needed to document the code can be drastically reduced. Comments should be limited to `helpwin` information and the documentation of tricks used in the code.

APPENDIX

C

Fourier Series

In this appendix, we describe the mathematical theory of ***Fourier series***. This will extend the presentation in Chapter 3 which stated the main result that *any periodic signal can be synthesized with a sum of harmonically related sinusoids,* even when the synthesis sum needs an infinite number of terms.[1] We use specific examples of pulse waves, triangular waves, and rectified sinusoids to show how the Fourier series representation is obtained and how each signal's spectrum is defined by the Fourier series. Finally, we focus on general properties of the Fourier series which not only extend the understanding of the frequency domain, but also make the job of signal analysis easier in practice.

As stated in Chapter 3, the two aspects of the Fourier theory are analysis and synthesis. Starting from $x(t)$ and calculating $\{a_k\}$ is called ***Fourier analysis***. As we will show, Fourier analysis of a continuous-time periodic signal is achieved by evaluating the following integral:

Fourier Analysis Integral

$$a_k = \frac{1}{T_0} \int_0^{T_0} x(t) e^{-j\omega_0 k t}\, dt \tag{C.1}$$

[1]There are some exceptions, but none of these would ever arise in the practice of engineering when processing actual signals.

where $\omega_0 = 2\pi/T_0 = 2\pi F_0$ is the fundamental radian frequency[2] and $T_0 = 1/F_0$ is the fundamental period of the periodic signal $x(t) = x(t + T_0)$ for all t.

The reverse process of starting from $\{a_k\}$ and generating a periodic signal identical to the signal $x(t)$ in (C.1) is called ***Fourier synthesis***. It is achieved with the following general formula for synthesis of periodic signals:

Fourier Synthesis Summation

$$x(t) = \sum_{k=-\infty}^{\infty} a_k e^{j\omega_0 kt} \tag{C.2}$$

When taken together, equations (C.1) and (C.2) comprise the ***Fourier series*** representation of the periodic signal $x(t)$. When the signal $x(t)$ is real, the complex amplitudes are *conjugate-symmetric* with respect to the index k (i.e., $a_{-k} = a_k^*$), and the synthesis formula becomes a sum of real sinusoids of the form

$$x(t) = A_0 + \sum_{k=1}^{\infty} A_k \cos(\omega_0 kt + \varphi_k) \tag{C.3}$$

where $A_0 = a_0$, and the amplitude and phase of the kth term come from the polar form, $a_k = \frac{1}{2}A_k e^{j\varphi_k}$. In other words, the condition $a_{-k} = a_k^*$ is sufficient for the synthesized waveform to be a *real* function of time.

C-1 Fourier Series Derivation

In Chapter 3, we gave several demonstrations that a sum of complex exponentials as in (C.2) can represent a wide range of interesting periodic signals, such as vowel-like signals, and we carried out the details of evaluating a Fourier integral for the periodic full-wave rectified sine (FWRS) wave. However, we did not explain why the Fourier integral in (C.1) finds the spectrum for any periodic signal. In this section, we show how to derive the coefficients for the harmonic sum in (C.2)—that is, we show how to go from the time waveform $x(t)$ to the Fourier coefficients a_k by using the ***Fourier series integral*** of (C.1) to perform Fourier analysis. This Fourier integral requires that we have a mathematical formula that defines $x(t)$ over one period. Several examples will be presented later to illustrate this point. On the other hand, if $x(t)$ is known only as a digital recording, then numerical methods such as those discussed in Chapter 8 will be applicable.

C-1.1 Fourier Integral Derivation

The derivation of the Fourier series integral formula (C.1) relies on a simple property of the complex exponential signal—the integral of a complex exponential over any number of complete periods is zero. In equation form,

[2]The fundamental frequency in hertz is F_0, so that we can use $f_k = kF_0$ to refer to the kth harmonic frequency. As a result, $f_1 = F_0$ and $f_0 = 0$.

$$\int_0^{T_0} e^{j(2\pi/T_0)kt}\,dt = 0 \qquad (k \neq 0) \tag{C.4}$$

where k is a *nonzero* integer. The integrand in (C.4) is a complex exponential whose frequency is $(2\pi/T_0)k$, so its period is T_0/k and the integral is taken over k periods. After substituting ω_0 for $2\pi/T_0$, here is the integration:

$$\int_0^{T_0} e^{j\omega_0 kt}\,dt = \left.\frac{e^{j\omega_0 kt}}{j\omega_0 k}\right|_0^{T_0} = \frac{e^{j(\omega_0 T_0)k} - 1}{j\omega_0 k} = 0 \qquad (k \neq 0)$$

The numerator is zero because $\omega_0 T_0 = 2\pi$ and $e^{j2\pi k} = 1$ for any integer k.

A key ingredient in the infinite series representation (C.2) is the form of the complex exponentials, all of which must repeat with the same period as the period of the signal $x(t)$, which is T_0. If we define $v_k(t)$ to be the complex exponential having frequency $\omega_0 k$, then

$$v_k(t) = e^{j\omega_0 kt} \tag{C.5}$$

Even though the minimum duration period of $v_k(t)$ would be T_0/k, the following shows that $v_k(t)$ repeats also with a period of T_0:

$$\begin{aligned} v_k(t + T_0) &= e^{j\omega_0 k(t+T_0)} = e^{j\omega_0 kt}\, e^{j(\omega_0 T_0)k} \\ &= e^{j\omega_0 kt}\, \overbrace{e^{j2\pi k}}^{1} = e^{j\omega_0 kt} = v_k(t) \end{aligned}$$

where again we have used $e^{j2\pi k} = 1$ for any integer k (positive or negative).

Next we can generalize the zero-integral property (C.4) of the complex exponential to involve two signals:

Orthogonality Property

$$\int_0^{T_0} v_k(t) v_\ell^*(t)\,dt = \begin{cases} 0 & \text{if } k \neq \ell \\ T_0 & \text{if } k = \ell \end{cases} \tag{C.6}$$

where the * superscript in $v_\ell^*(t)$ denotes the complex conjugate. Proving the orthogonality property is straightforward. We begin with

$$\int_0^{T_0} v_k(t) v_\ell^*(t)\,dt = \int_0^{T_0} e^{j\omega_0 kt} e^{-j\omega_0 \ell t}\,dt = \int_0^{T_0} e^{j\omega_0 (k-\ell)t}\,dt$$

There are two cases to consider for the last integral: when $k = \ell$ the exponent becomes zero, so the integral is

$$\int_0^{T_0} e^{j\omega_0(k-\ell)t}\,dt = \int_0^{T_0} e^{j0t}\,dt = \int_0^{T_0} 1\,dt = T_0$$

Otherwise, when $k \neq \ell$ the exponent is nonzero and we can invoke the zero-integral property in (C.4) to get

$$\int_0^{T_0} e^{j\omega_0(k-\ell)t}\,dt = \int_0^{T_0} e^{j\omega_0 mt}\,dt = 0$$

where $m = k - \ell \neq 0$.

The orthogonality property of complex exponentials (C.6) simplifies the rest of the Fourier series derivation. If we assume that the synthesis summation (C.2) is a valid representation of $x(t)$, then

$$x(t) = \sum_{k=-\infty}^{\infty} a_k e^{j\omega_0 kt} \tag{C.7}$$

and we can multiply both sides of (C.7) by the complex exponential $v_\ell^*(t)$ and integrate over the period T_0.

$$\int_0^{T_0} x(t)e^{-j\omega_0\ell t}\,dt = \int_0^{T_0} \left(\sum_{k=-\infty}^{\infty} a_k e^{j\omega_0 kt}\right) e^{-j\omega_0\ell t}\,dt \tag{C.8a}$$

$$= \sum_{k=-\infty}^{\infty} a_k \underbrace{\left(\int_0^{T_0} e^{j\omega_0(k-\ell)t}\,dt\right)}_{\text{nonzero only when } k=\ell} = a_\ell T_0 \tag{C.8b}$$

Notice that we are able to isolate one of the complex amplitudes (a_ℓ) in (C.8b) by applying the orthogonality property (C.6). The final analysis formula is obtained by dividing the left-hand side of (C.8a) and right-hand side of (C.8b) by T_0, giving an equation for a_ℓ. Since ℓ is an arbitrarily chosen index, we can replace ℓ with k to obtain the Fourier analysis integral previously asserted in (C.1)

A delicate step in the derivation occurs when going from (C.8a) to (C.8b) where the order of the infinite summation and the integration is interchanged. The validity of this manipulation depends on convergence properties of the infinite sum. If the sum is finite, there is no problem because the integral of a sum is the sum of the integrals of the individual terms in the sum. However, if the sum is infinite, as in the general theory of Fourier series, then special care must be taken. A sufficient condition can be stated: If $x(t)$ is a smooth function that is bounded and has only a finite number of discontinuities and zero crossings within one period, then the interchange is permissible and the Fourier coefficients are obtained by evaluating the Fourier integral as we have asserted.

The fact that every periodic signal, even a discontinuous wave, can be represented as the sum of an infinite number of sinusoids was one of the amazing claims in Fourier's famous thesis of 1807. Many years passed before mathematicians were able to develop a rigorous convergence proof based on mean-square convergence to support Fourier's claim. In this appendix, we will give more examples of Fourier series representations of periodic signals and discuss some of the important aspects of the mathematical theory underlying these representations.

EXERCISE C.1

Show that since $x(t) = x(t + T_0)$, the range of integration for (C.1) can be *any convenient interval* of length T_0. For example, break the integral into two integrals over $[-T_0/2, 0]$ and $[0, T_0/2]$ and then use a change of dummy variable of integration on the one over $[-T_0/2, 0]$ prior to combining back into the form (C.1) to show that

$$a_k = \frac{1}{T_0}\int_{-T_0/2}^{T_0/2} x(t)e^{-j\omega_0 kt}\,dt \tag{C.9}$$

EXERCISE C.2

Use (C.9) to show that if $x(t)$ is real and even, $x(-t) = x(t)$, then the Fourier series coefficients are real and $a_{-k} = a_k$.
Hint: Use Euler's relation to write $e^{-j\omega_0 kt} = \cos(\omega_0 kt) - j\sin(\omega_0 kt)$ in (C.9) and use the facts that $\cos(\cdot)$ is an even function and $\sin(\cdot)$ is an odd function. Also, the product of two even functions is even, and the product of an even function times an odd function is odd.

C-2 Examples of Fourier Analysis

We can represent *any periodic signal* by using a sum of complex exponential signals, if we constrain the *frequencies to be harmonics* as in (C.2). Wave shapes that do not look at all sinusoidal, such as pulse waves, triangular waves, and rectified sinusoids have a Fourier representation as a sum of complex exponentials, but with an infinite number of terms. In this section, the general formulas for their Fourier coefficients a_k are derived and shown to be relatively compact.

There are four concepts to keep in mind: (1) the *analysis* integral that extracts complex amplitudes $\{a_k\}$ from the signal $x(t)$, (2) the *spectrum* plot of a_k versus f, which is (3) the frequency-domain *representation* of the signal $x(t)$ as an infinite sum of harmonic sinusoids (complex exponentials), and (4) *approximate synthesis* of $x(t)$ with a finite number of sinusoids by truncating the infinite representation. These essential features of Fourier analysis and synthesis are depicted in Fig. C-1. We start with the signal defined in the time domain by a mathematical function $x(t)$ that is known to be periodic with period T_0. In order to determine the Fourier series coefficients, we need to know $x(t)$ over one period as a function that can be used in the Fourier analysis integral. If the Fourier integral can be evaluated to give a formula for the coefficients a_k, then, as depicted in the upper part of Fig. C-1, we have all the information required to plot the spectrum of the periodic signal $x(t)$ (i.e., we have obtained the frequency-domain representation of the signal).

One very important point about the spectrum is illustrated in Fig. C-1: All periodic signals with fundamental period T_0 have a spectrum that looks similar to the one in Fig. C-1. The only difference is the values of the coefficients a_k. Thus, whenever we have a spectrum that consists of harmonics, we know immediately that it is the spectrum of a periodic signal.

In Fig. C-1, we illustrate the four Fourier concepts enumerated above beginning with a full-wave rectified sine (FWRS) wave plotted at the bottom left of the figure. Its spectrum is depicted schematically at the top of the figure. The details of the Fourier analysis for the FWRS signal were presented in Section 3-5.2, but we might ask at this point, why do we need to determine the Fourier series since it would be very easy to write an equation that would define $x(t)$ over one period? In fact, we must write such an equation in order to begin the Fourier analysis calculation. The short answer is that the

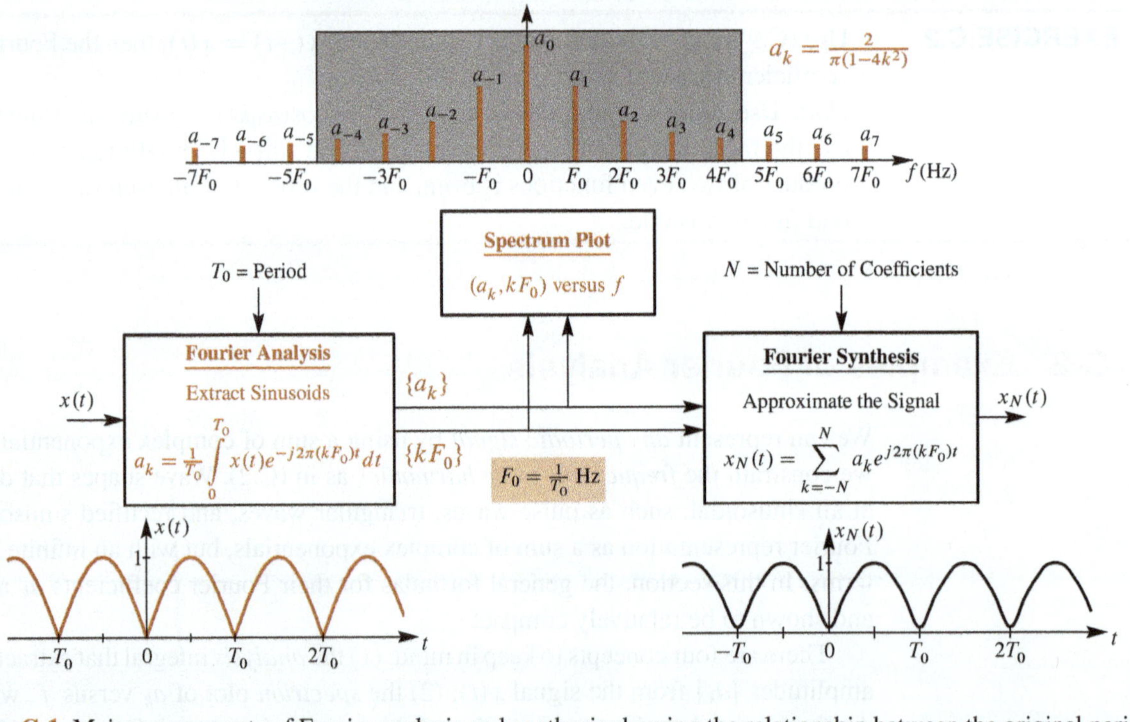

Figure C-1 Major components of Fourier analysis and synthesis showing the relationship between the original periodic signal $x(t)$, its spectrum, and the synthesized signal $x_N(t)$ that approximates the original.

frequency-domain representation of the signal as a sum of complex exponentials gives us different information that is not evident from either the equation defining one period or the plot of the waveform shown in Fig. C-1.

This brings to the fore an important point about the concept of Fourier synthesis. In general, the Fourier synthesis formula requires infinite limits as in (C.2), and we think of synthesis from the spectrum as a *mathematical representation* of the periodic signal $x(t)$. From this point of view, it is sufficient to know that the Fourier series for a signal exists, and we then use it to develop an understanding of the signal based on the spectrum and its properties, rather than the details of the time waveform.

Figure C-1 has a fourth component illustrated in the lower right-hand part of the figure. This part of the diagram suggests synthesizing a signal from the spectrum representation of the signal. In particular, an approximate signal can be synthesized according to the *finite* Fourier synthesis summation formula[3]

Finite Fourier Synthesis

$$x_N(t) = \sum_{k=-N}^{N} a_k e^{j\omega_0 kt} \tag{C.10}$$

DEMO
Fourier Series Finite Synthesis

When we regard the Fourier synthesis equation as a computational formula as in (C.10), we sum a (necessarily) finite number of the complex exponential terms evaluated for a range of values of t. We might want to do this simply to verify that the Fourier series does indeed sum up to a function that closely approximates the original function definition used to compute the coefficients a_k. We could also study the rate of convergence as we increase N toward infinity. From a more practical point of view, we may want to explore the effect of limited bandwidth on the detailed nature of the signal. This is suggested in Fig. C-1 by the shaded region superimposed on the spectrum. In the lower right corner of the figure, we see the waveform that results when all frequencies above $4F_0$ are eliminated from the original input signal. Thus, the resulting $x_4(t)$ might be thought of as the output of a filtering or absorption process that removes the high-frequency signal components that are *present in the original signal*. Figure C-1 shows the relationship between Fourier analysis and Fourier synthesis using representative plots for the FWRS wave case where the finite synthesis result with $N = 4$ (in the lower right corner) is an approximation to the original FWRS wave shown (in the lower left corner).

We shall now illustrate the various aspects of Fig. C-1 by considering several commonly used periodic signals. These examples will show that infinite limits are often required to represent signals that occur in practical applications.

LAB
Synthesis of Sinusoidal Signals

[3] If you have access to MATLAB, it is straightforward to write a *Fourier Synthesis Program* that implements (C.10). This MATLAB programming exercise is described in more detail in the lab projects devoted to music synthesis available on the Companion Website.

C-2.1 The Pulse Wave

A simple, but very common, example is the periodic pulse wave, which is defined for one period by

$$x(t) = \begin{cases} 1 & \text{for } 0 \le |t| < \tau/2 \\ 0 & \text{for } \tau/2 \le |t| \le T_0/2 \end{cases} \tag{C.11}$$

Figure C-2 shows a plot of this signal with the basic period being infinitely repeated with period T_0. It is called a pulse wave because it switches between one for τ seconds (the pulse width) and zero for the remainder of the period. Pulse signals are to be found throughout every digital hardware device, where one and zero signify ON (or TRUE), and OFF (or FALSE), respectively. In many applications where pulse waves occur, τ may be much less than the period T_0, which in turn may be very small (implying a high fundamental frequency). When $\tau = T_0/2$, the pulse signal is commonly called a *square wave* because it is ON for exactly half (50%) of its period. Often the terminology is extended to "50% duty cycle square wave."

To obtain the Fourier series representation of the pulse wave, we must derive a formula for the complex amplitudes a_k by using the Fourier analysis integral (C.1). First note that $x(t)$ is most conveniently defined over the interval $-T_0/2 \le t \le T_0/2$ rather than $0 \le t \le T_0$. Thus, it is easiest to use the form of the Fourier analysis integral (C.9) of Exercise C.1. After substituting $x(t)$ defined in (C.11) into the integral (C.9), we immediately recognize that the integral must be broken into three integrals to handle all the cases in the definition of $x(t)$, that is,

$$a_k = \frac{1}{T_0}\int_{-T_0/2}^{-\tau/2}(0)e^{-j\omega_0 kt}\,dt + \frac{1}{T_0}\int_{-\tau/2}^{\tau/2}(1)e^{-j\omega_0 kt}\,dt + \frac{1}{T_0}\int_{\tau/2}^{T_0/2}(0)e^{-j\omega_0 kt}\,dt \tag{C.12}$$

but the first and third integrals drop out because the signal $x(t)$ is zero for $\tau/2 \le |t| \le T_0/2$. Thus, we need only evaluate and simplify the integral from $-\tau/2$ to $\tau/2$ as follows:

$$\begin{aligned} a_k &= \left(\frac{1}{T_0}\right)\frac{e^{-j\omega_0 kt}}{-j\omega_0 k}\bigg|_{-\tau/2}^{\tau/2} \\ &= \left(\frac{1}{T_0}\right)\frac{e^{-j(2\pi/T_0)k(\tau/2)} - e^{-j(2\pi/T_0)k(-\tau/2)}}{-(j2\pi/T_0)k} = \frac{\sin(\pi k\tau/T_0)}{\pi k} \end{aligned} \tag{C.13}$$

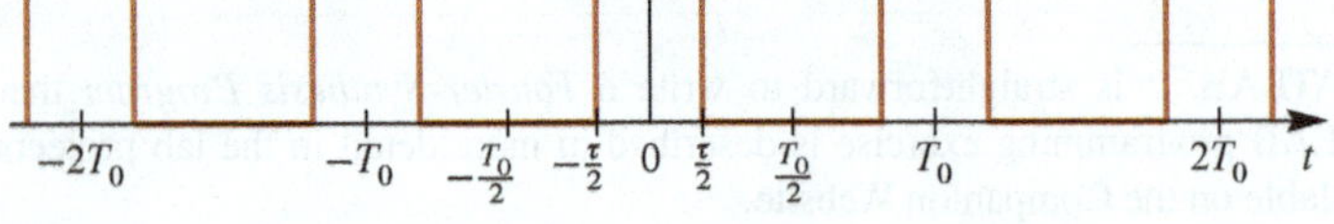

Figure C-2 Plot of a pulse wave signal with fundamental frequency $F_0 = 1/T_0$ and pulse width τ.

The formula in (C.13) is a valid expression for the Fourier coefficients of the pulse wave for all k except $k = 0$, where the formula gives an indeterminate form (0/0). Thus, we must determine the DC component directly by computing the average of $x(t)$ over one period. In order to do this, we return to (C.12) and set $k = 0$ before doing the integral, and we obtain

$$a_0 = \frac{1}{T_0}\int_{-\tau/2}^{\tau/2} (1)e^{-j0t}\,dt = \frac{1}{T_0}\int_{-\tau/2}^{\tau/2} (1)\,dt$$
$$= \frac{1}{T_0}\left(\frac{\tau}{2} - \left(-\frac{\tau}{2}\right)\right) = \frac{\tau}{T_0}$$

If we put this result together with (C.13), we obtain

$$a_k = \begin{cases} \dfrac{\sin(\pi k\tau/T_0)}{\pi k} & k = \pm 1, \pm 2, \pm 3, \ldots \\ \tau/T_0 & k = 0 \end{cases} \qquad \text{(C.14)}$$

Note that as defined in (C.11), the pulse wave is an even function of t. Therefore, as noted in Exercise C.2, the Fourier series coefficients a_k are real and they are an even function of k (i.e., $a_{-k} = a_k$).

EXAMPLE C-1 Square Wave

The *square wave* is the special case of the pulse wave, where $\tau = T_0/2$. In this case, the formula for the Fourier coefficients a_k becomes relatively simple: all the even-indexed coefficients (except for a_0) become zero.

$$a_k = \frac{\sin(\pi k(T_0/2)/T_0)}{\pi k} = \frac{\sin(\pi k/2)}{\pi k}$$

$$\Rightarrow \quad a_0 = \frac{1}{2},\ a_{\pm 1} = \frac{1}{\pi},\ a_{\pm 2} = 0,\ a_{\pm 3} = \frac{-1}{3\pi},\ a_{\pm 4} = 0,\ a_{\pm 5} = \frac{1}{5\pi}, \ldots$$

EXERCISE C.3 Show that the Fourier series synthesis summation for the pulse wave can be written as

$$x(t) = a_0 + \sum_{k=1}^{\infty} 2a_k \cos(\omega_0 kt)$$

where the a_k coefficients are given by (C.14).

C-2.1.1 Spectrum of a Pulse Wave

Figure C-3 shows the spectrum for three cases of the pulse wave whose Fourier coefficients are given by (C.14). For added clarity, the envelope of the Fourier coefficients,

$$a(f) = \frac{\sin(\pi f \tau)}{\pi f T_0} \tag{C.15}$$

is plotted with a thin line as if $a(f)$ is a continuous function of frequency. However, the Fourier series coefficients of the pulse wave are the values of $a(f)$ at discrete frequencies kF_0 (i.e., $a_k = a(kF_0)$). As shown in the plots of Fig. C-3, the envelope $a(f)$ in (C.15) has a "main lobe" centered at $f = 0$, where the function is relatively large compared to the "sidelobes" that surround the main lobe. The main lobe occupies the frequency interval between the lowest frequency zeros as in Fig. C-3, which occur when $\pi f \tau = \pm\pi$, or equivalently $f = \pm 1/\tau$. When τ is equal to T_0 divided by an integer as in the examples of Fig. C-3, the envelope of the Fourier coefficients has zeros at frequencies such that kF_0 is an integer multiple of $1/\tau$. For example, in Fig. C-3(a) where $\tau = T_0/2$, the Fourier coefficients are zero when kF_0 is an integer multiple of $1/\tau = 2/T_0 = 2F_0$. Likewise, when $\tau = T_0/4$ the coefficients are zero for integer multiples of $4F_0$, and so on.

The three examples in Fig. C-3 show that the magnitude of these a_k coefficients drops off as $1/k$, so they eventually become small as k increases. Even so, as defined in Section 3-2.5, the bandwidth of any pulse wave signal is infinite because a_k in (C.14) can be nonzero for large indices as k ranges over all the integers from $-\infty$ to $+\infty$. In

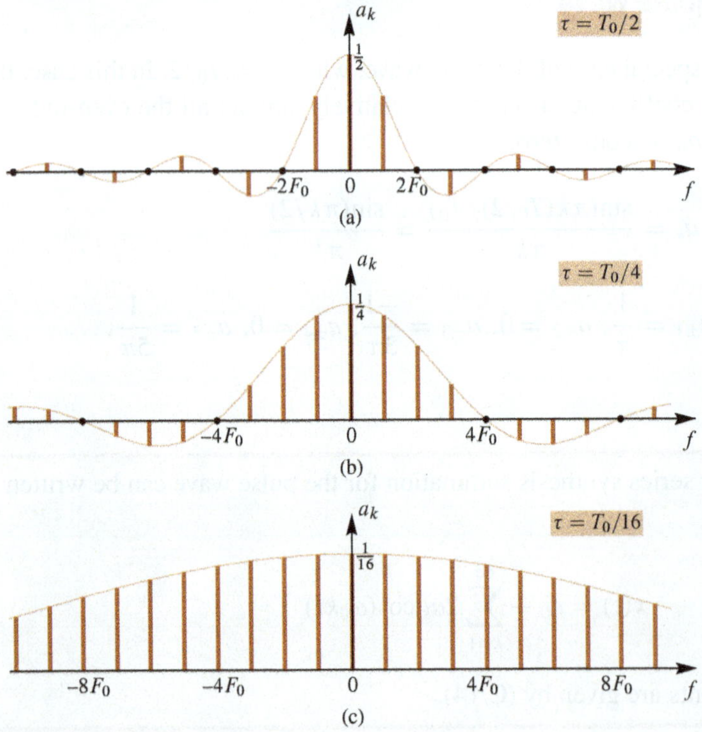

Figure C-3 Spectrum of a pulse wave for different pulse widths (holding T_0 fixed). (a) $\tau = T_0/2$ (square wave case) where the mainlobe extends from $-2F_0$ to $+2F_0 = 1/\tau$. (b) $\tau = T_0/4$ where the mainlobe extends from $-4F_0$ to $+4F_0 = 1/\tau$. (c) $\tau = T_0/16$ where the mainlobe extends from $-16F_0$ to $+16F_0 = 1/\tau$ and no sidelobes are visible. Only the range from $k = -9$ to $+9$ is shown, but the a_k coefficients range over $-\infty < k < \infty$.

Fig. C-3, we see that as the pulse width τ becomes smaller, the spectrum spreads out to higher frequencies (i.e., the relative size of higher frequency coefficients becomes larger). If we make the pulse width even narrower, the DC value decreases while the higher frequency coefficients become larger relative to the low frequency coefficients. In the limit of very narrow pulses, all the coefficients are equal but they have very small amplitude. This is illustrated in Fig. C-3(c) where the first zero coefficient would occur at $16F_0$, which is outside of the limits of the frequency scale shown.

C-2.1.2 Finite Synthesis of a Pulse Wave

Strictly speaking, the bandwidth of the pulse wave is infinite, but in a practical situation the signal might be transmitted through a communication channel or modified by a system that would remove frequencies above a certain limit, thereby reducing the bandwidth to a finite value. For example, suppose that the signal bandwidth is reduced to $5F_0$ so that all the frequencies above $5F_0$ have zero amplitude. The spectrum plots of Fig. C-3 show that for the case $\tau = T_0/2$, the omitted coefficients would have relatively small amplitudes. On the other hand, in the case $\tau = T_0/16$, the discarded coefficients (above $5F_0$) are about the same size as the low frequency coefficients. The effect of this sort of truncation on the time-domain signal is seen in Fig. C-4, which shows plots of

$$x_N(t) = \sum_{k=-N}^{N} a_k e^{j\omega_0 kt}$$

DEMO
Fourier Series Finite Synthesis

for $N = 5$ and a_k in (C.14). Instead of showing the effect of increasing N, Fig. C-4 shows the effect of shortening the pulse duration when the bandwidth is restricted to $5F_0$. For the widest pulse (Fig. C-4(a) with $\tau = T_0/2$), five harmonics of the fundamental are sufficient to produce a noticeably "flat" region where the pulse is "on." However, the time waveform in Fig. C-4(c) shows that the bandlimited signal for $\tau = T_0/16$ does not even reach the full pulse height. In all cases, we see oscillations during the on and off regions, and the sharp corners of the pulse are rounded off. Both of these features are due to the finite bandwidth of $5F_0$ Hz. In digital electronics, the oscillations are called "overshoot" and "undershoot," and the time to switch from off to on is called the "rise time." By increasing the bandwidth (increasing N), we can reduce the rise time, but the maximum overshoot will remain at about 0.09 for a unit amplitude pulse wave.[4]

In Fig. C-4(b), the pulse duration is shorter so that the pulse is rounded off; and in the case of $\tau = T_0/16$ in Fig. C-4(c), the original pulse shape is hard to discern. Indeed, in this case the bandlimited signal is broader than the ideal pulse and it fails to reach the desired pulse height. Again, increasing the bandwidth will lead to a signal that more faithfully reproduces the pulse signal—the rise time will be shorter but the oscillations will still be present. In general, we can conclude from these examples that very narrow, rapidly repeating pulses require very wide bandwidth for accurate reproduction.

[4]This overshoot is called the *Gibbs phenomenon*, and will be discussed in more detail in Section C-4.2.

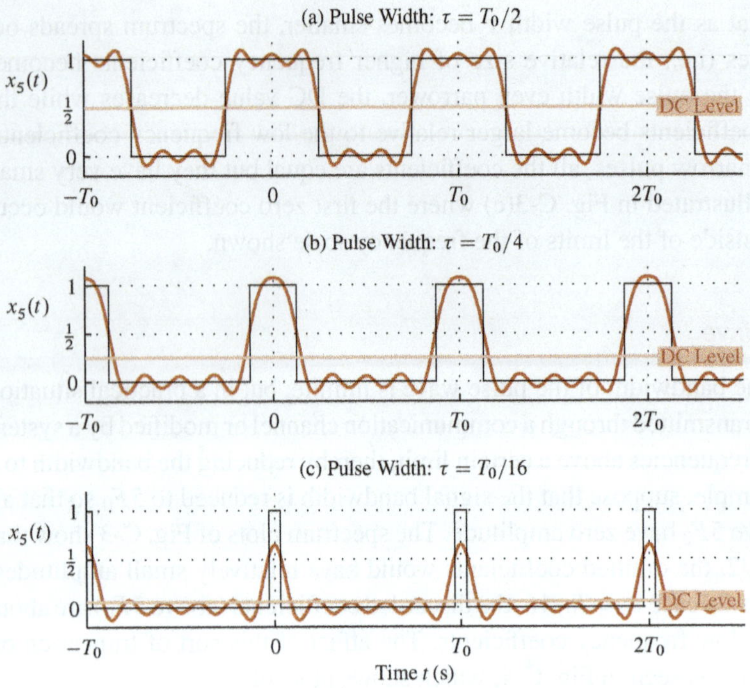

Figure C-4 Finite Fourier synthesis ($N = 5$) for different pulse widths. (a) $\tau = T_0/2$ (square wave case); DC level = 1/2. (b) $\tau = T_0/4$, DC level = 1/4. (c) $\tau = T_0/16$, DC level = 1/16.

EXERCISE C.4 Describe how the waveform $x_5(t)$ would look in the case of the pulse wave when $\tau = T_0/8$. If you have difficulty, you can synthesize $x_5(t)$ using (C.14) with this value of τ.

EXERCISE C.5 Use MATLAB to see how the plots of Fig. C-4 would look if N in the finite Fourier synthesis (C.10) was 21.

C-2.2 Triangular Wave

Another case that is also relatively simple to analyze is the triangular wave shown in Fig. C-5. The mathematical formula for the triangular wave consists of two segments. To set up the Fourier analysis integral, we choose the period from $-T_0/2$ to $+T_0/2$. The definition of the waveform over that period requires two line segments:

$$x(t) = \begin{cases} -2t/T_0 & \text{for } -T_0/2 \le t < 0 \\ 2t/T_0 & \text{for } 0 \le t < T_0/2 \end{cases} \tag{C.16}$$

The first line segment has a slope of $-2/T_0$, the second, $+2/T_0$. Unlike the pulse wave, the triangular wave is a continuous signal.

Now we invoke the Fourier integral (C.1) to derive a formula for the $\{a_k\}$ coefficients of the triangular wave. We might suspect from our earlier experience that the DC coefficient

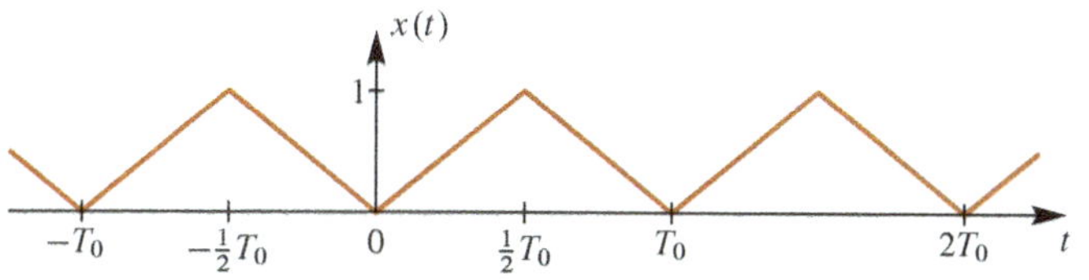

Figure C-5 Triangular wave with period T_0 s.

has to be found separately, so we address that case first. Using the Fourier analysis integral (C.1) and setting $k = 0$, we obtain

$$a_0 = \frac{1}{T_0}\int_0^{T_0} x(t)\,dt$$

If we recognize that the integral over one period is, in fact, the area under the triangle from 0 to T_0 in Fig. C-5, we get

$$a_0 = \frac{1}{T_0}(\text{area}) = \frac{1}{T_0}\left(\frac{T_0}{2}\right) = \frac{1}{2} \tag{C.17}$$

For the general case where $k \neq 0$, we must break the Fourier series analysis integral into two sections because the signal definition consists of two formulas:

$$a_k = \frac{1}{T_0}\int_{-T_0/2}^{0} (-2t/T_0)e^{-j\omega_0 kt}\,dt + \frac{1}{T_0}\int_0^{T_0/2} (2t/T_0)e^{-j\omega_0 kt}\,dt \tag{C.18}$$

The integrals are identical except for a minus sign and the limits of integration. However, integration by parts, or an integral table, or an on-line computer algebra system such as Wolfram|Alpha© would be needed to carry out the actual integrals. After integration by parts and some tedious algebraic steps, the integral for a_k can be written as

$$a_k = \frac{e^{-jk\pi} - 1}{\pi^2 k^2} \tag{C.19}$$

Since $e^{-jk\pi} = (-1)^k$, the numerator in (C.19) equals either 0 or -2, and we can write the following cases for a_k:

$$a_k = \begin{cases} \dfrac{-2}{\pi^2 k^2} & k = \pm 1, \pm 3, \pm 5, \ldots \\ 0 & k = \pm 2, \pm 4, \pm 6, \ldots \\ \dfrac{1}{2} & k = 0 \end{cases} \tag{C.20}$$

Note that in this case the coefficients are purely real and $a_{-k} = a_k$. This is true because the signal is real and also an even function of t. Also observe that, as in the case of the square wave, we cannot evaluate (C.19) for $k = 0$ to obtain the DC coefficient.

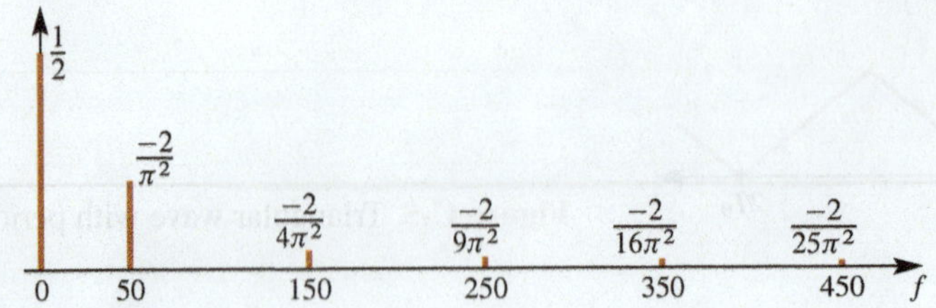

Figure C-6 One-sided spectrum of a triangular wave up to and including the 9th harmonic. The even harmonics at $f = 100$, 200, 300, and 400 Hz are zero. For negative frequencies, conjugate symmetry $a_{-k} = a_k^*$ holds.

EXERCISE C.6 Use integration by parts to show that

$$\int te^{-j\omega_0 kt}dt = e^{-j\omega_0 kt}\left(\frac{j\omega_0 kt + 1}{\omega_0^2 k^2}\right)$$

Then, starting from (C.18), derive the formula (C.20) for the Fourier series coefficients of the triangular wave.

C-2.2.1 Spectrum of a Triangular Wave

A plot of the spectrum[5] for the triangular wave is shown in Fig. C-6 for $F_0 = 50$ Hz. The complex amplitudes given in (C.20) fall off very rapidly, and only the odd-index harmonics are present for $k > 0$. From the spectrum in Fig. C-6 it is easy to write down a formula for the Fourier series representation of the triangular wave which is a time-domain signal. For each odd index k, there are two spectrum lines, one at $f = 50k$ and the other at $f = -50k$, which can be combined.

$$\frac{-2}{k^2\pi^2}e^{j2\pi(50k)t} + \frac{-2}{k^2\pi^2}e^{-j2\pi(50k)t} = \frac{-4}{k^2\pi^2}\cos(2\pi(50k)t) \tag{C.21}$$

The set of odd indices can be generated via $k = 2\ell - 1$ as $\ell = 1, 2, \ldots,$ so we can write

$$x(t) = \frac{1}{2} + \sum_{\ell=1}^{\infty} \frac{-4}{(2\ell-1)^2\pi^2}\cos(2\pi(50)(2\ell-1)t)$$

C-2.2.2 Finite Synthesis of a Triangular Wave

DEMO

Fourier Series Finite Synthesis

The ideal triangular wave in Fig. C-5 is a continuous signal, unlike the square wave which is discontinuous. Therefore, it is easier to approximate the triangular wave with a finite Fourier synthesis sum (C.10). Two cases are shown in Fig. C-7, for $N = 3$ and 9. The fundamental frequency is equal to $F_0 = 50$ Hz. In the $N = 9$ case, the approximation is nearly indistinguishable from the true triangularly shaped waveform but there is slight rounding at the top and bottom of the waveform. Adding harmonics for $N > 9$ will

[5] The plot of Fig. C-6 shows only the positive-frequency half of the spectrum plot. This is a simpler, yet sufficient, way to plot the spectrum of a real signal since the negative-frequency part can be constructed if needed by using the conjugate symmetry of the coefficients (i.e., $a_{-k} = a_k^*$).

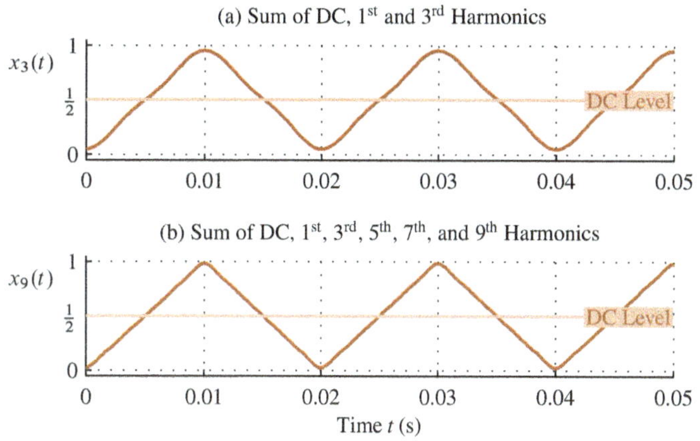

Figure C-7 Summing harmonic components for a 50 Hz triangular wave via finite Fourier synthesis (C.10). (a) $N = 3$. (b) Up to and including the 9th harmonic (i.e., $N = 9$). The constant DC level of 1/2 is indicated.

improve the synthesis near those points. Even the $N = 3$ case is a reasonably good approximation, despite using only DC and two sinusoidal terms. We can see the reason for this by examining the plot of the spectrum in Fig. C-6, which shows that the high-frequency components decrease in size very quickly for $k > 3$. This implies that the triangular wave can be transmitted more accurately through a bandlimiting system than can a pulse wave of the same fundamental frequency.

C-2.3 Half-Wave Rectified Sine

A rectified sine wave is easy to produce from a pure sinusoid using a simple electronic circuit. As discussed in Chapter 3, Section 3-5.2, the full wave rectified sine (FWRS) finds application in AC-to-DC power converters. A related case is the half-wave rectified sine (HWRS) signal shown in Fig. C-8 (i.e., the negative lobes of a sine wave have been clipped off). In electronics it is easy to do such clipping with a diode. The mathematical formula for the HWRS consists of two cases when defined over one period, $0 \le t \le T_0$:

$$x(t) = \begin{cases} \sin((2\pi/T_0)t) & \text{for } 0 \le t < T_0/2 \\ 0 & \text{for } T_0/2 \le t < T_0 \end{cases} \qquad \text{(C.22)}$$

The HWRS is a continuous signal, but its first derivative is discontinuous, because the slope changes abruptly at $t = 0, \pm T_0/2, \pm T_0, \ldots$, as shown in Fig. C-8.

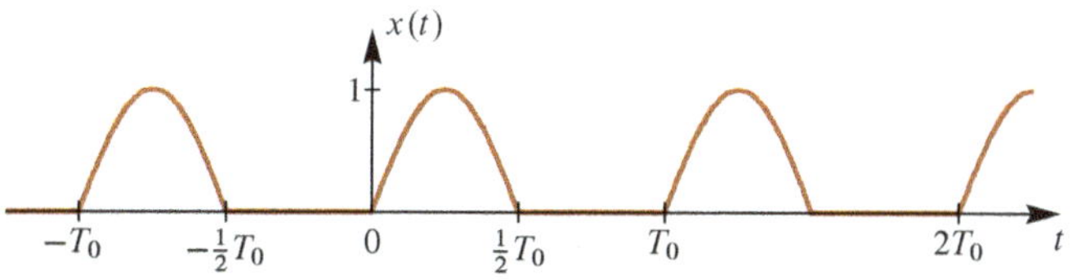

Figure C-8 Periodic half-wave rectified sine signal.

When $x(t)$ from (C.22) is used in the Fourier series analysis integral with $\omega_0 = (2\pi/T_0)$, we must break the integral into two sections because the signal definition consists of two pieces:

$$a_k = \frac{1}{T_0}\int_0^{T_0/2} \sin(\omega_0 t)e^{-j\omega_0 kt}\,dt + \cancelto{0}{\frac{1}{T_0}\int_{T_0/2}^{T_0} (0)e^{-j(2\pi T_0)kt}\,dt} \tag{C.23}$$

As in the FWRS case, we can use the inverse Euler relation for sine to write the remaining integral as the sum of two integrals

$$a_k = \frac{1}{2jT_0}\int_0^{T_0/2} e^{j\omega_0(1-k)t}\,dt + \frac{-1}{2jT_0}\int_0^{T_0/2} e^{-j\omega_0(1+k)t}\,dt \tag{C.24}$$

In (C.24) the special cases of $k = \pm 1$ must be handled separately by substituting $k = \pm 1$ before the integration is carried out. For example, when $k = 1$, we get

$$a_1 = \frac{1}{2jT_0}\int_0^{T_0/2} 1\,dt + \cancelto{0}{\frac{-1}{2jT_0}\int_0^{T_0/2} e^{-j\omega_0(2)t}\,dt} = \frac{1}{4j} = -j\frac{1}{4} \tag{C.25}$$

The second integral is zero because the period of the complex exponential $e^{-j\omega_0(2)t}$ is $T_0/2$ and (C.4) applies.

After integrating the complex exponentials in (C.24) and simplifying numerous complex-valued expressions, the final result for a_k can be written as

$$a_k = \begin{cases} \dfrac{1}{\pi(1-k^2)} & \text{for } k \text{ even} \\ \mp j\dfrac{1}{4} & \text{for } k = \pm 1 \\ 0 & \text{for } k = \pm 3, \pm 5, \ldots \end{cases} \tag{C.26}$$

Note that the DC component is $a_0 = 1/\pi \approx 0.3183$, which is exactly half the DC component of the FWRS. This is because the HWRS signal is zero over the last half of its period.

EXERCISE C.7 Starting from the integrals in (C.24), derive the formula in (C.26) where k is even for the Fourier series coefficients of the HWRS. Exploit complex exponential simplifications such as $e^{j2\pi k} = 1$, $e^{j\pi} = -1$, and so on.

C-2.3.1 Finite Synthesis of a Half-Wave Rectified Sine

DEMO
Fourier Series Finite Synthesis

Since the ideal HWRS signal in Fig. C-8 is a continuous signal, a finite Fourier sum approximation (C.10) can be expected to be better for a given number of coefficients than for the discontinuous square wave. Two cases are shown in Fig. C-9, for $N = 2$ and 8. In the $N = 8$ case, the approximation is very close to the original waveform, but small ripples are visible. Adding harmonics for $N > 8$ will improve the synthesis just a little bit by reducing the ripples. The $N = 2$ case exhibits noticeable ripples because it is only using DC and two sinusoidal terms.

It is possible to explain these differences by examining the spectrum, which will show that the relative magnitudes of the higher frequency components being dropped for $k > 2$ are quite small. A plot of the spectrum for the HWRS is shown in Fig. C-10 for $F_0 = 50$ Hz. The complex amplitudes given in (C.26) fall off very rapidly, and only the even-index harmonics are present for $k > 1$.

From the spectrum in Fig. C-10 it is easy to write down a formula for the Fourier series representation of HWRS waveform. For each index k, there are two spectrum lines, one at $f = 50k$ and the other at $f = -50k$, which can be combined. For example, when k is even and $k > 0$

$$\frac{1}{\pi(1-k^2)} e^{j2\pi(50k)t} + \frac{1}{\pi(1-k^2)} e^{-j2\pi(50k)t} = \frac{2}{\pi(1-k^2)} \cos(2\pi(50k)t) \qquad \text{(C.27)}$$

The set of even indices can be generated via $k = 2\ell$ as $\ell = 1, 2, \ldots$, so we can write

$$x(t) = \frac{1}{\pi} + \frac{1}{2}\sin(2\pi(50)t) + \sum_{\ell=1}^{\infty} \frac{2}{\pi(1-(2\ell)^2)} \cos(2\pi(50)(2\ell)t)$$

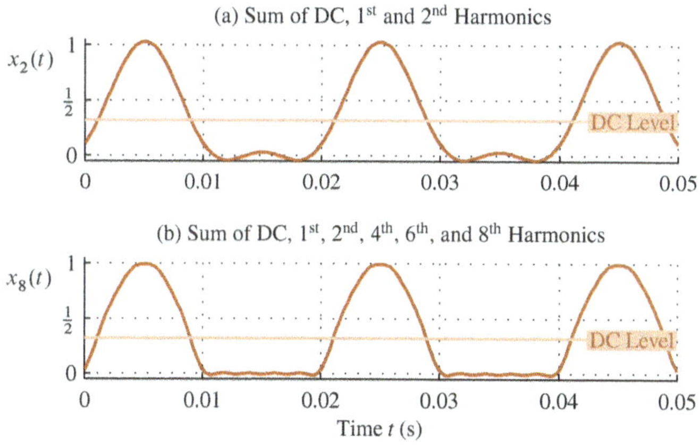

Figure C-9 Summing harmonic components for the half-wave rectified sine signal using finite Fourier synthesis (C.10). (a) $N = 2$. (b) Up to and including the 8th harmonic (i.e., $N = 8$). The fundamental frequency is $F_0 = 50$ Hz. The constant DC level of 0.3183 is indicated in both plots.

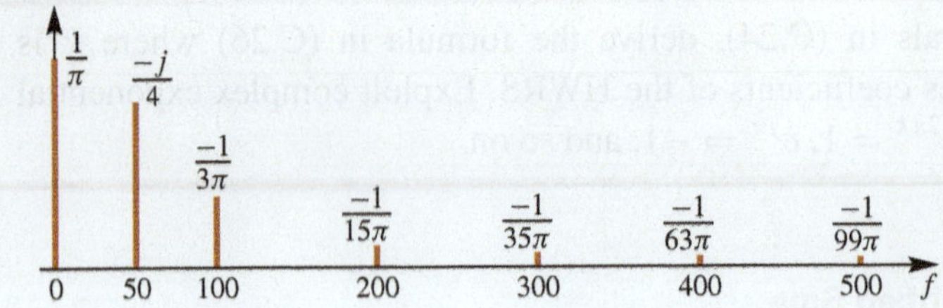

Figure C-10 One-sided spectrum of a half-wave rectified sine signal up to and including the 10th harmonic. The fundamental frequency is $F_0 = 50$ Hz. The odd harmonics at $f = 150$, 250, 350, and 450 Hz are zero. For negative frequencies, conjugate symmetry $a_{-k} = a_k^*$ holds.

It can be seen from the above analysis that half-wave rectification is another way to obtain a signal with a nonzero DC level from the sinusoidal signal that is available from the power company. The major difference from the FWRS signal is that the fundamental frequency of the HWRS output is half that of the FWRS fundamental. For the DC power supply application, this makes it somewhat more difficult to smooth out the ripples and achieve a constant output.

C-3 Operations on Fourier Series

In signal processing, when we manipulate signals we need to predict how the signal will change. One of the great advantages of the Fourier series representation of periodic signals is that operations on the signal $x(t)$ often have very simple corresponding operations on the Fourier coefficients. These relationships are often referred to as *properties* of the Fourier series representation. In Chapter 3, Section 3-3, we presented several properties of the spectrum of a sum of sinusoids. Since the Fourier series representation is the spectrum of a periodic signal, most of these properties must hold for the Fourier series. This spectrum must have harmonic spectrum lines, so the only complication that might arise is whether the operation would affect the harmonic nature of the spectrum lines, or equivalently, the periodicity of the signal. This section revisits these properties and gives several examples.

C-3.1 Scaling or Adding a Constant

A quick look at Section 3-3.1 reminds us that scaling is a simple property because multiplying a periodic signal by a scale factor (i.e., $\gamma x(t)$), will multiply all its Fourier series coefficients by the same scale factor (γ)

$$\gamma x(t) = \sum_{k=-\infty}^{\infty} (\gamma a_k) e^{j\omega_0 kt}$$

Likewise, we recall that when adding a constant only the DC component of the spectrum will change

$$x(t) + c = \underbrace{(a_0 + c)}_{\text{new DC value}} + \sum_{k \neq 0} a_k e^{j\omega_0 kt}$$

Neither of these operations will change the period of the signal, so the properties in Section 3-3.1 are true for the Fourier series.

EXAMPLE C-2 New Square Wave

Define a *bipolar* square wave as follows:

$$y(t) = \begin{cases} +1 & \text{for } 0 \le t < T_0/2 \\ -1 & \text{for } T_0/2 \le t < T_0 \end{cases}$$

This square wave can be related to the zero-one square wave of Example C-1 via $y(t) = 2(x(t) - 1/2)$. Subtracting $1/2$ from $x(t)$ will change only the DC value, then multiplying by two will double the size of all the Fourier coefficients. Since the Fourier coefficients of the zero-one square wave are

$$a_k = \begin{cases} 1/2 & \text{for } k = 0 \\ \dfrac{-j}{k\pi} & \text{for } k \text{ odd} \\ 0 & \text{for } k = \pm 2, \pm 4, \ldots \end{cases}$$

the Fourier coefficients $\{b_k\}$ for $y(t)$ are

$$b_k = \begin{cases} 2(a_0 - 1/2) = 0 & \text{for } k = 0 \\ 2a_k = 2\left(\dfrac{-j}{k\pi}\right) & \text{for } k \text{ odd} \\ 0 & \text{for } k = \pm 2, \pm 4, \ldots \end{cases}$$

C-3.2 Adding Signals

Another simple operation is addition, but there are two cases: If the two periodic signals have the same fundamental frequency, then the Fourier coefficients are summed

$$x(t) + y(t) = \sum_{k=-\infty}^{\infty} (a_k + b_k) e^{j\omega_0 k t}$$

where the Fourier coefficients of $x(t)$ and $y(t)$ are $\{a_k\}$ and $\{b_k\}$, respectively. There are special situations where ω_0 could change after the addition (see Problem P-C.11).

If the fundamental frequencies are different, the situation is more complicated because the sum signal might not even be periodic. If the sum signal is periodic, finding the new fundamental frequency for the sum involves a gcd (greatest common divisor) operation. Then using the new fundamental frequency, all the Fourier coefficients must be re-indexed prior to adding the Fourier coefficients.

C-3.3 Time-Scaling Property

A basic property of the Fourier series is that the Fourier coefficients are determined by the wave shape over one period, while the fundamental frequency is determined by the duration of the fundamental period. The operation of time-scaling, which is defined by the equation $y(t) = x(\alpha t)$, will change the length of each period but leave the wave shape the same. If $x(t)$ is a periodic signal with period T_0 and the scaling constant is $0 < \alpha < \infty$, then the period of $y(t)$ is T_0/α. In this case, we can determine the Fourier series coefficients of $y(t)$ by applying the time-scale operation to the Fourier series representation of $x(t)$. Thus, the following two expressions are equivalent:

$$x(\alpha t) = \sum_{k=-\infty}^{\infty} a_k e^{j(2\pi/T_0)k(\alpha t)} \tag{C.28a}$$

$$y(t) = \sum_{k=-\infty}^{\infty} a_k e^{j(2\pi/(T_0/\alpha))kt} \tag{C.28b}$$

We recognize (C.28b) as the Fourier series for a periodic function with period T_0/α, or fundamental frequency αF_0. Thus, if $\alpha < 1$, the period is greater than T_0 and the fundamental frequency is lower than F_0. When $\alpha > 1$, the opposite is true. The key observation is that the Fourier series coefficients of the time-scaled signal $x(\alpha t)$ are identical to those of the original signal $x(t)$, only the fundamental frequency changes.

EXERCISE C.8 Verify that the sum in (C.28b) produces a signal that satisfies $y(t + T_0/\alpha) = y(t)$, and, therefore, the period of $y(t) = x(\alpha t)$ is T_0/α if the period of $x(t)$ is T_0.

EXAMPLE C-3 Time-Scaling Pulse Waves

For pulse waves, we have already derived a formula (C.14) for the Fourier coefficients, a_k, and we can easily show that the time-scaling property is true for all pulse signals with the same ratio between pulse duration and period, τ/T_0. The formula (C.14) for the Fourier coefficients of a pulse wave is repeated below

$$a_k = \begin{cases} \dfrac{\sin(\pi k(\tau/T_0))}{\pi k} & k = \pm1, \pm2, \pm3, \ldots \\ \tau/T_0 & k = 0 \end{cases}$$

Since τ and T_0 always appear together as the ratio τ/T_0, the a_k coefficients depend only on the ratio.

For example, when the period is 50 ms and the pulse width is 10 ms, the ratio $\tau/T_0 = 0.2$. Thus, we can determine the Fourier coefficients for all cases where

$\tau/T_0 = 0.2$ by using the Fourier analysis integral for a 0.2 s pulse repeated with a period of 1 s. The following integral is for a pulse length of τ' with a period of 1 s

$$a_k = \int_{-\tau'/2}^{\tau'/2} e^{-j2\pi kt} dt$$

so the limits on the integral are $\pm\tau'/2 = \pm(\tau/T_0)/2 = \pm 0.1$ for this case. This integral has fewer symbols to complicate the integration and subsequent algebraic manipulations. It would be ideal for use with symbolic analysis programs such as Wolfram|Alpha.

EXERCISE C.9 Type in the following code to Wolfram|Alpha and interpret the result:

```
int_(-tau/2)^(tau/2) e^(-i 2 pi k t) dt
```

C-3.4 Time-Shifting Property

In Section 3-3.3, the time-shifting property of the spectrum was derived. Since the Fourier series defines a harmonic spectrum, it must obey the same property. If we form a new periodic signal $y(t)$ by time shifting $x(t)$, then there is a simple change in the Fourier coefficients $\{a_k\}$, which are the spectrum of $x(t)$. If the Fourier coefficients of $y(t)$ are $\{b_k\}$, then the time-shifting property of the spectrum implies

Time Shifting Property

$$y(t) = x(t - \tau_d) \quad \longleftrightarrow \quad b_k = a_k e^{-j\omega_0 k \tau_d}$$

Thus, time-shifting $x(t)$ will multiply its Fourier coefficients a_k by a complex exponential.

EXAMPLE C-4 Delayed Triangular Wave

Consider a triangular wave $x(t)$ with period $T_0 = 0.04$ s, as depicted in the top plot of Fig. C-11. The Fourier series coefficients a_k for the triangular wave $x(t)$ are given by (C.20) and the period is $T_0 = 0.04$ s, so $\omega_0 = 50\pi$. Therefore, the Fourier series coefficients of the delayed signal $y(t) = x(t - 0.01)$ are $b_k = a_k e^{-j(50\pi)k(0.01)} = a_k e^{-j\pi k/2} = (-j)^k a_k$ or

$$b_k = (-j)^k a_k = \begin{cases} \dfrac{j2(-1)^{k+1}}{\pi^2 k^2} & k = \pm 1, \pm 3, \pm 5, \dots \\ 0 & k = \pm 2, \pm 4, \pm 6, \dots \\ 1/2 & k = 0 \end{cases} \tag{C.29}$$

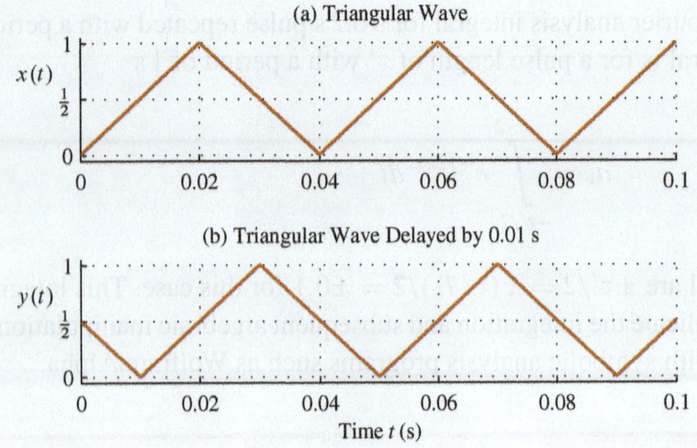

Figure C-11 Illustration of time delay. (a) Triangular wave with period $T_0 = 0.04$ s. (b) Triangular wave delayed by 0.01 s.

For convenience in making the plot, the waveforms of Fig. C-11 were created with the finite sum in (C.10) using $N = 35$ and a_k from (C.20) and b_k from (C.29). The errors at the slope discontinuities are so small as to be unnoticeable on the scale of plotting. Thus, the figure gives further evidence that the Fourier series approximation converges uniformly with large N if the signal is continuous.

C-3.5 Differentiation Property

In Section 3-3.4, if we form a new signal $y(t)$ by taking the derivative of $x(t)$, then $y(t)$ is also periodic with the same period and its Fourier coefficients can be obtained as follows:

Differentiation Property

$$y(t) = \frac{d}{dt}x(t) \quad \longleftrightarrow \quad b_k = (j\omega_0 k)a_k$$

Assuming that the derivative operator can be applied to each term in the Fourier series for $x(t)$, we get

$$y(t) = \frac{d}{dt}x(t) = \sum_{k=-\infty}^{\infty} a_k \frac{d}{dt}e^{j\omega_0 kt} = \sum_{k=-\infty}^{\infty} \underbrace{(j\omega_0 k)a_k}_{b_k} e^{j\omega_0 kt}$$

Thus, differentiating $x(t)$ will multiply its Fourier coefficients a_k by $(j\omega_0 k)$.

EXAMPLE C-5 Differentiated Triangular Wave

Consider a triangular wave $x(t)$ as depicted in the plot of Fig. C-12(a). The derivative is the slope of the linear segments, but is discontinuous at $t = 0, 0.02, 0.04, \ldots$; hence, the derivative is a square wave as shown in Fig. C-12(b). In general, the Fourier series coefficients a_k for the triangular wave $x(t)$ are given by (C.20). When the period is $T_0 = 0.04$ s, $\omega_0 = 2\pi/(0.04) = 50\pi$, so the Fourier series coefficients of the

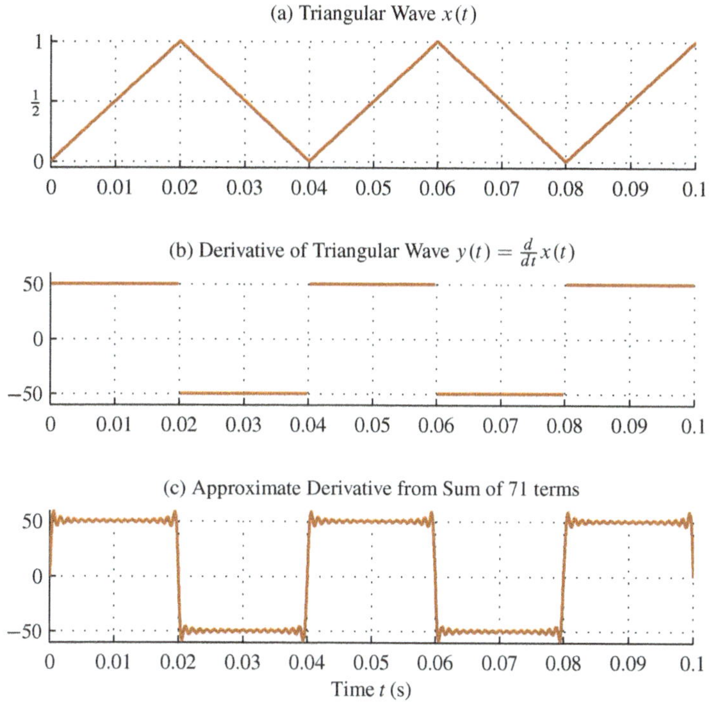

Figure C-12 Illustration of time differentiation. (a) Triangular wave. (b) Derivative of the triangular wave. (c) Finite Fourier sum approximation using b_k for $-35 \le k \le 35$.

differentiated signal $y(t)$ are $b_k = (j50\pi k)a_k$, or

$$b_k = (j50\pi k)a_k = \begin{cases} 50\dfrac{(-2j)}{\pi k} & k = \pm1, \pm3, \pm5, \dots \\ 0 & k = \pm2, \pm4, \pm6, \dots \\ 0 & k = 0 \end{cases} \tag{C.30}$$

These are the Fourier series coefficients for a square wave that varies between +50 and −50. The DC value is zero because the multiplier $(j50\pi k)$ is zero for $k = 0$. The plot in Fig. C-12(c) was constructed using (C.30) and the finite synthesis sum (C.10) with $N = 35$. Like the pulse waveform synthesis in Fig. C-3, we see ripples near the discontinuities. Away from the points of discontinuous slope, however, the approximation converges to +50 and −50, which are the slopes of the triangular wave in the first and last half of the period, respectively.

C-3.6 Frequency-Shifting Property and Multiplying by a Sinusoid

If you multiply a periodic signal by a complex exponential, the Fourier series components will move in frequency, but their values will be unchanged. From Section 3-3.5 we expect frequency shifting of the spectrum lines (up or down). When multiplying by a sinusoid, which is the sum of two complex exponentials, the spectrum will be the sum of two frequency-shifted spectra, one shifted up and the other down. Figure C-13 presents an example of multiplying a periodic pulse wave $x(t)$ by a sinusoid at twelve times the

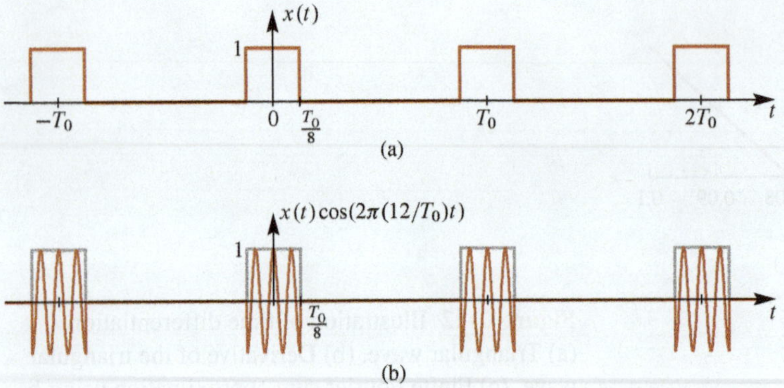

Figure C-13 Illustration of sinusoidal multiplication. (a) Pulse wave with $\tau = T_0/4$. (b) After multiplication by a cosine wave whose frequency, $12F_0$, is 12 times the fundamental frequency of the pulse wave.

fundamental frequency of $x(t)$. The pulse wave has Fourier series representation using the coefficients $\{a_k\}$,

$$x(t) = \sum_{k=-\infty}^{\infty} a_k e^{j\omega_0 kt} \tag{C.31a}$$

and the cosine wave can be written as the sum of two complex exponentials

$$\cos(12\omega_0 t) = \frac{1}{2}e^{j12\omega_0 t} + \frac{1}{2}e^{-j12\omega_0 t} \tag{C.31b}$$

Using the Fourier representation, the product is

$$y(t) = x(t)\cos(12\omega_0 t) = \left(\sum_{k=-\infty}^{\infty} a_k e^{j\omega_0 kt}\right)\left(\frac{1}{2}e^{j12\omega_0 t} + \frac{1}{2}e^{-j12\omega_0 t}\right) \tag{C.31c}$$

The product signal, $y(t)$, will be periodic since T_0 is a period of the cosine wave. Thus, the fundamental frequency of $y(t)$ will also be F_0, and $y(t)$ has a Fourier series with coefficients that will be denoted as $\{b_k\}$.

$$y(t) = \sum_{k=-\infty}^{\infty} b_k e^{j\omega_0 kt} \tag{C.31d}$$

After we expand (C.31c) into two summations, we can identify the frequency shifting of the spectrum of the pulse wave Fourier series, which is shown in Fig. C-14(a). This is illustrated by (C.31e) and Figs. C-14(b) and (c). The first sum in (C.31e) is the spectrum in Fig. C-14(b); the second sum is Fig. C-14(c).

$$\sum_{k=-\infty}^{\infty} b_k e^{j\omega_0 kt} = \frac{1}{2}\sum_{k=-\infty}^{\infty} a_k e^{j\omega_0 (k+12)t} + \frac{1}{2}\sum_{k=-\infty}^{\infty} a_k e^{j\omega_0 (k-12)t} \tag{C.31e}$$

In order to combine the sums, we must change the index of summation so the complex exponential is the same in both, which gives

$$\sum_{k=-\infty}^{\infty} b_k e^{j\omega_0 kt} = \frac{1}{2}\left(\sum_{\ell=-\infty}^{\infty} a_{\ell-12} e^{j\omega_0 (\ell)t} + \sum_{\ell=-\infty}^{\infty} a_{\ell+12} e^{j\omega_0 (\ell)t}\right) \tag{C.31f}$$

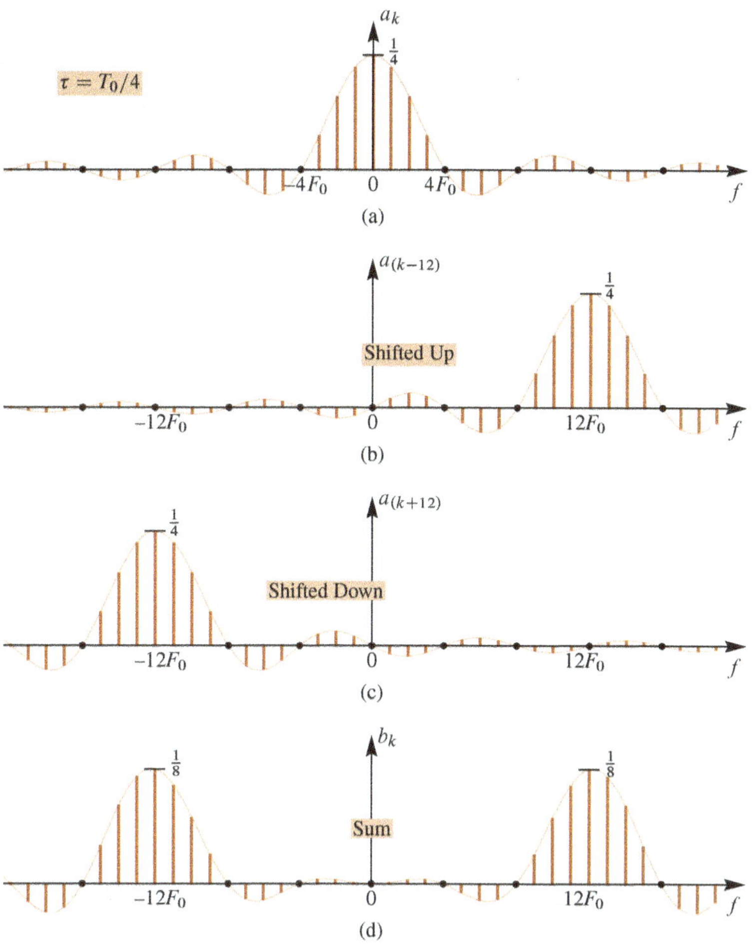

Figure C-14 Frequency shifting for a periodic pulse wave. (a) Spectrum of a pulse wave with $\tau = T_0/4$. (b) Spectrum for the pulse wave multiplied by $e^{+j12\omega_0 t}$ which shifts the spectrum in (a) up to be centered at $12\omega_0 = 2\pi(12F_0)$. (c) Spectrum for the pulse wave multiplied by $e^{-j12\omega_0 t}$ which shifts the spectrum down. (d) Spectrum for the pulse wave multiplied by $\cos(12\omega_0 t)$.

Finally, we see that the Fourier series coefficients $\{b_k\}$ for $y(t)$ are

$$b_k = \frac{a_{k-12} + a_{k+12}}{2} \qquad \text{for all } k \tag{C.32}$$

This spectrum for (C.32) is shown in Fig. C-14(d), where the two shifted copies in Figs. C-14(b) and (c) have been added together after scaling by $1/2$. In general, the addition of the two shifted spectra involves a complex phasor addition, but this case is simple because the cosine has zero phase and the $\{a_k\}$ coefficients of the pulse wave are real.

C-4 Average Power, Convergence, and Optimality

Parseval's Theorem for the Fourier series establishes an equivalence between the sum of the squares of the Fourier coefficients and the signal power. In effect, a spectrum plot of the coefficients $\{|a_k|^2\}$ can be interpreted as the distribution of signal power versus frequency.

Therefore, when the Fourier series of a signal has only a few large coefficients, we can determine what fraction of the signal power is contained in those frequency components. Likewise, when we truncate the Fourier series and synthesize a signal from a finite Fourier sum like (C.10), we can estimate the approximation error by summing $|a_k|^2$ for the coefficients being dropped. This concept underlies audio compression methods such as MP3, and also JPEG image compression, which achieve low bit rates by operating on a frequency domain representation of the signal.

The *average power* of the periodic signal $x(t)$ is the total energy[6] of one period of $x(t)$ divided by the period.

$$P = \frac{1}{T_0}\int_0^{T_0} |x(t)|^2 dt. \tag{C.33}$$

The average power is a convenient measure of the size (or strength) of the periodic signal.

EXAMPLE C-6 Average Power of a Sinusoid

For the case where $x(t) = A\cos(\omega_0 t + \varphi)$, the average power integral

$$P_s = \frac{1}{T_0}\int_0^{T_0} |A\cos(\omega_0 t + \varphi)|^2 dt$$

can be evaluated directly, by using the trigonometric identity $|\cos\theta|^2 = \cos^2\theta = \frac{1}{2}(1 + \cos 2\theta)$ to simplify the integrand

$$P_s = \frac{1}{T_0}\int_0^{T_0} \tfrac{1}{2}A^2 dt + \underbrace{\frac{1}{T_0}\int_0^{T_0} \tfrac{1}{2}A^2\cos(2\omega_0 t + 2\varphi)dt}_{\text{integral over two periods is zero}} = \tfrac{1}{2}A^2$$

Note that the average power depends only on the amplitude of the sinusoidal signal.

The result in Example C-6 can be related to the Fourier series coefficients if we recall that Euler's formula leads directly to the Fourier series of one sinusoid, i.e.,

$$A\cos(\omega_0 kt + \varphi) = \tfrac{1}{2}Ae^{j\varphi}e^{j\omega_0 kt} + \tfrac{1}{2}Ae^{-j\varphi}e^{-j\omega_0 kt} \tag{C.34}$$

---

[6]The "total energy" of a signal is defined as

$$E = \int_{-\infty}^{\infty} |x(t)|^2 dt$$

This terminology comes from physics where if $x(t)$ were a velocity, the kinetic energy would be proportional to the square of velocity, or x^2. The total energy of a periodic function would be infinite, but the average power, being an integral over 0 to T_0, is generally finite.

The Fourier series has only two nonzero coefficients, $a_k = \frac{1}{2}Ae^{j\varphi}$ and $a_{-k} = \frac{1}{2}Ae^{-j\varphi}$. Now consider a single complex exponential signal $x_k(t) = a_k e^{j\omega_0 kt}$ as would appear in a general Fourier series representation. Since $|x_k(t)| = |a_k|$, its average power is

$$P_k = \frac{1}{T_0}\int_0^{T_0} |x_k(t)|^2 dt = \frac{1}{T_0}\int_0^{T_0} |a_k|^2 dt = |a_k|^2$$

so the average power of a single complex exponential is its magnitude squared. Adding the average powers of the two complex exponentials in (C.34) turns out to give an alternate way to obtain the average power of the sinusoid:

$$P_s = |a_k|^2 + |a_{-k}|^2 = |\tfrac{1}{2}Ae^{j\varphi}|^2 + |\tfrac{1}{2}Ae^{-j\varphi}|^2 = \tfrac{1}{2}A^2 \tag{C.35}$$

This is not a coincidence. It is a special case of a more general result that holds when the Fourier series has more than two coefficients. The general result given below is known as Parseval's Theorem, and it applies to any periodic signal $x(t)$ that has a Fourier series:

Parseval's Theorem

$$P = \frac{1}{T_0}\int_0^{T_0} |x(t)|^2 dt = \sum_{k=-\infty}^{\infty} |a_k|^2 \tag{C.36}$$

EXERCISE C.10 Use Parseval's Theorem to determine the average power of the signal

$$x(t) = 2\cos(200\pi t) + \sin(400\pi t)$$

whose period is 10 ms.

EXERCISE C.11 Explain why the average power of the signal $x(t) = 2\sin(120\pi t) + \cos(240\pi t)$ is the same as the average power of that of the signal in Exercise C.10.

EXERCISE C.12 Use Parseval's Theorem to complete the proof of the famous formula:

$$\frac{\pi^2}{M} = \sum_{k=1}^{\infty} \frac{1}{k^2}$$

by finding the numerical value of the integer M. In the history of mathematics, this was known as the Basel problem, which was solved by Euler in 1735.
`http://en.wikipedia.org/wiki/Basel_problem`
Hint: use Parseval's Theorem with the square wave and its Fourier series coefficients.

C-4.1 Derivation of Parseval's Theorem

It is relatively straightforward to demonstrate the validity of Parseval's Theorem. The average power of a periodic signal can be expressed as

$$P = \frac{1}{T_0}\int_0^{T_0} |x(t)|^2 dt = \frac{1}{T_0}\int_0^{T_0} x(t)x^*(t)dt \tag{C.37}$$

Recall that the conjugate of a sum is equal to the sum of the conjugates. If the Fourier series representation of $x(t)$ exists, then the complex conjugate of the periodic signal can be represented by the Fourier series

$$x^*(t) = \left(\sum_{k=-\infty}^{\infty} a_k e^{j\omega_0 kt}\right)^* = \sum_{k=-\infty}^{\infty} a_k^* e^{-j\omega_0 kt} \tag{C.38}$$

Substituting (C.38) into (C.37) and interchanging the order of integration and summation gives the desired result:

$$P = \frac{1}{T_0}\int_0^{T_0} x(t)\left(\sum_{k=-\infty}^{\infty} a_k^* e^{-j\omega_0 kt}\right) dt$$

$$= \sum_{k=-\infty}^{\infty} a_k^* \underbrace{\left(\frac{1}{T_0}\int_0^{T_0} x(t)e^{-j\omega_0 kt}dt\right)}_{=a_k} = \sum_{k=-\infty}^{\infty} a_k^* a_k = \sum_{k=-\infty}^{\infty} |a_k|^2 \tag{C.39}$$

Therefore, we have demonstrated that the average power of a periodic signal is equal to the sum of the average powers of the individual complex exponential signals that comprise the Fourier series representation of the periodic signal.[7]

C-4.2 Convergence of Fourier Synthesis

We can think of the finite Fourier synthesis sum (C.10) as making an approximation to the true signal, i.e.,

$$x(t) \approx x_N(t) = \sum_{k=-N}^{N} a_k e^{j\omega_0 kt}$$

[7] It should be noted that implicit in the above demonstration of the validity of the Parseval Theorem, is that the frequencies are harmonically related so that the term within the large parentheses on the second line of the derivation of (C.39) is, in fact, the Fourier integral for obtaining a_k from $x(t)$. An alternate derivation would use the Fourier series representation for both $x(t)$ and $x^*(t)$ and then invoke the orthogonality of the harmonically related complex exponentials to eliminate integrals involving different frequencies.

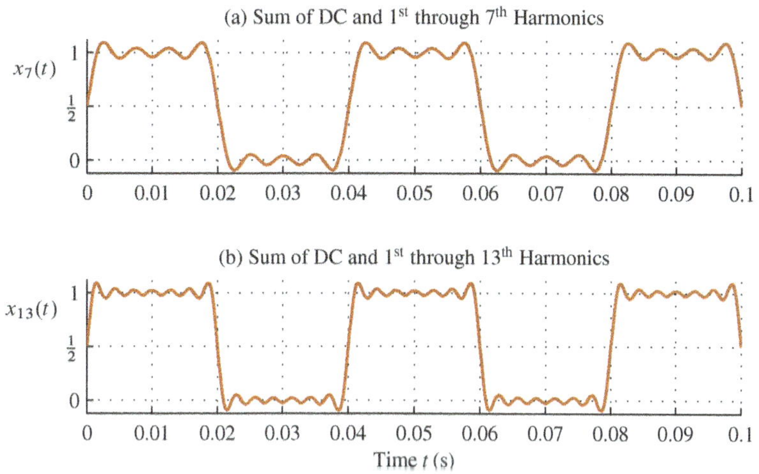

Figure C-15 Summing harmonic components via the finite Fourier synthesis sum (C.10). (a) $N = 7$. (b) $N = 13$.

DEMO

Fourier Series Finite Synthesis

In fact, our discussion so far suggests that with enough terms in the sum, the approximation will be perfect (i.e., $x_N(t) \to x(t)$, as $N \to \infty$). This is illustrated by Fig. C-15, which shows the square wave approximations for $N = 7$ and $N = 13$.

One way to assess the quality of the approximation is to define an error signal, $e_N(t) = x(t) - x_N(t)$, that measures the difference between the true signal and the signal that would be synthesized with N terms. Then we can quantify the error by measuring a feature of the error. For example, a commonly used feature is the maximum magnitude of the difference over one period, which is called the *worst-case error*.

$$E_{\text{worst}} = \max_{t \in [0, T_0)} |x(t) - x_N(t)| \tag{C.40}$$

For the square wave, $|e_N(t)|$, the absolute value of the error, is shown in Fig. C-16 for $N = 7$ and $N = 13$. Figure C-16 was generated with period $T_0 = 0.04$ s by using the Fourier coefficients from Example C-1 in a MATLAB script to generate a plot of the absolute error. Since the ideal square wave is discontinuous, the worst-case error is always half the size of the jump in the waveform precisely at the discontinuity point. The more interesting feature of the absolute error comes from examining the size of the "ears" seen in Fig. C-15 on either side of the discontinuity. In Fig. C-16, this *overshoot* error is found on either side of the large peaks of the error: in $e_7(t)$ at $t \approx 0.0025$ s, with size 0.092; in $e_{13}(t)$ the overshoot peak occurs at $t \approx 0.00143$ s, and is of height 0.09. In fact, it has been proven that as N increases, the overshoot peak moves closer to the discontinuity, but the peak height remains at about 9% of the size of the discontinuity. (For these examples, the size of the discontinuity jump is one.) The mathematical analysis of this approximation error was first published by J. Willard Gibbs[8] in 1898, and is now referred to as the *Gibbs phenomenon* in the theory of Fourier series.

If we perform a similar error analysis for the triangular wave, or the rectified sines, the worst-case error is well-behaved because these signals are continuous. The worst-case

[8]Gibbs received the first American doctorate awarded in engineering from Yale in 1863.

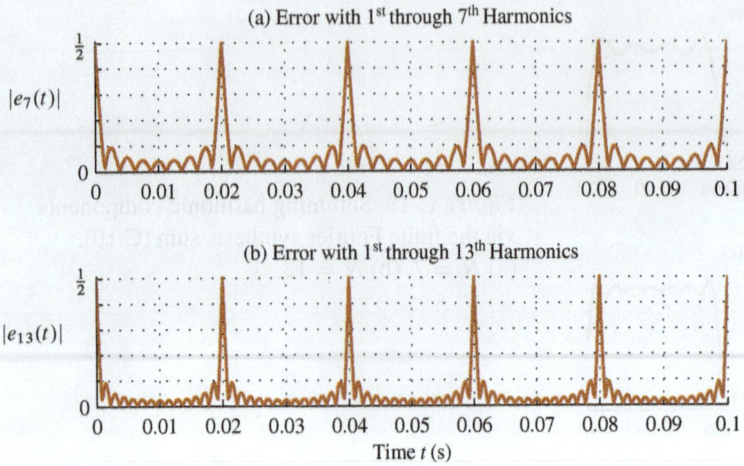

Figure C-16 Error magnitude when approximating a square wave with a sum of harmonic components $x_N(t)$ via finite Fourier synthesis (C.10). (a) $N = 7$. (b) $N = 13$.

error will converge to zero as $N \to \infty$ (i.e., there is no Gibbs phenomenon). For the $N = 3$ and $N = 9$ approximations of the triangular wave, the worst-case errors can be measured by making a zoomed version of Fig. C-7. When these errors are measured, the result is 0.0497 for $N = 3$ and 0.0202 for $N = 9$. Since there is no discontinuity involved, these worst-case errors get smaller as N increases.

EXERCISE C.13 Carry out a measurement of the worst-case error for the full-wave rectified sine signal when approximating with N terms for $N = 3$ and $N = 9$.

C-4.3 Minimum Mean-Square Approximation

Worst-case error only measures point-wise convergence of the approximation to the ideal signal, but it is not the only way to measure the size of the approximation error difference. In fact, to prove convergence of the Fourier series for all signals including the square wave, another error measure has been developed. This is the mean-square error, where the measurable feature of the approximation error is an averaging integral of the squared error over the whole period. The mean-square error is the average power in the error signal $e_N(t) = x(t) - x_N(t)$, which is denoted E_N and defined as

$$E_N = \frac{1}{T_0} \int_0^{T_0} |x(t) - x_N(t)|^2 \, dt$$

where $x(t)$ is a periodic signal and $x_N(t)$ is any finite Fourier series written as

$$x_N(t) = \sum_{k=-N}^{N} \alpha_k e^{j\omega_0 kt} \tag{C.41}$$

Here the coefficients α_k are not necessarily equal to the coefficients a_k derived from $x(t)$ by doing the Fourier series analysis integral. The question then is the following: If we want to minimize the error E_N, how should we choose the finite set of coefficients α_k for $-N \leq k \leq N$ to best approximate the periodic signal $x(t)$, which we know is represented exactly by the *infinite* Fourier series with coefficients a_k? To answer this question, we begin by writing

$$\begin{aligned} e_N(t) &= x(t) - x_N(t) \\ &= \sum_{k=-\infty}^{\infty} a_k e^{j\omega_0 kt} - \sum_{k=-N}^{N} \alpha_k e^{j\omega_0 kt} = \sum_{k=-\infty}^{\infty} (a_k - \alpha_k) e^{j\omega_0 kt} \end{aligned} \tag{C.42}$$

where, by definition, $\alpha_k = 0$ for $|k| > N$. Now using Parseval's Theorem, we can express the mean-square approximation error as

$$E_N = \frac{1}{T_0} \int_0^{T_0} |x(t) - x_N(t)|^2 \, dt = \sum_{k=-\infty}^{\infty} |a_k - \alpha_k|^2 \tag{C.43}$$

Since $\alpha_k = 0$ for $|k| > N$, the error E_N can be written as

$$E_N = \sum_{k=-N}^{N} |a_k - \alpha_k|^2 + \sum_{|k|>N} |a_k|^2 \tag{C.44}$$

Equation (C.44) provides the answer to our question. Since all the individual terms in the sums on the right-hand side are positive, it follows that if we choose $\alpha_k = a_k$ for $-N \leq k \leq N$, then the first sum is zero and the second sum over all the Fourier coefficients, where $|k| > N$ will be the residual mean-square error of the approximation. Thus, we can state that the choice $\alpha_k = a_k$ for $|k| \leq N$ minimizes the mean-square approximation error for a finite number of terms. Since all the terms in the second sum are positive, we conclude that the mean-square approximation error always decreases as we add more terms (increase N), because we have fewer terms remaining in (C.44). Thus, the infinite Fourier series converges in the sense that the mean-square approximation error goes to zero if we include all the terms in the Fourier series. Therefore, by using the mean-square error measure, the theory of Fourier series can be made mathematically rigorous even for discontinuous signals like the square wave.

Convergence in the mean-square sense may be a somewhat surprising result if we recall the Gibbs behavior shown in Fig. C-16. However, if we increase N to a large value, the plot of the approximation error magnitude will retain the shape of those in Fig. C-16, but the oscillations will be very fast over the entire period and very small except in a narrow vicinity around the discontinuity. We can already see this from a comparison of $|e_7(t)|$ and $|e_{13}(t)|$. We can say intuitively that in the limit of large N, all of the error occurs at the single discontinuity point and the integral of the squared error approaches zero. Note, however, that with this interpretation, the worst-case error is still half the size of the discontinuity, but it occurs only at one point on each period of the waveform.

EXERCISE C.14 Make a plot of $|e_{100}(t)|$ over one period of the square wave and compare it to the plots in Fig. C-16. In order to obtain a plot that shows the behavior of the approximation around the discontinuity, it will be necessary to evaluate the error signal at thousands of time instants across one period.

C-5 The Spectrum in Pulsed-Doppler Radar Waveform Design

Radar is a system for using electromagnetic wave propagation to measure the distance from a fixed antenna to another (possibly moving) object. The design of signals for use in radar is an example where the spectrum provides mathematical insight about trade-offs in the selection of time-domain parameters of waveforms. The frequency-shifted pulse wave of Fig. C-13 is the basis for the pulsed-Doppler radar waveform used in many radars. The same waveform is used in the biomedical application of ultrasound to measure blood flow. This waveform enables simultaneous measurement of range and velocity, and is needed in systems where a single transducer is used to transmit and receive the signals. This section describes how the spectrum can be used to understand the radar waveform design problem.

C-5.1 Measuring Range

The word RADAR is an acronym for RAdio Detection And Ranging. The simplest radar measurement is range to a target (or reflector). The basic principle of range measurement is illustrated by Fig. C-17. A short pulse is transmitted through the atmosphere, it reflects off of an object, and returns to the receiving antenna. Range is determined by measuring the time elapsed from the beginning of the pulse until the reflected pulse arrives back at the sending location. In radar systems, a single antenna is used both to transmit and to receive the signal.[9] Figure C-18 shows that the transmitted radar signal must be very short if the transmit and receive antenna is shared, because the signal travels at the speed of light, $c = 3 \times 10^8$ m/s,

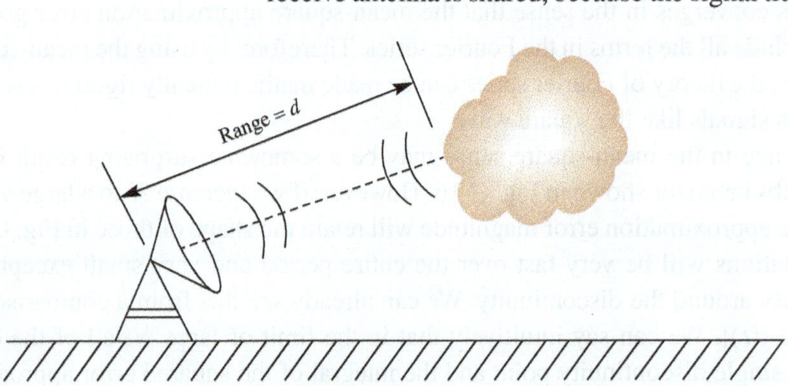

Figure C-17 Illustration of radar system for tracking rain clouds.

[9] Ultrasound systems use high-frequency acoustic transducers.

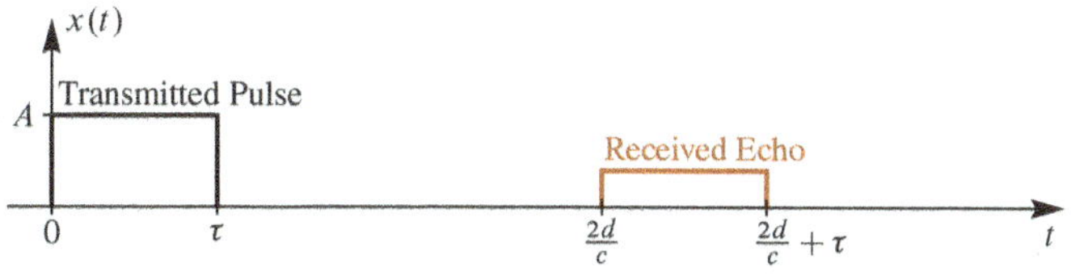

Figure C-18 Illustration of pulsed radar signals: transmitted pulse and received echo delayed by the two-way travel time.

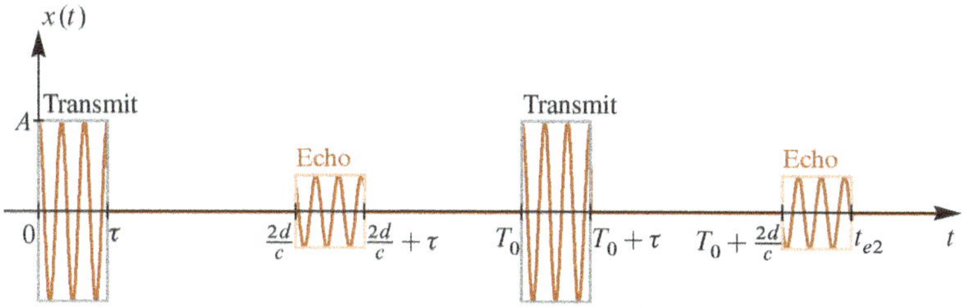

Figure C-19 Pulsed high-frequency radar transmitted signal $r(t)$ and received echoes. The pulse repetition interval is T_0 s. The radar frequency is F_r Hz.

and we must be able to distinguish between the transmitted pulse and its echo from the target. For example, if we want to detect a close reflecting target at 1500 m range, then the two-way travel time (out to the target and back) is

$$\tau_d = 2\left(\frac{d}{c}\right) = 2\left(\frac{1500}{3 \times 10^8}\right) = 1000 \times 10^{-8} = 10\,\mu\text{s}$$

In order to receive the returned pulse at this time without confusion, the transmitted pulse signal must be off (zero) as shown in Fig. C-18, so its duration τ would have to be less than 10 μs. On the other hand, a more distant target at 120 km would have a two-way travel time of 800 μs, so the close target distance at 1500 m imposes a much tighter constraint on the duration of the transmitted pulse.

In Fig. C-18, the transmitted pulse is shown as a simple on–off waveform. However, practical radar systems transmit periodic pulsed sine wave signals of the form

$$r(t) = p(t)\cos(2\pi F_r t + \varphi) \tag{C.45}$$

where $p(t)$ is the on/off periodic pulse signal as in Fig. C-13(a). Such a signal is shown in Fig. C-19 along with the received echoes from the pulses. As suggested by Fig. C-14(d), the spectrum of the periodic sinusoidal pulse signal $r(t)$ in (C.45) is centered on F_r. By adjusting the center frequency F_r of the sinusoidal pulses, the radar signal spectrum can be positioned at frequencies where electromagnetic transmission is most efficient. Each pulse "samples in range" the environment along a direction determined by the antenna orientation. By sending repeated pulses and rotating the antenna, it is possible to scan a surrounding region of space. As we will see below, the periodic sinusoidal pulse waveform also allows the tracking of the velocity of targets.

C-5.2 Measuring Velocity from Doppler Shift

The measurement of velocity is based on the Doppler-shift principle, which is experienced in everyday life, for example, as the changing sound of a horn on a moving train. When the train is coming toward the listener the frequency of the sound from the horn goes up; after the train passes and moves away from the listener, the frequency drops. In radar, the same frequency shift happens when the reflecting target is moving toward, or away from, the radar antenna. Doppler shift is the basis of police radar used to catch speeders on a highway, and also weather radar which measures wind velocity from reflections off of rain drops and other wind-blown particles.

To understand the Doppler phenomenon, it is helpful to begin by assuming that the transmitted radar signal is the continuous-wave (not pulsed) sinusoid

$$s(t) = A\cos(2\pi F_r t + \varphi) \tag{C.46}$$

As depicted in Fig. C-17, the transmitted signal propagates from the radar antenna and is reflected by a target such as a moving rain cloud at nominal range d_0 and moving with velocity v relative to the radar antenna (which is also the receiving antenna). In other words, the target distance in Fig. C-17 varies with time as $d(t) = d_0 - vt$, where it is assumed for simplicity that v is positive in the radial direction from the target to the receiving antenna. Assuming that the velocity v is small compared to the speed of electromagnetic propagation $c = 3 \times 10^8$ m/s, the roundtrip time delay from the antenna to the target and back to the antenna is $\tau_d = 2(d_0 - vt)/c$. Therefore, the continuous-wave signal received back at the antenna is

$$\hat{s}(t) = B\cos(2\pi F_r(t - \tau_d) + \varphi) = B\cos(2\pi F_r(1 + 2v/c)t - 4\pi d_0 F_r/c + \varphi)$$

where B denotes the (usually much smaller) amplitude of the returned sinusoidal signal. Thus, the return signal has frequency $F_r + F_d = F_r(1+2v/c)$ and the amount of frequency shift is seen to depend on the *radial velocity* of the target relative to the velocity of signal propagation—that is, the Doppler frequency shift F_d is

$$F_d = 2\frac{v}{c}F_r \text{ Hz} \tag{C.47}$$

where v is the radial velocity of the target, and F_r is the radio frequency used by the radar. Because c is very large compared to velocities of even supersonic aircraft, the Doppler shift is small compared to the typical radar frequency F_r. For example, for a 1 GHz radar, wind in a rain storm at 30 m/s ($\approx$ 67 mph) will produce a Doppler shift of

$$F_d = \frac{2(30)}{3 \times 10^8}10^9 = 200 \text{ Hz}$$

The model for the Doppler phenomenon is easy to state in the frequency domain. If we assume that the radar signal (or train horn) is a pure sinusoid, then its spectrum consists of two lines at $\pm F_r$. When the Doppler shift is positive, $F_d > 0$, the spectrum lines are shifted up to $\pm(F_r + F_d)$; for negative Doppler, $F_d < 0$, the lines at $\pm(F_r + F_d)$ are shifted down. The radar receiver must be able to measure the amount of frequency shift, and then the velocity can be obtained from (C.47) via

$$v = \frac{1}{2}\frac{c}{F_r}F_d \text{ m/s} \tag{C.48}$$

Because the frequency shift is extremely small compared to F_r, accurate measurement of target velocity relies on detecting very small movement of a spectrum line from F_r to $F_r \pm F_d$.

Although we asserted that the spectrum consists of two lines, this is only true if the transmitted time signal is an infinitely long sinusoid, $A\cos(2\pi F_r t + \varphi)$, which is useless for measuring range where a short pulse (e.g., $\tau = 10\,\mu\text{s}$) is needed for detecting the time delay τ_d. The clever solution to this dilemma is the frequency-shifted periodic pulse signal shown in Fig. C-19. In radar, this signal is called the pulsed-Doppler waveform.

C-5.3 Pulsed-Doppler Radar Waveform

Recall that the frequency-shifted sinusoidal signal being transmitted is

$$r(t) = p(t)\cos(2\pi F_r t + \varphi) \qquad -\infty < t < \infty \tag{C.49}$$

where $p(t)$ is a periodic on/off signal with pulse duration τ and pulse repetition period $T_0 = 1/F_0$. Although it is not a required condition, it simplifies our discussion here to assume that $F_r = MF_0$. Thus, the signal $r(t)$ will be exactly periodic with period T_0, and the spectrum of $r(t)$ will have an infinite set of lines at frequencies kF_0 with k an integer. This is similar to the spectrum shown in Fig. C-14(d), and is depicted for the radar case in Fig. C-20(a), where only the spectrum lines near the peak at $f = F_r$ are shown.

At the radar antenna, the transmitted and received signals must not overlap in time. In order that the returned pulse be distinct from the transmitted pulse in each period, the two-way travel time to the nearest target must be greater than the pulse width τ. The period of the pulse wave must also obey a range constraint; the two-way travel time to the farthest target must be within the time interval as shown in Fig. C-19. Thus, we require $T_0 \geq \tau + 2d_{\text{max}}/c$ where τ is the pulse length and d_{max} is the maximum range. For example, if we pick $T_0 = 1$ ms and $\tau = 10\,\mu\text{s}$, then targets between 1500 m and 148 km will cause echoes within the period as depicted in Fig. C-19.

If we assume that the time duration of $r(t)$ is infinite, then the transmitted radar signal consists of an infinite number of sinusoidal pulses spaced with period T_0. Since $r(t)$ is a periodic signal with $T_0 = 1$ ms, it has a Fourier series and its spectrum consists of lines with a spacing of $F_0 = 1000$ Hz as depicted in Fig. C-20(a).[10] When $F_r = 1$ GHz, the spectrum will have lines at $f = F_r \pm 1000\ell$ Hz, with the largest amplitude lines being clustered near the radar's frequency F_r Hz. With this spectrum we can detect a limited range of frequency shifts because a Doppler shift will move the spectrum line at F_r Hz to $F_r + F_d$. As long as the frequency shift up or down does not get too close to the neighboring harmonic lines at $F_r \pm 1000$ Hz, there is no ambiguity about the amount of frequency shift. With a 1000 Hz spacing of harmonics, the maximum unambiguous Doppler shift is ± 500 Hz. Converting to velocity, this would be ± 75 m/s (or ≈ 201 mph). Notice that the period T_0 controls both the maximum range and the maximum unambiguous velocity interval. Making T_0 smaller increases the unambiguous velocity interval while shortening the maximum range.

[10]Incidentally, if $F_r = 1$ GHz, the spectrum line at $f = F_r$ is, in this case, the millionth harmonic.

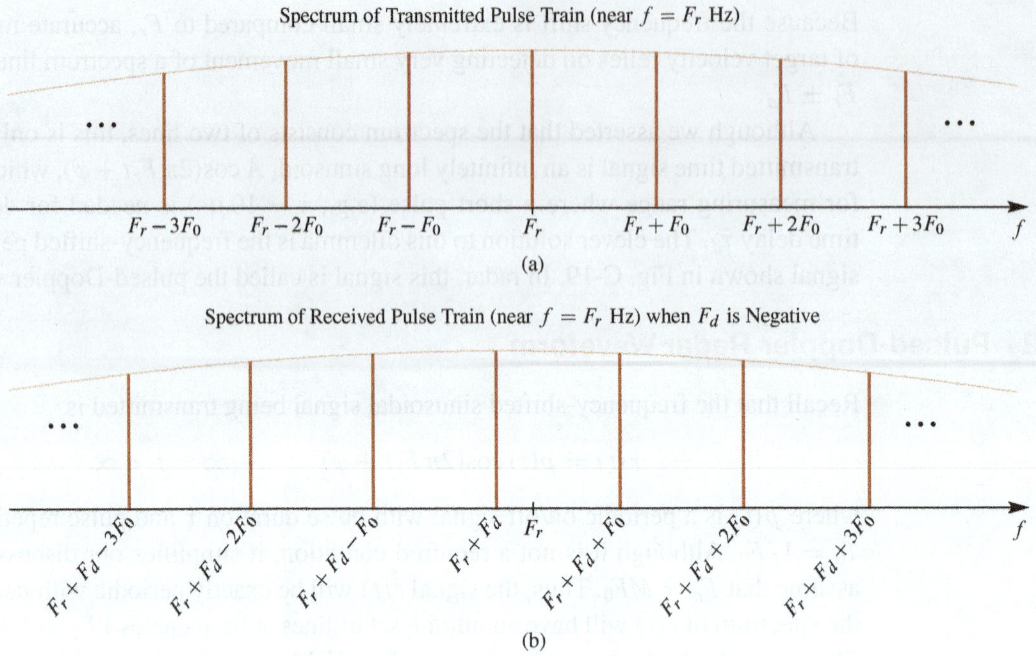

Figure C-20 Spectra of (a) periodic pulse wave frequency shifted to $f = F_r$ Hz (assuming $F_r = MF_0$) and then (b) Doppler shifted down by F_d Hz (i.e., F_d is negative). Only the spectrum lines near $f = F_r$ Hz are shown. When the number of pulses is finite, the lines become Dirichlet shapes with finite width inversely proportional to the number of pulses.

C-5.4 Measuring the Doppler Shift

In a practical radar system, the signal length cannot be infinite, so the pulsed-Doppler waveform would have to be truncated to a finite number of pulses (N_p). As shown in Chapter 8, this changes the spectrum by broadening the harmonic lines to have a Dirichlet shape with nonzero width. The width of the Dirichlet peaks impacts the frequency shift measurement which must detect the movement of the spectrum line from F_r to $F_r + F_d$; the narrower the Dirichlet width, the easier the measurement. In Chapter 8, we also show that the width of the spectrum lines is inversely proportional to the total signal duration $N_p T_0$. Thus, by using more pulses in the finite pulsed-Doppler waveform we can make the spectrum lines as narrow as needed. For example, with 50 pulses and a one millisecond pulse repetition interval, the width of the measured spectrum lines would be 20 Hz, and it would be easy to detect frequency shifts of that size. When converted to velocity in a 1 GHz radar, a Doppler shift of 20 Hz is 3 m/s, so the measurements would be accurate to within ± 1.5 m/s.

The processing done in the radar receiver to compute the frequency shift would require spectrum analysis with the DFT—a topic that is treated in Chapter 8. Our main goal here has been to describe how a sufficiently complicated waveform can be used simultaneously for two different measurements. The spectrum representation has been central to our discussion of the many trade-offs that arise in the design of that waveform.

C-6 Problems

NOTE

Problems with Solutions

The problems in this section address the major concepts of Fourier Series presented in this appendix. On the Companion Website there are many related homework problems which can be used for review and practice. Some of them have detailed solutions.

P-C.1 A periodic signal $x(t) = x(t+T_0)$ is described over one period $-T_0/2 \le t \le T_0/2$ by the equation

$$x(t) = \begin{cases} 1 & \text{for } |t| < t_c \\ 0 & \text{for } t_c < |t| \le T_0/2 \end{cases}$$

where $t_c < T_0/2$.

(a) Sketch the periodic function $x(t)$ over the time interval $-2T_0 < t < 2T_0$ for the case $t_c = T_0/4$.

(b) Determine the DC coefficient a_0.

(c) Determine a formula for the Fourier series coefficients a_k in the finite Fourier representation (3.37). Your final result should depend on t_c and T_0.

(d) Sketch the spectrum of $x(t)$ for the case $\omega_0 = 2\pi(100)$ and $t_c = T_0/4$. Use a frequency range from $-10\omega_0$ to $+10\omega_0$.

(e) Sketch the spectrum of $x(t)$ for the case $\omega_0 = 2\pi(100)$ and $t_c = T_0/10$. Use a frequency range from $-10\omega_0$ to $+10\omega_0$.

(f) From your results in (d) and (e), what do you conclude about the relationship between t_c and the relative size of the high-frequency components of $x(t)$? When t_c decreases, do the high-frequency components get bigger or smaller relative to the size of a_0?

P-C.2 The periodic waveform in Fig. P-C.2 has the property of *half-wave symmetry* (i.e., the last half of the period is the negative of the first half). More precisely, signals with half-wave symmetry have the property that

$$x(t + T_0/2) = -x(t) \qquad -\infty < t < \infty \tag{C.50}$$

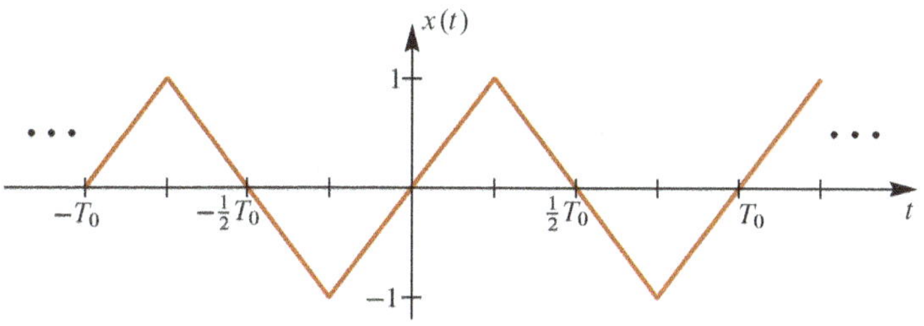

Figure P-C.2

In this problem we will show that this condition has an interesting effect on the Fourier series coefficients for the signal.

(a) Suppose that $x(t)$ is a periodic signal with half-wave symmetry and is defined over half a period by

$$x(t) = t \qquad \text{for } 0 \le t < T_0/2$$

where T_0 is the period of the signal. Plot this periodic signal for $-T_0 \le t \le T_0$.

(b) Prove that the DC coefficient a_0 is zero for any periodic signal having half-wave symmetry. *Hint:* Split the integral for a_0 into two parts.

(c) Prove that all the even indexed Fourier series coefficients are zero for a signal with half-wave symmetry (i.e., $a_k = 0$ if k is an even integer).

P-C.3 A signal $x(t)$ is periodic with period $T_0 = 8$. Therefore, it can be represented as a Fourier series of the form

$$x(t) = \sum_{k=-\infty}^{\infty} a_k e^{j(2\pi/8)kt}$$

It is known that the Fourier series coefficients for this representation of a particular signal $x(t)$ are given by the integral

$$a_k = \frac{1}{8}\int_{-4}^{0} (4+t)e^{-j(2\pi/8)kt}\,dt$$

(a) In the integral expression for a_k above, the integrand, the period, and the limits define the signal $x(t)$. Determine an equation for $x(t)$ that is valid over one period.

(b) Using the result from part (a), draw a plot of $x(t)$ over the range $-8 \le t \le 8$ s. Label your plot carefully.

(c) Determine a_0, the DC value of $x(t)$.

P-C.4 A periodic signal $x(t)$ with a period $T_0 = 10$ is defined over one period, $0 \le t \le 10$ as

$$x(t) = \begin{cases} 0 & 0 \le t \le 5 \\ 2 & 5 < t \le 10 \end{cases}$$

(a) Sketch the periodic function $x(t)$ over the time interval $-10 \le t \le 20$.

(b) Determine the DC coefficient of the Fourier series, a_0.

(c) Use the Fourier analysis integral (3.27) to find a_1, the *first* Fourier series coefficient (i.e., a_k for $k = 1$).

(d) If we add a constant value of one to $x(t)$, we obtain the signal $y(t) = 1 + x(t)$ with $y(t)$ given over one period by

$$y(t) = \begin{cases} 1 & 0 \le t \le 5 \\ 3 & 5 < t \le 10 \end{cases}$$

This signal can also be represented by a Fourier series, but with different coefficients:

$$y(t) = \sum_{k=-\infty}^{\infty} b_k e^{jk\omega_0 t}$$

Explain how b_0 and b_1 are related to a_0 and a_1. *Note*: You should not have to evaluate any new integrals explicitly to answer this question.

P-C.5 The following signal is the product of a sinusoid and a finite Fourier series:

$$x(t) = \sin(7t)\left(\sum_{k=-3}^{3} \frac{3}{2+jk}\, e^{jkt}\right)$$

(a) Explain why the term in parentheses is real-valued.

(b) Use the frequency shifting property to make a sketch of the spectrum of the signal defined above.

(c) Since the signal $x(t)$ is real, verify that the spectrum is conjugate symmetric.

P-C.6 Suppose that a periodic signal is defined as

$$q(t) = \cos(3\pi t)\sum_{k=-2}^{2}\left(\frac{j5}{k+j1.5}\right)e^{j8\pi kt}$$

(a) Prove that the signal $q(t)$ is real-valued.

(b) Determine the fundamental period of $q(t)$.

(c) Determine the DC value of $q(t)$.

(d) Make a table listing all the nonzero Fourier coefficients of the signal $q(t)$.

(e) Make a spectrum plot of this signal showing only the positive frequency components. Label all the complex amplitudes.

P-C.7 Consider the periodic function $x(t)$ plotted in Fig. P-C.7.

(a) Find the DC value a_0 and the other Fourier coefficients a_k for $k \neq 0$ in the Fourier series representation of $x(t)$.

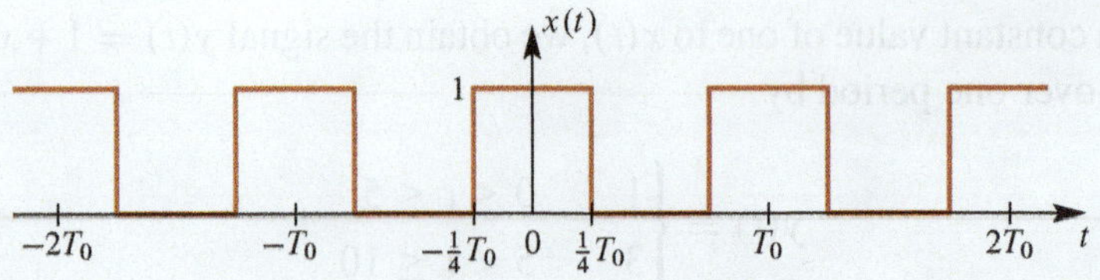

Figure P-C.7

(b) Define a new signal as $y(t) = 2x(t - T_0/2)$, and then sketch the waveform of $y(t)$. Use the scaling and time shifting properties to write down the Fourier series coefficients b_0 and b_k for $k \neq 0$ for the periodic signal $y(t)$ without evaluating any integrals.

P-C.8 Suppose that a periodic signal is defined over one period as:

$$x(t) = \begin{cases} 7 & \text{for } 0 \le t \le 5 \\ 3 & \text{for } 5 < t < 8 \end{cases}$$

(a) Assume that the period of $x(t)$ is 8 s. Plot $x(t)$ over the range $-10 \le t \le 10$ s.

(b) Determine the DC value of $x(t)$ from areas, or from the Fourier series integral.

(c) Write the Fourier integral expression for the coefficient a_k in terms of the specific signal $x(t)$ defined above. Set up all the specifics of the integral (i.e., limits of integration and the integrand).

(d) Determine a general expression for the Fourier series coefficients, $\{a_k\}$, by evaluating the Fourier integral expression. Simplify the expression into polar form, so that the magnitude and phase are explicit.

(e) It might be true that every fifth, or every eighth, coefficient is zero for $k > 0$. For example, $a_5 = 0$, $a_{10} = 0$, $a_{15} = 0$, etc.; or, $a_8 = 0$, $a_{16} = 0$, $a_{24} = 0$, etc. Determine which speculation is true.

P-C.9 Make plots of the spectra for both the FWRS and HWRS signals. Use the complex amplitudes from (3.34) and (C.26) and assume that $T_1 = 0.02$ s for the full-wave signal and $T_0 = 0.02$ s for the half-wave signal. The result should be different fundamental frequencies.

P-C.10 It turns out that the Fourier series coefficients for the signal $y(t) = x(\alpha t)$, which has period T_0/α, are exactly the same as the Fourier series coefficients for the signal $x(t)$, which has period T_0. This is the time-scaling property of the Fourier series.

(a) For the pulse wave (C.11), the Fourier series coefficients are given by (C.14). Suppose that $p(t)$ is a periodic pulse wave with $\tau = 0.02$ s, where $T_0 = 0.1$ s is the period. Make a sketch of $q(t) = p(t/5)$; in the process, determine the period and pulse width of $q(t)$.

(b) From the known Fourier series (C.14) compare the expressions for the Fourier series coefficients of $p(t)$ and $q(t)$. Manipulate the expressions to show equality.

(c) Prove that the time-scaling property is true. The following Fourier analysis integral gives the Fourier series coefficients of the time-scaled signal $x(\alpha t)$:

$$b_k = \frac{1}{T_0/\alpha} \int_0^{T_0/\alpha} x(\alpha t) e^{-j(2\pi/(T_0/\alpha))kt} dt$$

By changing the dummy variable of integration to $u = \alpha t$, show that the right-hand side is the same as the Fourier analysis integral for $x(t)$ (i.e., $b_k = a_k$).

P-C.11 If a periodic signal has a Fourier series where all the odd-indexed coefficients are zero, then the fundamental frequency has been chosen incorrectly. Suppose that $x(t)$ where $f_0 = 3$ Hz and the Fourier coefficients are

$$a_k = \begin{cases} j12/k & \text{for } k = \pm 1, \pm 2, \pm 4, \pm 6 \\ 0 & \text{otherwise} \end{cases}$$

(a) Sketch the spectrum for $x(t)$.

(b) Define $y(t) = x(t) + 24\sin(6\pi t)$. Sketch the spectrum for $y(t)$.

(c) Explain why the fundamental frequency of $y(t)$ is not 3 Hz.

(d) Determine the correct fundamental frequency for $y(t)$, and also redefine the Fourier coefficients of $y(t)$ for the new f_0.

P-C.12 Suppose that a periodic signal $x(t)$ is defined by the plot in Fig. P-C.12 (only the section $-8 \le t \le 8$ is shown):

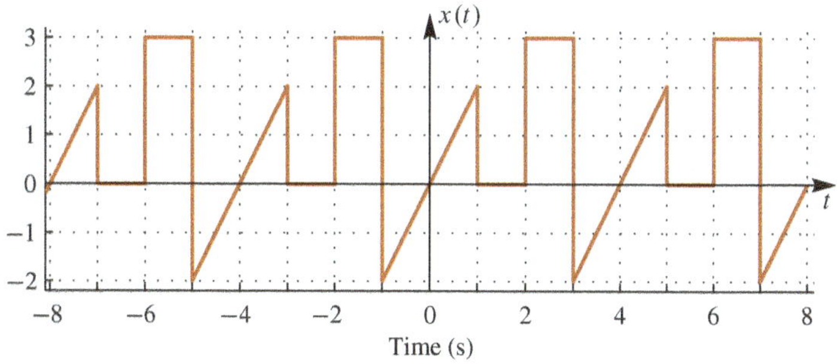

Figure P-C.12

(a) Determine the period T_0 of $x(t)$.

(b) Determine the fundamental frequency of $x(t)$ in Hz.

(c) Determine the DC value of $x(t)$.

(d) Write the Fourier integral expression for the coefficient a_k in terms of the specific signal $x(t)$ defined above. Set up all the specifics of the integral(s) including the limits of integration, but do not evaluate any integral. All parameters in the integral(s) should have numeric values.
Note: more than one integral might be needed.

P-C.13 The following MATLAB code defines vectors xt and zt, which correspond to a finite segment of the periodic signals $x(t)$ and $z(t)$:

```
tt = -1000:0.1:1000;     %- in seconds
xt = zeros(size(tt));
for k = [-3,-1,0,1,3]      %- loop on k for only these indices
   xt = xt + (1 + abs(k) - j*k)*exp(j*0.02*pi*k*tt);
end
xt = real(xt);            %- x(t) should be real-valued
zt = xt - 3.2 + 8*cos(0.06*pi*tt - pi);
```

(a) Determine the fundamental period of the signal $x(t)$ defined in the MATLAB code above.

(b) The signal $x(t)$ has a Fourier series with coefficients $\{a_k\}$, defined by the MATLAB code. Determine the numerical values of the $\{a_k\}$ coefficients for $|k| \leq 3$ in *polar form.*

(c) The new signal defined for $z(t)$ is also periodic with a Fourier series, so it can be expressed in the following Fourier series with new coefficients $\{b_k\}$:

$$z(t) = \sum_{k=-3}^{3} b_k \, e^{j0.02\pi kt}$$

Determine the numerical values of the $\{b_k\}$ coefficients for $|k| \leq 3$ in *polar form.*

P-C.14 Shown in Fig. P-C.14 are spectrograms for two periodic signals, $x_1(t)$ and $x_2(t)$. *The frequency axis has units of Hz.*

(a) The spectrum line at $f = 2000$ Hz is the same N-th harmonic of both signals; determine N.

(b) The two signals have the same fundamental frequency. Determine the fundamental period T_0 of each signal using *accurate estimates* from the spectrograms.

(c) One of these signals is a 50% duty-cycle square wave; recall the harmonic structure of its Fourier series. Determine which signal fits this description and explain your answer.

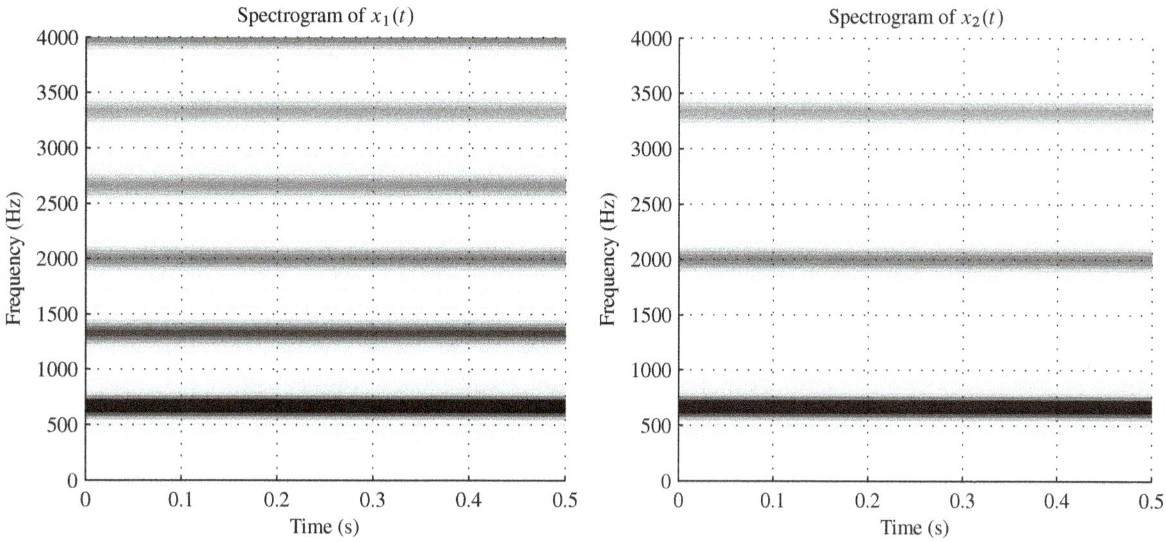

Figure P-C.14

P-C.15 The properties of the Fourier series can be used to avoid actual integration of the Fourier integral.

(a) Determine the Fourier series coefficients of a full-wave rectified cosine (FWRC) signal $x(t) = |\cos(2\pi t)|$ by using the time-shifting property and the known Fourier series coefficients for the full-wave rectified sine, $s(t) = |\sin(2\pi t)|$, which are

$$a_k = \frac{2}{\pi(1 - 4k^2)}$$

(b) Starting from the known Fourier series coefficients of the half-wave rectified sine (HWRS) given in (C.26), determine the Fourier series coefficients of a full-wave rectified sine (FWRS) signal by using the additive and time shifting properties.

(c) For the pulse wave (C.11), the Fourier series coefficients are given by (C.14). Suppose that $p(t)$ is a periodic pulse wave with $\tau = 0.02$ s, where $T_0 = 0.1$ s is the period. If the Fourier series coefficients of $p(t)$ are denoted by a_k, define a new signal $q(t)$ by changing the sign of every other coefficient, i.e.,

$$q(t) = \sum_{k=-\infty}^{\infty} (-1)^k a_k e^{j\omega_0 kt}$$

Make a plot of $q(t)$. Recall that $(-1)^k$ can be written as a complex exponential.

P-C.16 The Fourier series coefficients of the triangular wave in Fig. P-C.7 can be obtained without doing integration of the Fourier analysis integral. The key is to use the differentiation property.

(a) If the triangular wave is $x(t)$, make a sketch of the derivative signal $y(t) = \frac{d}{dt}x(t)$.

(b) One way to determine the Fourier series coefficients of $y(t)$ is by invoking the differentiation property assuming that the Fourier coefficients of $x(t)$ are a_k. If the Fourier coefficients of $y(t)$ are b_k, give an equation for b_k in terms of a_k.

(c) The result of part (a) should be a scaled square wave with zero DC. Thus, it is relatively easy to write down the Fourier series coefficients of $y(t)$ from the pulse wave results.

(d) Equate the results of parts (b) and (c) and solve for a_k to find the Fourier series coefficients of the triangular wave in Fig. P-C-2.

APPENDIX

D

Laboratory Projects

This appendix contains a list of the laboratories that are currently available on the Companion Website. Each of the labs corresponds to material primarily from one chapter, although some of the lab projects use concepts from several chapters. There are two types of lab descriptions: lab projects which are longer and lab exercises which are short enough to be done during a typical 2–3 hour supervised lab session. Tables D-1 and D-2 summarize 25 labs available on the Companion Website with a cross reference to the primary chapter(s) involved in each lab. The structure of the labs is as follows:

Overview: Each lab starts with a brief review of the key concepts to be studied and implemented.

Pre-Lab: The Pre-Lab section consists of a few exercises that introduce some simple MATLAB commands that will be needed in the lab. These should be easy enough that students can finish them successfully with little or no outside help.

Warm-up: The warm-up section consists of more simple exercises that introduce the MATLAB functionality needed for the Exercises or Project of that lab. The warm-up is intended to be completed during a *supervised lab time*, so that students can ask questions of an expert. In our own use of these experiments, we have implemented

Table D-1 Laboratory Projects on the Companion Website.

Lab	*Subject*	*Chapter*
P-1	Introduction to MATLAB	2
P-2	Complex Exponentials: Multipath	2
P-3	Complex Exponentials: Direction Finding	2
P-4	AM and FM Sinusoidal Signals	3
P-5	Synthesis of Sinusoidal Signals (Speech or Music)	3
P-6	Shepard Scale Synthesis	3
P-7	FM Synthesis for Musical Instruments	3
P-8	Digital Images: A/D and D/A	4
P-9	Sampling, Convolution, and FIR Filtering	5
P-10	Edge Detection in Images: UPC decoding	5
P-11	Hearing Test via Frequency Response	6
P-12	Frequency Response: Bandpass and Nulling Filters	6
P-13	Encoding and Decoding Touch-Tone Signals	7 and 9
P-14	Octave Band Filtering	9
P-15	Cochlear Implant Filter Bank Simulation	9 or 10
P-16	PeZ—The z, n, and $\hat{\omega}$ Domains	10

Table D-2 Short Laboratory Exercises on the Companion Website.

Lab	*Subject*	*Chapter*
S-1	Complex Exponentials: Source Localization	2
S-2	Complex Exponentials: Direction of Arrival	2
S-3	Complex Exponentials: Beamforming with Phasors	2
S-4	Convolution and FIR Filtering	5
S-5	DLTI GUI and Nulling Filters	6
S-6	FIR Filter Design	7
S-7	Spectrogram: AM and FM Signals	3 or 8
S-8	Spectrogram: Harmonic Lines and Chirp Aliasing	4 or 8
S-9	PeZ: Remove Interference from EKG Signals	10

a system in which the laboratory instructor must verify the appropriate steps by initialing an ***Instructor Verification*** sheet for a few key steps. Examples of this verification sheet are included in the lab writeups on the Companion Website.

Exercises: The bulk of the work in each lab consists of exercises that require some MATLAB programming and plotting. All of the exercises are designed to illustrate theoretical ideas presented in the text. Furthermore, we have included numerous processing examples that involve real signals such as speech, music, and images. The short labs have only Exercises and Homework.

Projects: In some cases, the labs require implementations that are so large and complicated that it is no longer fair to call them exercises. In labs such as FM Synthesis for Musical Instruments, Synthesis of Sinusoidal Signals (Speech or Music), Encoding and Decoding Touch-Tone Signals, and others, the problem statement is more like that of a design project. In these cases, students should be given some flexibility in creating an implementation that satisfies a general objective. In addition, many individual parts must be completed to make the whole project function correctly.

Homework: In the short labs, there are homework assignments that are extensions of of the exercises. These could be assigned as unsupervised work that complements the main body of the lab.

NOTE
Install DSP First Toolbox for MATLAB

These labs can be done with a recent version of MATLAB. Since a few of the labs require M-files that are not part of the standard MATLAB release, we also provide a package of *DSP First* M-files containing functions developed for this book. The Companion Website gives a link to instructions for installing the DSP First toolbox.

Index